# International Conference on Automotive and Electronics Technologies 2014



Automotive Simulation World Congress 2014 and ANSYS Electronics Simulation EXPO 2014

Tokyo, Japan
9 - 10 October 2014

ISBN: 978-1-5108-4103-1

**Printed from e-media with permission by:**

Curran Associates, Inc.
57 Morehouse Lane
Red Hook, NY 12571

proceedings.com

**Some format issues inherent in the e-media version may also appear in this print version.**



Printed by Curran Associates, Inc. (2017)



ANSYS Inc.
Southpointe
2600 ANSYS Drive
Canonsburg, PA 15317
USA

Phone: 844.Go.ANSYS
844.462.6797
Fax: 724.514.9494

www.ansys.com

**Additional copies of this publication are available from:**

Curran Associates, Inc.
57 Morehouse Lane
Red Hook, NY 12571 USA
Phone: 845-758-0400
Fax: 845-758-2633
Email: curran@proceedings.com
Web: www.proceedings.com

# Keynote

Interpreter Session (headphone icon)

## October 9, 2014 - Thursday -morning-

ANSYS Electronics Simulation Expo

Automotive Simulation World Congress

| Time | Session |
|---|---|
| 9:30 - 9:50 | **1K-1** (Interpreter Session)<br>Welcome Address ..... N/A<br>Gianmaria Bruni<br>Ferrari AF Corse Team, Italy<br>2013 World Champion of the WEC Championship |

| Time | Automotive Simulation World Congress | ANSYS Electronics Simulation Expo |
|---|---|---|
| 10:00 - 10:50 | **1K-A2** (Interpreter Session) | **1K-B2** (Interpreter Session) |
| | Welcome and Introduction<br>Dr. Sandeep Sovani<br>ANSYS, Inc. ..... N/A | Welcome and Introduction<br>Sudhir Sharma<br>ANSYS, Inc. ..... N/A |
| | Towards a Predictive and Comprehensive Engine Simulation Methodology<br>Dr. Dirk Linse<br>BMW Group, Germany ..... 1 | Computational Electromagnetics Has Changed My Life<br>Dr. Zoltan Cendes<br>Life Fellow, USA ..... N/A |

| Time | Session |
|---|---|
| 11:00 - 12:20 | **1K-3** (Interpreter Session)<br>CAE utilization and promotion in DENSO -from now and future-<br>Dr. Shigeru Akaike<br>DENSO CORPORATION, Japan ..... 15<br>Addressing Complexity with Comprehensive Simulation<br>Dr. Swaminathan Subbiah<br>ANSYS, Inc. ..... 27 |
| 12:20 - 13:30 | Lunch |

## October 10, 2014 - Friday -morning-

ANSYS Electronics Simulation Expo

Automotive Simulation World Congress

| Time | Session |
|---|---|
| 9:30 - 10:40 | **2K-1** (Interpreter Session)<br>Welcome and Updates of yesterday<br>Dr. Sandeep Sovani and Sudhir Sharma<br>ANSYS, USA ..... N/A<br>ANSYS Systems Solutions for Innovative Automotive Systems<br>Eric Bantegnie<br>Esterel Technologies .... N/A |
| 11:00 - 12:20 | **2K-2** (Interpreter Session)<br>Future vision for EVs and HVs<br>Shoichi Sasaki<br>Limited Liability Company SASAKI Motor Technology Research Institute ..... N/A<br>Automotive industry transformation from simulation, to device evolution, to the datacenter<br>Frank Soqui<br>Intel Corporation USA ..... N/A |
| 12:20 - 13:30 | Lunch |

# Guest speaker

**Gianmaria Bruni**
Ferrari AF Corse Team, Italy
2013 World Champion of the WEC Championship

**Dr. Dirk Linse**
BMW Group, Germany
Development Engineer, "Simulation Combustion"

**Dr. Shigeru Akaike**
DENSO CORPORATION, Japan

**Dr. Zoltan Cendes**
Life Fellow

**Eric Bantegnie**
Esterel Technologies
President & CEO

**Shoichi Sasaki**
Limited Liability Company SASAKI Motor Technology Research Institute

**Frank Soqui**
Intel Corporation, USA
General Manager, Technical Compute Group

# Rooms for Keynote sessions

**Please find the room for each Keynote session.**

| Room | October 9 | October 10 |
|---|---|---|
| A | 1K-1 1K-A2 1K-3 | 2K-1 2K-2 |
| B | 1K-B2 | |

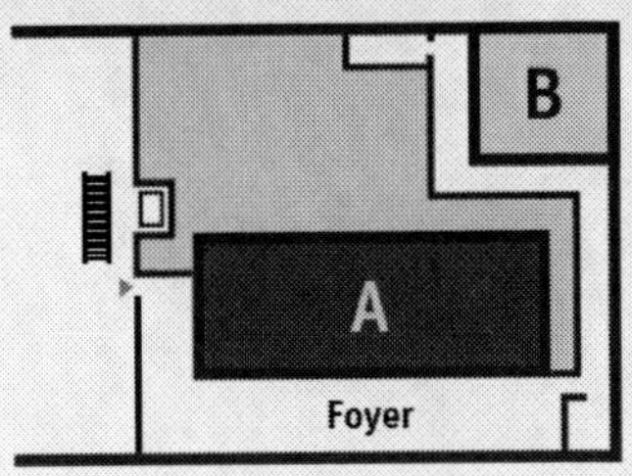

# Program October 9, 2014 - Thursday -afternoon-

Interpreter Session

ANSYS Electronics Simulation Expo

Automotive Simulation World Congress

| Time | Red<br>Powertrain (Interpreter Session) | Blue<br>Aerodynamics, Underhood (Interpreter Session) | Purple<br>Body, Chassis, Interior | Green<br>Motors & Power Electronics (Interpreter Session) | Yellow<br>RF, Microwave, & Communications | Orange<br>IC Power and Reliability |
|---|---|---|---|---|---|---|
| 13:30<br>\|<br>14:00 | 1A-1<br>Aeroacoustics modeling of Fan ..... N/A<br>Himanshu Agrawal<br>Cummins Research and Technology India Ltd. Pune | 1B-1<br>Prediction of Narrow-band noise using LES ..... N/A<br>Katsutomo Kanai<br>Honda R&D Co., Ltd. Automobile R&D Center, Japan | 1C-1<br>Breakthrough in Brake NVH and Wear Simulation ..... N/A<br>Vinay Carpenter<br>ANSYS India | 1D-1<br>New Electromagnetic-Thermal Coupled Analyses for Electrical Machine Design ..... 168<br>Dr. Marius Rosu<br>ANSYS, Inc. | 1E-1<br>Practice of IEC standards, IC-EMC modeling method and measuring method (Activities of the JEITA Semiconductor EMC Subcommittee) ..... N/A<br>Seiji Hayashi<br>Japan Electronics and Information Technology Industries Association | 1F-1<br>Advancing Chip-Package-System Convergence through a Comprehensive RTL to Silicon Methodology ..... N/A<br>Aveek Sarkar<br>ANSYS, Inc. |
| 14:05<br>\|<br>14:35 | 1A-2<br>SCR mixing pipe optimization and numerical models validation ..... N/A<br>Dr. Zhiguo Zhao<br>Tenneco, China | 1B-2<br>Aero-Vibro-Acoustics for Wind Noise Application ..... 65<br>Ashok Khondge<br>ANSYS India | 1C-2<br>Radiator Fan-Shroud assembly Design for Low NVH ..... N/A<br>Ashok Patidar<br>VE Commercial Vehicles Ltd (A Volvo Group and Eicher Motors Joint Venture) | 1D-2<br>The Influence of Outbreak Magnetic Field by a Superconductor Coil for Surrounding Devices and Environment .... 183<br>Takahiro Koga<br>ANSYS Japan K.K. | 1E-2<br>EMI prediction method for SiC inverter by the modeling of structure and the accurate model of power device ..... N/A<br>Sari Maekawa<br>Toshiba Corporation | 1F-2<br>Measurements and Simulation of RF Substrate Noise Coupling in Wireless Communication ICs . N/A<br>Makoto Nagata<br>Kobe University |
| 14:50<br>\|<br>15:20 | 1A-3<br>Higher automotive simulation productivity in ANSYS Fluent 15.0 with NVIDIA GPUs .....41<br>Bhushan Desam, Ph.D.<br>Nvidia Corporation, USA | 1B-3<br>Vehicle Aerodynamic Shape optimization with integration of morphing and optimization technologies ..... N/A<br>Swatantra Kumar Singh<br>Tata Motors, India | 1C-3<br>Engine-Gasket HPC Supension NV Analysis using Linear Perturbation ..... 124<br>Toru Hiyake<br>ANSYS Japan K.K. | 1D-3<br>Comprehensive Solution for Nissan LEAF Motor Performance Prediction ..... 200<br>Dr. David Staton<br>Motor Design Ltd. | 1E-3<br>A new design approach using Sparameter to achieve noise reduction for power-device-mounted-board ..... N/A<br>Akihiro Tanaka<br>RITA Electronics, Ltd. | 1F-3<br>Achieving Dynamic Voltage Drop Sign-off in Automotive Designs ..... N/A<br>Soenke Grimpen<br>Infineon, Germany |
| 15:25<br>\|<br>15:55 | 1A-4<br>Total Exhaust Simulation ..... N/A<br>Saurabh Srivastava<br>ANSYS India | 1B-4<br>Underhood Thermal Management ..... 77<br>Hamid Ghazialam<br>ANSYS, Inc. | 1C-4<br>Overcoming the Barriers Imposed by Geometry -Introduction of SpaceClaim- ..... 136<br>Ichiro Kawabata<br>SpaceClaim Japan K.K. | 1D-4<br>High-Performance Real-Time Simulation based on ETAS HiL Test System and ANSYS Maxwell FEA Model ..... 213<br>Dr. Marius Rosu<br>ANSYS, Inc. | 1E-4<br>PCB design and simulation technique reducing Common-mode voltage ..... N/A<br>Atsushi Nakamura<br>Ultimate Technologies Incorporated | 1F-4<br>Efficient On-chip ESD Protection Design and Verification Methodology using EDA Tools ..... N/A<br>Yoko Otsuka<br>Renesas System Design Co., Ltd |
| 16:10<br>\|<br>16:40 | 1A-5<br>Numerical Simulation on Burst Containment Hyperthermal Highspeed Experiment of Turbine Housing ..... N/A<br>Xiaowei Du<br>Honeywell, China | 1B-5<br>How to boost Fluent Adjoint using RBF Morph ..... 98<br>Marco Evangelos, Biancolini<br>RBF Morph, Italy | 1C-5<br>Resin Transfer Molding Simulation for Woven Fabric Composite Materials ..... 145<br>Arata Hayashigaki<br>CYBERNET SYSTEMS CO., LTD., Japan | 1D-5<br>High frequency, high power magnetic design with Maxwell 3D – from geometry creation to component optimization ..... 221<br>Dr. Jenna Pollock<br>Tesla Motors, Inc., USA | 1E-5<br>Magnetic field analysis of contactless power transmission system on moving via electromagnetic induction ..... N/A<br>Tetsuya Takura<br>Tohoku Institute of Technology | 1F-5<br>A Novel Approach for On-Die Regulator Power Network Modeling and Analysis ..... N/A<br>Teik Wah Lim<br>Altera Corporation, India |
| 16:45<br>\|<br>17:15 | 1A-6<br>Calculation of heat transfer characteristics in the exhaust system of an automotive ..... 59<br>Itsuhei Kohri<br>Tokyo City University, Japan | 1B-6<br>Numerical study of DrivAer body model with add-on under-floor cover and rear spoiler using Ansys Fluent .. N/A<br>Aniruddha Joshi<br>Tata Consultancy Services Limited India | 1C-6<br>Press forming simulation of textile composite material ..156<br>Koji Yamamoto<br>CYBERNET SYSTEMS CO.,LTD., Japan | 1D-6<br>MEMS Multiphysics Simulations in ANSYS Workbench ..... 234<br>Metin Ozen, Ph.D.<br>Ozen Engineering Inc. | 1E-6<br>Analysis of electromagnetic wave propagation in the Automobile ..... N/A<br>Shinji Fukui<br>NIPPON SOKEN, INC. | 1F-6<br>Power Noise Analysis for Mixedsignal Integrated Circuits .. N/A<br>Fumihiro Minami<br>Toshiba Corporation Semiconductor & Storage Products Company |

## Reception

17:30-19:00

After all the sessions, a reception will be held at the foyer and the room indicated with the blue color. We hope this will be a good opportunity for you to exchange information with one another.

# Program October 10, 2014 - Friday -afternoon-

Interpreter Session

ANSYS Electronics Simulation Expo

Automotive Simulation World Congress

| Time | Red<br>Powertrain (Interpreter Session) | Green<br>Aerodynamics, Underhood (Interpreter Session) | Orange<br>Engines | Blue<br>Automotive Electronics (Interpreter Session) | Yellow<br>RF, Microwave, & Communications | Purple<br>MBD & Embedded System Development |
|---|---|---|---|---|---|---|
| 13:30<br>\|<br>14:00 | 2A-1<br>Measurement and Prediction technology of cooling capability for Hybrid drivetrain components ..... 256<br>Tadashi Yamada<br>TOYOTA MOTOR CORPORATION, Japan | 2B-1<br>Leveraging SGI's UV and VizServer platform to enable multiple CAE users to run ANSYS software with remote visualization in a data center .... 322<br>Tony DeVarco<br>SGI, USA | 2C-1<br>CFD simulations with chemical kinetics for advanced engine combustion .....380<br>Tsujimura Taku<br>National Institute of Advanced Industrial Science and Technology | 2D-1<br>Case study of automotive ESD analysis ..... 446<br>Hongkeun Kim<br>SL Corporation, Korea | 2E-1<br>Non-contact power transmission technology on rotary or slide system using the electric field coupling ..... N/A<br>Kenichi Harakawa<br>ExH Corporation | 2F-1<br>MBD solution covering from system design to verification by real-time simulation for automotive systems ..... N/A<br>Kosuke Konishi<br>IDAJ Co, Ltd. Japan |
| 14:05<br>\|<br>14:35 | 2A-2<br>Evaluation method for Pressure Loss Using Energy Dissipation..266<br>Hiroshi Okumura<br>DENSO TECHNO Co.,Ltd. Japan | 2B-2<br>An example of low speed competition car development using CFD ..... 332<br>Tetsuya Fujimoto<br>Sophia University, Japan | 2C-2<br>Numerical Investigation of Soot Formation in Gasoline Direct Injection Engines ..... 396<br>Jun Hashimoto<br>Oita University, Japan | 2D-2<br>Vehicle EMC simulation and prediction ..... N/A<br>Xue Yue, Ph.D.<br>CSR Qingdao Sifang Co.,ltd., China | 2E-2<br>Maxwell-3D analysis of small capacity contactless power transfer system applied AGV ..... N/A<br>Kiyotaka Fuji<br>Fellow Energy Co.,Ltd./ Kyushu Institute of Technology Adjunct professor | 2F-2<br>An Application of SCADE Suite to the Design of Embedded Software for a Line-Tracing Robot<br>- For Participating in the Embedded Technology Software Design Robot Contest - ..... N/A<br>Shuji Takahashi<br>Gunma University |
| 14:50<br>\|<br>15:20 | 2A-3<br>RBF Morph mesh morphing ACT extension for ANSYS Mechanical ..... 270<br>Marco Evangelos, Biancolini<br>RBF Morph, Italy | 2B-3<br>Aerodyanamics Development in SUBARU ..... N/A<br>Yusuke Oki<br>FUJI HEAVY INDUSTRIES LTD., Japan | 2C-3<br>A Study of Autoignition in an Incylinder Engine by using Numerical Analyses and Optical Measurements ..... 408<br>Akira Iijima<br>Nihon University, Japan | 2D-3<br>Simulation of stray magnetic couplings in EMC* filters for automotive power electronic devices.<br>(* Electromagnetic Compatibility) .....456<br>Takashi Masuzawa<br>DENSO CORPORATION, Japan | 2E-3<br>Thermal design optimization considering the solar radiation and heat convection ..... N/A<br>Toshihiro Tsujimura<br>Toshiba Corporation Semiconductor & Storage Products Company | 2F-3<br>Prospects for Control Development Using Model-Based Requirement Decomposition ..... N/A<br>Junichi Ichihara<br>AZAPA Co. LTD |
| 15:25<br>\|<br>15:55 | 2A-4<br>CAEBAT Battery Thermal Management Project by General Motors, ANSYS and Esim ..... 293<br>Sandeep Sovani<br>ANSYS, Inc. | 2B-4<br>Rain Water Management ..... 351<br>Padmesh Mandloi<br>ANSYS India | 2C-4<br>Study of Combustion Chamber Design for High Compression Ratio ..... 422<br>Munemasa Hashimoto<br>NEW A.C.E INSTITUTE CO., LTD. | 2D-4<br>Application of the simulation technology for the development of automotive product<br>Higashitani Hiroshi .. 467<br>Panasonic Corporation Automotive & Industrial Systems Company, Japan | 2E-4<br>Spatial Design in Wireless Power Transfer with Resonance field ..... N/A<br>Tatsuya Hosotani<br>Murata Manufacturing Co., Ltd. | 2F-4<br>Problem applying Model-baseddevelopment in ISO26262 ..... N/A<br>Aoki Yuho<br>SGS Japan Inc. |
| 16:10<br>\|<br>16:40 | 2A-5<br>Process Integration and Comprehensive Solution Paradigms: Engine Combustion, Conjugate Heat Transfer, Durability and Fatigue ..... 303<br>Laz Foley<br>ANSYS, Inc. | 2B-5<br>Study on modeling of aerodynamic sound by using both ANSYS and acoustic solver, WAON ..... N/A<br>Masaaki Mori<br>CYBERNET SYSTEMS CO.,LTD., Japan | 2C-5<br>Design of components for high temperatures and thermo-mechanical fatigue (TMF) ..... N/A<br>Vinay Carpenter<br>ANSYS India | 2D-5<br>HEV/EV full system design validation including components and circuit ..... N/A<br>Liu Yan<br>Chery New Energy Automotive Technology Co.Ltd., China | 2E-5<br>Highly Accurate and Robust Automotive Radar System Design Using a Combined Electromagnetic Field Solver/Circuit Simulator Design Methodology.<br>Markus Kopp<br>ANSYS, Inc. ..... N/A | 2F-5<br>System Modeling and Simulation:Trends and Enablers for Electrified Vehicle Systems ..... N/A<br>Kazuhiro Kawamura<br>ANSYS Japan K.K. |
| 16:45<br>\|<br>17:15 | 2A-6<br>ANSYS Customization Solutions .....313<br>Padmesh Mandloi<br>ANSYS India | 2B-6<br>ANSA-TGrid: A Common Platform for Automotive CFD Preprocessing ..... 365<br>Mohammad "Peyman" Davoudabadi, Ph.D.<br>ANSYS, Inc. | 2C-6<br>ANSYS Strategy for Internal Combustion Engine Simulations ..... 433<br>Ellen Meeks<br>ANSYS, Inc. | 2D-6<br>Improving mechanical reliability of PC Board/Package ..... N/A<br>Kazuhiro Matsuda<br>ANSYS Japan K.K. | | 2F-6<br>The Future of Embedded Software Validation ..... N/A<br>Casey Soileau<br>National Instruments Japan Corporation |

# Sponsor



## Special Sponsor

CYBERNET

In both the automotive industry and electronics industries, taking measures against aerodynamic and vibration noise is becoming essential to improve the product quality. In addition, the approach to the composite and polymeric materials has become an increasingly important issue in order to ensure the strength and to reduce the weight of the product. At Cybernet booth, we will introduce various case studies of Structural Analysis using ANSYS, Acoustic Analysis and Composite Solution.

## Platinum Sponsor

intel

Headquartered in Santa Clara, Calif., Intel designs and manufactures a variety of essential technologies including microprocessors, chipsets and the additional hardware, software and related services that together serve to accelerate your exploration and discovery. Intel and ANSYS have a long history of collaboration, resulting in ANSYS software solutions designed and optimized for high-performance computing on Intel architecture. ANSYS and Intel software developers continue to work closely to tune and optimize ANSYS codes using the Intel Software Development Tools Suite to enable you to affordably move your ideas from concept to production faster than ever before.

## Gold Sponsors

GOMPUTE
A GRIDCORE COMPANY

Gompute delivers comprehensive solutions for High Performance Computing, in-house, in-the-cloud or both. With over 10 years'of experience with the ANSYS product portfolio, providing solutions and services to the IT and Engineering communities, we deliver a collaborative and highly productive work environment for geographically distributed engineering teams. Combined with Gompute On-Demand, we ensure that you have the option to either scale out or completely host your HPC environment, ranging from 1 -> 1000's of cores in a Gompute owned Super Computing Centre. To learn more on how we work with HPC, visit www.gompute.com.

sharc

Sharc develops and provides revolutionary meshing and post-processing software for the most extreme CFD workflows. Using advanced proprietary data formats, innovative rendering engines and unique meshing algorithms, Sharc is able to reduce time from CAD to visualization by orders of magnitude using the extreme meshing software, Harpoon and the industry leading visualization suite of software Periscope. Our technical staff provide customization and tailoring of the software to customer specific requirements, streamlining the engineering process even further.

## Sliver Sponsors

NVIDIA. hp | SCSK The Partner for Success ZUKEN. FUJITSU sgi

## Exhibitors

Empowered by Innovation NEC

BETA CAE Systems SA

ESTECO

Automotive Simulation World Congress
October 9-10, 2014

# TOWARDS A PREDICTIVE AND COMPREHENSIVE ENGINE SIMULATION METHODOLOGY.

Dirk Linse

BMW
GROUP

# IN-CYLINDER FLOW SIMULATION OF A BMW FOUR-CYLINDER GASOLINE ENGINE.

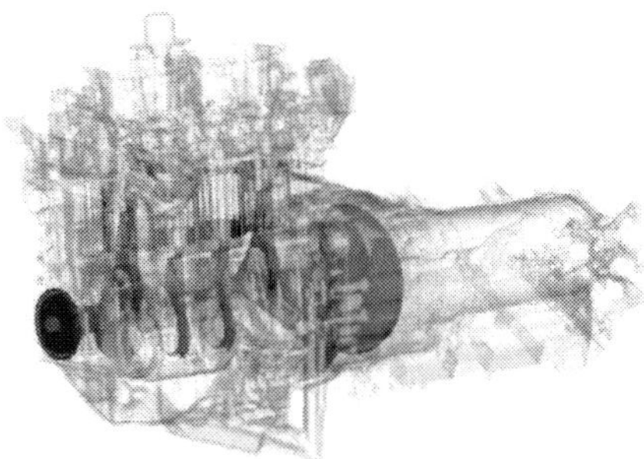

Page 2

# OVERVIEW OF THE BMW IC-ENGINE WORKFLOW USING AN ADVANCED MESHING STRATEGY.

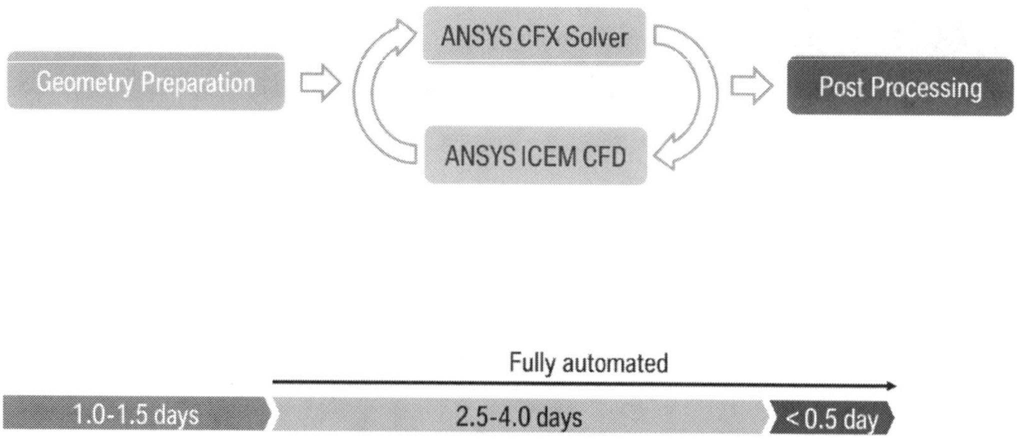


Page 3

# WE USE MESHES THAT ARE COMPRISED OF TETRAEDER, PRISM, AND HEXAEDER ELEMENTS.

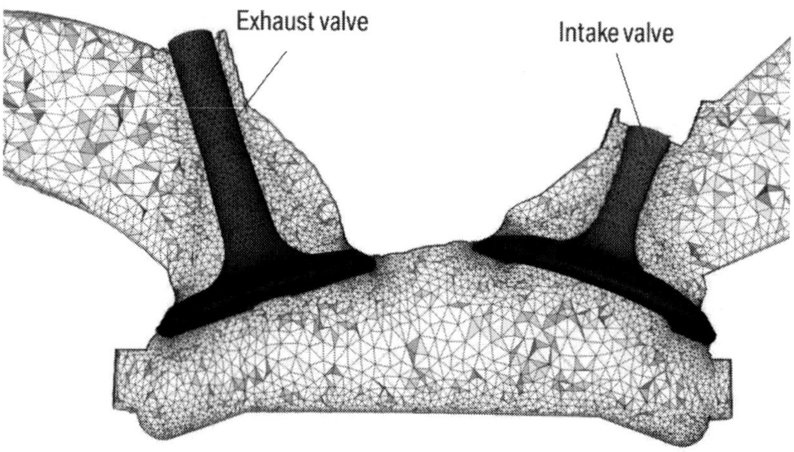


Page 4

## CONTINUOUS BOUNDARY LAYERS AND LOCAL REMESHING INCREASE ACCURACY AND ROBUSTNESS.

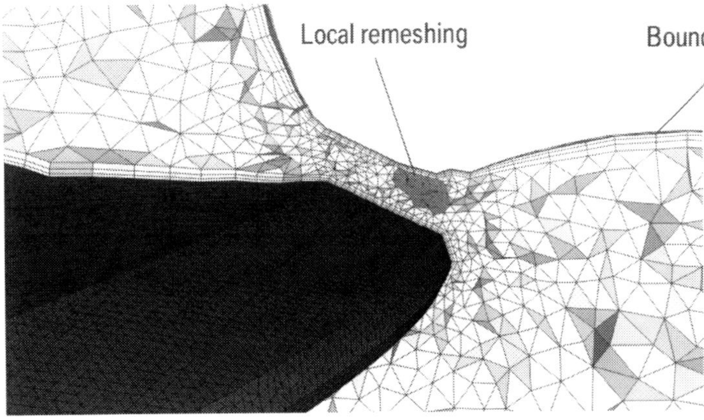


Page 5

## FOR SMALL VALVE LIFTS A HEXAEDER MESH IS USED TO RESOLVE THE INTAKE VALVE GAP.

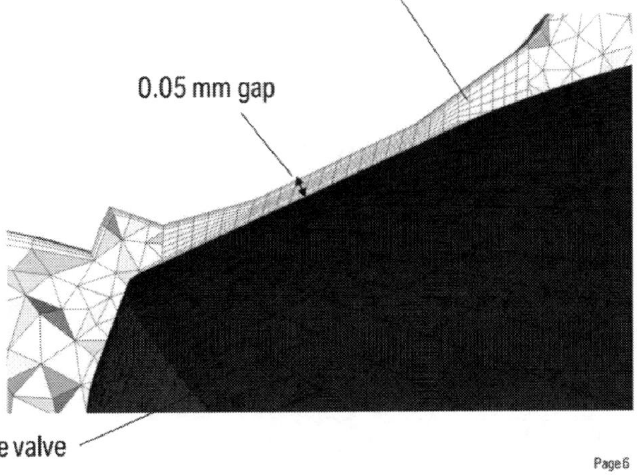


Page 6

## INTAKE VALVE GAP IS SMALLER THAN THE THICKNESS OF A HAIR.

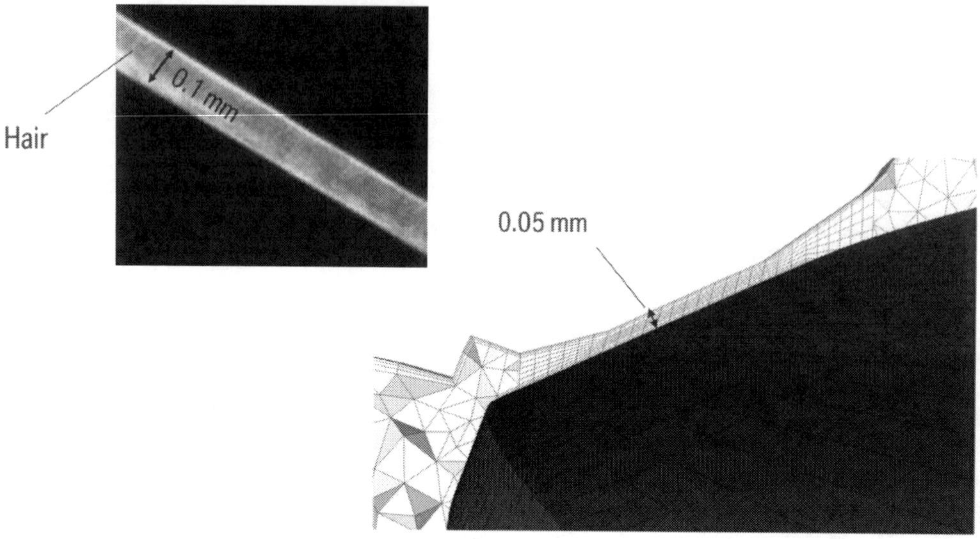


Page 7

## VARIOUS COMPLEX PHENOMENA NEED TO BE CONSIDERED FOR IN-CYLINDER FLOW SIMULATIONS.

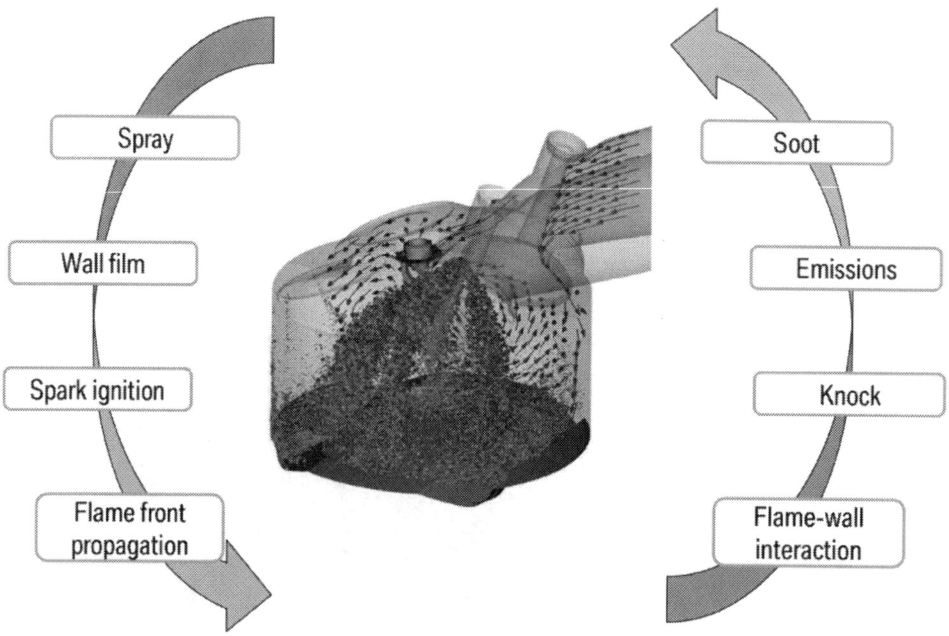


Page 8

## SOOTING DIFFUSION FLAMES DUE TO INCREASED FUEL-WETTING OF THE PISTON.

Intake port side

Exhaust port side

Page 9

We need predictive models!

Page 10

## SPARK IGNITION MODEL TAKES INTO ACCOUNT SPARK DYNAMICS, CHEMISTRY, AND DIFFUSION EFFECTS.

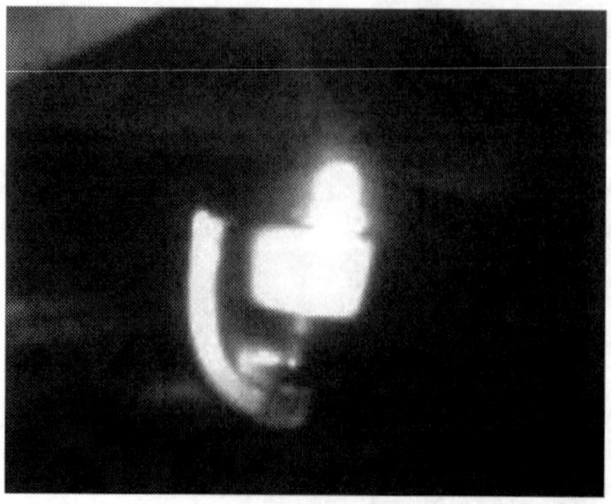

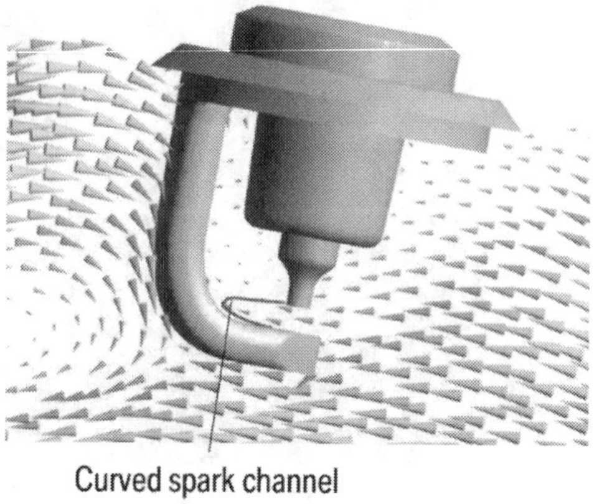


Page 11

## STATISTICAL KNOCK MODEL PROVIDES THE LOCATION OF AUTO-IGNITION AND ITS PROBABILITY.

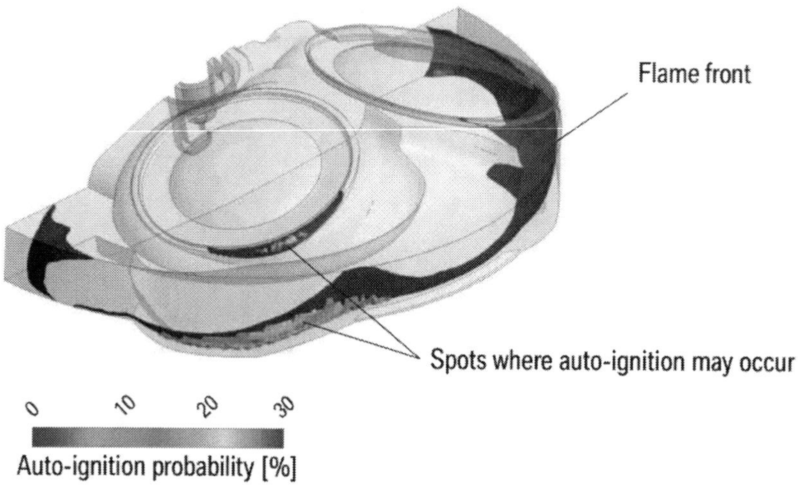


Page 12

Combine simulation and experiment!

Page 13

## VISUALIZATION OF A PRE-IGNITION EVENT USING HIGH-SPEED ENDOSCOPY.

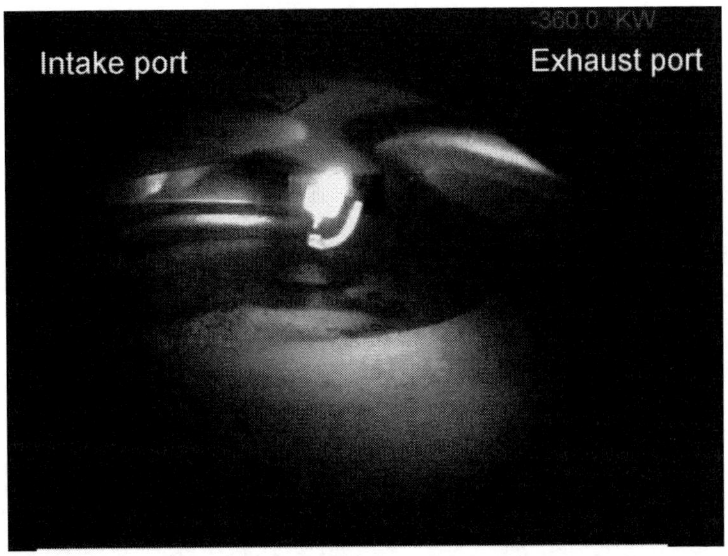


Page 14

Predict with confidence!

Page 15

# CURRENT AND FUTURE APPLICATIONS FOR THE SIMULATION OF INTERNAL COMBUSTION ENGINES.

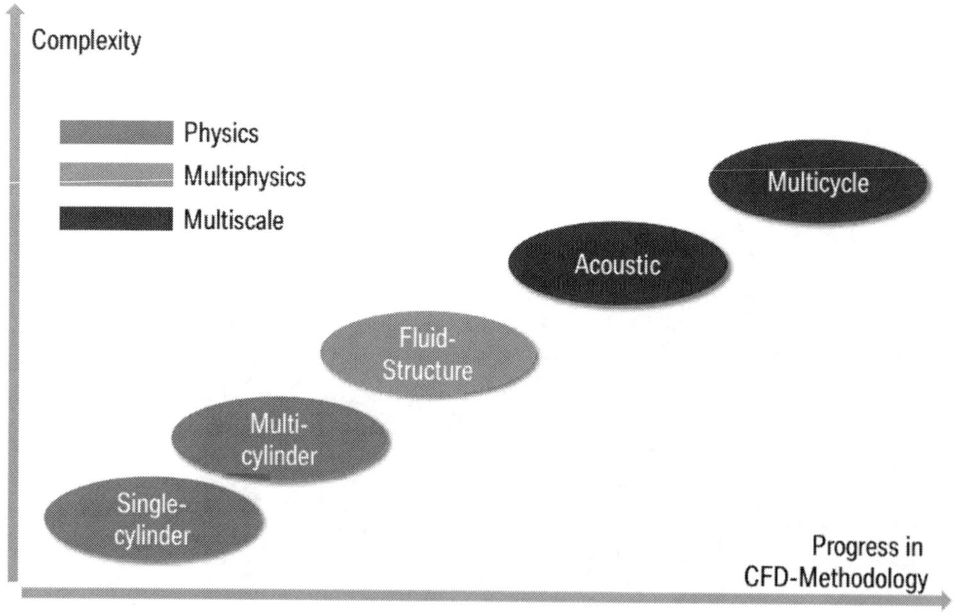


Page 16

# CURRENT AND FUTURE APPLICATIONS FOR THE SIMULATION OF INTERNAL COMBUSTION ENGINES.

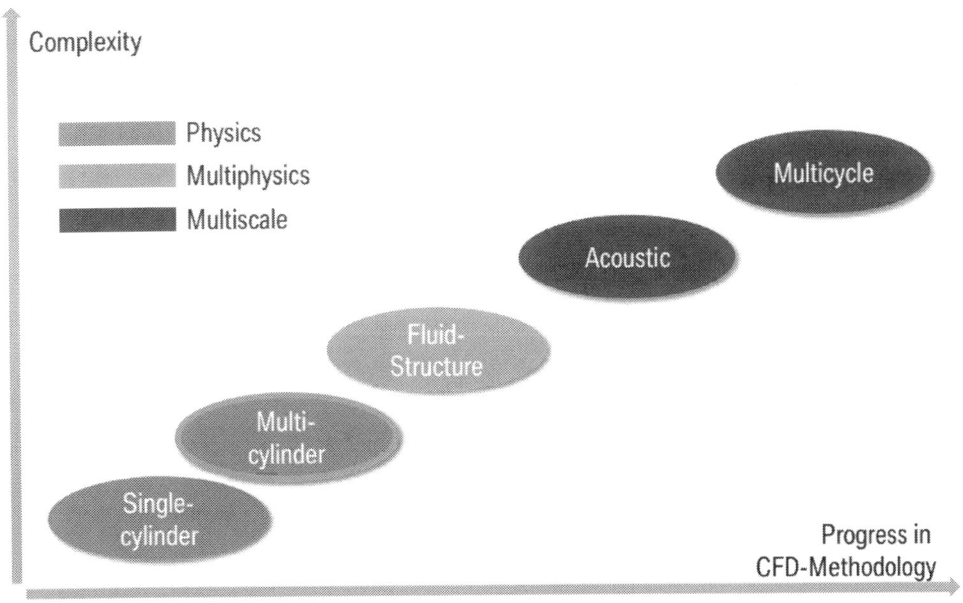


Page 17

# BMW TURBOCHARGED 3-CYLINDER ENGINE. SINGLE HOUSING FOR MANIFOLD AND TURBINE.

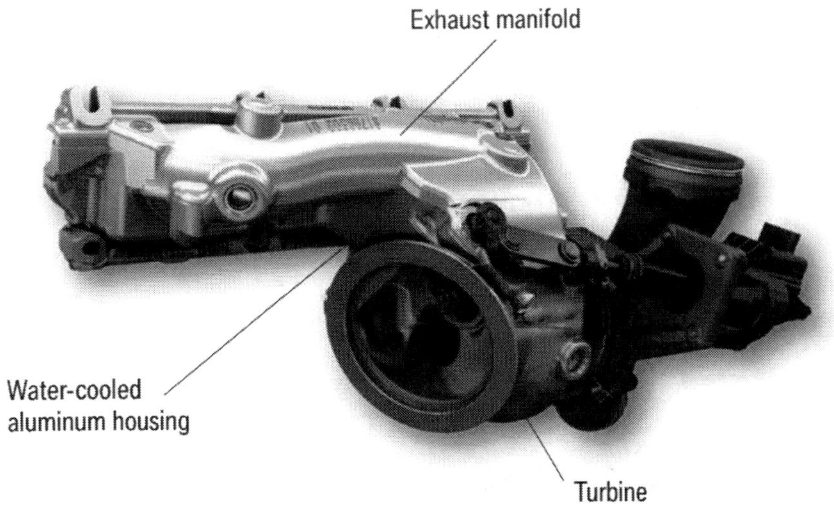


Page 18

## CURRENT AND FUTURE APPLICATIONS FOR THE SIMULATION OF INTERNAL COMBUSTION ENGINES.

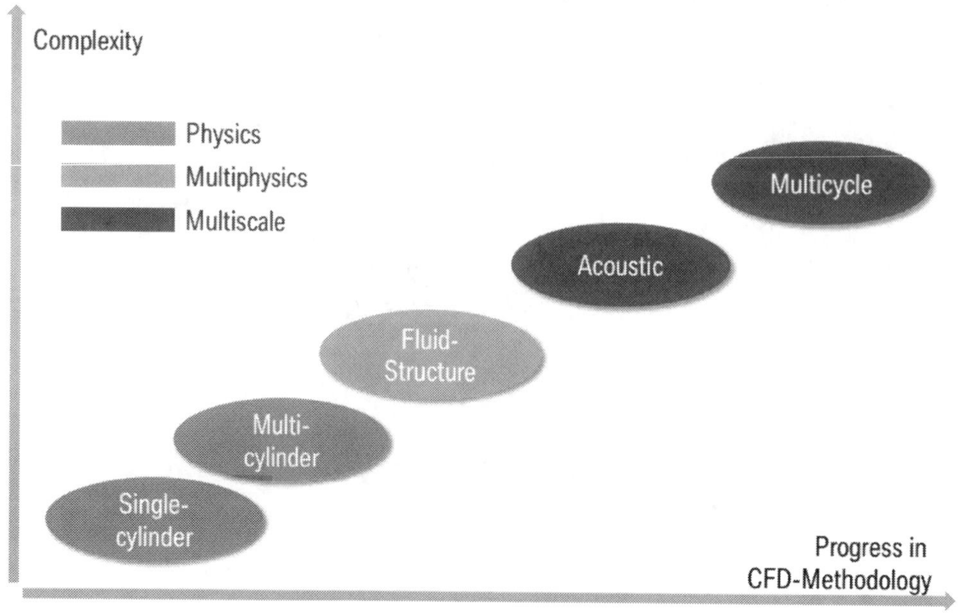


Page 19

## WASTEGATE PLATE IS SUBJECT TO HIGH THERMAL AND DYNAMICAL STRESSES AND MAY START TO OSCILLATE.

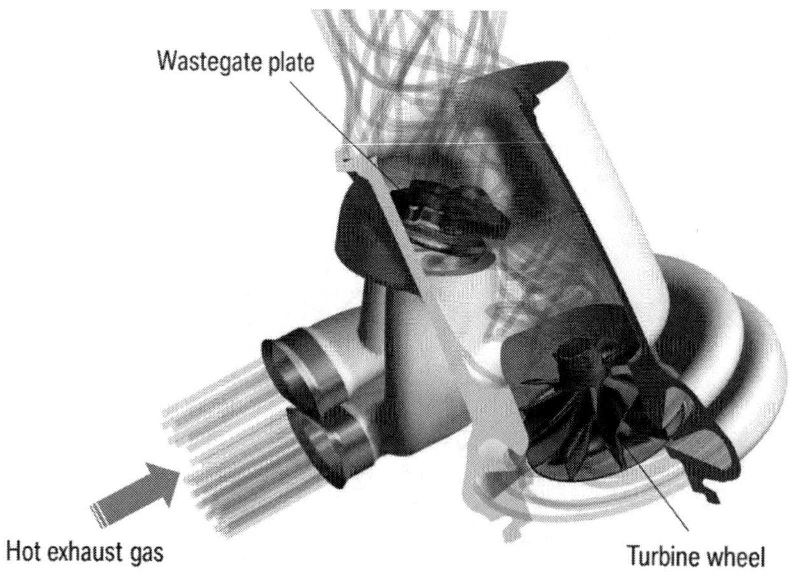


Page 20

## SIMULATION OF A RATTLING WASTEGATE PLATE USING 2-WAY FLUID-STRUCTURE INTERACTION.

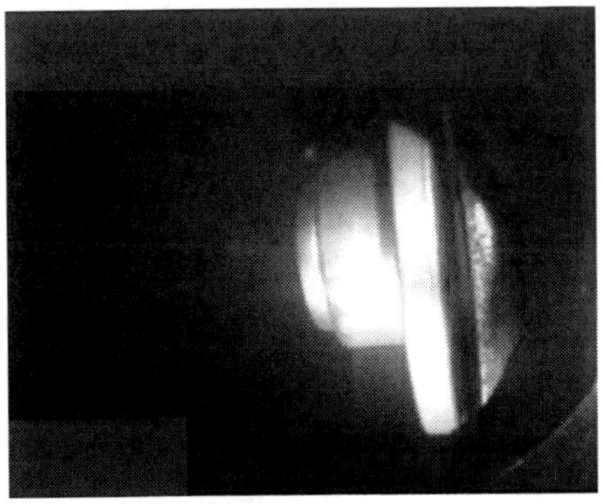

Experiment

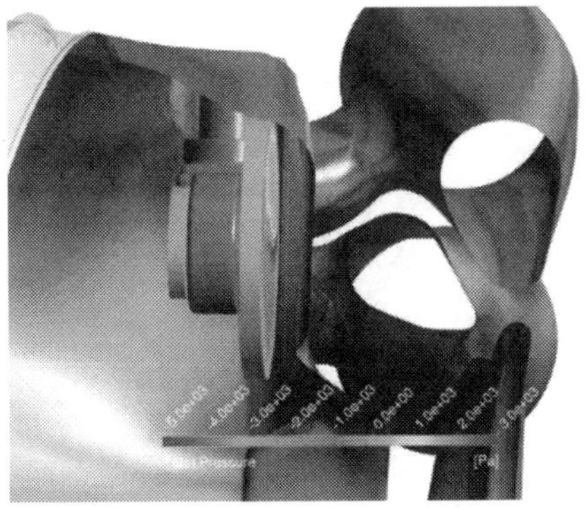

Simulation

Page 21

## CURRENT AND FUTURE APPLICATIONS FOR THE SIMULATION OF INTERNAL COMBUSTION ENGINES.

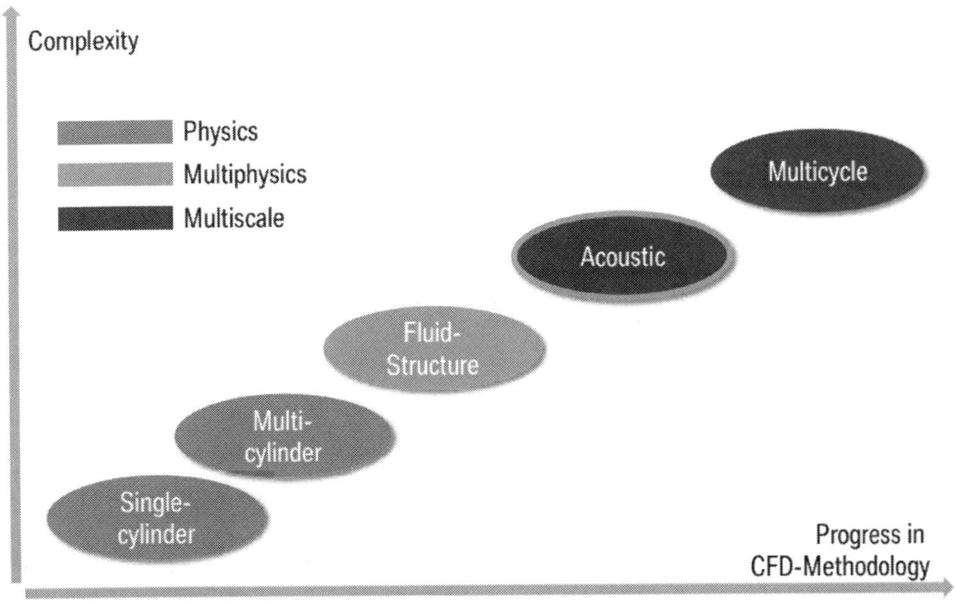


Page 22

# SIMULATION OF JET NOISE OF EXHAUST SYSTEMS USING SCALE-RESOLVING TURBULENCE MODELS.

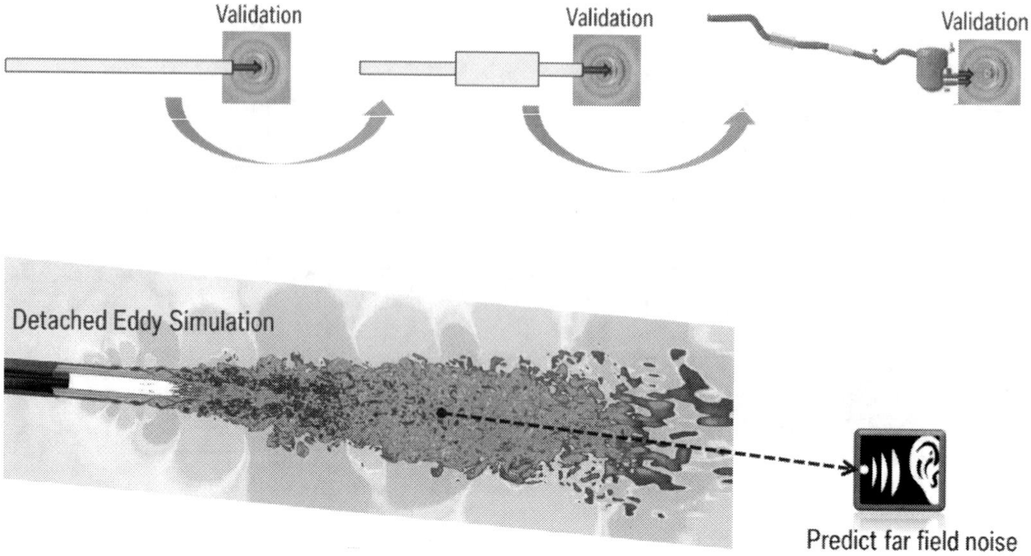


Page 23

# CURRENT AND FUTURE APPLICATIONS FOR THE SIMULATION OF INTERNAL COMBUSTION ENGINES.

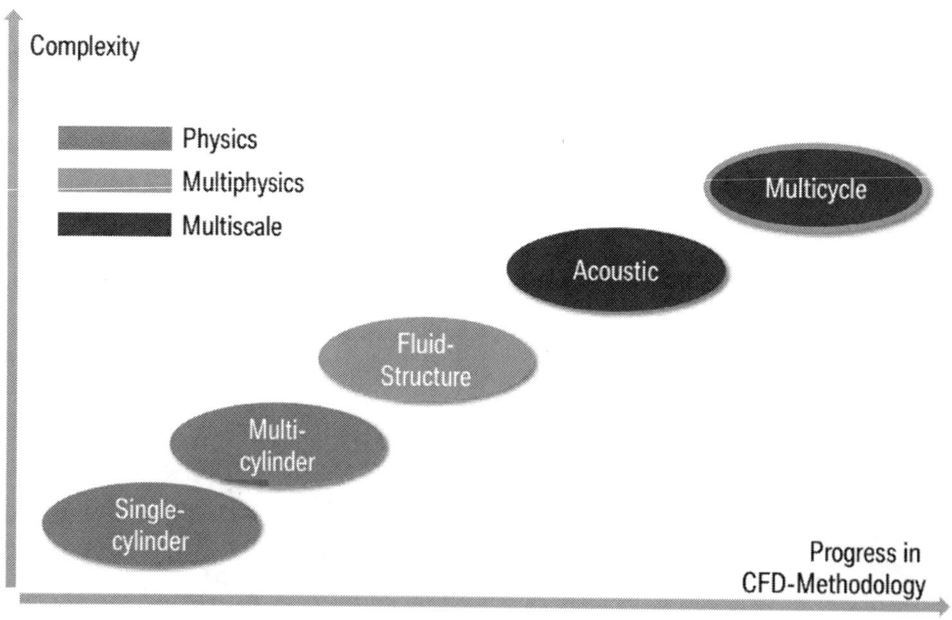


Page 24

## LARGE-EDDY-SIMULATION OF IN-CYLINDER FLOW PHENOMENA IS STILL AN ACTIVE AREA OF RESEARCH.

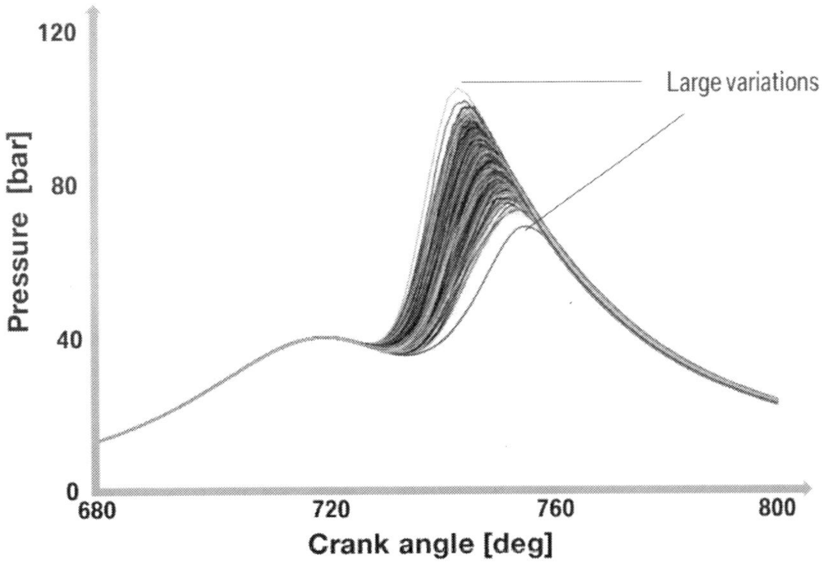


Page 25

## CURRENT AND FUTURE APPLICATIONS FOR THE SIMULATION OF INTERNAL COMBUSTION ENGINES.

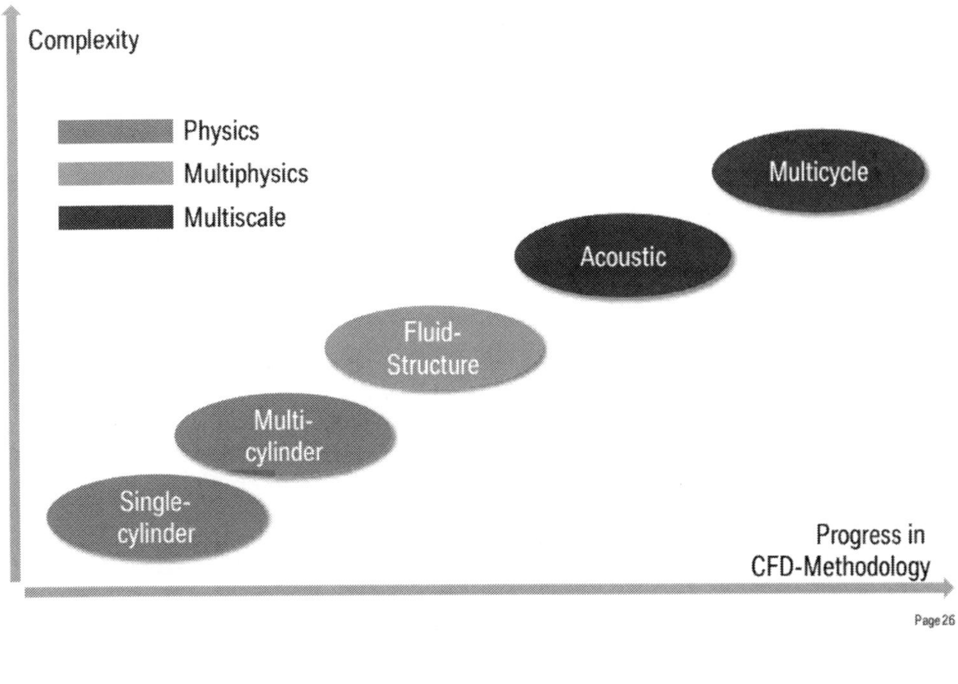


Page 26

Push it to the limit!

Page 27

# CAE Utilization and Promotion in DENSO — Past, Present and Future

1. Corporate profile
2. CAE utilization and promotion
   1) Expand CAE
   2) Enhance CAE
   3) Embed CAE
3. Summary

Shigeru AKAIKE
Digital Engineering Dept., DENSO CORPORATION




## 1-1. Corporate Profile

| | |
|---|---|
| **Established** | December 16, 1949 |
| **Capital** | 187.4 billion yen (US$1.8 billion) |
| **Net sales** | |
| Consolidated basis | 4,095.9 billion yen (US$39.8 billion) |
| Non-consolidated basis | 2,490.8 billion yen (US$24.2 billion) |
| **Net income** | |
| Consolidated basis | 287.4 billion yen (US$ 2.8 billion) |
| Non-consolidated basis | 219.7 billion yen (US$ 2.1 billion) |
| **Employees** | |
| Consolidated basis | 139,842 |
| Non-consolidated basis | 38,581 |
| **Consolidated subsidiaries** (Japan 62, North America 26, Europe 35, Asia/Oceania 56, South America/Others 6) | 185 |
| **Affiliates under the equity method** (Japan 13, North America 4, Europe 3, Asia/Oceania 11, South America/Others 2) | 33 |

*There are two non-consolidated subsidiaries.

Notes:
U.S. dollar amounts have been translated, for convenience only, at the rate of 102.92 yen = US$1,
the approximate exchange rate prevailing on March 31, 2014. Billion is used in the American sense of one thousand million.

/ as of March 31, 2014

## 1-2. Automotive Products



### Environment

Hybrid and electric vehicle components,
gasoline engine management system,
diesel engine management system,
starter, alternator, radiator, etc.

### Safety

Sensing technologies for driving assist systems,
actuator & computers for antilock brake system (ABS) /
electronic stability control (ESC),
adaptive front-lighting system (AFS),
airbag sensors & electronic control units,
periphery monitoring system, instrument cluster,
rain sensor for automatic windshield wiper, etc.

### Comfort & Convenience

Car air-conditioning system,
air conditioner for buses, air purifiers,

Car navigation system,
electronic toll collection system (ETC),
remote security system,
remote touch controller, smart key,
advanced vehicle operation system (AVOS), etc.

## 1-3. Organization



Board of Directors

President & CEO
Executive Vice Presidents

| | |
|---|---|
| | Audit |
| | Business Innovation |
| | New Business Promotion |
| Corporate Center | Corporate Planning, Accounting, Corporate Communications, Legal, IT, Secretarial, General Admin., Human Resources, Health Care Promotion, External Affairs |
| Engineering Research & Development Center | Technology Planning, Engineering Development Promotion, Intellectual Property, Design, R&D, DE |
| Production Promotion Center | Quality Control, Safety & Environment, Production Engineering, Materials Engineering, Machinery & Tools, Prototype Manufacturing, Construction, Production Control, Logistics |
| Purchasing Group | Purchase |
| Sales and Marketing Group | Original Equipment Manufacturers Sales, Global Business Planning, Technical Service |
| Powertrain Control Systems Business Group | Please refer to the next page |
| Electronic Systems Business Group | |
| Thermal Systems Business Group | |
| Information & Safety Systems Business Group | |
| Aftermarket Business Unit | Development and sales of aftermarket products |

/ as of January. 1, 2014

## 1-4. Business Group System



**DENSO is organized into four business groups to strengthen global management.**

| | |
|---|---|
| Powertrain Control Systems Business Group | Development and production of gasoline and diesel engine control systems and related products, hybrid and electric car products, powertrain products, and power supply and starting system parts such as alternators and starters |
| Electronic Systems Business Group | Development and production of semiconductor sensors, microelectronic devices including ICs (integrated circuits), and electronic products including engine control computers |
| Thermal Systems Business Group | Development and production of air conditioning systems for cars and buses, truck refrigeration units, air purifiers and related air conditioning products, radiators, and cooling systems |
| Information & Safety Systems Business Group | Development and production of car navigation systems, ITS (intelligent transport system) products such as electronic toll collection, telematics, driving safety products including sensor computers for airbags, body ECUs (engine control units), and meters |

| | |
|---|---|
| **Goals** | Strengthen links among projects for greater systemization and mobility.<br>Encourage rapid decision-making.<br>Reinforce management in order to link projects from domestic and overseas group companies. |




## 2. CAE Promotion in DENSO



An EVP of DENSO comments:

- Previously, CAE was a tool for limited groups of specialists.
- But it has gradually been promoted as
  1) Presentation tool for sales expansion ⇨ **Expand**
  2) Engineering tool for new idea breeding ⇨ **Enhance**
  3) Alternative tool to physical prototyping, though it is not yet perfect. ⇨ **Embed**

# Definition of a CAE Engineer

| Simulator Utilization | Modelling know-how/experience | Theoretical background |
|---|---|---|
| Use of a single engineering product | | |
| Customization for a single physics | Less | Concept of CAE tech. |
| Customization for single/multiple physics | | Theory of CAE tech. |
| Development for single/multiple physics | More | Theory of the latest CAE tech. |

CAE Expert

CAE Specialist






## CAE promotion target as of Jan. '97

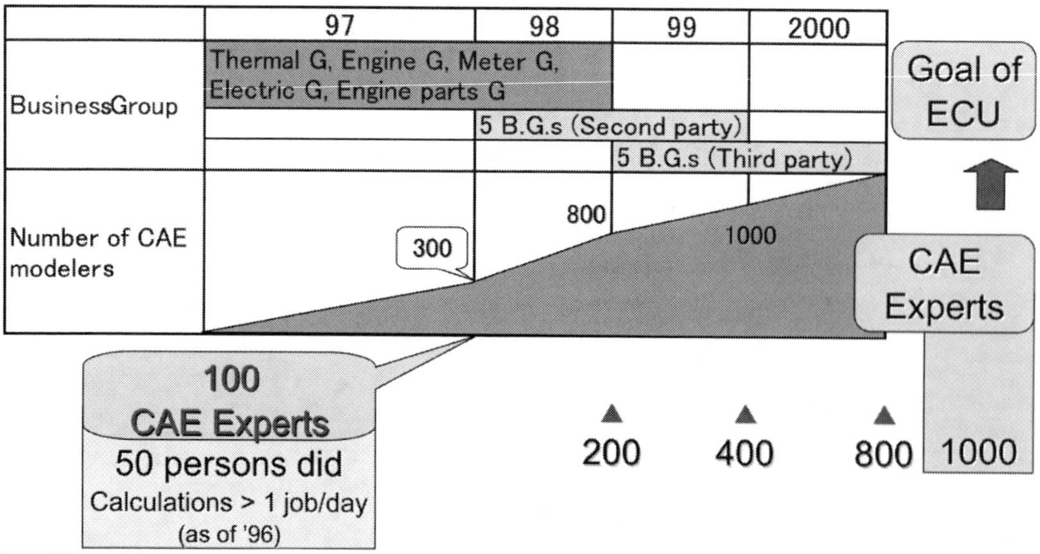

## 2-1. Expand CAE in DENSO (Organization for CAE Promotion)



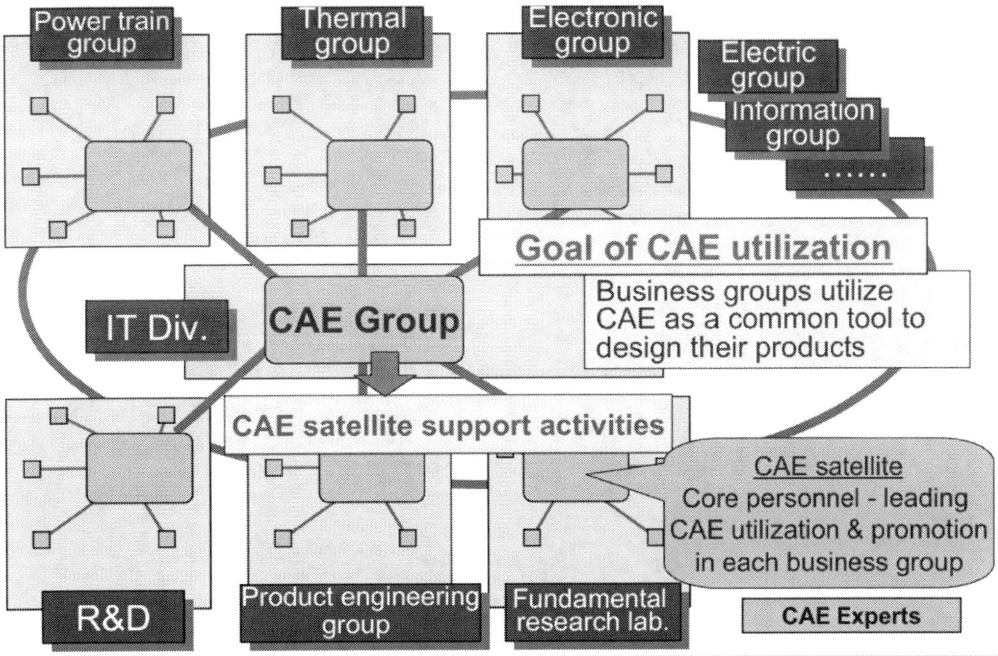





## 2-2. Enhance CAE in DENSO (Adopted Software)



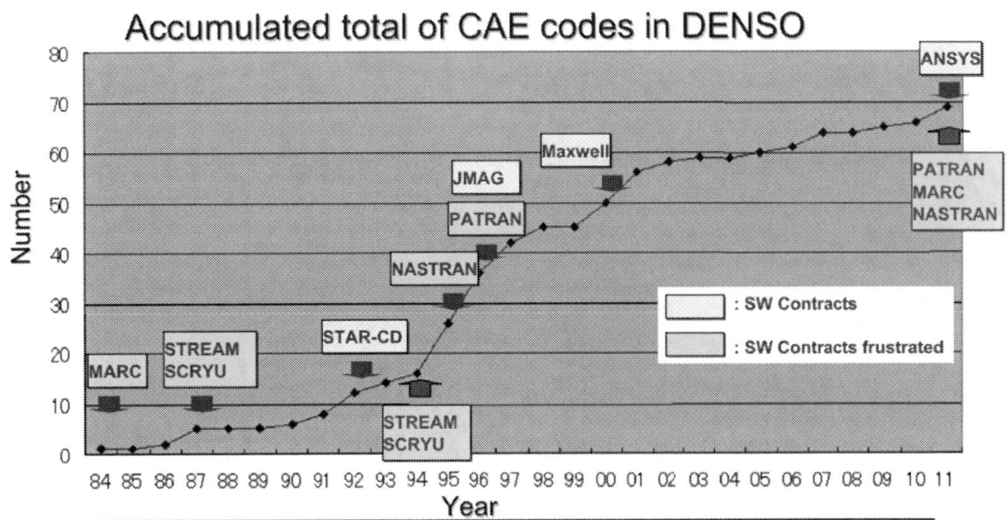


**(i) Instead of developing in-house CAE codes, we adopt commercial codes and improve them internally.**

**(ii) We adopted many CAE codes in the 1990s (probably in pace with many new developments by software venders).**

## 2-2. Enhance CAE in DENSO (Problems & Challenges)



**What is the problem?**

Investment depends on the corporate performance. CAE investment is especially affected by the economic trends because CAE is treated as R&D expenditure/investment in DENSO.

➢When the corporate income increases:

We adopt the most suitable CAE code for our requirements, which has lead to the present situation with many kinds of software and thousands of licenses in DENSO.

➢When the corporate income decreases:

A good opportunity to cut down on software cost, because everyone accepts reduction/adjustment of software licenses

1) Need to merge similar software codes
   → Benchmarking to identify their basic functionalities
2) Need to replace simulators with more effective ones
   → Log analysis for the operation rate of each code

**Challenge we faced:**

We had two core strength analysis codes. We started benchmarking to compare their functionalities.




## 2-2. Enhance CAE in DENSO (Benchmarking)



**Request sent to ANSYS top management (Nov. 2, 2009)**

*Our relationship & expectation*

**Let me express my thanks to everyone from you and your Japanese partner**, for their collaboration with us.

This year we, together with you and your Japanese partner, have challenged to reduce the total CAE-license costs, by merging structural-CAE softwares into yours.

**We will continue our try to merge softwares into your products,** especially for multi-physics analyses, such as flow-vibration, structure-electric magnetic, etc. **We expect your products to gain more ability** in those areas.

**We expect to keep good relationship with you and your Japanese partner** to develop new CAE function, in order to enhance the product development capability at DENSO

CEO Mr. C

**Request sent to Company M top management (Oct 23, 2009)**

**Our relationship & expectation**

**Let me express my thanks to Mr. David and everyone from your company,** for their collaboration with us .

This year we, together with you, have challenged to reduce the total CAE-license costs, by merging structural-CAE softwares into your products. **We have switched some licenses into your products.** I appreciate you supports to the project.

**We will continue our try to merge softwares into your products,** except for multi-physics analyses, such as flow-vibration, structure-electric magnetic, etc. **We expect your products to gain more ability** in those areas, in order to replace other products.

**We expect to keep good relationship with you** to develop new CAE function, in order to enhance the product development capability at DENSO

CEO Mr. D

## 2-2. Enhance CAE in DENSO (Benchmark Results)



- Company M could not accommodate our requests

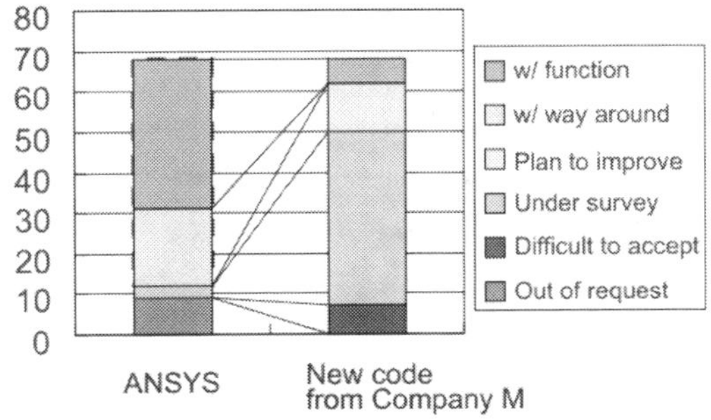


Answers from ANSYS & Company M

- Large part of users requested adoption of ANSYS

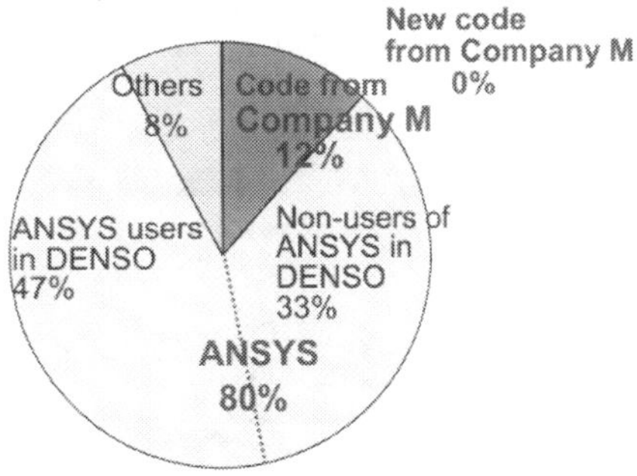


Questionnaire results from the users

**We selected ANSYS as the core strength analysis code in 2011**

Hey now!!
Support of ANSYS is OK or not ?

⇦ **Reliable information from my human network**




## 2-2. Enhance CAE in DENSO (Summary)



- **Functionality: Periodic benchmarking,**
  **not only before, but also after installation**
  ANSYS ⇔ NASTRAN, MARC, ABAQUS
  STAR-CCM+ ⇔ Fluent, IconCFD, SCRYU/TETRA
  JMAG ⇔ Maxwell
  DUFUL ⇔ RecurDyn, ADAMS
  VA-One ⇔ Seam3D
  :
- **Support: Reliable information from my human network**
  in one's own company
  in one's company group
  in the business group
  in the industry
  :
  MRB (Management Review Board), TRB
- **Cost**
- **Contract (global)**

# 2-3. Strategy to Embed CAE (Road Map Activities)



**Technology management strategies**

**1st layer: Society/regulations**
Social trends, restrictions, and automobile industry

*Expected period: until 2030*

**1.5th layer: Automobiles & automotive society**
Future vehicles and future society (customer value)

**2nd layer: Systems**
System functionalities needed in the future

*Adjustment*

**3rd layer : Products**
DENSO products that can meet the needs of the future

*Adjustment*

**4th layer : Technology**
Elemental technologies and core technologies needed to realize the products

Adjustment

Agreeing on what kind of "customer value" we will provide to contribute to the automotive society.

**Company-wide cross-functional road map members (in the country and from overseas)**

**Discussion with carmakers and research institutions**
**Creation of region-specific road maps**

# 2-3. Strategy to Embed CAE (DENSO CAE Trends and Direction)



## Trends

1. EV/HV analysis needs will increase.
2. Multiphysics analysis needs will increase.
   Kinematics-vibration-noise analysis will be especially required to address environmental problems (turbomachinery, compressors, belt-drives, etc.).
3. Material design CAE has to be bolstered against China risk (for example, rare earths & metals).

## Direction : Tackle challenges to accomplish the World-First (CAE)

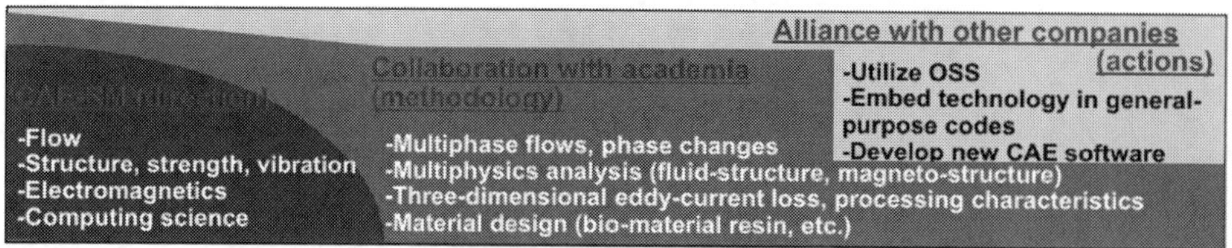

## 2-3. Strategy to Embed CAE
### (Collaboration & Alliance Significantly Improves CAE Competence)



**Industry-Academia**

Reflect basic theories into an effective phase

Analyses at universities (items to be embedded):
- Flow by the particle method
- High-precision electromagnetic field
- Polymeric heat-transfer characteristics
- Metal-to-metal joints
- Magnetic particle compression

**Industries (BP)**

Embed the theories/methods into CAE codes
- Prometech
- ANSYS
- JSOL
- CD Adapco
- SGI

Awareness to academic works

Collaboration

Compare theories / calculations with phenomena
- Testing
- Grasping phenomena
- Reviewing theories

CAE Dept.
BG, R&D
Prod. Promotion

DENSO

Increased presence of DENSO CAE

Alliance

**Demonstrate distinguished CAE competence by embedding the world-first technology into software**




## 2-3. Strategy to Embed CAE
### (CAE Global Organization in DENSO)



**Digital Engineering Dept.**

Cooperation
- Strategic technology promotion
- New technology development
- Technology development
- Deployment & support

Enhance CAE technology with business groups by solving problems in development of typical products

**B.G. CAE Team**

Develop product-specific CAE simulators for engineers

Engineer

Utilize CAE simulators in the design process

Number of personnel (plan)

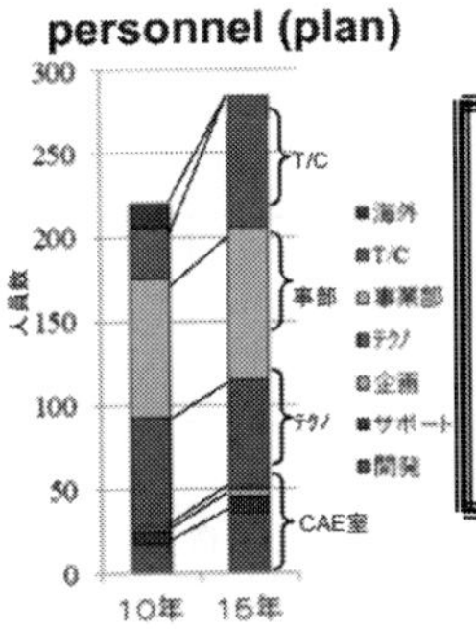


Entrustment

**TECHNO**

Strengthen cooperation

TECHNO CAE engineers develop their ability at DE Dept. (starting Apr. 2011)

Education
Increased efficiency

Commission

**O S T/C**

Train engineers from overseas T/Cs at HQs to enable joint development of CAE technology in the near future

**Develop company-wide CAE competence**

## 2-3. Strategy to Embed CAE
### (Current CAE Responsibilities in Global DENSO)



| | DE Dept. | B.G. CAE Team | DENSO TECHNO | DMVN |
|---|---|---|---|---|
| Technology strategic planning | ◎ | | | |
| Advanced technology R&D | ◎ | | | |
| Technology development | ◎ | ○ | | |
| Deployment & support | ◎ | ○ | | |
| Application analysis | | ○ | Dom. ◎ | |
| Repeat analysis | | | Dom. ◎ | PT ◎ |

Added-value ↑

← Urgency, difficulty

1. **DENSO TECHNO is mainly responsible for domestic application & repeat analyses.**
2. **DENSO MANUFACTURING VIETNAM (DMVN) is responsible for repeat analysis from DENSO Japan Power train B.G.**




## 2-3. Strategy to Embed CAE
### (Future CAE Responsibilities in global DENSO)



| | DE Dept. | B.G. CAE Team | DENSO TECHNO | DMVN |
|---|---|---|---|---|
| Technology strategic planning | ◎ | | | |
| Advanced technology R&D | ◎ | | | |
| Technology development | ◎ | ○ | → ○ | ◌ |
| Deployment & support | → ○ | ○ | ◎ | |
| Application analysis | | ○ | Dom. ◎ | → OS ◎ |
| Repeat analysis | | | ◎ → ○ | ◎ |

Added-value ↑

← Urgency, difficulty

1. **DENSO TECHNO will take more responsibilities for higher-level CAE tasks.**
2. **DMVN will address global DENSO's application & repeat analyses.**
3. **DMVN will collaborate with DE dept. to develop CAE Tech.**

## 3. Summary



- We have been promoting CAE in DENSO focusing on three points:
  1st Expanding
  2nd Enhancing
  3rd Embedding

- We have come to leverage CAE in place of physical prototyping (though it is not yet perfect).
- To leverage CAE as a perfect alterative to physical prototyping, we will have to attack multiphysics analysis vigorously.
- To further develop CAE competence, we will have to strengthen the global CAE (DE) network.

- We would like to maintain our strong Strategic Partnership with ANSYS, Inc.
  - by sharing our visions and goals
  - through on-going and future joint projects for the world-first CAE
- We have appreciated the support and proposals by ANSYS.
  Proposals beyond our expectations would be greatly welcome!




## Closing Remarks



# My mottos

- **People who work hard can achieve almost anything.**
- **And those people who try hard will impress and delight others**
- **But those who persevere will truly gain the support from others**

**Meanwhile,**
**You might think hard work would realize your dream.**
**That is reverie and arrogance.**
**Taichi YAMADA, a Japanese screenwriter (Nikkei ASSOCIE, February 19th 2008)**

**If you climb higher, you can see the horizon!!**

DENSO

**Thank You**

DENSO ANSYS Automotive Simulation World Congress 2014

2014 **Automotive Simulation** World Congress

2014 **ANSYS Electronics** Simulation Expo

# Addressing Complexity with Comprehensive Simulation

Dr. S. Subbiah
V.P. of Global Product Operations
ANSYS, USA

---

## Automotive and electronics industries are increasingly intertwined due to these trends



**Dramatically increasing Software & Electronics content**

- 10-100 M software lines of code in modern vehicles
- The share of electronics in total vehicle cost is projected to grow to 40% in 2015

**Fuel Efficiency & Emissions Reduction**

- Hybrid/Electric Vehicles, and Fuel Cell
- Optimizing of Engines, Aerodynamics, etc with smart controls

**Power Efficiency, Miniaturization & Integration**

- Performance-per-watt becoming a design-in decision
- Mixed-signal content

**Proliferation of Wirelessly Connected Devices**

- Sensor loaded products ushering in the Internet of Things
- 11 Terabytes of mobile data per month by 2017

## Active Safety

2014 Automotive Simulation World Congress
October 9-10 . Tokyo, Japan

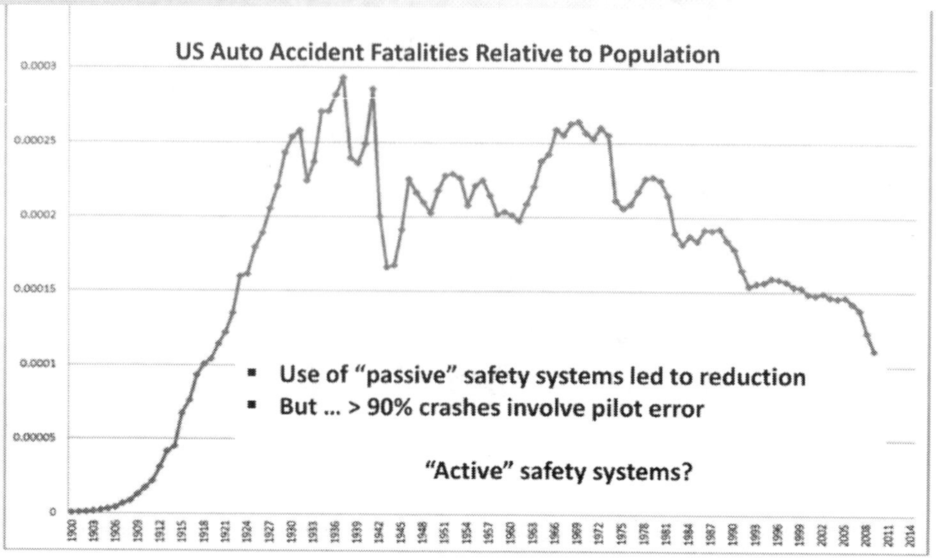


## Connectivity

2014 Automotive Simulation World Congress
October 9-10 . Tokyo, Japan

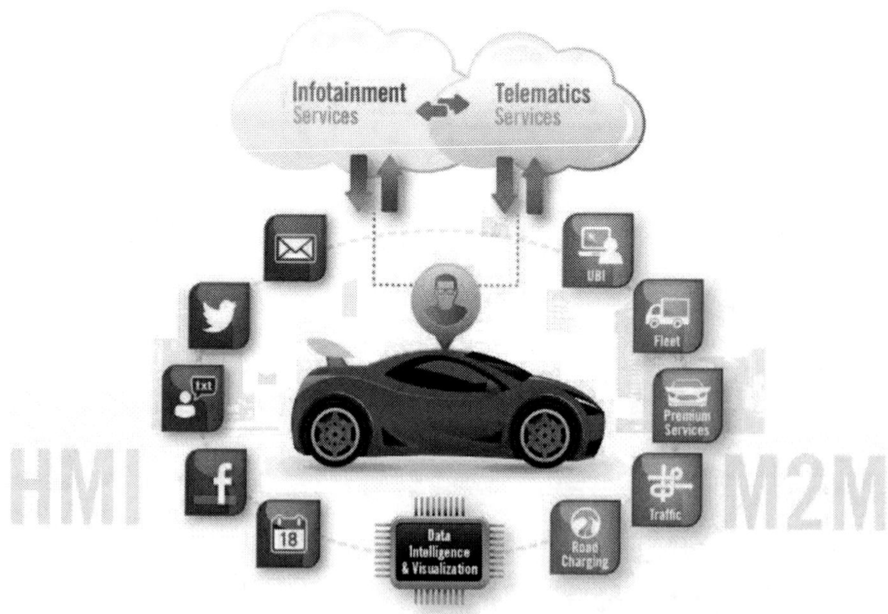

## Connectivity
### System and Component Design Challenges



**Interference Avoidance**

System Design:

- EMI / EMC Compliance
- Design, placement and immunity

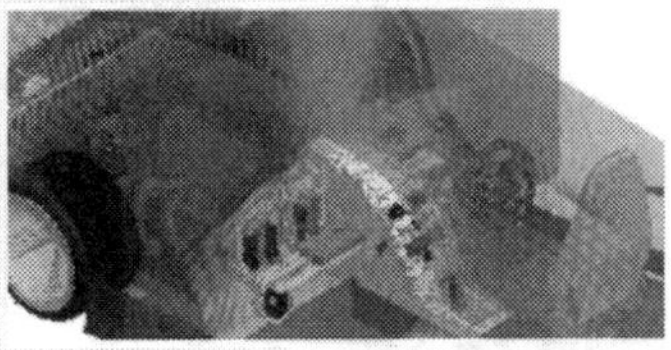

Component Design:

- Feature / Mixed-signal integration
- Noise prediction, isolation planning

Single chip with multipl analog and digital functic

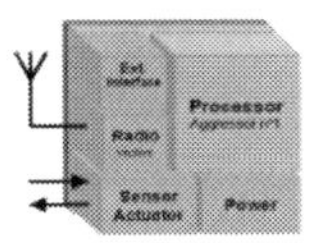


SoC Architecture

Noise coupling → Spikes in FM Spectrum impacts audio quality and performance

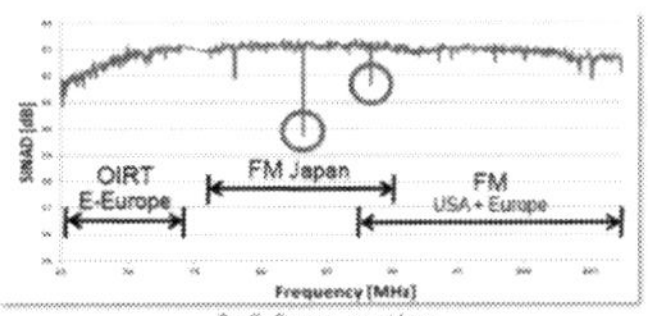


SoC Cross-section

---

2014 Automotive Simulation World Congress
October 9-10 . Tokyo, Japan

## Comprehensive simulation that considers all aspects of a system is necessary to tackle complexity

**IC Engine and Control System**

- Controller Software
- Network and Communication
- Chip and Board Design
- Air Flow
- Combustion
- Coolant Flow and Boiling
- Conjugate Heat Transfer
- Structural Strength
- Thermal Structural Durability
- Electromagnetic Sensors
- Electromagnetic Starter/Alternator

Reference: Automotive Sensor Technology, Book by John Turner

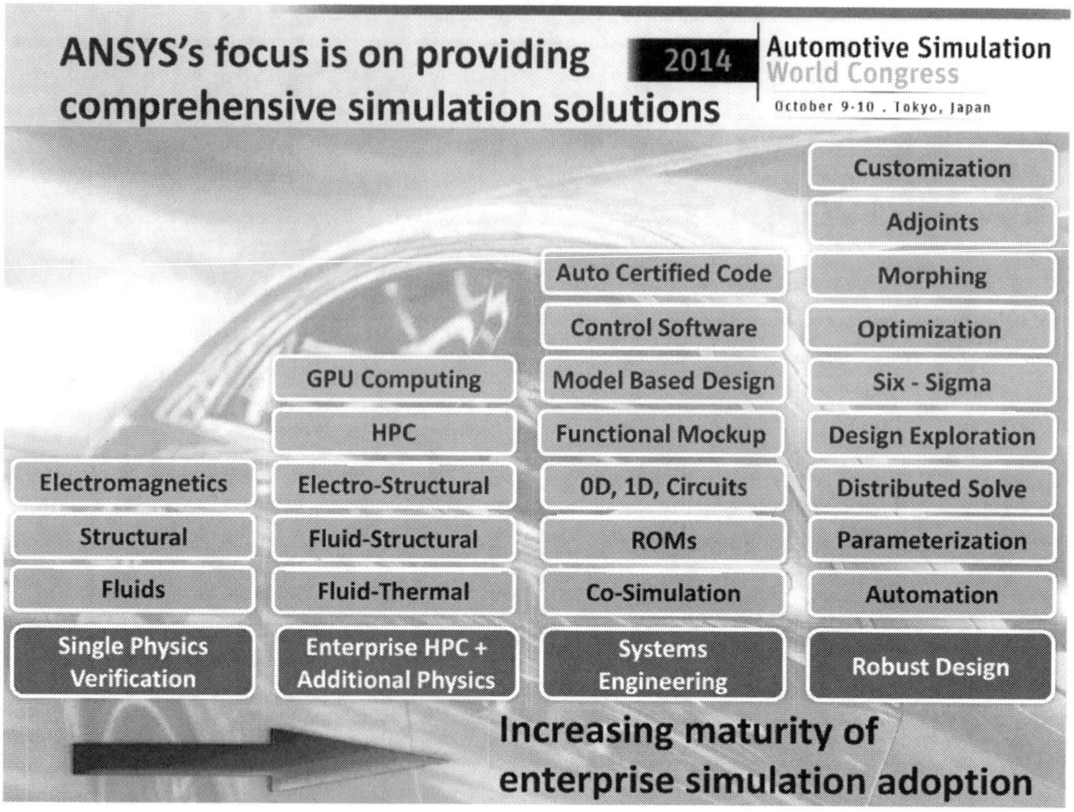
ANSYS's focus is on providing comprehensive simulation solutions
2014
Automotive Simulation World Congress
October 9-10 . Tokyo, Japan
Customization
Adjoints
Auto Certified Code
Morphing
Control Software
Optimization
GPU Computing
Model Based Design
Six - Sigma
HPC
Functional Mockup
Design Exploration
Electromagnetics
Electro-Structural
0D, 1D, Circuits
Distributed Solve
Structural
Fluid-Structural
ROMs
Parameterization
Fluids
Fluid-Thermal
Co-Simulation
Automation
Single Physics Verification
Enterprise HPC + Additional Physics
Systems Engineering
Robust Design
Increasing maturity of enterprise simulation adoption

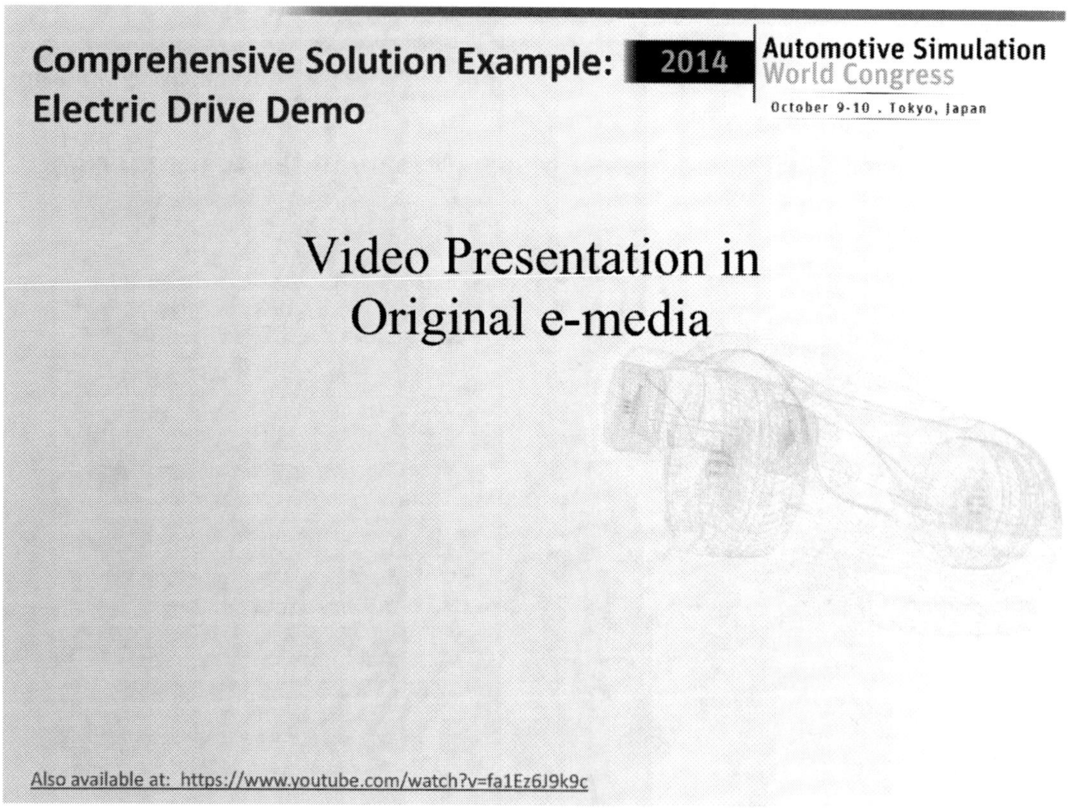
Comprehensive Solution Example: Electric Drive Demo
2014
Automotive Simulation World Congress
October 9-10 . Tokyo, Japan
Video Presentation in Original e-media
Also available at: https://www.youtube.com/watch?v=fa1Ez6J9k9c

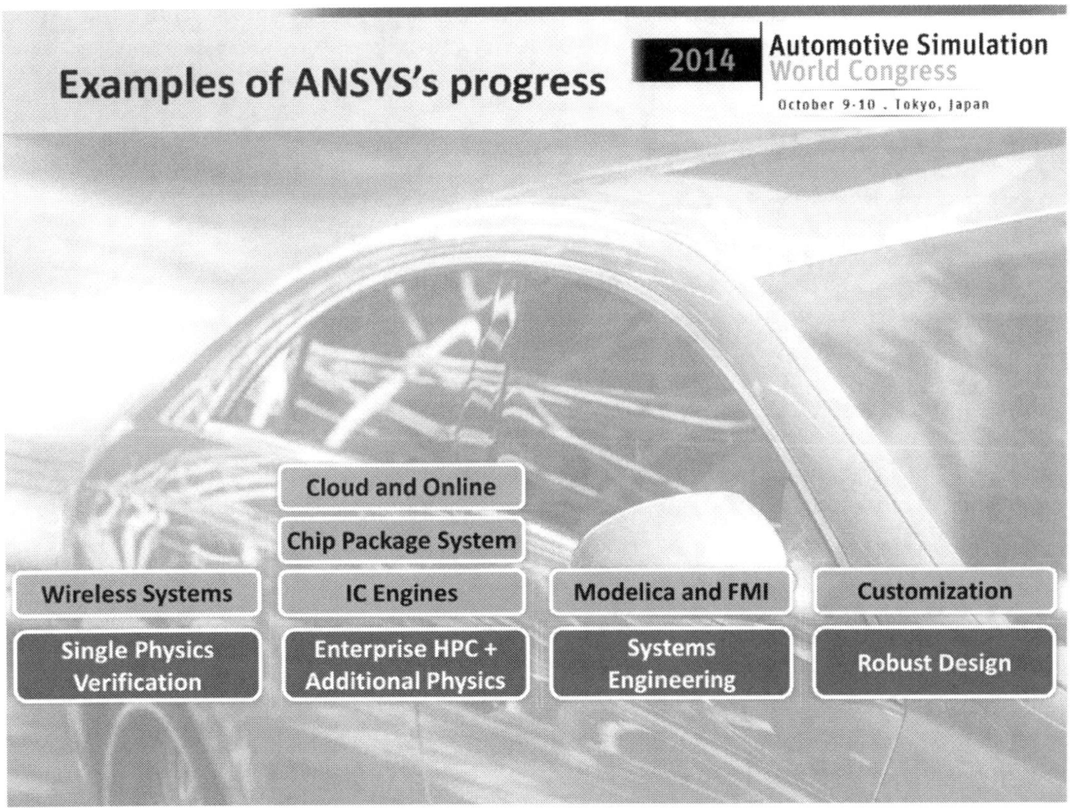
Examples of ANSYS's progress
2014
Automotive Simulation
World Congress
October 9-10 . Tokyo, Japan
Cloud and Online
Chip Package System
Wireless Systems
IC Engines
Modelica and FMI
Customization
Single Physics Verification
Enterprise HPC + Additional Physics
Systems Engineering
Robust Design

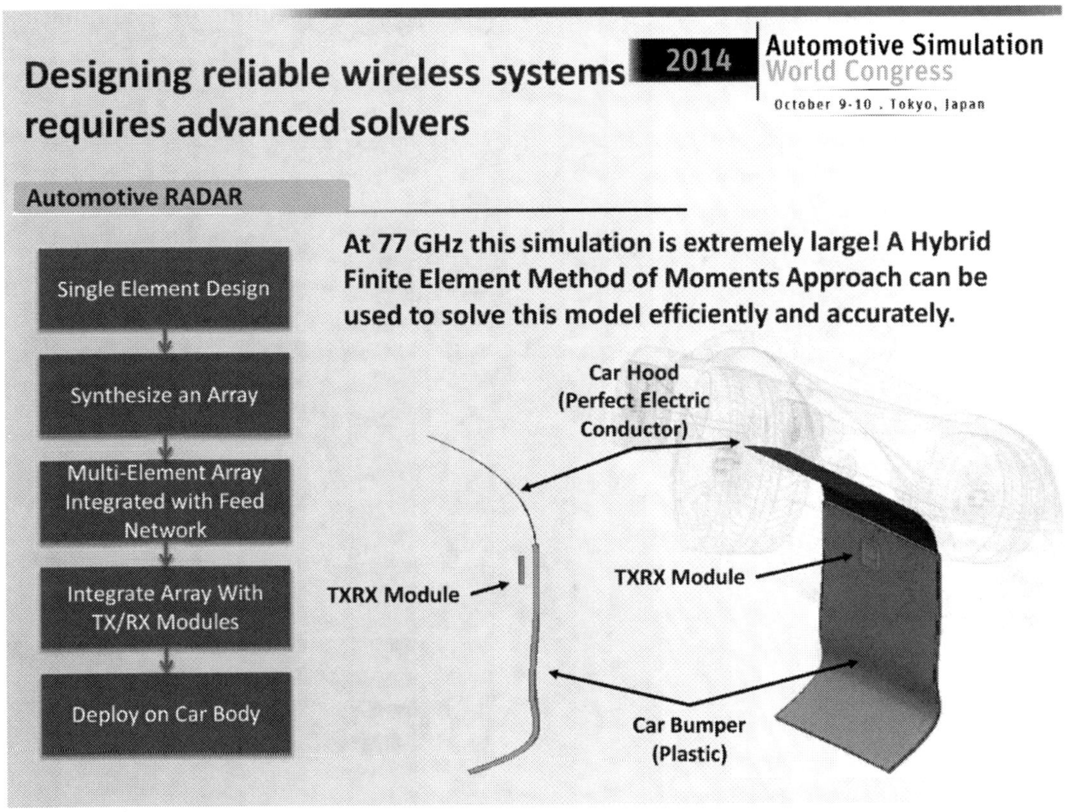
Designing reliable wireless systems requires advanced solvers
2014
Automotive Simulation
World Congress
October 9-10 . Tokyo, Japan
Automotive RADAR
Single Element Design
Synthesize an Array
Multi-Element Array Integrated with Feed Network
Integrate Array With TX/RX Modules
Deploy on Car Body
At 77 GHz this simulation is extremely large! A Hybrid Finite Element Method of Moments Approach can be used to solve this model efficiently and accurately.
Car Hood
(Perfect Electric
Conductor)
TXRX Module
TXRX Module
Car Bumper
(Plastic)

## ANSYS now provides a complete solution for designing automotive RADAR

2014 Automotive Simulation World Congress
October 9-10 . Tokyo, Japan

**Best Practices**

**Geometry Capture and Cleanup**
- Model 3D geometry with ease using SpaceClaim
- ANSYS acquired SpaceClaim in 2014

**Antenna Design and Simulation**
- Perform RF/Microwave design using HFSS
- Predict and mitigate EMI/EMC issues early

**Combined Electromagnetics + Systems**
- Use hybrid electromagnetic solvers
- Use hybrid electromagnetic circuit silmulations

**Thermal and Vibration Analysis**
- Predict and improve product robustness
- Perform impact and vibration testing through Explicit Dynamics and ANSYS Mechanical

## Improved accuracy of autonomous vehicle long range RADARs

2014 Automotive Simulation World Congress
October 9-10 . Tokyo, Japan

**Full antenna array design including obstructions**

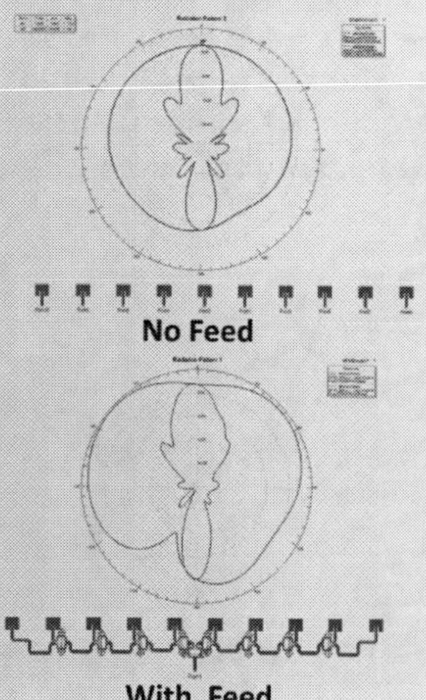


"Low" Antenna Pattern    "High" Antenna Pattern

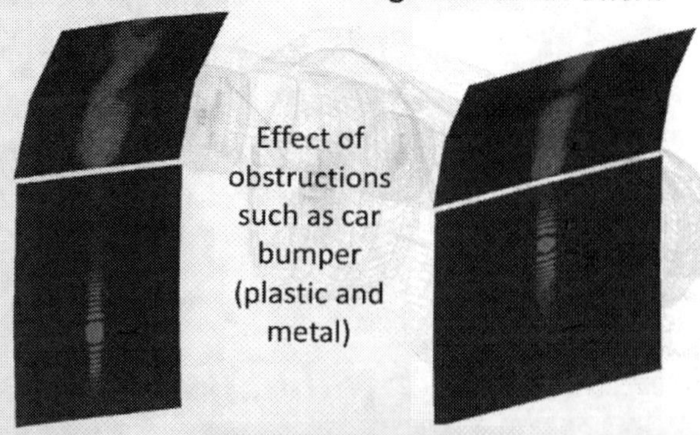


Highly Accurate and Robust Automotive Radar System Design
2E-5 Markus Kopp

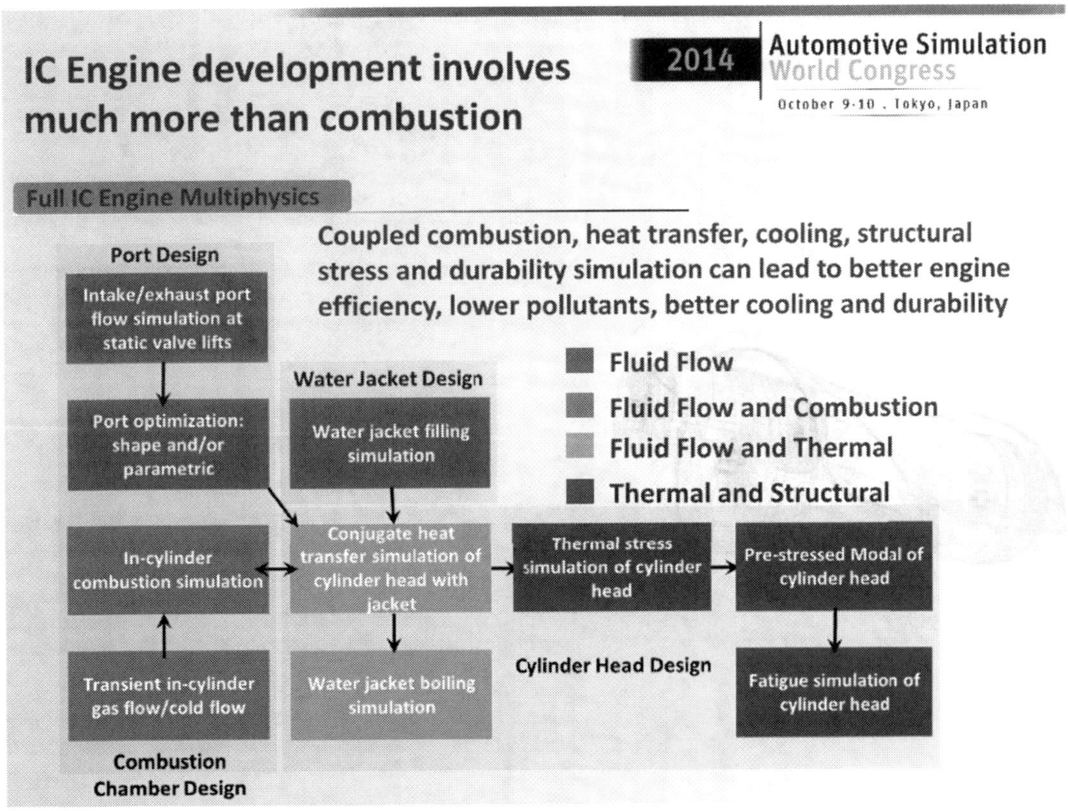
IC Engine development involves much more than combustion
2014 Automotive Simulation World Congress
October 9-10 . Tokyo, Japan
Full IC Engine Multiphysics
Coupled combustion, heat transfer, cooling, structural stress and durability simulation can lead to better engine efficiency, lower pollutants, better cooling and durability
Fluid Flow
Fluid Flow and Combustion
Fluid Flow and Thermal
Thermal and Structural
Port Design
Intake/exhaust port flow simulation at static valve lifts
Port optimization: shape and/or parametric
Water Jacket Design
Water jacket filling simulation
In-cylinder combustion simulation
Conjugate heat transfer simulation of cylinder head with jacket
Thermal stress simulation of cylinder head
Pre-stressed Modal of cylinder head
Transient in-cylinder gas flow/cold flow
Water jacket boiling simulation
Cylinder Head Design
Fatigue simulation of cylinder head
Combustion Chamber Design

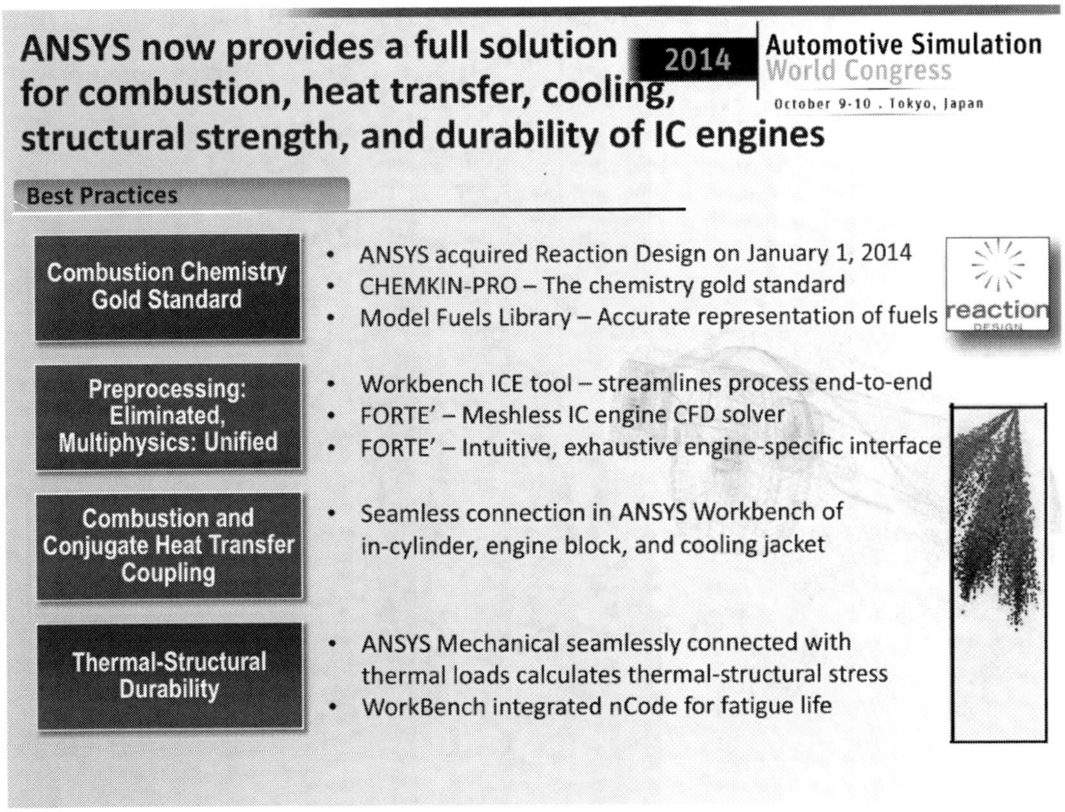
ANSYS now provides a full solution for combustion, heat transfer, cooling, structural strength, and durability of IC engines
2014 Automotive Simulation World Congress
October 9-10 . Tokyo, Japan
Best Practices
Combustion Chemistry Gold Standard
• ANSYS acquired Reaction Design on January 1, 2014
• CHEMKIN-PRO – The chemistry gold standard
• Model Fuels Library – Accurate representation of fuels
reaction
Preprocessing: Eliminated, Multiphysics: Unified
• Workbench ICE tool – streamlines process end-to-end
• FORTE' – Meshless IC engine CFD solver
• FORTE' – Intuitive, exhaustive engine-specific interface
Combustion and Conjugate Heat Transfer Coupling
• Seamless connection in ANSYS Workbench of in-cylinder, engine block, and cooling jacket
Thermal-Structural Durability
• ANSYS Mechanical seamlessly connected with thermal loads calculates thermal-structural stress
• WorkBench integrated nCode for fatigue life

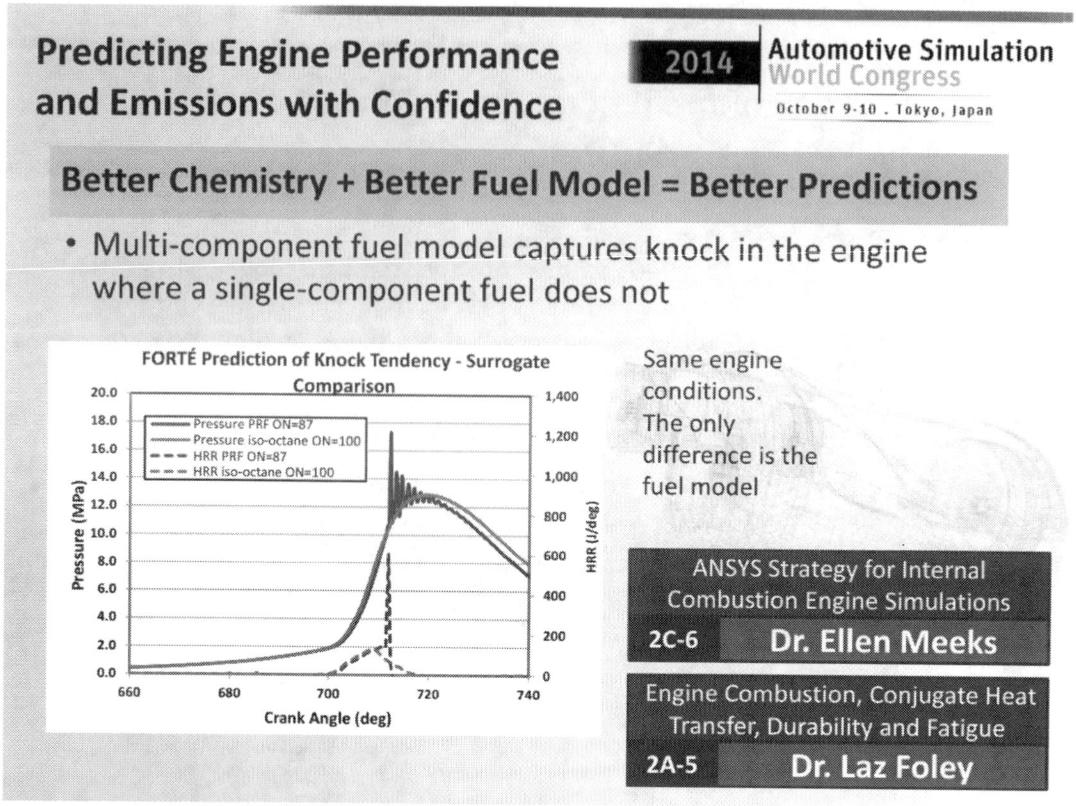
Predicting Engine Performance and Emissions with Confidence
2014 Automotive Simulation World Congress
October 9-10 . Tokyo, Japan
Better Chemistry + Better Fuel Model = Better Predictions
• Multi-component fuel model captures knock in the engine where a single-component fuel does not
FORTÉ Prediction of Knock Tendency - Surrogate Comparison
Pressure PRF ON=87
Pressure iso-octane ON=100
HRR PRF ON=87
HRR iso-octane ON=100
Pressure (MPa)
HRR (J/deg)
Crank Angle (deg)
Same engine conditions. The only difference is the fuel model
ANSYS Strategy for Internal Combustion Engine Simulations
2C-6 Dr. Ellen Meeks
Engine Combustion, Conjugate Heat Transfer, Durability and Fatigue
2A-5 Dr. Laz Foley

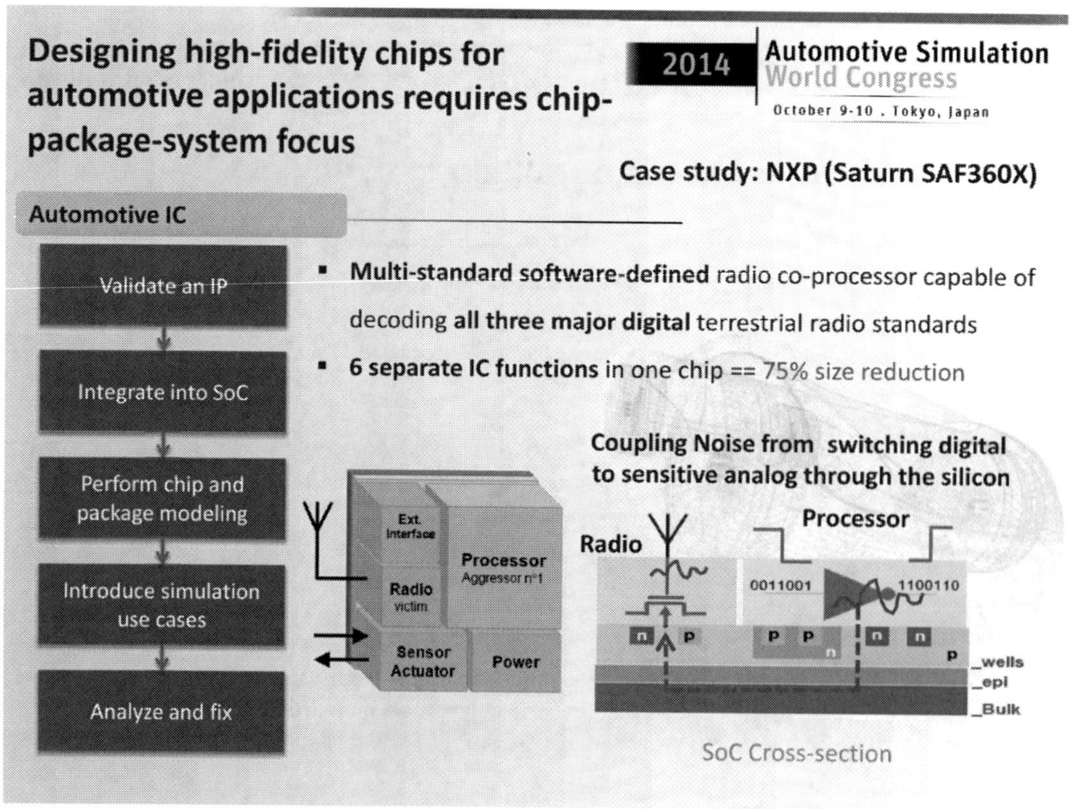
Designing high-fidelity chips for automotive applications requires chip-package-system focus
2014 Automotive Simulation World Congress
October 9-10 . Tokyo, Japan
Case study: NXP (Saturn SAF360X)
Automotive IC
Validate an IP
Integrate into SoC
Perform chip and package modeling
Introduce simulation use cases
Analyze and fix
▪ Multi-standard software-defined radio co-processor capable of decoding all three major digital terrestrial radio standards
▪ 6 separate IC functions in one chip == 75% size reduction
Ext. Interface
Processor
Aggressor n°1
Radio
victim
Sensor Actuator
Power
Coupling Noise from switching digital to sensitive analog through the silicon
Radio
Processor
0011001
1100110
wells
epi
Bulk
SoC Cross-section

## ANSYS solutions for RTL + IP + SoC + Package modeling enables comprehensive noise coupling analysis

2014 Automotive Simulation World Congress
October 9-10 . Tokyo, Japan

**Best Practices**

**IP Validation and Modeling using Totem**
- Validate analog and memory IP EM/IR down to transistor level
- Create models for SoC validation

**RTL Optimization using PowerArtist**
- Predict early power numbers
- Analyze and reduce power to meet target

**Package Modeling using RedHawk-CPA**
- Use electromagnetic solvers and connect package models
- Review chip-package design issues simultaneously

**Chip + Package Co-analysis in RedHawk**
- Perform chip + package + silicon substrate analysis
- Use test-vectors or VectorLess methods

## ANSYS solutions for RTL + IP + SoC + Package modeling enabled comprehensive noise coupling analysis

2014 Automotive Simulation World Congress
October 9-10 . Tokyo, Japan

Noise coupling → Spikes in FM Spectrum
Impacts audio quality and performance

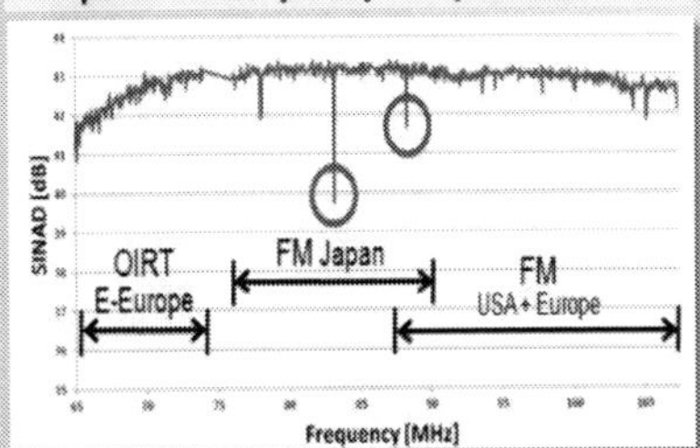


**Result Analysis**

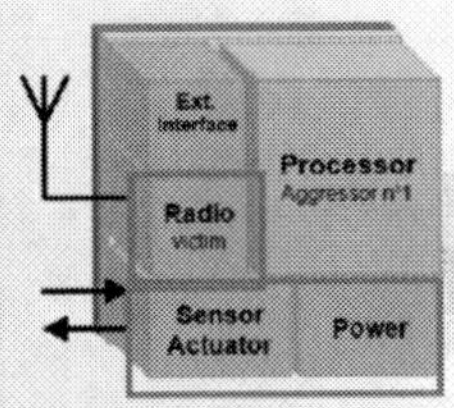


**Design Changes**

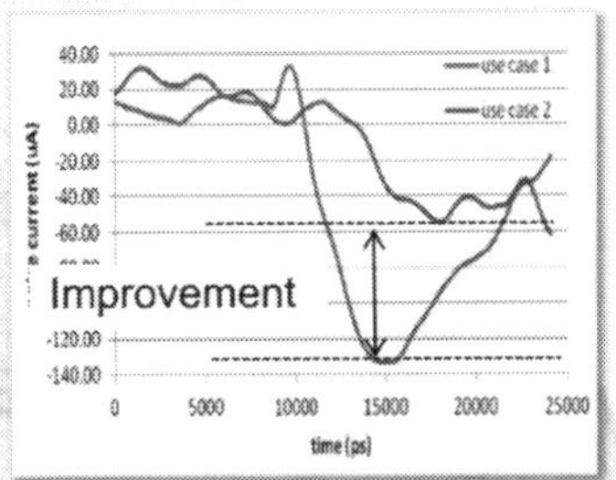


Noise Current Amplitude versus time

**Fix Validation**

**Advancing Chip-Package-System Convergence through a Comprehensive RTL to Silicon Methodology**

**1F-1** | **Aveek Sarkar**

## Globalization And HPC Scale-up Are Driving A Paradigm Shift In The Deployment Of Simulation

2014 Automotive Simulation World Congress
October 9-10 . Tokyo, Japan

**HPC, Cloud & Collaboration**

**HPC infrastructure and data is evolving to centralized and consolidated resources, with users remote, global, and mobile**

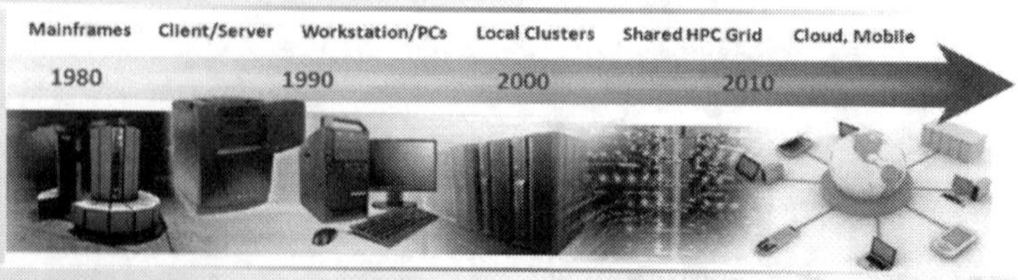


**Driven by:**

- **Scaled up demand for high performance computing (HPC) to support higher fidelity simulation results and design exploration**
- **Operational efficiency and asset optimization**
- **Need for collaboration and data sharing across geographically distributed users**
- **Increasing focus on simulation IP protection and process traceability**

## Moving Simulation To The Datacenter (or Cloud) Requires A Focus On End-user Process And Data

2014 Automotive Simulation World Congress
October 9-10 . Tokyo, Japan

**Best Practices**

**Consolidate HPC and data**
- Operational efficiency
- Avoid moving data
- Enable collaboration

**Enable effective remote access**
- Beyond batch processing
- Remote 3D graphics & interactive use
- Mobile tools for job management

**Manage the data**
- For IP protection and compliance
- For efficient re-use and retrieval
- In context of PLM

**Software asset management**
- Common tools vs. point solutions
- Shared centralized licensing
- Consolidated business engagement

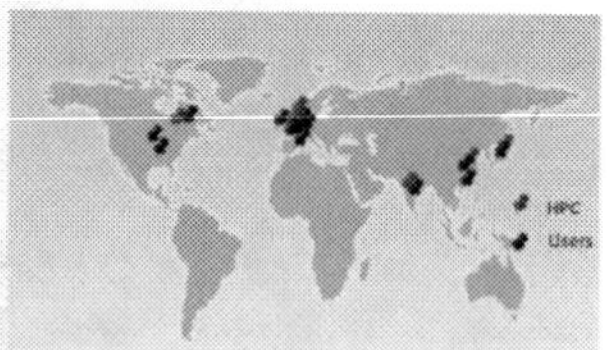


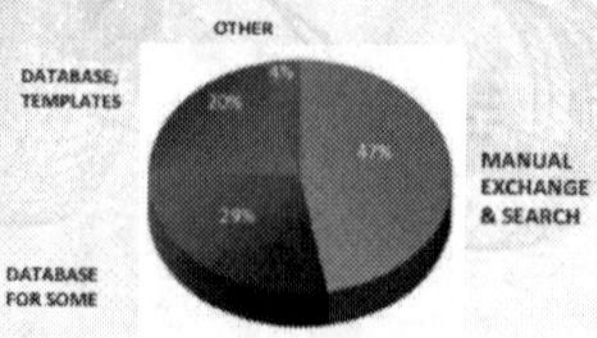


Management of Simulation
Data Source: CPDA

## ANSYS Is Building An Open Network Of Cloud Partners, Developing Best-Practice Simulation Cloud

2014 Automotive Simulation World Congress
October 9-10 . Tokyo, Japan

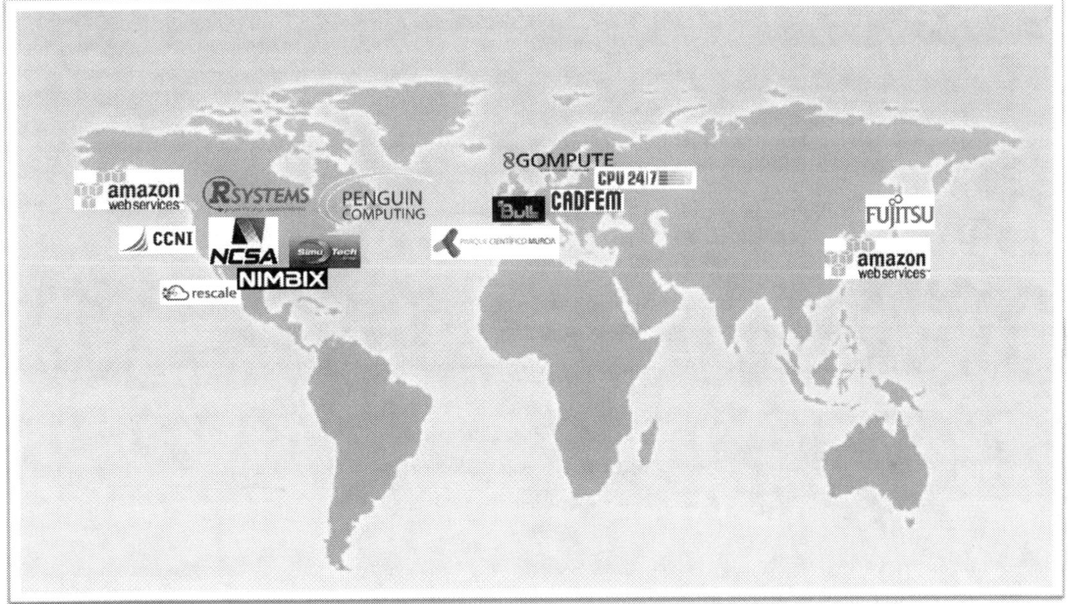


## Modeling a complex system requires integration of diverse physical models from multiple sources

2014 Automotive Simulation World Congress
October 9-10 . Tokyo, Japan

### Modelica and FMI

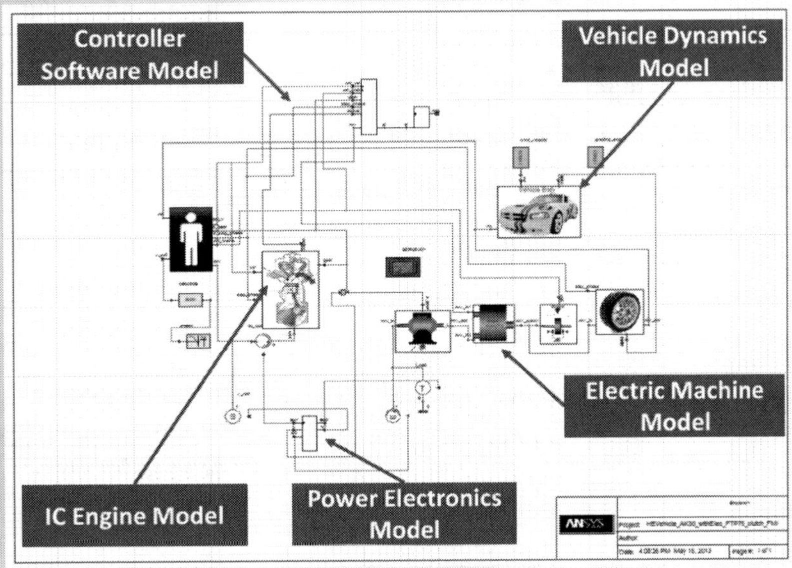

## ANSYS Simplorer has been developed into a "backbone" for integrating and delivering a broadly applicable system modeling solution

2014 Automotive Simulation World Congress
October 9-10 . Tokyo, Japan

**Best Practices**

**Multidomain Physical Systems Modeler**

- ANSYS Simplorer: an integrated environment for modeling, simulating, and analyzing multi-domain physical systems

**Connection with 2D-3D Solvers**

- ANSYS Simplorer connects with ANSYS 2D-3D solvers for various physics for accurate component representation
- Reduced Order Models (ROMs)

**Connections with 3rd Party Tools**

- Connection with tools such as Matlab, Simulink
- Connection with Modelica and FMI Libraries

**Embedded Software and Control System Design**

- Embedded software modeling, verification and life-cycle management
- Certified automated code generation

## ANSYS announced partnership with Modelon

ANSYS + Modelon

2014 Automotive Simulation World Congress
October 9-10 . Tokyo, Japan

ANSYS Simplorer system modeling solution now provides

- Access to Modelon's Modelica Compiler
- Modelon's best-in-class Modelica libraries with powerful physical plant modeling capabilities
- Modelon's service and training

Solutions for Innovative Automotive Systems
2K-2 Eric Bantegnie

## Capturing specific workflows, knowledge and processes is key to deploying simulation to large designer groups

2014 Automotive Simulation World Congress
October 9-10 . Tokyo, Japan

### Customization

**Process automation reduces CAE time at Whirlpool by 60%**

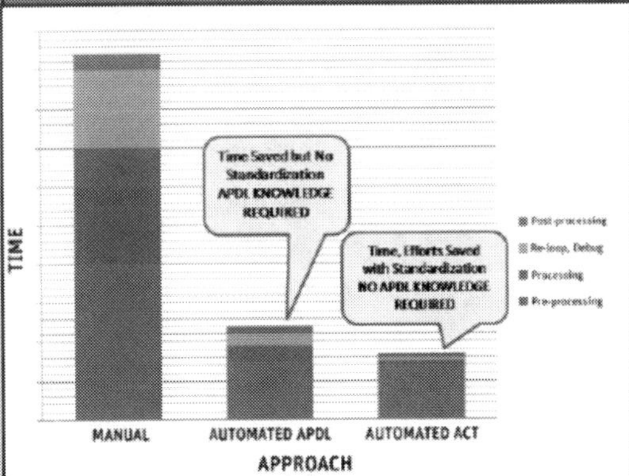


Customization makes CAE software
- Easy to use
- More industry specific
- Captures custom workflows
- Captures best practices
- Creates vertical applications
- Brings standardization

And delivers numerous benefits
- Improves productivity
- Standardization enhances reliability and quality
- Improves knowledge retention
- Reduces training needs

## ANSYS has increased focus on customization and is delivering several powerful customization tools

2014 Automotive Simulation World Congress
October 9-10 . Tokyo, Japan

### Best Practices

**Journaling and Scripting**
- Python scripting in ANSYS Workbench
- E.g. interaction with Microsoft Excel
- Journaling and Scripting in individual solvers, e.g. APDL

**Wizards and Templates**
- Streamlined UI to guide user through complete simulation process
- Adoption of best practices

**ACT (Application Customization Toolkit)**
- The Application Customization Toolkit (ACT) is the tool for customizing the Mechanical interface within ANSYS Workbench

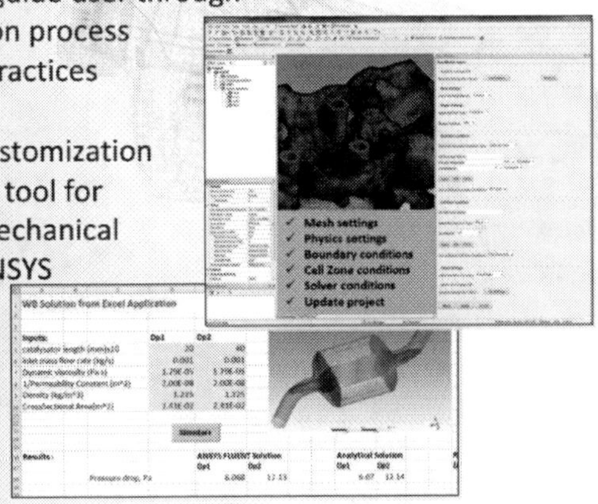

## Customization Example: Automotive HVAC Extension

2014 Automotive Simulation World Congress
October 9-10 . Tokyo, Japan

- Complete setup in DM using ACT objects
- Smart and Rapid object creation
  - Room, Person, Blocks, Fans, Openings and Vents
- Support for CAD Import and definition of bodies
- Visual representation in DM similar to Mechanical
- Full automation in DM. (*No need to visit Meshing, Fluent and Post*)

**Extension Manager**

**ACT Tool Bar in DM**

Room | Block | Fan | Person | Mesh | Post & Solution

**ACT Objects in DM Tree**

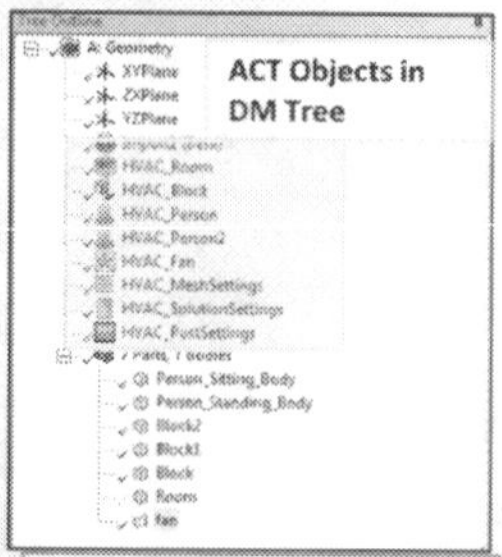

**Rapid Model Creation**
(Model Created Under 5 minutes)

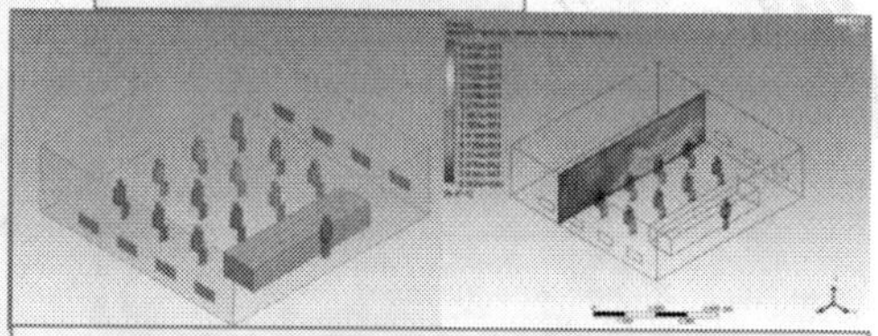

**Automated Report and Images**

Customization for Streamlining of Fluid and Mechanical Simulations for Designers

2A-6 Padmesh Mandloi

## Thank you

2014 Automotive Simulation World Congress
October 9-10 . Tokyo, Japan

Wednesday, December 10, 2014 | 2014 Automotive Simulation World Congress | 28

# Higher automotive simulation productivity in ANSYS Fluent 15.0 with NVIDIA GPUs


**Bhushan Desam, Ph.D.**
**NVIDIA Corporation**


## AGENDA

- GPUs in Enterprise Computing
- Business Challenges in Product Development
- GPU Computing for CAE Applications
- GPU Acceleration in ANSYS Fluent 15.0
- NVIDIA GPU Performance on Automotive Applications
- NVIDIA Products & Solutions for ANSYS Simulations
- Value of GPU-accelerated Simulations

2

NVIDIA Enterprise Group
Visualization, Accelerated Computing & Virtualization
NVIDIA
QUADRO
Revolutionizing Design & Visualization
TESLA
Accelerating Momentum in HPC and Big Data Analytics
GRID
Enabling End-to-End Enterprise Virtualization
3

NVIDIA
TESLA
Accelerating Computing
GPUs enable tremendous breakthroughs by simply enabling us to do more, faster
ANSYS, as a technology leader, leverage GPUs to accelerate engineering simulation for better design
The world's top10 energy-efficient supercomputers use NVIDIA GPUs according to TheGreen500 list
4

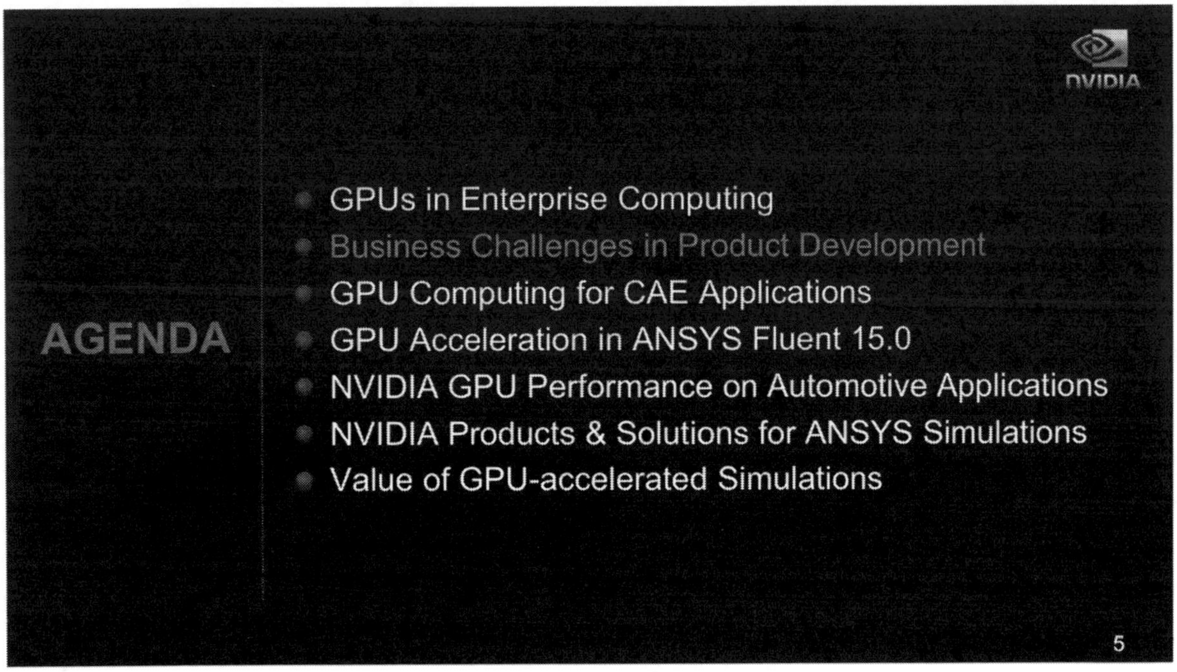
NVIDIA
AGENDA
GPUs in Enterprise Computing
Business Challenges in Product Development
GPU Computing for CAE Applications
GPU Acceleration in ANSYS Fluent 15.0
NVIDIA GPU Performance on Automotive Applications
NVIDIA Products & Solutions for ANSYS Simulations
Value of GPU-accelerated Simulations
5

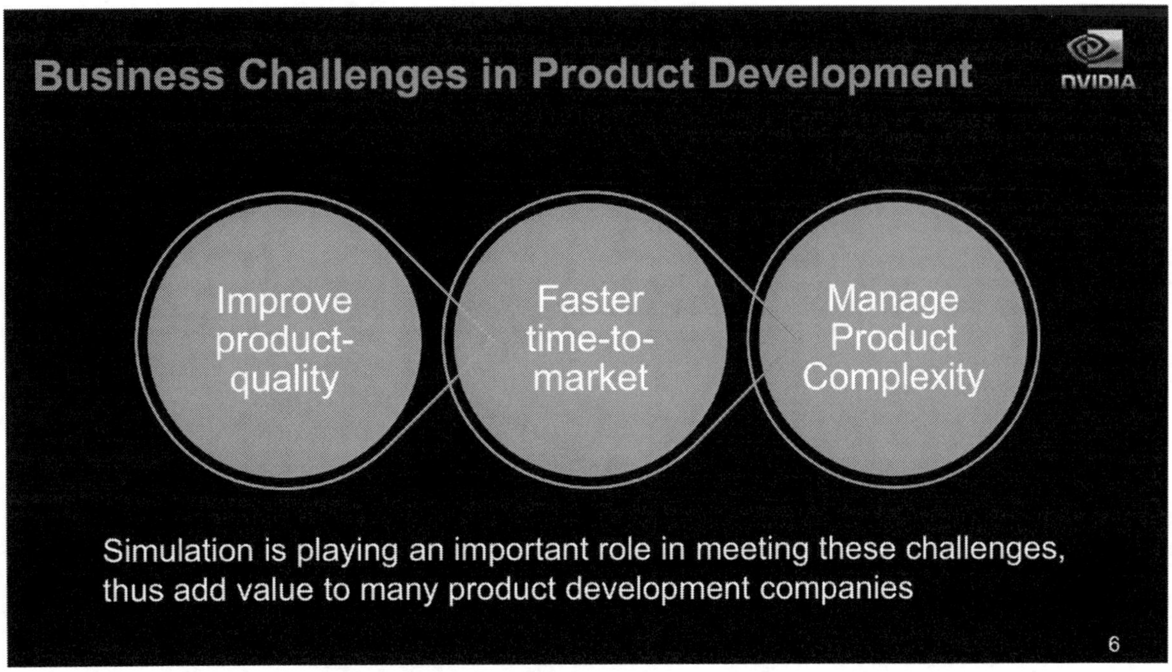
Business Challenges in Product Development
NVIDIA
Improve product-quality
Faster time-to-market
Manage Product Complexity
Simulation is playing an important role in meeting these challenges, thus add value to many product development companies
6

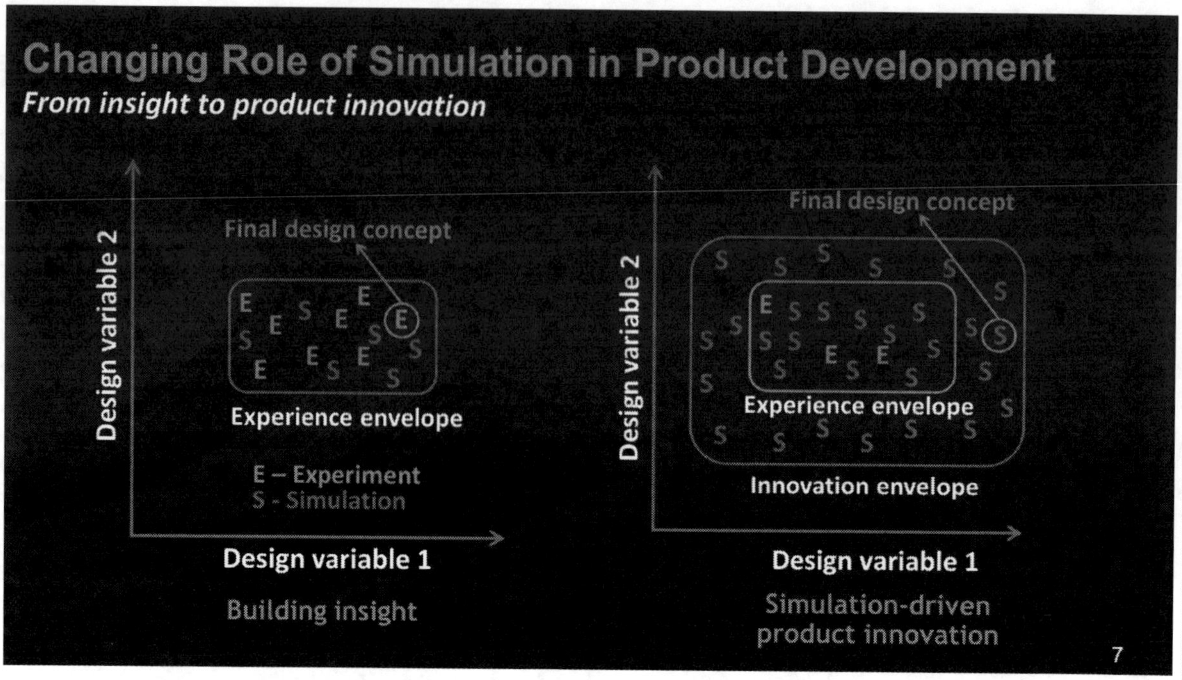
Changing Role of Simulation in Product Development
From insight to product innovation
Design variable 2
Final design concept
Experience envelope
E – Experiment
S - Simulation
Design variable 1
Building insight
Design variable 2
Final design concept
Experience envelope
Innovation envelope
Design variable 1
Simulation-driven product innovation
7

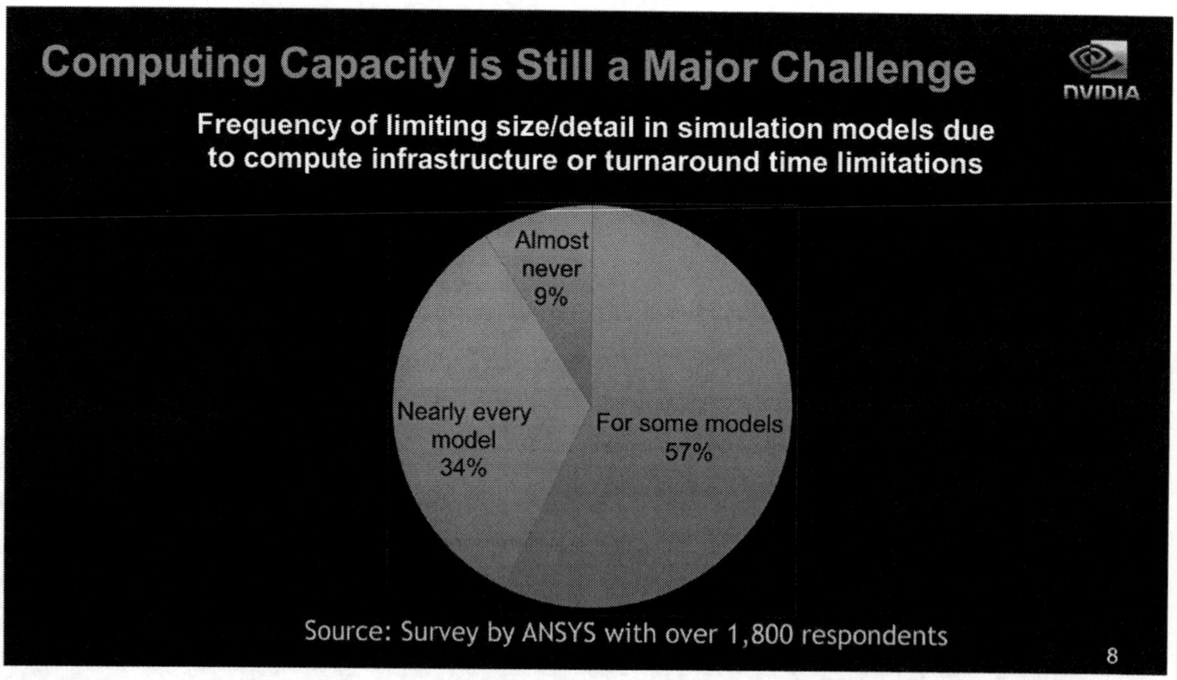
Computing Capacity is Still a Major Challenge
NVIDIA
Frequency of limiting size/detail in simulation models due to compute infrastructure or turnaround time limitations
Almost never 9%
Nearly every model 34%
For some models 57%
Source: Survey by ANSYS with over 1,800 respondents
8

Hardware trends in HPC
NVIDIA
Recent advancements have revolutionized the computational speed available on the desktop
Multi-core processors
GPGPU (general-purpose computing on graphics processing units)
Large amounts of RAM
Solid state derives (SSD)
NVIDIA TESLA
6X
more flops compared to the latest CPU
NVIDIA GPU
9

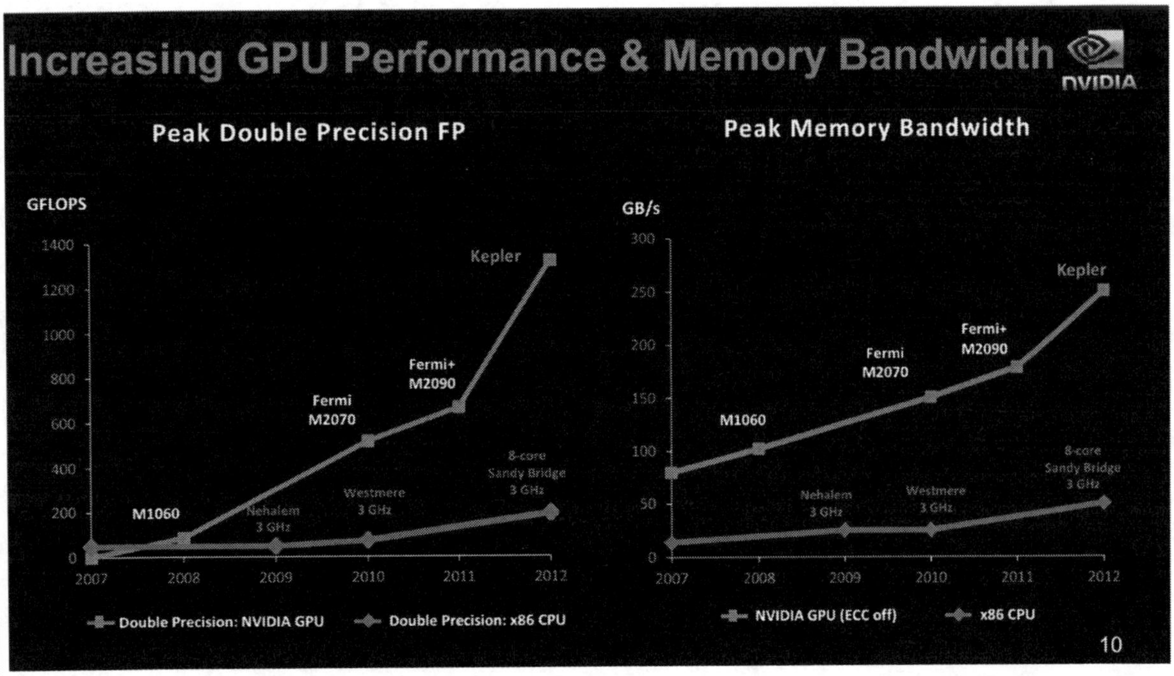
Increasing GPU Performance & Memory Bandwidth
NVIDIA
Peak Double Precision FP
GFLOPS
1400
1200
1000
800
600
400
200
0
2007
2008
2009
2010
2011
2012
Kepler
Fermi+
M2090
Fermi
M2070
M1060
Nehalem
3 GHz
Westmere
3 GHz
8-core
Sandy Bridge
3 GHz
Double Precision: NVIDIA GPU
Double Precision: x86 CPU
Peak Memory Bandwidth
GB/s
300
250
200
150
100
50
0
2007
2008
2009
2010
2011
2012
Kepler
Fermi+
M2090
Fermi
M2070
M1060
Nehalem
3 GHz
Westmere
3 GHz
8-core
Sandy Bridge
3 GHz
NVIDIA GPU (ECC off)
x86 CPU
10

AGENDA

- GPUs in Enterprise Computing
- Business Challenges in Product Development
- GPU Computing for CAE Applications
- GPU Acceleration in ANSYS Fluent 15.0
- NVIDIA GPU Performance on Automotive Applications
- NVIDIA Products & Solutions for ANSYS Simulations
- Value of GPU-accelerated Simulations

11

## Basics of GPU Computing

- **GPU is an accelerator attached to an x86 CPU**
- **GPU acceleration is user-transparent**
  - Jobs launch and complete without additional user steps
- **Schematic of a CPU with an attached GPU accelerator**
  - CPU begins/ends job, GPU manages heavy computations

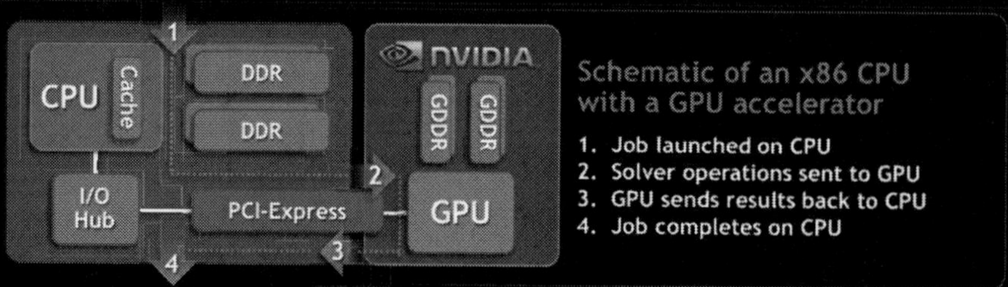


12

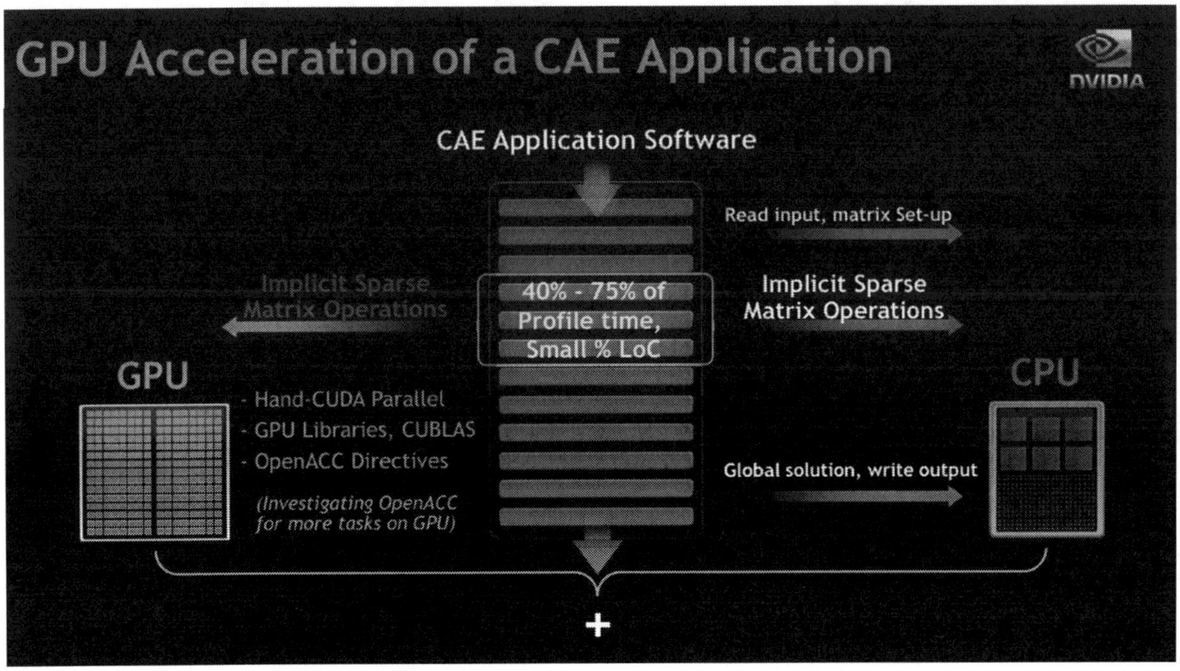
GPU Acceleration of a CAE Application
NVIDIA
CAE Application Software
Read input, matrix Set-up
Implicit Sparse Matrix Operations
40% - 75% of Profile time, Small % LoC
Implicit Sparse Matrix Operations
GPU
CPU
- Hand-CUDA Parallel
- GPU Libraries, CUBLAS
- OpenACC Directives
(Investigating OpenACC for more tasks on GPU)
Global solution, write output
+

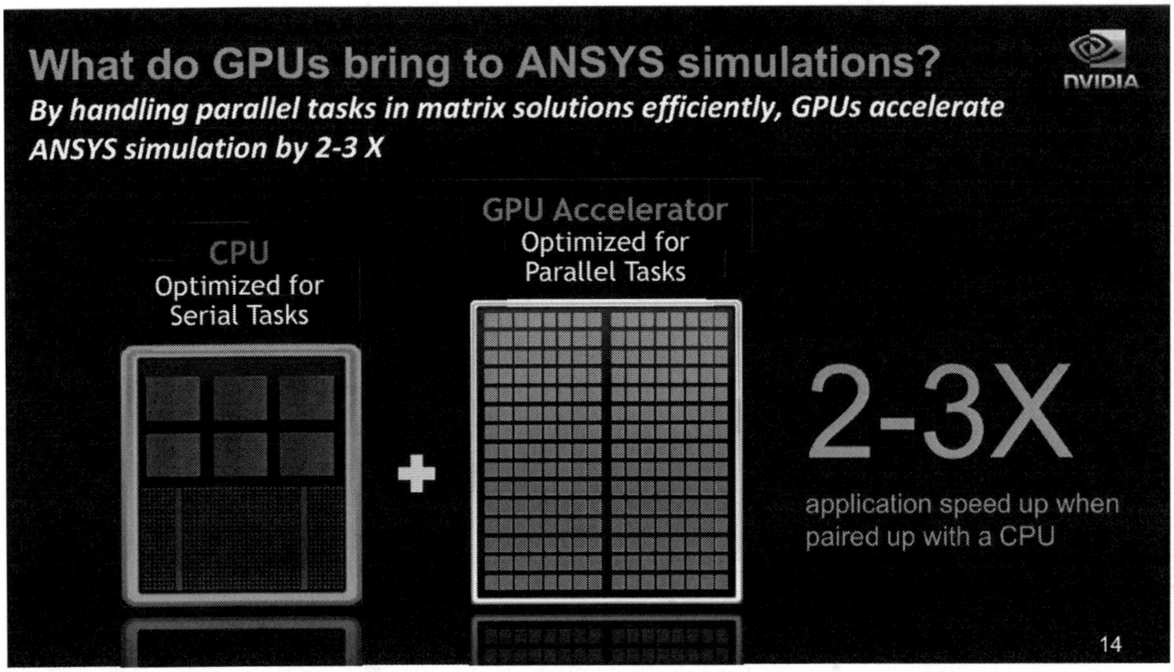
What do GPUs bring to ANSYS simulations?
NVIDIA
By handling parallel tasks in matrix solutions efficiently, GPUs accelerate ANSYS simulation by 2-3 X
CPU
Optimized for Serial Tasks
GPU Accelerator
Optimized for Parallel Tasks
2-3X
application speed up when paired up with a CPU
14

**AGENDA**

- GPUs in Enterprise Computing
- Business Challenges in Product Development
- GPU Computing for CAE Applications
- GPU Acceleration in ANSYS Fluent 15.0
- NVIDIA GPU Performance on Automotive Applications
- NVIDIA Products & Solutions for ANSYS Simulations
- Value of GPU-accelerated Simulations

15

## How to Enable NVIDIA GPUs in Fluent 15.0?

Windows:

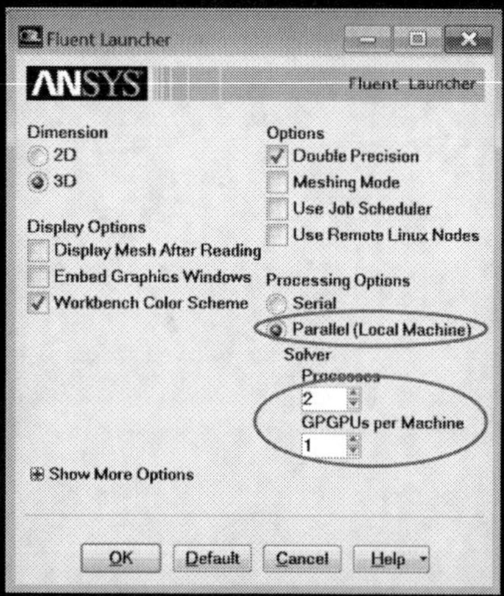


Linux:

fluent 3ddp -g -ssh –t2 -gpgpu=1 -i journal.jou

Cluster specification:

*nprocs* = Total number of fluent processes

*M* = Number of machines

*ngpgpus* = Number of GPUs per machine

Requirement 1

$nprocs \mod M = 0$

Same number of solver processes on each machine

Requirement 2

$\left(\frac{nproc}{M}\right) \mod ngpgpus = 0$

No. of processes should be an integer multiple of GPUs

16

## CPU-GPU configurations

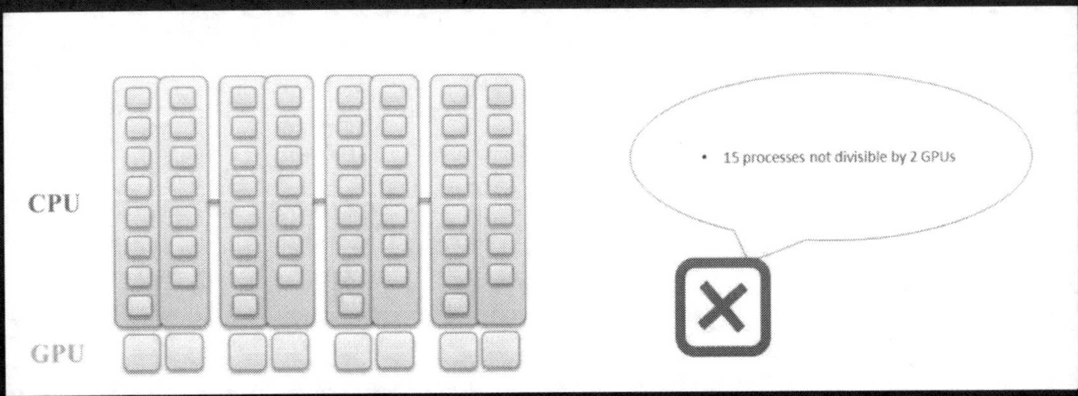


17

## ANSYS 15.0 HPC licenses (new)

- Treats each GPU socket as a CPU core, which significantly increases simulation productivity from HPC licenses
- All ANSYS HPC products unlock GPUs in 15.0, including HPC, HPC Pack, HPC Workgroup, and HPC Enterprise products.

GPU support in ANSYS – Licensing comparison

| License type | ANSYS 14.5 | ANSYS 15.0 |
|---|---|---|
| HPC per core licenses | None | 1 HPC license – 1 GPU<br>2 HPC licenses – 2 GPUs<br>'N' HPC licenses – 'N' GPUs |
| HPC pack | 1 HPC pack – 1 GPU<br>2 HPC packs – 4 GPUs | 1 HPC pack – 4 GPUs<br>2 HPC packs – 16 GPUs |

18

## GPU Acceleration in Fluent 15.0

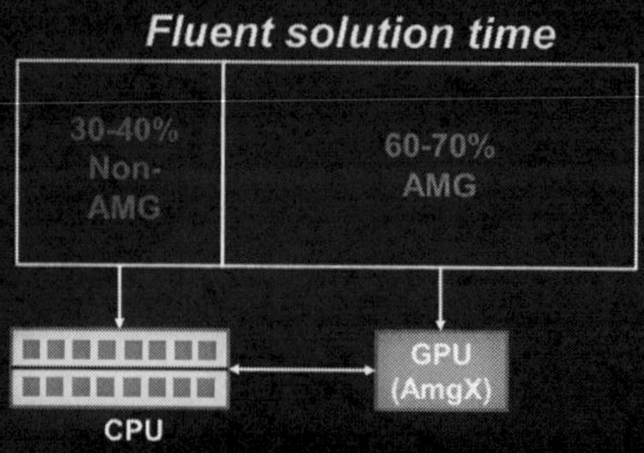


- In flow problems, pressure-based coupled solver typically spends 60-70% time in AMG where as segregated solver only spends 30-40% of the time in AMG
- Higher AMG times are ideal for GPU acceleration, thus coupled-problems benefit from GPUs

19

## Fluent Speed-up from GPU acceleration

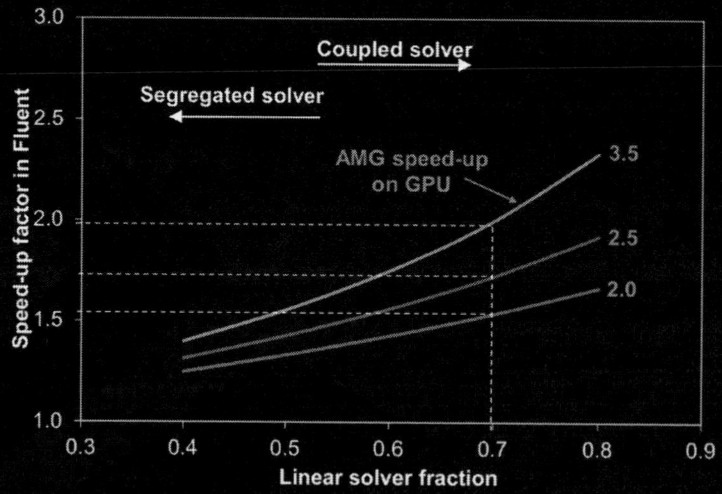


20

## AGENDA

- GPUs in Enterprise Computing
- Business Challenges in Product Development
- GPU Computing for CAE Applications
- GPU Acceleration in ANSYS Fluent 15.0
- NVIDIA GPU Performance on Automotive Applications
- NVIDIA Products & Solutions for ANSYS Simulations
- Value of GPU-accelerated Simulations

21

## Automotive Applications

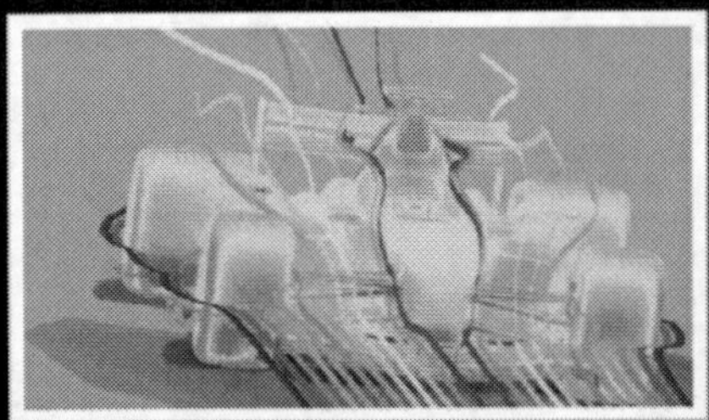

External aerodynamics

In-cylinder flow analysis

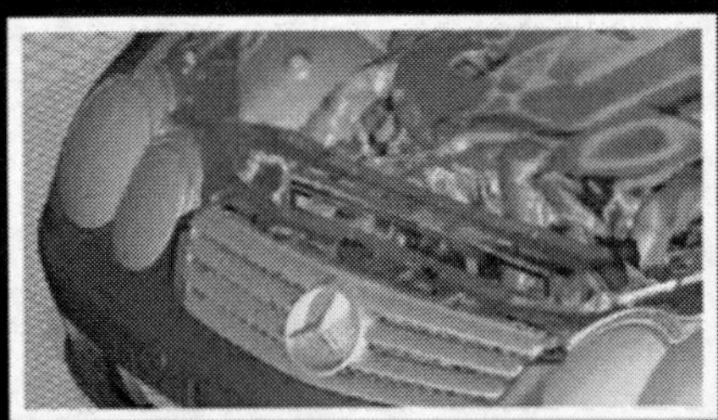

Underhood thermal analysis

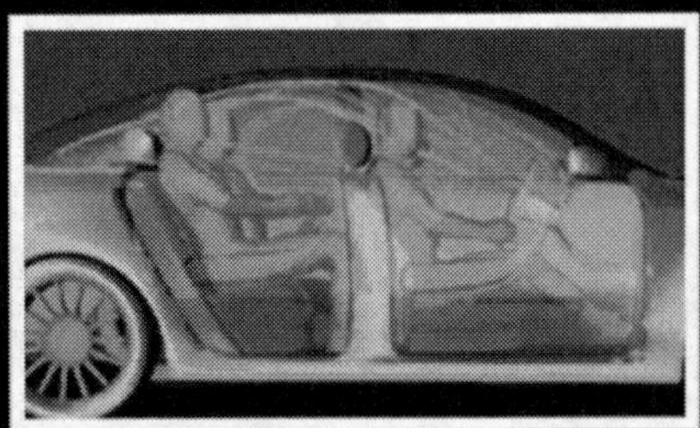

Passenger comfort analysis

22

# GPU Acceleration of Water Jacket Analysis

*ANSYS Fluent 15.0 performance on pressure-based coupled Solver*

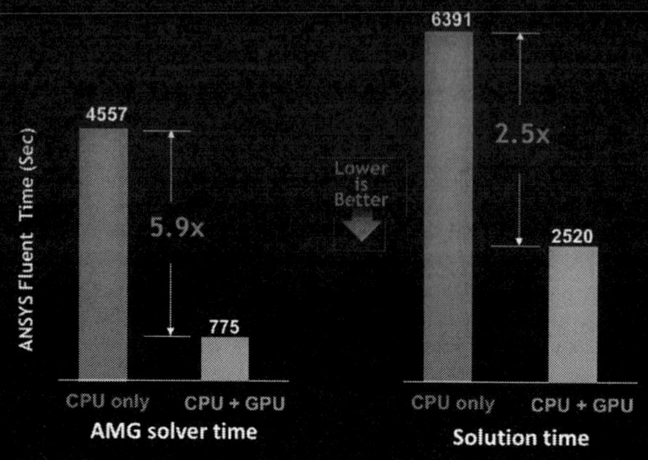


**Water jacket model**

- Unsteady RANS model
- Fluid: water
- Internal flow
- CPU: Intel Xeon E5-2680; 8 cores
- GPU: 2 X Tesla K40

**NOTE: Times for 20 time steps**

23

# ANSYS Fluent 15.0 power consumption study

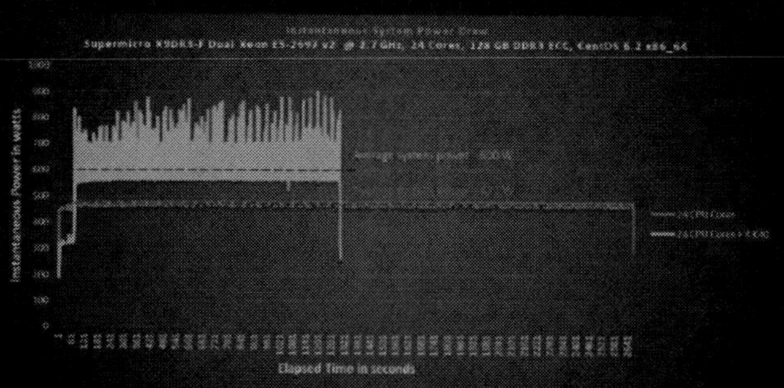

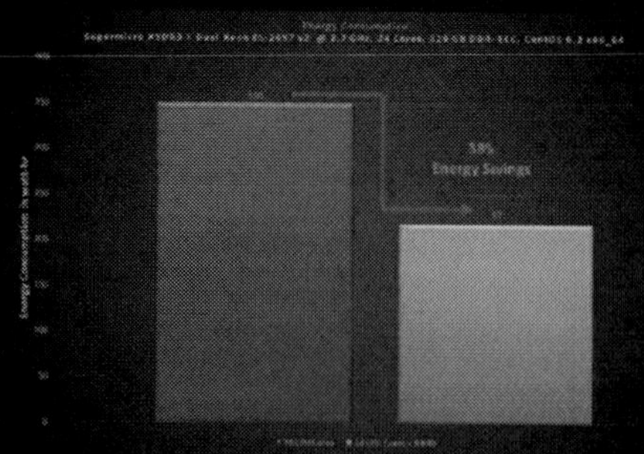

- Adding GPUs to a CPU-only node resulted in 2.1x speed up while reducing energy consumption by 38%

24

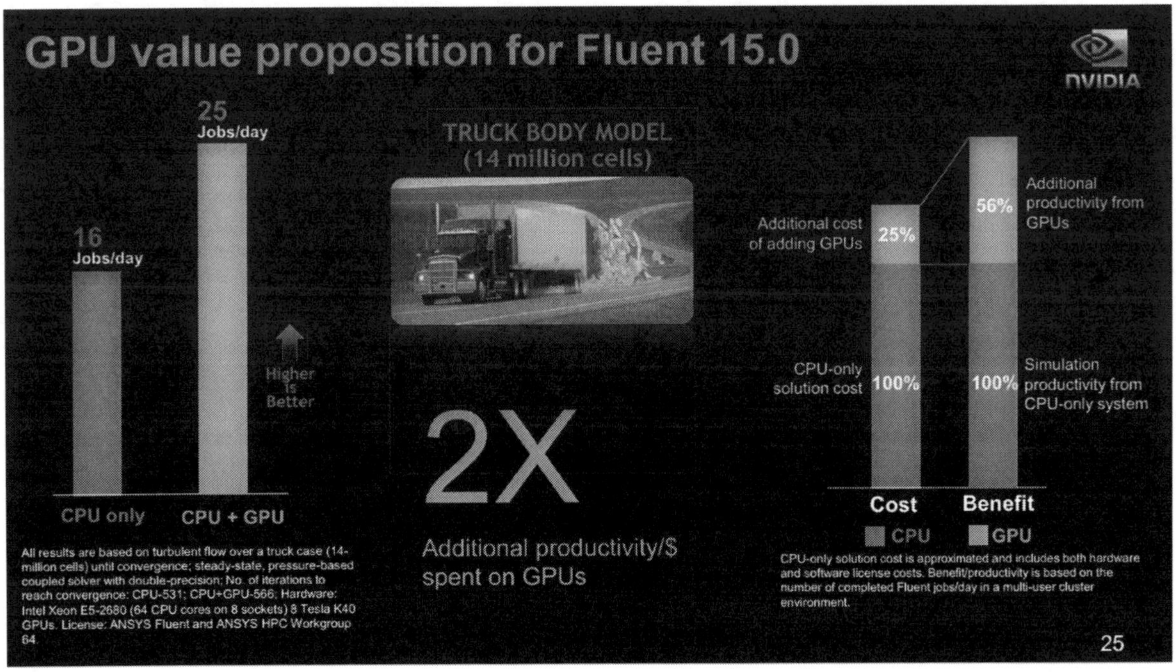
GPU value proposition for Fluent 15.0
NVIDIA
25
Jobs/day
16
Jobs/day
Higher
is
Better
CPU only
CPU + GPU
All results are based on turbulent flow over a truck case (14-million cells) until convergence; steady-state, pressure-based coupled solver with double-precision; No. of iterations to reach convergence: CPU-531; CPU+GPU-566; Hardware: Intel Xeon E5-2680 (64 CPU cores on 8 sockets) 8 Tesla K40 GPUs. License: ANSYS Fluent and ANSYS HPC Workgroup 64.
TRUCK BODY MODEL
(14 million cells)
2X
Additional productivity/$ spent on GPUs
Additional cost of adding GPUs
25%
CPU-only solution cost
100%
56%
Additional productivity from GPUs
100%
Simulation productivity from CPU-only system
Cost
Benefit
CPU
GPU
CPU-only solution cost is approximated and includes both hardware and software license costs. Benefit/productivity is based on the number of completed Fluent jobs/day in a multi-user cluster environment.
25

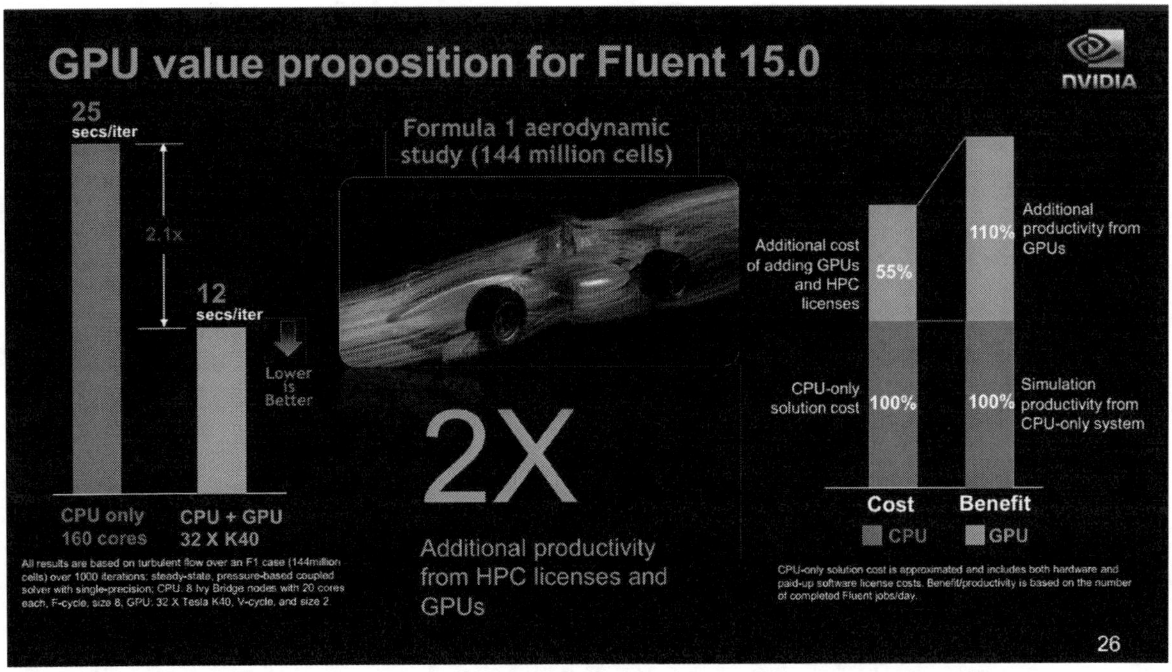
GPU value proposition for Fluent 15.0
NVIDIA
25
secs/iter
2.1x
12
secs/iter
Lower
is
Better
CPU only
160 cores
CPU + GPU
32 X K40
All results are based on turbulent flow over an F1 case (144million cells) over 1000 iterations; steady-state, pressure-based coupled solver with single-precision; CPU: 8 Ivy Bridge nodes with 20 cores each, F-cycle, size 8; GPU: 32 X Tesla K40, V-cycle, and size 2.
Formula 1 aerodynamic study (144 million cells)
2X
Additional productivity from HPC licenses and GPUs
Additional cost of adding GPUs and HPC licenses
55%
CPU-only solution cost
100%
110%
Additional productivity from GPUs
100%
Simulation productivity from CPU-only system
Cost
Benefit
CPU
GPU
CPU-only solution cost is approximated and includes both hardware and paid-up software license costs. Benefit/productivity is based on the number of completed Fluent jobs/day.
26

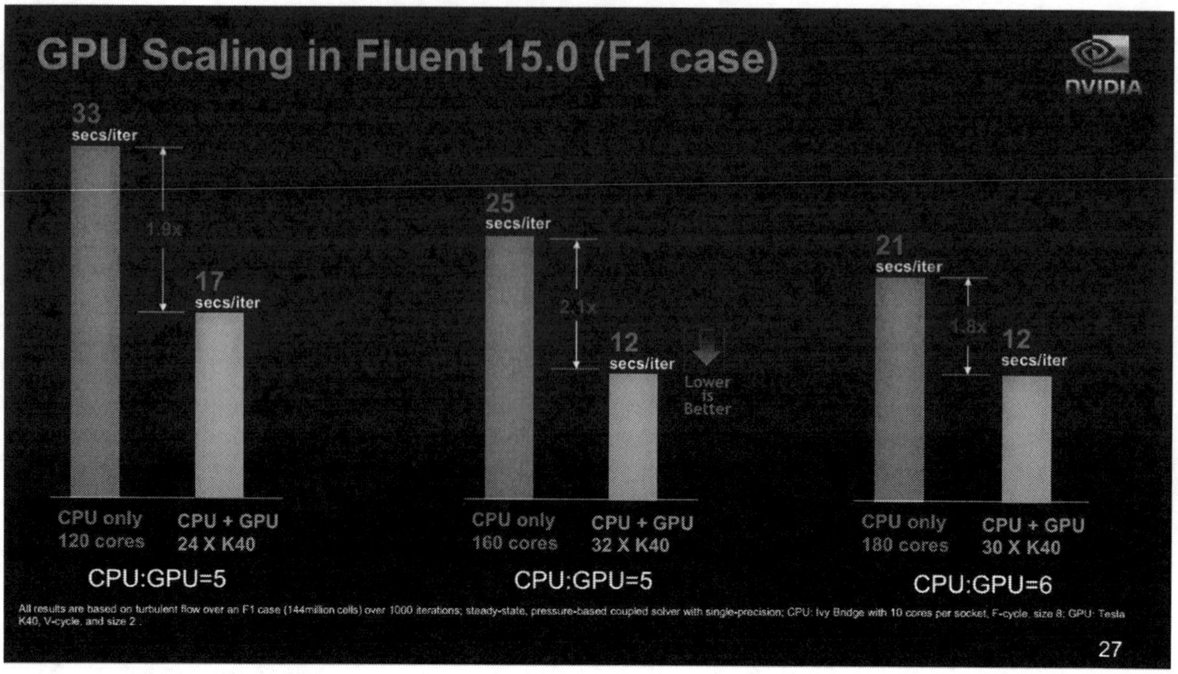
GPU Scaling in Fluent 15.0 (F1 case)
NVIDIA
33
secs/iter
1.9x
17
secs/iter
25
secs/iter
2.1x
12
secs/iter
Lower
is
Better
21
secs/iter
1.8x
12
secs/iter
CPU only
120 cores
CPU + GPU
24 X K40
CPU only
160 cores
CPU + GPU
32 X K40
CPU only
180 cores
CPU + GPU
30 X K40
CPU:GPU=5
CPU:GPU=5
CPU:GPU=6
All results are based on turbulent flow over an F1 case (144million cells) over 1000 iterations; steady-state, pressure-based coupled solver with single-precision; CPU: Ivy Bridge with 10 cores per socket, F-cycle, size 8; GPU: Tesla K40, V-cycle, and size 2.
27

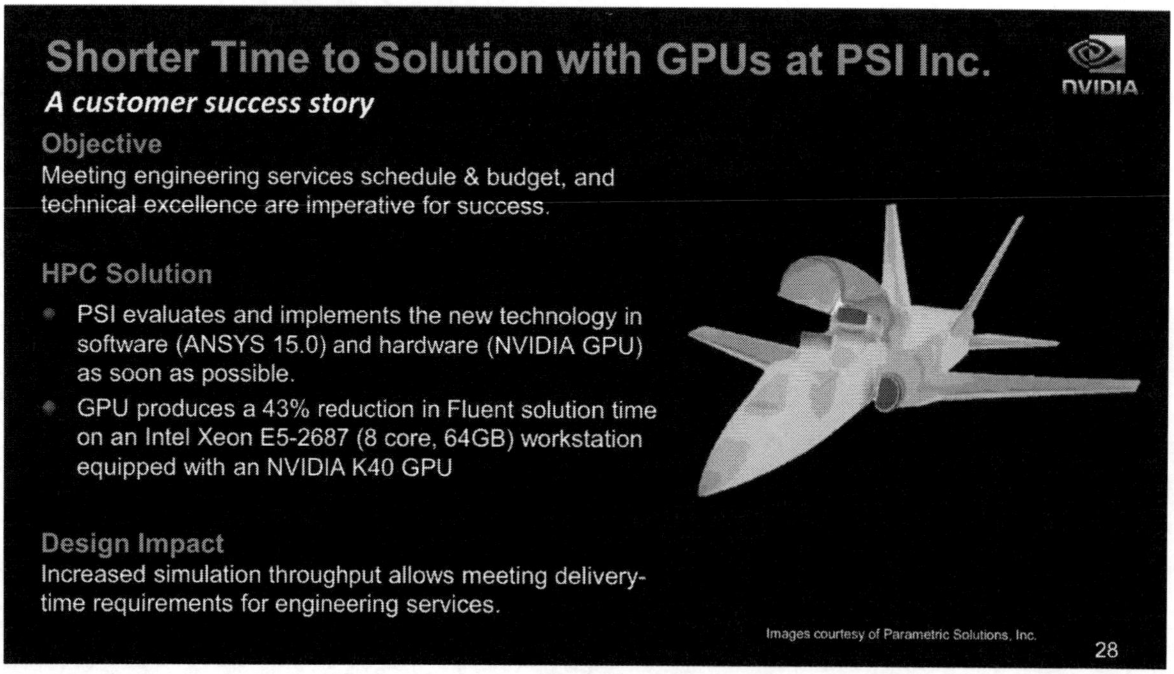
Shorter Time to Solution with GPUs at PSI Inc.
NVIDIA
A customer success story
Objective
Meeting engineering services schedule & budget, and technical excellence are imperative for success.
HPC Solution
PSI evaluates and implements the new technology in software (ANSYS 15.0) and hardware (NVIDIA GPU) as soon as possible.
GPU produces a 43% reduction in Fluent solution time on an Intel Xeon E5-2687 (8 core, 64GB) workstation equipped with an NVIDIA K40 GPU
Design Impact
Increased simulation throughput allows meeting delivery-time requirements for engineering services.
Images courtesy of Parametric Solutions, Inc.
28

# ANSYS Fluent 15.0 Resources

- ANSYS solution web page at NVIDIA – http://www.nvidia.com/ansys
- GPU user guide for ANSYS Fluent 15.0 available at - http://www.ansys.com/Resource+Library/Technical+Briefs/Accelerating+ANSYS+Fluent+15.0+Using+NVIDIA+GPUs
- Previously recorded ANSYS IT webcast series webinar titled "How to Speed Up ANSYS 15.0 with NVIDIA GPUs", which is available at http://www.ansys.com

29

# Optimizing AmgX performance

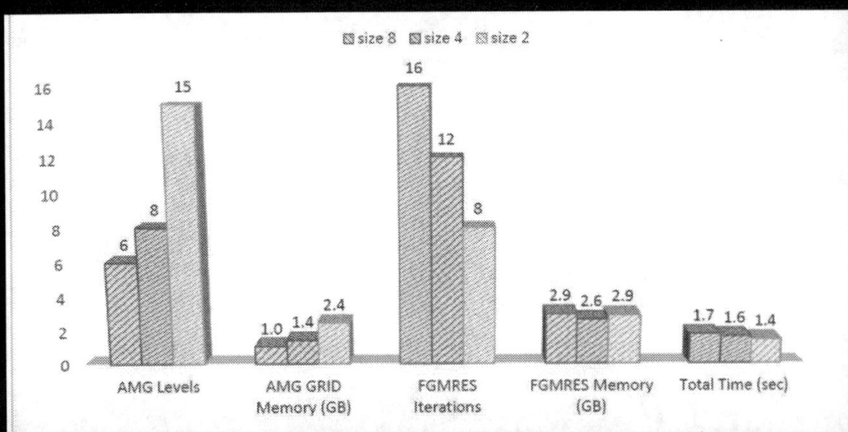

GPU User Guide for ANSYS Fluent

30

NVIDIA

# AGENDA

- GPUs in Enterprise Computing
- Business Challenges in Product Development
- GPU Computing for CAE Applications
- GPU Acceleration in ANSYS Fluent 15.0
- NVIDIA GPU Performance on Automotive Applications
- NVIDIA Products & Solutions for ANSYS Simulations
- Value of GPU-accelerated Simulations

31

NVIDIA

## NVIDIA Kepler family GPUs for CAE simulations

K20 (5 GB)
K20X (6 GB)
K40 (12 GB)

K6000 (12 GB)

32

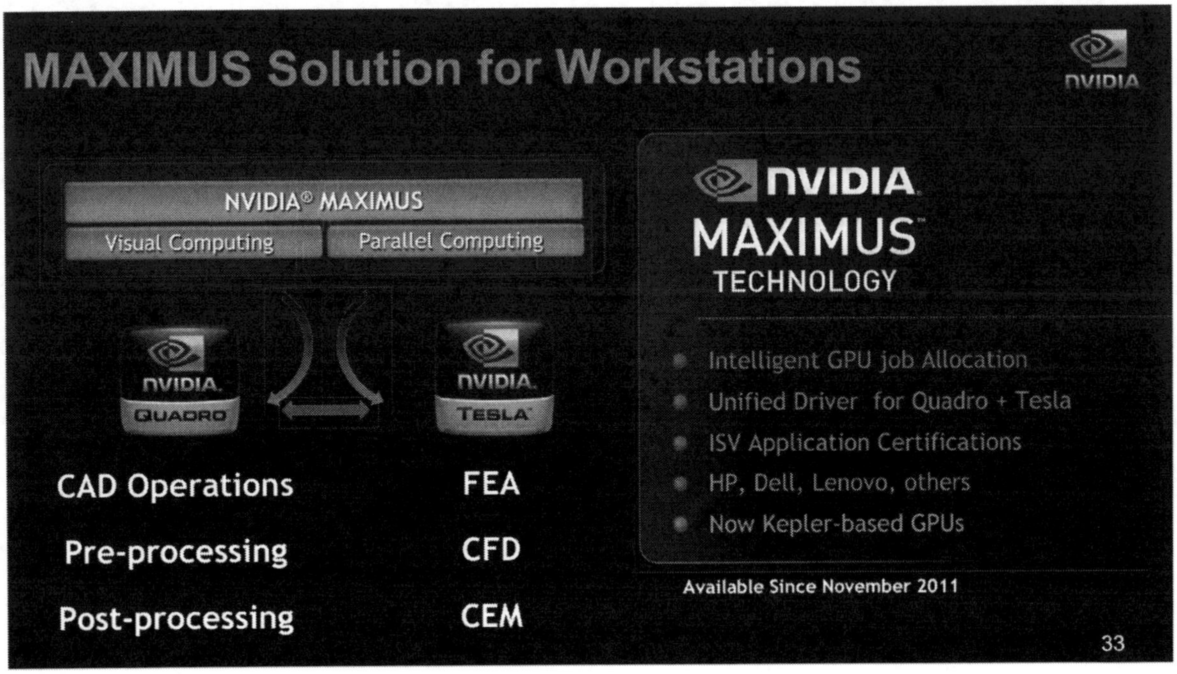
MAXIMUS Solution for Workstations
NVIDIA® MAXIMUS
Visual Computing
Parallel Computing
NVIDIA QUADRO
NVIDIA TESLA
CAD Operations
Pre-processing
Post-processing
FEA
CFD
CEM
NVIDIA MAXIMUS TECHNOLOGY
Intelligent GPU job Allocation
Unified Driver for Quadro + Tesla
ISV Application Certifications
HP, Dell, Lenovo, others
Now Kepler-based GPUs
Available Since November 2011
33

AGENDA
GPUs in Enterprise Computing
Business Challenges in Product Development
GPU Computing for CAE Applications
GPU Acceleration in ANSYS Fluent 15.0
NVIDIA GPU Performance on Automotive Applications
NVIDIA Products & Solutions for ANSYS Simulations
Value of GPU-accelerated Simulations
34

## Benefits of GPU-accelerated simulations

***More simulations in the same amount of time or same number of simulations in less amount of time***

**Improve product-quality**

More design points can be analyzed for better-quality products without slipping project schedules

**Faster time-to-market**

Simulation times can be cut into half, thus shorter product development times

**Complex simulations**

Mesh sizes can be doubled or advanced models can be used without increasing simulation times

35

ANSYS
2014 Automotive Simulation World Congress
on October 9-10, 2014, at JP Tower Hall and Conference, Tokyo

# Calculation of heat transfer properties in the exhaust system of an automotive

**Itsuhei KOHRI**

Tokyo City University

## ABSTRACT

Saving time to develop a new vehicle is one of the essential motivations by predicting heat environment in the engine compartment and around exhaust system with Computational Fluid Dynamics (CFD). However, it is not always practical because large calculations are needed to solve the heat and mass flow field including internal flow of the exhaust system with full-3D calculation, and also around the vehicle body simultaneously. To reduce the time and the cost, the coupled system of 1D with 3D has been developed to predict heat environment in the engine room and around the exhaust system. This method made it possible to evaluate heat environment on various driving conditions by calculating the heat generation and transfer characteristics of the catalyst. In the previous phase of the study, the characteristics for 1D-model was given by experiments. However, the procedure depending on the experiment was not always practical at the first stage of the vehicle development, where the prototype was not prepared.

In this study, the heat and mass flow field in the exhaust system is calculated with 1D-calculations in spite of experiments. Actually, various problems to be solved for 1D-3D simulation have been still remained. One of the principal problems is the treatment of the pulsating flow of the exhausted gas. Basically, in case of this, unsteady calculation of URANS will be required. However, it takes much time for the unsteady calculation to analyze heat transfer characteristics of the exhaust system. Then, we studied applicability of the steady state calculation of SRANS. At the first step, we discussed fundamental features of SRANS regarding heat transfer properties by comparing with URANS. Secondary, the basic investigation on URANS was performed with simplified 2D model. Finally, the potential and the limitation of SRANS calculation have been clarified.

---

Professor,
Department of Mechanical Systems Engineering

1-28-1 Tamazutsumi, Setagaya-ku, Tokyo 158-8557, Japan

ANSYS
2014 Automotive Simulation World Congress
on October 9-10, 2014, at JP Tower Hall and Conference, Tokyo

## Calculation of heat transfer properties in the exhaust system of an automotive

自動車排気系の伝熱特性計算

**Itsuhei KOHRI**
Tokyo City University

TOKYO CITY UNIVERSITY

---

## Outline

- Background and Purpose
  - Purpose and Goal of this study
  - Previous results : 1D-3D simulation
- 1st Step : Simulation of real exhaust system
  - Real exhaust system model
  - Unsteady characteristics of heat transfer
- 2nd Step : Basic investigation on transient calculation
  - Simple exhaust manifold model
  - Prediction of thermal characteristic
  - Discus about unsteady and steady.
- Conclusion

TOKYO CITY UNIVERSITY 2

---

## BACKGROUND AND PURPOSE

TOKYO CITY UNIVERSITY 3

---

## Background

- Thermal management in the engine compartment
  - To control heat balance of engine cooling system
  - To control temperature of various parts
    → to prevent heat damage
- Design of cooling system in the engine compartment
  - Empirical and/or experimental procedure
  - CFD calculation --- partial/hybrid simulation → full simulation
- Prediction of the surface temperature of parts with CFD
  - Publications concerning thermal management with CFD
    - Wang etc. 2010, JSAE } Full 3D ⇨ These methods require large computational resources.
    - Tanaka etc. 2012,JSAE
  - Attitude for prediction of temperature with CFD
    - More precise calculation to follow accuracy
    - Smaller and simpler calculation aiming economical/effective design

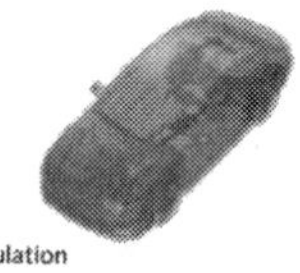

TOKYO CITY UNIVERSITY 4

---

## Early attempt on cooling simulation

- Recent approach of "Smaller and simpler simulation"
  - In this study, authors will follow "Smaller and simpler calculation"
    - I.Kohri et.al. 2011, ASME-JSME-KSME
    - K.Ishikawa et.al. 2012, JSAE --- 1D-3D hybrid
- Previous authors' works (1D-3D hybrid simulation)
  - Heat flow inside exhaust system: 1D-CFD } coupling
  - Heat flow outside exhaust system: 3D-CFD
  - Computational procedure
    - 1D-calculation and 3D-calculation are processed alternately.
    - Heat flux "$Q_r$" is output of 1D which is input of 3D.
    - Surface temperature "$T_w$" is output of 3D which is input of 1D

| 1D-simulation (UDF) | | 3D-simulation (Fluent) |
|---|---|---|
| ・Equation of Continuity<br>・Momentum equation<br>・State equation<br>・Energy equation<br>・Heat transfer equation | $Q_r$ →<br>← $T_w$ | ・Equation of Continuity<br>・RANS equation<br>・Heat transfer equation<br>・Thermal radiation equation |

TOKYO CITY UNIVERSITY 5

---

- Calculative condition
  - 1D-simulation
    - Steady, compressible flow
    - Heat generation in catalyst is considered
    - Heat transfer properties are given by experiments.
  - 3D-simulation
    - Steady (SRANS), incompressible, RNG
    - All tetrahedral, 9million of cells
- Results of early attempt
  - The simulation predicted surface temperature of typical component with order of 15°C accuracy in engine compartment.
  - This method has been proved as practical for vehicle development.

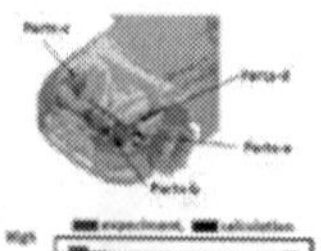

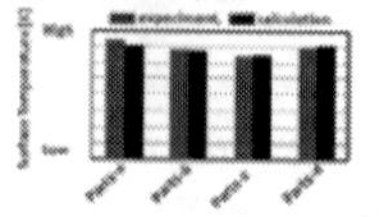


TOKYO CITY UNIVERSITY 6

## Purpose of the study

- Problems remained in above 1D-3D simulation to be solved
  - Heat transfer property should be expected for arbitrary systems
    - Since heat transfer property is determined by experiment for individual system, it is not available to employ 1D-3D simulation for arbitrary system.
  - To discuss fundamental features of SRANS regarding heat transfer properties by comparing with URANS.
    - Individual property is calculated with local 3D simulation inside of the exhaust system, as substitute for the experiment.
    - The property obtained by local simulation is given to 1D simulation.
  - To clarify potential and limitation of SRANS calculation, which is the goal of this investigation.
- Concrete procedure in this study
  - We focused on "clarify the potential of SRANS".
    - 1st step : Simulation of real exhaust system
    - 2nd step : Basic Investigation On Transient Calculation ← *Main content*

TOKYO CITY UNIVERSITY 7

## SIMULATION OF REAL EXHAUST SYSTEM

TOKYO CITY UNIVERSITY 8

## Past results prior to this study

- Past results regarding analysis in exhaust system.
  - Sugiura : Comparing SRANS with URANS
    - Straight pipe model whose geometry is different from the current model
    - Pulsation induced reverse flow on unsteady state to increase HTC
    - However, SRANS does not gives remarkably different HTC from that by time-averaged value of URANS
  - Park : Analysis with URANS
    - Geometry is similar to the current model
    - Pulsation induced both reverse flow and vortex shedding which are expected to enhance the diffusion of the fluid near the wall
    - No comparison with SRANS
- Problems remained to be solved
    - To clarify the effects of both reverse flow and vortex shedding on heat transfer property for the current model

TOKYO CITY UNIVERSITY 9

## Calculation condition

- Specification of CFD for URANS and SRANS

Calculation condition

| | | | |
|---|---|---|---|
| - | | URANS | SRANS |
| - | | Case1 | Case2 |
| Governing equation | | RANS equation | |
| | | Continuity equation | |
| | | Energy /heat transfer equation | |
| Fluid | | Compressible | |
| State | | Unsteady | Steady |
| Turbulence Model | | RNG k-ε model | |
| Boundary layer | | Std. wall function | |
| Time Step [s] | | 0.0001 | - |
| Wall | Ambient Temperature [K] | 373 | |
| | Heat Transfer Coefficient [W/m²] | 20 | |

- Grid structure

Mesh information

| | |
|---|---|
| Cell structure | All Tetrahedral |
| Total number of cells | 1.8 million |

TOKYO CITY UNIVERSITY 10

- Transient condition
  - Mass flow rate of each port

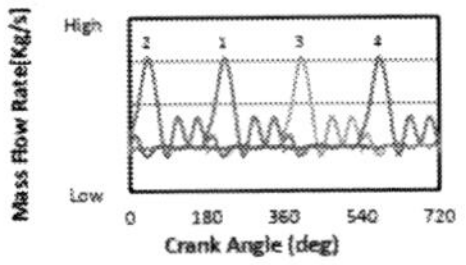


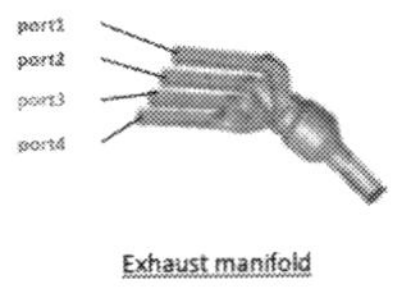


Exhaust manifold

  - Gas temper at inlet of each port

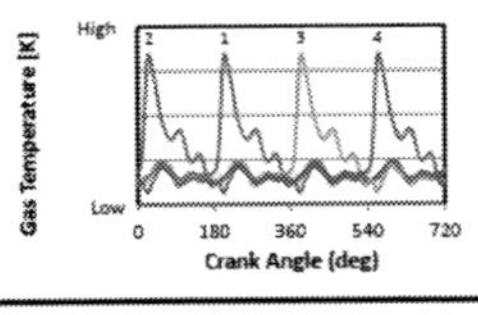


TOKYO CITY UNIVERSITY 11

## URANS vs. SRANS

- Comparison between URANS and SRANS
  - Averaged flow velocity
    - Flow field of SRANS is different from that of time-averaged URANS.

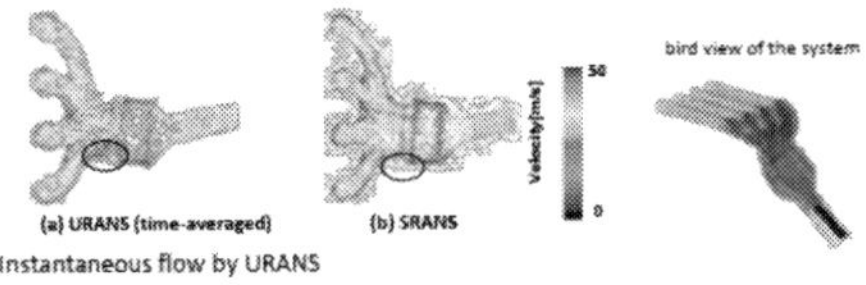


  - Instantaneous flow by URANS

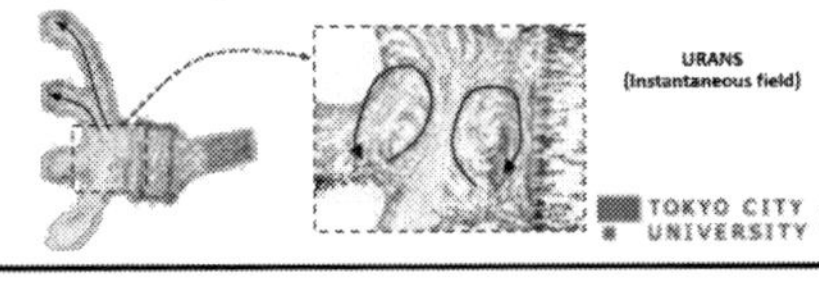


TOKYO CITY UNIVERSITY 12

- Heat transfer coefficient (HCT)
  - URANS predicts twice higher HTC than SRANS.

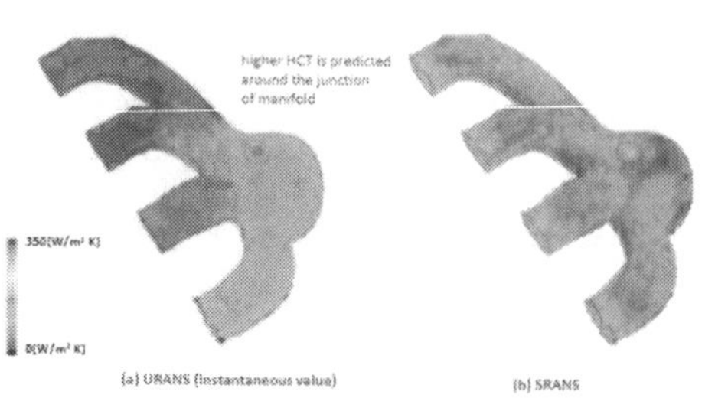


(a) URANS (instantaneous value) (b) SRANS

TOKYO CITY UNIVERSITY (13)

## Problems in SRANS simulation

- Problems in SRANS simulation
  - Certainly, SRANS simulation gives different HTC from URANS.
  - URANS needs longer computational time than SRANS
    - URANS : several weeks
    - SRANS : a few hours
    - As far as using URANS, it spends a lot of time to evaluate several cases on heat transfer characteristics.
  - Whereas, we have much successful experiences with SRANS in this field.
    - What is the essential problem involved in this problem?
- Optimal choice of URANS and SRANS
  - Needs some fundamental investigation with simplified exhaust system mode to find the essential mechanism which caused the difference.

TOKYO CITY UNIVERSITY (14)

## BASIC INVESTIGATION ON TRANSIENT CALCULATION

TOKYO CITY UNIVERSITY (15)

## Purpose of the investigation in this chapter

- Purpose of the investigation in this chapter
  - To evaluate the effect of vortex on heat property.
    - The simplified exhaust model is employed to analyze heat property and flow in a convenient.
    - Simplified model is able to extract the typical flow(i.e. vortex, swirl and impinging flow) inside exhaust system.

TOKYO CITY UNIVERSITY (16)

## Simplified model

- Simplified mode of exhaust manifold
  - 2-dimensional model for light calculations

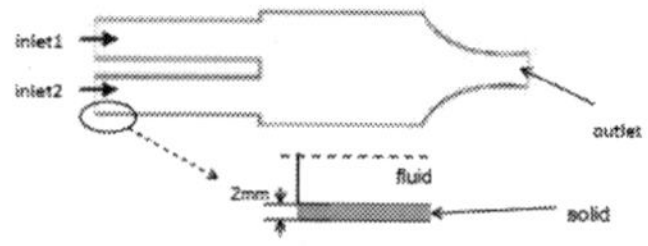


- Grid structure
  - All tetrahedral in solid and fluid region
  - Total number of cells is 0.1 millions

TOKYO CITY UNIVERSITY (17)

## Boundary condition

- Calculation condition in this study

| - | | Case1 | Case2 |
|---|---|---|---|
| Fluid | | Compressible | |
| State | | Unsteady | Steady |
| Turbulence Model | | SST k-ω model | |
| Total Number of Cells | | 0.1 Million | |
| Time Step [s] | | 0.001 | - |
| Inlet | Mass Flow Rate[K/s] | 0.1 – 0.8 [1)] | 0.5 |
| | Temperature[K] | 1120 | Approx. 1120 [2)] |
| Outer surface | Outside Temperature[K] | 1000 [3)] | |
| | Heat Transfer Coefficient[W/m² K] | 20 [4)] | |

- 1) time averaged mass flow rate is adjusted for both cases
- 2) time averaged enthalpy of the entering gas is adjusted for both cases
- 3) assuming as a heat protector be attached
- 4) HTC between outside air and outer surface of the model

TOKYO CITY UNIVERSITY (18)

- Pulsating flow condition
  - Inlet condition of SRANS is identical with time-averaged Inlet condition of URANS
    - Averaged values of mass flow rate, enthalpy and turbulent quantities at inlet are adjusted for both calculations
    - Inlet temperature of SRANS is determined for inlet enthalpy to be equal to that of URANS

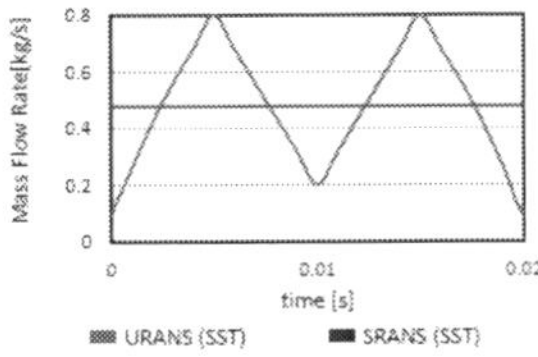


TOKYO CITY UNIVERSITY (19)

## Flow structure inside pipe

- Averaged flow field
  - URANS predicted higher velocity near wall than SRANS
  - Flow structure was difference between URANS and SRANS
    - URANS predicts transient vertices and swirl flow generating by pulsating flow.

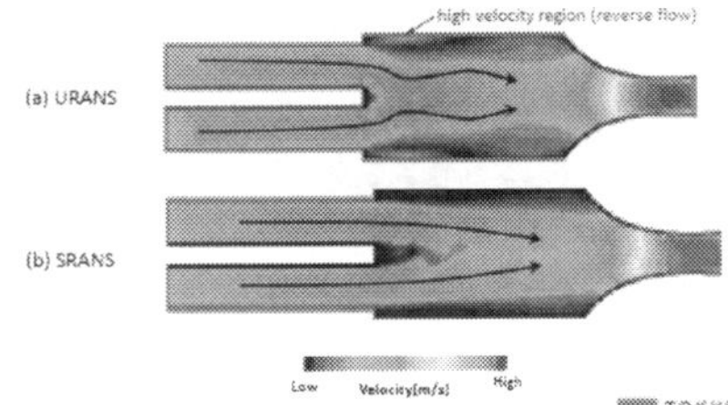


TOKYO CITY UNIVERSITY (20)

- Instantaneous flow field
  - Fluctuation of velocity field
  - Large vortices are generated alternately.
  - Large vortices entrains mass and heat which enhances heat transfer.

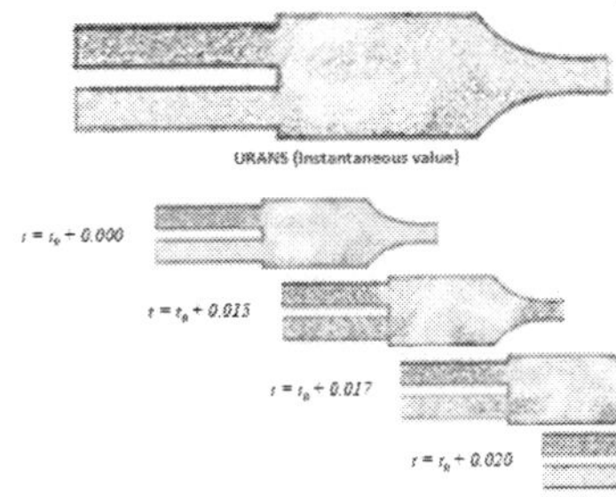


TOKYO CITY UNIVERSITY (21)

- Comparison of velocity near wall
  - Boundary layer profile is different.
  - Reverse flow is significant for URANS.
  - Velocity gradient near the wall is steeply for URANS than SRANS.

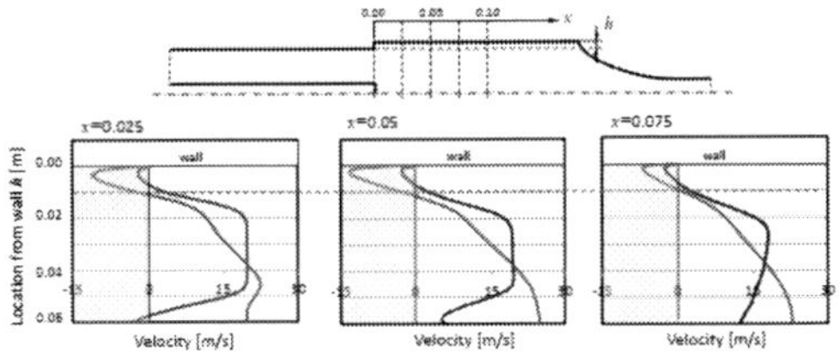


TOKYO CITY UNIVERSITY (22)

## Influence of pulsating flow on heat transfer

- What effect does pulsating flow has heat transfer?
  - HTC by URANS predicts twice higher than that by SRANS at the junction where pulsating flow generates vortices.
  - Motion of vortices promotes increase in heat transfer at inside surface.

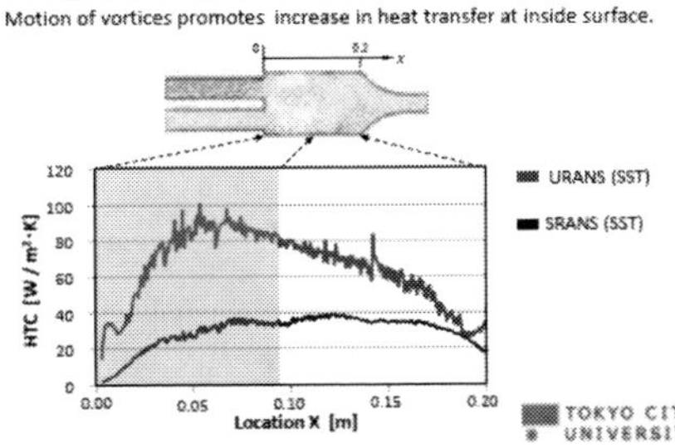


TOKYO CITY UNIVERSITY (23)

## Comparison of heat transfer

- Comparison of heat transfer
  - SRANS predicted 30% lower heat flux than URANS

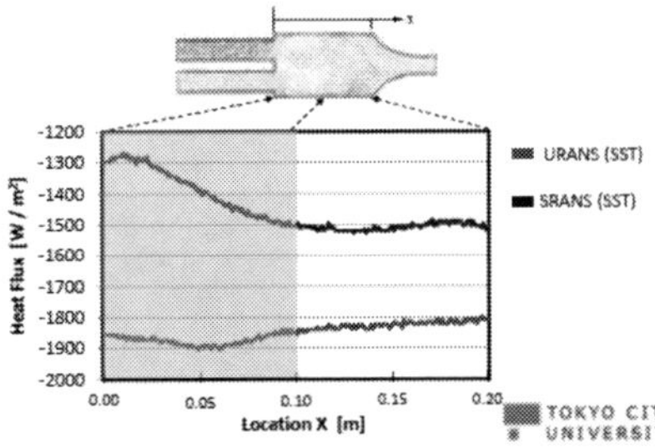


TOKYO CITY UNIVERSITY (24)

- Comparison of gas temperature
  - Temperature downstream of back step by SRANS is remarkable lower than that by URANS.
  - The difference is due to the effects of the generation of a large vortex. The vortex entrains heat so that its heat transfer is enhanced.

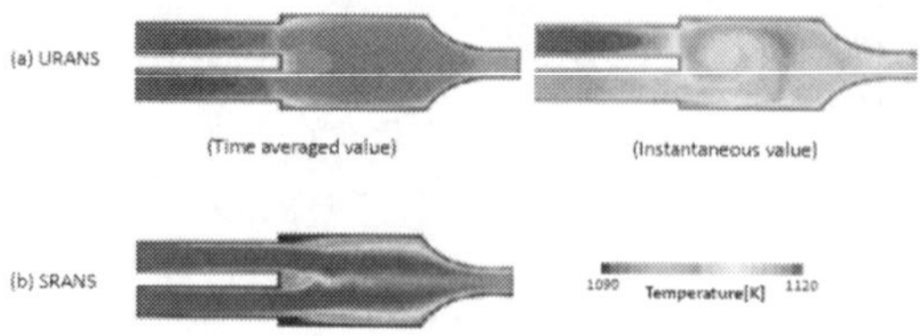


TOKYO CITY UNIVERSITY 25

## SUMMARY

TOKYO CITY UNIVERSITY 26

## Conclusions

- URANS and SRANS
  - For fluctuating phenomenon, SRANS does not always give average values. There is possibility that it will make a serious mistake.
  - Of course, for instance, SRANS sometimes gives practical averaged value for fluctuation phenomenon, such as oscillating single pipe flow and so on.
  - However, for pulsating flow with two inlet pipes, where inflow enters alternately, SRANS gives significantly different average values.
  - In case of that , at least, URANS is required. LES will be better.
- Choice of solution
  - If you have enough time and computational resources for your analysis, URANS is recommended without hesitation.
  - When you have to save time for your mission, if the transient phenomenon is mainly governed by turbulent diffusion, e.g. in the boundary layer or the shear layer, SRANS is applicable.
  - However, if the transient phenomenon contains large vortex structures which governs heat and/or mass transfer, URANS should be adopted, but not SRANS.

TOKYO CITY UNIVERSITY 27

2014 Automotive Simulation World Congress

# Aero-Vibro Acoustics For Wind Noise Application

David Roche and Ashok Khondge

ANSYS, Inc.

## Outline

1. Wind Noise
2. Problem Description
3. Simulation Methodology
4. Results
5. Summary

## Aerodynamics Noise Generation

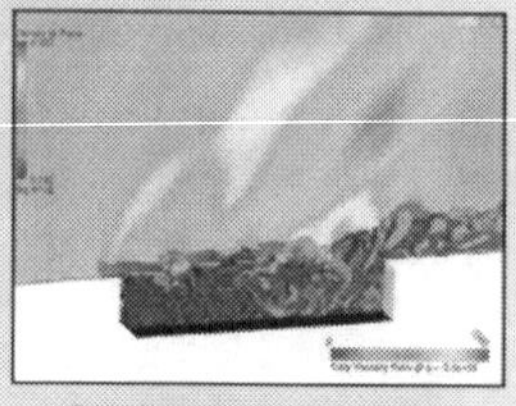

Cavity Resonance

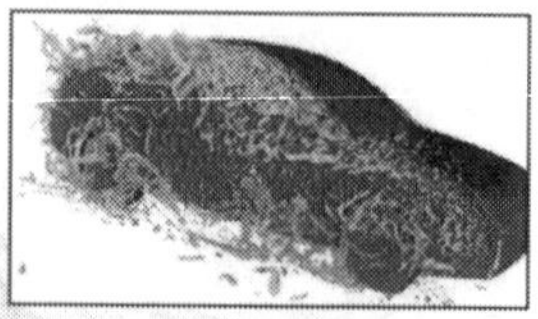

A-Pillar Vortex

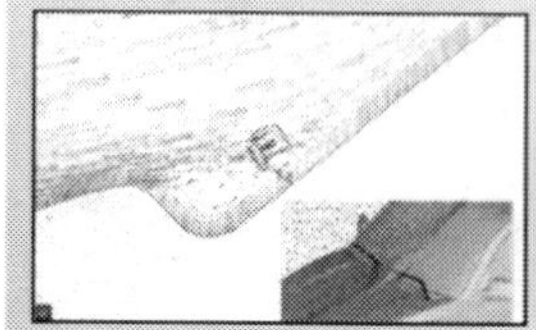

Flow separation/ Vortex shading

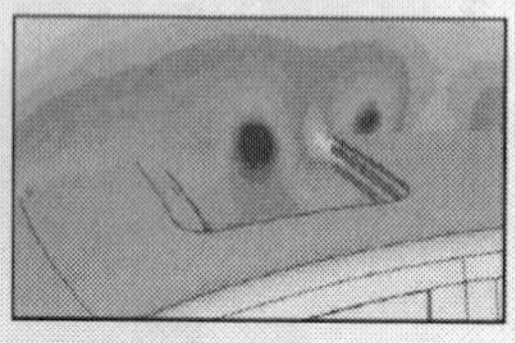

Cavity Resonance

Flow separation/ vortex shading

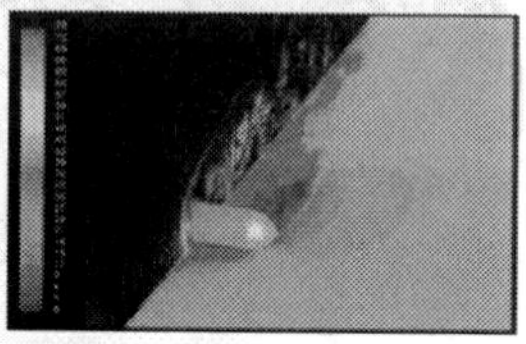

Flow separation/Vortex shading





## Wind Noise

- **High Frequency (> 500 Hz) Noise generated by head wind and perceived inside automotive cabin @ highway driving speeds**
- **It is cost beneficial to address wind noise "UPFRONT" in Design Process**
  1. **Highest degree of freedom to change exterior design causing wind noise**
  2. **Avoids expensive late countermeasures**

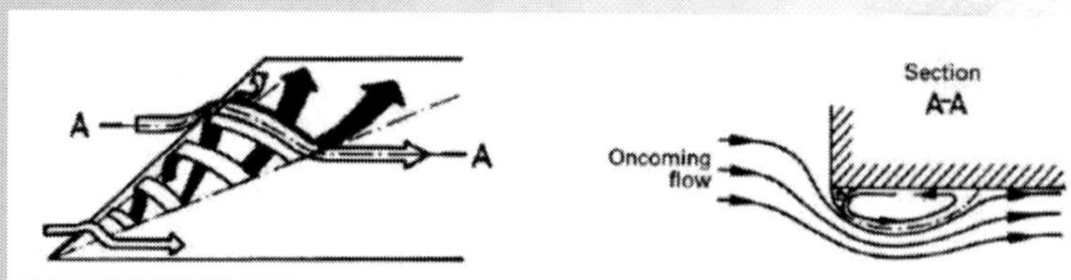


[1]

2014 Automotive Simulation World Congress
October 9-10 . Tokyo, Japan

## Problem Description

- Demonstrate Aero-vibro-acoustics coupling to predict noise at the interior of the Hyundai simplified model [2]
- HSM Model – Simplified model released by Hyundai Kia Motors for 2013 KSNVE Open Benchmark[2]

**Sound Source** Transient Separated flow at A-pillar → **Transfer Path** [Side glass, windshield] → **Receiver** Ear of a driver

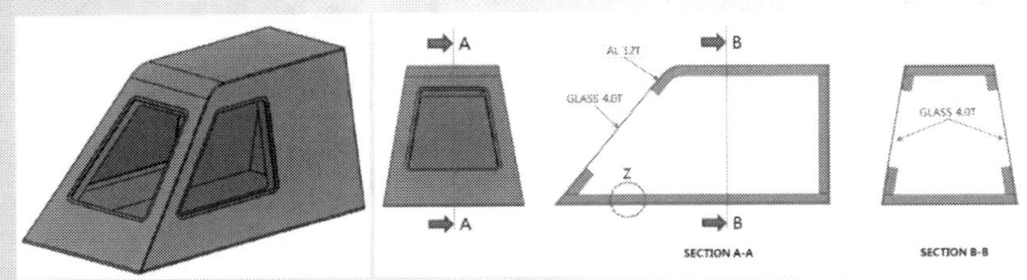




2014 Automotive Simulation World Congress
October 9-10 . Tokyo, Japan

## Simulation Methodology [1]

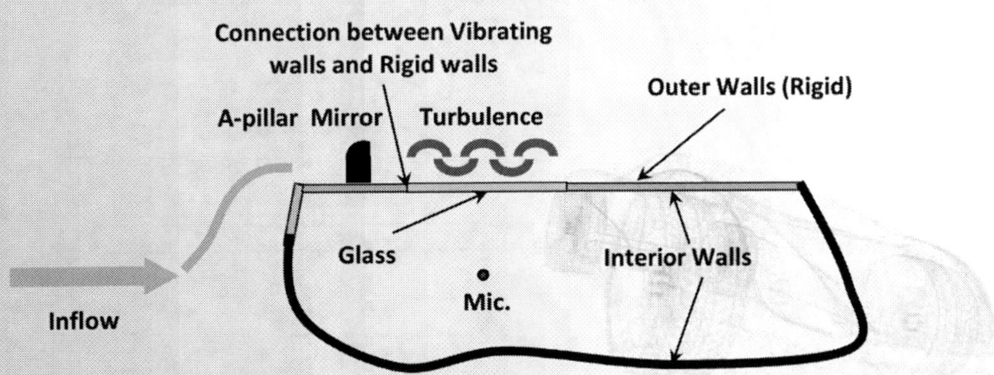


External CFD Model – Transient Flow → Vibrating Surfaces (Side Glass, Windshield) → Acoustics Model (Car Interior)

CFD modeling

Vibroacoustics Modeling

# Simulation Methodology [2]

Solve Transient CFD → Time → Freq. domain transform → Mapping Freq. Domain Pressure Loading → Solve "Vibro-Acoustics" Model

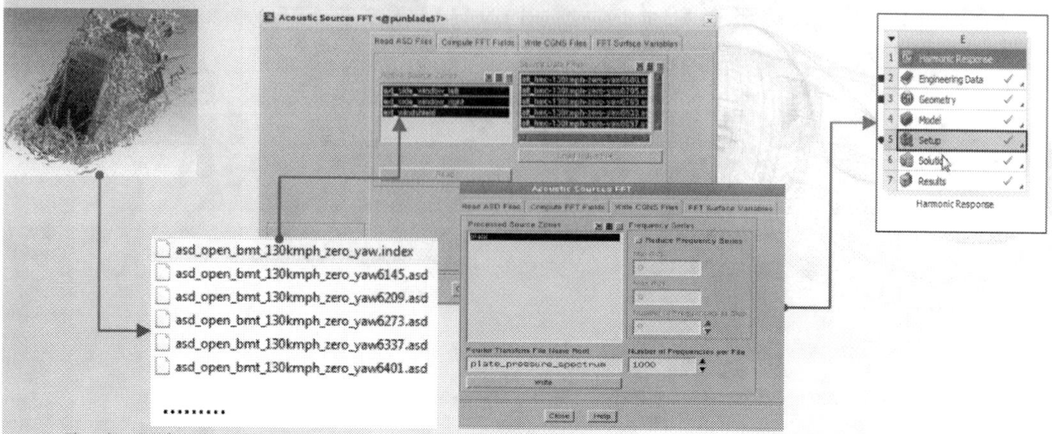






# Workflow Studied

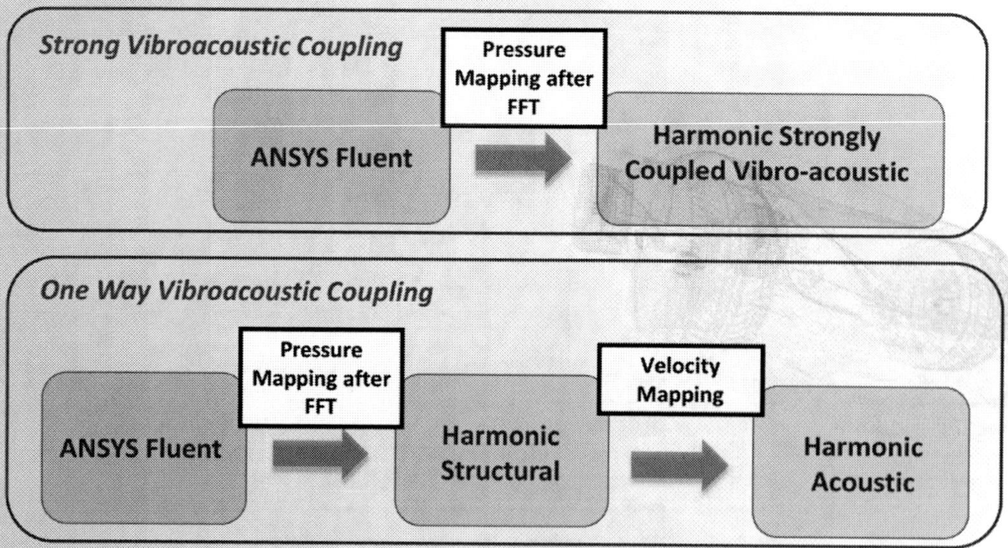

# Test Measurement Points

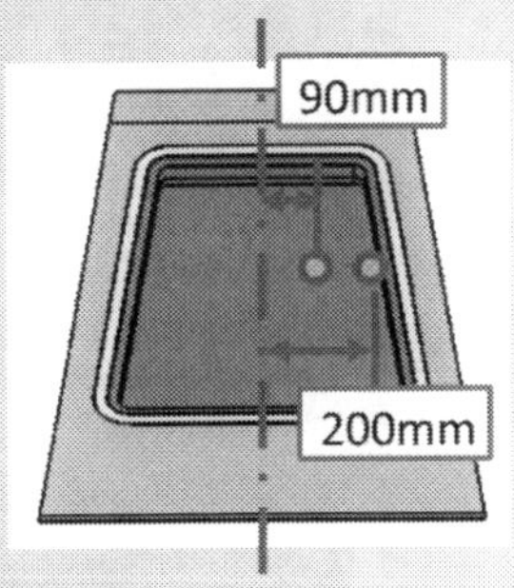


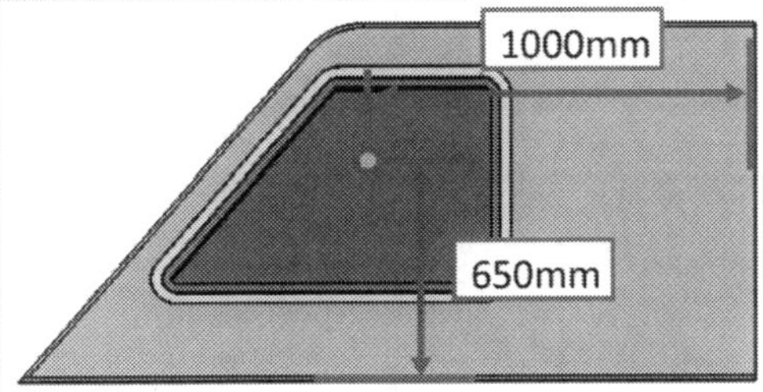


Interior Microphone Locations

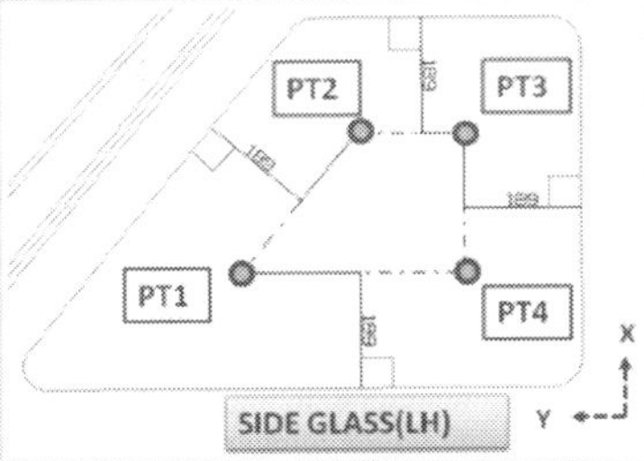


Accelerometers mounted on LH Side Glass



2014 Automotive Simulation World Congress
October 9-10 . Tokyo, Japan

# Transient CFD Modeling

- CFD Domain consists of External HSM surfaces, side glasses and windshield & wind tunnel boundaries
- Configuration Studied : HSM – 0 deg. Yaw
- – Nozzle Inlet : Velocity Inlet : 130 kmph [Profile]
- – Tunnel Outlet : Pressure Outlet (Gauge Pressure = 0 Pa)
- – Tunnel Inlet : Pressure Inlet (Gauge Tot. Pressure = 0 Pa)
- – Tunnel Top, floor, side, wall BC (No-Slip)

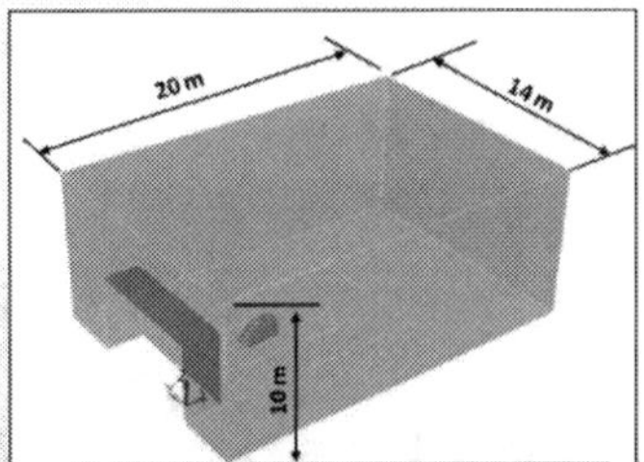


CFD Domain

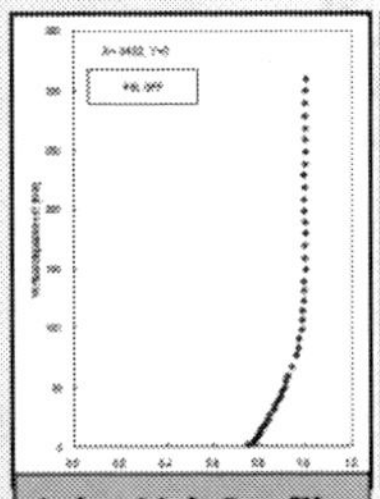

Inlet Vel. Profile

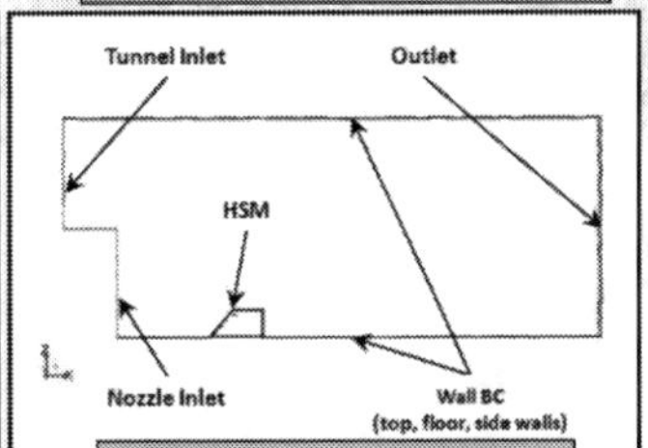


Boundary Conditions

# Mesh Details

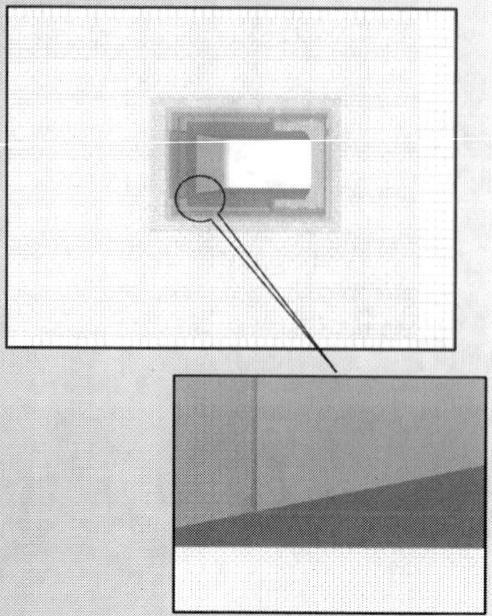

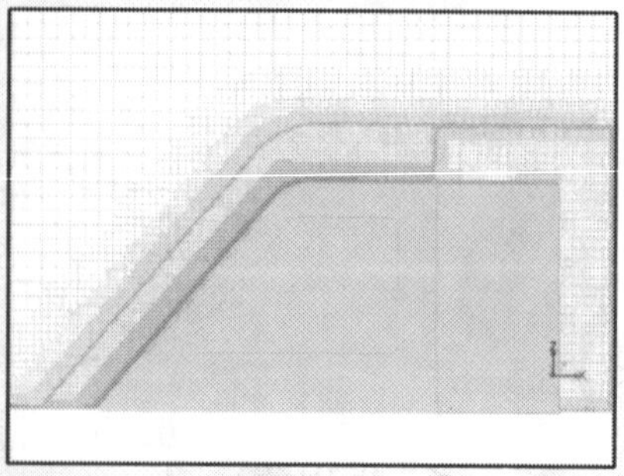

- Total Cell Count = 45 Million
- Prims Cell Count = 23 Million
- No of Prism Layers = 12
- First Prism Layer Height = 0.05 mm
- Surface Mesh Size : A-pillar 1.5 to 2.0 mm, Side glasses, windshield = 2.0 to 3.0 mm





# Solver Settings : Transient CFD Simulation

1. Solver : ANSYS Fluent, Pressure Based, Double Precision, Transient , Gradient – Green Gauss Node Based
2. Transient Formulation : 2nd Order Implicit, Time Step Size = 2e-5 s
3. Material - Air as Ideal Gas
4. Turbulence Model Steady State : SST K-Omega
5. Turbulence Model Transient : DDES – SST K-Omega
6. Pressure Velocity Coupling – SIMPLEC
7. Spatial Discretization
   - Pressure : Second Order, Density : Second Order Upwind
   - Momentum : Bounded Central Differencing
   - TKE & Specific Dissipation Rate : Second Order Upwind
   - Energy : Second Order Upwind

## Solution Procedure

1. Run Steady State Simulation using SST K-Omega Turbulence Model
2. Steady State Simulation Solver Settings – Pressure Based Coupled Solver
3. Switch to Transient Simulation, Use Second Order Temporal Discretization
4. Switch to DDES SST K-Omega Turbulence Model
5. Run initial transient simulation to achieve dynamic steady state
6. Run Final transient simulation [time step size 2e-5, no. of iterations per time step = 8]
7. Export the ASD Data on Wind-shield, side-glass surfaces at every time step
8. Perform Surface Acoustics FFT to transform source data into Frequency Domain





## Transient Flow Field

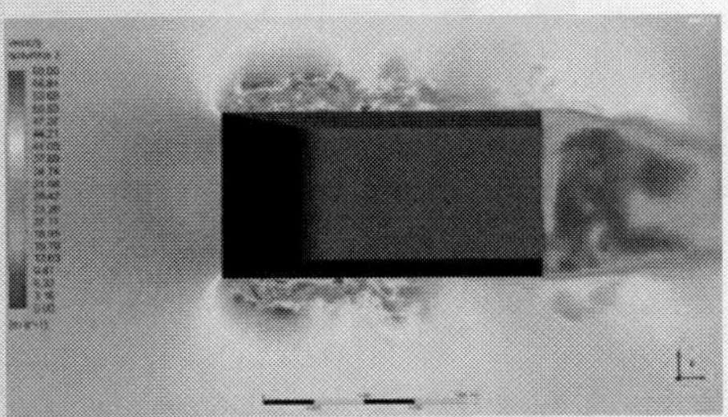

Velocity Contours at Z = 0.5 m

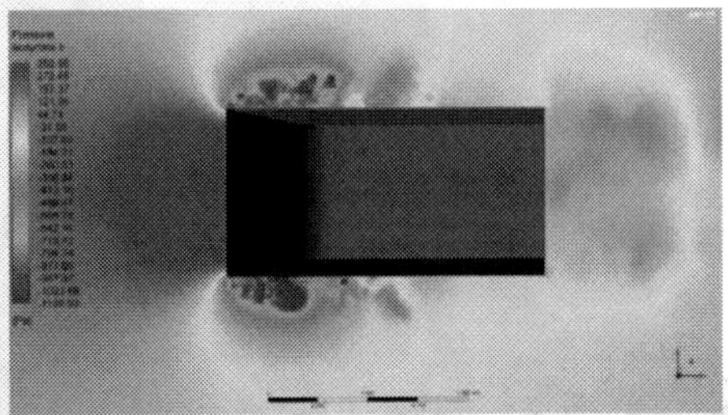

Pressure Contours at Z = 0.5 m

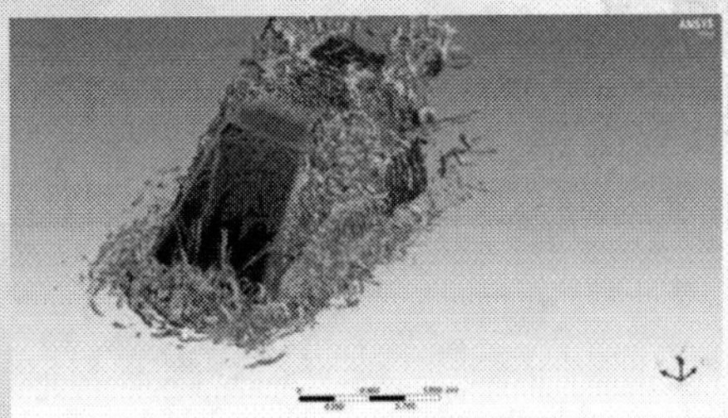

Iso-surface of Q-Criterion colored by velocity magnitude

## Source Data : Transformation Time → Frequency Domain

**Acoustic Source FFT is a Beta Feature in R 16.0**

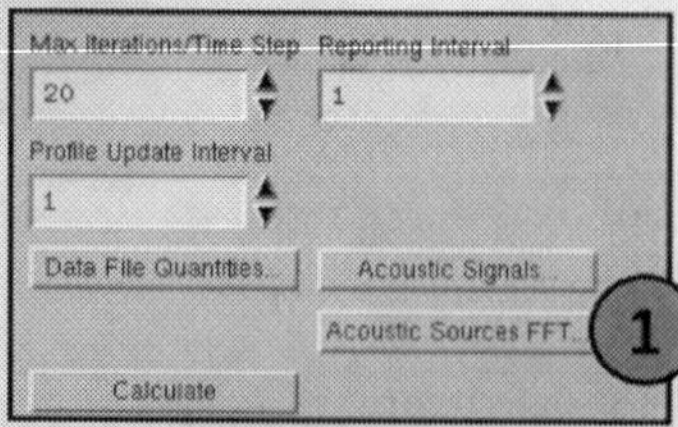


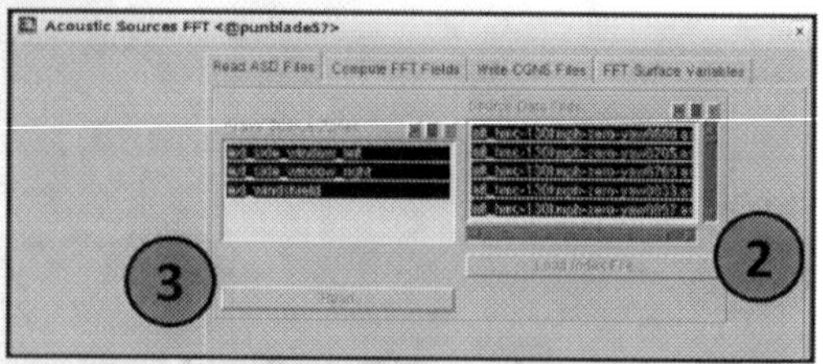


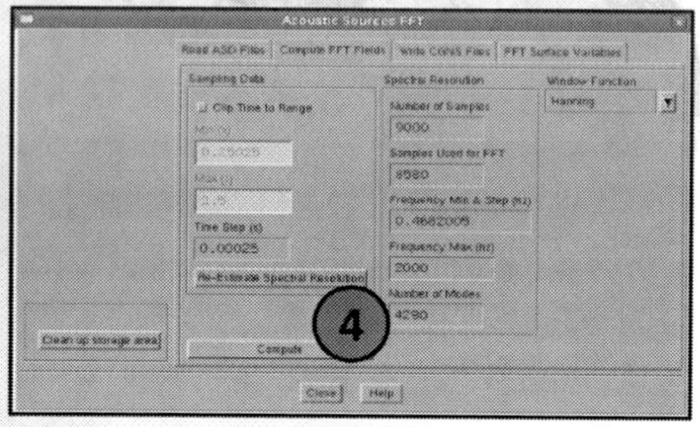


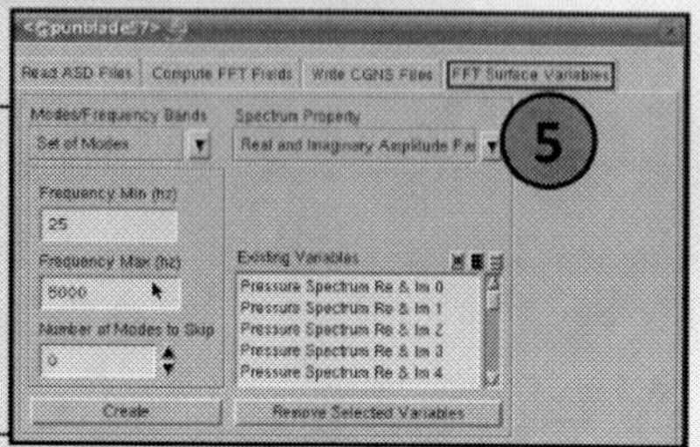


1. Set Modes (Real /Img.)
2. Octave (SPL)
3. 1/3rd Octave (SPL)
4. Constant Band (SPL)





## Surface dB Map

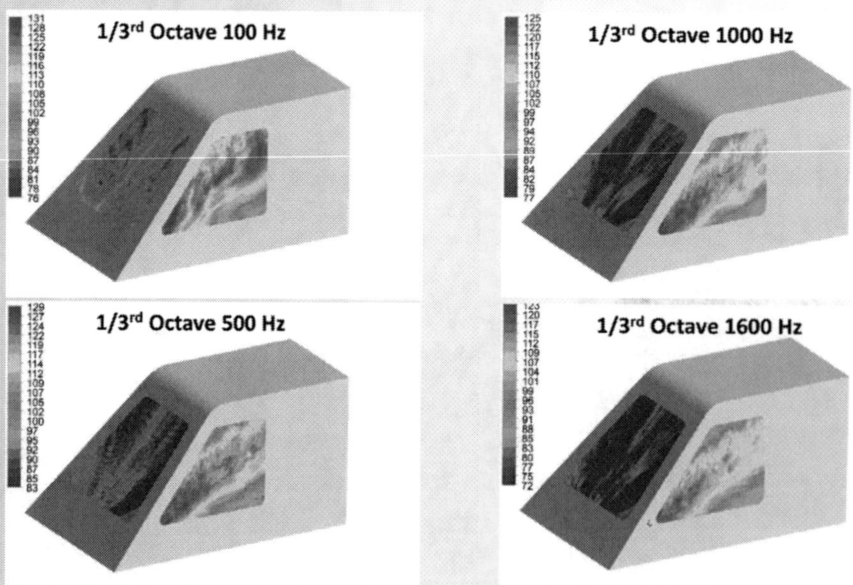

2014 Automotive Simulation World Congress
October 9-10 . Tokyo, Japan

# Acoustics Pressure Loading in Freq. Domain

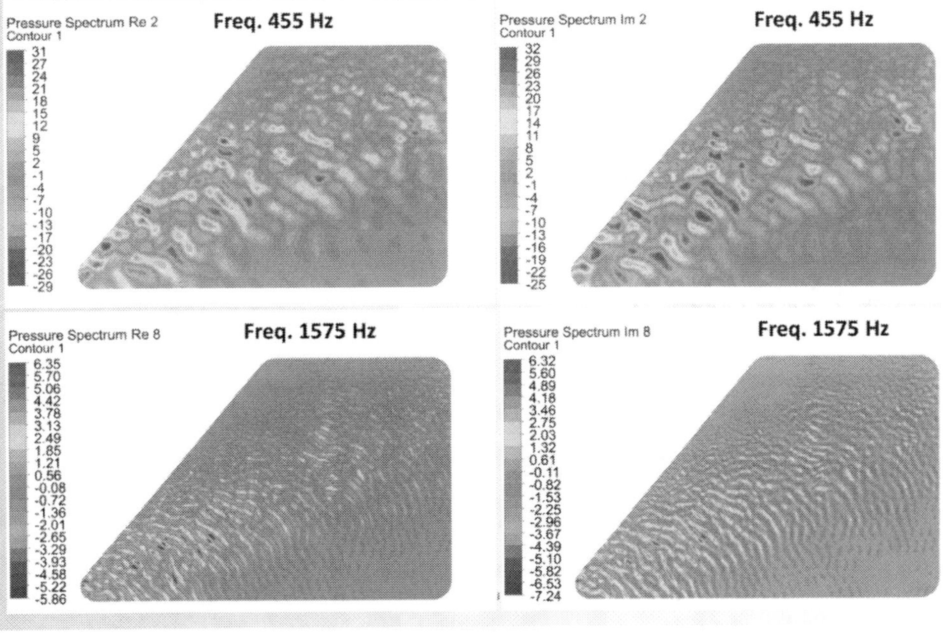


2014 Automotive Simulation World Congress
October 9-10 . Tokyo, Japan

# Acoustics Source Mapping

ANSYS Fluent | **Mapping for 100 Hz** | ANSYS Mechanical

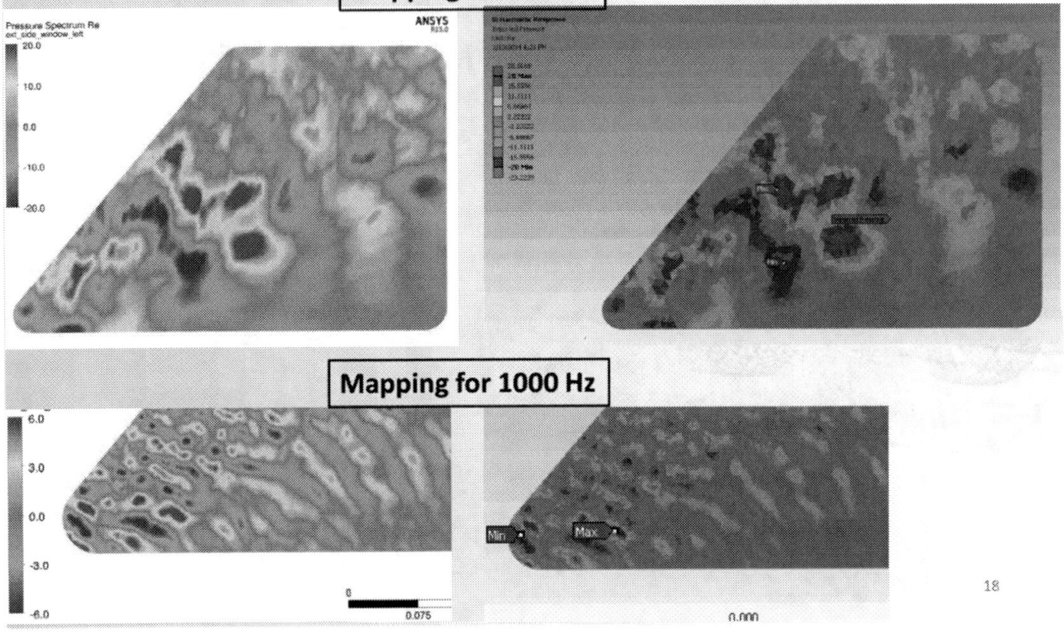


18

# Vibroacoustics Modeling

## Structural Material Properties

| PROPERTIES | GLASS | AL 6061 | HEAVY LAYER |
|---|---|---|---|
| Thickness (mm) | 4 | 12 | 1 |
| Density (kg/m³) | 2500 | 2700 | 2000 |
| Young's Modulus (GPa) | 70 | 69 | 0.04 |
| Poisson's ratio | 0.22 | 0.33 | 0.45 |

| PROPERTIES | Air | Foam |
|---|---|---|
| Mass Density (kg/m³) | 1.2 | 1.2 |
| Sound Speed (m/s) | 343 | 343 |
| Fluid Resistivity (N s /m⁴) | 6.83E+16 | |
| Porosity | 0.879 | |
| Tortuosity | 3.31 | |
| Viscous Length (m) | 9.483e-10 | |
| Thermal Length (m) | 1.2174e-10 | |

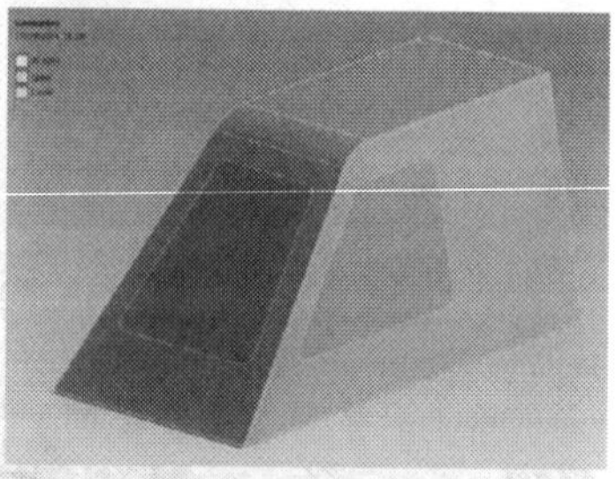

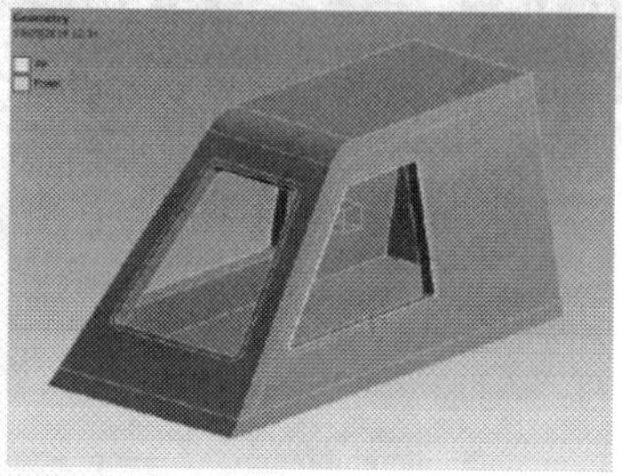





# Vibroacoustics Modeling

**Strong Coupling:**

- **Full Vibroacoustics harmonic analysis from 50 to 1000 Hz (117 frequencies)**

**Weak Coupling:**

- **Full structural harmonic analysis from 50 to 1000 Hz (117 frequencies)**
- **Full acoustic harmonic analysis from 50 to 1000 Hz (117 frequencies)**

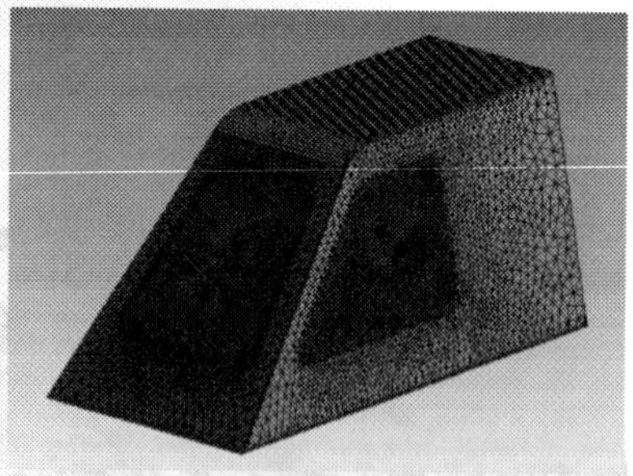

## Results : Acceleration vs Frequency

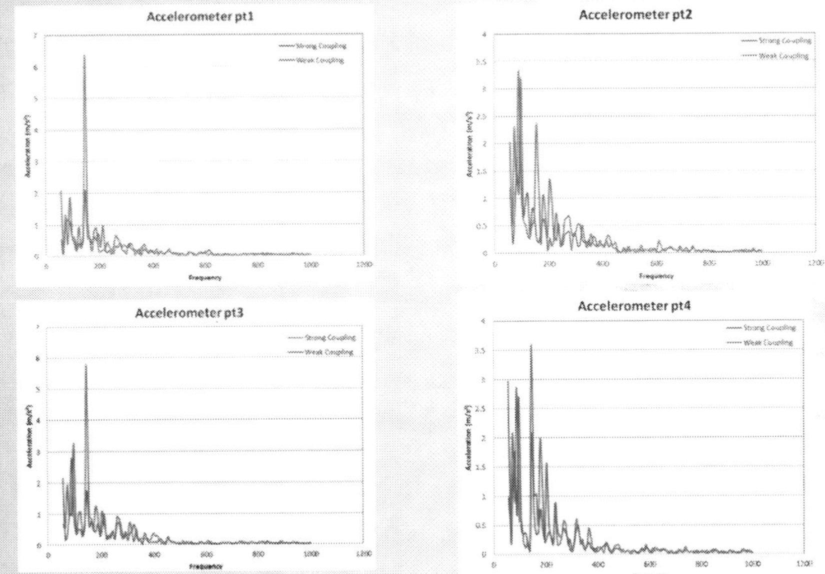






## Results : SPL(dB) vs 1/3$^{rd}$ Octave Freq. @ Interior Microphone

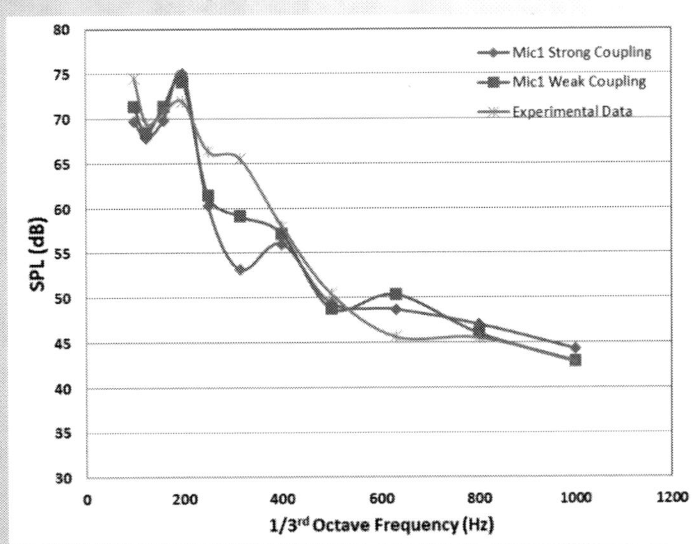

## Summary

1. Aero and Vibroacoustics coupling is demonstrated using transient CFD and Vibroacoustics modeling
2. ANSYS Fluent solver is used for transient aeroacoustics simulation
3. A Vibroacoustics simulation is done using ANSYS Mechanical using two approaches
   - Strong Vibroacoustics coupling approach
   - Weak Vibroacoustics coupling approach
4. Simulation Results are fairly in good agreement with Test Data
5. Differences between strong and weak Vibroacoustics coupling are observed
   - It doesn't seem possible to neglected the effects of the acoustics cavity on the deformation of the structure

ANSYS

Realize Your Product Promise™

# Underhood Thermal Modeling Using ANSYS Fluent – Models and Latest Advances

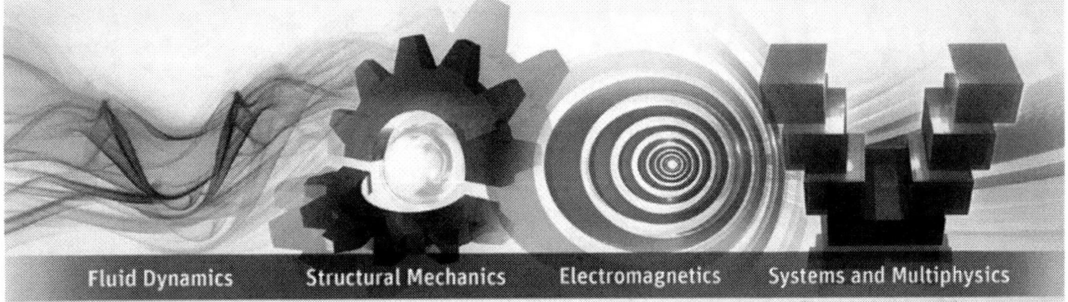



***Hamid Ghazialam***

***Principal Technical Services***


ANSYS

## Complete Automotive CFD

**Aerodynamics Aeroacoustics**

**Underhood Thermal Management**

**ANSYS CFD**

**Cabin, HVAC**

**Powertrain**

  October 6, 2014

## Complete Automotive CFD

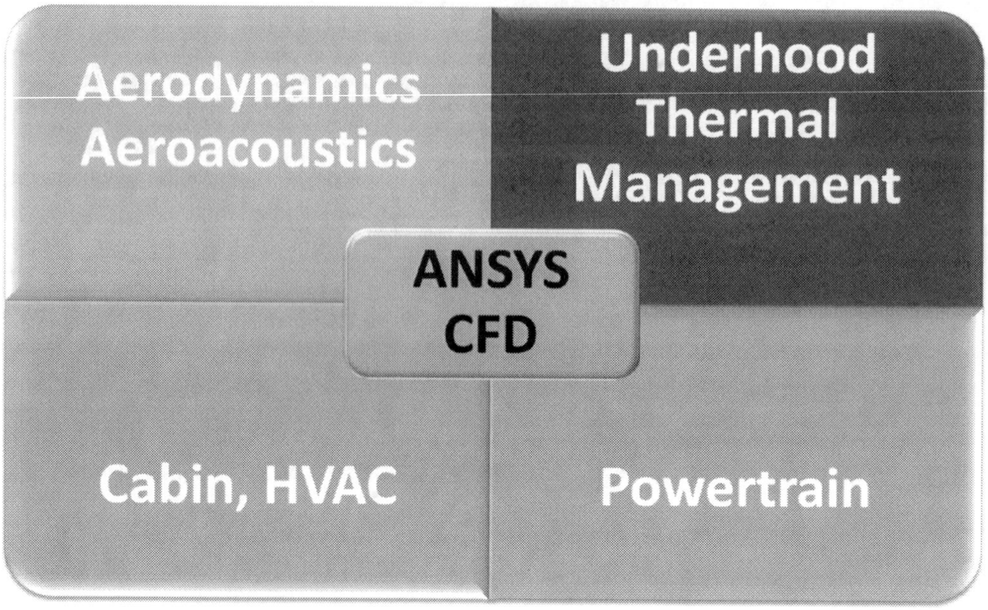




## Outline

**Underhood Simulation Using ANSYS Fluent**

- ❑ Introduction
- ❑ Model Overview
- ❑ Advances in Solver Capabilities
- ❑ Advances in Automatic Mesh Generation
- ❑ Summary and Conclusions
- ❑ Appendix

## ANSYS Introduction

- **Type of UTM simulation**
  - Front-End Air Flow (FEAF)
  - Full Thermal
  - Loal Thermal
- **Industries**
  - Automotive
  - Off-Highway
  - Truck and Bus
  - Aircraft
  - Military
- **Challenges**
  - Complex geometry
  - Dirty CAD
  - Large models

- **Road conditions simulated**
  - **Vehicle moving at constant speed**
    - Steady-state
  - **Constant speed → Idle**
    - Steady-state
  - **Constant speed → Soak**
    - Transient
  - **Warm up → Grade → Soak**
    - Transient
  - **Warm up → Grade → Idle**
    - Transient
  - **City-drive**
    - Transient

## ANSYS Outline

**Underhood Simulation Using ANSYS Fluent**

- Introduction
- Challenges
- Model Overview
- Advances in Solver Capabilities
- Advances in Automatic Mesh Generation
- Summary and Conclusions
- Appendix

## Models ANSYS Fluent Offers

- ❑ Steady-state as well as efficient transient flow solvers
  - Pressure-Based Coupled Solver
- ❑ Ability to automate the case setup and solution process
- ❑ Robust parallel processing and outstanding scalability
- ❑ Porous media (for heat exchangers and grills)
- ❑ Dual Cell and Macro based Heat Exchanger models
- ❑ Fan models (MRF, plane fan, and sliding mesh)
- ❑ Fast View-Factor Based Radiation Models
  - 45 minutes on 80 million cell using 256 cores!
- ❑ Exhaust Skin Temperatures
- ❑ Automated Post-Processing

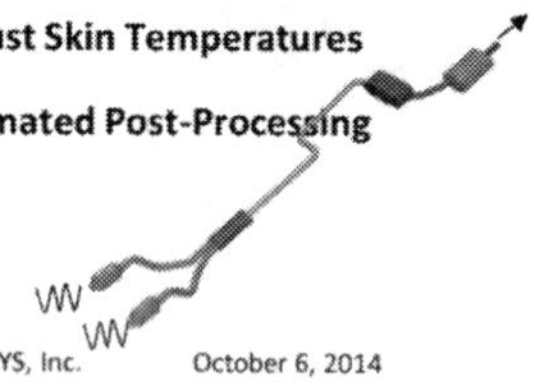

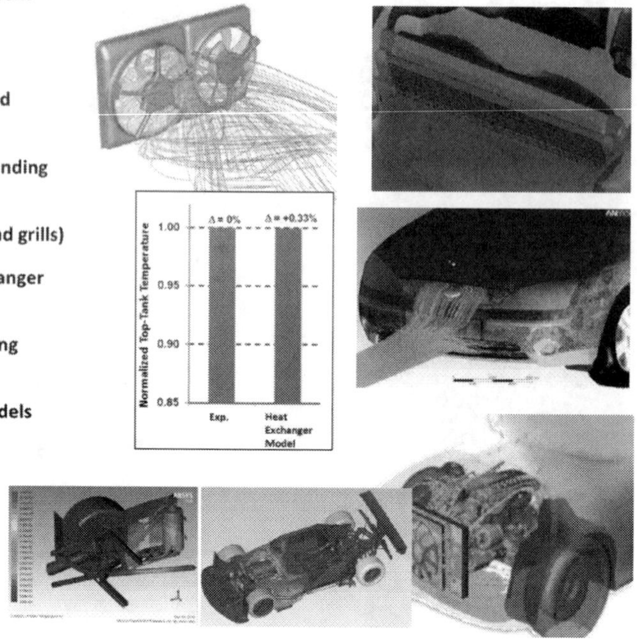




## Outline

### Underhood Simulation Using ANSYS Fluent

- ❑ Introduction
- ❑ Model Overview
- ❑ Advances in Solver Capabilities
- ❑ Advances in Automatic Mesh Generation
- ❑ Summary and Conclusions
- ❑ Appendix

# Multi-Layer Shell Conduction Models

- Allows conduction in planar direction
- Thin shield is meshed as a single-surface
- Accounts conduction at junctions
- User Inputs: material, thickness, heat generation (if any)
  - Thermal conductivity can be bi-axial
- Two Models

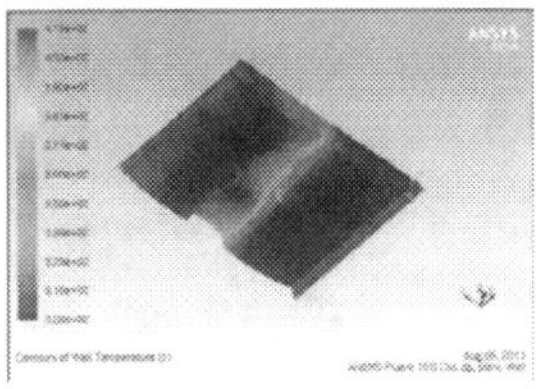

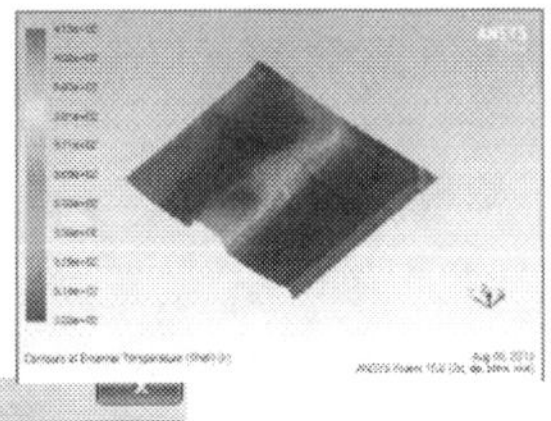

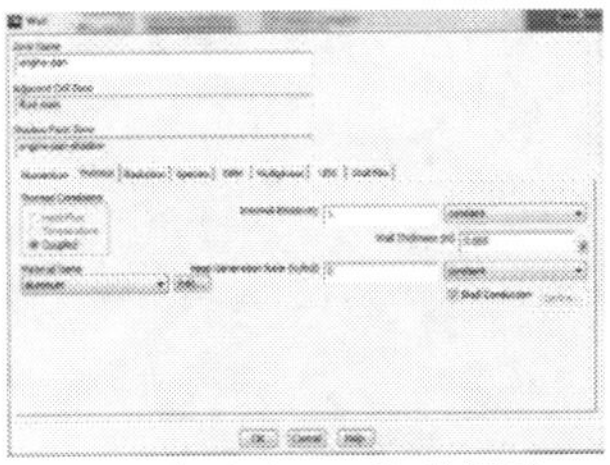

*Single Layer Shell (SLS)*

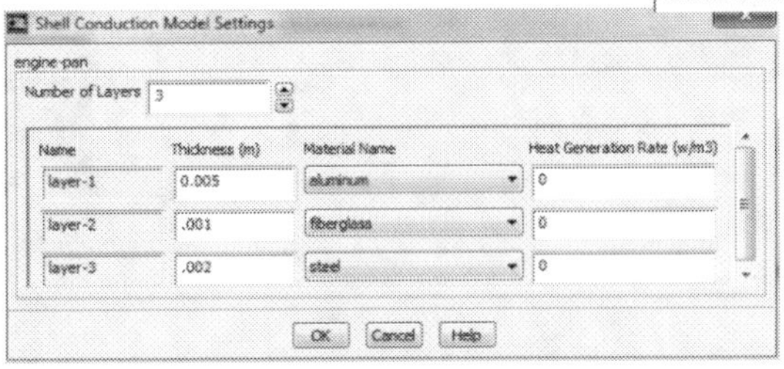


New: *Multi-Layer Shell (MLS)*



# Fluid-Solid Explicit Mapping

- With the mapping technique we can totally avoid conformal mesh, which can significantly reduce the time it takes to generate the mesh.

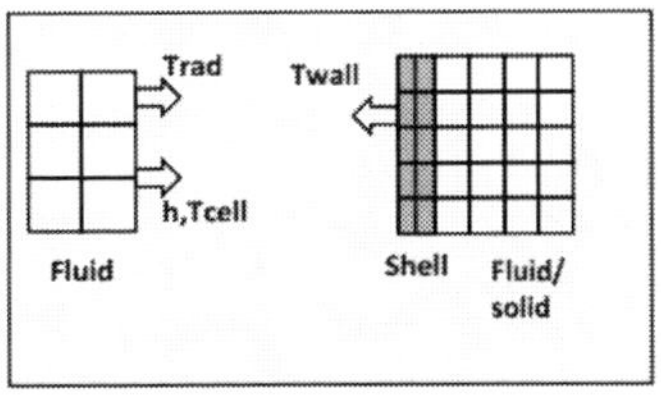


Fluid/Shell Mapping

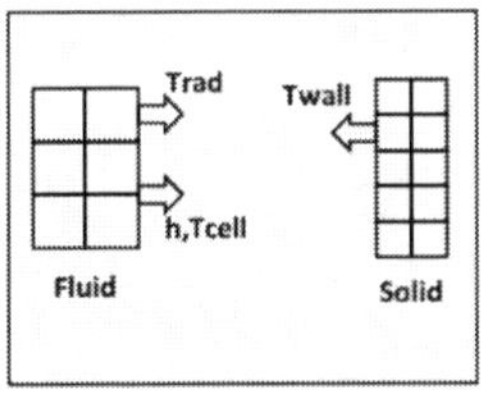


Fluid/Solid Mapping

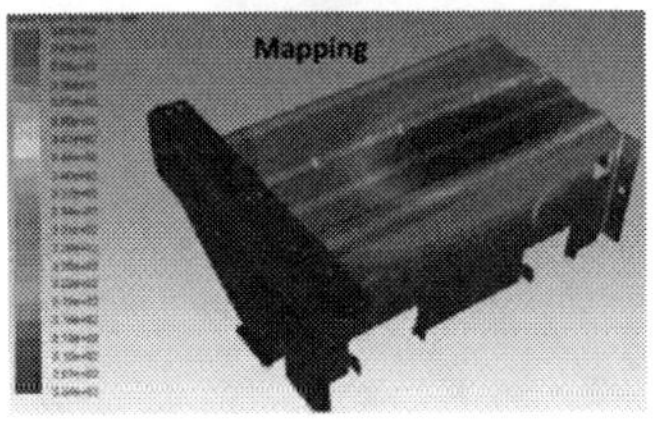


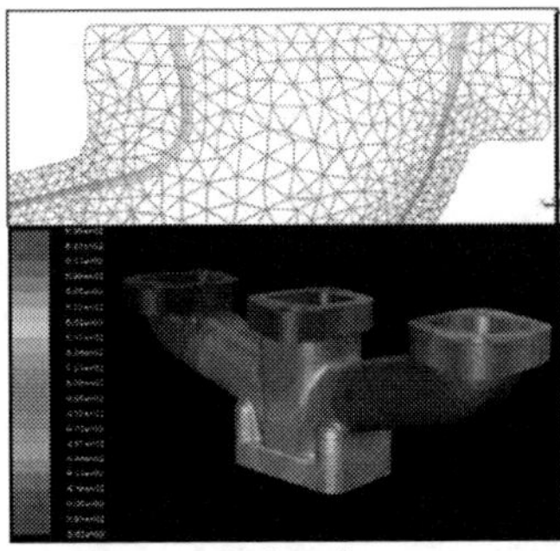

Conformal

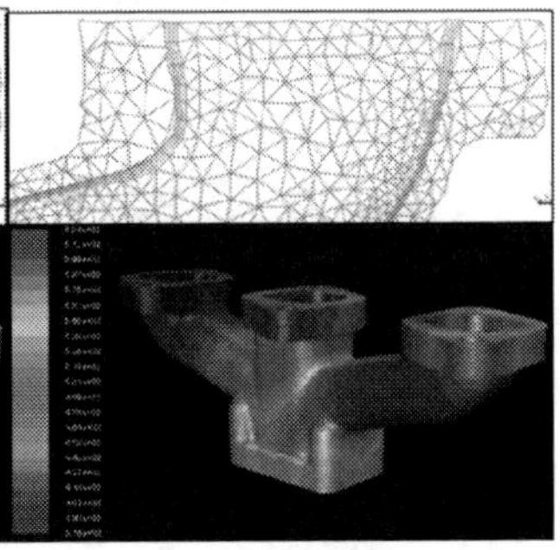

Mapping

## Fast and Accurate Modeling Full Vehicle Transient Soaking and Other Road Conditions

- CPU Effort: 30 Hrs. wall-clock, 48 CPUs, 12 million cells, 60 minutes physical time

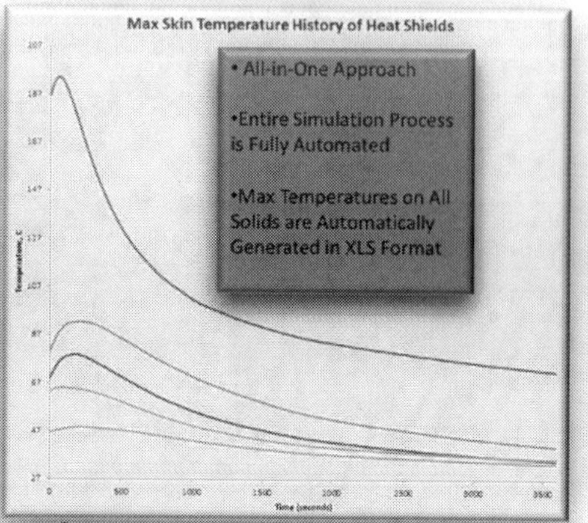


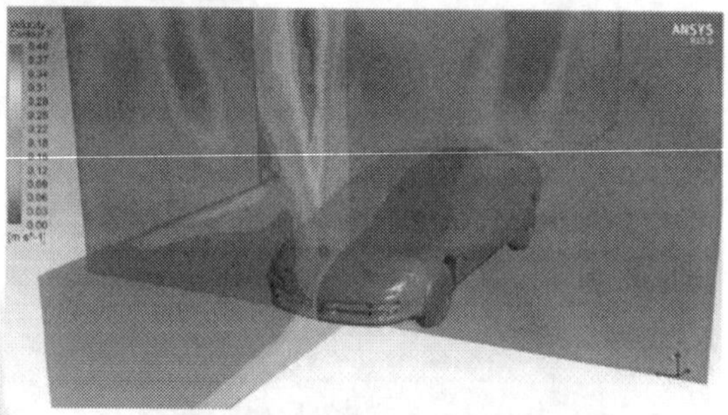

Animations Generated in CFD Post

## Outline

**Underhood Simulation Using ANSYS Fluent**

- Introduction
- Model Overview
- Advances in Solver Capabilities
- Advances in Automatic Mesh Generation
- Summary and Conclusions
- Appendix

## Introduction to the Wrapper Scripts

- **The entire wrapping process is fully automated using scripts**
- **Three scripts**
  - **AdvWrapNPrisms_R15R16_v500_Main.bin**
    - This is the main scheme that defines all the functions
  - **AdvWrapNPrisms_R15R16_v500_UserInputs.scm**
    - This is the user input scheme. The user only modifies this file.
  - **AdvWrapNPrisms_R15R16_v500_Run.scm**
    - This is the run scheme that call the functions defined in above



## Case 1: Application of Script on a Dirty Engine

**Step 1: Import geometry**

**Step 2: create Box**

- It will create object

**Step 3: define object for engine**

- Note: Automatic hole detection (skinning) will be applied on the engine, thus need to assign it to a separate object.

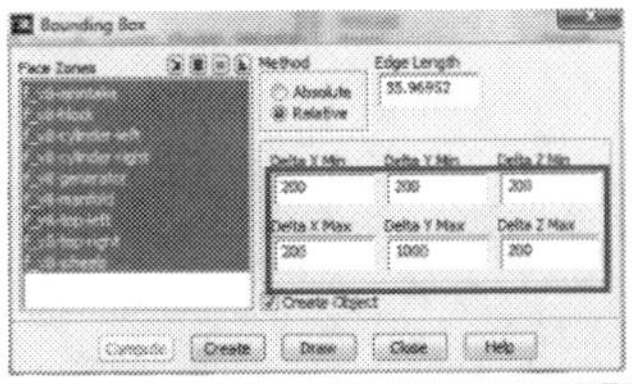

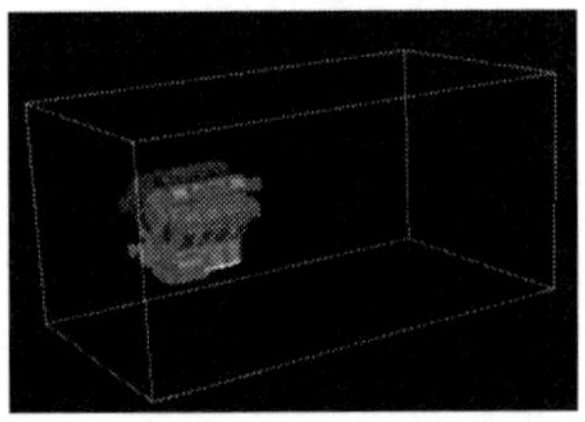

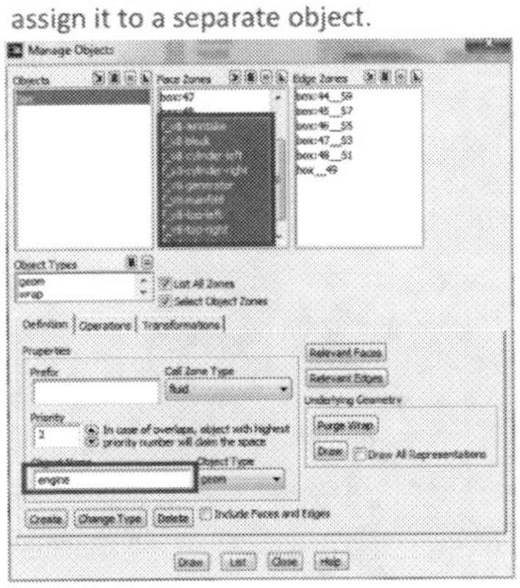

# Case 1: Application of Script on a Dirty Engine

**Step 5: extract features for object engine**

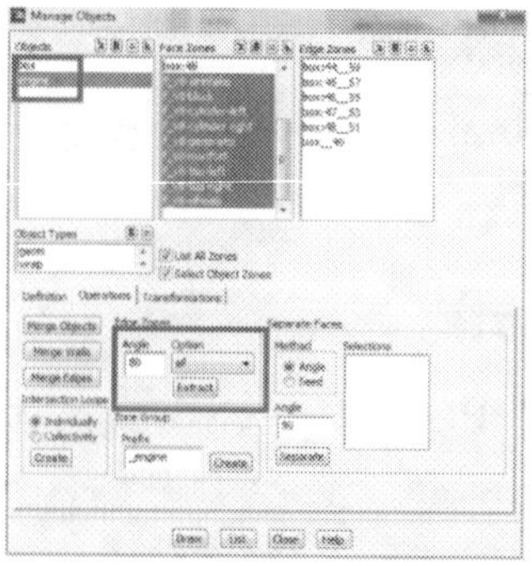

**Step 6: define material point for the main fluid region**

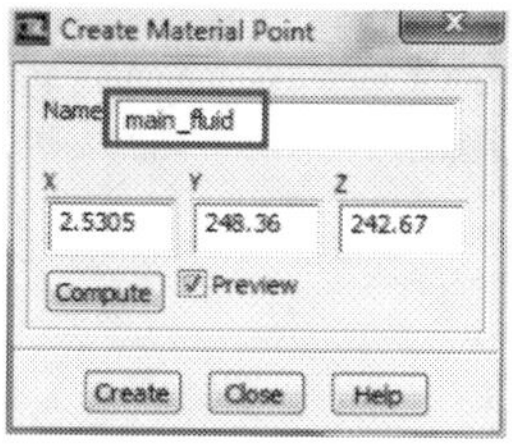


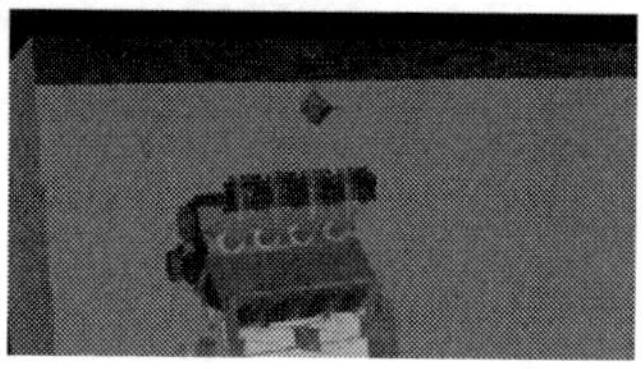



# Case 1: Application of Script on a Dirty Engine

**Step 7: Define curvature, proximity, and soft SF and save as engine.msh.gz**

- Note: Max allowable tet size = global Max

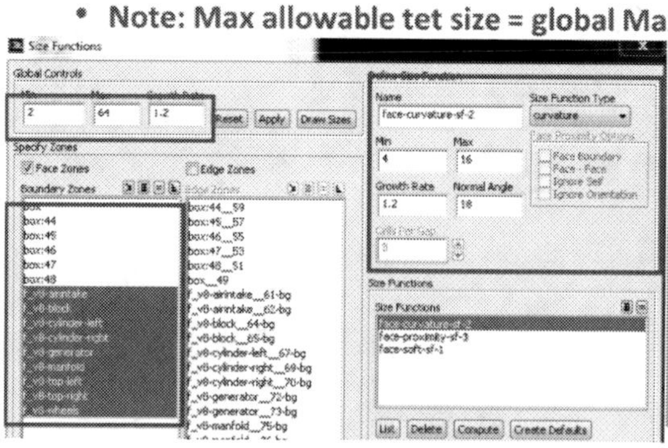

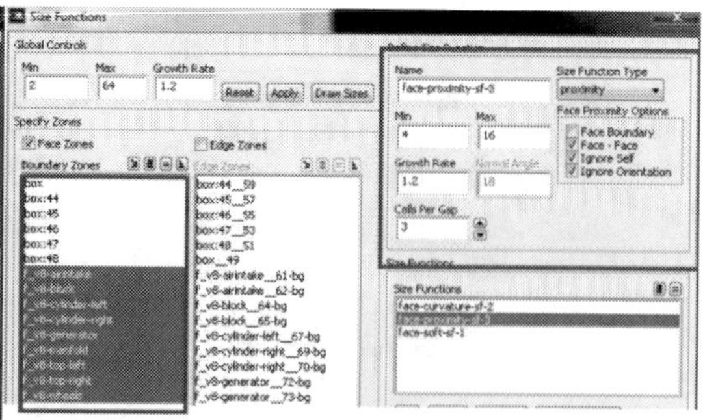

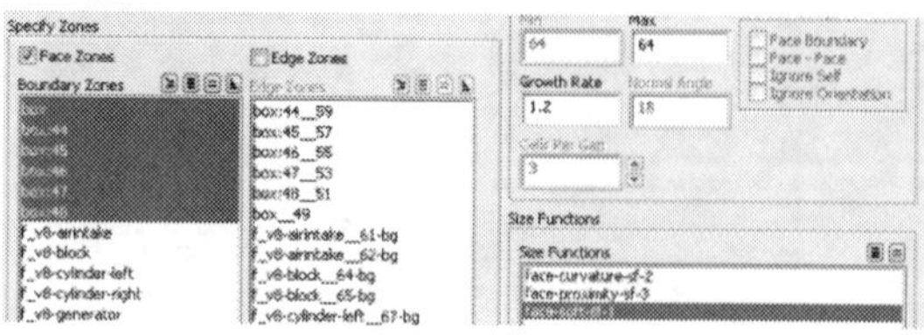

## Case 1: Application of Script on a Dirty Engine

Step 8: Edit user input scheme as shown on the right

Step 9: Run AdvWrapNPrisms_R15R16_v500_Run.scm

It will produce engine-final.msh.gz and print out total meshing time (8.6 minutes):

```
(define input_geometry "engine")
(define compute_size_field "yes")
(define extract_intersected_feature_edges "no")
(define dirty_objects '(("engine" 1 16 skin)))
(define live_regions '("main_fluid"))
(define additional_coarsening_factor_and_max_size '((1.1 1000) (1.5 24)))
(define prism_layer_method "aspect_ratio")
(define PrismLayers 3)
(define PrismAspectRatio 5)
(define PrismLayerGrothRate 1.2)
(define volume_fill_type "tet")
(define tet_resolution_factor 1.5)
```

```
Computed SF in ----------------------------------0.8 minutes (0.013333333 hours).
Automatic closing of holes in -----------------1.05 minutes (0.0175 hours).
Wrapped all live regions in -------------------0.41666667 minutes (0.0069444444 hours).
Imprinted all wrappers in ---------------------0.9 minutes (0.015 hours).
Coarened all to target size in ----------------0.6 minutes (0.01 hours).
Surface Improved in ---------------------------0.38333333 minutes (0.0063888889 hours).
PostTet in ------------------------------------0.95 minutes (0.015833333 hours).
Fully conformal prism generated in ------------1.9666667 minutes (0.032777778 hours).
Volume fill in ---------------------------------1.0166667 minutes (0.016944444 hours).
Rezoned in ------------------------------------1.3 minutes (0.021666667 hours).
==========================================================
T O T A L  M E S H I N G  T I M E: ------------9.3833333 minutes (0.15638889 hours)
```

  October 6, 2014

## Case 1: Application of Script on a Dirty Engine

**High quality tet + prism generated from all faces, all conformal, fully automated. There are no pyramids or non-conformal interfaces!**

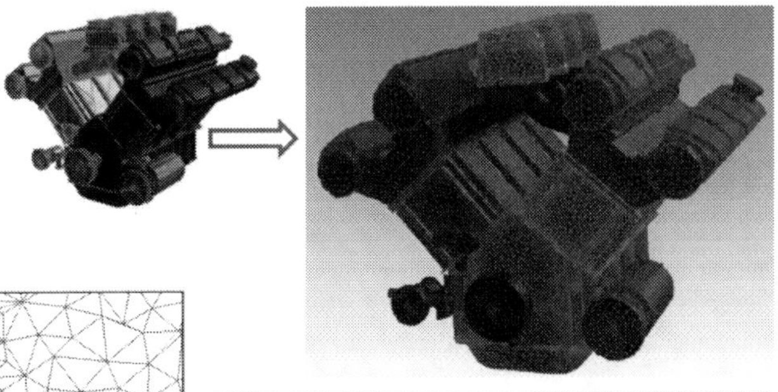

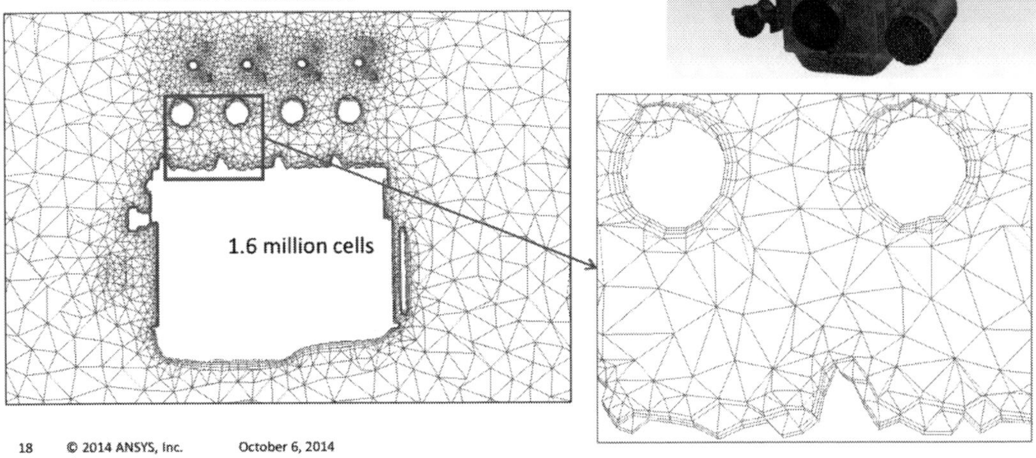


  October 6, 2014

## Case 1: Application of Script on a Dirty Engine

- Change to hexcore in the user input on the right and re-run.

(define volume_fill_type "hexcore")

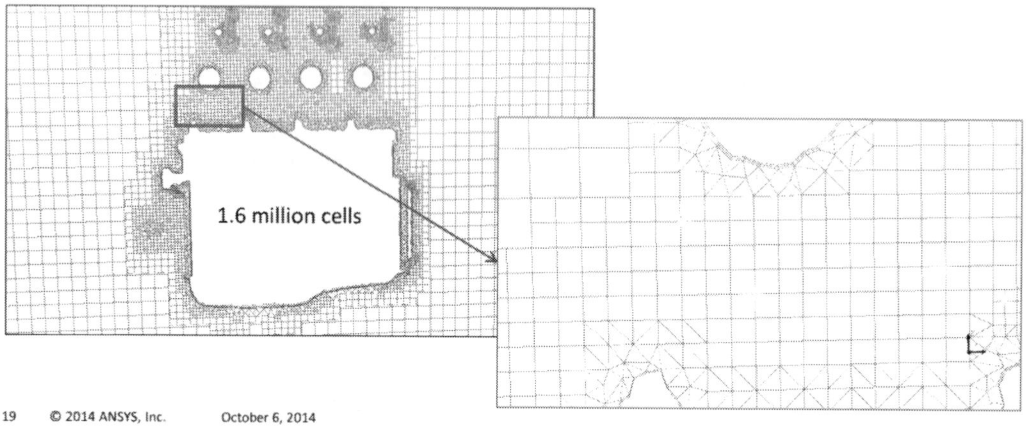




## Case 1: Application of Script on a Dirty Engine (BOI)

- Create 2 boxes for body-of-influence (BOI)
  - No need to create objects
  - Surfaces can also be imported
- Add "boi" size functions for those boxes
- Save
- Re-run script

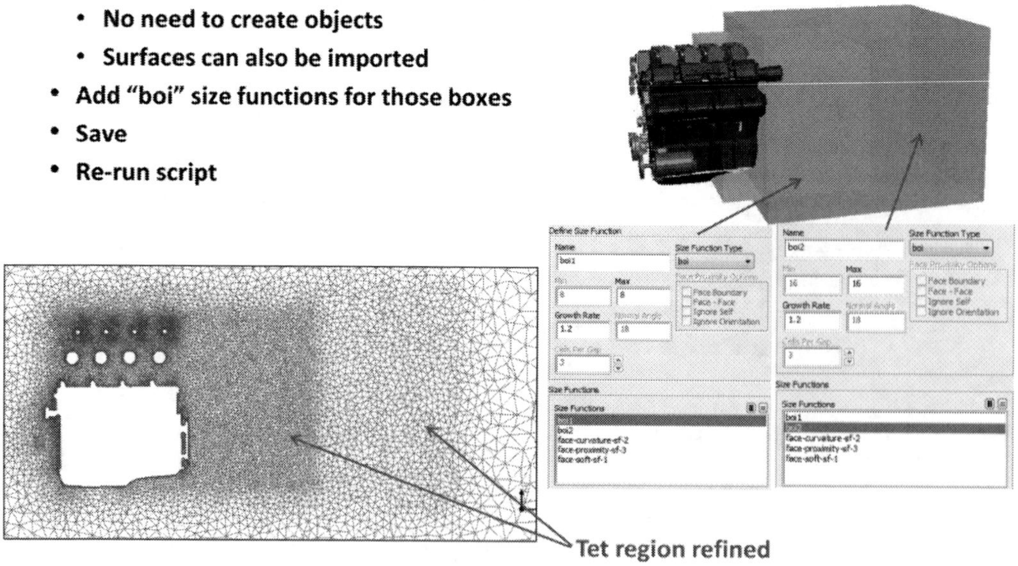

## Case 1: Application of Script on a Dirty Engine – Additional Prism Layer on Select Zones

Extra layers can be generated on select zones (e.g. __p-*)

```
;====local prism layer settings
(define list_of_zones_with_extra_layers "__p-*")
(define local_prism_layer_method "aspect_ratio")
(define local_prism_layers 7)
(define local_last_ratio_percentage 40)
(define local_prism_layer_first_height 0.2592)
(define local_prism_aspect_ratio 20)
(define local_prism_layer_growth_rate 1.2)
```

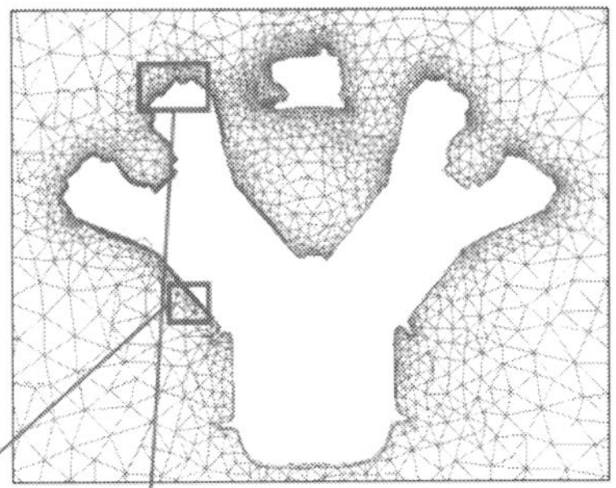

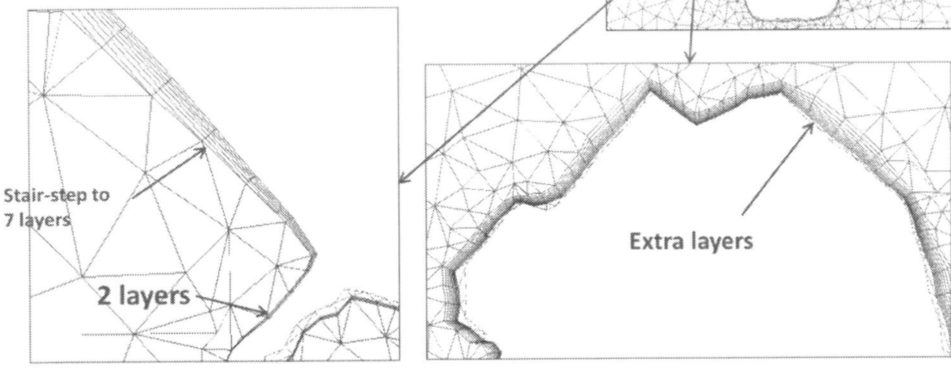


  October 6, 2014

## Case 1: Application of Script on a Dirty Engine – Feature Resolution Factor

**Feature resolution factor can be used to determine the degree in which the geometry is cleaned.**

(define dirty_objects '(("engine" 1 16 skin)))

(define dirty_objects '(("engine" 4 16 skin)))

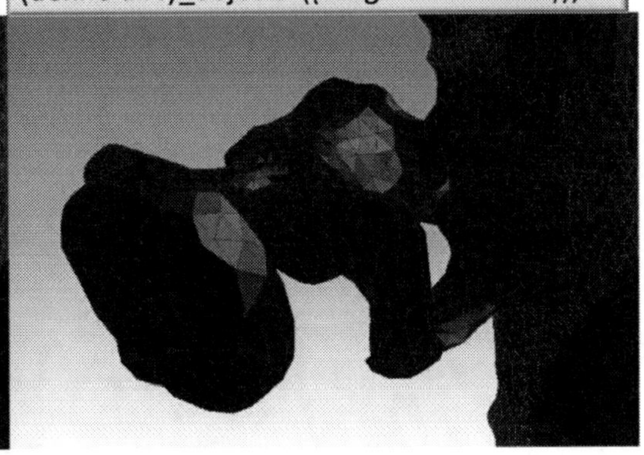

# Case 2: Application of Script on a Dirty FEAF

Step 1: Import / read faceted geometry

Step 2: Create box

Step 3: Create Heat Exchanger volume mesh and copy + triangulate its quad sides.

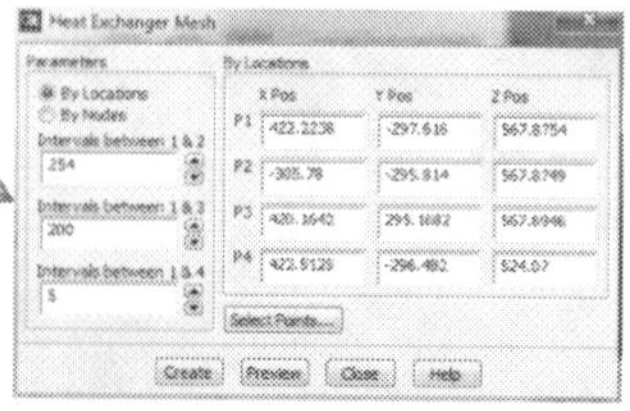

Step 4: Create cylinder around the fan

Step 5: Create objects

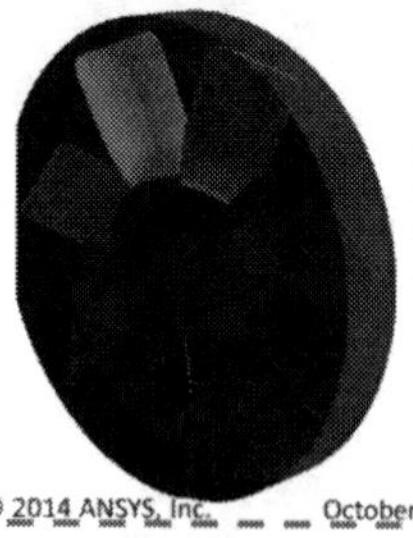

Note: the triangulated quad side is part of the object.

Needs skinning

  October 6, 2014

# Case 2: Application of Script on a Dirty FEAF

**Step 6: extract features**

**Step 7: define SF**

**Step 8: define material points**

- One for main fluid region and
- One for the MRF region

**Step 9: Edit user input file**

```
(define input_geometry "feaf")
(define compute_size_field "yes")
(define extract_intersected_feature_edges "yes")
(define dirty_objects '(("eng" 1 32 skin)))
(define live_regions '("main_fluid" "fan_fluid"))
(define coarsening_method "standard_coarsening")
(define additional_coarsening_factor_and_max_size '((1.1 1000) (1.5 24)))
(define prism_layer_method "aspect_ratio")
(define PrismLayers 3)
(define prism_layer_first_height 1)
(define PrismAspectRatio 10)
(define PrismLayerGrothRate 1.2)
(define volume_fill_type "tet")
(define tet_resolution_factor 1.5)
```

**Step 10: Execute run script**

```
Computed SF in ------------------------------4.2166667 minutes (0.070277778 hours).
Automatic closing of holes in ---------------7.0166667 minutes (0.11694444 hours).
Wrapped all live regions in -----------------3.4333333 minutes (0.057222222 hours).
Imprinted all wrappers in -------------------10.6 minutes (0.17666667 hours).
Coarened all to target size in --------------4.3333333 minutes (0.072222222 hours).
Surface Improved in -------------------------2.7 minutes (0.045 hours).
PostTet in ----------------------------------3.9666667 minutes (0.066111111 hours).
Fully conformal prism generated in ----------7.6833333 minutes (0.12805556 hours).
Volume fill in -------------------------------3.0666667 minutes (0.051111111 hours).
Rezoned in ----------------------------------5.0666667 minutes (0.084444444 hours).
==============================================================
T O T A L  M E S H I N G  T I M E: ----------52.083333 minutes (0.86805556 hours)
```

  October 6, 2014

## ANSYS Case 2: Application of Script on a Dirty FEAF

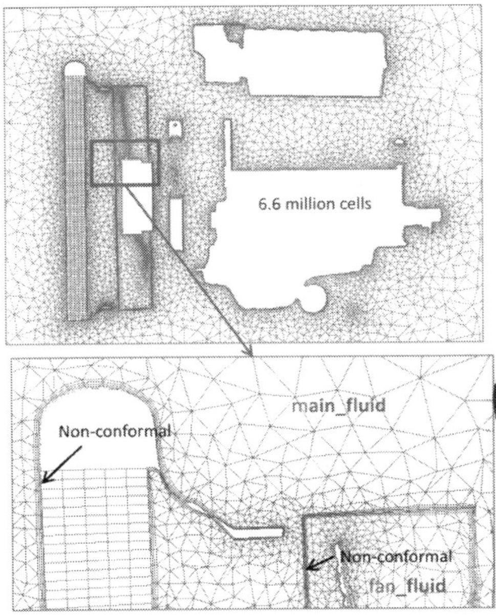

## ANSYS Case 3: Automatic Gap Closure

Gaps in external aero surfaces

Unwanted gaps are common in complex geometry.
The script will automatically close gaps before the main wrap.

```
(define objects_with_gaps '(("ext" 12 0.375)))
```

**Max size to cover all gaps**

**~ 4 times smaller than min gap to be closed**

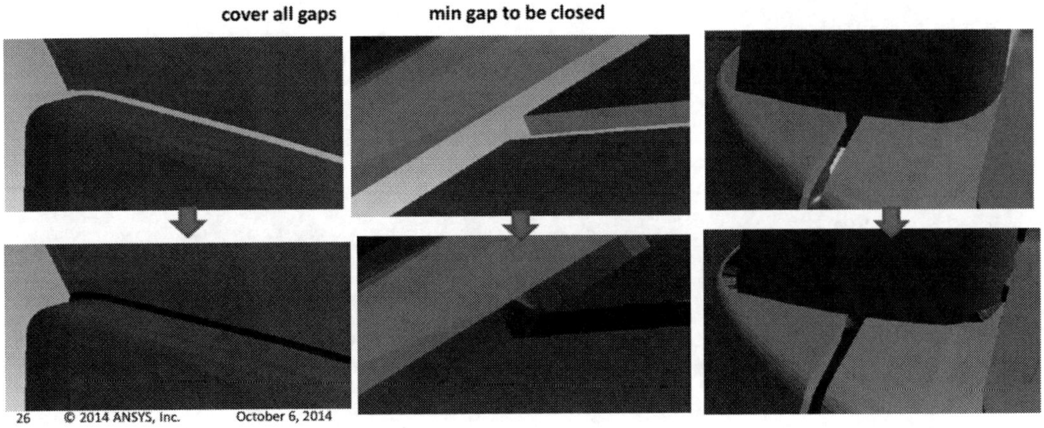

## Automated Meshing with Explicit Mapping

1. Import CAD in SCDM
2. (*Automated*) Export STL into Fluent Meshing
3. (*Automated*) Setup and run script to obtain main fluid region with 2 prism layers
4. (*Automated*) In parallel create solid meshes
5. (*Automated*) Merge main fluid with solid
6. (*Automated*) Generate contact pairs
7. (*Automated*) Setup solid2solid NCI and fluid2solid mapping
8. Run case with radiation and CHT

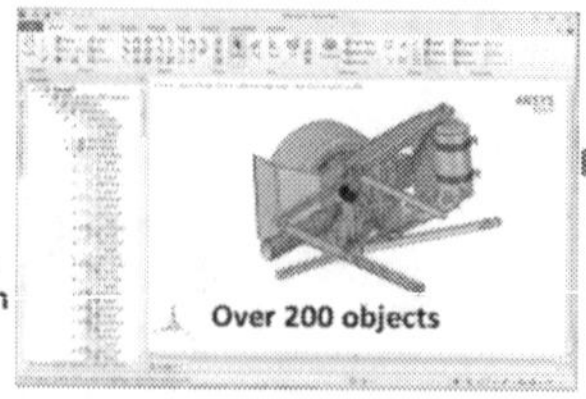

Solid Meshing (automatic) – ~2.5 CPU-hrs

Main fluid

Run Script in batch (automatic) – 1 CPU-hr

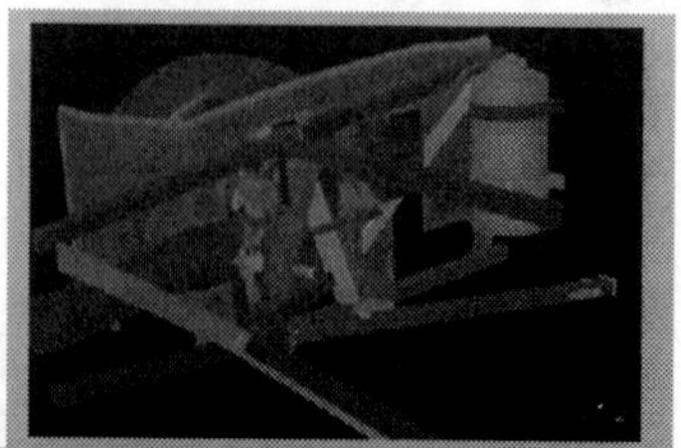

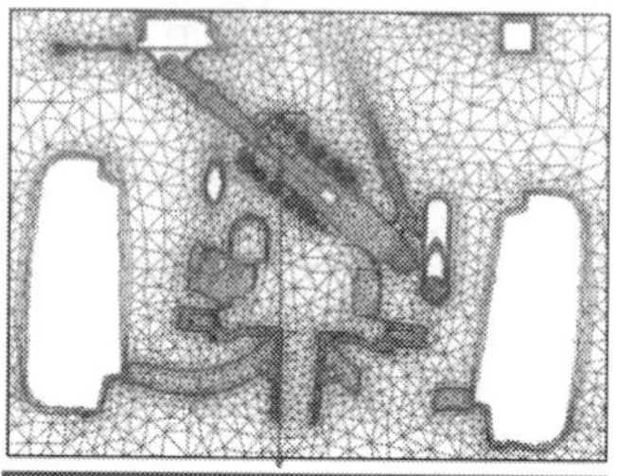

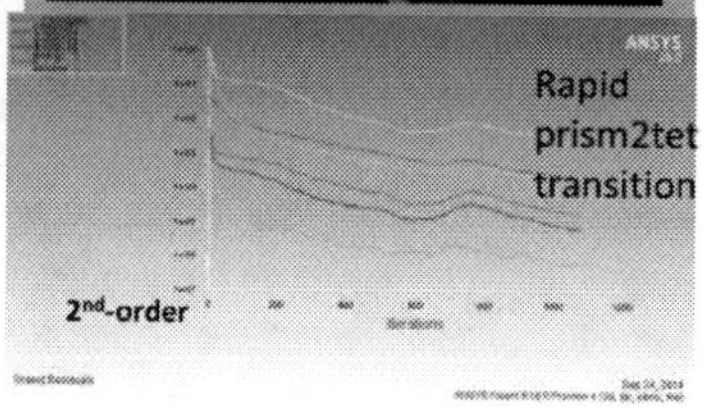


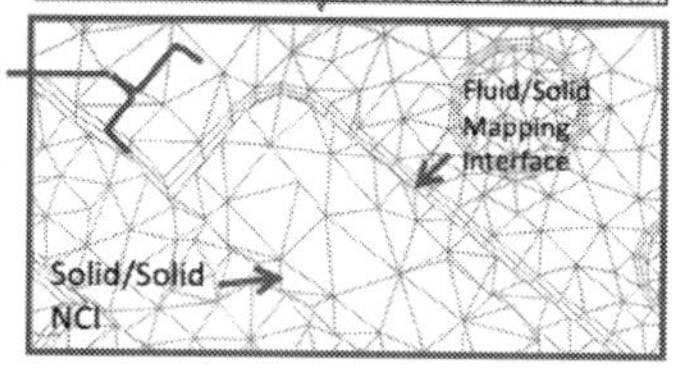


## Automated Meshing with Explicit Mapping

**To speed up solid meshing, we can run multiple sessions simultanously, meshing chunks of solids per core (this will be scripted).**

**208 solids:**

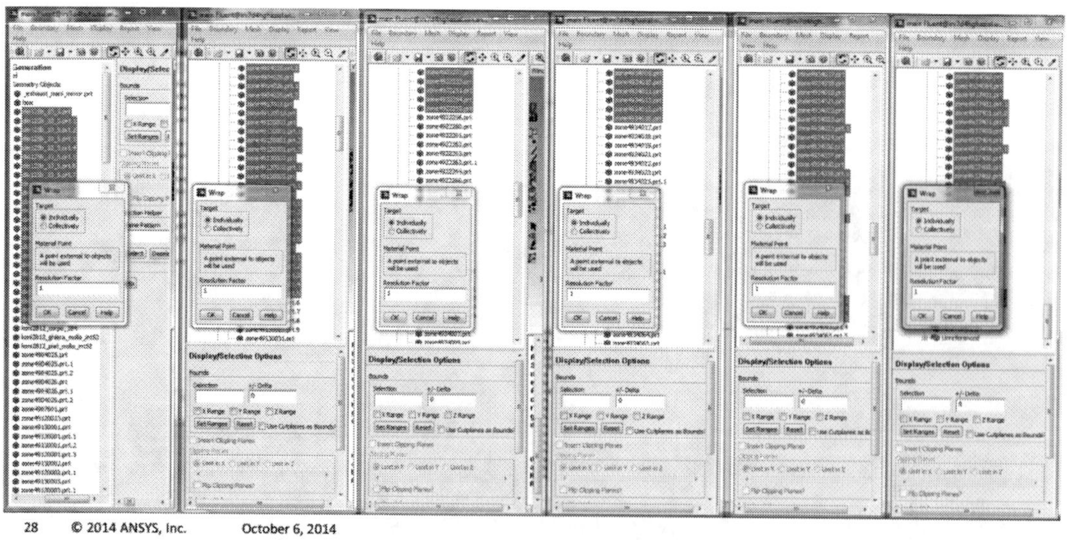

# ANSYS An-Isotropic Y+ Adaption

1. Solve for flow with 3 layers
2. (*Automatic*) Perform an-isotropic adaption based on y+ (~2)
3. Improve volume mesh if needed
4. Perform final solution with radiation, mapping, etc.

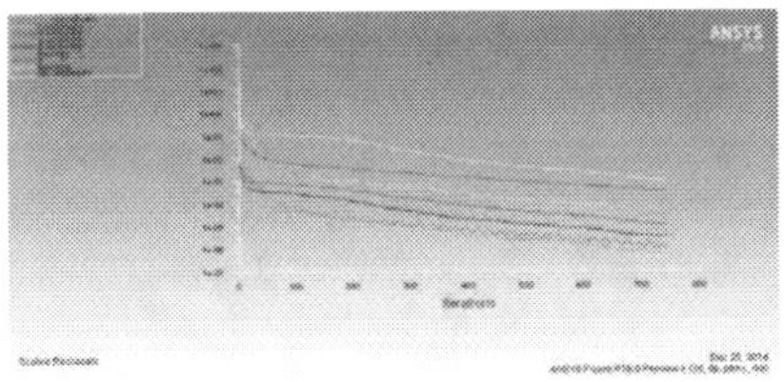


Excellent Convergence on Adapted Mesh

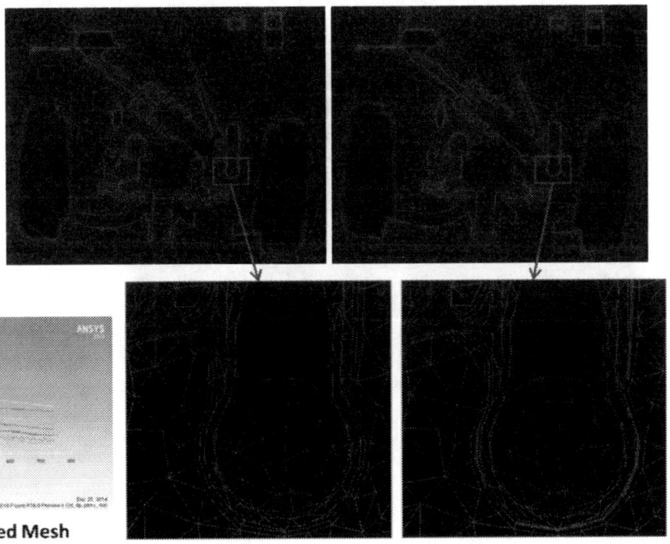

3 layers (5.80 million total cells)

Adapted (max 23 layers!) (7.96 million total cells)



ANSYS

## ANSYS Imprinted Surface Creation

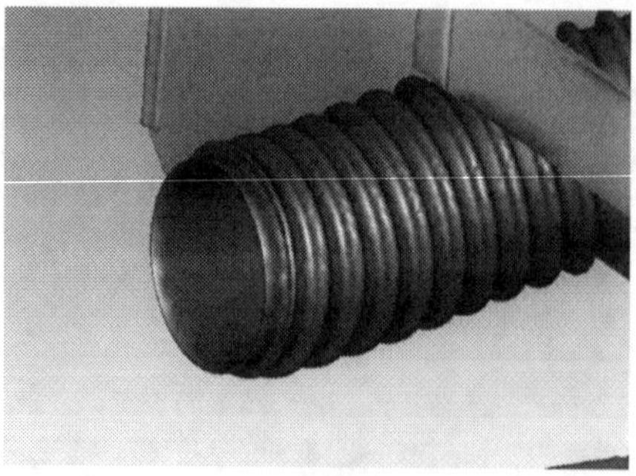

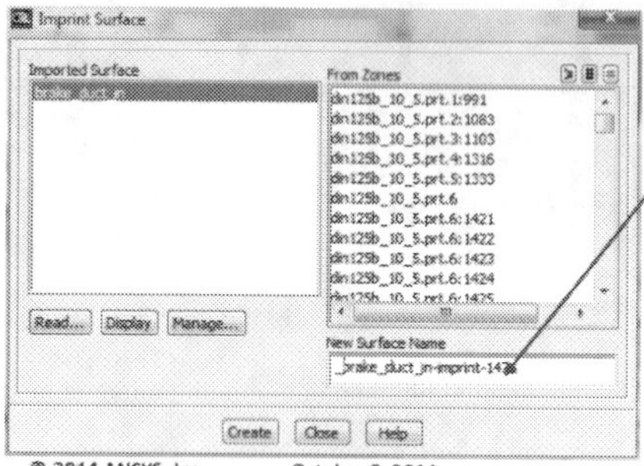


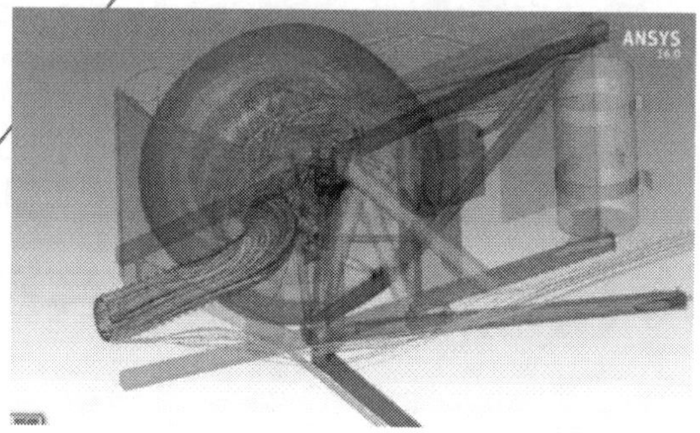




## ANSYS Outline

**Underhood Simulation Using ANSYS Fluent**

- ❑ Introduction
- ❑ Model Overview
- ❑ Advances in Solver Capabilities
- ❑ Advances in Automatic Mesh Generation
- ❑ Summary and Conclusions
- ❑ Appendix

## Summary

- ANSYS CFD is fully capable to accurately predict fluid flow and heat transfer in automotive underhood application for different road conditions
- We are working to improve and reduce the turnaround time:
  - We introduced schemes that will – starting from CAD - generate high-quality meshes faster, and more automated from extremely complex and dirty geometry.
  - New mapping technique in R16 allows us to include many solids for CHT highly automated.
- Automatic continuous prism layer generation and being able to adapt the mesh to obtain mesh-independent result on such complex and dirty geometry with many contributing solids is state-of-the-art.



## Outline

### Underhood Simulation Using ANSYS Fluent

- Introduction
- Model Overview
- Advances in Solver Capabilities
- Advances in Automatic Mesh Generation
- Summary and Conclusions
- Appendix

## Mapping Goal: To be Able to Converge with Poor Geometry Resolution

- In spite of the fact that nuts-n-bolts are not resolved enough, all though conduction path maybe broken, but solution still converges.
- As is generally the case, for accurate CFD predicitons, geometry has to be accurately represented and enough resolution to be given.

Since the bolt is not important (but was included by mistake), the geometry is purposely not well resolved.



## Mapping Goal: To be Able to Converge with Poor Geometry Representation

Solid-solid conduction takes place only if the two solids "touching" within a tolerance. If a solid overlaps beyond that tolerance, conduction will be broken. But it will still converge!

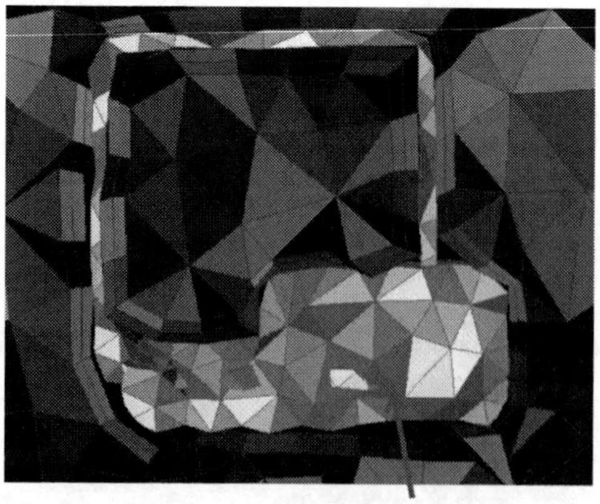

Solid is not important but was included by mistake!

# Accuracy of a Single Layer Tet for Thin Baffles

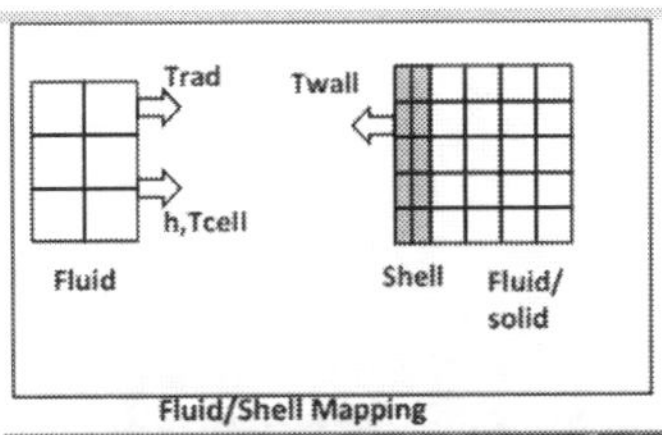


Fluid/Shell Mapping

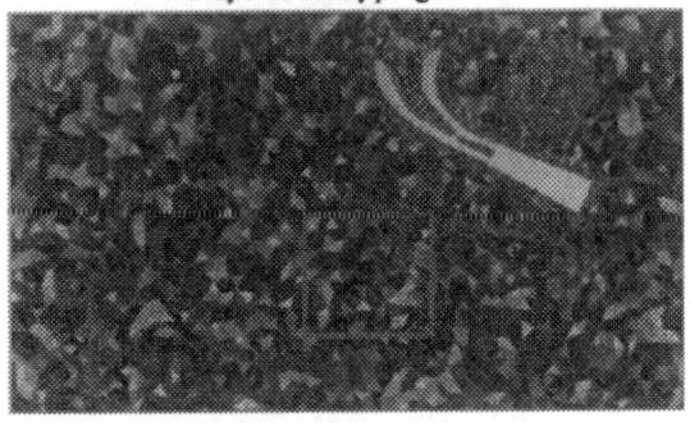

- Coarse tet mesh can be used to mesh the baffles as solids
  - What is the accuracy level?

Baffle is meshed as a single layer tet

Baffle is not meshed but shell conduction is applied on one side

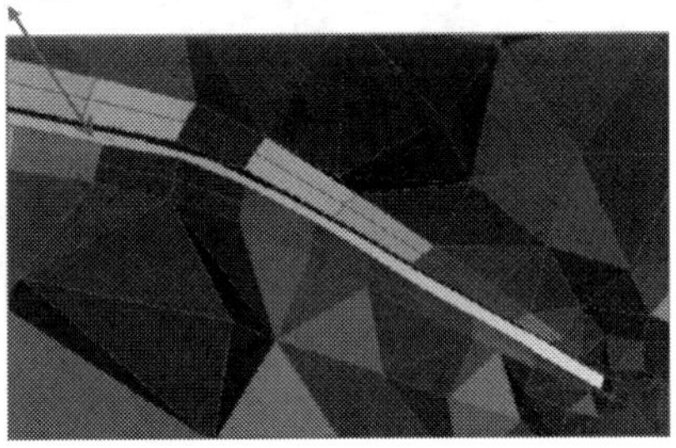

Fluid/Solid Mapping

Fluid/Shell Mapping



# Accuracy of a Single Layer Tet for Thin Baffles

**For this steel baffle structgure, a single tet layer compares well with multi-layer shell**

1-layer Shell on Baffle

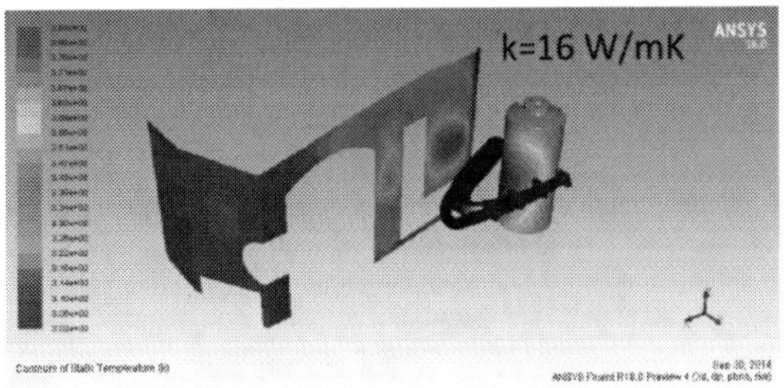


Tet on Baffle

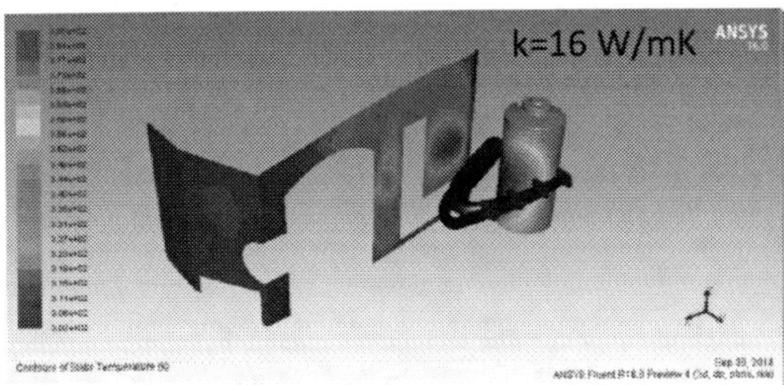


4-layer Shell on Baffle

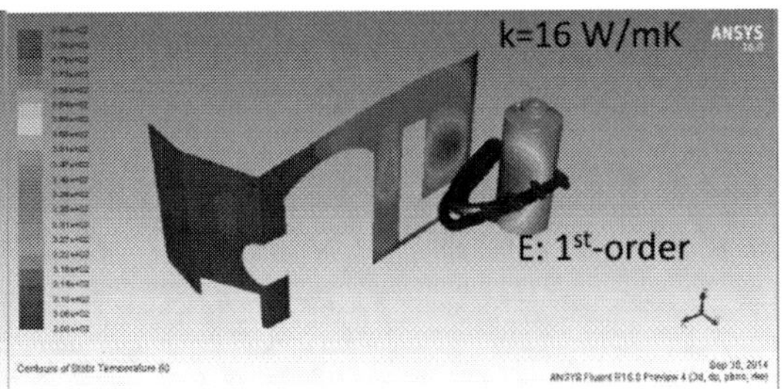

# Automated Meshing and Mapping

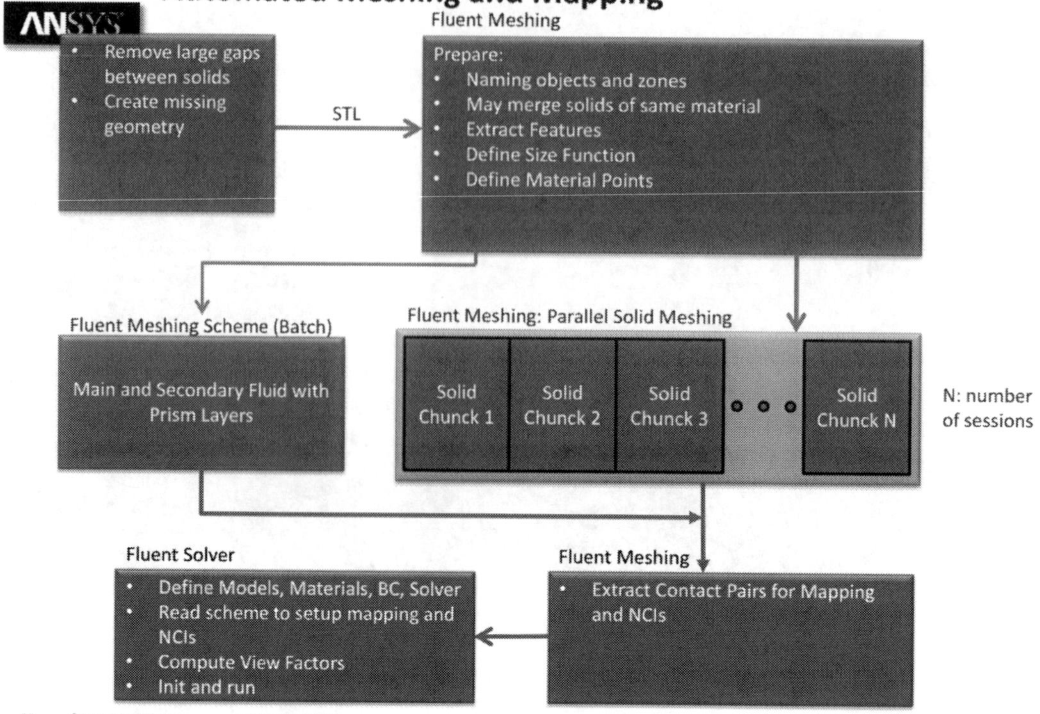


39  October 6, 2014

ANSYS

# Appendix – Additional Utilities For UTM

❑ **Quick Case setup**

- ***UH_Wall_Panel_Scheme_Utility_CAF.scm***
  - Write/Read wall boundary condition in EXCEL format
- ***Cellzone_Panel_Scheme_Utility_1.scm***
  - Write/Read settings for MRF cell zones in EXCEL format
- ***Export-Material-Properties.scm***
  - Write material library in EXCEL format
- ***Read-Material-Setup.scm***
  - Read material settings for cell zones in EXCEL format

40  October 6, 2014

**Source code will be shared.*

## Appendix – Additional Utilities For UTM

- **Automatic Post-processing**
  - ***Postprocessing_complete_tool_ver7.scm***
    - *Write out information such as Min. Max. and Average Temperature on fluid cell zones, solid cell zones and wall face zones in EXCEL format*
    - *Create and export temperature contour plots on solid zones*
  - ***plot-t-on-shells.scm***
    - *Create and export temperature contour plots on shell conduction zones*
  - ***take-plane-cut.scm*** (Edit and give inputs before read in)
    - *Create and export temperature and velocity magnitude contour plots, velocity vector plot on selective x,y and z plane cuts*
  - ***get-wall-adjacent-cell.scm*** (Edit and give inputs before read in)
    - *Create and export static temperature and HTC contour plots on selective components*
  - ***export-wall-temperature.scm***
    - *Write Max. Temperature on solid walls and shell conduction walls VS. flow time during transient simulation*
  - ***get-flow-adj-cell.scm*** (Edit and give inputs before read in)
    - *Report mass and energy flow rate on selective face zones in normal and opposite normal direction.*

41  October 6, 2014

(rbf-morph)™
Welcome to the World of Fast Morphing!

2014 Automotive Simulation World Congress

# RBF Morph software
## How to boost Fluent Adjoint using RBF Morph

**Prof. Marco Evangelos Biancolini**
*University of Rome Tor Vergata*
info@rbf-morph.com
biancolini@ing.uniroma2.it

Tor Vergata

(rbf-morph)™

ASWC 2014
October 2014 - Tokyo, Japan

www.rbf-morph.com
RBF Morph, an ANSYS Inc. Partner
ANSYS

---

(rbf-morph)™
Welcome to the World of Fast Morphing!

## Outline

2014 Automotive Simulation World Congress

- Company Introduction
- RBF Morph Software Line
- Ongoing Researches
- Industrial Applications
- Fluent Adjoint Coupling

ASWC 2014
October 2014 - Tokyo, Japan

www.rbf-morph.com
RBF Morph, an ANSYS Inc. Partner
ANSYS

# Company Introduction



RBF Morph is a pioneer and world-leading provider of numerical morphing techniques and solutions conceived to efficiently handle shape optimization studies concerning most challenging industrial applications.
We are an independent software-house and vendor. Our main product is **RBF Morph**™, that is a unique morpher that combines a very accurate control of the geometrical parameters with an extremely fast mesh smoothing properly designed to be integrated in advanced computational optimization procedures.

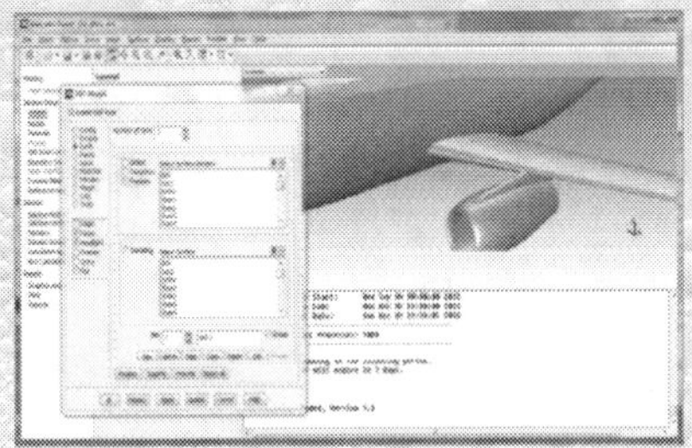

The **RBF Morph** tool is currently available in the market mainly as add-on of the CFD commercial code ANSYS® Fluent®.





# Company Introduction



The **RBF Morph** tool had its inception in 2008 as on-demand solution for a Formula 1 top team. The need was a novel technology able to change the shape of large CFD numerical models as fast as possible. The final result had been so good that the technology was packaged in a commercial software product and launched onto the market.

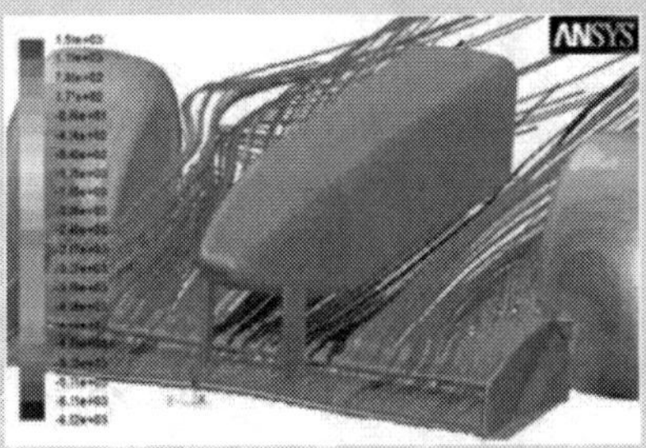


At present, Dr. Marco Evangelos Biancolini is the unique owner of the **RBF Morph** technology and, as Director, avails himself of the collaboration of several experts for the deliver of products and services.

(rbf-morph)™ Welcome to the World of Fast Morphing!

## Company Introduction

2014 Automotive Simulation World Congress

- Morphing-based numerical tools and services
- RBF Morph Milestones
  - ✓ 2008: tool implementation for Formula 1 top team consultancy activity
  - ✓ 2009: founded in Italy
  - ✓ 2009: Software Partner of ANSYS
  - ✓ 2009: at EASC **RBF Morph** won the *Most Advanced Approach Award Most Innovative Approach using Simulation Methods*
  - ✓ 2011: strategic partnership with Tor Vergata University (Rome)
  - ✓ 2012: OEM partner of ANSYS
  - ✓ 2013: beneficiary of an FP7 AAT Project RBF4AERO
  - ✓ 2013: at ASWC **RBF Morph** awarded for the *Best use of HPC*
  - ✓ 2013: Partner of Enginsoft
  - ✓ 2014: beneficiary of FP7 Project RIBES
  - ✓ 2014: beneficiary of FP7 Fortissimo

ANSYS

ENGIN SOFT

easco EUROPEAN AUTOMOTIVE SIMULATION CONFERENCE

2013 Automotive Simulation World Congress

ASWC 2014
October 2014 - Tokyo, Japan

Tor Vergata

www.rbf-morph.com ANSYS
RBF Morph, an ANSYS Inc. Partner

(rbf-morph)™ Welcome to the World of Fast Morphing!

## Knowledge Galaxy

2014 Automotive Simulation World Congress

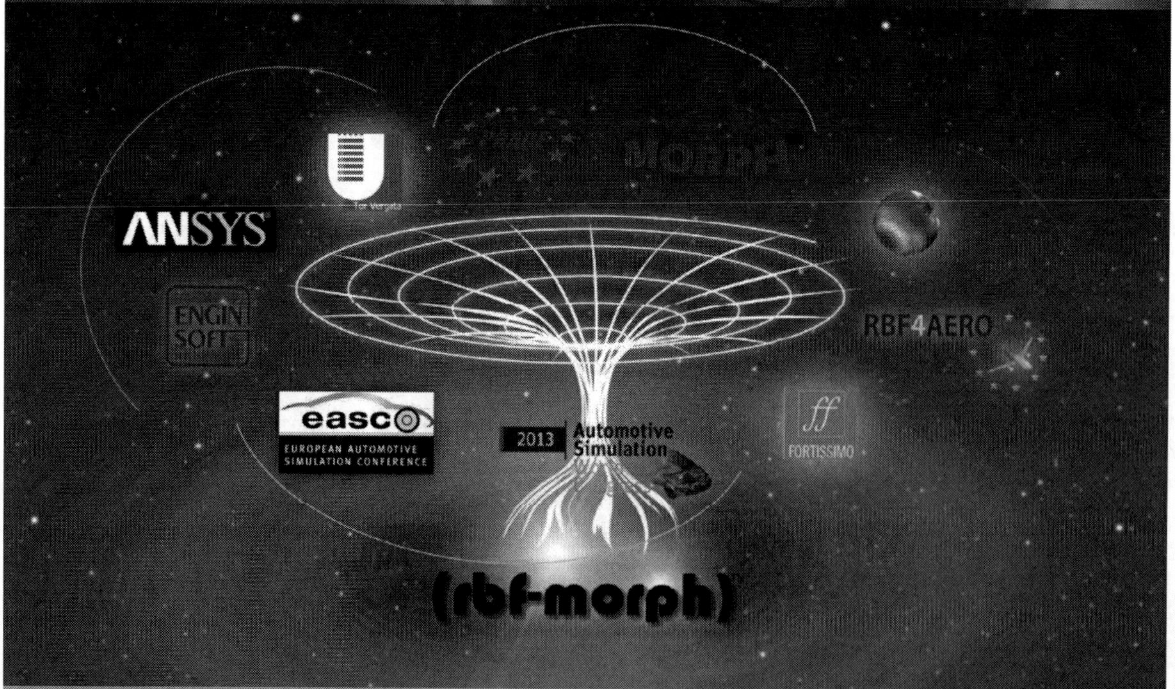


ASWC 2014
October 2014 - Tokyo, Japan

Tor Vergata

www.rbf-morph.com ANSYS
RBF Morph, an ANSYS Inc. Partner

(rbf-morph)™

Welcome to the World of Fast Morphing!

2014 Automotive Simulation World Congress

# RBF Morph software line

ASWC 2014

October 2014 - Tokyo, Japan

www.rbf-morph.com

RBF Morph, an ANSYS Inc. Partner

(rbf-morph)™ 

Welcome to the World of Fast Morphing!

2014 Automotive Simulation World Congress

# Mesh Morphing with RBF

- A system of **Radial Basis Functions** is used to **fit a solution** for the mesh movement/**Morphing**, from a list of **source points** and their **displacements**.
- The RBF problem definition does not depend on the mesh
- Radial Basis Function interpolation is used to derive the displacement in **any location** in the space, each component of the displacement is interpolated:

$$\begin{cases} v_x = s_x(\mathbf{x}) = \sum_{i=1}^{N} \gamma_i^x \phi(\|\mathbf{x} - \mathbf{x}_{k_i}\|) + \beta_1^x + \beta_2^x x + \beta_3^x y + \beta_4^x z \\ v_y = s_y(\mathbf{x}) = \sum_{i=1}^{N} \gamma_i^y \phi(\|\mathbf{x} - \mathbf{x}_{k_i}\|) + \beta_1^y + \beta_2^y x + \beta_3^y y + \beta_4^y z \\ v_z = s_z(\mathbf{x}) = \sum_{i=1}^{N} \gamma_i^z \phi(\|\mathbf{x} - \mathbf{x}_{k_i}\|) + \beta_1^z + \beta_2^z x + \beta_3^z y + \beta_4^z z \end{cases}$$

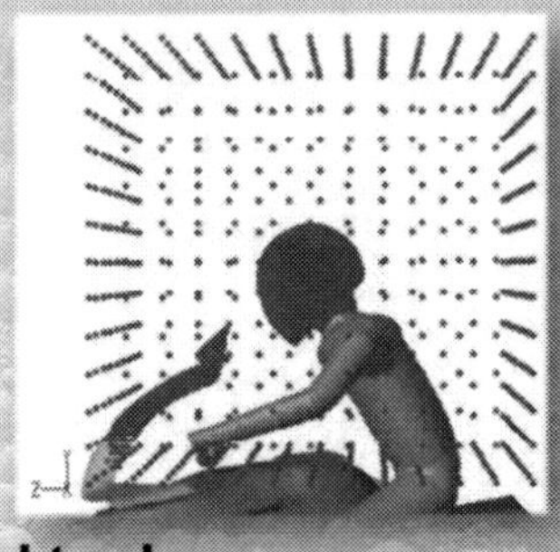

- RBF are recognized as on of the **best mathematical tool** for mesh morphing. The main issue is about performances required for the solution of **large dataset**.

ASWC 2014

October 2014 - Tokyo, Japan

www.rbf-morph.com

RBF Morph, an ANSYS Inc. Partner

# RBF Morph software line

- HPC RBF **general purposes** library (state of the art algorithms, parallel, GPU). This is the numerical kernel of our software. **Millions** of RBF centers can be fitted in a short time.
- Awarded mesh morphing software available as an add-on for **ANSYS Fluent** CFD solver
- **Stand alone** morphing software + smoothing commands for different mesh formats
- ANSYS Mechanical **ACT module** (first beta version already working since June 2014)

# Fluent add-on

- **Add on** fully integrated within **Fluent** (GUI, TUI & solving stage), **Workbench** and **Adjoint Solver** NEW!
- **Mesh-independent** RBF fit used for surface mesh morphing and volume mesh smoothing
- **Parallel** calculation allows to morph **large size** models (many millions of cells) in a short time
- Management of **every kind of mesh** element type (tetrahedral, hexahedral, polyhedral, etc.)
- Support of the **CAD re-design** of the morphed surfaces
- **Multi fit** makes the Fluent case truly parametric (only 1 mesh is stored)
- **Precision**: exact nodal movement and exact feature preservation (**RBF** are better than **FFD**)

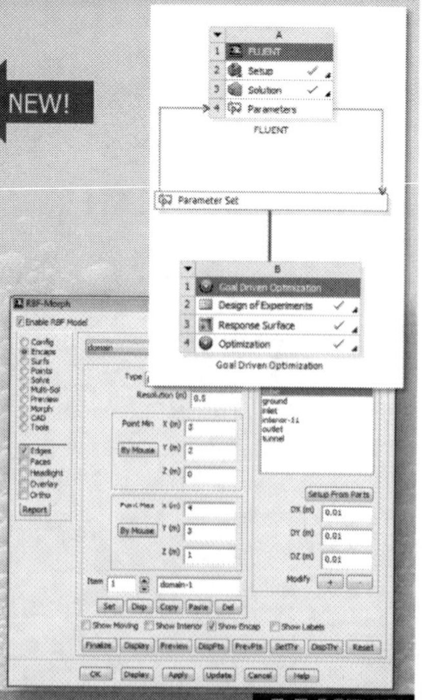

# (rbf-morph)™ ACT module for Mechanical

Welcome to the World of Fast Morphing!

Automotive Simulation World Congress

- Deeply integrated in ANSYS Mechanical: same look & feel, same interaction logic
- Nested in the usual Mechanical tree as an added object, shares its scoping tools for **geometrical** and **mesh elements** selections
- Written in **python** and **xml**, uses external RBF Morph core libraries
- Child **hierarchical logic** for complex morphings (two steps, three steps, ..., n steps setups)

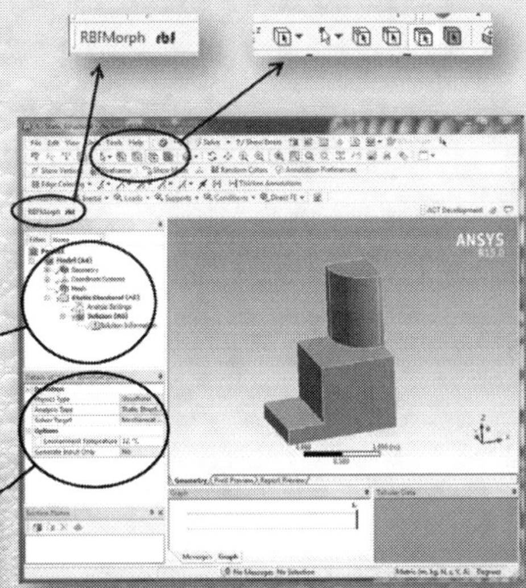


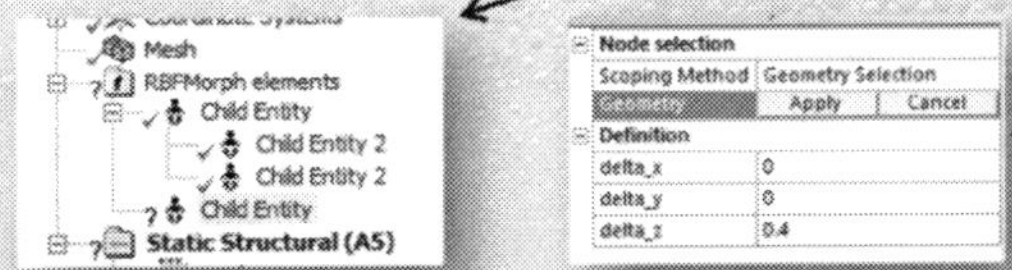


ASWC 2014
October 2014 - Tokyo, Japan

Tor Vergata

ANSYS

www.rbf-morph.com
RBF Morph, an ANSYS Inc. Partner

# (rbf-morph)™ RBF Morph Stand Alone

Welcome to the World of Fast Morphing!

Automotive Simulation World Congress

- RBF solutions are fully compatible and **exchangeable** between add-on and standalone versions
- Support for STL and CGNS file formats. Selected morphed surfaces can be exported in STL format and **back to CAD** is possible via STEP files
- **Add-on-like** interface
- **Solver independent** process currently supports many mesh formats
- Functions **scriptable** via tcl
- Global supported bi-harmonic functions and $C^0$, $C^2$, $C^4$ compact supported functions available

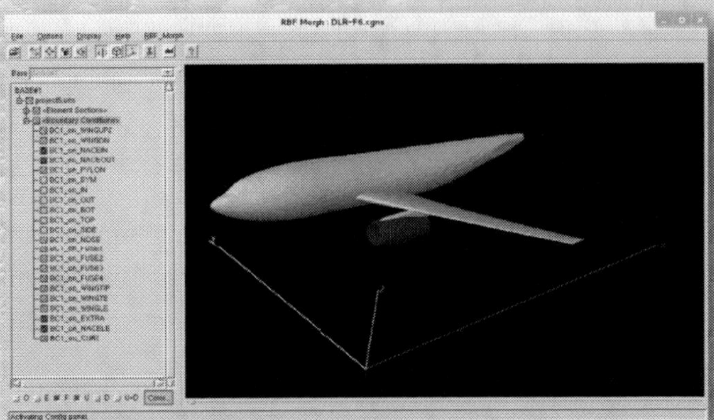


ASWC 2014
October 2014 - Tokyo, Japan

Tor Vergata

ANSYS

www.rbf-morph.com
RBF Morph, an ANSYS Inc. Partner

(rbf-morph)™ Welcome to the World of Fast Morphing!

2014 Automotive Simulation World Congress

# Ongoing RBF Morph Researches

ASWC 2014
October 2014 - Tokyo, Japan

Tor Vergata

www.rbf-morph.com
RBF Morph, an ANSYS Inc. Partner

---

(rbf-morph)™ Welcome to the World of Fast Morphing!

## Ongoing RBF research

2014 Automotive Simulation World Congress

- RBF Morph and Adjoint coupling: Adjoint sculpting, Adjoint preview, Augmented DOE
- **STL** targeting, **CAD** controlled surfaces
- **Mesh to CAD** features
- Mapping of **magnetic** and **pressure** loads
- Interpolation of **hemodynamic** flow fields acquired *in vivo*
- Strain and **stress calculation** (experimental data, coarse FEM, isostatic lines)

ASWC 2014
October 2014 - Tokyo, Japan

Tor Vergata

www.rbf-morph.com
RBF Morph, an ANSYS Inc. Partner

## (rbf-morph)™ RBF4AERO EU Project

Welcome to the World of Fast Morphing!

2014 Automotive Simulation World Congress

- "Innovative Benchmark Technology for Aircraft Engineering Design and Efficient Design Phase Optimisation" –

  **ACP3-GA-2013-605396**

- **www.rbf4aero.eu**

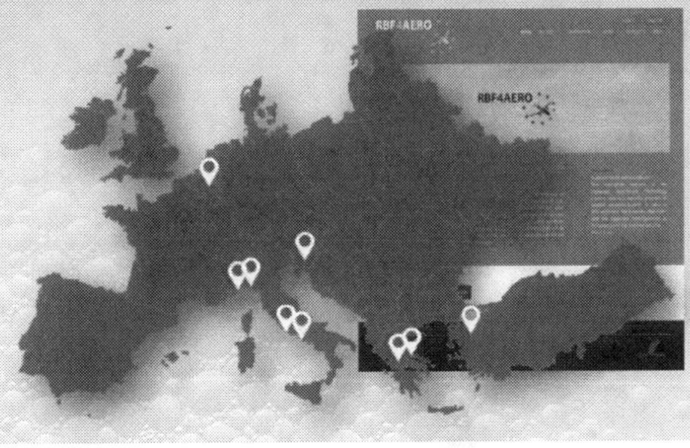


RBF AERO

D'APPOLONIA | PIPISTREL | Tor Vergata | TEI | VON KARMAN INSTITUTE | IAC



## (rbf-morph)™ RIBES EU Project

Welcome to the World of Fast Morphing!

2014 Automotive Simulation World Congress

- **R**adial basis functions at fluid **I**nterface **B**oundaries to **E**nvelope flow results for advanced **S**tructural analysis

  **JTI-CS-2013-GRA-01-052**

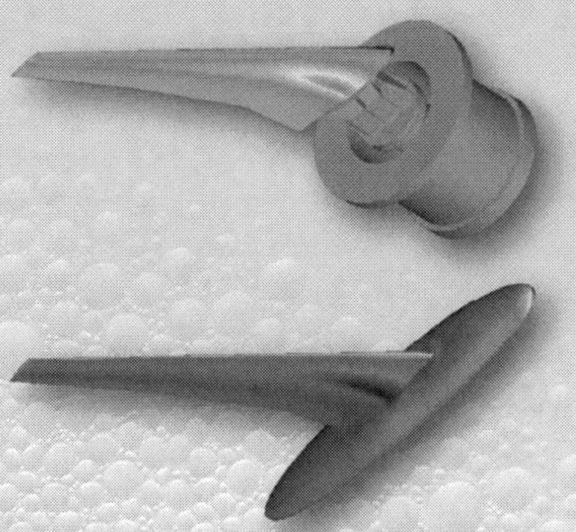

RIBES

CLEANSKY | Tor Vergata

## (rbf-morph)™ Fortissimo EU Project

Welcome to the World of Fast Morphing!

2014 Automotive Simulation World Congress

- **F**actories **O**f the Future **R**esources, **T**echnology, **I**nfrastructure and **S**ervices for **SI**mulation and **MO**delling
- Approved experiment: "Virtual Automatic Rapid Prototyping Based on Fast Morphing on HPC Platforms"

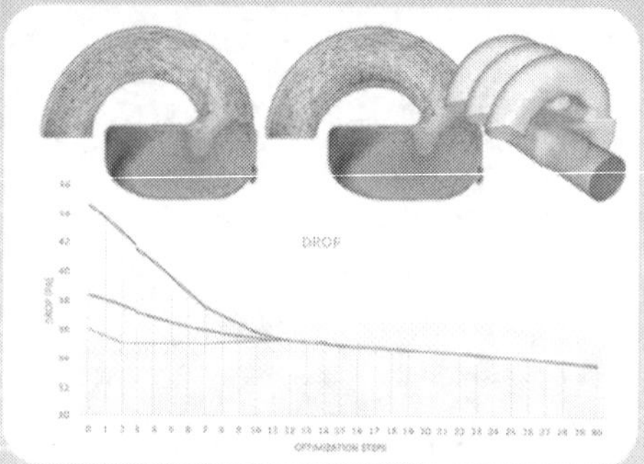

FORTISSIMO

CINECA SCAI SuperComputing Applications and Innovation

Tor Vergata

(rbf-morph)™

Welcome to the World of Fast Morphing!

2014 Automotive Simulation World Congress

# Industrial Applications

(rbf-morph)™
Welcome to the World of Fast Morphing!
2014 Automotive Simulation World Congress
Motorbike Windshield (Bricomoto, MRA)
(rbf-morph)™
Welcome to the World of Fast Morphing!
BRICO MOTO
MRA
MOTORCYCLE WINDSHIELDS
www.rbf-morph.com
ASWC 2014
October 2014 - Tokyo, Japan
www.rbf-morph.com
ANSYS
RBF Morph, an ANSYS Inc. Partner

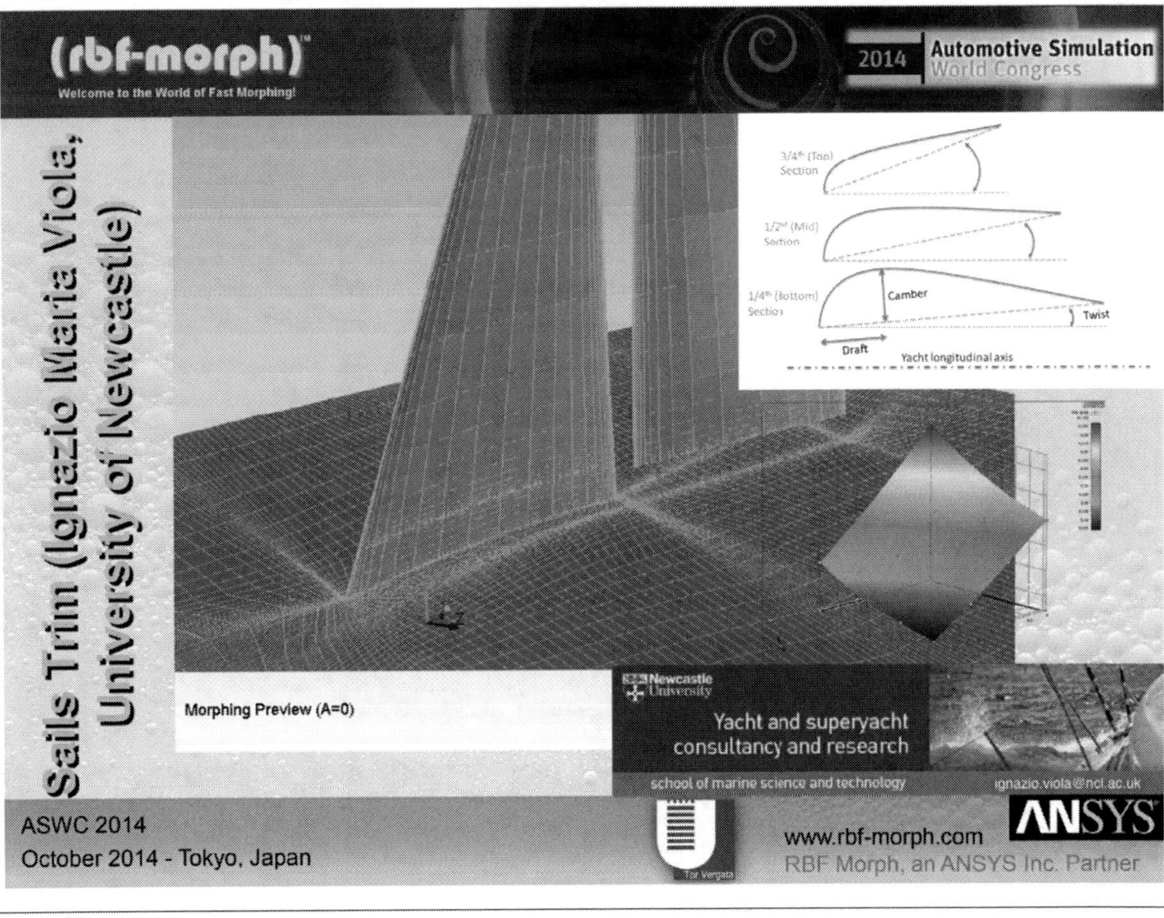
(rbf-morph)™
Welcome to the World of Fast Morphing!
2014 Automotive Simulation World Congress
Sails Trim (Ignazio Maria Viola, University of Newcastle)
Camber
Twist
Draft
Yacht longitudinal axis
Morphing Preview (A=0)
Newcastle University
Yacht and superyacht consultancy and research
school of marine science and technology
ignazio.viola@ncl.ac.uk
ASWC 2014
October 2014 - Tokyo, Japan
www.rbf-morph.com
ANSYS
RBF Morph, an ANSYS Inc. Partner

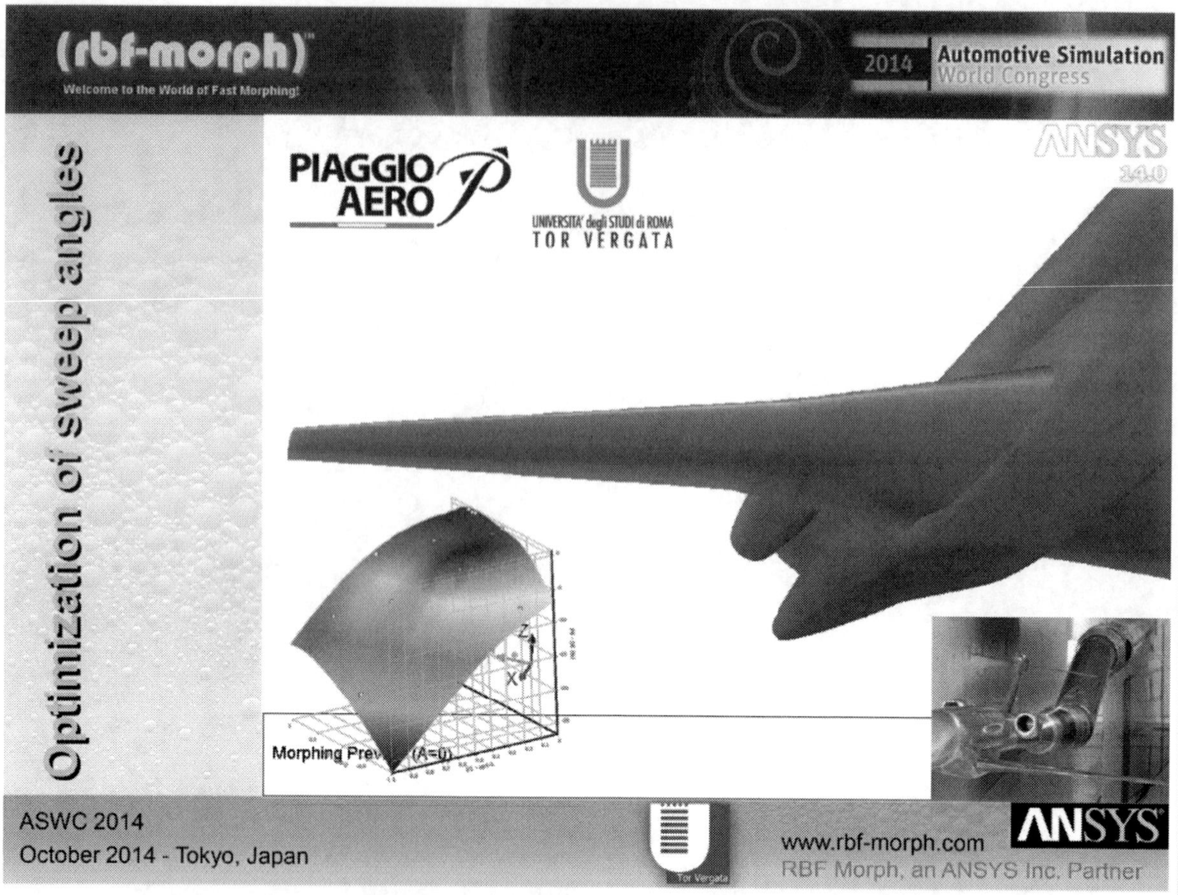
(rbf-morph)™
Welcome to the World of Fast Morphing!
2014
Automotive Simulation
World Congress
ANSYS
14.0
Optimization of sweep angles
PIAGGIO AERO
UNIVERSITA' degli STUDI di ROMA
TOR VERGATA
ASWC 2014
October 2014 - Tokyo, Japan
Tor Vergata
www.rbf-morph.com
RBF Morph, an ANSYS Inc. Partner
ANSYS

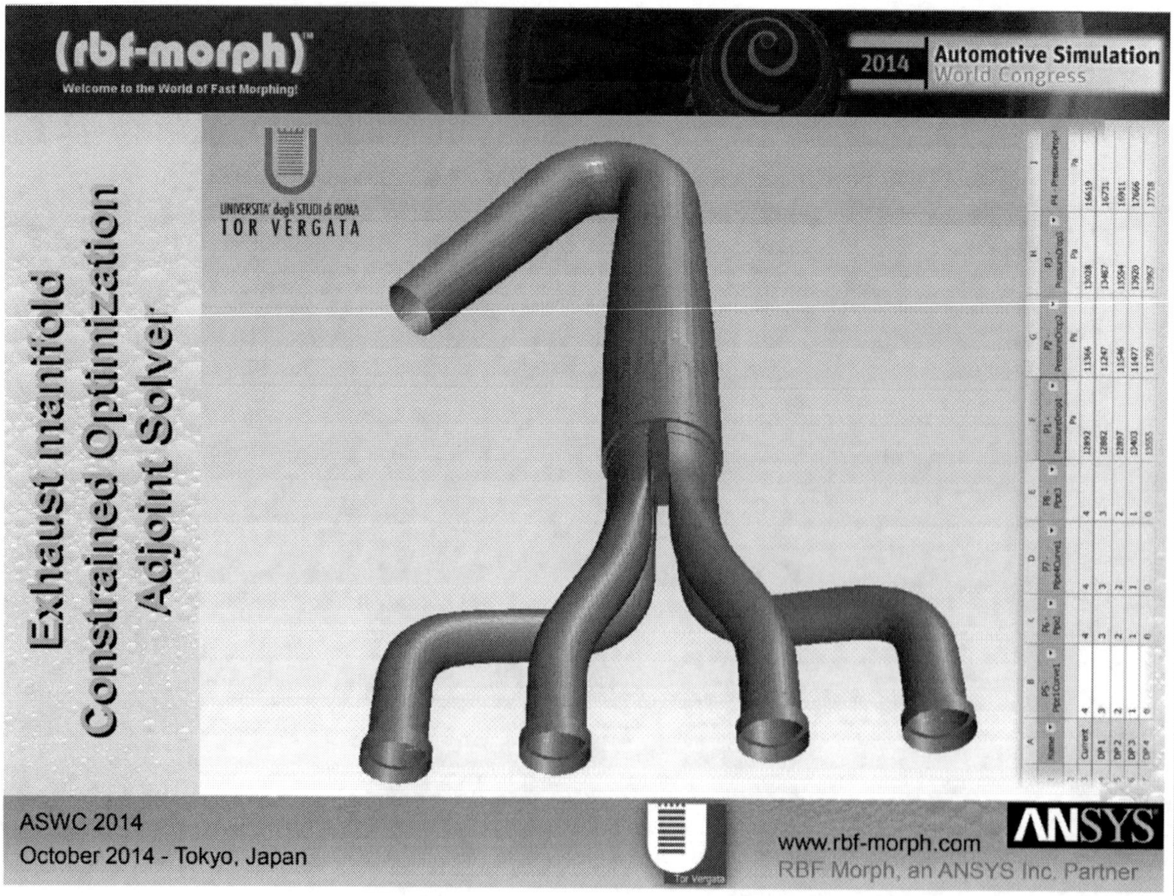
(rbf-morph)™
Welcome to the World of Fast Morphing!
2014
Automotive Simulation
World Congress
Exhaust manifold
Constrained Optimization
Adjoint Solver
UNIVERSITA' degli STUDI di ROMA
TOR VERGATA
ASWC 2014
October 2014 - Tokyo, Japan
Tor Vergata
www.rbf-morph.com
RBF Morph, an ANSYS Inc. Partner
ANSYS

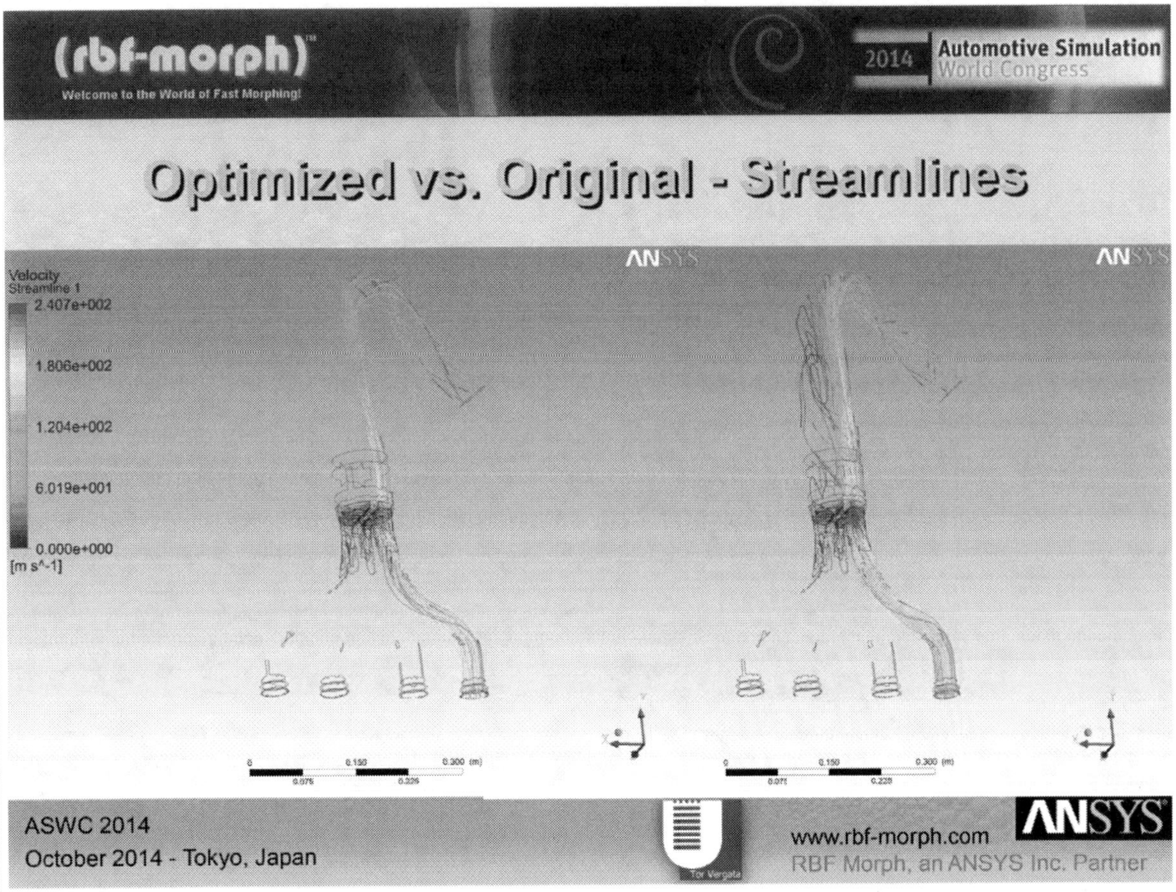
(rbf-morph)™
Welcome to the World of Fast Morphing!
2014 Automotive Simulation World Congress
Optimized vs. Original - Streamlines
Velocity
Streamline 1
2.407e+002
1.806e+002
1.204e+002
6.019e+001
0.000e+000
[m s^-1]
ANSYS
ASWC 2014
October 2014 - Tokyo, Japan
www.rbf-morph.com
RBF Morph, an ANSYS Inc. Partner
ANSYS

(rbf-morph)™
Welcome to the World of Fast Morphing!
2014 Automotive Simulation World Congress
MIRA Reference car
(MIRA ltd)
MIRA
MIRA Reference Car
Shape Optimisation using RBF-Morph
Smarter Thinking
ASWC 2014
October 2014 - Tokyo, Japan
www.rbf-morph.com
RBF Morph, an ANSYS Inc. Partner
ANSYS

(rbf-morph)™
Welcome to the World of Fast Morphing!

2014 Automotive Simulation World Congress

# 50:50:50 Project Volvo XC60 (Ansys, Intel, Volvo)

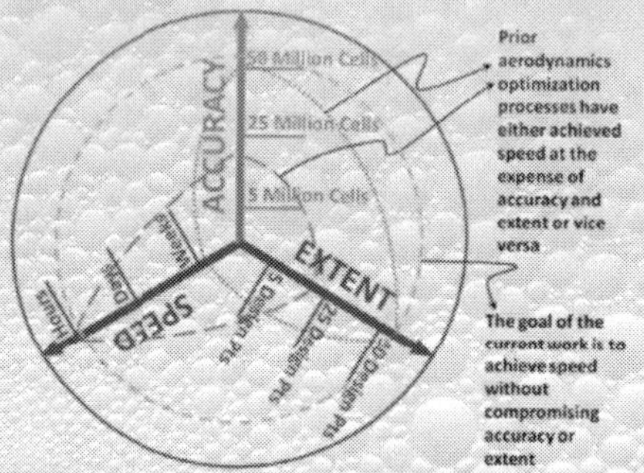


ASWC 2014
October 2014 - Tokyo, Japan

www.rbf-morph.com
RBF Morph, an ANSYS Inc. Partner

ANSYS

(rbf-morph)™
Welcome to the World of Fast Morphing!

# HPC Workflows

2014 Automotive Simulation World Congress

| Case | Motorbike windshield | Reference car | Sedan | Hull | **Volvo XC60** | Sails | DLR F6 | IR5 |
|---|---|---|---|---|---|---|---|---|
| **Organization** | MRA/UTV | MIRA | ANSYS | Leeds | **ANSYS** | New Castle / UTV | MorphLab/ UTV | Dallara |
| **Year** | 2009 | 2010 | 2011 | 2011 | **2012** | 2013 | 2013 | 2013 |
| **#Mcells** | 1,5 | 5,2 | 6 | 0,3 | **50** | 1,5 | 14 | 80 |
| **mesh type** | tets | poly | tets | hexa | **tets** | hexa | tets | tets |
| **#par** | 3 | 3 | 2 | 8 | **4** | 4 | 8 | 5 |
| **#design** | 45 | 27 | 9 | 45 | **50** | 100 | 81 | 1 |
| **RS Tool** | modeF | Mathcad | DX | DX | **DX** | DX/ Mathcad | DX | FSI |
| **ncores** | 4 | 2 | 12 | 4 | **240** | 16 | 16 | 256 |
| **RUN (hr)** | 48 | 300 | 24 | 45 | **50** | 26 | 102 | 1 |
| **Time to set-up one par (hr)** | 1,5 | 2,5 | 2 | 1 | **2** | 2 | 1 | 2 |
| **Time to set-up (hr)** | 4,5 | 7,5 | 4 | 8 | **8** | 8 | 8 | 8 |
| **Serial time one design (hr)** | 4,27 | 22,22 | 32,00 | 4,00 | **240,00** | 4,16 | 20,15 | 256,00 |
| **Serial time one design (hr/Mcells)** | 2,84 | 4,27 | 5,33 | 13,33 | **4,80** | 2,77 | 1,44 | 3,2 |

ASWC 2014
October 2014 - Tokyo, Japan

www.rbf-morph.com
RBF Morph, an ANSYS Inc. Partner

ANSYS

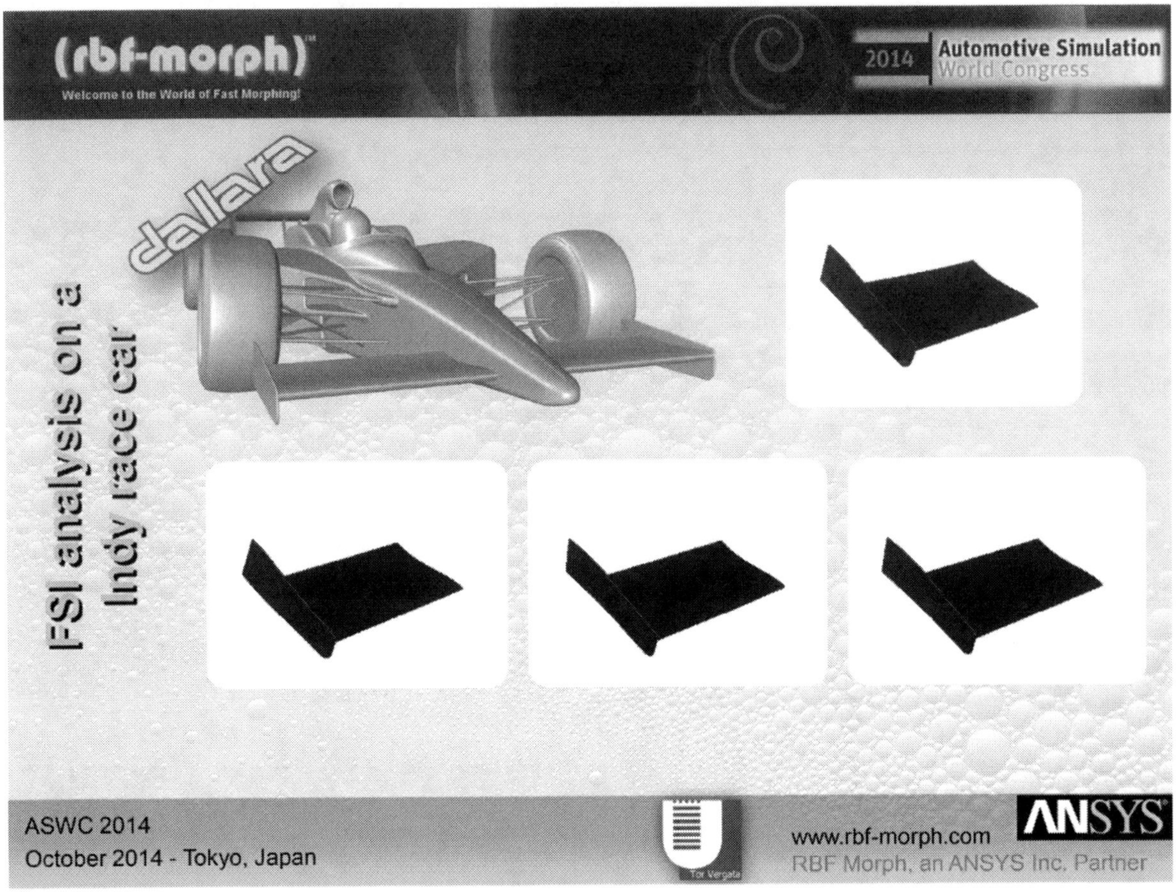


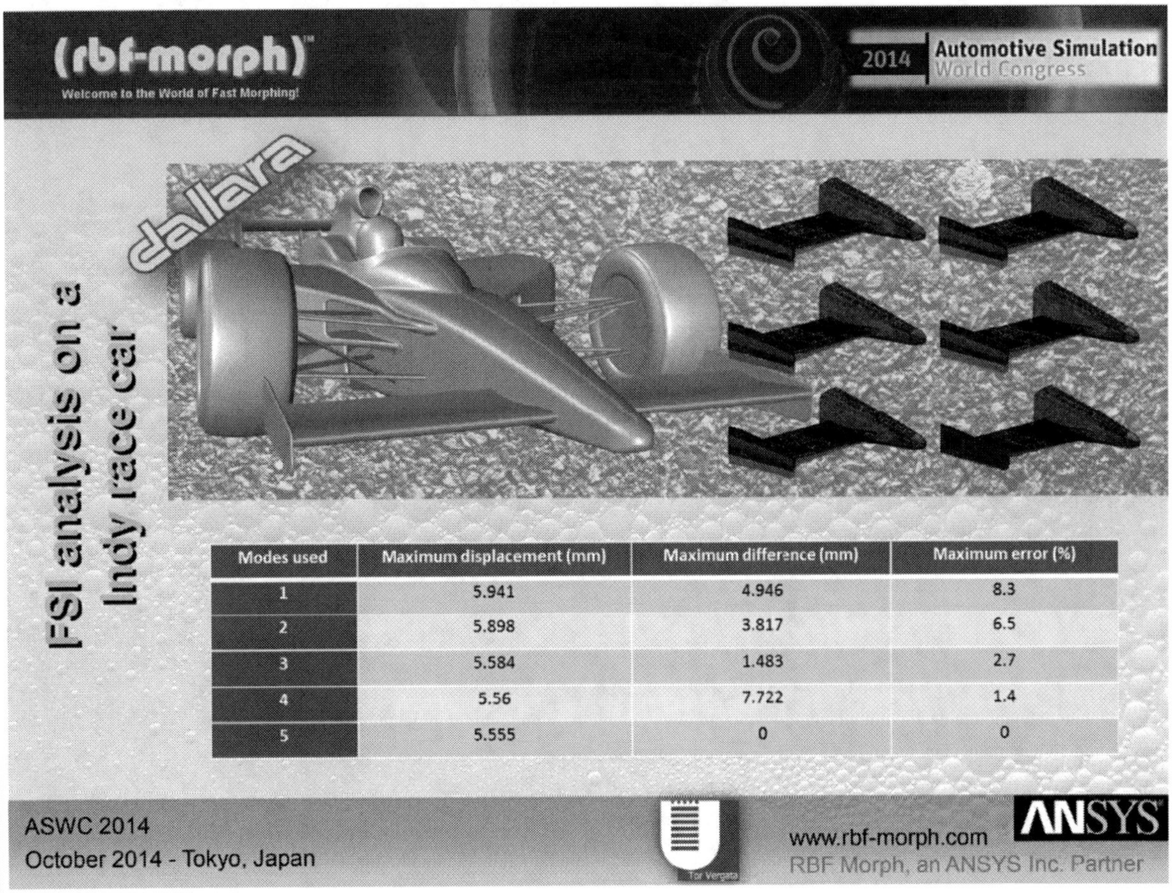


| Modes used | Maximum displacement (mm) | Maximum difference (mm) | Maximum error (%) |
|---|---|---|---|
| 1 | 5.941 | 4.946 | 8.3 |
| 2 | 5.898 | 3.817 | 6.5 |
| 3 | 5.584 | 1.483 | 2.7 |
| 4 | 5.56 | 7.722 | 1.4 |
| 5 | 5.555 | 0 | 0 |

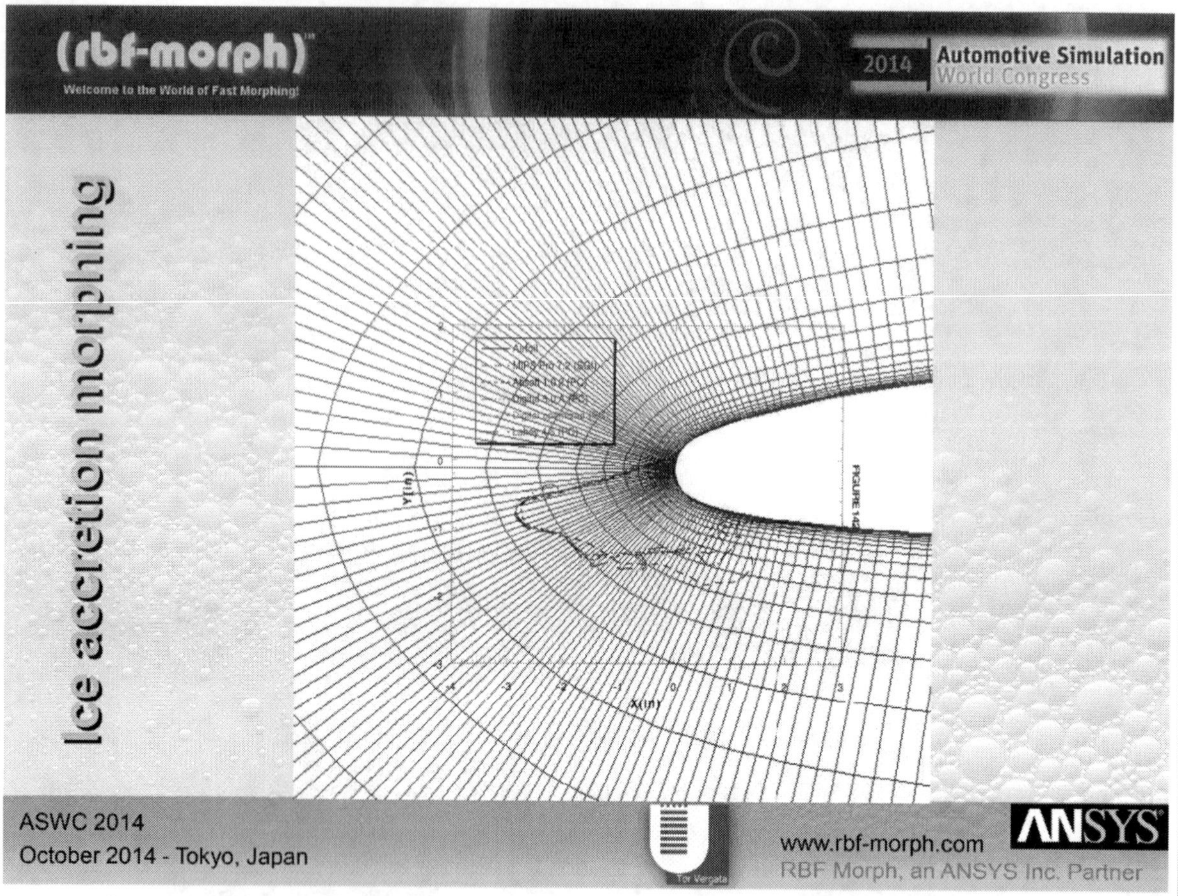
(rbf-morph)™
Welcome to the World of Fast Morphing!
2014
Automotive Simulation
World Congress
Ice accretion morphing
Y(in)
X(in)
ASWC 2014
October 2014 - Tokyo, Japan
www.rbf-morph.com
ANSYS
RBF Morph, an ANSYS Inc. Partner

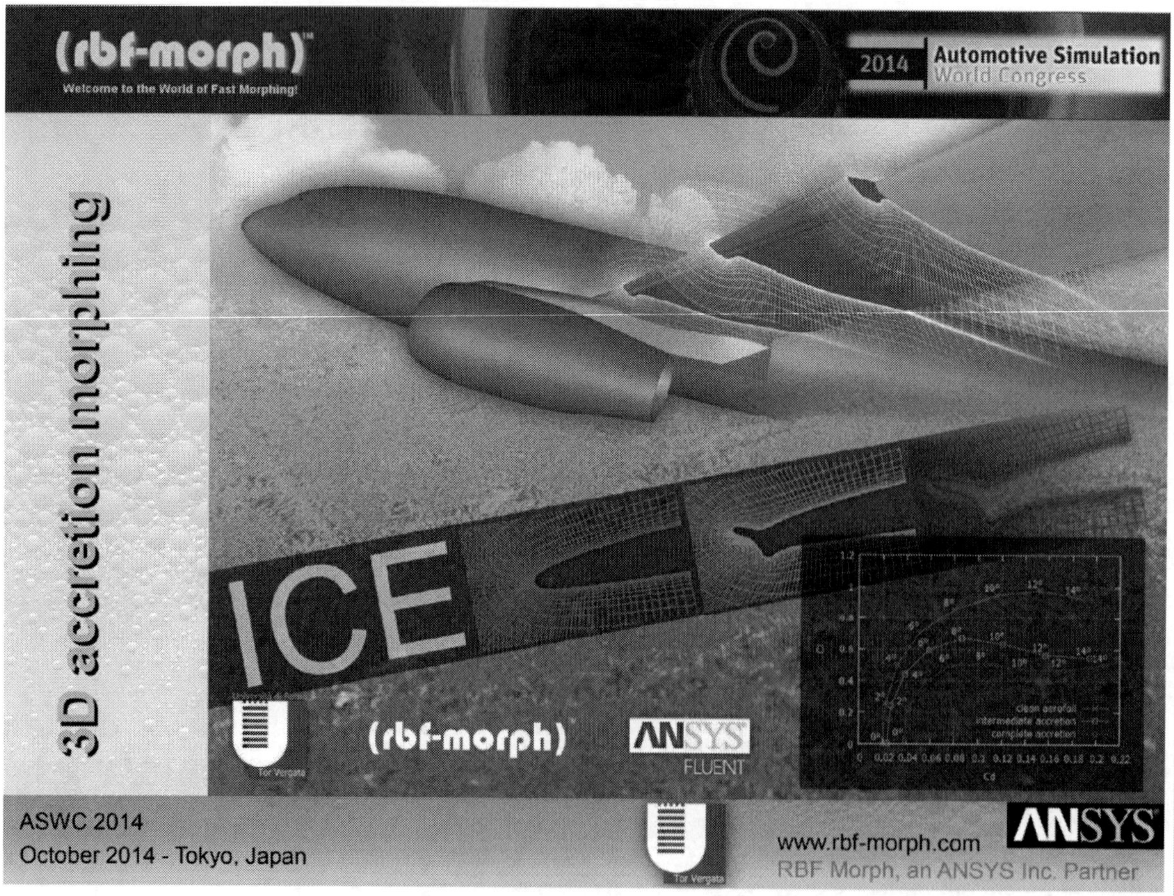
(rbf-morph)™
Welcome to the World of Fast Morphing!
2014
Automotive Simulation
World Congress
3D accretion morphing
ICE
(rbf-morph)
ANSYS
FLUENT
clean aerofoil
intermediate accretion
complete accretion
Cd
ASWC 2014
October 2014 - Tokyo, Japan
www.rbf-morph.com
ANSYS
RBF Morph, an ANSYS Inc. Partner

(rbf-morph)™
Welcome to the World of Fast Morphing!
2014
Automotive Simulation
World Congress
Fluent Adjoint Coupling
ASWC 2014
October 2014 - Tokyo, Japan
Tor Vergata
www.rbf-morph.com
ANSYS
RBF Morph, an ANSYS Inc. Partner

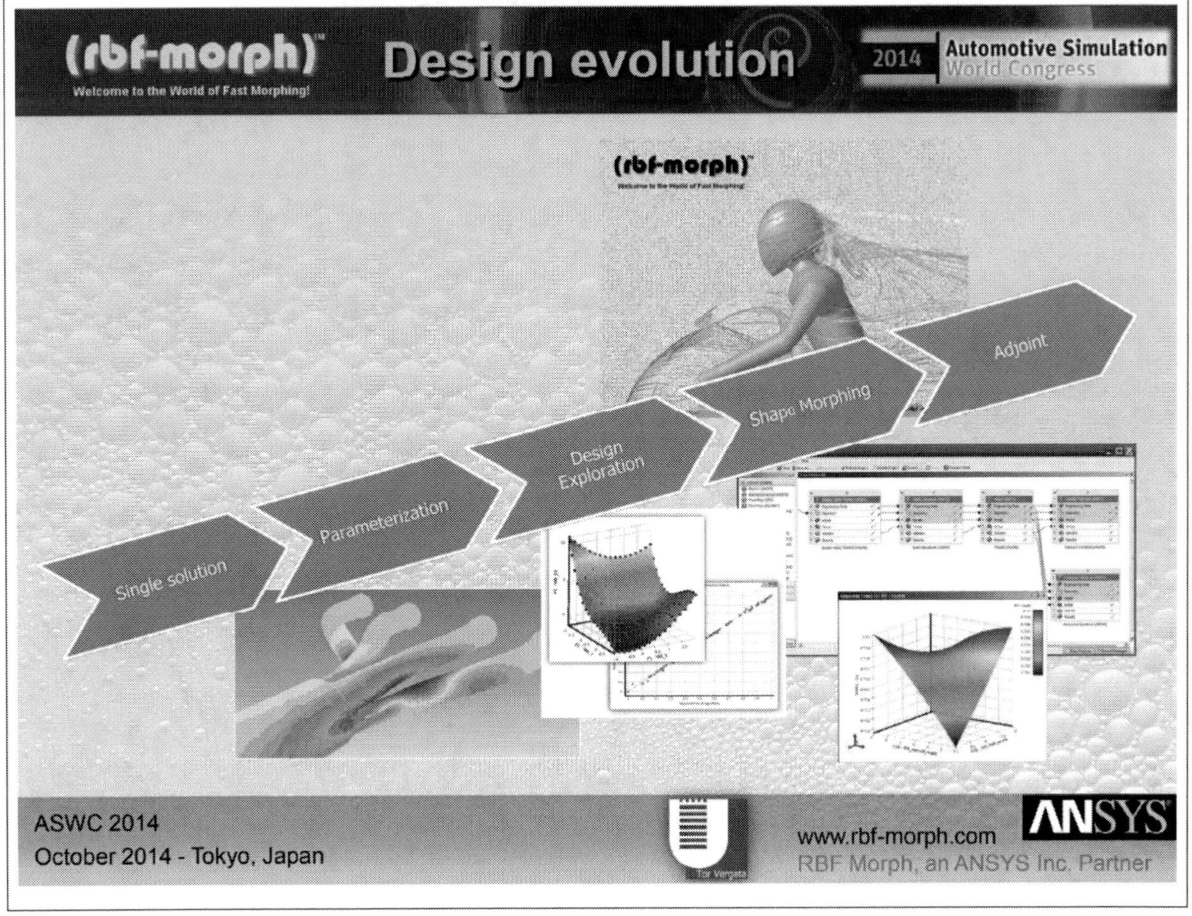
(rbf-morph)™
Welcome to the World of Fast Morphing!
Design evolution
2014
Automotive Simulation
World Congress
Single solution
Parameterization
Design Exploration
Shape Morphing
Adjoint
ASWC 2014
October 2014 - Tokyo, Japan
Tor Vergata
www.rbf-morph.com
ANSYS
RBF Morph, an ANSYS Inc. Partner

## (rbf-morph)™ Adjoint Key Ideas

Welcome to the World of Fast Morphing!

2014 Automotive Simulation World Congress

- An **adjoint solver** allows to compute the **derivatives** of an engineering quantity with respect to the positions of all the nodes of the mesh.
- Post-process the **adjoint solution allows** to get
  - Shape Sensitivity
  - Contour & Vector plots of sensitivities
- Coupling with a **mesh morpher** allows to full exploit sensitivity data
  - An **FFD morpher** comes with the adjoint solver
  - **RBF Morph** is an excellent companion to take advantage of adjoint sensitivity data!

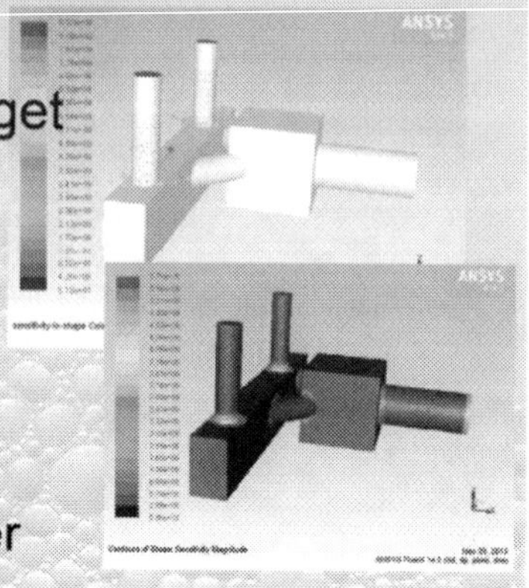

ASWC 2014
October 2014 - Tokyo, Japan

Tor Vergata

www.rbf-morph.com
RBF Morph, an ANSYS Inc. Partner
ANSYS

## (rbf-morph)™ Possible strategies

Welcome to the World of Fast Morphing!

2014 Automotive Simulation World Congress

- **Adjoint sculpting**
- **Adjoint preview** (Sensitivity analysis with imposed shapes)
- **Augmented-DOE**

ASWC 2014
October 2014 - Tokyo, Japan

Tor Vergata

www.rbf-morph.com
RBF Morph, an ANSYS Inc. Partner
ANSYS

## (rbf-morph)™ Adjoint Sculpting

Welcome to the World of Fast Morphing!

2014 Automotive Simulation World Congress

- **Adjoint shape** can be captured by RBF Morph
- The morphing action can be **accurately** controlled (Box, Surfs, Points...)
- Multiple "sculpted" shape modifications can be **combined**
- Optimized shape can be exported **back to CAD**
- Shape data can be transferred to a different model (the method is meshless)

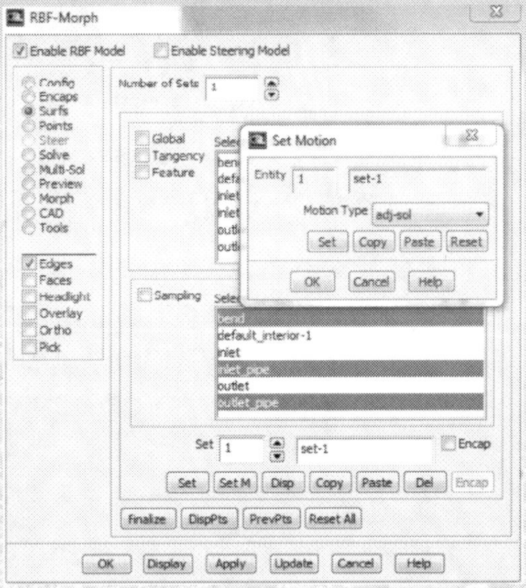


ASWC 2014
October 2014 - Tokyo, Japan

www.rbf-morph.com
RBF Morph, an ANSYS Inc. Partner

## (rbf-morph)™ Adjoint preview

Welcome to the World of Fast Morphing!

2014 Automotive Simulation World Congress

- **Nodal sensitivities** are used to compute the sensitivity with respect to **shape parameters**
- A **single adjoint** solution can be used to analyze the observable change introduced by **many** imposed shape variations
- The **computational cost** is unaffected by the number of shape variations analyzed!
- **Shape variations** can be easily defined and blended together using **RBF Morph**
- Derivatives are available at a **given state** of shape parameters (i.e. after morphing)

ASWC 2014
October 2014 - Tokyo, Japan

www.rbf-morph.com
RBF Morph, an ANSYS Inc. Partner

# Augmented DOE

- Adjoint derivatives with respect to shape parameters allows to **enrich** the **response surface** with tangent data
- Each full computed points of DOE (CFD+adjoint) comes with the value of the **response** and all the **local derivatives**
- **Example**: with 9 shape parameters10 augmented DOE runs produce the same information of 100 CFD runs. The cost of the adjoint update is twice of simple CFD so a 5x speed-up can be achieved.

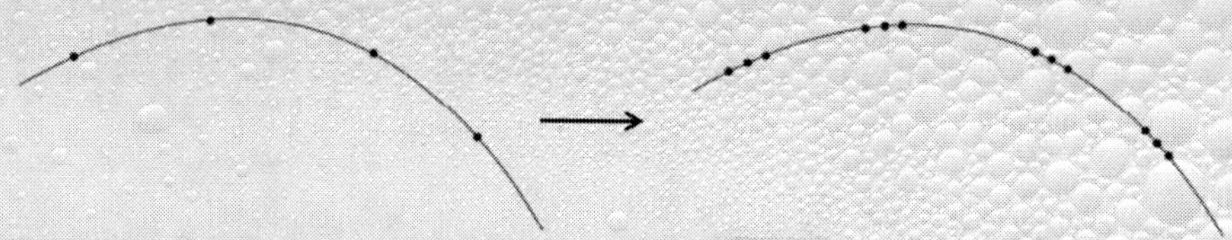





# Fluent Adjoint Coupling Examples

## (rbf-morph)™ Adjoint Self Sculpting

Welcome to the World of Fast Morphing!

2014 Automotive Simulation World Congress

- Manifold system
- Shape information coming from the **Fluent Adjoint** morpher are used to control the shape
- A single adjoint (baseline shape) is used to define shape modifications in **two locations**
- New shape modifications can be **combined** as usual

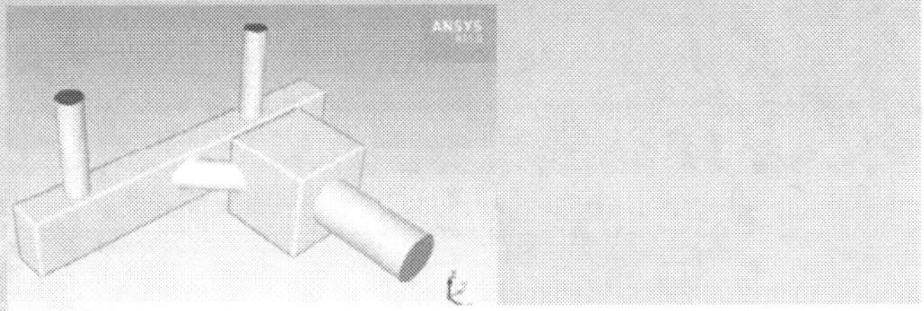

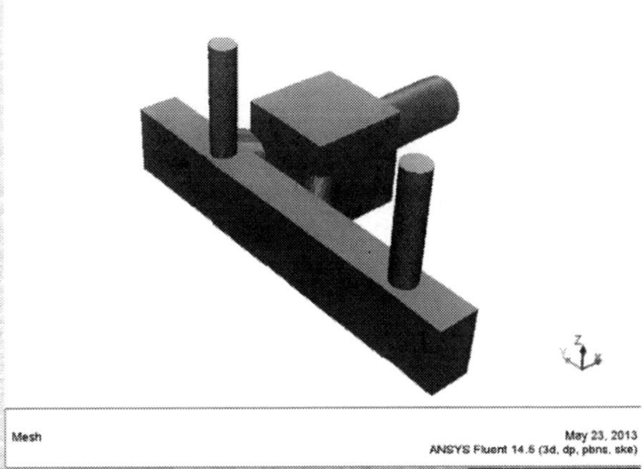


ASWC 2014
October 2014 - Tokyo, Japan

Tor Vergata

www.rbf-morph.com
RBF Morph, an ANSYS Inc. Partner

ANSYS

## (rbf-morph)™ Adjoint Self Sculpting

Welcome to the World of Fast Morphing!

2014 Automotive Simulation World Congress

- Obtained shapes are used to update the original **CAD**
- Both **runners** are sculpted using the same solution data

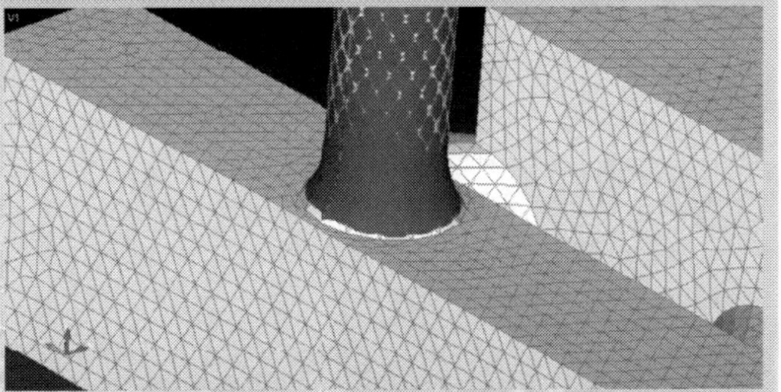

ASWC 2014
October 2014 - Tokyo, Japan

Tor Vergata

www.rbf-morph.com
RBF Morph, an ANSYS Inc. Partner

ANSYS

## (rbf-morph)™ Adjoint Self Sculpting

Welcome to the World of Fast Morphing!

2014 Automotive Simulation World Congress

- 90 deg bend optimization
- New shape is sculpted using **adjoint data**
- Original geometry (2 cylinders and a torus) is transformed in **NURBS**
- **NURBS** are morphed using the **back to CAD** tool of RBF Morph

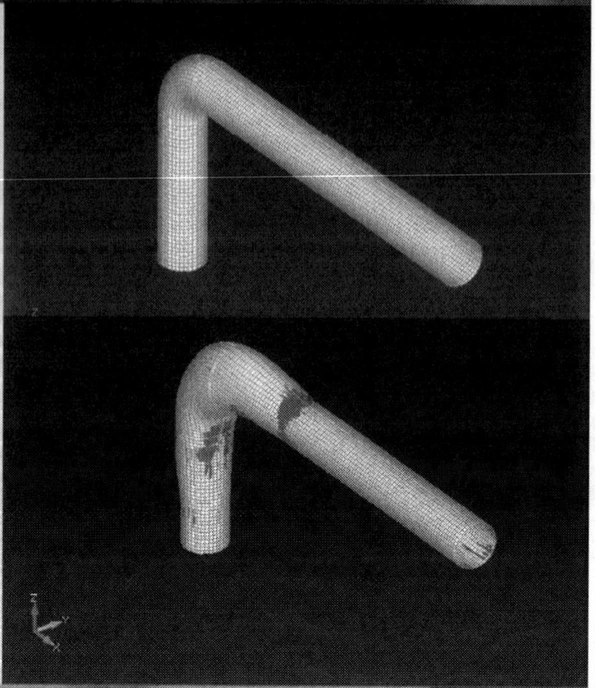

ASWC 2014
October 2014 - Tokyo, Japan

www.rbf-morph.com
RBF Morph, an ANSYS Inc. Partner

ANSYS

## (rbf-morph)™ Adjoint preview

Welcome to the World of Fast Morphing!

2014 Automotive Simulation World Congress

- A cube immersed in a wind tunnel is made parametric using **3 shapes**
- Steepest descent gradient method allows to reduce **drag** by **16.7%**

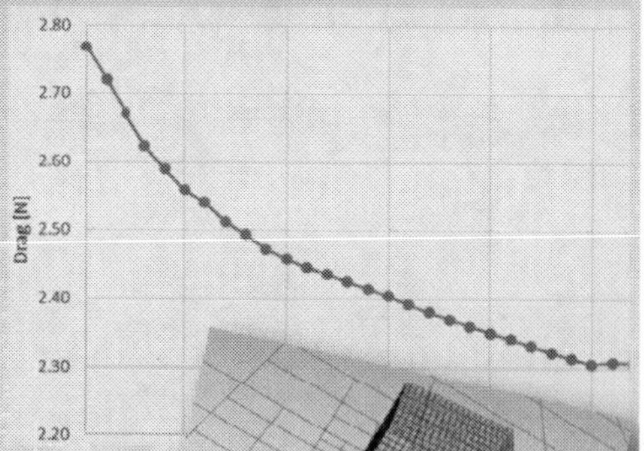


$$\alpha_j = -\beta \frac{dI}{dx_j}$$

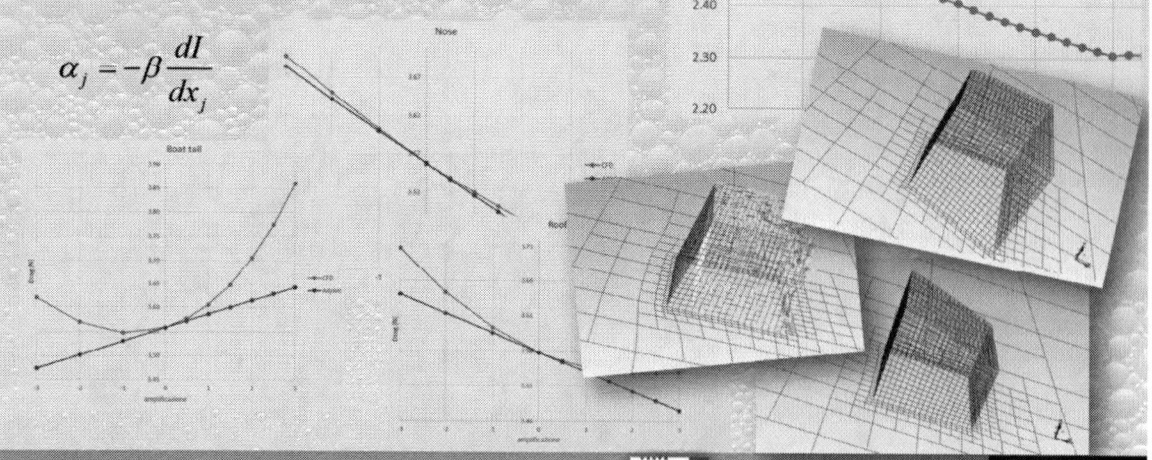


ASWC 2014
October 2014 - Tokyo, Japan

www.rbf-morph.com
RBF Morph, an ANSYS Inc. Partner

ANSYS

## (rbf-morph)™ Engine Air box

Welcome to the World of Fast Morphing!

2014 Automotive Simulation World Congress

- **32 shape parameters** are used to control the geometry of the plenum and of the three runners
- Original **packaging** constraints are preserved
- Each complete run consists in 3 CFD + adjoint simulation (one for pressure drop of each runner)
- Design **objectives**: minimum pressure drop, uniform pressure drop.
- Design **constraints**: regular shape of runners.

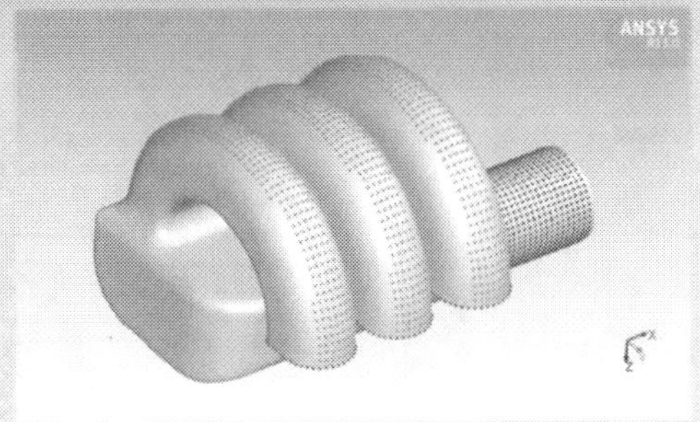


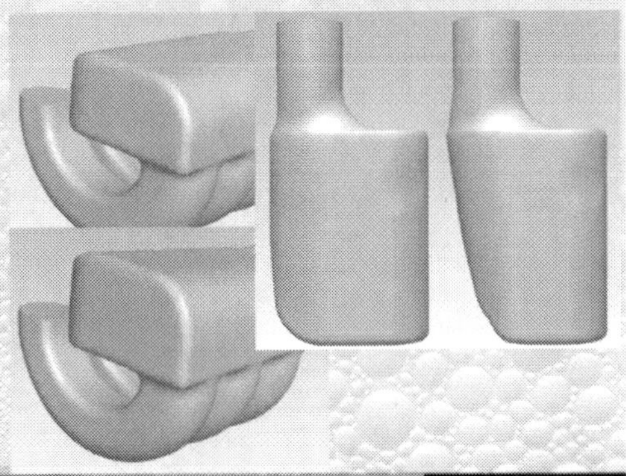

ASWC 2014
October 2014 - Tokyo, Japan

www.rbf-morph.com
RBF Morph, an ANSYS Inc. Partner

## (rbf-morph)™ Engine Air box

Welcome to the World of Fast Morphing!

2014 Automotive Simulation World Congress

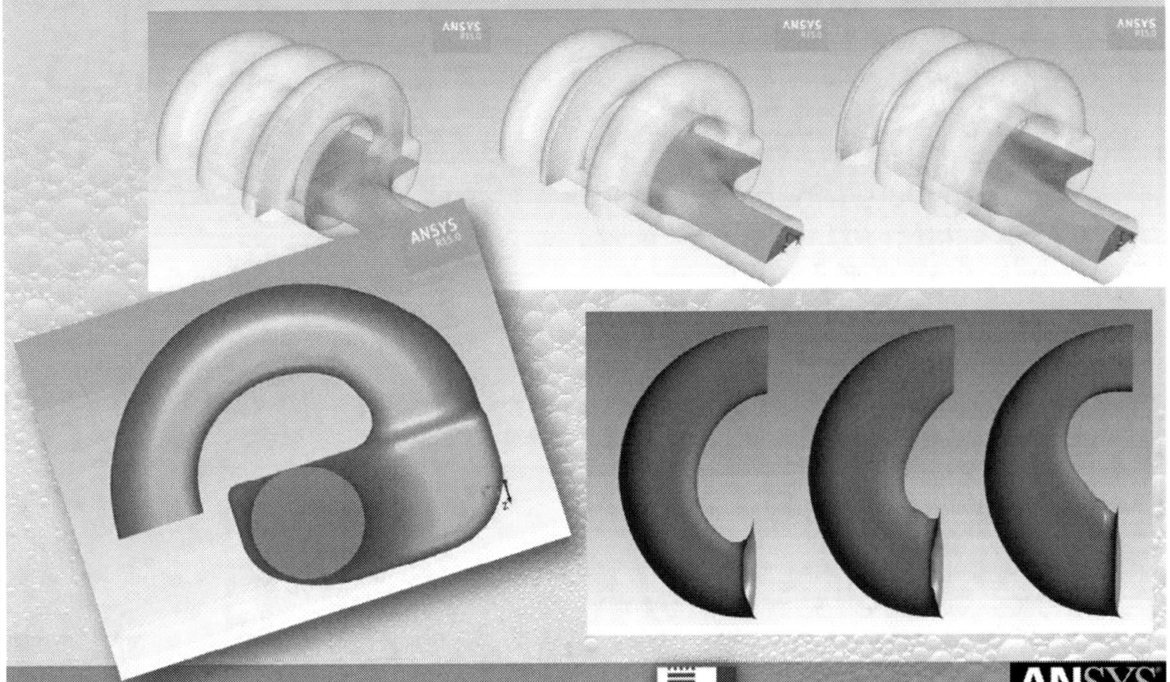


ASWC 2014
October 2014 - Tokyo, Japan

www.rbf-morph.com
RBF Morph, an ANSYS Inc. Partner

## (rbf-morph)™ Engine Air box

Welcome to the World of Fast Morphing!

2014 Automotive Simulation World Congress

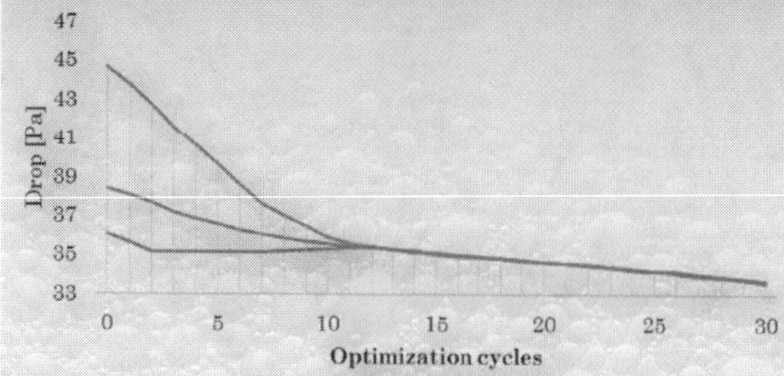


| | Mean pressure Drop [Pa] | Unbalance |
|---|---|---|
| Baseline | 39.7 | 12.45% |
| Optimized | 33.635 | 0.12% |
| Reduction | 15.3% | 99.0% |

- Obtained shape allows to get a **15.3%** reduction of pressure drop and uniform distribution.

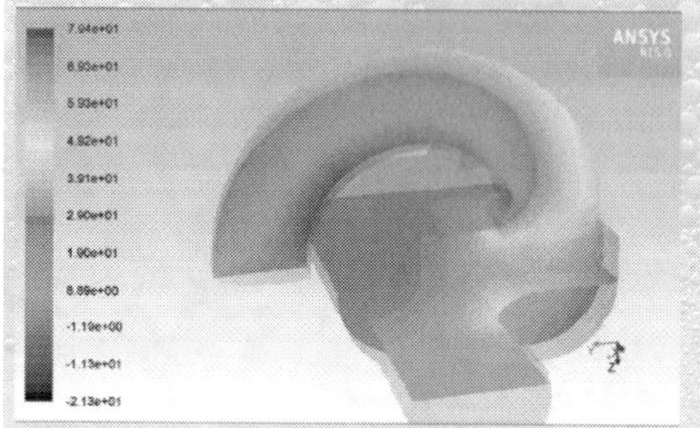


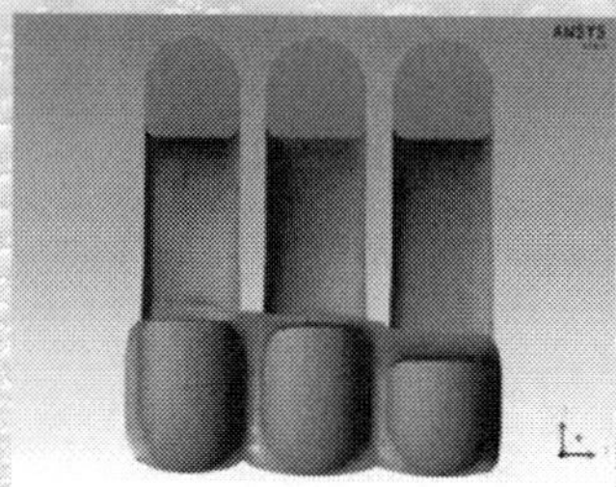



## (rbf-morph)™ Sedan car

Welcome to the World of Fast Morphing!

2014 Automotive Simulation World Congress

- **3 shape parameters** are used to control the shape of the rear part (tail, roof, back window)
- Each complete run consists in a **CFD + adjoint** simulation (drag and its 3 derivatives are obtained)
- The **steepest descent** method is used to update the 3 shape parameters at each design cycle.
- Mesh size **3.6 millions** cells.

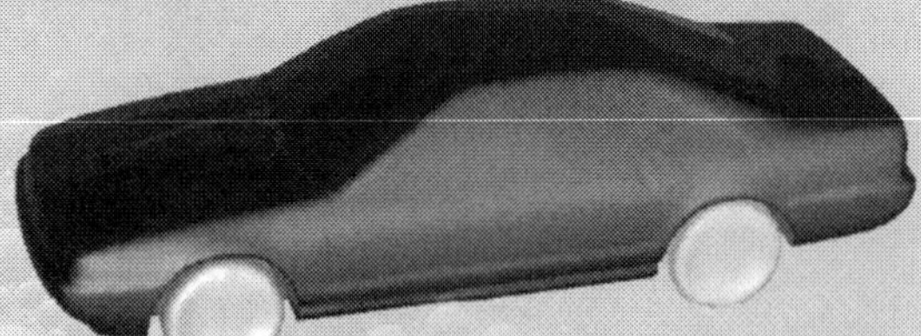

$$\alpha_j = -\beta \frac{dI}{dx_j}$$

(rbf-morph)
Welcome to the World of Fast Morphing!

# Sedan car

2014 Automotive Simulation World Congress

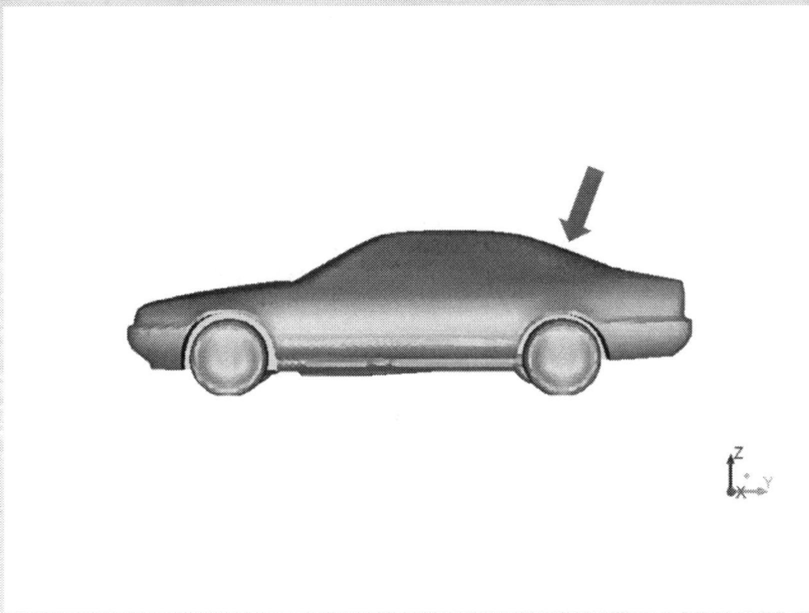

ASWC 2014
October 2014 - Tokyo, Japan

Tor Vergata

www.rbf-morph.com
RBF Morph, an ANSYS Inc. Partner

(rbf-morph)
Welcome to the World of Fast Morphing!

Sedan car

2014 Automotive Simulation World Congress

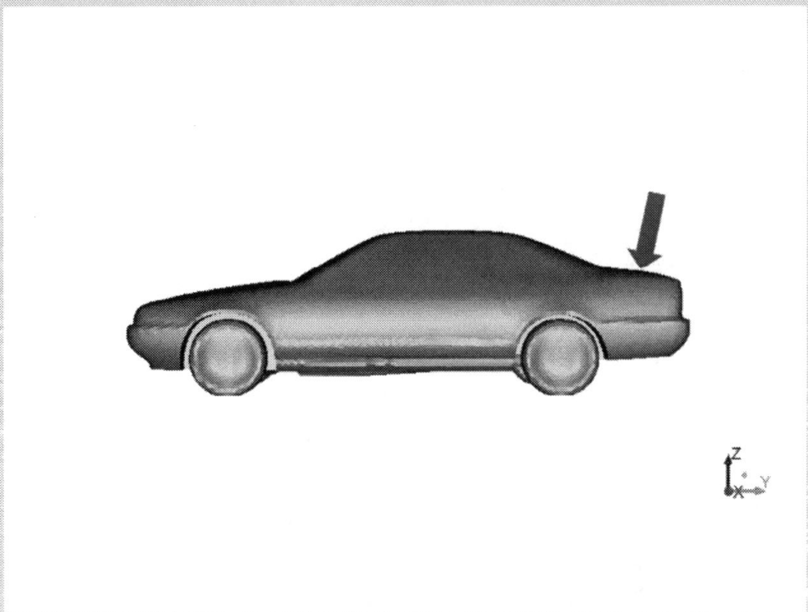

ASWC 2014
October 2014 - Tokyo, Japan

Tor Vergata

www.rbf-morph.com
RBF Morph, an ANSYS Inc. Partner

# Sedan car

(rbf-morph)™ Welcome to the World of Fast Morphing!

2014 Automotive Simulation World Congress

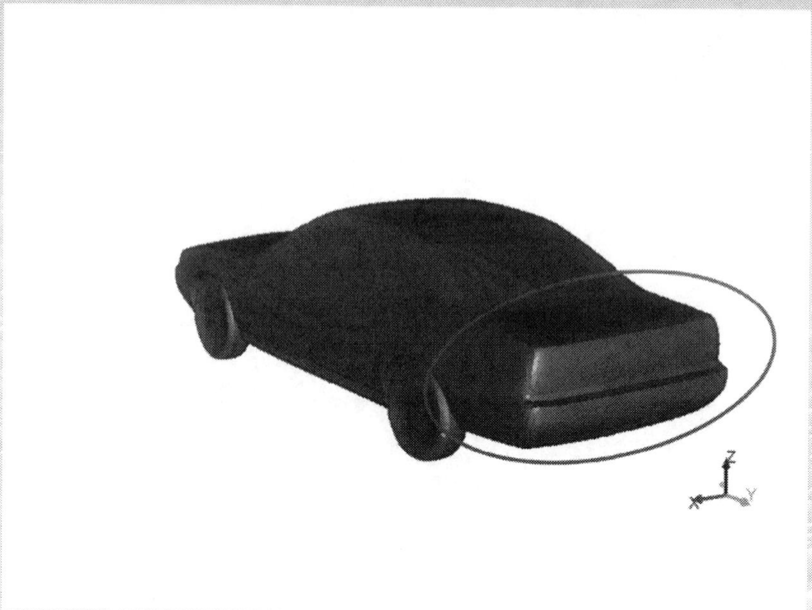

ASWC 2014
October 2014 - Tokyo, Japan

www.rbf-morph.com
RBF Morph, an ANSYS Inc. Partner

ANSYS

# Sedan car

(rbf-morph)™ Welcome to the World of Fast Morphing!

2014 Automotive Simulation World Congress

- A **3.13%** drag reduction is achieved after 33 cycles
- With just 3 parameters a **standard DOE** has comparable costs.

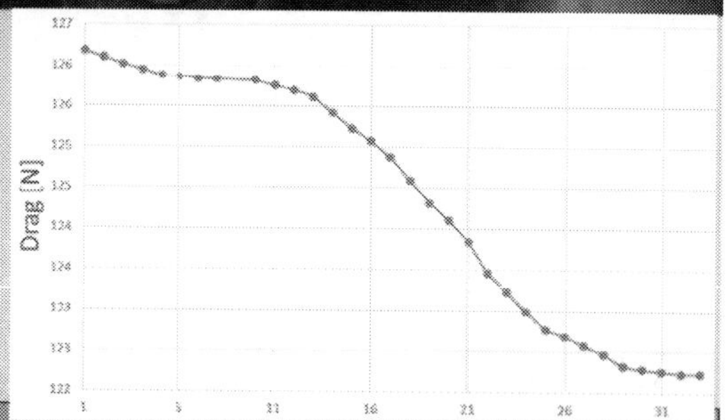

ASWC 2014
October 2014 - Tokyo, Japan

www.rbf-morph.com
RBF Morph, an ANSYS Inc. Partner

ANSYS

(rbf-morph)™ Welcome to the World of Fast Morphing!

## Conclusions

2014 Automotive Simulation World Congress

- RBF Morph has been successfully coupled with **adjoint solver** of Fluent.
- Advanced mesh morphing allows to effectively enable **adjoint self sculpting**. Flow solution is used to define new shapes.
- RBF Morph allows to **pick, mix and reuse** self sculpted shapes.
- RBF Morph allows to **bring back to CAD** sculpted shapes.
- A **shape parametric** CFD model can be defined using ANSYS Fluent and RBF Morph (standard shapes or self sculpted ones can be mixed)
- Local **derivatives** with respect to shape parameters can be readily computed using the **adjoint preview** tool of RBF Morph.
- The effectiveness of RBF Morph + Fluent adjoint coupling has been demonstrated facing **challenging industrial applications**

ASWC 2014
October 2014 - Tokyo, Japan

www.rbf-morph.com
RBF Morph, an ANSYS Inc. Partner

---

(rbf-morph)™ Welcome to the World of Fast Morphing!

2014 Automotive Simulation World Congress

ご清聴ありがとうございました!

Prof. Marco Evangelos Biancolini

E-mail: info@rbf-morph.com
Web: www.rbf-morph.com

Goo.gl/1svYd
linkedin.com/company/rbf-morph
Twitter.com/RBFMorph
youtube.com/user/RbfMorph

ASWC 2014
October 2014 - Tokyo, Japan

www.rbf-morph.com
RBF Morph, an ANSYS Inc. Partner

2014 Automotive Simulation World Congress

# Engine-Gasket HPC trial and Suspension NVH analysis using Linear_Perturbation

ANSYS Japan
Mechanical BU
Toru Hiyake

## Agenda

2014 Automotive Simulation World Congress

- Topic 1 : Engine-Gasket Nonlinear Analysis
  - Legacy Data Import
  - Gasket modeling
  - Bolt Pretension modeling
  - HPC Performance

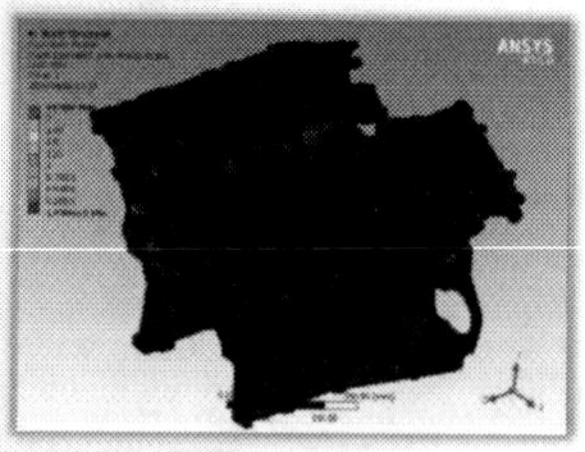

- Topic 2 : Leaf-Spring Suspension Dynamic Analysis
  - Tire Model
  - Leaf Spring Model
  - Suspension Assy Model
  - Linear Perturbation

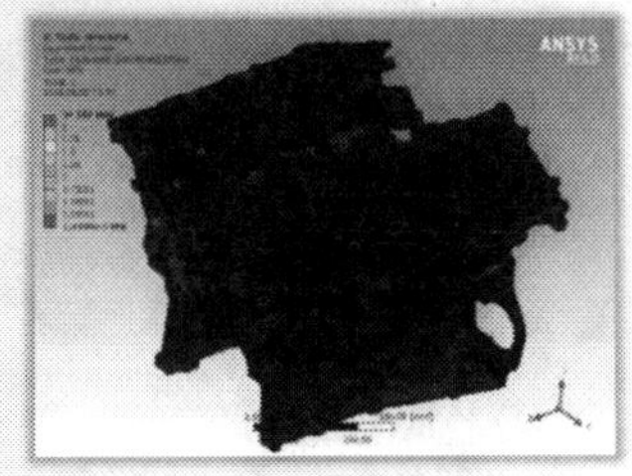

# Topic 1 : Engine-Gasket Nonlinear Analysis





< Topic 1 : Engine-Gasket Nonlinear Analysis >

## Legacy Data Import

- **Data Import in Mechanical using WB tool(External Model)**

Legacy FE Data (.cdb file)

↓

ANSYS Mechanical

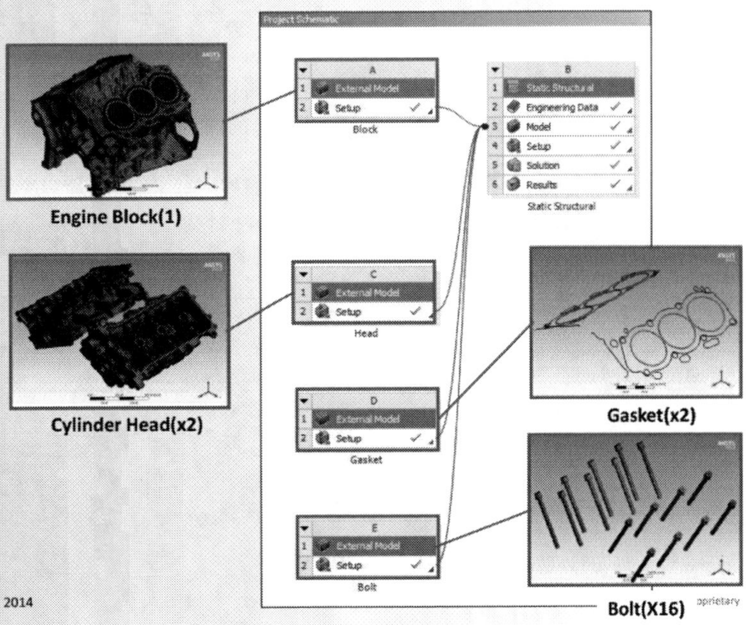

< Topic 1 : Engine-Gasket Nonlinear Analysis >

## Legacy Data Import

2014 Automotive Simulation World Congress

- **Engine Block(1) + Cylinder Head(x2) + Gasket(x2) + Bolt(x16)**

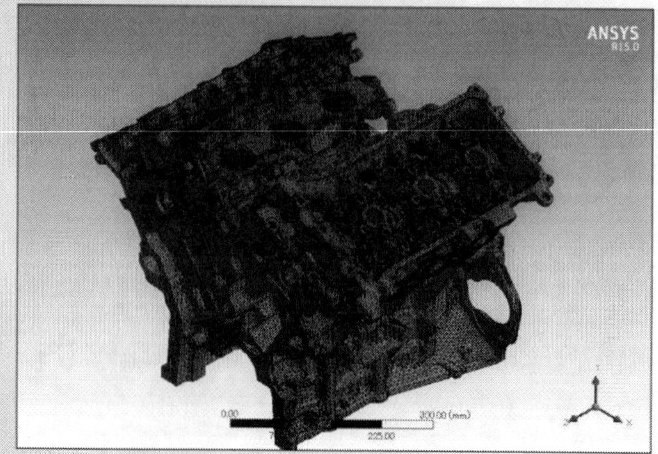


| | |
|---|---|
| Elements | : 2,017,780 |
| Nodes | : 3,379,929 |
| Total DOF | : 10,130,299 |





< Topic 1 : Engine-Gasket Nonlinear Analysis >

## Gasket Modeling : Element Setting

2014 Automotive Simulation World Congress

- **ANSYS Gasket Elements**
  - Inter192-195
- **Change Element Type**
  - From Solid185 to Inter195
    - Solid185 is one order structural element(original element in .cdb file)
    - Inter195 is one order gasket element
  - Using Command Object

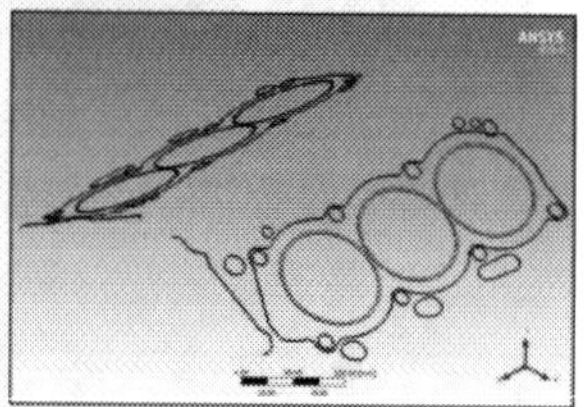

Gasket

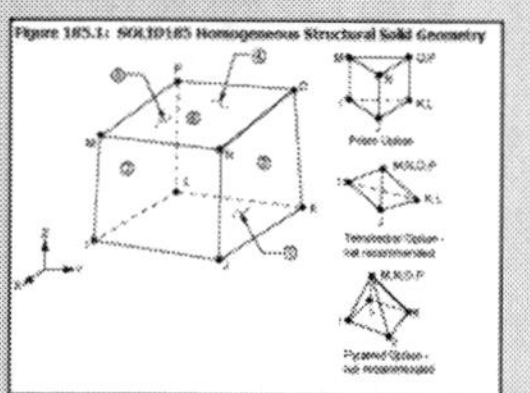

Solid185 element

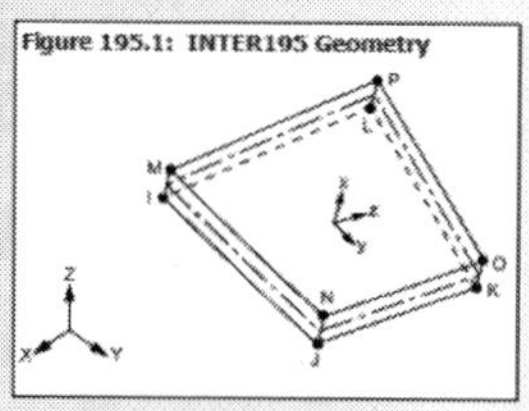


Inter195 element

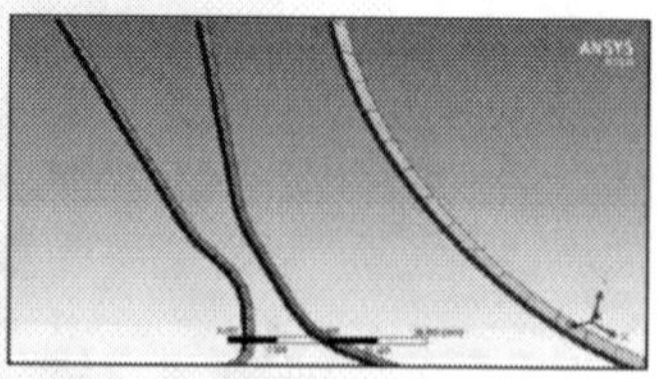

Gasket Mesh

Command Object

< Topic 1 : Engine-Gasket Nonlinear Analysis >

## Gasket Modeling : Material Setting

2014 Automotive Simulation World Congress

- **Gasket Material**
  - The gasket material is usually under compression and is highly nonlinear.
  - The material also exhibits quite complicated unloading behavior when compression is released.

- **WB Engineering Data**
  - ANSYS support Gasket material

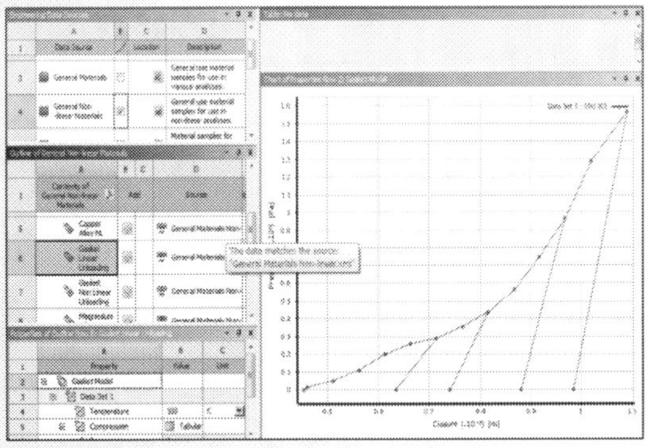

Gasket material in WB Engineering Data

 October 2, 2014 

< Topic 1 : Engine-Gasket Nonlinear Analysis >

## Boundary Conditions : Bolt Model

2014 Automotive Simulation World Congress

- **This boundary condition applies a pretension load to a cylindrical face, to bodies, typically to model a bolt under pretension.**
  - Easy Bolt modeling
- **In Mechanical , user can set this boundary condition easily.**
  - Pattern copy using object generator

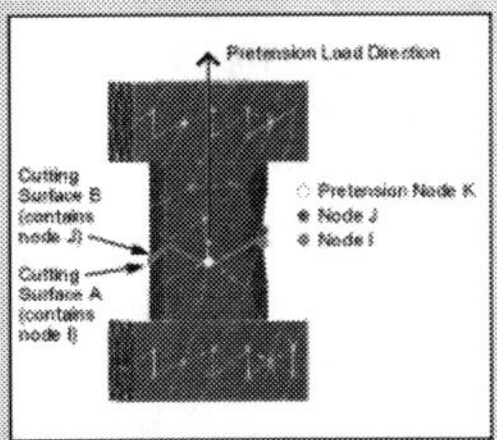


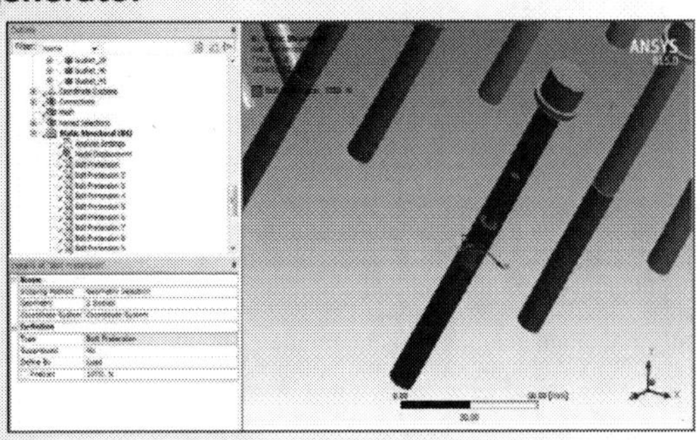

Bolt Pretension Setting in Mechanical

 October 2, 2014 

< Topic 1 : Engine-Gasket Nonlinear Analysis >

## Boundary Conditions : Constraints

2014 Automotive Simulation World Congress

- Direct Node Selection for displacement's constraint.

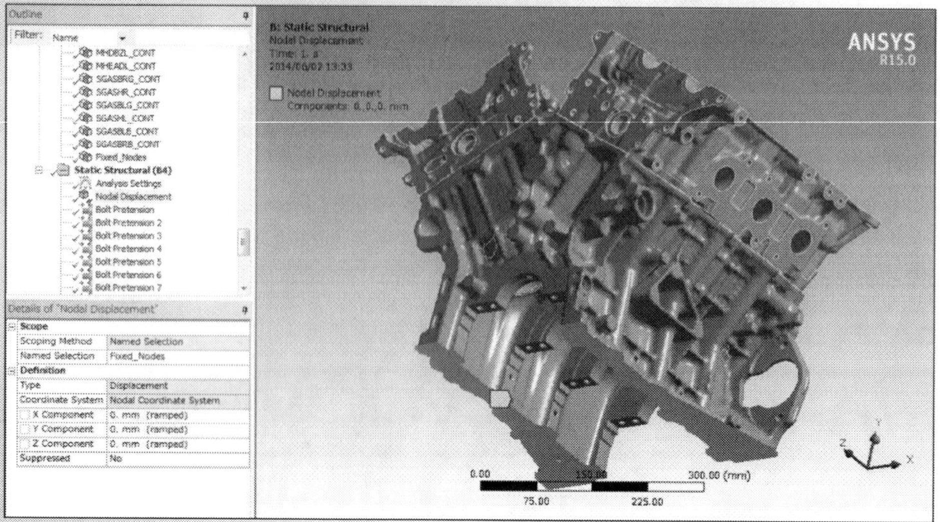


Boundary Condition Setting in Mechanical



< Topic 1 : Engine-Gasket Nonlinear Analysis >

## Solver Setting

2014 Automotive Simulation World Congress

- Analysis Type
  - Non-Linear Statics

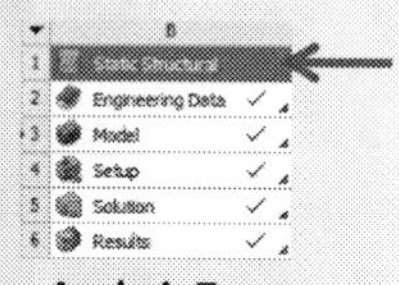


Analysis Type

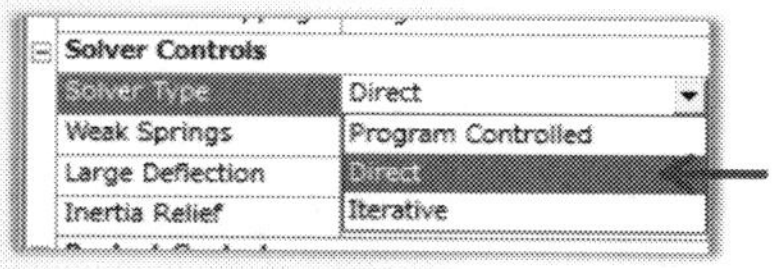


Solver Type

- Solver Type
  - Sparse Direct Solver
  - PCG Iterative Solver

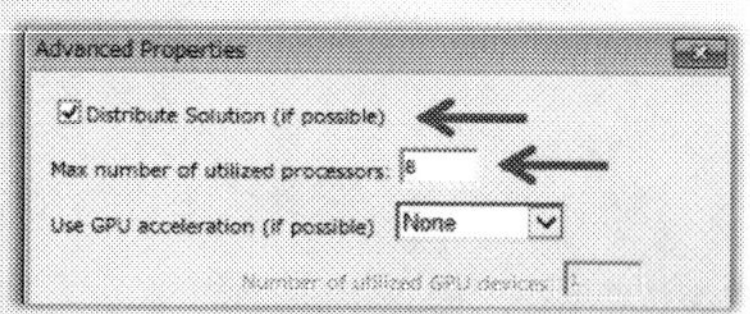


HPC Setting

- HPC Setting
  - Parallel Type
    - SMP(Shared Memory Parallel)
    - DMP(Distribute Memory Parallel)
  - Number of Cores
    - 2,4,8,12

| Solver Type | Parallel Type | Cores |
|---|---|---|
| Sparse | SMP | 2,4,8,12 |
| | DMP | |
| PCG | SMP | |
| | DMP | |

Setting Combination

< Topic 1 : Engine-Gasket Nonlinear Analysis >

# Hardware Information

2014 Automotive Simulation World Congress

- DELL T5610
  - Type A1 & A2 & A3

- DELL T7600
  - Type B

| | CPU | Memory | HD Drive | OS |
|---|---|---|---|---|
| Type A1 | 8 Core x 2(16 Core) , 2.6GHz | 64GB , 1866MHz | 3TB , 7200RPM | Win7 |
| Type A2 | 8 Core x2(16 Core) , 2.6GHz | 64GB , 1866MHz | 256GB , SSD | Win7 |
| Type A3 | 8 Core x2(16 Core) , 2.6GHz | 64GB , 1866MHz | 420GB, Fusion-io | Win7 |
| Type B | 6 Core x 2(12 Core) , 2.3GHz | 256GB , 1600MHz | 2TB , 7200RPM | Win8 |



< Topic 1 : Engine-Gasket Nonlinear Analysis >

# Analysis Requirements

2014 Automotive Simulation World Congress

- Sparse Direct Solver Information

```
SPARSE MATRIX DIRECT SOLVER.
 Number of equations =    10130299,    Maximum wavefront =    930
 Memory allocated for solver =              13335.623 MB
 Memory required for in-core =             106084.285 MB
 Optimal memory required for out-of-core =   5468.802 MB
 Minimum memory required for out-of-core =   1613.289 MB
```

← 100G Byte!!

| H/W | Solver Type | Memory Mode | .LN09 |
|---|---|---|---|
| Type A | Direct | Optimal Out-of-Core | 95GByte |
| | PCG | In-Core | - |
| Type B | Direct | In-Core | - |
| | PCG | In-Core | - |

- Nonlinear EQUILIBRIUM ITERATION
  - Total Number : 8

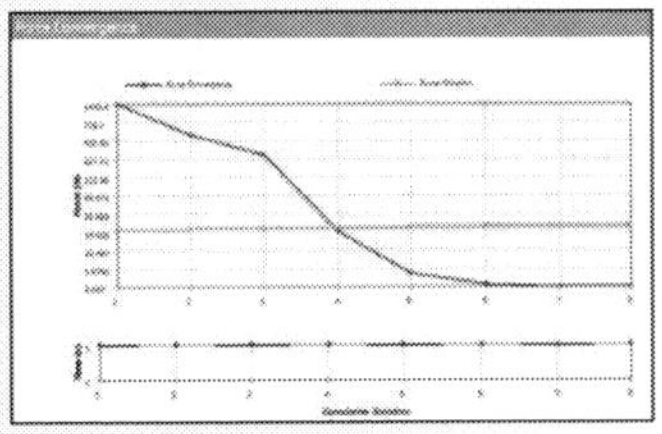

Number of Iteration

< Topic 1 : Engine-Gasket Nonlinear Analysis >

# Results

2014 Automotive Simulation World Congress

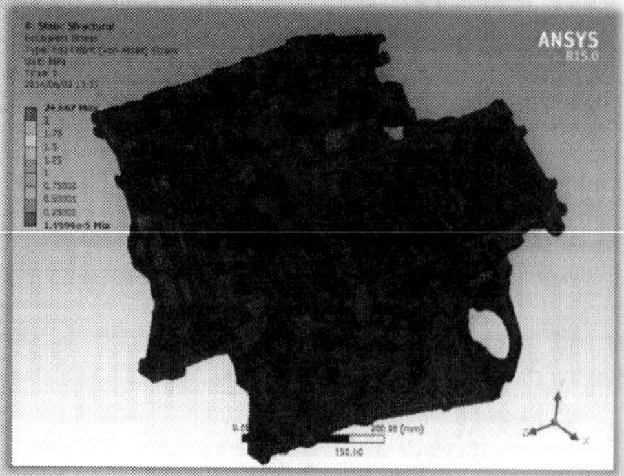


Mises Stress

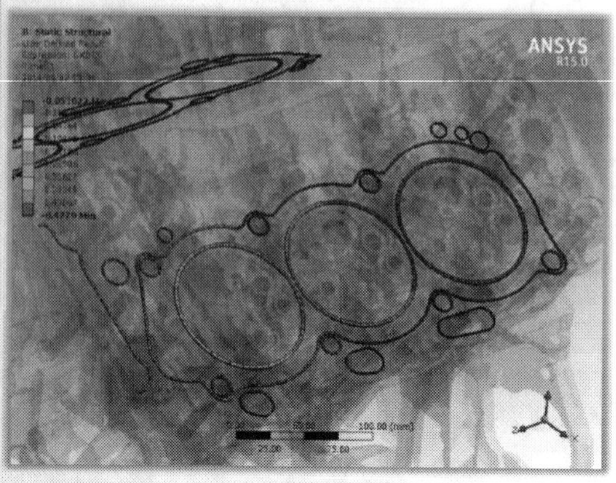


Gasket Disp.



< Topic 1 : Engine-Gasket Nonlinear Analysis >

# Results : Hardware v.s. Solver

2014 Automotive Simulation World Congress

- **Hardware** : HD , SSD , Fusion-io , Large_MEM
- **Solver** : Direct_SMP , Direct_PCG , PCG_SMP , PCG_DMP
- **Number of Cores** : 2(Default)

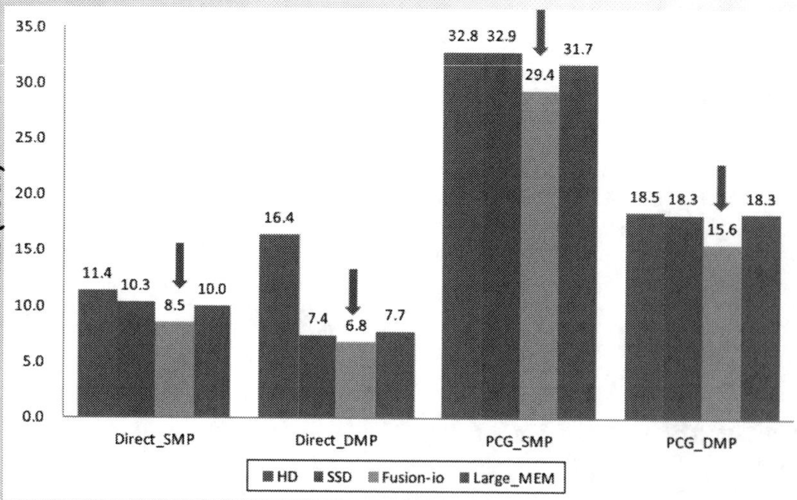

< Topic 1 : Engine-Gasket Nonlinear Analysis >

## Results : Parallel v.s. Solver

2014 Automotive Simulation World Congress

- Hardware : SSD
- Solver : Direct_SMP , Direct_PCG , PCG_SMP , PCG_DMP
- Number of Cores : 2 , 4 , 8 , 12

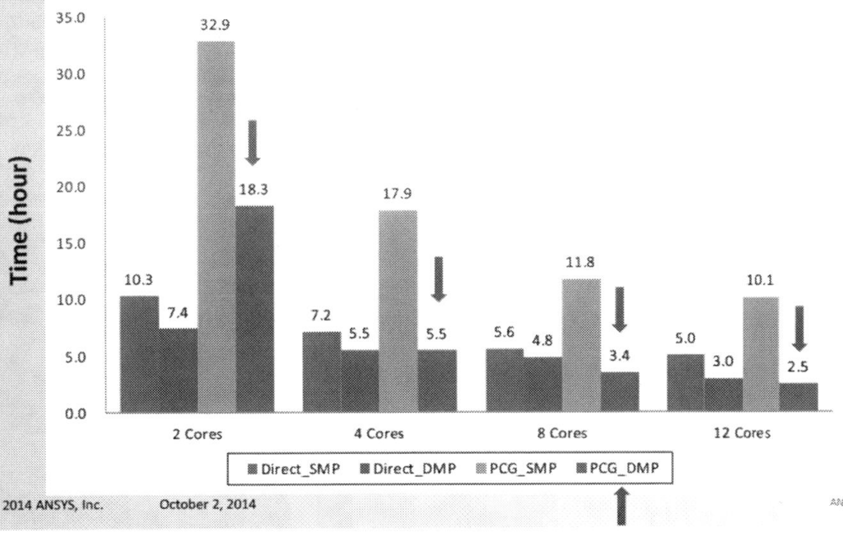


 October 2, 2014 

2014 Automotive Simulation World Congress

## Topic 2 : Leaf-Spring Suspension Dynamic Analysis

 October 2, 2014 

< Topic 2 : Leaf-Spring Suspension Dynamic Analysis >

# Process Flowchart

2014 Automotive Simulation World Congress

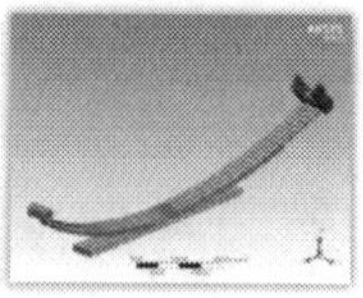

- **Workbench Schematic**

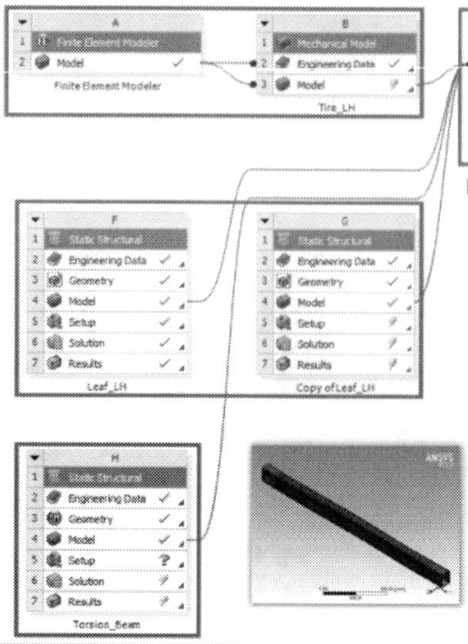

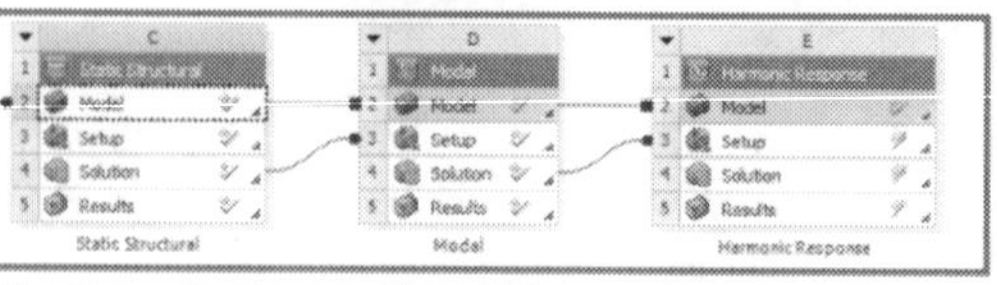


**Nonlinear Statics** **Eigen Value** **Frequency Response**

→

**Linear Perturbation**

< Topic 2 : Leaf-Spring Suspension Dynamic Analysis >

# *Tire Modeling*

2014 Automotive Simulation World Congress

- **Modeling Keyword**
  - **Hyper Elastic Material**
  - **Reinforce Elements**
  - **Hydrostatic Fluid Pressure**
  - **Contact**

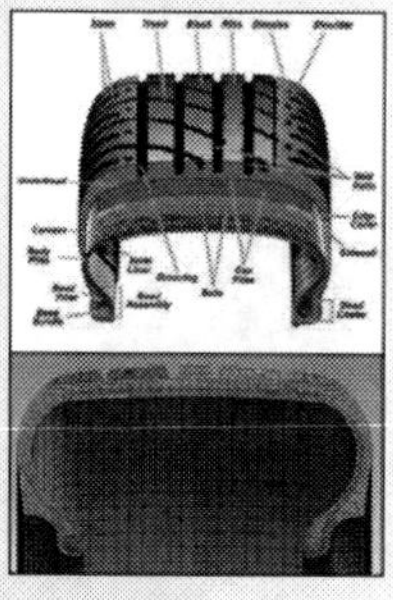

**Section**

**Tire & Wheel**

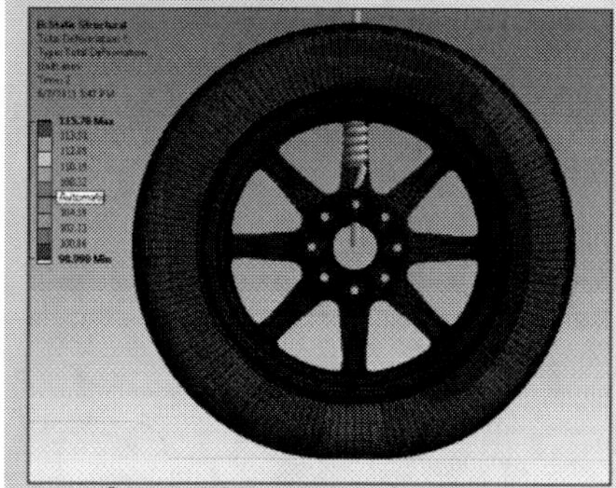

**Deformation**

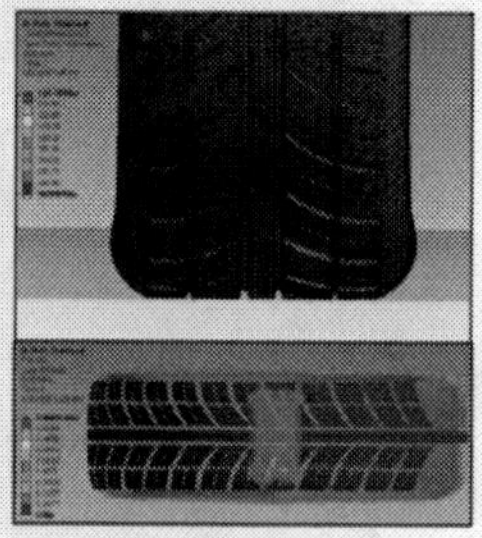

**Foot Print**

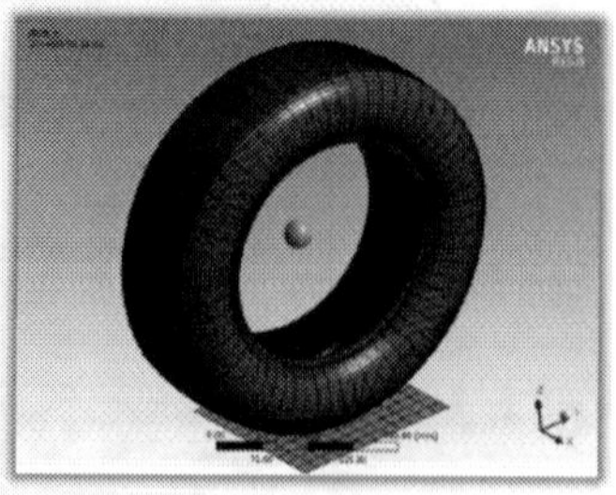

**Tire model for NVH Analysis**

< Topic 2 : Leaf-Spring Suspension Dynamic Analysis >

2014 Automotive Simulation World Congress

## Leaf Spring Modeling

- Modeling Keyword
  - Large Deformation
  - Contact
  - Bushing

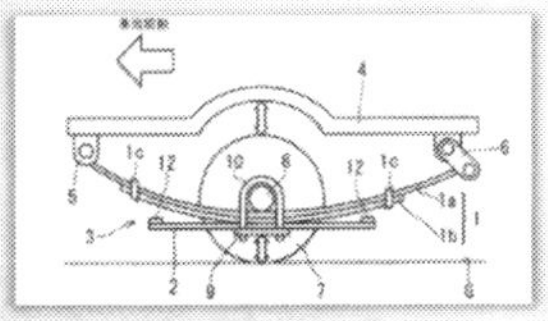

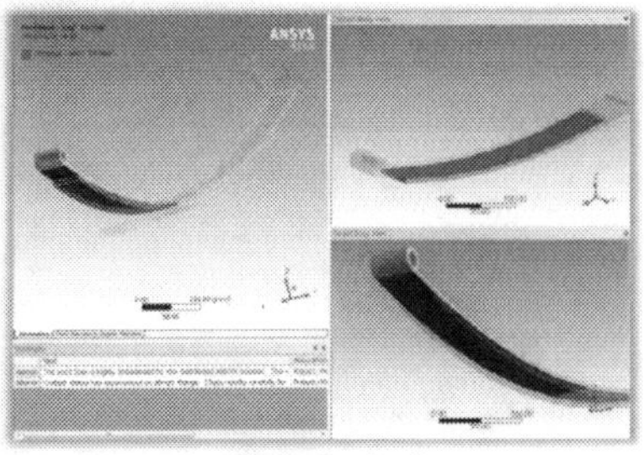

Plate Contact

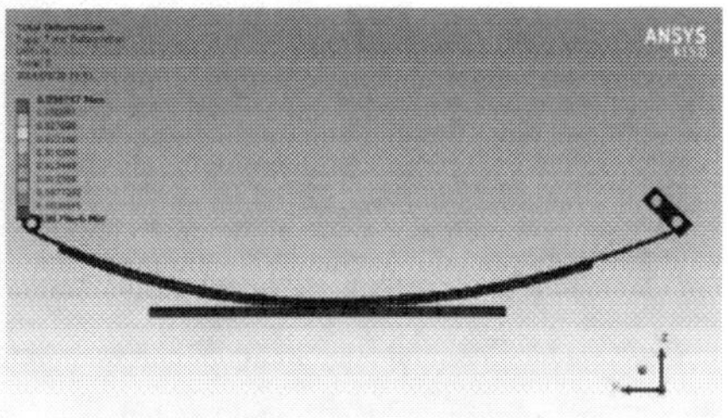

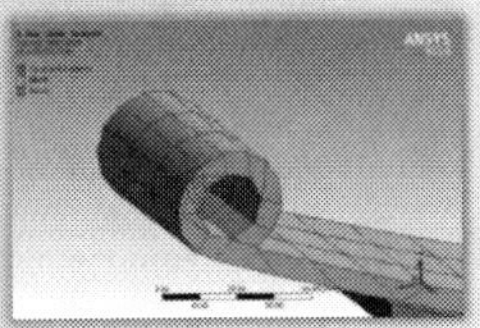

Front Bush

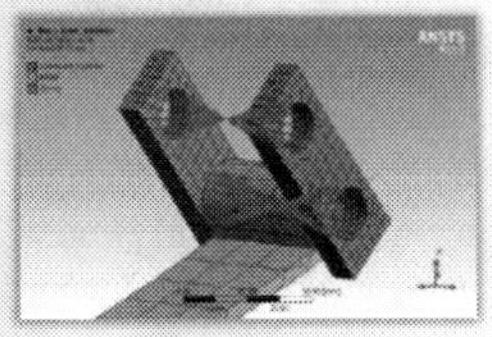

Rear Bush x 2

Deformation(Animation)



< Topic 2 : Leaf-Spring Suspension Dynamic Analysis >

2014 Automotive Simulation World Congress

## Suspension Modeling

- Boundary Conditions
  - Tire Inner Pressure
    - 0.2MPa
  - Contact between Tire and Ground
    - Penalty Method
    - Friction : 0.2
  - Contact between Plates
    - Augmented Lagrange Method
    - Friction : 0.2
  - Enforced Displacement
    - Ground Area
    - Z direction : +40mm

Displacement

< Topic 2 : Leaf-Spring Suspension Dynamic Analysis >

# *Suspension Analysis*

2014 Automotive Simulation World Congress

- **Non linear Statics**
  - Elapsed Time : 506(s)

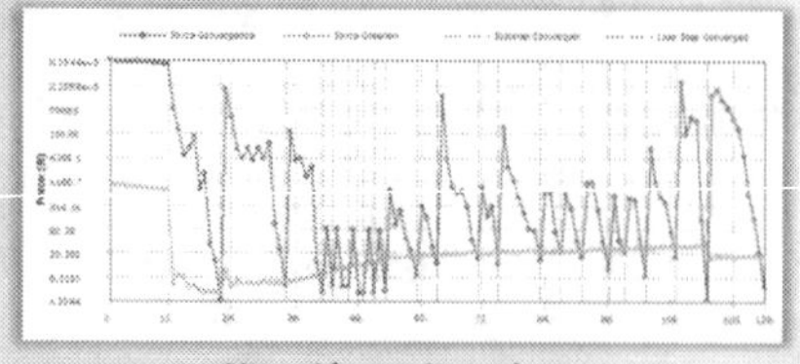

Non_Linear Iterations

Displacement

Equivalent Elastic Strain(Animation)

Equivalent Stress



< Topic 2 : Leaf-Spring Suspension Dynamic Analysis >

# *Suspension Analysis*

2014 Automotive Simulation World Congress

- **Eigen Value Analysis after Static Analysis**
  - Linear Perturbation Process
    - Initial Stress
    - Last deformation & Stiffness

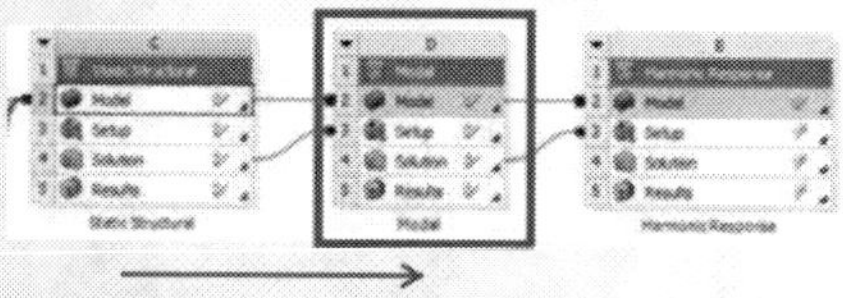

Mode 2 : 16Hz(Animation)

Mode 4 : 20Hz(Animation)

< Topic 2 : Leaf-Spring Suspension Dynamic Analysis >

## *Suspension Analysis*

2014 Automotive Simulation World Congress

- **Frequency Response Analysis after Eigenvalue Analysis**
  - Modal FRF Creation

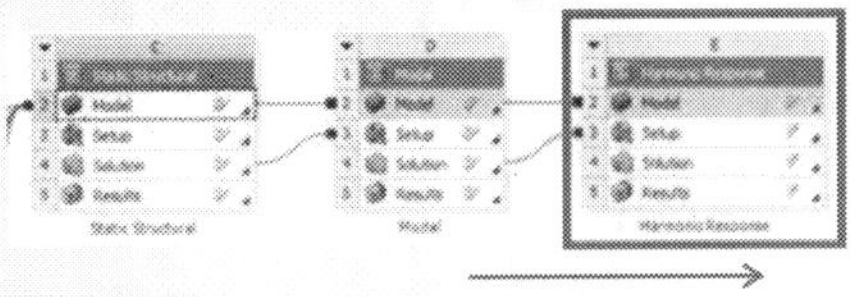

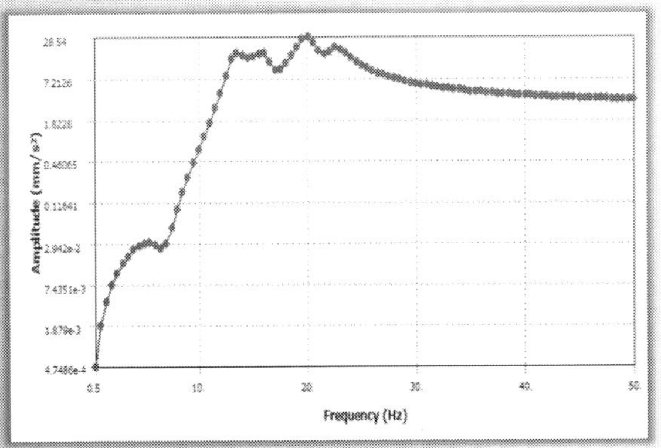


Frequency Response Function

Input Point (Under Leaf +Z)

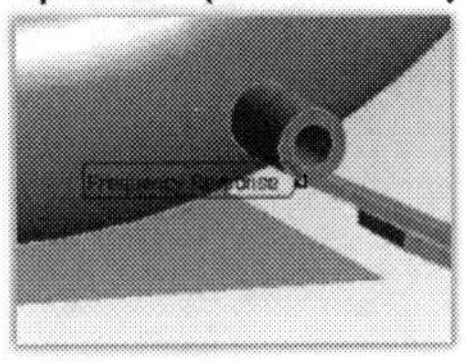

Output Point (Leaf Tip +Y)



## Conclusion

2014 Automotive Simulation World Congress

- **ANSYS HPC performance**
  - ANSYS's solver performance is continuing to be improved.
    - SMP/DMP
    - Direct/Iterative
    - HPC
- **ANSYS Non_Linear Staics & Dynamics Analysis**
  - ANSYS has many nonlinear technology.
    - Gasket
    - Rubber
    - Contact
    - Reinforce
    - Hydrostatic Fluid
    - Etc...
  - Statics & Dynamics combination analysis(Linear perturbation) is very easy using WB.

2014 Automotive Simulation World Congress

# Overcoming the Barriers Imposed by Geometry - Introduction of SpaceClaim -

Ichiro Kawabata, SpaceClaim Japan K.K.

2014 Automotive Simulation World Congress

## About SpaceClaim

- **Multi-Purpose Modeler**
  - Create New Models or Import, Repair and Edit files
  - Works with any CAD data
  - Exceptionally easy to learn and inexpensive to deploy
- **SpaceClaim is the best tool for many engineering tasks**
  - Lowest TCO in the history of 3D
  - Another Tool in your toolbox

## How does SpaceClaim do it?

### It's all about the user interface

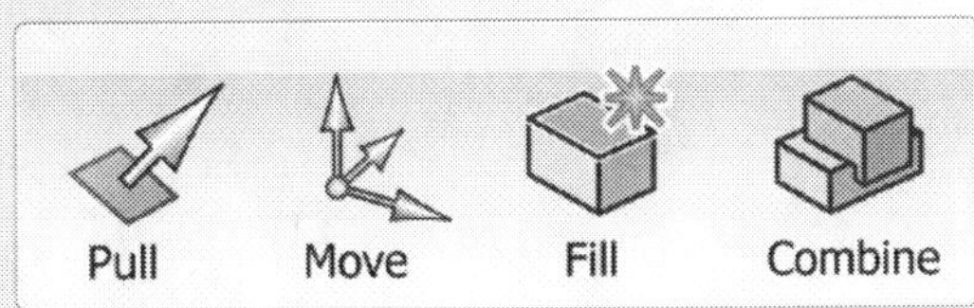


- Simple and easy-to-understand tools
- Fast and immediate feedback
- No need to understand model history



2014 Automotive Simulation World Congress

## Where are customers using SpaceClaim

- Design
  - 2D Replacement, Brainstorming, Concept Modeling, Simulation Driven Design
- CAE
  - Model Preparation, Simplification and Design Optimization
- Manufacturing
  - Fixtures, Tooling, Process Optimization
- Bid Modeling
  - Real-Time Collaboration, 3D Specifications, faster quotes

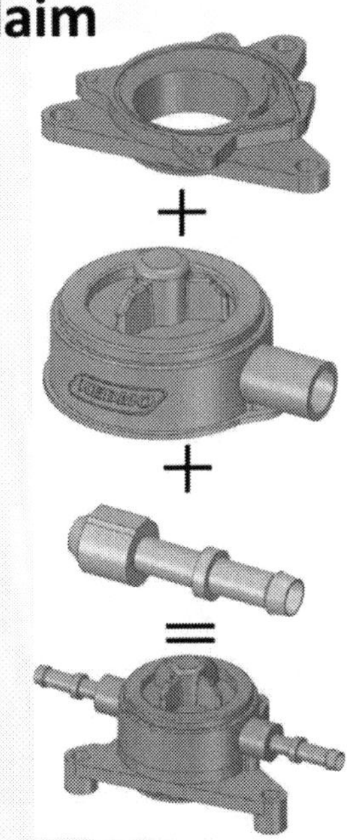

2014 Automotive Simulation World Congress

## SpaceClaim Corporation

- **Founded in late 2005 by industry leaders**
  - Founders at leading CAD companies PTC and SolidWorks
- **Headquartered in the USA, with regional offices in Japan, UK, Germany, France and Denmark**
  - >150 OEM and reseller partners around the world
  - Software translated into 13 languages
- **May, 2014 acquired by ANSYS, Inc.**

Monday, October 06, 2014 2014 Automotive Simulation World Congress 5

2014 Automotive Simulation World Congress

## Selected Corporate Customers

- Samsung
- Toyota Motor Corporation
- NASA
- General Dynamics
- Microsoft
- Rolls-Royce
- Boeing
- Siemens medical
- Medtronic
- Moog
- TE connectivity
- Cummins
- GE
- LG
- Harris
- Ford
- Bosch
- Eaton Powering Business Worldwide
- Johnson & Johnson
- P&G
- Whirlpool
- Alstom
- The Bold Look of Kohler
- Bombardier
- Raytheon
- Lockheed Martin
- Volvo
- Lotus
- Johnson Electric
- Huisman
- John Deere
- Schlumberger
- Pratt & Whitney
- Bell Helicopter
- U.S. Army
- Renault Sport F1
- Honeywell
- Jacobs
- Varian semiconductor equipment

"We can create and modify 3D models in SpaceClaim with a speed not achievable in other software tools." – Borg Warner

Monday, October 06, 2014

2014 Automotive Simulation World Congress

# CAE - BorgWarner

BorgWarner

- Tier 1 automotive supplier
- Analysis team had no access to 3D modeling tools
  - Company uses at least three major CAD systems to work with suppliers
  - CAE users are CAE experts
  - Need to change designs in ways that conflict with modeling intent
- SpaceClaim lets CAE team make the changes they need, including prep for analysis in ANSYS

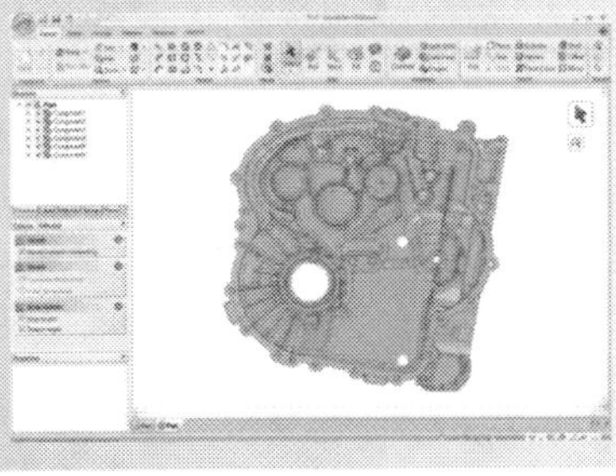

2014 Automotive Simulation World Congress 7

2014 Automotive Simulation World Congress

# CAE - Kranunion

- Uses SpaceClaim to prepare data for simulation in ANSYS
- Combines data from SolidWorks and other CAD, CAE, and 2D

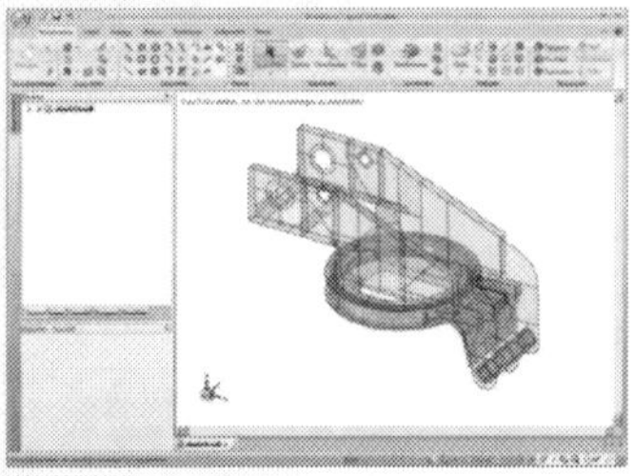

**"It used to take us a good two days to prepare a geometry for simulation from the bottom up with key points, lines and surfaces to the final model. Today, it takes around an hour and a half to import the 3D model and prepare it in SpaceClaim."**

*Reimar Herrschuh,*
*Calculation Engineer*

KIROW

Monday, October 06, 2014 2014 Automotive Simulation World Congress 8

## CAE - SENZ Umbrellas

- Make umbrellas that don't invert in heavy wind
- Outsource detailed design to manufacturing partner in China
  - They design the rough shape
  - Manufacturer sends them prototype of manufactured part
  - They approve and go into production
  - They never see or care about the CAD files!

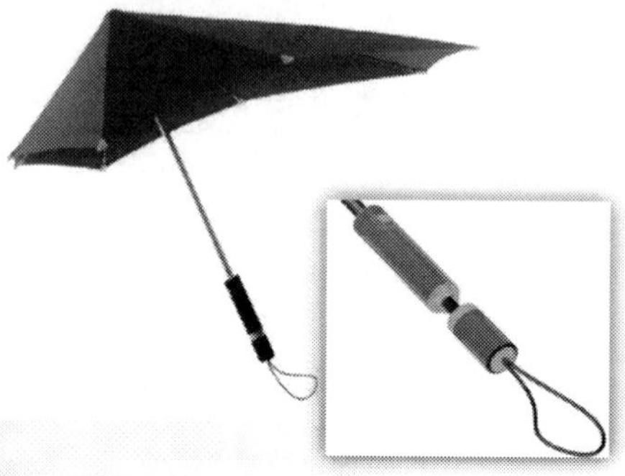

*"SpaceClaim does the job perfectly for the concept phase."*

Gerwin Hoogendoorn, CEO

senz°





## CAE - Engrana

- Simulation preparation challenge
  - Preparing customer-supplied geometries for analysis
- Using SpaceClaim for design optimization
- Incorporating SpaceClaim into the workflow at ENGRANA

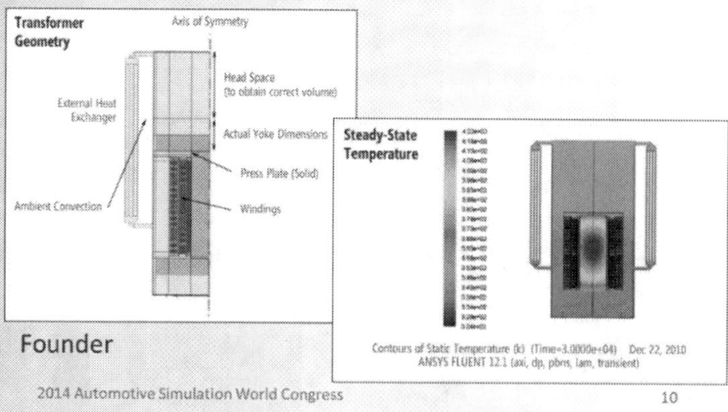


*"I especially appreciate the intuitive nature of SpaceClaim. The tool figures out what I want to do before I tell it – SpaceClaim almost reads my mind."*

Bill Wangard, Ph.D. Founder

## CAE and Modeling Mesh Data – Automotive

- **Global automotive OEM needed to reverse engineer the structural qualities of competitive chassis components**
  - Only shell (thin-walled) elements required
  - Previous process took three weeks in Altair Hyperworks
  - Goal: create geometry in one week in SpaceClaim
- **CAD-experienced contractor used SpaceClaim**
  - In two weeks, learned SpaceClaim and created the geometry
  - Scan data –> STL model –> Solid model –> Catia

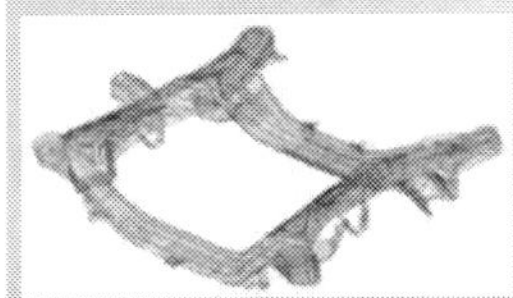

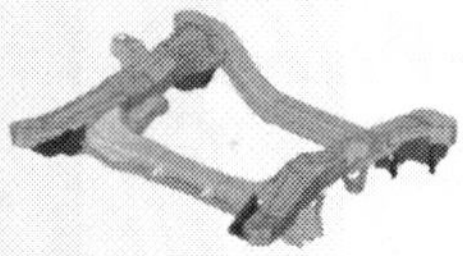

2014 Automotive Simulation World Congress

## Automotive

- **All engineers train CatiaV5 long time (a few weeks).**
  - Type A: Many engineers use CatiaV5. (45%)
  - Type B: Some engineers previously used CatiaV5. (Currently not) (25%)
  - Type C: Some engineers cannot use CatiaV5. Asked another engineers modeling (30%)
- **After 1 day training of SpaceClaim**
  - Type A: use CatiaV5, also SpaceClaim
  - Type B: use SpaceClaim
  - Type C: use SpaceClaim
- **Especially, CAE engineers propose their ideas to Designer.**

## Low TCO, Compelling ROI

- SpaceClaim advantages:
  - Immediate productivity
  - Reduced training cuts cost
    - > 50%
- <u>>70% of users report no need for classroom training</u>
- Typical reported payback period for $4k purchase is approx. 2 months
- <u>Average reported productivity gain is 5.8x</u>

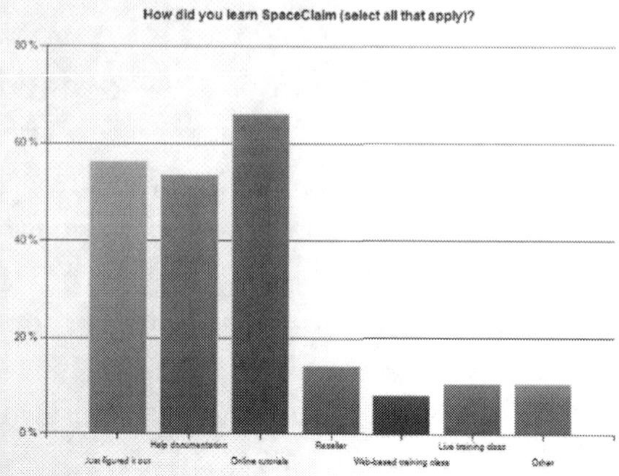


**"We've been using SpaceClaim for about 17 months and have more than recouped our investment in time saved, error reduction, and faster time-to-market. SpaceClaim is extremely easy to learn. Even our intern, <u>without training, can be productive</u> in just two weeks. You can't beat that."**

*- Michael Lieblich, FAA Designated Engineering Representative, AeroNautique*





## SC2014 new features

- Performance – 30% faster model load & multi-threaded import
- Pinned reference geometry: points, axes, planes that update with model changes to position and shape
- Dynamic interference
- UI changes/new tools: Insert tab, Measurement, power selection of curves/std features
- Excel plugin for equation-linked driving dimensions

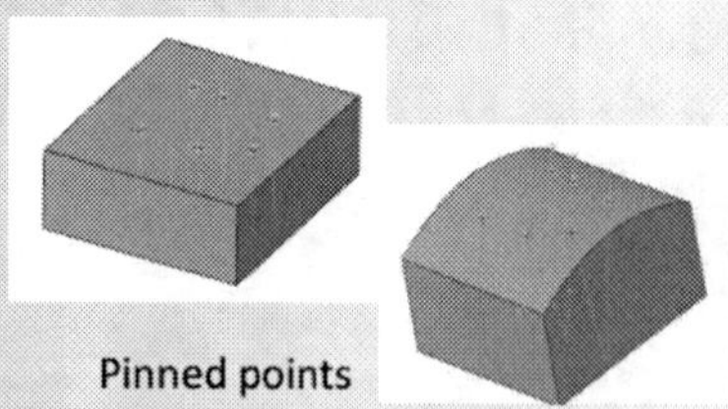

Pinned points and axes

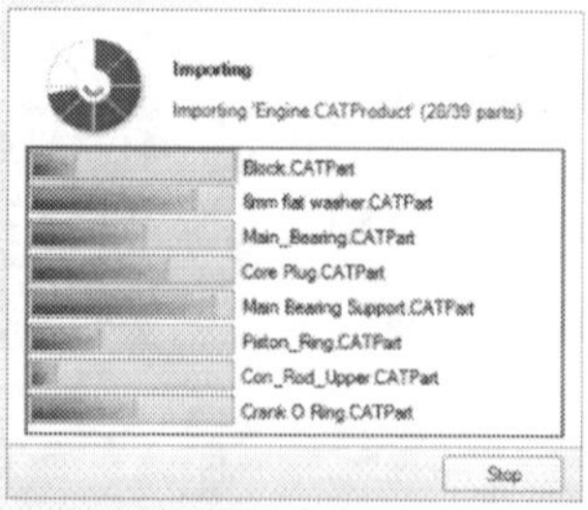

# SC2014 new features (CAE)

- Surface simplification tool
- Driving dimensions from mass properties
- Support for baffle geometry via new Shared Topology 'merge' option

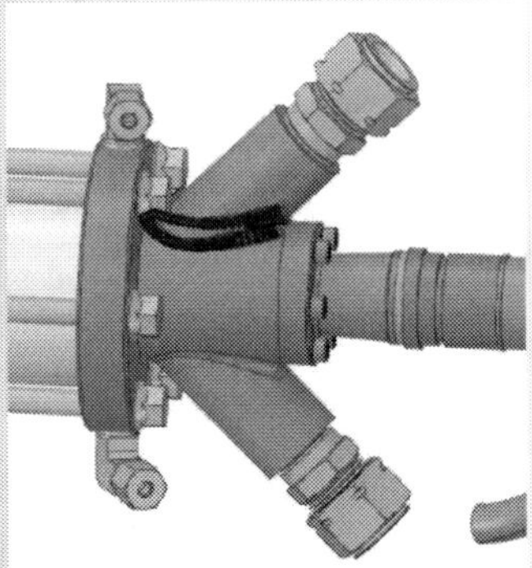

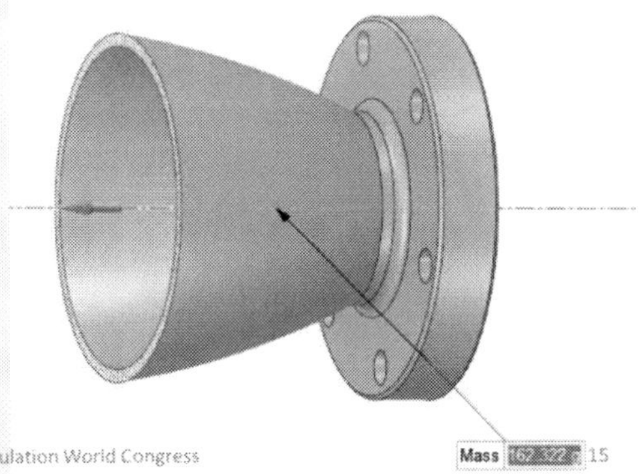




2014 Automotive Simulation World Congress

# ANSYS 15.0 Support for Non-manifold Surface Topology

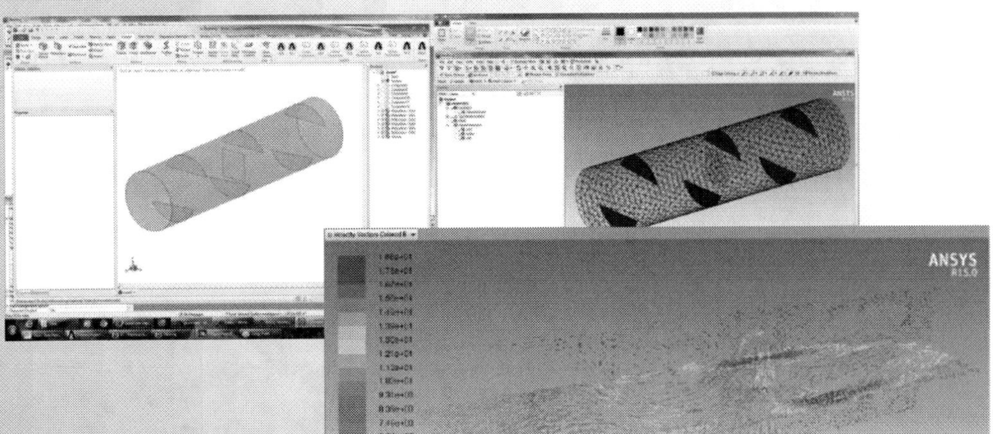

## Summary

- **Engineer benefit from using the best tool for the job**
  - Lead in the application of 3D direct modeling to enable product teams to better innovate and collaborate
- **SpaceClaim demonstrates a measurable return on investment**
  - Model preparation for simulation, rapid prototyping, and CAM
  - Concept modeling





## Thank You

SpaceClaim direct editing is really the way to go for most solid modeling chores. I truly believe it is the future. I say this as SolidWorks's seventh customer (user since 1995). I love the streamlined approach to tools that is based on the notion of moving and push/pull commands

*- Jeff Sand, Senior Design Officer, BRD MOTORCYCLES*

TECH BRIEFS 2012 Product of the Year

DesignNews Best New Products Award WINNER

SPACECLAIM CORPORATION

# Resin Transfer Molding Simulation for Woven Fabric Composite Materials

**Cybernet Systems Co., LTD,**
**Mechanical CAE Division**
**Arata Hayashigaki**

CYBERNET SYSTEMS CO., LTD.

## Carbon Fiber Reinforced Plastic (CFRP) vs. "Mass Production in Transportation"

*From project start (1960' s)*

*30 years passed !*

**1994 Airplane**

http://www.carbonfiber.gr.jp/field/craft.html

**Boeing 777**

*20 years passed !*

**2014 Vehicle (in Japan)**

http://www.extremetech.com/extreme/162582-bmw-i3-will-bmws-new-ev-finally-be-the-breakthrough-for-carbon-fiber-cars

**BMW i3 (Compact EV)**

**Price : ¥ 5 millions !**

*Strategy toward cost-cutting?*

 

# Merits of CFRP

## Specific Weight

Fe : Al : Glass Fiber : Carbon Fiber : Plastic
= 7.8 : 2.7 : 2.5 : 1.8 : 1.3

**Steel wt. : CFRP wt. = 4 : 1**

**Replace "body-shell" & "frame"
(bonnet, door, trunk lid, ...)
→ Most effective weight saving !**

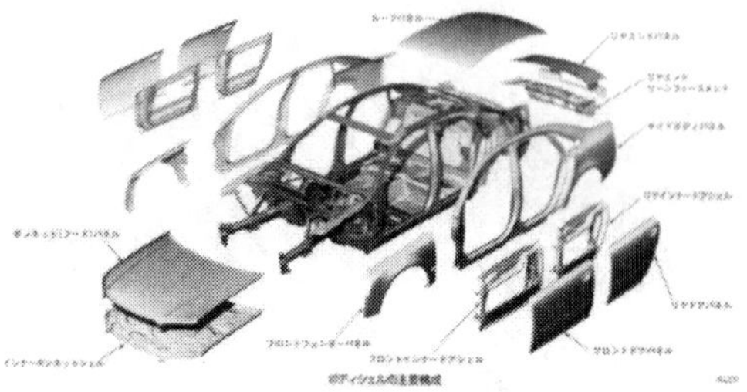

http://www.weblio.jp/content/

## Strength

**Specific Strength; St : CF = 1 : 10**
**Specific Elasticity; St : CF = 1 : 7**

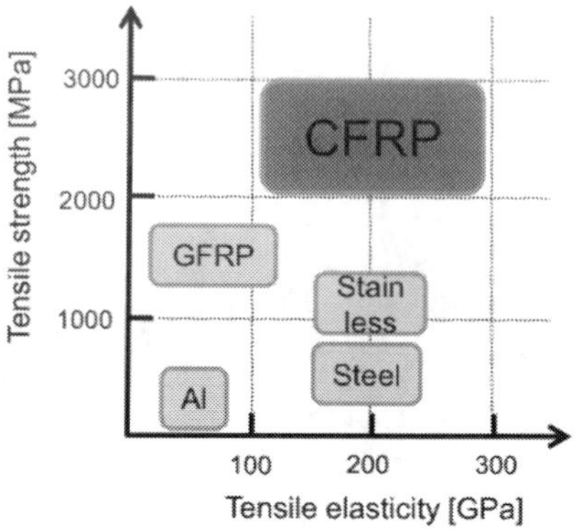






# Demerits of CFRP

## Cost

1. **Raw material expense**
2. **Production cost**

| | Raw Mat. (yen/kg) | After molding (yen/kg) |
|---|---|---|
| Fe Alloy | 80 | 100 |
| Al Alloy | 600 | 800 |
| Mg Alloy | 3,000 | 6,000 |
| Fiber(PAN) | 3,000 | 5,000~10,000 |

**CFRP production cost**

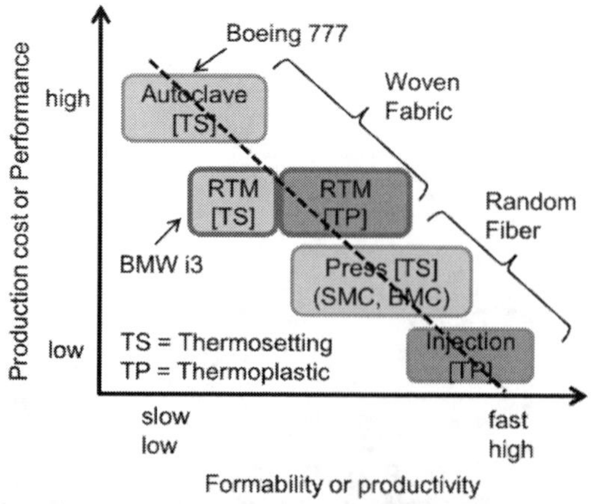

# Agenda

1. Analytic System Linkage in "RTM" Simulation

2. Testing Analysis Examples

3. Summary

 


## 1. Analytic System Linkage in "RTM" Simulation

### Resin Transfer Molding (RTM)

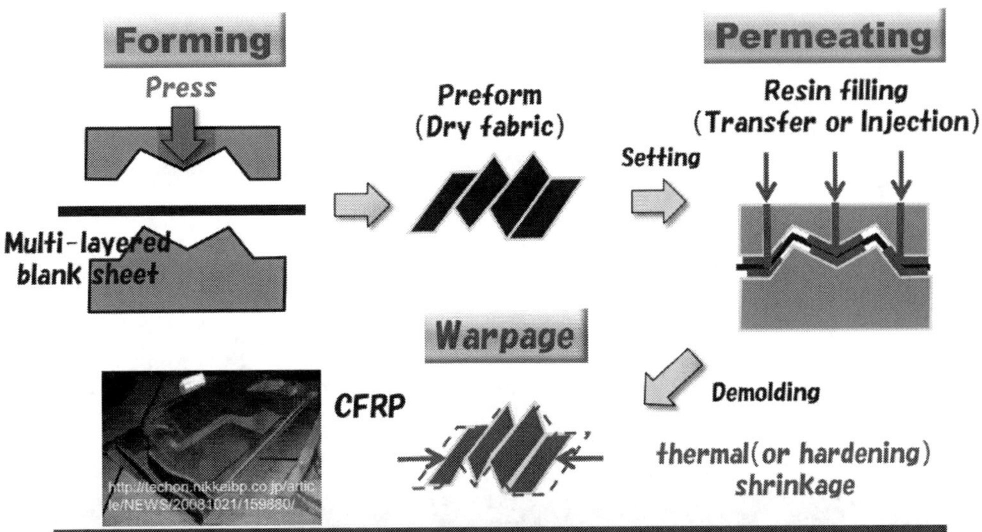


 

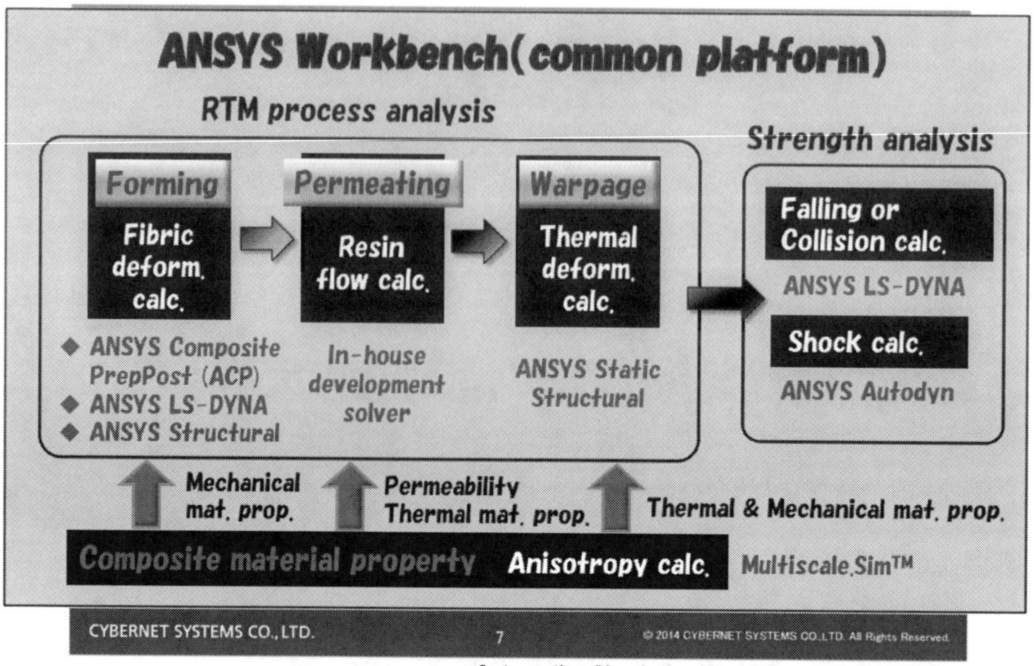


Automotive Simulation World Congress 2014 in Tokyo

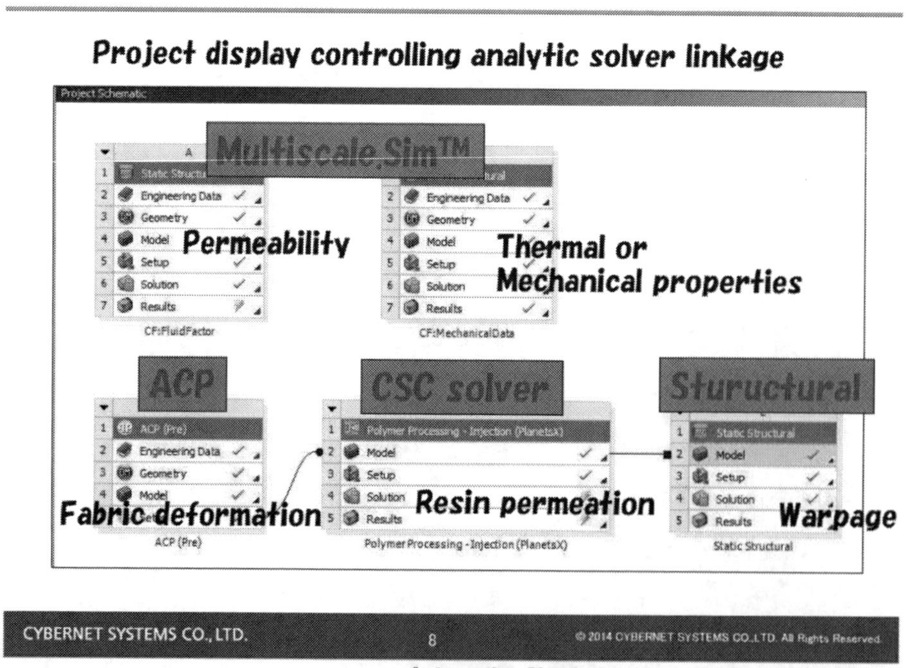


Automotive Simulation World Congress 2014 in Tokyo

## What' s homogenization method in Multiscale.Sim™ ?

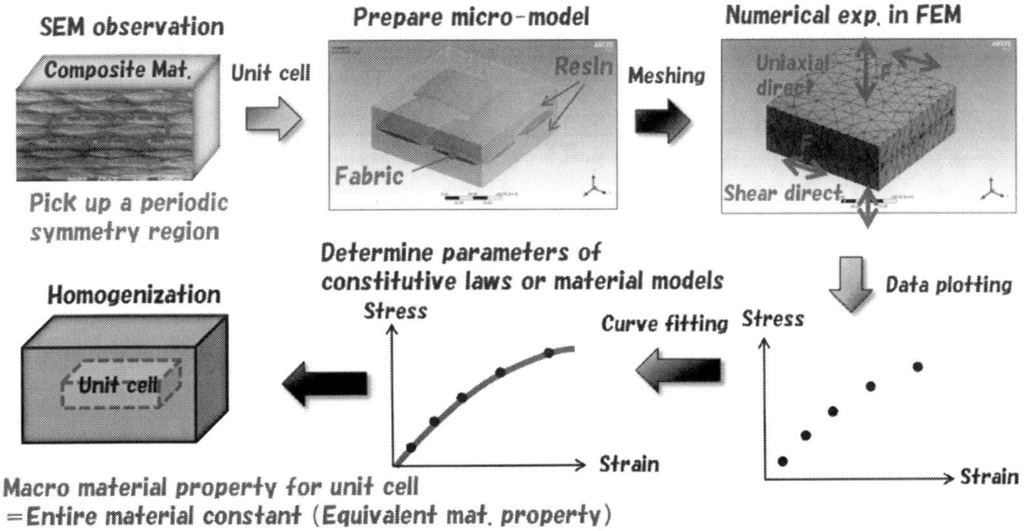


 



## What' s draping method in ANSYS Composite PrepPost(ACP) ?

① Deformation analysis based on simple geometrical interpretation (without considering blank-sheet' s stiffness).

→ Location of folds after forming

→ Optimum form of blank sheet before forming

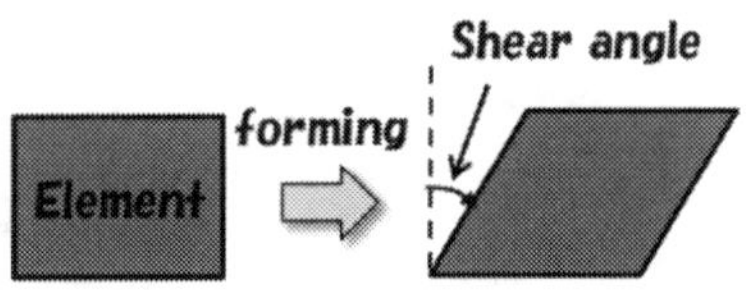


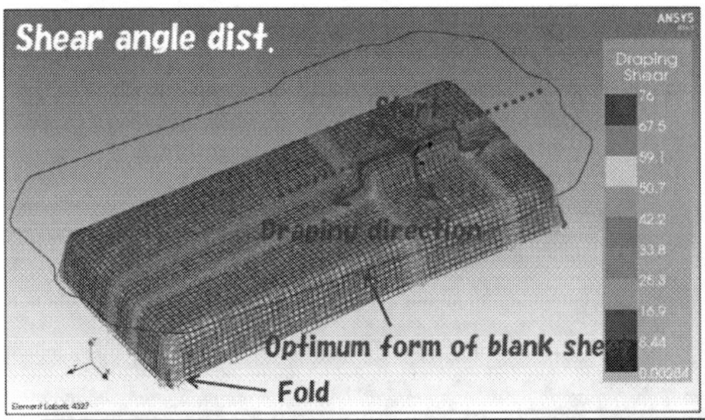


When changing a draping direction,

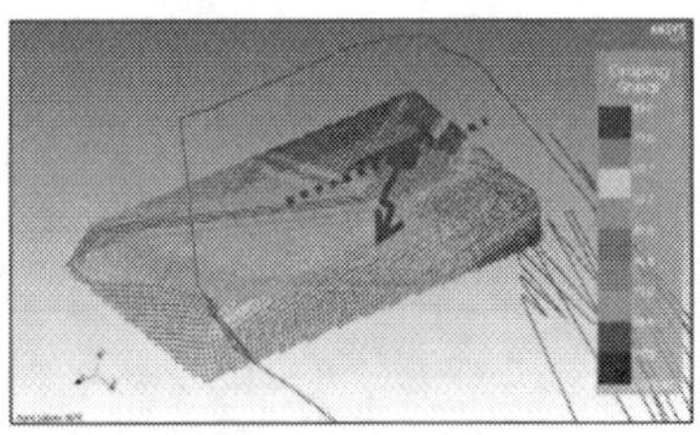

 

## ② Predict changing principal directions of fabric

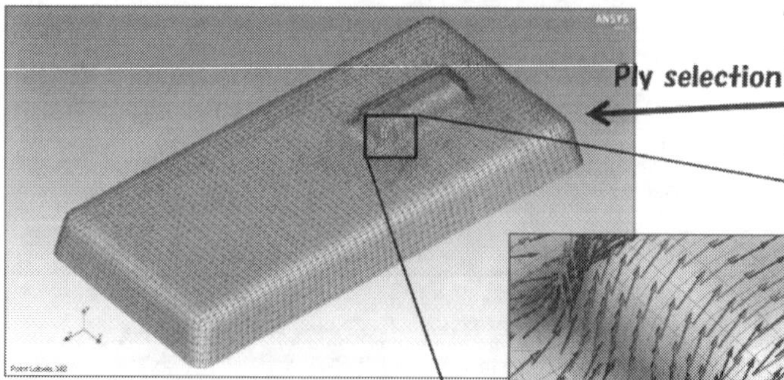


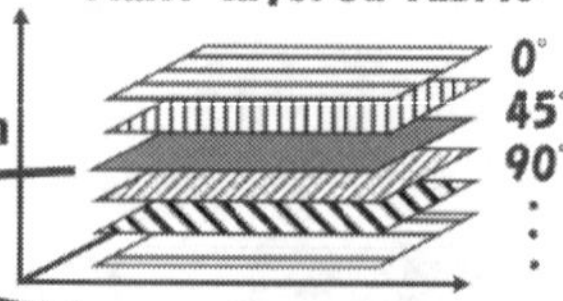


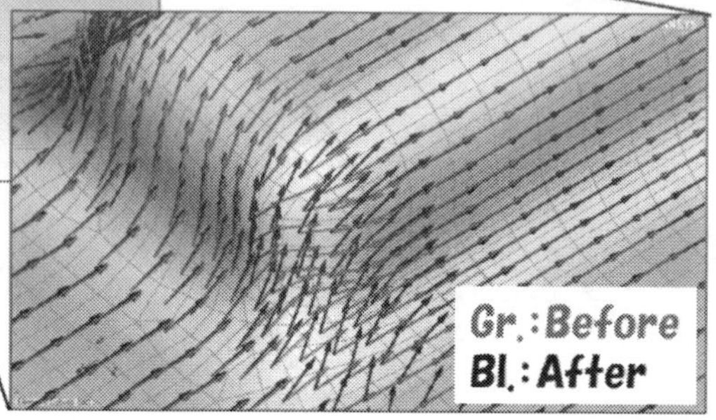


From variation of principal direction, recalculate anisotropic material property after forming.

 



## 2. Testing Analysis Examples

Example: Door trim

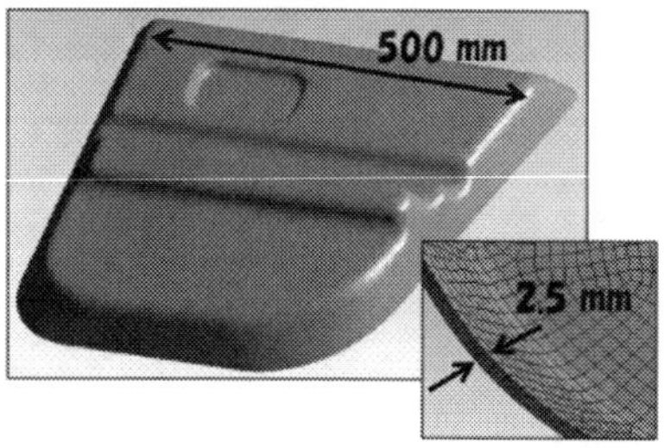


Micro mat. analysis (M.Sim™)

+

Principal direction analysis of dry fabric after forming (ACP)

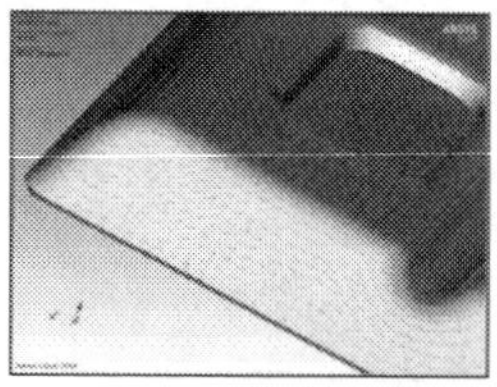

Permeability coefficient to dry fabric before forming

Permeation analysis (with homogenized macro material values)

Permeability coeff. after forming, Linear expansion coeff. & elastic modulus & thermal conductivity of CFRP after forming

Warp analysis(Structural)

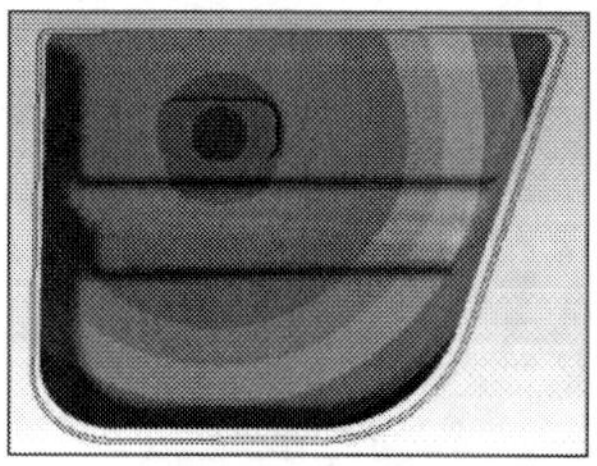

Volume shrinkage

Linear exp. coeff. & elastic modulus of CFRP

Darcy's law

 

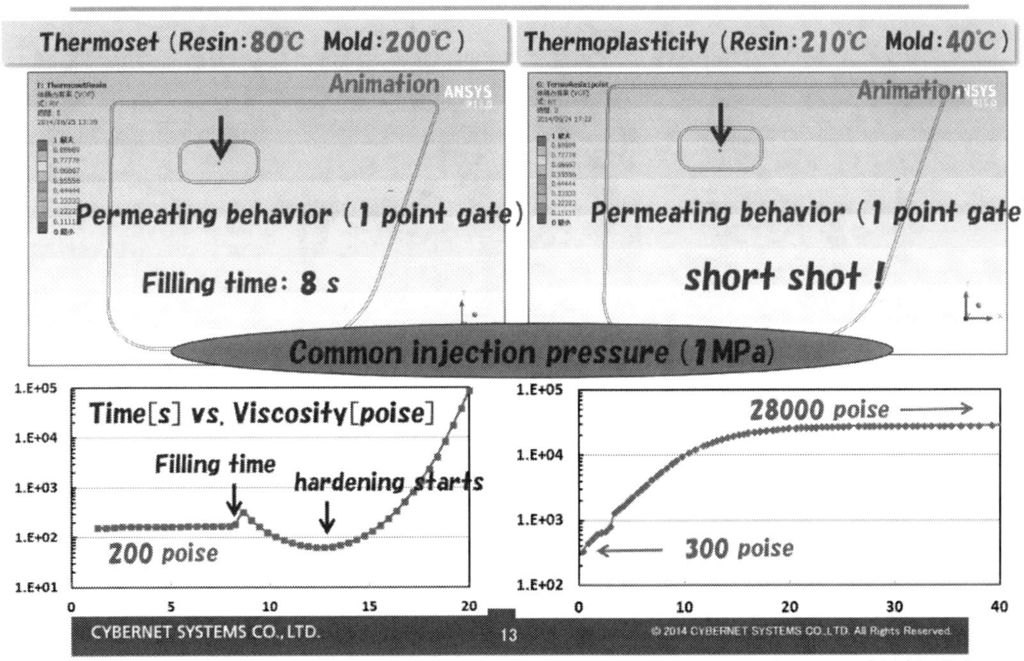


Automotive Simulation World Congress 2014 in Tokyo

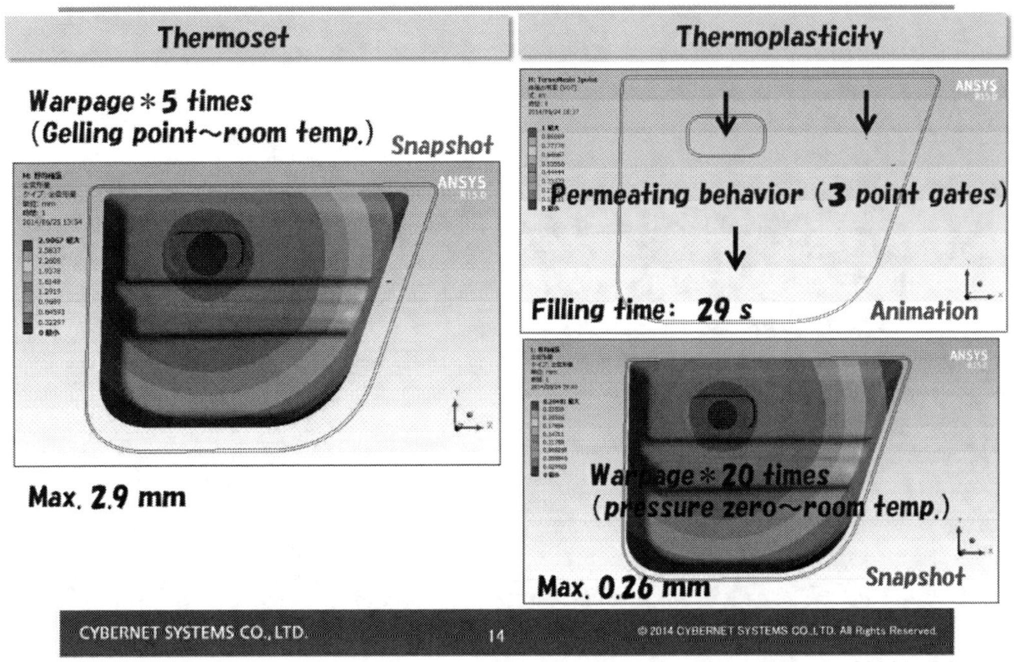


Automotive Simulation World Congress 2014 in Tokyo

## Close-up [Relation between folds and permeability]

Shear angle after forming

Arrival time of flow (thermoplastic+3 point gates)

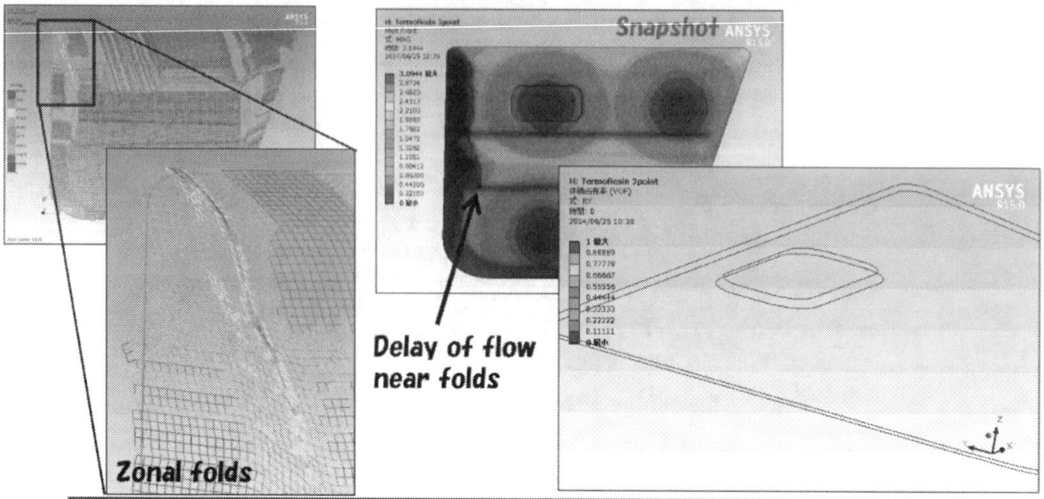


 



## Close-up [Possibility of air trap (void) generation]

Marking output for void regions
(< size **0.03 cc**)

Thermoplastic+3 point gates

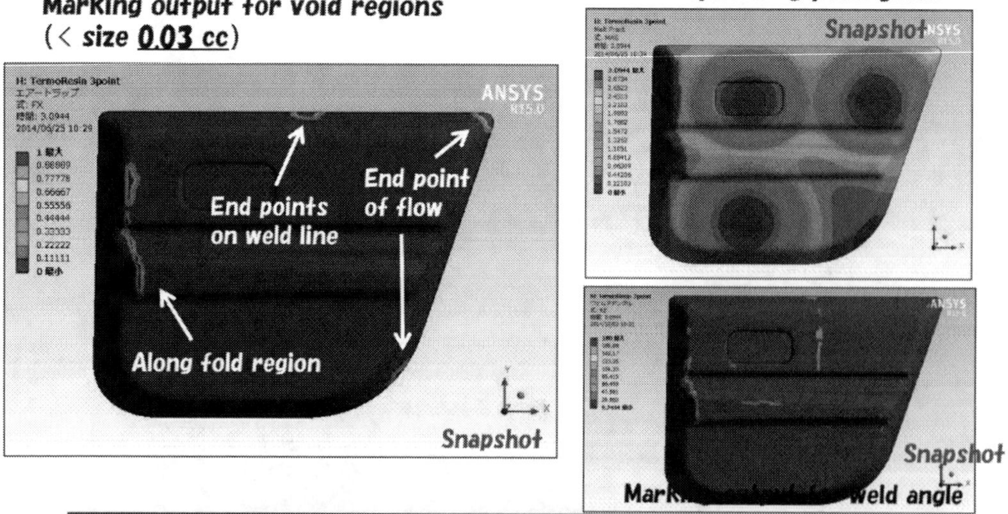


 

## 3. Summary

① An integrated environment of RTM simulation is constructed on ANSYS workbench, in which dedicated analytic solvers are assigned to each molding process of RTM.

② A closing system composed of softwares only, because Multiscale.Sim™ can give necessary material property values for all molding processes without relying on experimental measurements.

③ A CAE system to develop a fast & high-performance molding process for the purpose of "VEHICLE" mass production.

 



# BacKup Slides

 

## What's a solution toward CFRP cost-cutting?

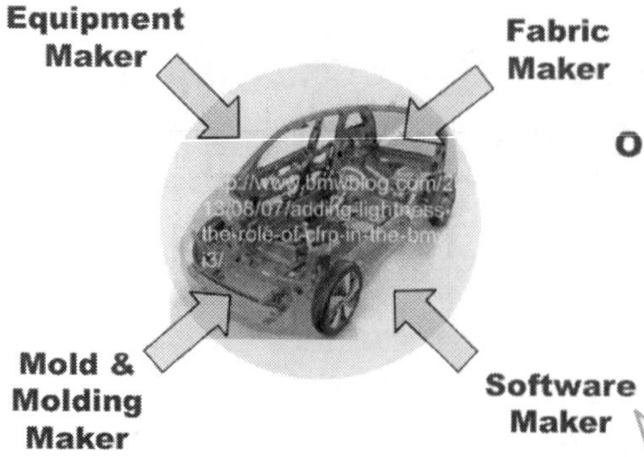


**Our proposal for CAE system linkage**

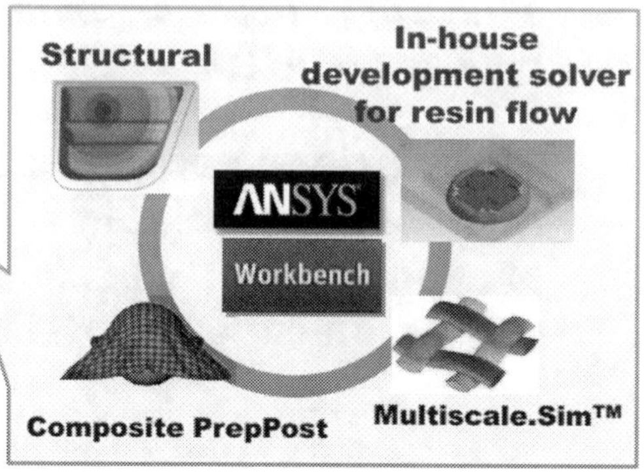


1. *Cost-cutting of fabric*
2. *Selection of a fast & high-performance molding process by means of effective utilization of CAE systems*
3. *Mass production of equipment & mold*

 


*··· available for RTM*

## *Equivalent property values calculable by Multiscale.Sim™*

### *[Linear type]*

| *Material prop.* | *Constitutive laws* |
|---|---|
| *Elastic modulus* | *$E_{ij}$: Complete or Orthogonal anisotropy* |
| *Thermal conductivity* | *$\lambda_{ij}$: Complete or Orthogonal anisotropy* |
| *Linear expansion coefficient* | *$\alpha_{ij}$: Orthogonal anisotropy* |
| *Permeability coefficient* | *$K_{ij}$: Orthogonal anisotropy* |
| *Density* | *$\rho$ : Isotropy* |
| *Specific heat* | *c: Isotropy* |

### *[Non-linear type]*

| *Nonlinear materials* | *Material models* |
|---|---|
| *Elastic-plasticity* | *Anisotropic Hill* |
| *Creep* | *Anisotropic Hill* |
| *Visco-elasticity* | *Anisotropic generalized Maxwell* |
| *Hyper-elasticity* | *Kaliske's anisotropic strain energy density function* |
| *Damage* | *Anisotropic damage model* |

CYBERNET SYSTEMS CO., LTD.  
Automotive Simulation World Congress 2014 in Tokyo

## Raw material expense

**Fe : Fiber = 1 : 40**
**Al : Fiber = 1 : 5**

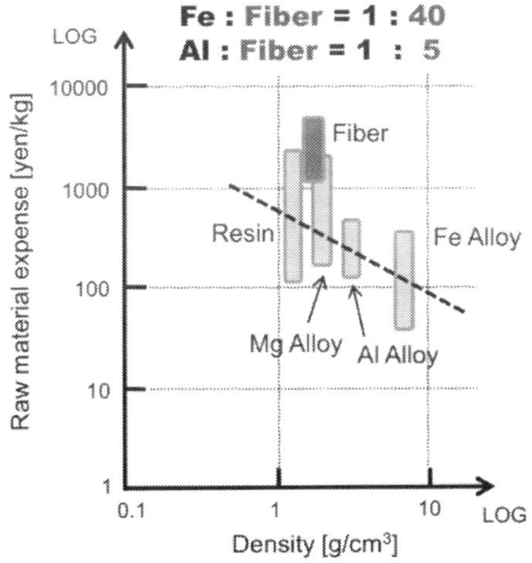


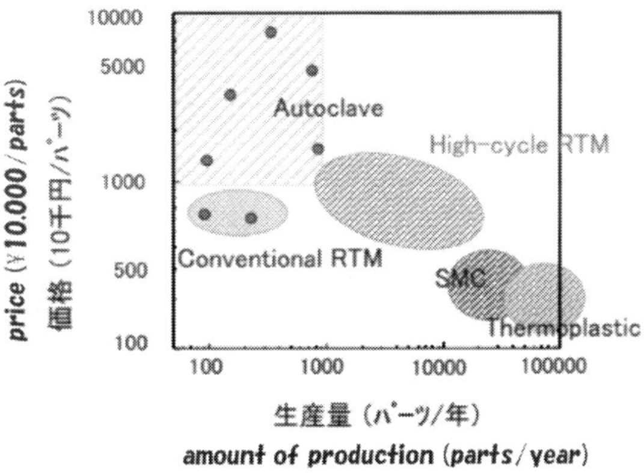


Reference cited:
http://www.nisri.jp/jisedai/docs/lecture_20131010_umemoto.pdf

CYBERNET SYSTEMS CO., LTD. 21 

**Automotive Simulation World Congress 2014 in Tokyo**

*Multiscale.Sim is a trademark of Cybernet Systems Co., Ltd. in Japan and other countries.*

*ANSYS and any and all ANSYS, Inc. brand, product, service and feature names, logos and slogans are registered trademarks or trademarks of ANSYS, Inc. or its subsidiaries in the United States or other countries.*

*All other brand, product, service and feature names or trademarks are the property of their respective owners.*

CYBERNET SYSTEMS CO., LTD.

**Automotive Simulation World Congress 2014 in Tokyo**

CYBERNET SYSTEMS CO., LTD.

2014 Automotive Simulation World Congress
October 9-10, Tokyo, Japan

# Composite Press Forming Simulation of Dry Textile Fabric Preform.

CYBERNET SYSTEMS CO.,LTD.
Mechanical CAE Division
Koji Yamamoto

[redacted]

CYBERNET SYSTEMS CO., LTD.

## Agenda

- Motivation
  - Technical issues for composite material analysis
  - Resin Transfer Molding(RTM) Process
- Analytical Approach for RTM Process
  - ANSYS Composite Prepost (Drape Sim.) & Transfer Molding Analysis
  - LS-DYNA (Explicit FEM Press Sim.) & Transfer Molding Analysis
- Multiscale analysis add on tool for ANSYS Workbench.
  - Homogenization and Localization
- Press Simulation for Dry Fabric.
  - Material Model.
  - Identification material constants by Multiscale Analysis.
- Transfer Molding Simulation for Dry Fabric
  - How to consider fiber orientation and fineness of preform.
- Summary




ANSYS

CYBERNET SYSTEMS CO., LTD.

## Motivation 1 : Technical Issues for Composite

- Characteristic of composite materials
  - ➢ Light and Hard.
  - ➢ Concurrency production for material and molding.
  - ➢ Airplane industry is preceding to apply.
- Key point to reduce vehicle weight
  - ➢ Apply to primary structure.
    - Body
    - Panel (Hood, Roof)
  - ➢ Mass production process.
  - ➢ Simulation technique.

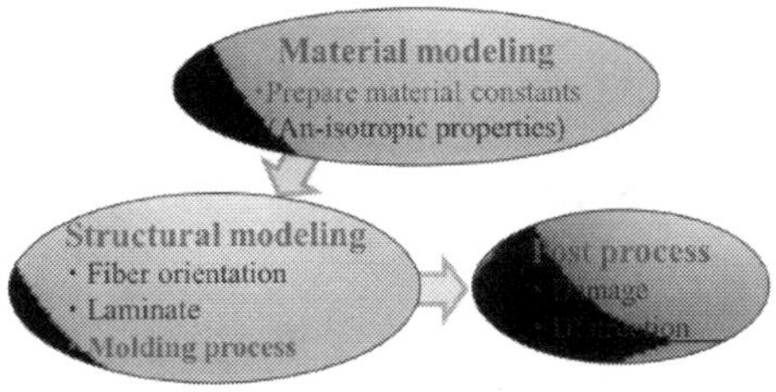


Technical issues for simulation

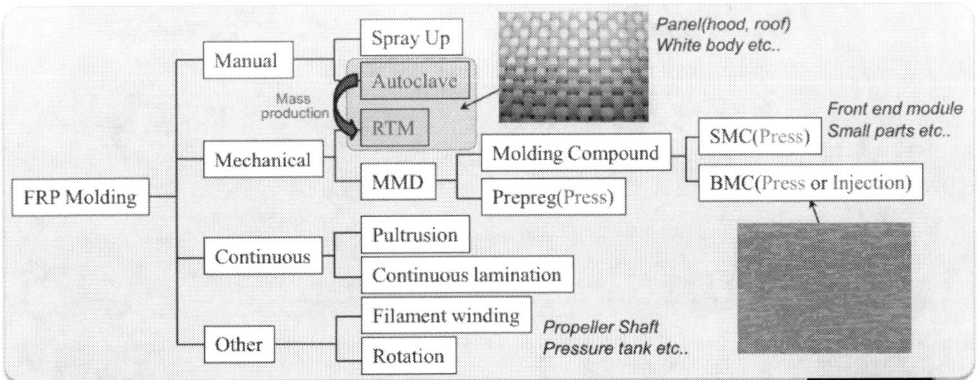


 3 ANSYS

CYBERNET SYSTEMS CO., LTD.

## Motivation 2 : RTM Process

### RTM = Resin Transfer Molding

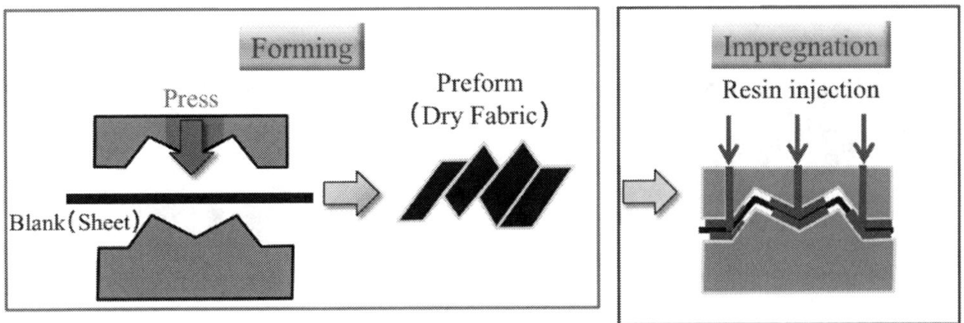


- Important factor for product quality.
  - ➢Forming
    - Fiber coarseness and fineness
    - Fiber orientation
    - Wrinkle
  - ➢Impregnation
    - Void
    - Warping
    - Residual stress

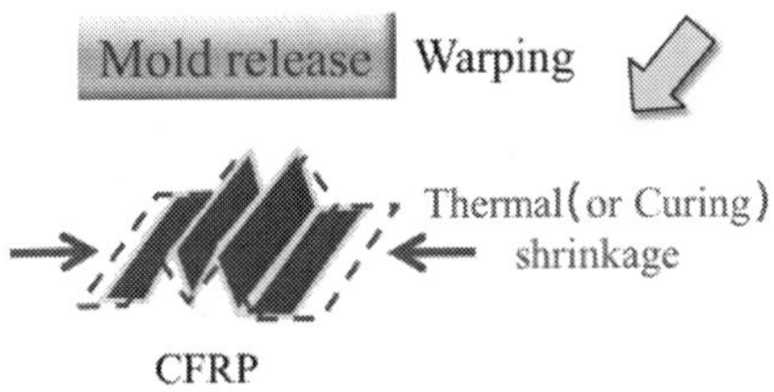


 4 ANSYS

CYBERNET SYSTEMS CO., LTD.

## Analytical Approach for RTM Process

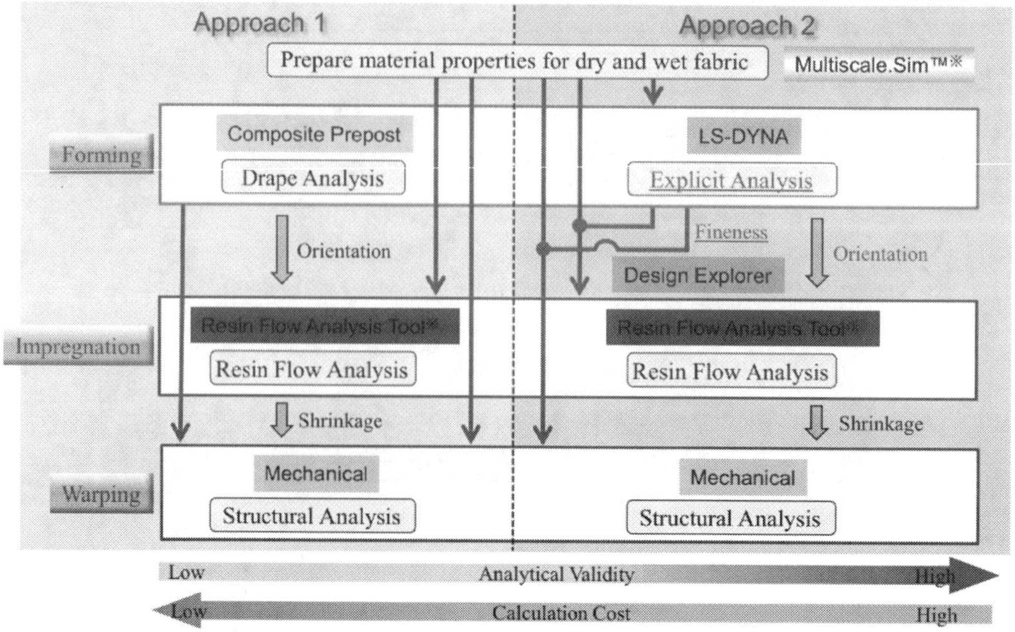


※ CYBERNET has developed original ANSYS add-on tool for Multiscale analysis and resin flow analysis.

 5

ANSYS

CYBERNET SYSTEMS CO., LTD.

## Deformation of Dry Fabric during Forming Process

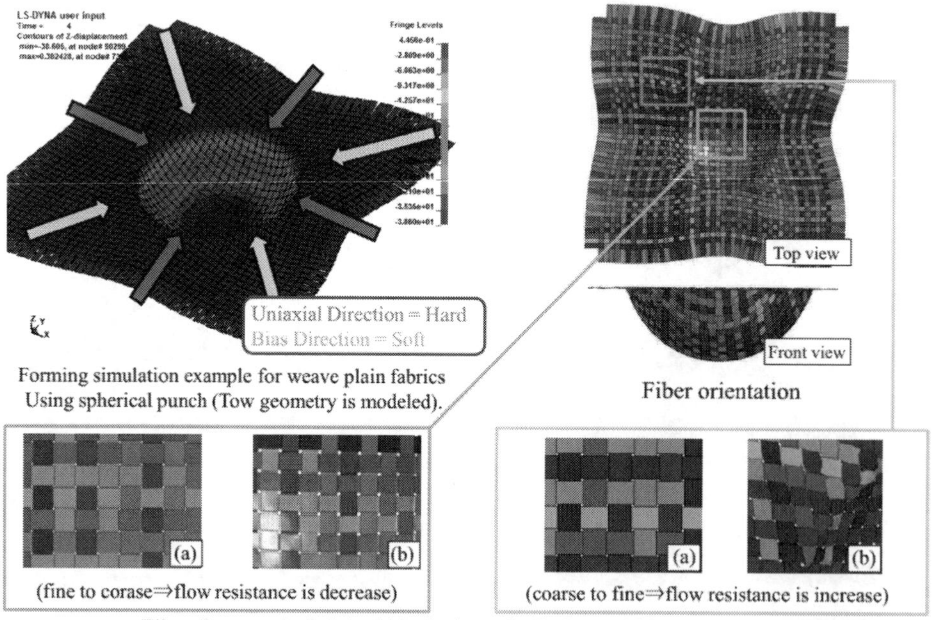


Fiber fineness (a:Original blank shape, b: Deformed blank shape)

 6

ANSYS

CYBERNET SYSTEMS CO., LTD.

## Drape Simulation vs. Explicit Simulation

We have to consider the material properties of dry fabric to perform press forming analysis with accuracy. Multiscale analysis is very useful to reduce time cost to prepare the an-isotropic material constants.

Comparison of properties for weave styles※

| Property | Plain | Twill | Satin | Basket | Leno | Mock Leno |
|---|---|---|---|---|---|---|
| Good Stability | *** | *** | ** | ** | ***** | *** |
| **Good Drape** | ** | **** | ***** | *** | * | ** |
| Symmetrical | **** | *** | * | *** | * | **** |
| Low Crimp | ** | *** | **** | ** | ***** | ** |

(*****=excellent, ****=good, ***=acceptable, **=poor, *=very poor)

Drape properties are strongly depended on weave styles.

Drape Properties
≒
Shear modulus

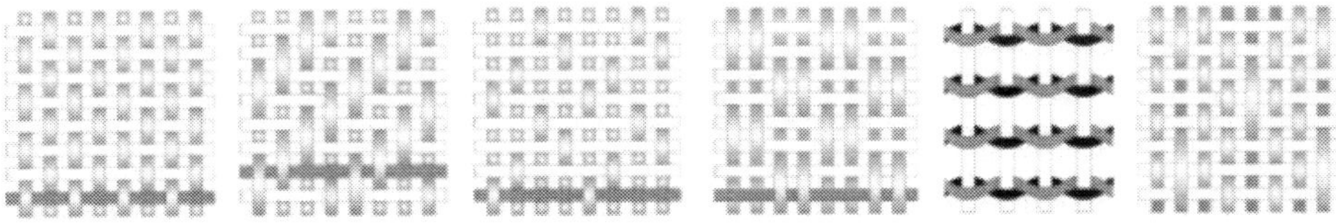

Plain Twill Satin Basket Leno Mock Leno

Typical weave styles

※ Reference "Woven Fabrics in Polymeric Composite Materials" (http://www.azom.com/article.aspx?ArticleID=1091)

 7 ANSYS

CYBERNET SYSTEMS CO., LTD.

## Multiscale Analysis Overview

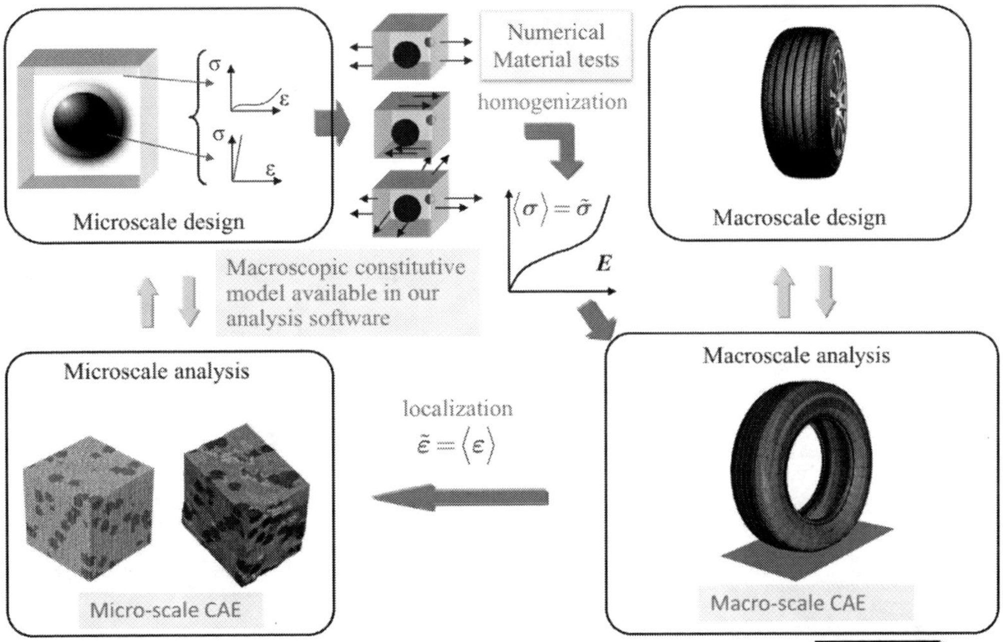


 8 ANSYS

**CYBERNET SYSTEMS CO.,LTD.**

## Development organization

Multiscale.Sim™ ※is an ANSYS add-in tool which is based on the homogenization method and developed by Industry-university cooperation projects of four organizations that have been established since 2007.

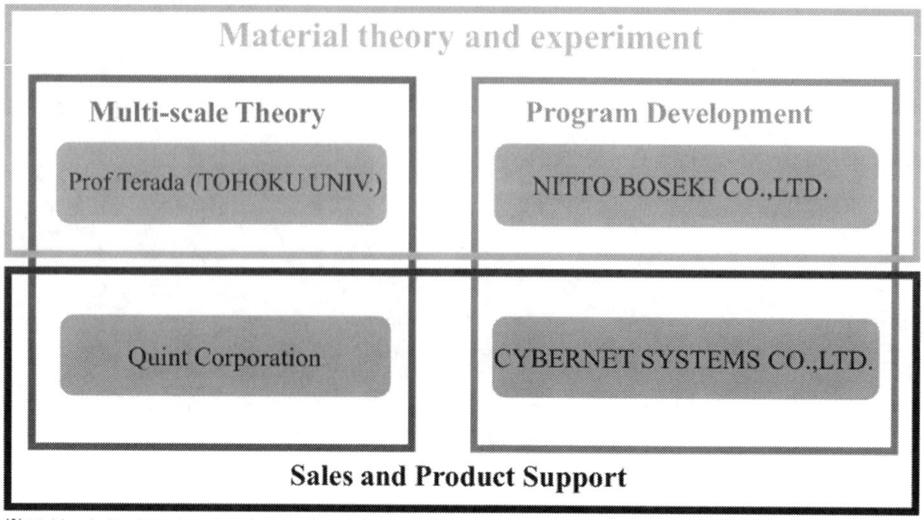


※ Multiscale.Sim is trademark in Japan and applying in Global at now.

 9

ANSYS

**CYBERNET SYSTEMS CO.,LTD.**

## Homogenization: Micro-/macro-scopic boundary value problems

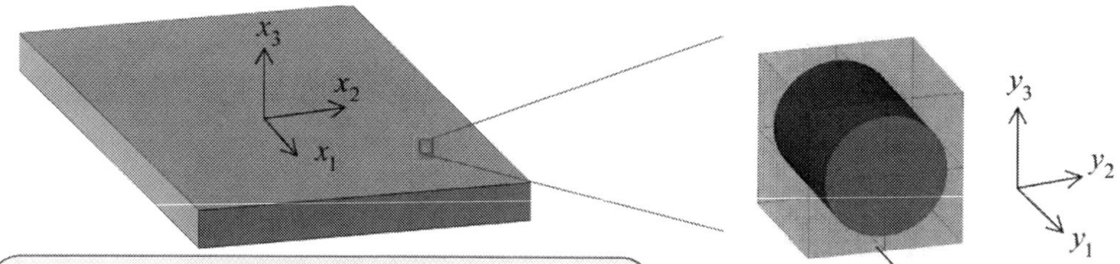


**Macro-scale governing equation**

- Equilibrium eqn. $\partial_x^{\mathrm{T}} \boldsymbol{\Sigma} + \bar{\boldsymbol{B}} = \boldsymbol{0}$
- Displ.-stain relation $\boldsymbol{E} = \partial_x \boldsymbol{U}$
- Constitutive eqn. (Anisotropic elasticity)

$$\begin{Bmatrix} \Sigma_1 \\ \Sigma_2 \\ \Sigma_3 \\ \Sigma_4 \\ \Sigma_5 \\ \Sigma_6 \end{Bmatrix} = \begin{bmatrix} C_{11} & C_{12} & C_{13} & C_{14} & C_{15} & C_{16} \\ & C_{22} & C_{23} & C_{24} & C_{25} & C_{26} \\ & & C_{33} & C_{34} & C_{35} & C_{36} \\ & & & C_{44} & C_{45} & C_{46} \\ & \text{sym.} & & & C_{55} & C_{56} \\ & & & & & C_{66} \end{bmatrix} \begin{Bmatrix} E_1 \\ E_2 \\ E_3 \\ E_4 \\ E_5 \\ E_6 \end{Bmatrix}$$

$$\boldsymbol{E} = \frac{1}{|Y|} \int_Y \varepsilon \, dY$$

$$\boldsymbol{\Sigma} = \frac{1}{|Y|} \int_Y \boldsymbol{\sigma} \, dY$$

**Micro-scale governing equation**

$$\begin{cases} \partial_y^{\mathrm{T}} \boldsymbol{\sigma} = \boldsymbol{0} \\ \varepsilon = \partial_y \boldsymbol{u} \\ \quad = \boldsymbol{E} + \partial_y \boldsymbol{u}^* \\ \sigma = D\varepsilon \end{cases}$$

 10

ANSYS

## Homogenization: Numerical material testing (NMT)

$$E=\begin{Bmatrix}E_1\\E_2\\E_3\\E_4\\E_5\\E_6\end{Bmatrix} \Rightarrow \begin{Bmatrix}\Sigma_1\\\Sigma_2\\\Sigma_3\\\Sigma_4\\\Sigma_5\\\Sigma_6\end{Bmatrix}=\begin{bmatrix}C_{11}&C_{12}&C_{13}&C_{14}&C_{15}&C_{16}\\&C_{22}&C_{23}&C_{24}&C_{25}&C_{26}\\&&C_{33}&C_{34}&C_{35}&C_{36}\\&&&C_{44}&C_{45}&C_{46}\\&\text{sym.}&&&C_{55}&C_{56}\\&&&&&C_{66}\end{bmatrix}\begin{Bmatrix}E_1\\E_2\\E_3\\E_4\\E_5\\E_6\end{Bmatrix} \Rightarrow \begin{Bmatrix}\Sigma_1\\\Sigma_2\\\Sigma_3\\\Sigma_4\\\Sigma_5\\\Sigma_6\end{Bmatrix}=\Sigma$$

*Calculation of macroscopic elastic constants by NMT*

Volume average of microscopic straina

$$E=\frac{1}{|Y|}\int_Y \varepsilon\, dY$$

Input: macroscopic strain

**Micro-scale FE analysis**

$$\varepsilon=\partial_y u = E+\partial_y u^*$$

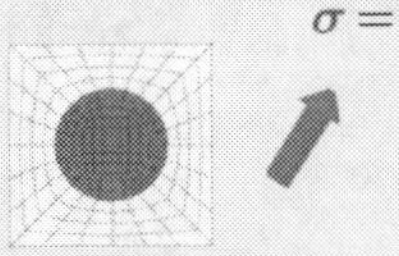

$$\sigma = D\varepsilon$$

$$\partial_y^{\mathrm{T}}\sigma=0$$

Volume average of microscopic stress

$$\Sigma=\frac{1}{|Y|}\int_Y \sigma\, dY$$

Output: macroscopic stress





ANSYS





## Homogenization: Numerical material testing (NMT)

$$\begin{Bmatrix}\Sigma_1\\\Sigma_2\\\Sigma_3\\\Sigma_4\\\Sigma_5\\\Sigma_6\end{Bmatrix}=\underbrace{\begin{bmatrix}C_{11}&C_{12}&C_{13}&C_{14}&C_{15}&C_{16}\\&C_{22}&C_{23}&C_{24}&C_{25}&C_{26}\\&&C_{33}&C_{34}&C_{35}&C_{36}\\&&&C_{44}&C_{45}&C_{46}\\&\text{sym.}&&&C_{55}&C_{56}\\&&&&&C_{66}\end{bmatrix}}_{\text{Unknown parameters}}\begin{Bmatrix}E_1\\E_2\\E_3\\E_4\\E_5\\E_6\end{Bmatrix}$$

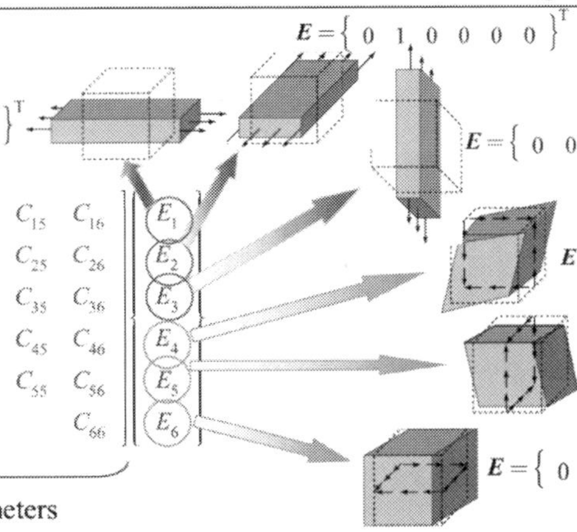


Homogenization (Volume averaging)

$$\Sigma=\frac{1}{|Y|}\int_Y \sigma\, dY=\begin{Bmatrix}\Sigma_1\\\Sigma_2\\\Sigma_3\\\Sigma_4\\\Sigma_5\\\Sigma_6\end{Bmatrix}_{E=\{0\ 0\ 0\ 0\ 0\ 1\}^{\mathrm{T}}}=\begin{bmatrix}C_{11}&C_{12}&C_{13}&C_{14}&C_{15}&C_{16}\\C_{12}&C_{22}&C_{23}&C_{24}&C_{25}&C_{26}\\C_{13}&C_{23}&C_{33}&C_{34}&C_{35}&C_{36}\\C_{14}&C_{24}&C_{34}&C_{44}&C_{45}&C_{46}\\C_{15}&C_{25}&C_{35}&C_{45}&C_{55}&C_{56}\\C_{16}&C_{26}&C_{36}&C_{46}&C_{56}&C_{66}\end{bmatrix}\begin{Bmatrix}0\\0\\0\\0\\0\\1\end{Bmatrix}=\begin{Bmatrix}C_{16}\\C_{26}\\C_{36}\\C_{46}\\C_{56}\\C_{66}\end{Bmatrix}$$





ANSYS

CYBERNET SYSTEMS CO., LTD.

## Multiscale.Sim™ GUI for Workbench

Multiscale.Sim™ GUI menu for Workbench 14.5 and 15.0 has been developed by using ANSYS Customization Toolkit. Mechanical APDL GUI is also implemented by UIDL.

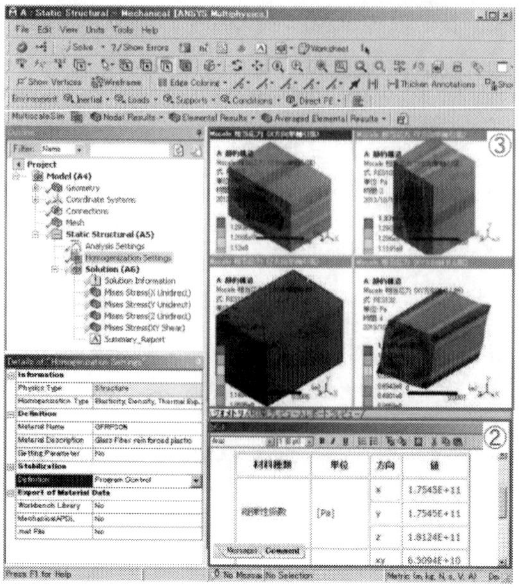

■ **Main function**

① Model creation templates (DM)

② Store the homogenization results to engineering database.

③ Contour and vector plot for numerical material testing.

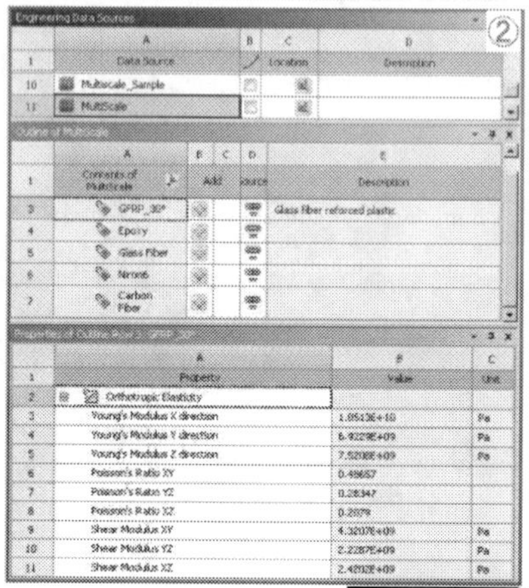





CYBERNET SYSTEMS CO., LTD.

## Supported Material Models on Multiscale.Sim™

■ Useful function for RTM simulation.

- Non-Linear numerical material testing. (However, dry fabric model is not supported.)
- Homogenization function to identify seepage coefficient using "darcy's law".

| Analysis Field | Analysis Type | Material Models | Material anisotropy |
|---|---|---|---|
| Structural | Linear | Elastic Coefficient | Anisotropic & Orthotropic |
| | | Density | - |
| | | Thermal Expansion | Orthotropic |
| | Non-Linear | Elasto-Plastic | Orthotropic (Hill)※2 |
| | | Visco-Elastic | Anisotropic(Anisotropic maxwell)※1,2 |
| | | Hyper-Elastic | Orthotropic (Kariske)※2 |
| | | Creep | Orthotropic (Hill)※2 |
| Thermal | Linear | Thermal Conduction | Anisotropic & Orthotropic |
| | | Specific Heat | - |
| Fluid | Linear | Seepage coefficient | Anisotropic & Orthotropic |
| Electromagnetic | Linear | Permittivity | Anisotropic & Orthotropic |
| | | Permeability | Anisotropic & Orthotropic |
| | | Electrical Resistivity | Anisotropic & Orthotropic |

※1 Orthotropic visco-elastic model is implemented by "usermat" subroutine.

※2 Curve fitting function is implemented by "userop" subroutine

CYBERNET SYSTEMS CO., LTD.

## How about the Material Property of Dry Fabric?

Dry fabric material property for pure shearing mode is evaluated by numerical material testing.

Analysis Conditions

| Condition | Detail |
|---|---|
| Fiber | Carbon |
| Weave style | Plane |
| Volume Fraction | 35.0 [Vol.%] |
| Testing | Pure Shearing (in plane) |

Material properties for carbon tow

| Condition | Detail |
|---|---|
| ρ | 1.44 [g/cm3] |
| E1 | 99.1 [GPa] |
| E2 | 7.4 [GPa] |
| G12 | 2.5 [GPa] |
| G23 | 5.0 [GPa] |
| v12 | 0.2 [-] |
| v23 | 0.2 [-] |
| μ | 0.0 [-] |

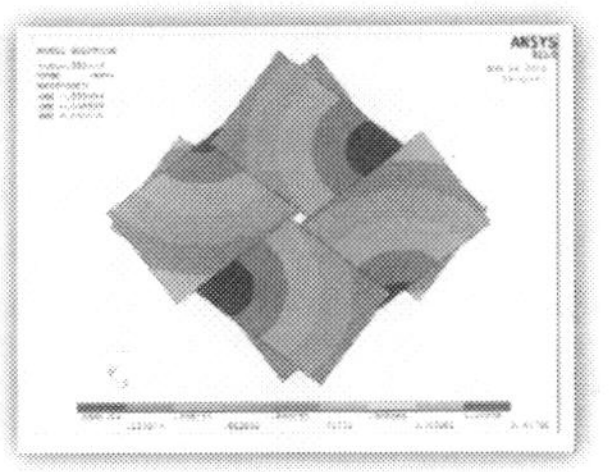

Deformation shape during pure shearing numerical material testing

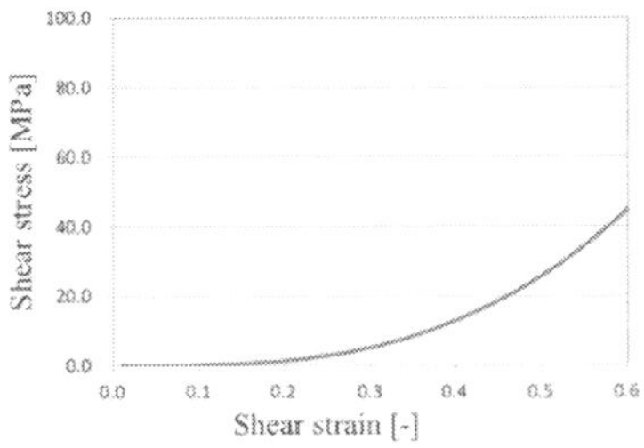


Macroscopic material response

Good correlation

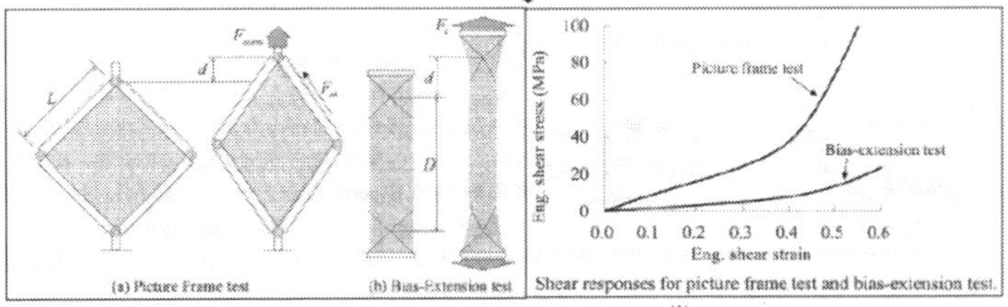


Results of reference paper※

※Reference : Journal of the Society of Materials Science, Japan, Vol.63, No.5, pp.380-385, May.2014





ANSYS

CYBERNET SYSTEMS CO., LTD.

## Material Model for Dry Fabric of LS-DYNA

- Characteristic of dry fabric material property
  - Shear modulus is increased drastically by shear locking.
  - Other deformation mode show linear behavior.
- *MAT_MACROMECHANICS_DRY_FABRIC

$$[C]^{-1} = \begin{bmatrix} 1/E_1 & -\nu_{12}/E_1 & -\nu_{12}/E_1 & 0 & 0 & 0 \\ -\nu_{12}/E_1 & 1/E_2 & -\nu_{23}/E_2 & 0 & 0 & 0 \\ -\nu_{12}/E_1 & \nu_{23}/E_2 & 1/E_2 & 0 & 0 & 0 \\ 0 & 0 & 0 & 1/\mu G_{12} & 0 & 0 \\ 0 & 0 & 0 & 0 & 1/\mu G_{23} & 0 \\ 0 & 0 & 0 & 0 & 0 & 1/\mu G_{12} \end{bmatrix}$$

Discount factor

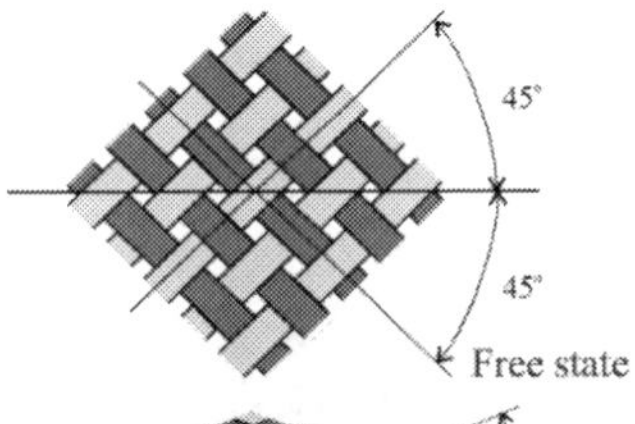


Free state

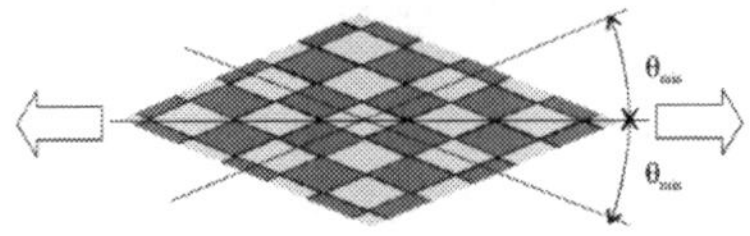

Locking state by shearing

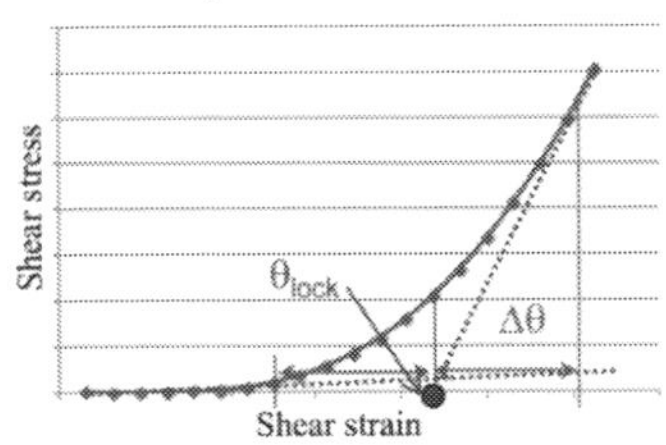


Rough estimating procedure

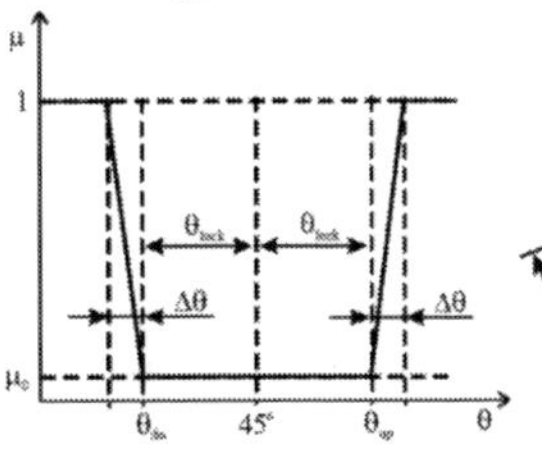


Discount factor as a function of braid angle.

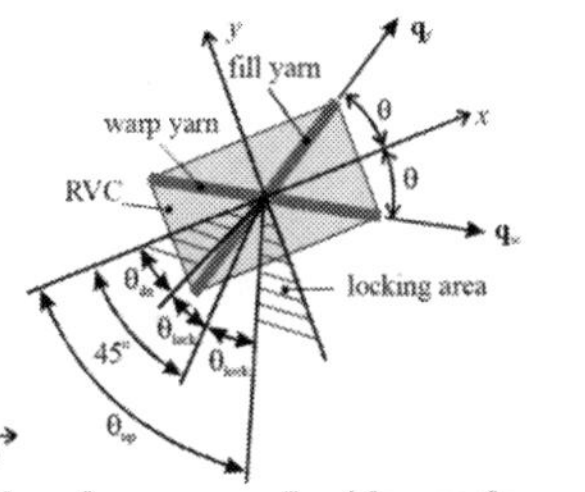


Locking angle





ANSYS

CYBERNET SYSTEMS CO., LTD.

## Curve Fitting to Identify Material Constants

Material constants are identified by fitting the numerical material testing results using GA optimization algorism. Initial value of optimization is estimated from stress-strain curve of shear testing as mentioned previous slide.

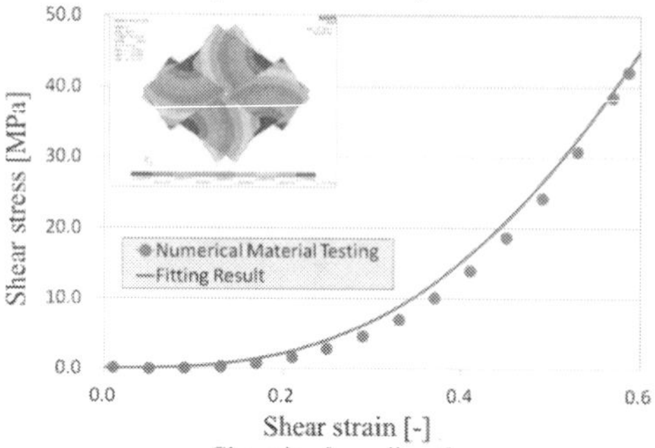


Shear in plane direction

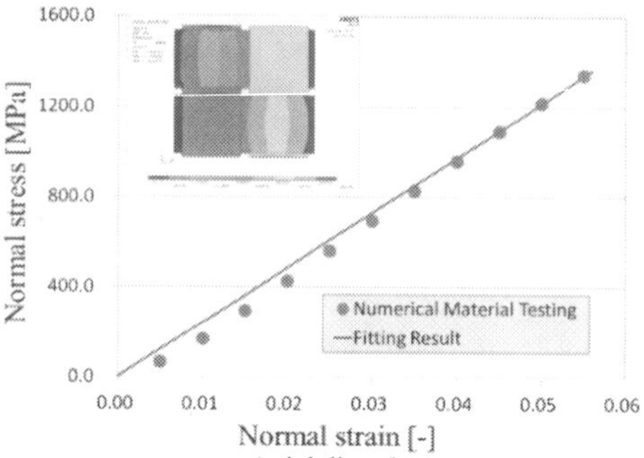


Axial direction

Curve fitting results for each testing mode.

Identified parameter for MICROMECHANICAL_DRY_FABRIC.

| mid | ro | e1 | e2 | g12 | g23 | v12 | v23 |
|---|---|---|---|---|---|---|---|
| 1 | 1440 | 99.1e9 | 1.0e9 | 2.5e9 | 5e9 | 0.2 | 0.2 |
| xt | thi | thl($\theta_{lock}$) | bfi | bwi | dscf | cnst | altr($\Delta\theta$) |
| 1.0 | 45.0 | 30.0 | 3.65 | 3.65 | 1000.0 | 0.02 | 5.0 |



17

ANSYS

CYBERNET SYSTEMS CO., LTD.

## Validation for Homogenized Model

Analysis result for homogenized model is validated by comparing direct model.

Blank sheet is modeled as shell163 element. k file is created by "LS-DYNA Export" component systems in workbench environment and modified the fabric material properties using text editor.

- Analysis conditions
  - ·Press Velocity 5000.0[mm/sec]
  - ·Material Model
    Punch : *MAT_RIGID
    Die : *MAT_RIGID
    Tow : *MAT2※1
    Weave : *MAT235※2
- Model Statistics

| | Direct | Homogenized |
|---|---|---|
| Nodes | 117595 | 16314 |
| Elements | 129857 | 42501 |

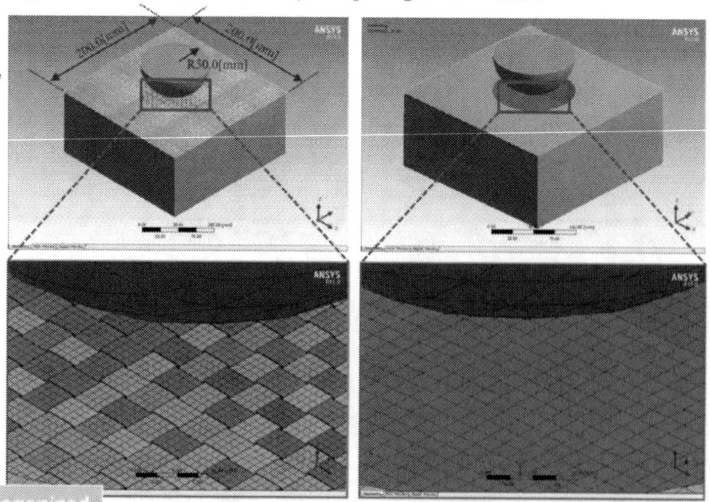


Direct Model    Homogenized Model

※1 : ORTHOTROPIC_ELASTIC (for Direct Model)
※2 : MICROMECHANICS_DRY_FABRIC (for Homogenized Model)



18

ANSYS

CYBERNET SYSTEMS CO., LTD.

## Validation for Homogenized Model

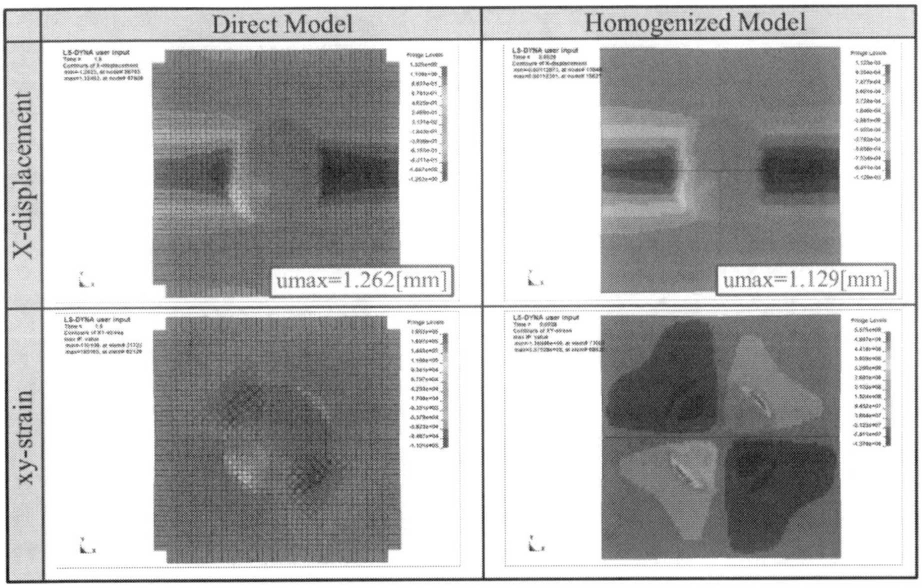


➢Calculation cost { Direct model : 5h31min
Homogenized model : 0h22min





ANSYS

CYBERNET SYSTEMS CO., LTD.

## Drape Simulation vs. Explicit FEM Simulation

There are two modeling approach for press forming simulation.

- 1. Drape sim. of ANSYS Composite Prepost
  - ➢ Shear angle is estimated using energy algorithm assuming shear modulus distributed uniformly.

$$\min E = \frac{1}{2} G\gamma^2 = \sum_{i=1}^{4} \cos^2 \alpha_i$$

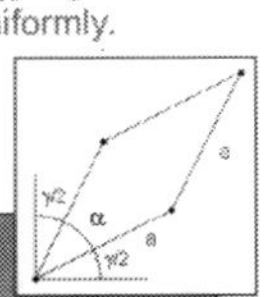

- 2. Explicit FEM Simulation LS-DYNA and so on.
  - ➢ Deformation properties are estimated using FEM taking into account nonlinear properties of shear modulus.

$$[M]\{\ddot{u}_n\} + [C]\{\dot{u}_n\} + [K]\{u_n\} = \{F_n\}$$

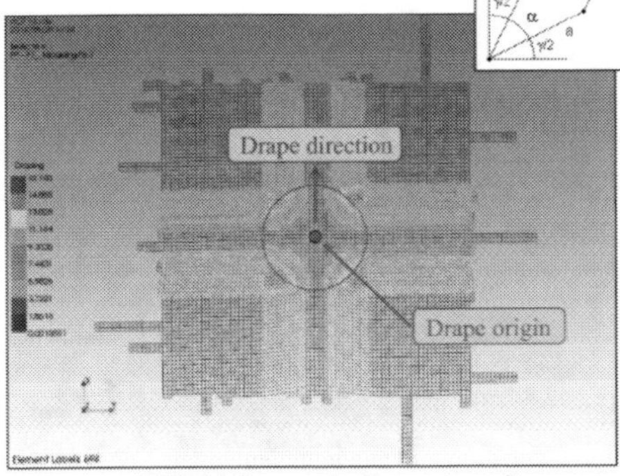


ANSYS Composite Prepost result

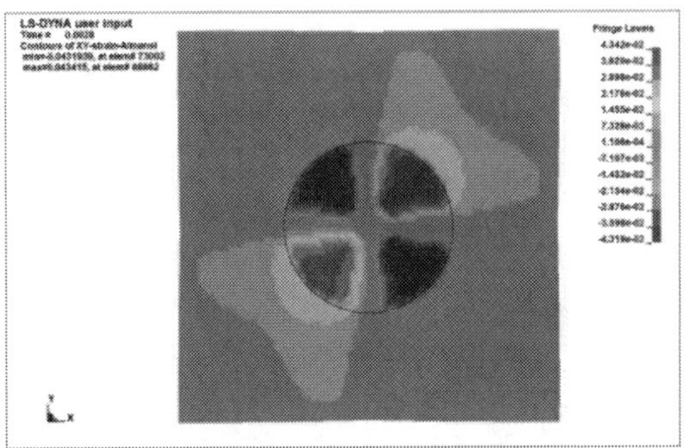

LS-DYNA result (Homogenized model)

Shear angle distribution for press forming by spherical punch
(Dot line means outline of punch : Pushing depth = 15.0[mm])





ANSYS

CYBERNET SYSTEMS CO., LTD.

## Fiber Reorientation by shear deformation

Element motion is composed of stretch and rigidly moving component.

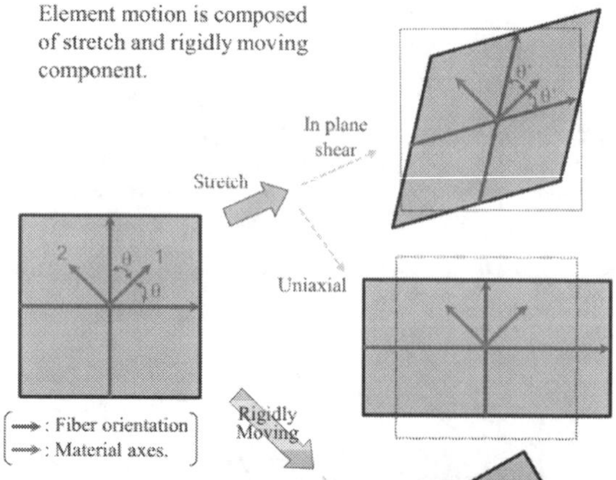


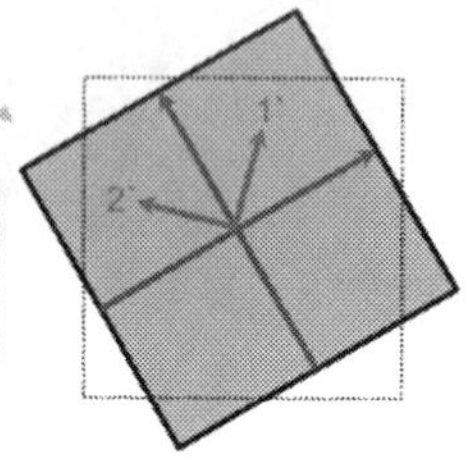


Influential factor for reorientation

| | In-plane Shear | Uniaxial | Rigidly Moving |
|---|---|---|---|
| Fiber Ori. → | ○ | × | × |
| Material Axes → | × | × | ○ |

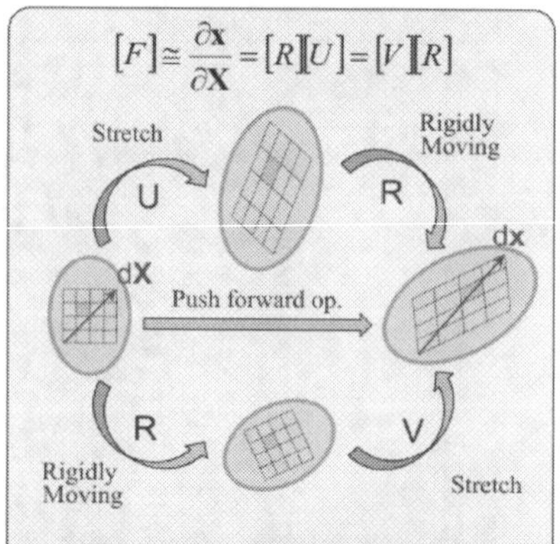


Deformation gradient tensor [F]

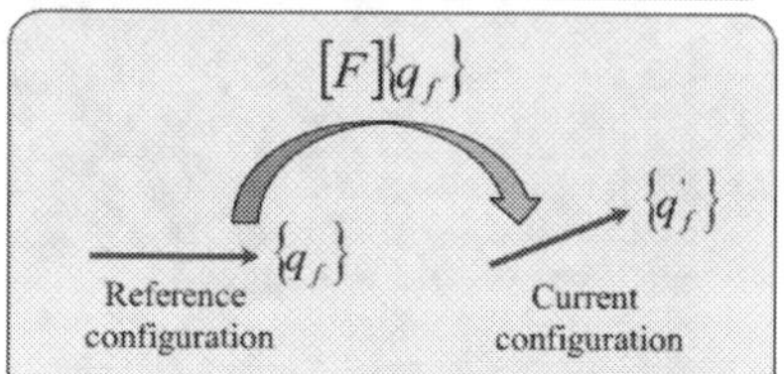


Push forward operation





CYBERNET SYSTEMS CO., LTD.

## Coupled analysis : Structural to Resin Flow

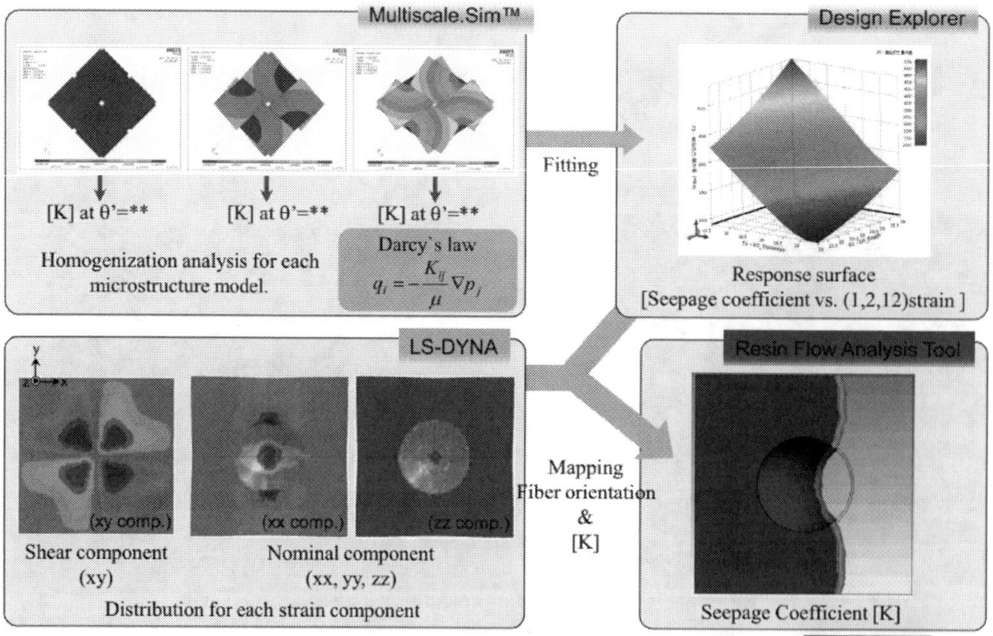

CYBERNET SYSTEMS CO., LTD.

## Summary

**Purpose : Development explicit press simulation system for RTM process.**

- Forming process
  - Macro Analysis
    - Explicit press simulation by LS-DYNA.
    - Dry fabric microstructure is homogenized
  - Micro Analysis (=Predict material properties)
    - Predict material properties for dry fabric using homogenization method of Multiscale.Sim™.
    - Material constants for dry fabric constitutive law is identified by GA algorithm.
  - Post process
    - Fiber Reorientation
      - Extraction from deformation gradient tensor.
    - Fiber fineness
      - Response surface of seepage coefficient vs. strain component

- Future plan -

- Impregnation process
  - Resin transfer simulation taking into account seepage coefficient distribution.
- Mold release process
- Verification & Validation





ANSYS

CYBERNET SYSTEMS CO., LTD.



ANSYS

ANSYS

Realize Your Product Promise®

# New Electromagnetic-Thermal Coupled Analyses for Electrical Machine Design

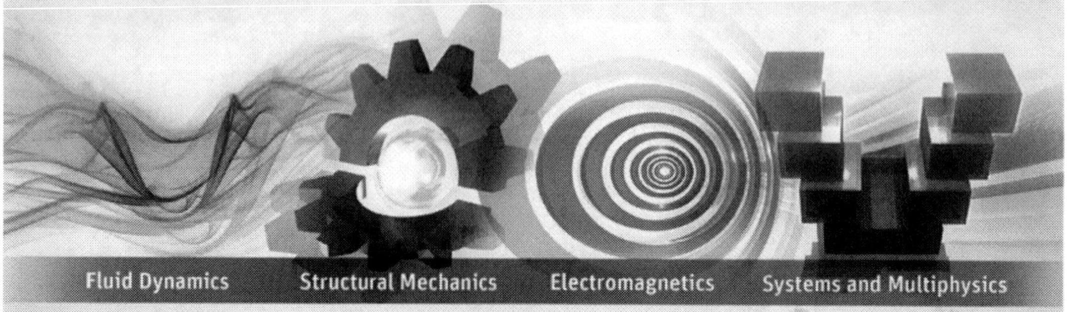

Fluid Dynamics | Structural Mechanics | Electromagnetics | Systems and Multiphysics

**Marius Rosu, PhD**

**EM Lead Product Manager, ANSYS Inc.**

ANSYS

## Acknowledgments

Manoj Nagulapally - ANSYS CFD Expert, USA

Mike Slack - ANSYS CFD Expert, UK

Yang Hu - ANSYS Electromagnetics, USA

Eddie Chong - MDL CFD Expert, UK

2  October 2, 2014 ANSYS Confidential

# Outline

**Integrated Motor Design Flow**

**Magnetic and Thermal Sizing Solution**
- **Case Study - PM Machine Design**

**Hybrid CFD Solution**
- **Case Study – IPM Motor Design**

**Electromagnetic - CFD Coupled Solution**
- **Case Study - Traction EV Motor Design**



# Motor Design Flow

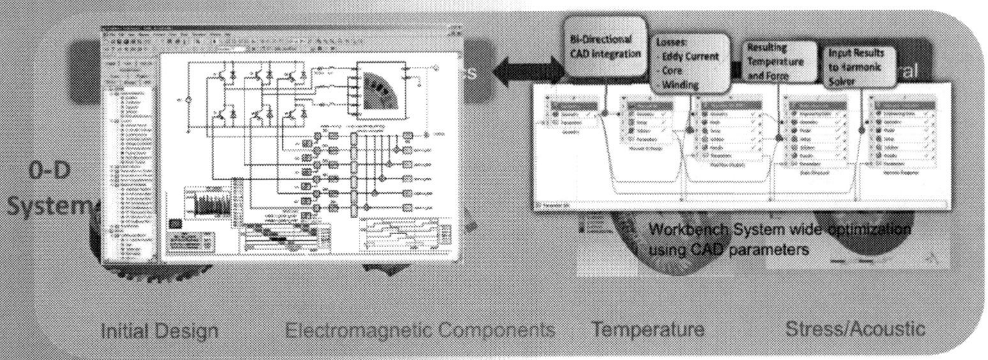

ANSYS

# Outline

Integrated Motor Design Flow

Magnetic and Thermal Sizing Solution
- Case Study - PM Machine Design

Hybrid CFD Solution
- Case Study – IPM Motor Design

Electromagnetic - CFD Coupled Solution
- Case Study - Traction EV Motor Design



ANSYS

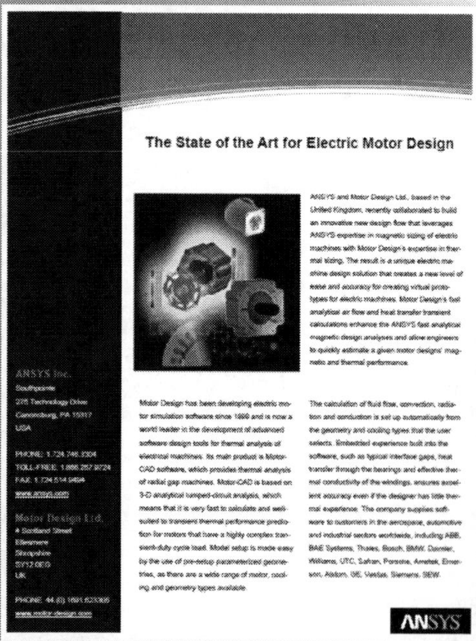

The State of the Art for Electric Motor Design

ANSYS and Motor Design Ltd., based in the United Kingdom, recently collaborated to build an innovative new design flow that leverages ANSYS expertise in magnetic sizing of electric machines with Motor Design's expertise in thermal sizing. The result is a unique electric machine design solution that creates a new level of ease and accuracy for creating virtual prototypes for electric machines. Motor Design's fast analytical air flow and heat transfer transient calculations enhance the ANSYS fast analytical magnetic design analyses and allow engineers to quickly estimate a given motor designs' magnetic and thermal performance.

Motor Design has been developing electric motor simulation software since 1999 and is now a world leader in the development of advanced software design tools for thermal analysis of electrical machines. Its main product is Motor-CAD software, which provides thermal analysis of radial gap machines. Motor-CAD is based on 3-D analytical lumped-circuit analysis, which means that it is very fast to calculate and well-suited to transient thermal performance prediction for motors that have a highly complex transient-duty cycle load. Model setup is made easy by the use of pre-setup parameterized geometries, as there are a wide range of motor, cooling and geometry types available.

The calculation of fluid flow, convection, radiation and conduction is set up automatically from the geometry and cooling types that the user selects. Embedded experience built into the software, such as typical interface gaps, heat transfer through the bearings and effective thermal conductivity of the windings, ensures excellent accuracy even if the designer has little thermal experience. The company supplies software to customers in the aerospace, automotive and industrial sectors worldwide, including ABB, BAE Systems, Thales, Bosch, BMW, Daimler, Williams, UTC, Safran, Porsche, Ametek, Emerson, Alstom, GE, Vestas, Siemens, SEW.

ANSYS Inc.
Southpointe
275 Technology Drive
Canonsburg, PA 15317
USA

Motor Design Ltd.
4 Scotland Street
Ellesmere
Shropshire
UK

ANSYS

**ANSYS Inc. and Motor Design Ltd.**

Collaborated to build an innovative new design flow that leverages,
- ANSYS expertise in magnetic sizing of electric machines
- Motor Design expertise in thermal sizing of electric machines

## First-Pass Integration with ANSYS Software

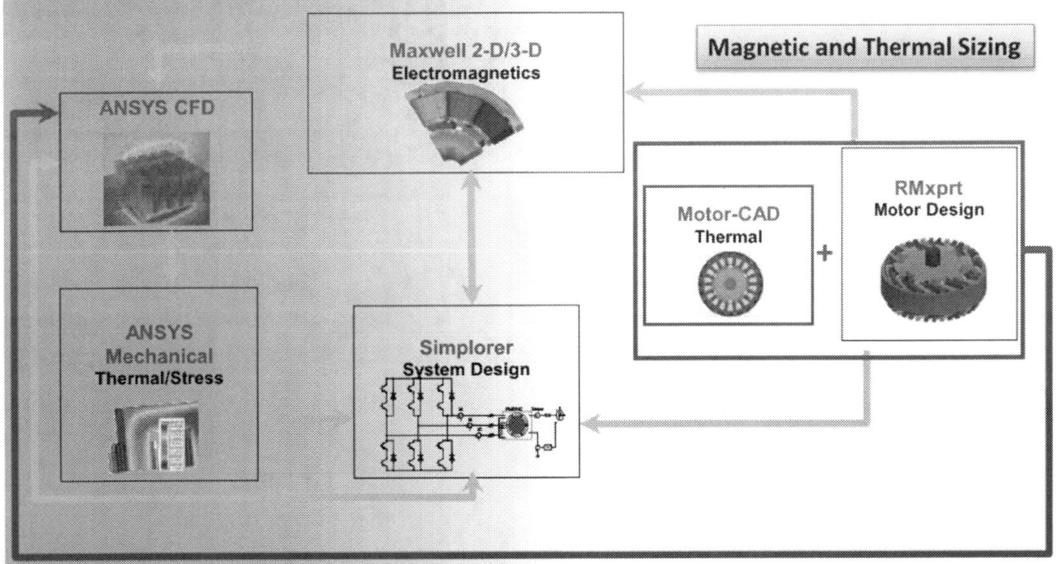




## Magnetic and Thermal Design Sizing Solution

- Template-based electrical machine design
- Fast analytical calculations for popular electric machine topologies
- One-click setup for Maxwell 2D and 3D finite element Solutions
- Network analysis dedicated to thermal analysis
- Heat transfer data calculated automatically
- Detailed cooling systems & facilitates optimisation

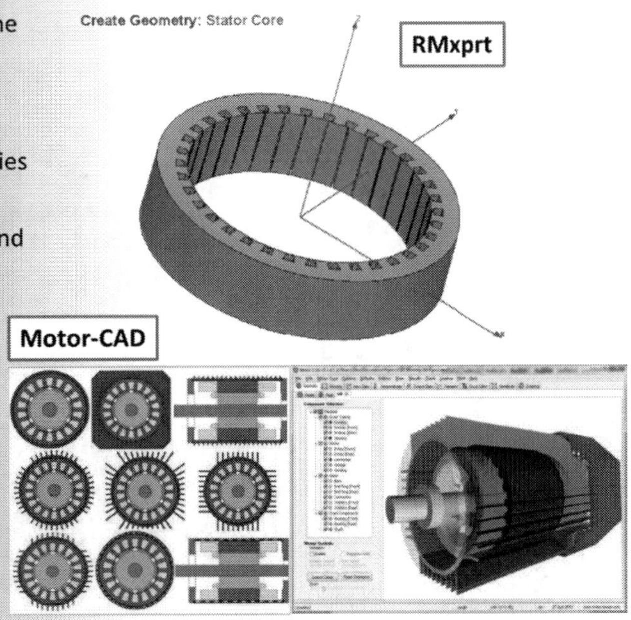

ANSYS
RMxprt - Magnetic Sizing
Adjust-Speed Synchronous Machine
Brushless Permanent-Magnet DC Motor
Claw-Pole Alternator
DC Machine
Generic Rotating Machine
Line-Start Permanent-Magnet Synchronous Motor
Permanent-Magnet DC Motor
Single-Phase Induction Motor
Switched Reluctance Motor
Three-Phase Induction Motor
Three-Phase Non-Salient Synchronous Machine
Three-Phase Synchronous Machine
Universal Motor
9
© 2013 ANSYS, Inc.
October 2, 2014
ANSYS Confidential

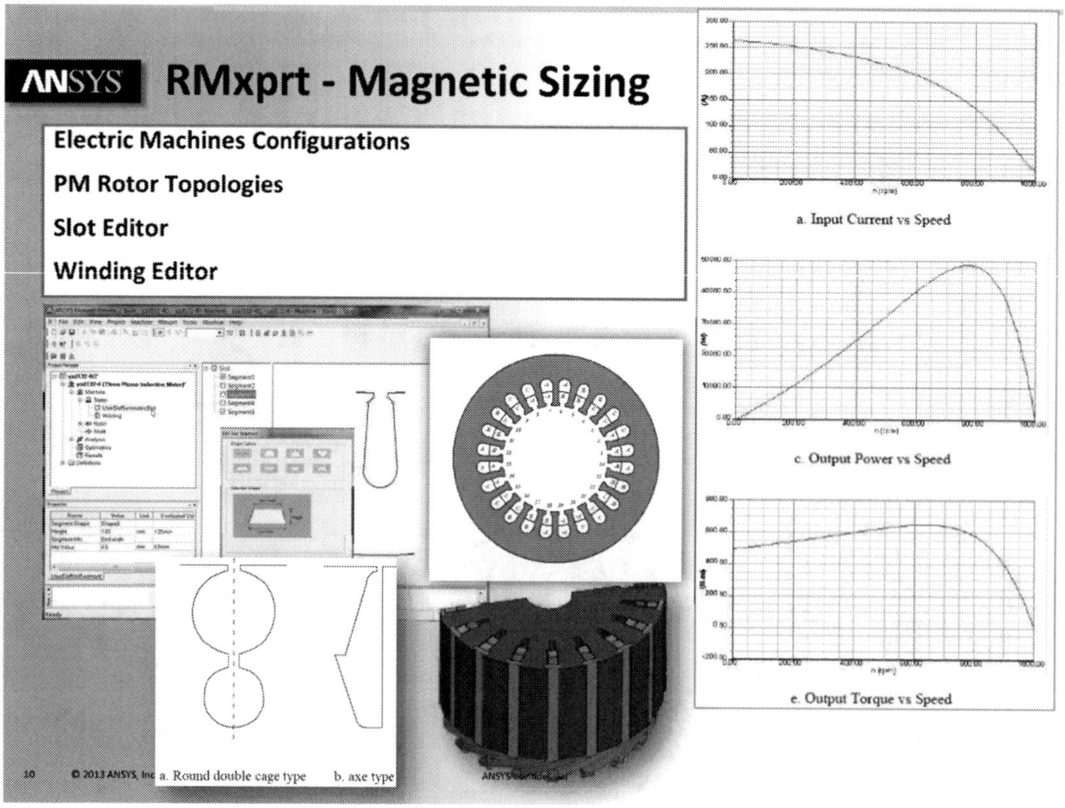
ANSYS
RMxprt - Magnetic Sizing
Electric Machines Configurations
PM Rotor Topologies
Slot Editor
Winding Editor
a. Input Current vs Speed
c. Output Power vs Speed
e. Output Torque vs Speed
a. Round double cage type
b. axe type
10
© 2013 ANSYS, Inc

# ANSYS Housing in RMxprt – New Capability

- Machine
  - Stator
  - Rotor
  - Shaft
  - Housing
    - Fin
    - Duct
    - Front Cap
    - Rear Cap
    - Foot
    - Flange

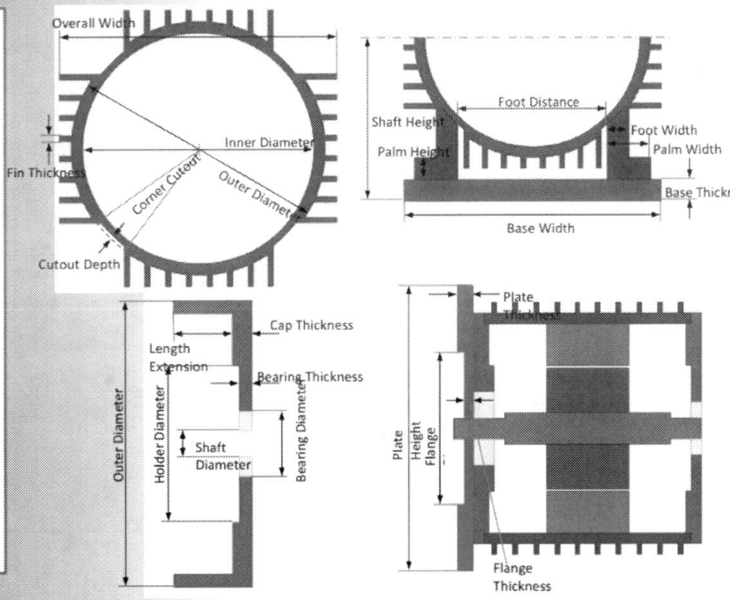


11 © 2013 ANSYS, Inc. October 2, 2014 ANSYS Confidential

# ANSYS New UDP for Thermal Field Analysis

- Housing
  - Four different profiles
  - Six different fin types
  - Three different duct types
- EndCaps

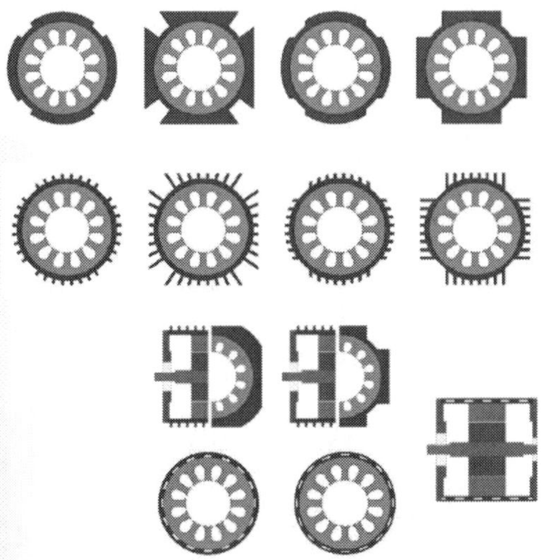

12 © 2013 ANSYS, Inc. October 2, 2014 ANSYS Confidential

## Motor-CAD – Thermal Sizing

**Includes proven models for an extensive range of cooling types:**

- TENV: Totally enclosed non-ventilated
- TEFC: Totally enclosed fan cooled
- Through Ventilation
- TE with Internal Circulating Air
- Open end-shield cooling
- Water Jackets
- Submersible cooling
- Wet Rotor & Wet Stator cooling
- Spray Cooling
- Direct conductor cooling

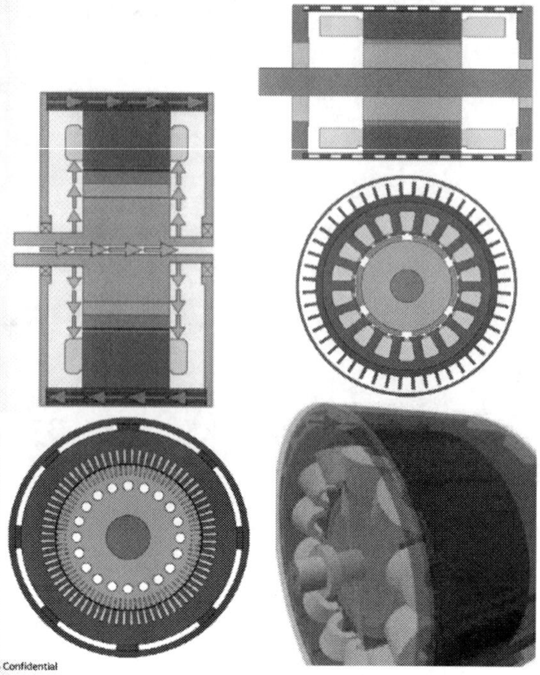



## Motor-CAD – Thermal Analysis

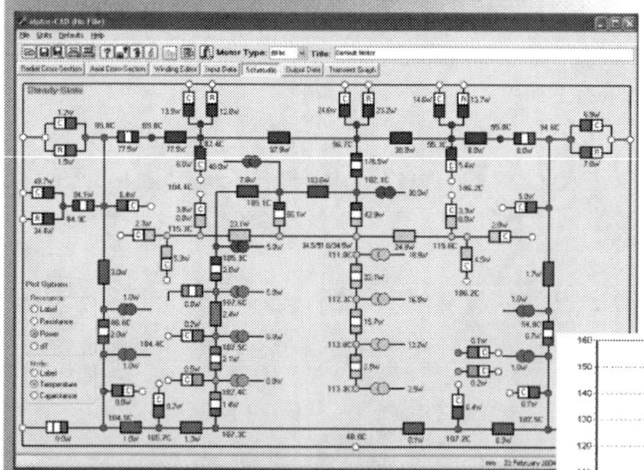

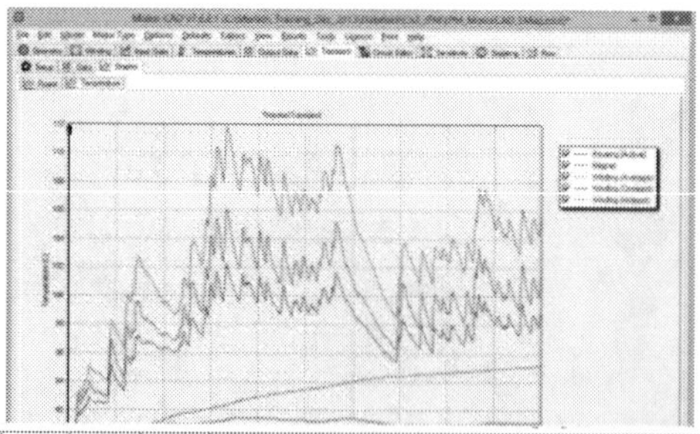

- Analytical lumped circuit analysis
- Combined Thermal and Flow network analyses
- Steady state thermal analysis
- Complex transient duty cycle analysis

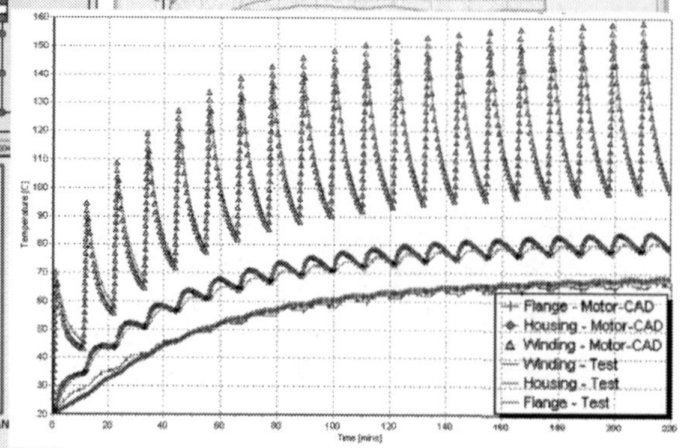

# Motor-CAD Integrated Thermal FEA Solver

Motor-CAD fully integrated FEA thermal solver allows fast calibration of analytical model giving best compromise of accuracy & calculation speed

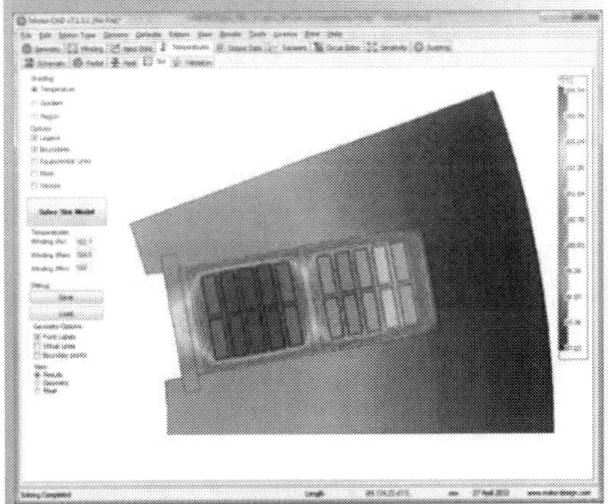

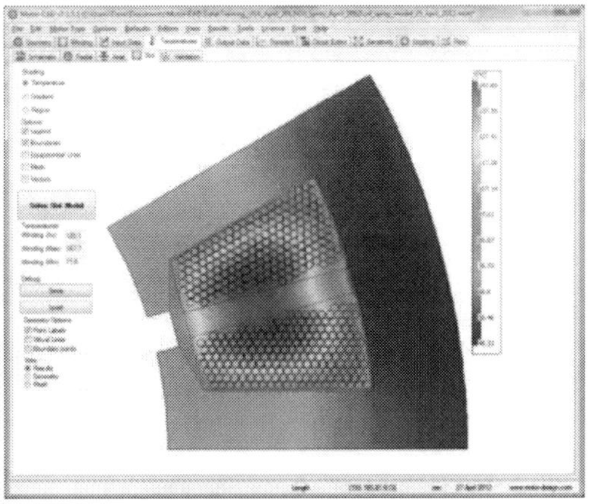

15 © 2013 ANSYS, Inc. October 2, 2014 ANSYS Confidential

# Case Study - PM Machine Design

## RMxprt - Motor CAD

- Electromagnetic performance calculated in RMxprt
- Default housing in RMxprt has axial fins which are suitable for TEFC cooling
- Geometry and power losses are exported to Motor-CAD

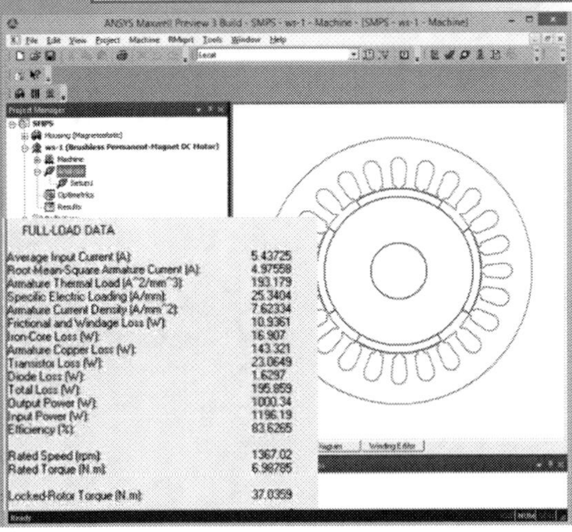


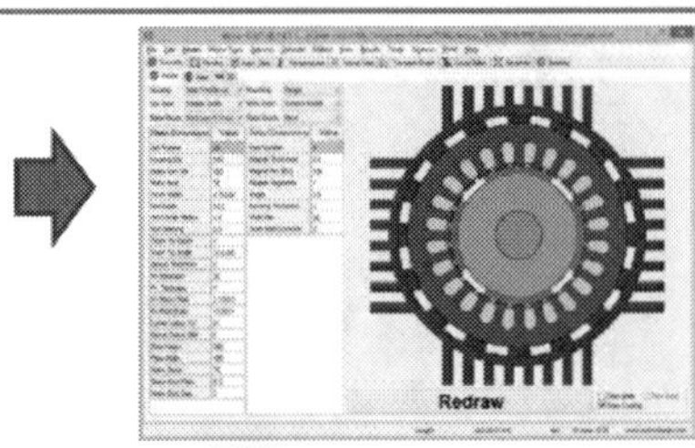


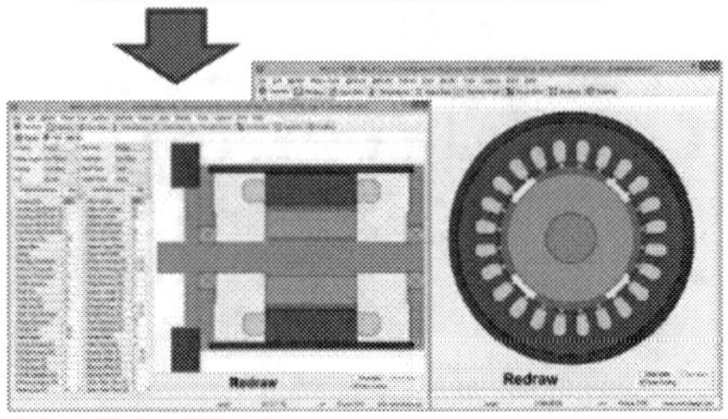

## ANSYS Case Study - PM Machine Design

### RMxprt - Motor CAD

- Winding also exported from RMxprt
- Motor-CAD updates winding to account for wire size available, enamel on wire and slot liner available

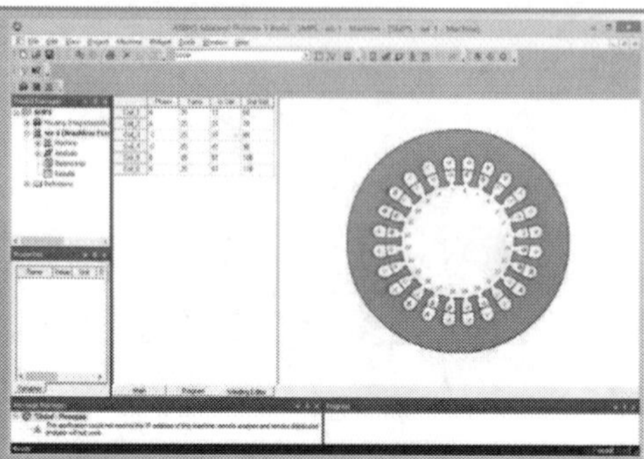

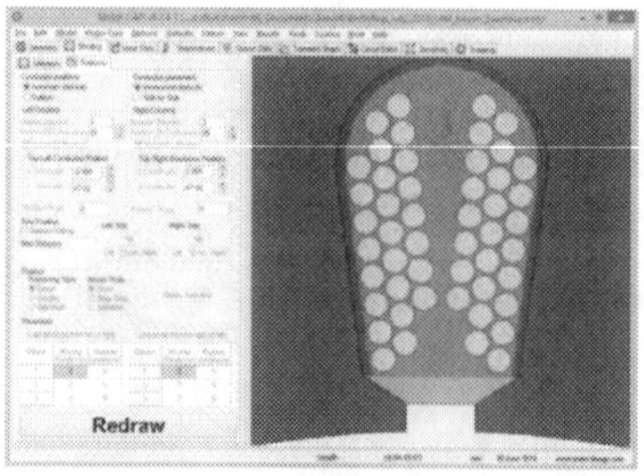


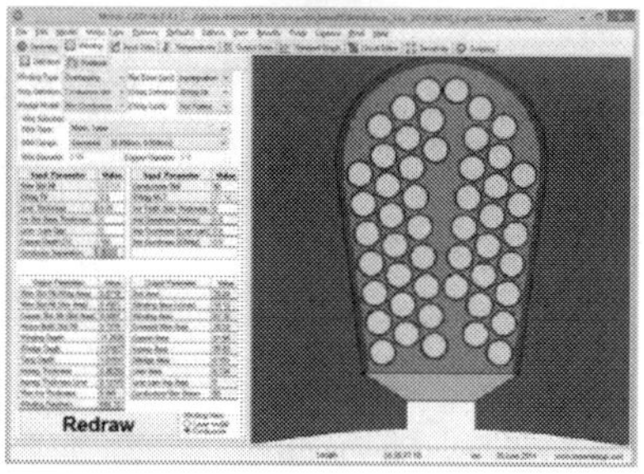




## ANSYS Case Study - PM Machine Design

**Steady-State Thermal**
- Instantaneous simulation

**Repetitive Duty Cycle Analysis**
- Rapid simulation

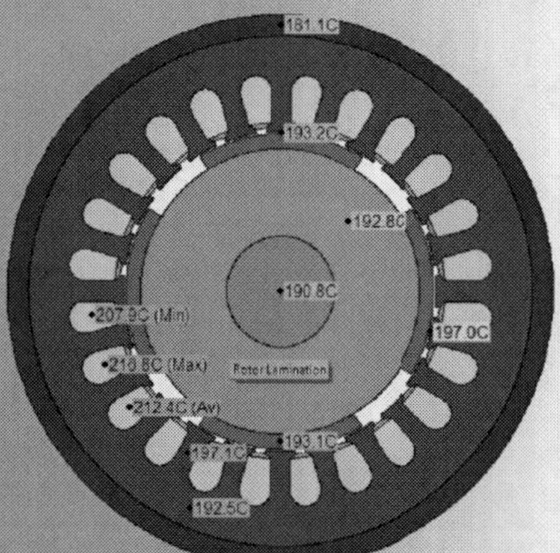


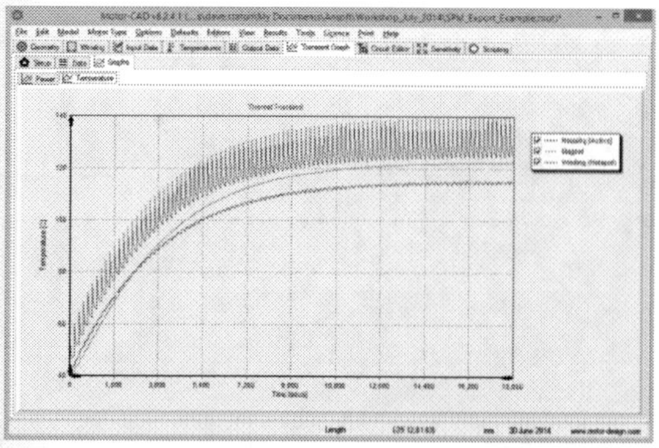

## ANSYS Case Study - PM Machine Design

**Robust Design Analysis on Thermal Sizing**
- Parametric analysis
- Sensitivity analysis

**Selected Design**

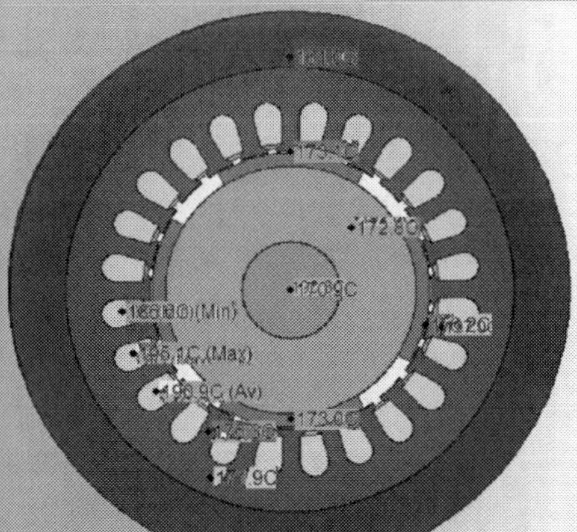

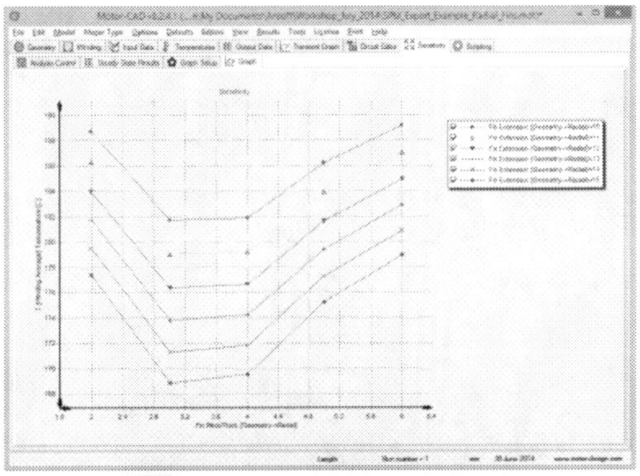

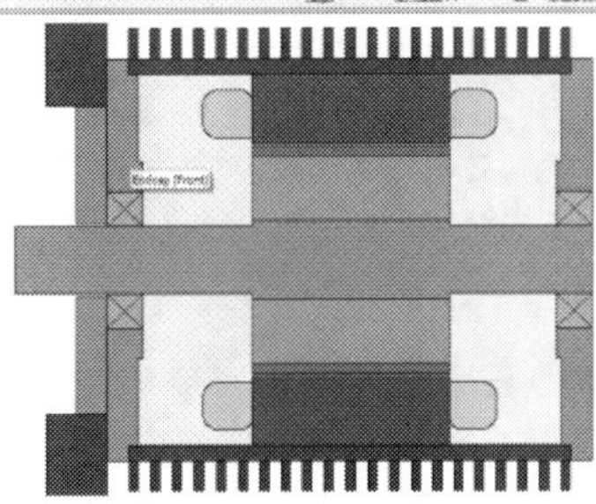



## ANSYS Outline

**Integrated Motor Design Flow**

**Magnetic and Thermal Sizing Solution**
- **Case Study - PM Machine Design**

**Hybrid CFD Solution**
- **Case Study – IPM Motor Design**

**Electromagnetic - CFD Coupled Solution**
- **Case Study - Traction EV Motor Design**

## ANSYS Motor-CAD Thermal Network

- Equivalent thermal network of a machine can be generated based on the machine information (geometry, materials, cooling types, heat sources, etc.)
- Convective heat transfer is predicted using heat transfer correlations
- The accuracy of the convection is only as good as the appropriateness and validity of the correlations used in the calculation, e.g. flow over fins, housing water jacket, etc.

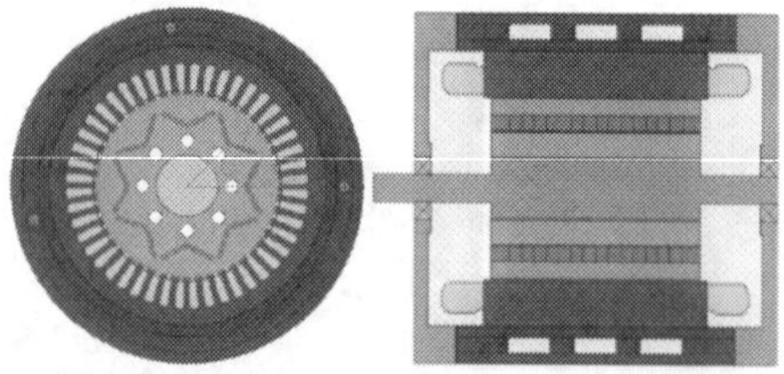

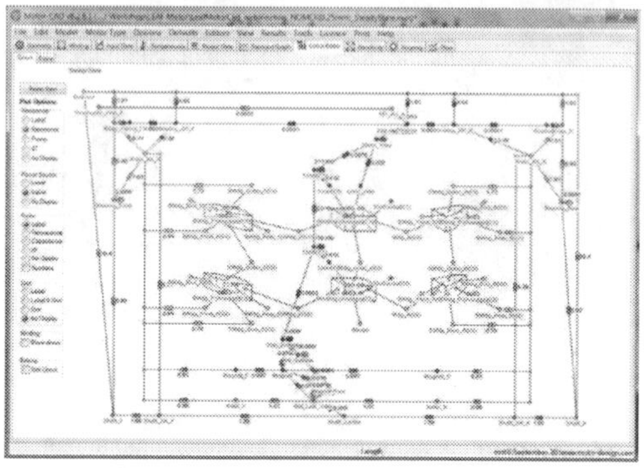



## ANSYS Hybrid Model : CFD Domain Definition

Motor – Thermal-Flow Network Equivalence

- Motor-CAD thermal network is imported into ANSYS Icepak
- ANSYS Icepak network replaces the thermal resistances between the housing and ambient with a CFD domain
- Inside the housing, the machine is modelled using Motor-CAD thermal network, whereas the convective heat transfer outside the housing is modelled using an Icepak CFD domain
- The thermal network and the CFD domain are coupled together through multi-faceted cylinder housing interfaces

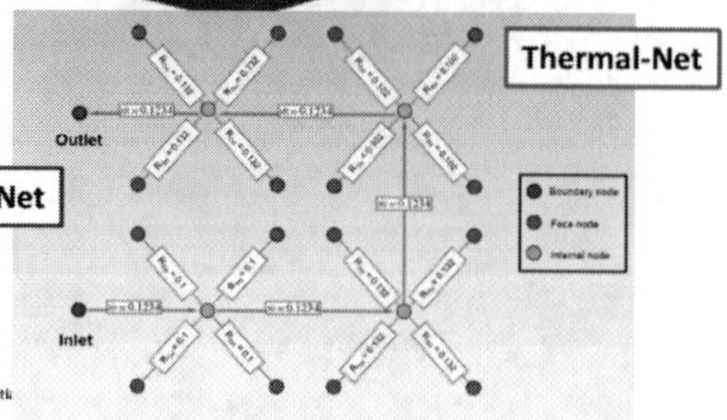


For the current application only thermal network is used

## Detailed and Hybrid Models: Comparison

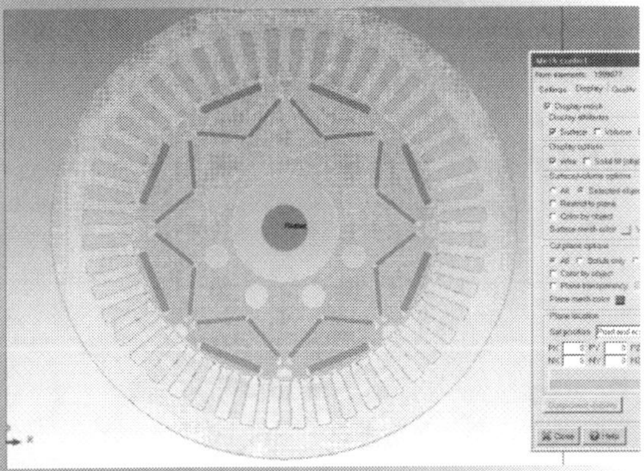

Detailed Model : 2M elements

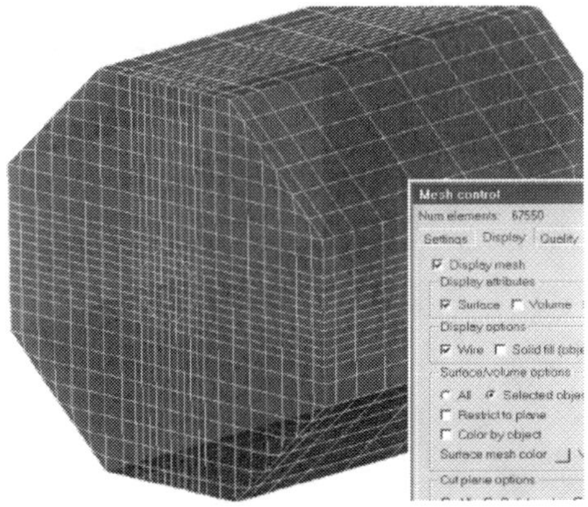

Network Model : 67k elements

**3D Natural Convection Simulation on Hybrid Model**
- Turbulent flow model
- Radiative heat transfer model
- Fully parallelized solver
- Converged CFD solution in ~ 2 minutes



## Hybrid Model: CFD Simulation Results

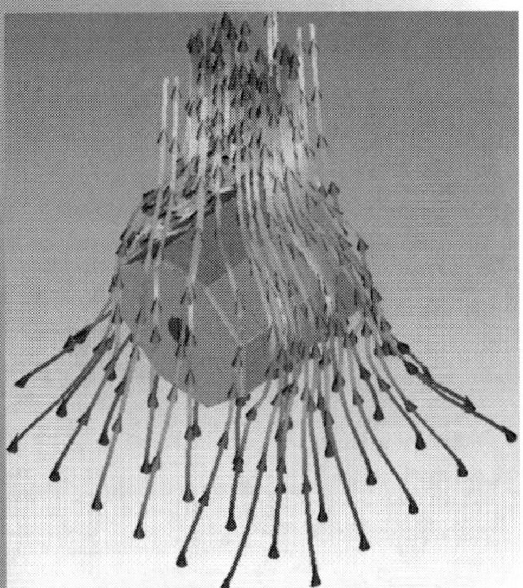

Flow tracks around network

| Section | Value | Mean |
|---|---|---|
| All | Temperature (C) | 93.4868 |
| Housing_1 | | 93.107 |
| Housing_2 | | 93.038 |
| Housing_3 | | 93.114 |
| Housing_4 | | 93.039 |
| Housing_5 | | 93.108 |
| Housing_6 | | 93.038 |
| Housing_7 | | 93.108 |
| Housing_8 | | 93.037 |
| Housing_OH_F_1 | | 93.123 |
| Housing_OH_F_2 | | 93.075 |
| Housing_OH_F_3 | | 93.128 |
| Housing_OH_F_4 | | 93.076 |
| Housing_OH_F_5 | | 93.124 |
| Housing_OH_F_6 | | 93.075 |
| Housing_OH_F_7 | | 93.123 |
| Housing_OH_F_8 | | 93.075 |
| Housing_OH_R_1 | | 93.123 |
| Housing_OH_R_2 | | 93.075 |
| Housing_OH_R_3 | | 93.128 |
| Housing_OH_R_4 | | 93.076 |
| Housing_OH_R_5 | | 93.124 |
| Housing_OH_R_6 | | 93.075 |
| Housing_OH_R_7 | | 93.123 |
| Housing_OH_R_8 | | 93.075 |
| ECap_F | | 94.258 |
| ECap_R | | 94.258 |
| Shaft_F | | 116.624 |
| Shaft_R | | 116.627 |
| Housing_F (0.0 W) | | 93.194 |
| Housing_R (0.0 W) | | 93.194 |
| Stator_Yoke (734.1 W) | | 109.943 |
| Tooth_outer (193.8 W) | | 118.517 |
| Stat_Surface (5.0 W) | | 124.299 |
| Rot_Surface (5.0 W) | | 126.546 |
| Magnet (4.669 W) | | 126.584 |
| Rot_Lam_Yoke (34.89 W) | | 127.567 |
| Shaft_Centre (0.0 W) | | 127.044 |
| Shaft_OH_F (0.0 W) | | 120.898 |
| Shaft_OH_R (0.0 W) | | 120.899 |
| Rotor_ES_F (0.0 W) | | 111.549 |
| Rotor_ES_R (0.0 W) | | 111.55 |
| HFI_WJ_A (0.0 W) | | 66.268 |
| Magnet_Pole (0.0 W) | | 126.594 |
| Bearing_Front (2.0 W) | | 110.716 |
| Bearing_Rear (2.0 W) | | 110.717 |
| Tooth_inner (193.8 W) | | 124.198 |
| Rotor_F (0.0 W) | | 117.873 |
| Rotor_R (0.0 W) | | 117.874 |

Nodal temperature report

# Outline

**Integrated Motor Design Flow**

**Magnetic and Thermal Sizing Solution**

- **Case Study - PM Machine Design**

**Hybrid CFD Solution**

- **Case Study – IPM Motor Design**

**Electromagnetic - CFD Coupled Solution**

- **Case Study - Traction EV Motor Design**



# More Extensive FEA/CFD

Automated ANSYS couplings speed up creation of 2D/3D FEA and CFD models to make more detailed calculations

- 3D electromagnetic effects
- CFD to check/calibrate flow calculations
- AC loss in winding

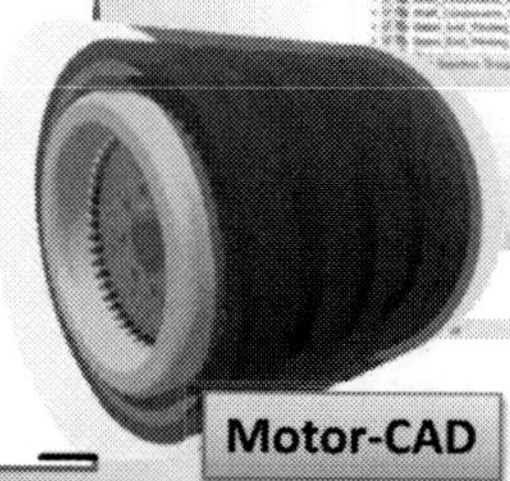


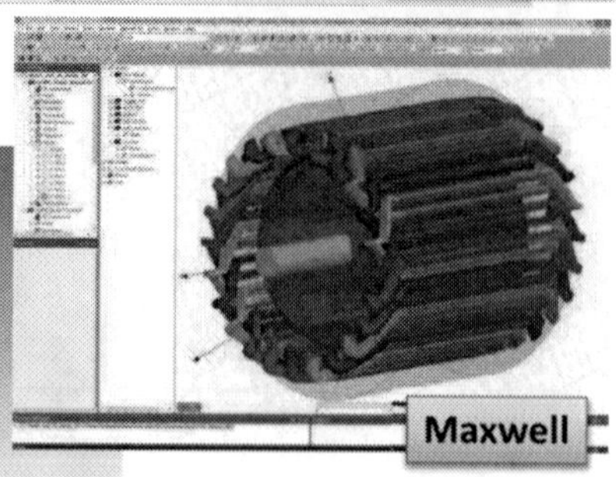


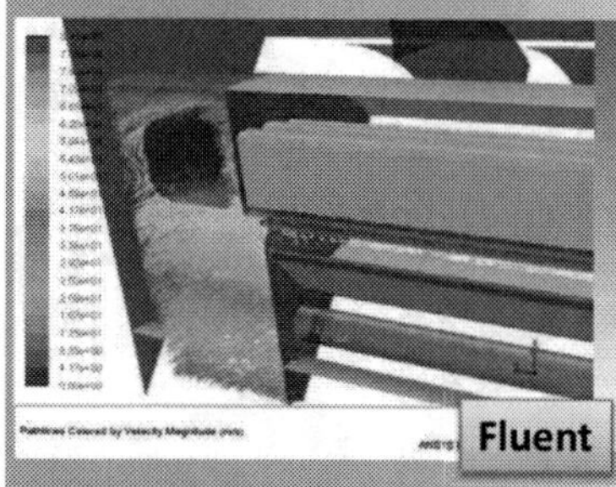

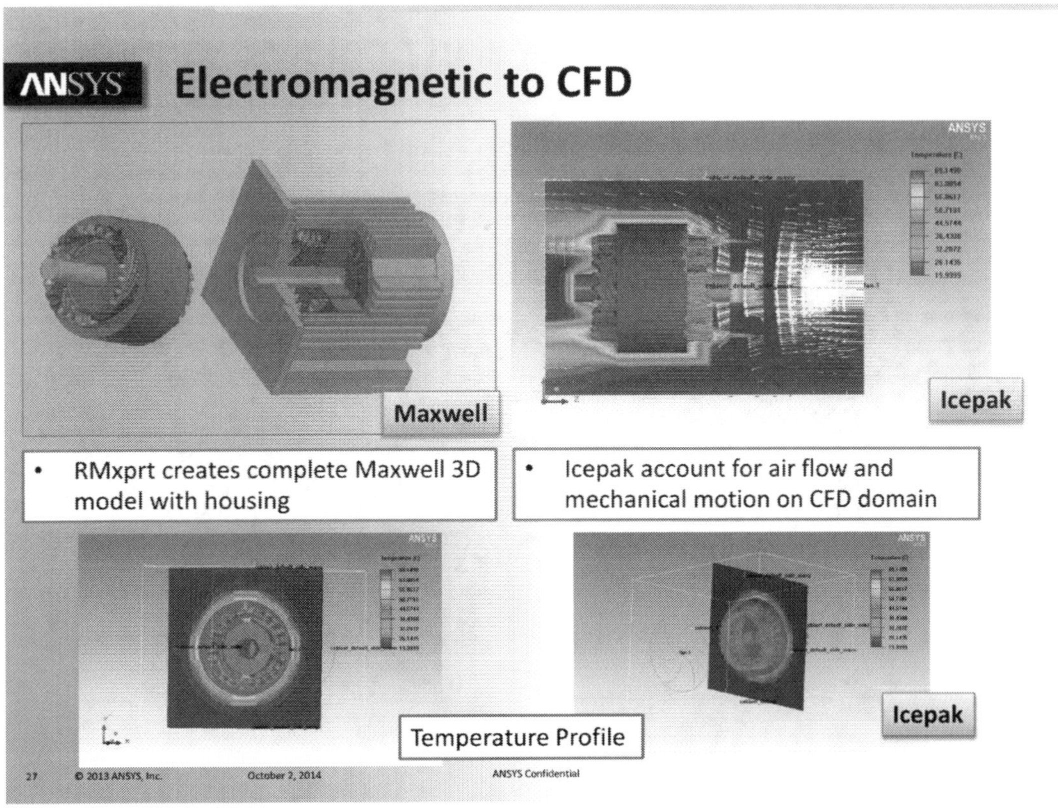
ANSYS
Electromagnetic to CFD
Maxwell
Icepak
RMxprt creates complete Maxwell 3D model with housing
Icepak account for air flow and mechanical motion on CFD domain
Temperature Profile
Icepak
27
© 2013 ANSYS, Inc.
October 2, 2014
ANSYS Confidential

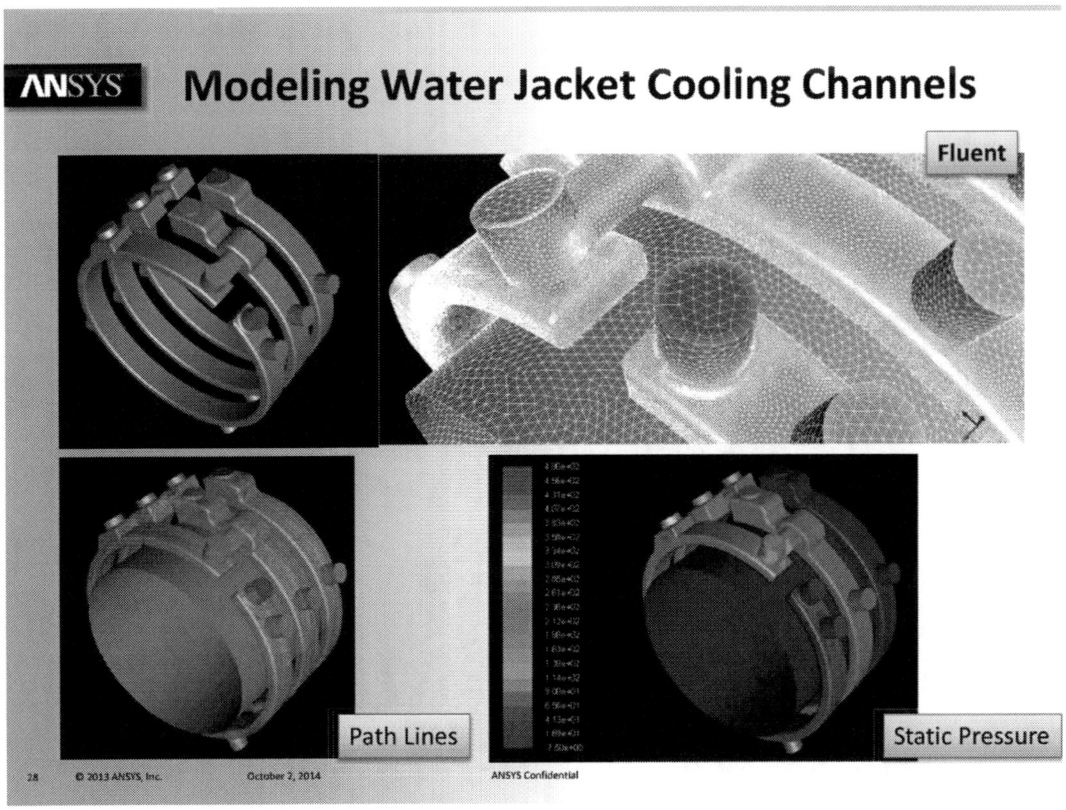
ANSYS
Modeling Water Jacket Cooling Channels
Fluent
Path Lines
Static Pressure
28
© 2013 ANSYS, Inc.
October 2, 2014
ANSYS Confidential

## Modeling Water Jacket Cooling Channels

Distribution of coolant using different channel configurations

Fluent

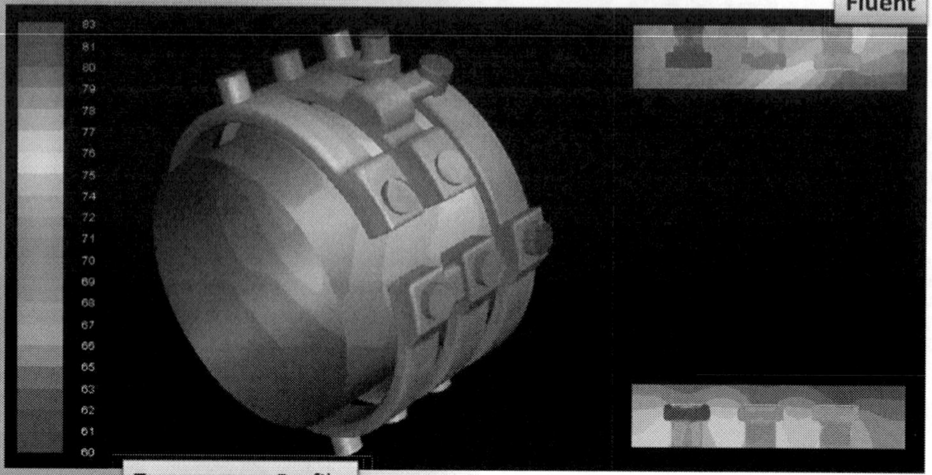

Temperature Profile

29 © 2013 ANSYS, Inc. October 2, 2014 ANSYS Confidential

## Summary

Comprehensive Electromagnetic and Thermal coupled design concepts are considered at different level of design adoption

- Sizing design - Rapid Virtual Prototyping where thousands of "what if's" can be performed in seconds;
- Hybrid design - Fast investigation of various cooling systems using thermal network coupled with CFD domain
- Physics-based design - Electromagnetic-CFD coupled solution for detailed analysis

30 © 2013 ANSYS, Inc. October 2, 2014 ANSYS Confidential

2014 ANSYS Electronics Simulation Expo

# The Influence of Outbreak Magnetic Field by a Superconducting Magnet for the surrounding devices and environment

High Energy Accelerator Research Organization(KEK) / Tesuya Yokoo, Ph.D
Tohoku University / Kenji Ohoyama, Dr.Sci
ANSYS Japan / Takahiro Koga

---



## Agenda

1. Background
2. Motivation
3. Electromagnetics Analysis
4. Vector Hysteresis Modeling
5. Simulation Model
6. Simulation Results
7. Conclusion

2014 ANSYS Electronics Simulation Expo
October 9-10 . Tokyo, Japan

## Background

**Tohoku University** and **KEK** joint project

3

2014 ANSYS Electronics Simulation Expo
October 9-10 . Tokyo, Japan

## Background

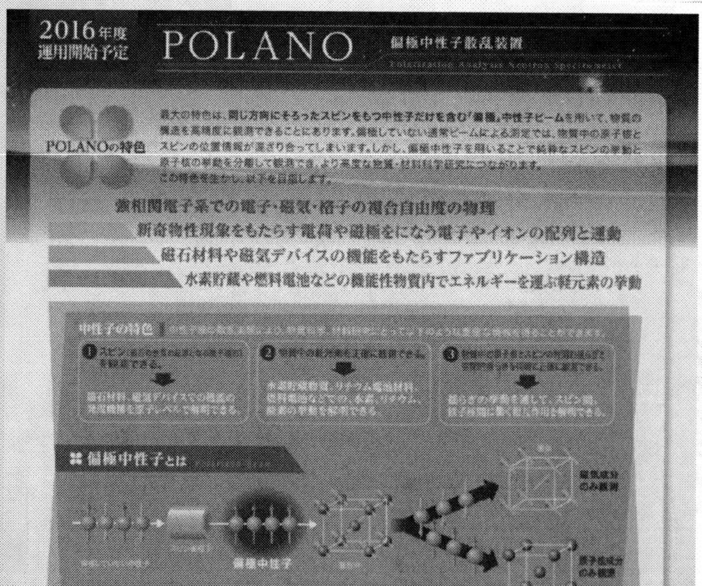


4

# Background

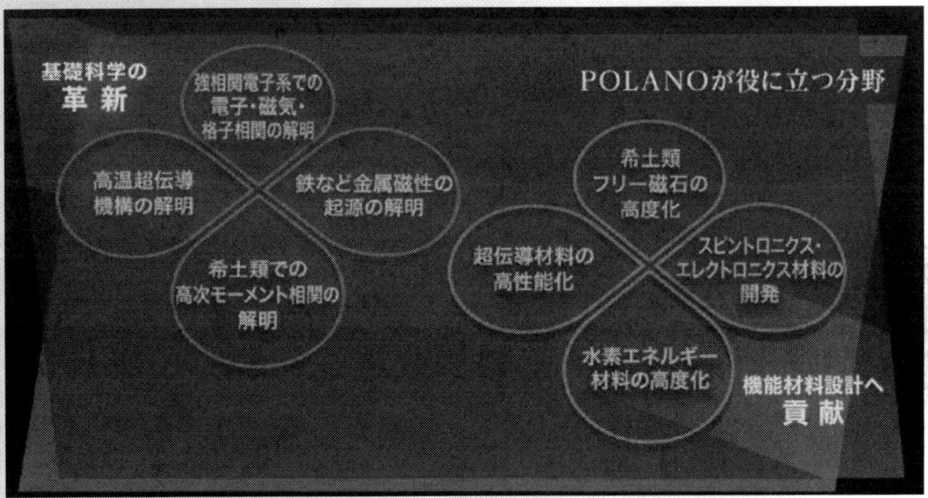




2014 ANSYS Electronics Simulation Expo
October 9-10 . Tokyo, Japan

# Background

2014 ANSYS Electronics Simulation Expo
October 9-10 . Tokyo, Japan

## Motivation

- For the under constructing large-scale experimental devices and laboratory, the outbreak magnetic field by the Superconducting Magnet (7 tesla) may adversely affect the environment/system.
- Perform visualization by electromagnetic simulation for remanence magnetization.
- Precision Goal: 10 tesla ~ 10uT order (100,000~0.1 gauss)

* Earth magnetism: 46uT (Japan)

7

2014 ANSYS Electronics Simulation Expo
October 9-10 . Tokyo, Japan

## Remanence Magnetization

### Hysteresis Loop

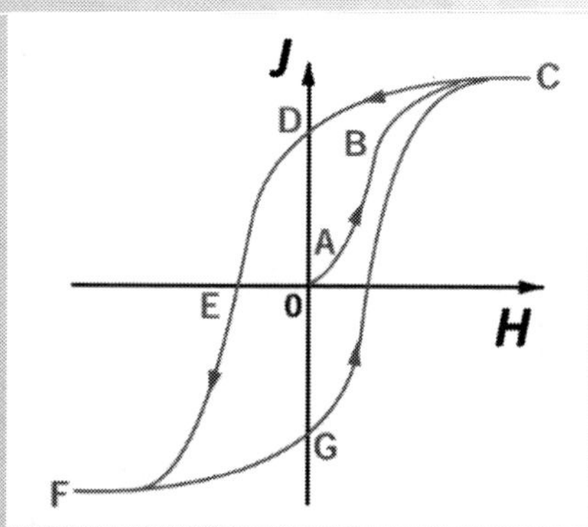


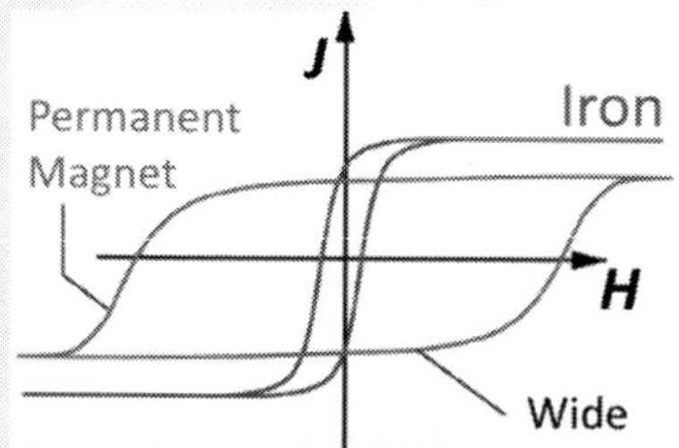


Soft/Hard Magnetic Material

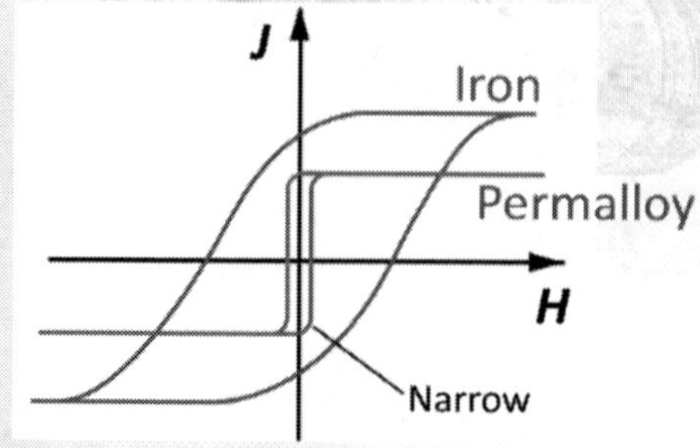


Soft Magnetic Material

8

## Electromagnetics Analysis

### ANSYS Maxwell

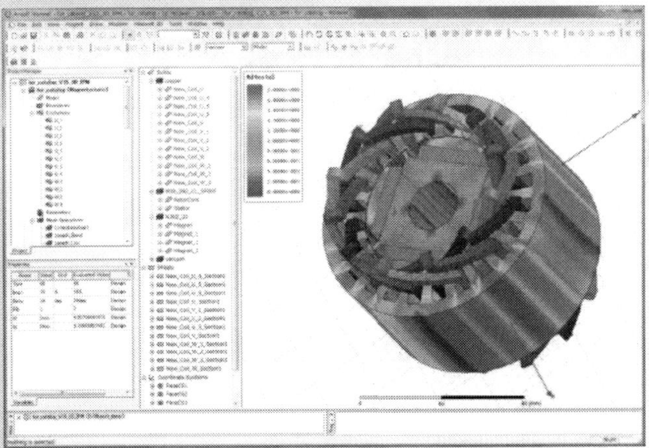

**Applications**
- ✓ Motor
- ✓ Actuator
- ✓ Coil, Inductor
- ✓ Transformer
- ✓ RFID
- ✓ Induction Heater
- ✓ Permanent Magnet
- ✓ Sensor
- ✓ Capacitor
- ✓ Electrode





## Vector Hysteresis Modeling

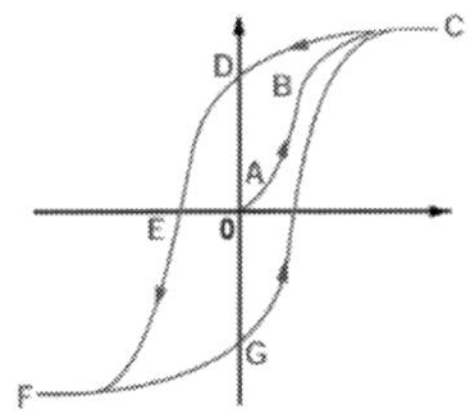


- ANSYS Maxwell has a capability to handle major/minor BH hysteresis loop simulation by transient solver
- Remanence magnetic field at soft magnetic material

Soft magnetic core

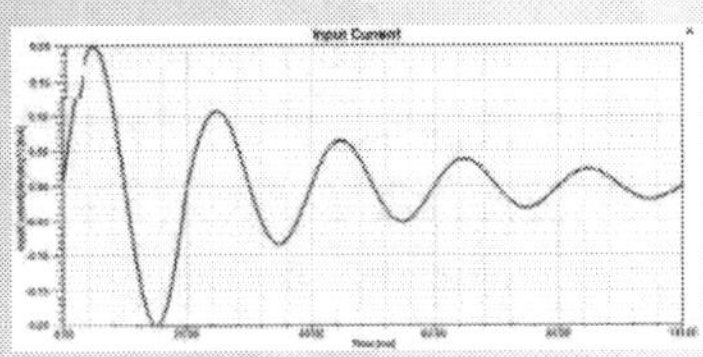


Input Current waveform

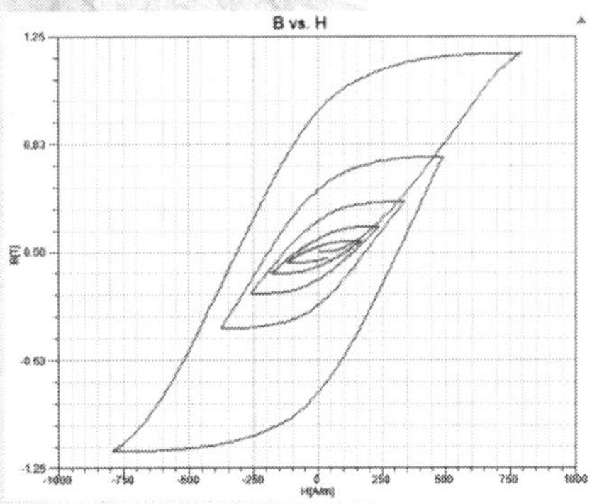


Tracking BH Operating point

## Simulation Model

- Magnetic Analysis
  - Superconducting Magnet
  - Magnetic Devices (Beam Line Devices)
  - Laboratory Model



2014 ANSYS Electronics Simulation Expo
October 9-10 . Tokyo, Japan

## Superconducting Magnet

- Target Point: 6.83 tesla

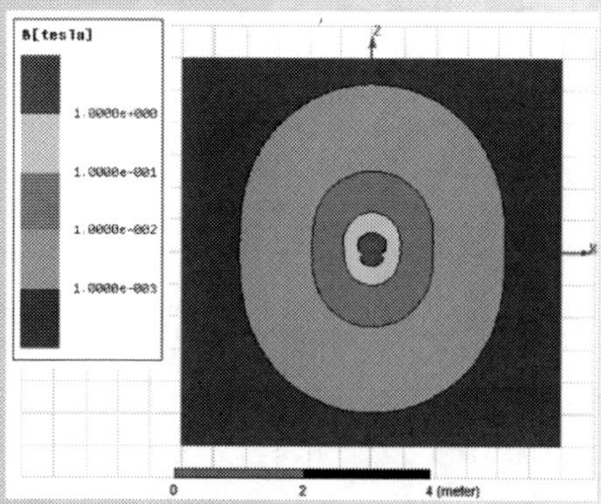

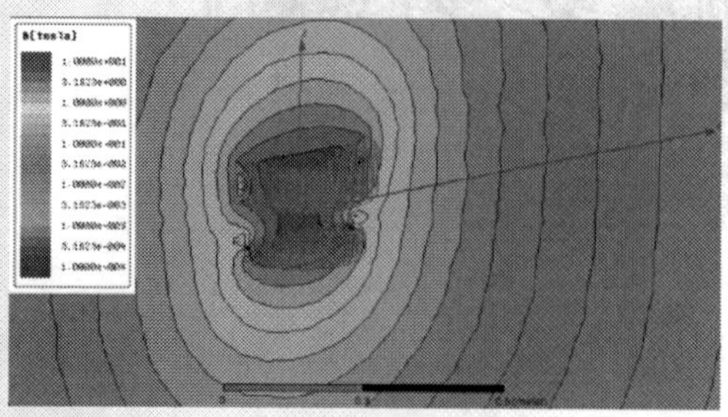

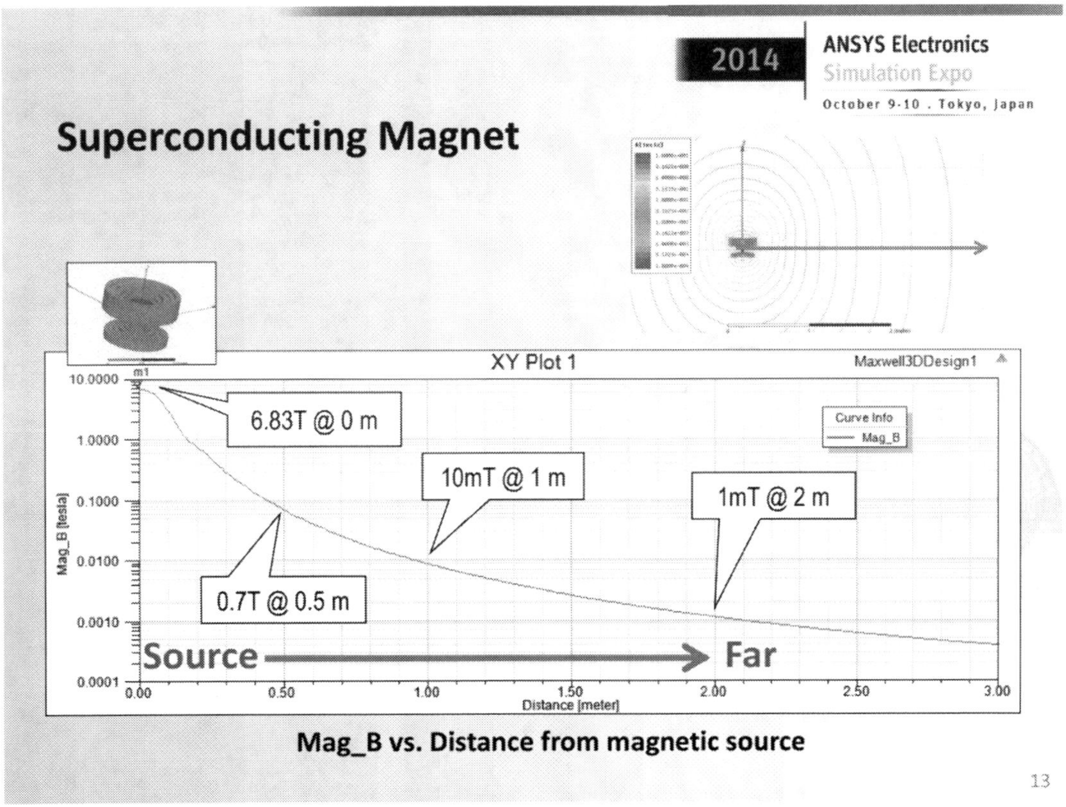
2014
ANSYS Electronics
Simulation Expo
October 9-10 . Tokyo, Japan
Superconducting Magnet
XY Plot 1
Maxwell3DDesign1
Curve Info
Mag_B
6.83T @ 0 m
10mT @ 1 m
1mT @ 2 m
0.7T @ 0.5 m
Source
Far
Mag_B [tesla]
Distance [meter]
Mag_B vs. Distance from magnetic source
13

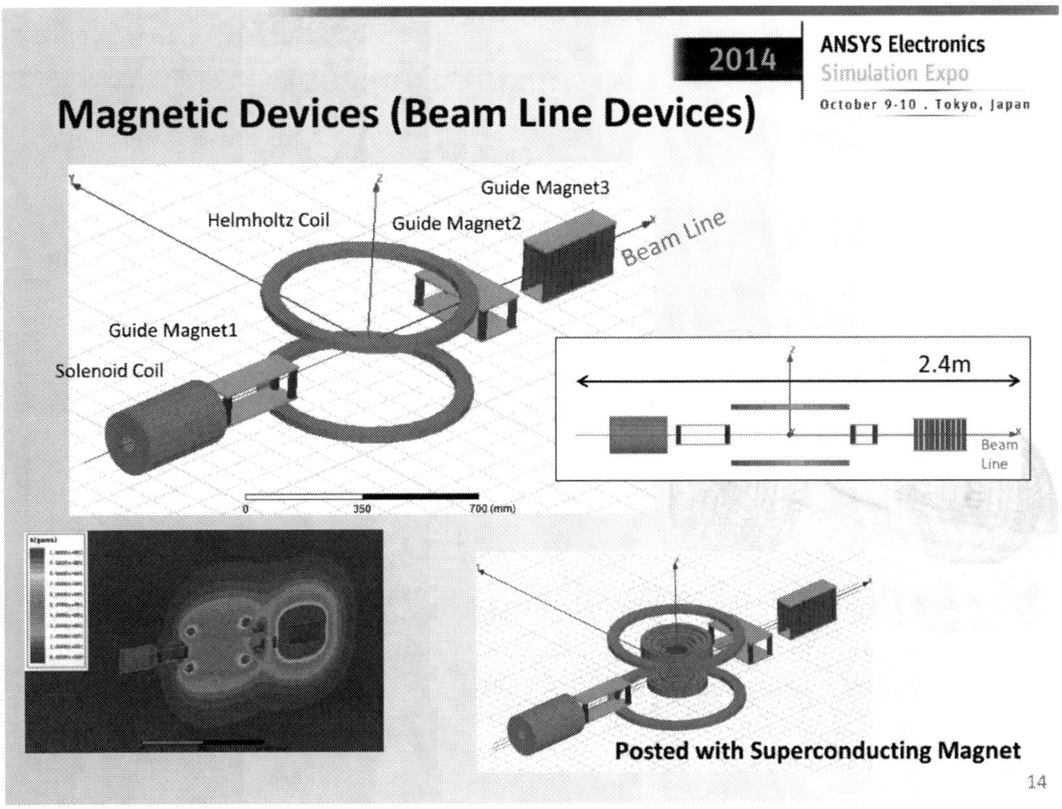
2014
ANSYS Electronics
Simulation Expo
October 9-10 . Tokyo, Japan
Magnetic Devices (Beam Line Devices)
Guide Magnet3
Helmholtz Coil
Guide Magnet2
Beam Line
Guide Magnet1
Solenoid Coil
2.4m
Beam Line
0
350
700 (mm)
Posted with Superconducting Magnet
14

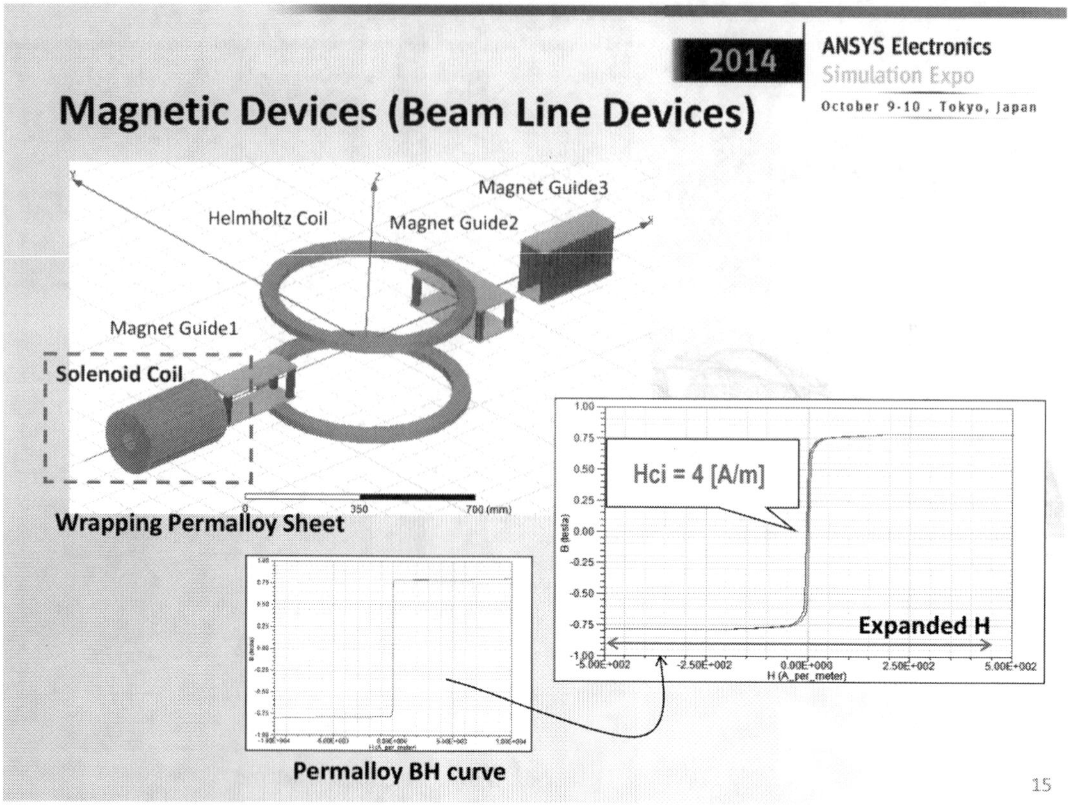
2014
ANSYS Electronics
Simulation Expo
October 9-10 . Tokyo, Japan
Magnetic Devices (Beam Line Devices)
Magnet Guide3
Helmholtz Coil
Magnet Guide2
Magnet Guide1
Solenoid Coil
0
350
700 (mm)
Wrapping Permalloy Sheet
Hci = 4 [A/m]
Expanded H
H (A_per_meter)
B (tesla)
Permalloy BH curve
15

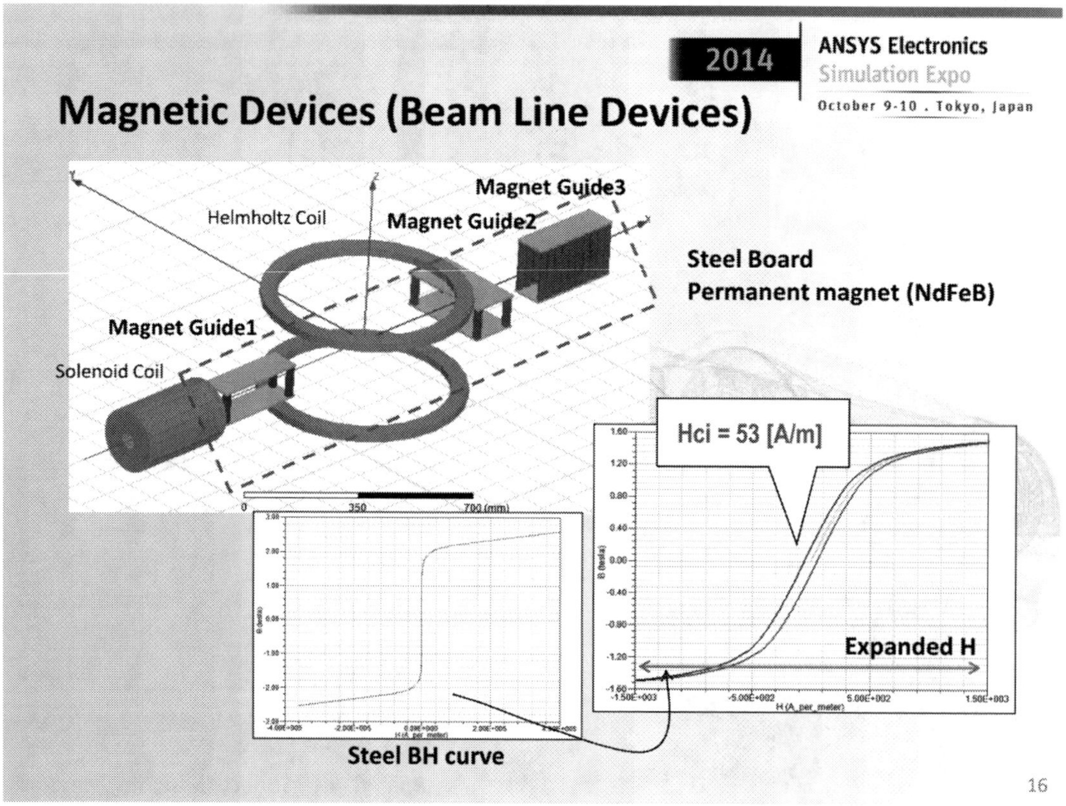
2014
ANSYS Electronics
Simulation Expo
October 9-10 . Tokyo, Japan
Magnetic Devices (Beam Line Devices)
Magnet Guide3
Helmholtz Coil
Magnet Guide2
Steel Board
Permanent magnet (NdFeB)
Magnet Guide1
Solenoid Coil
0
350
700 (mm)
Hci = 53 [A/m]
Expanded H
H (A_per_meter)
B (tesla)
Steel BH curve
16

2014 ANSYS Electronics Simulation Expo
October 9-10 . Tokyo, Japan

# Laboratory Model

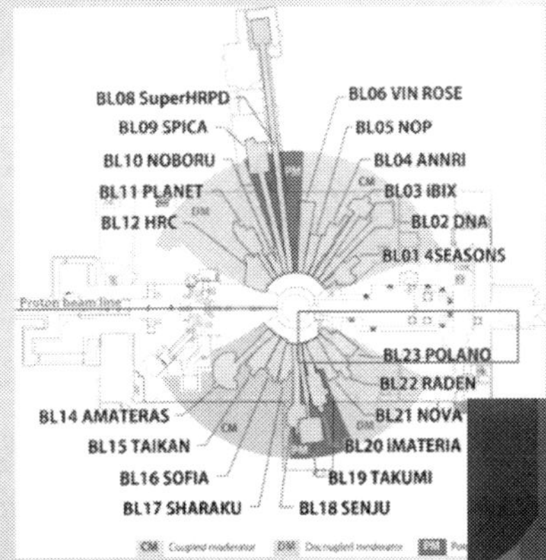


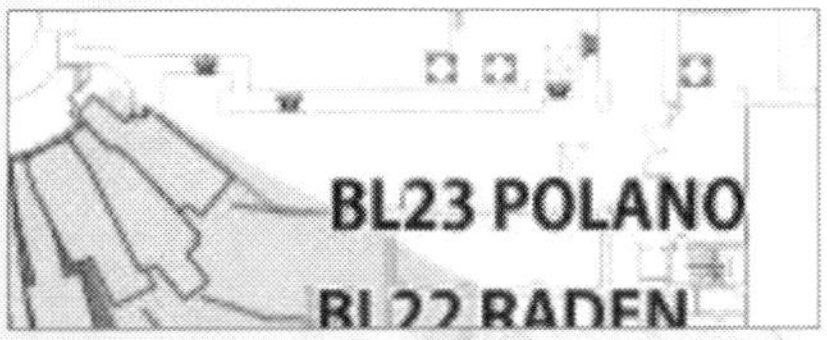


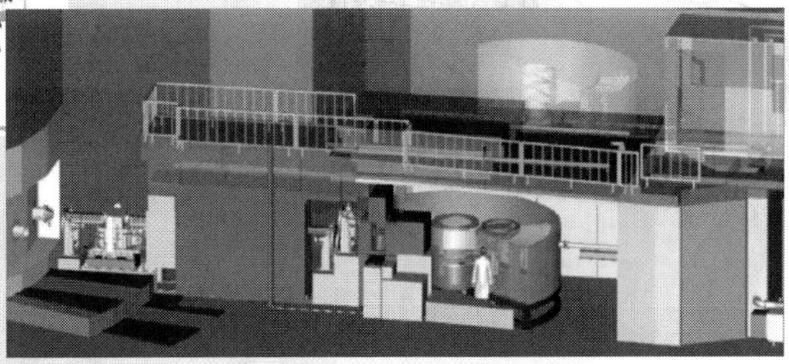

http://nc-imr.imr.tohoku.ac.jp/instruments.html
http://j-parc.jp/researcher/MatLife/ja/instrumentation/ns.html

17

2014 ANSYS Electronics Simulation Expo
October 9-10 . Tokyo, Japan

# Laboratory Model

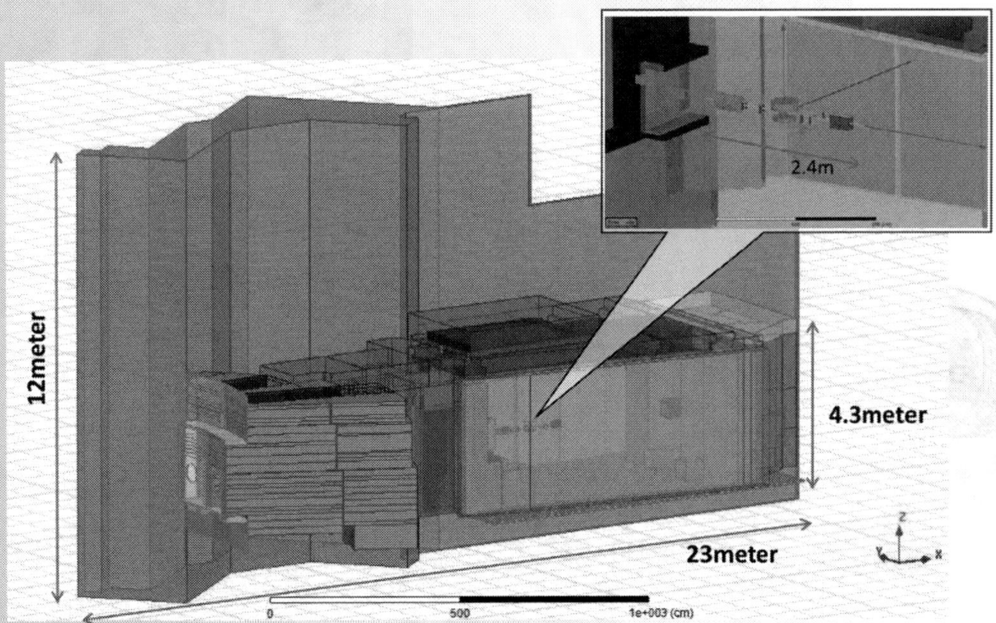


18

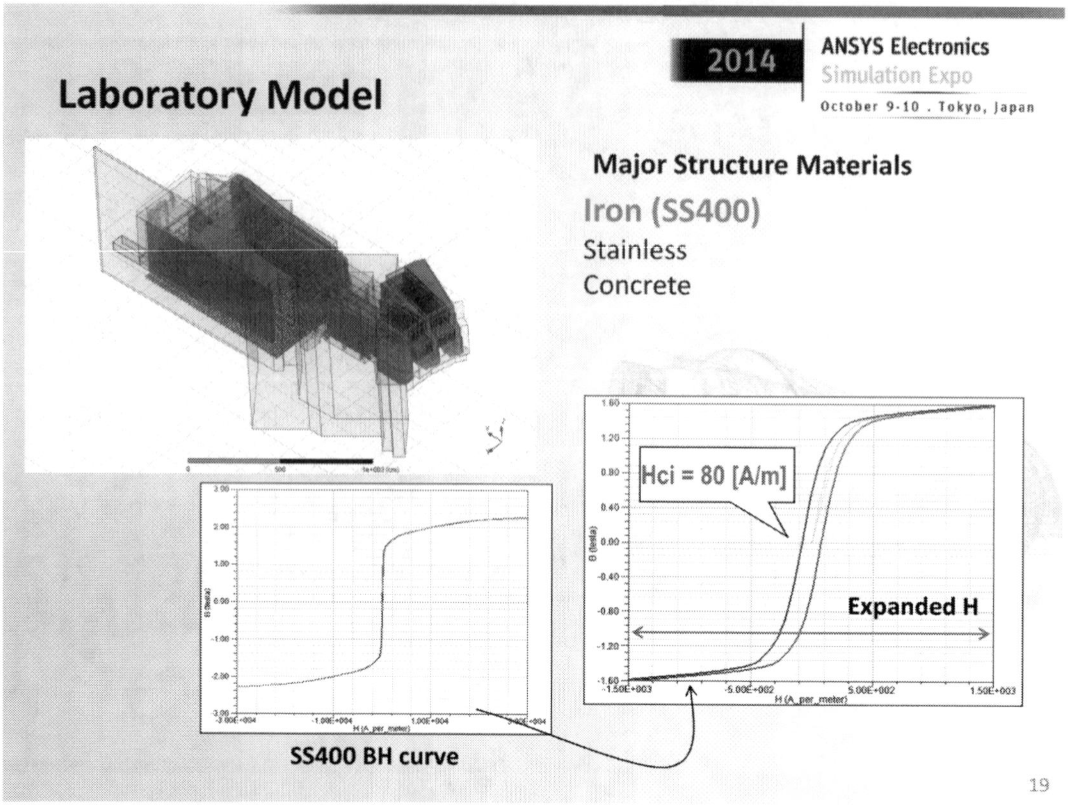
2014
ANSYS Electronics
Simulation Expo
October 9-10 . Tokyo, Japan
Laboratory Model
Major Structure Materials
Iron (SS400)
Stainless
Concrete
Hci = 80 [A/m]
Expanded H
SS400 BH curve
19

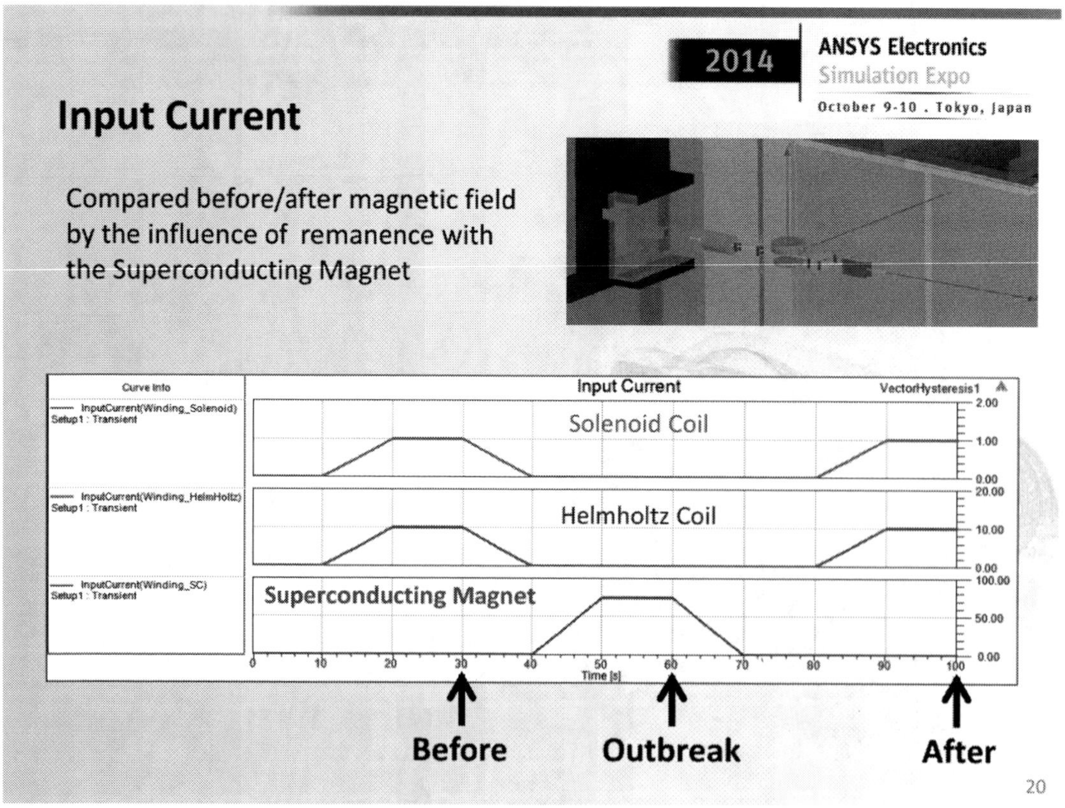
2014
ANSYS Electronics
Simulation Expo
October 9-10 . Tokyo, Japan
Input Current
Compared before/after magnetic field
by the influence of remanence with
the Superconducting Magnet
Curve Info
Input Current
Solenoid Coil
Helmholtz Coil
Superconducting Magnet
Time [s]
Before
Outbreak
After
20

2014 ANSYS Electronics Simulation Expo
October 9-10 . Tokyo, Japan

## Simulation Results

- B Vector plot on Beam Line
- Magnetic Field around Beam Line devices
- Magnetic Field at Laboratory

21

2014 ANSYS Electronics Simulation Expo
October 9-10 . Tokyo, Japan

## B Vector Plot on Beam Line

- Before/After Bx, By, Bz plot
- Devices: No input current

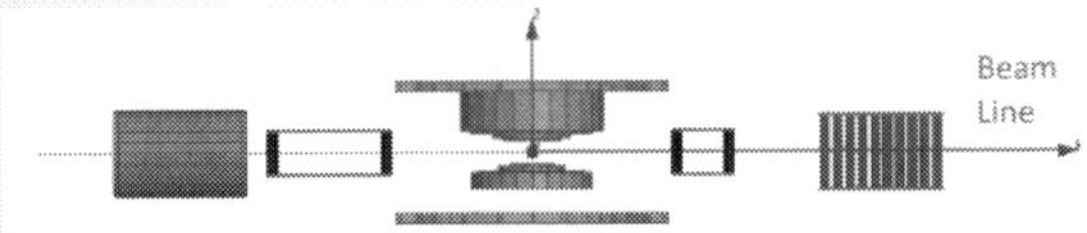


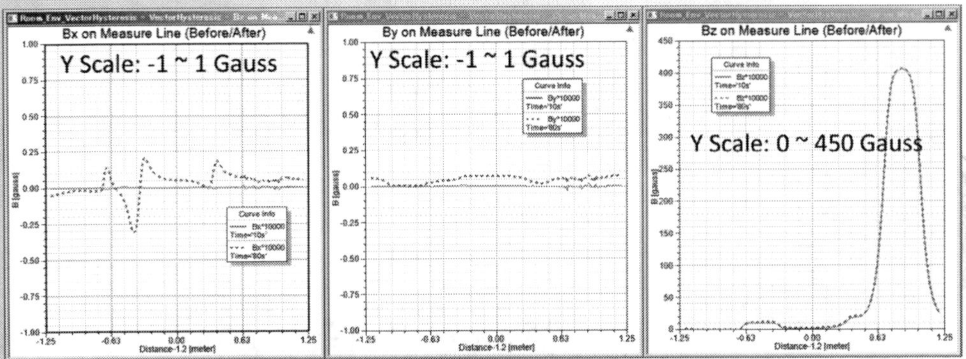


| Bx | By | Bz |
|---|---|---|
| Max. Delta: 0.31 Gauss | Max. Delta: 0.071 Gauss | Max. Delta: 2.68 Gauss |

( ━━ : Before / ▪▪▪▪▪ : After)

22

2014 ANSYS Electronics Simulation Expo
October 9-10 . Tokyo, Japan

## B Vector Plot on Beam Line

- Before/After Bx, By, Bz plot
- Devices: No input current

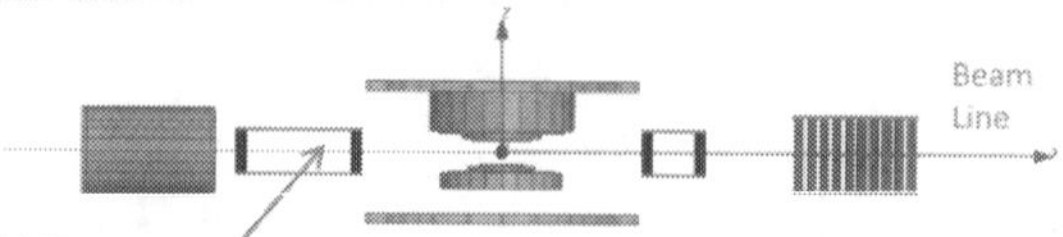


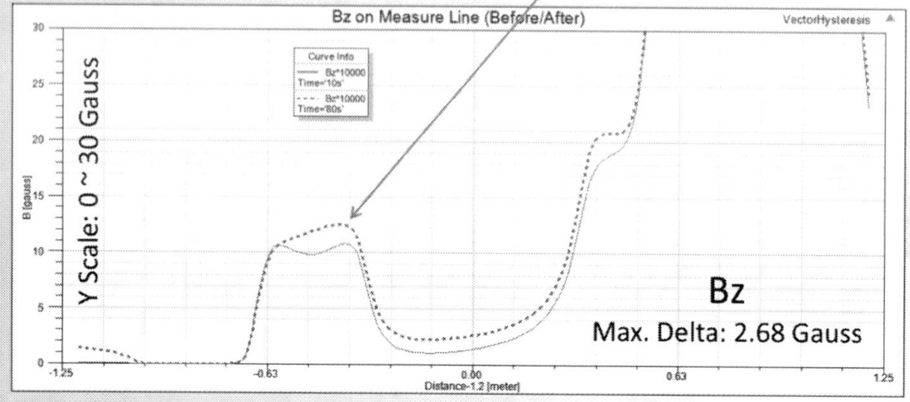


( ▬▬ : Before ／ ▪▪▪▪▪ : After)

23

2014 ANSYS Electronics Simulation Expo
October 9-10 . Tokyo, Japan

## Magnetic Field around Beam Line devices

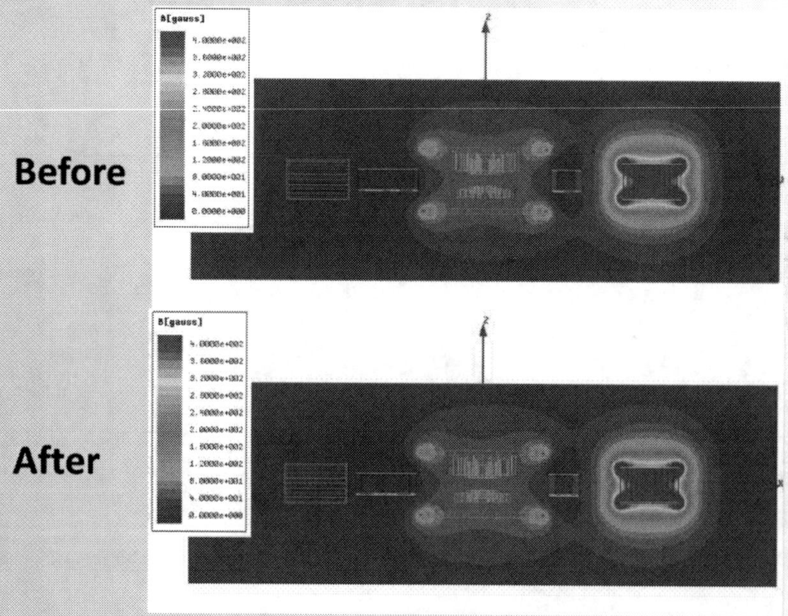


24

2014 ANSYS Electronics Simulation Expo
October 9-10 . Tokyo, Japan

## Magnetic Field around Beam Line devices

**Before**

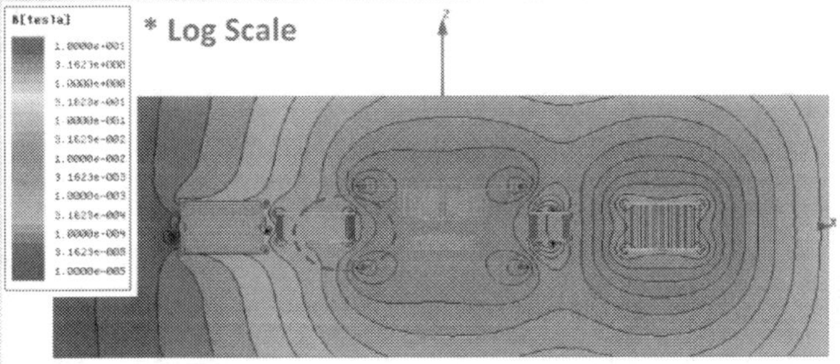


**After**

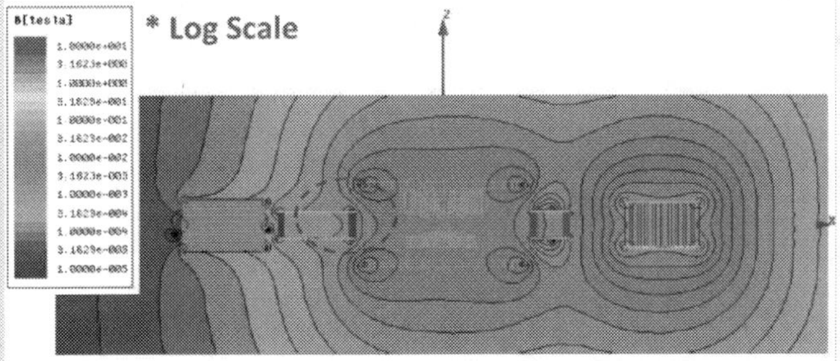


25

2014 ANSYS Electronics Simulation Expo
October 9-10 . Tokyo, Japan

## Magnetic Field around Beam Line devices

**Outbreak**

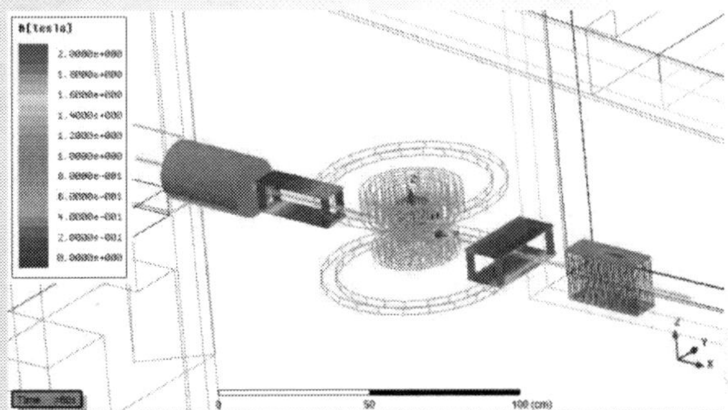

**After**

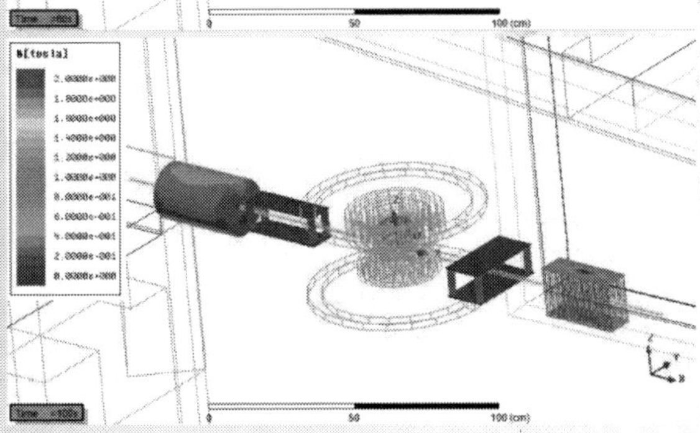

26

2014

ANSYS Electronics
Simulation Expo
October 9-10 . Tokyo, Japan

## Magnetic Field around Beam Line devices

### B Vector Field

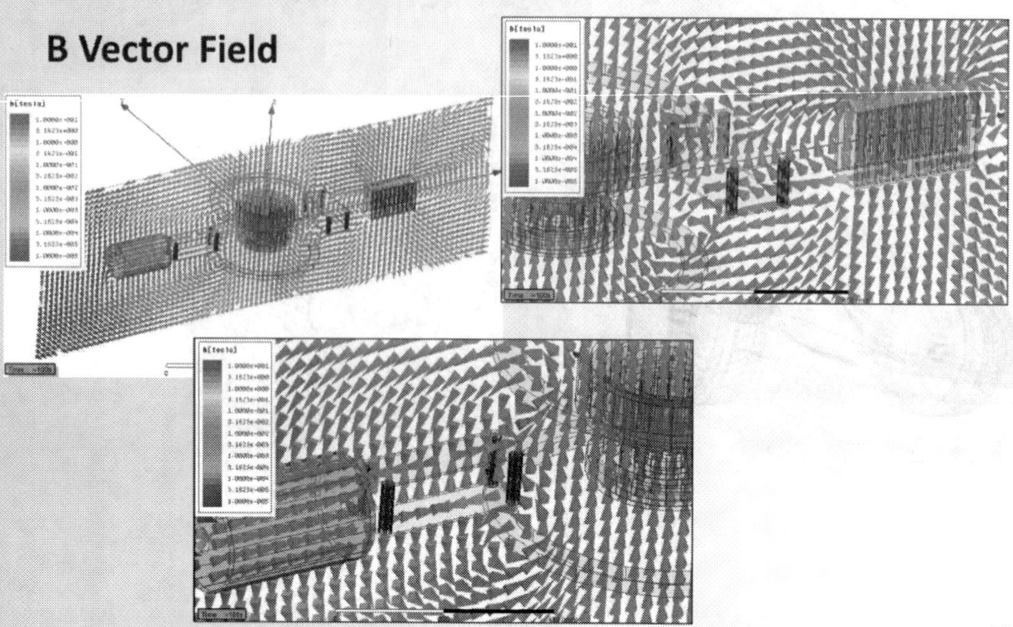

27

---

2014

ANSYS Electronics
Simulation Expo
October 9-10 . Tokyo, Japan

## Magnetic Field Laboratory

### Structure under the floor

Outbreak

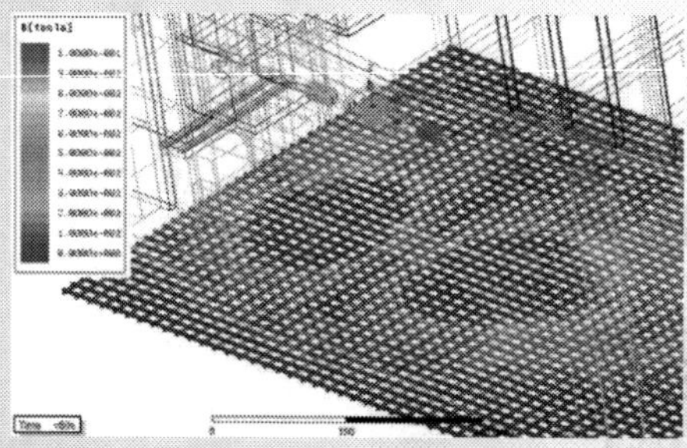

After

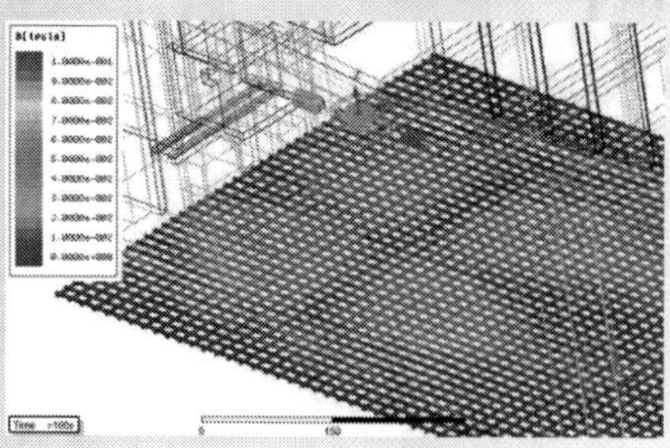

28

2014 ANSYS Electronics Simulation Expo
October 9-10 . Tokyo, Japan

## Magnetic Field Laboratory

### Side Wall

Outbreak

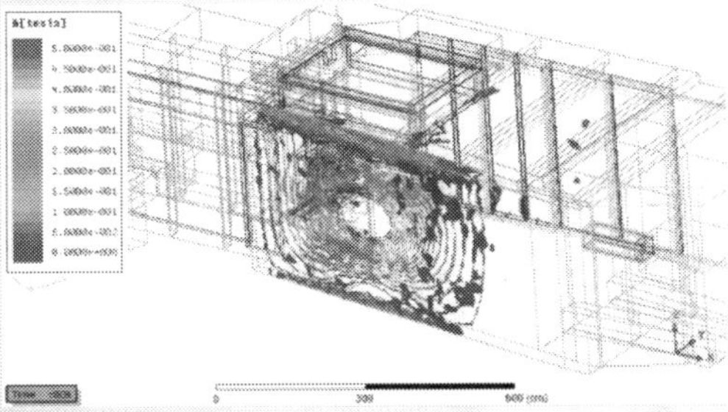

After

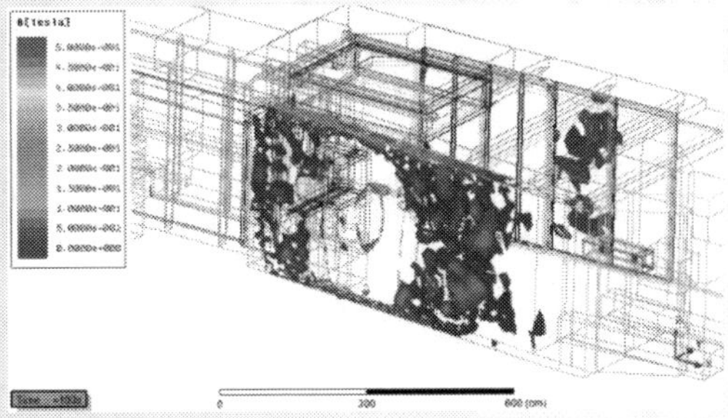

29

## Magnetic Field Laboratory

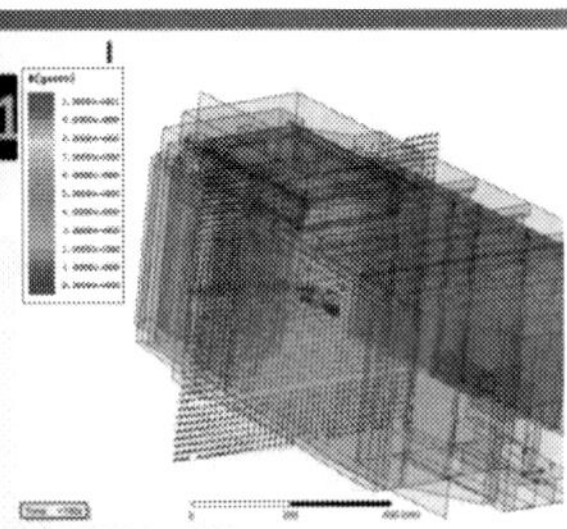

### YZ Cross Planes

Scale ~100 Gauss

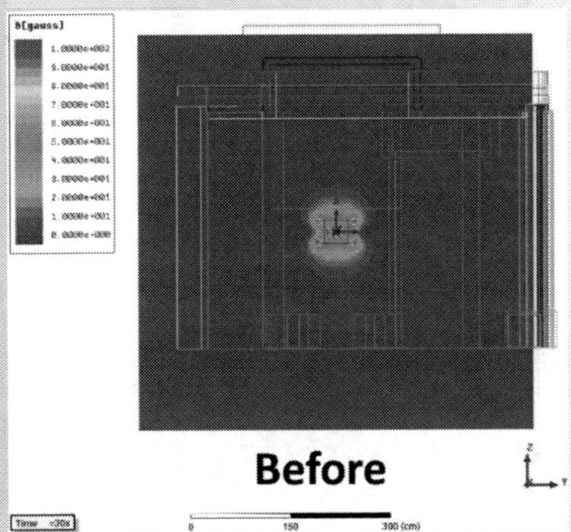


Scale ~100 Gauss

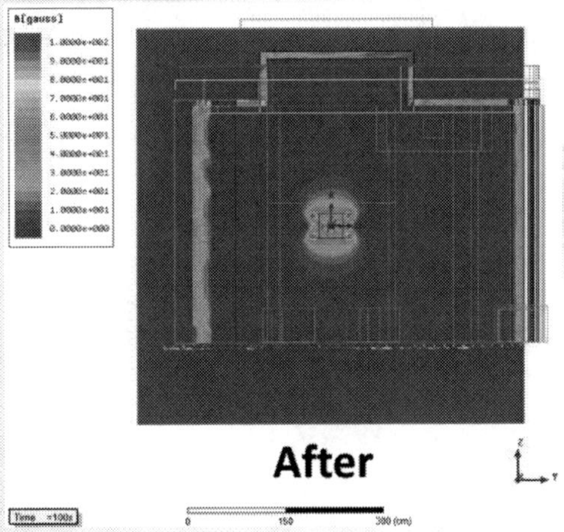


30

2014 ANSYS Electronics Simulation Expo
October 9-10 . Tokyo, Japan

# Magnetic Field Laboratory

## XY, XZ Cross Planes (After)

Scale ~100 Gauss

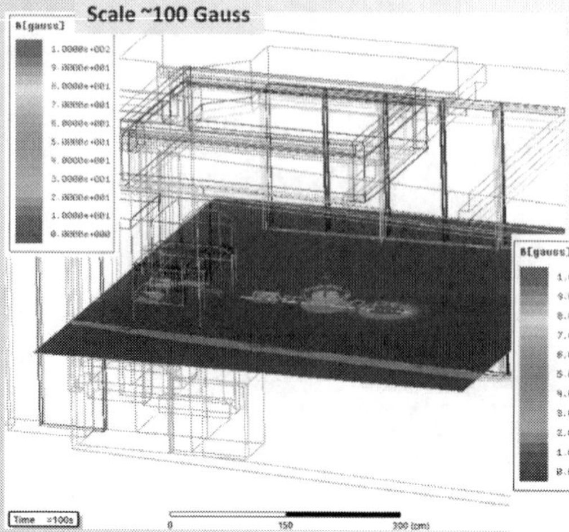


After

Scale ~10 Gauss

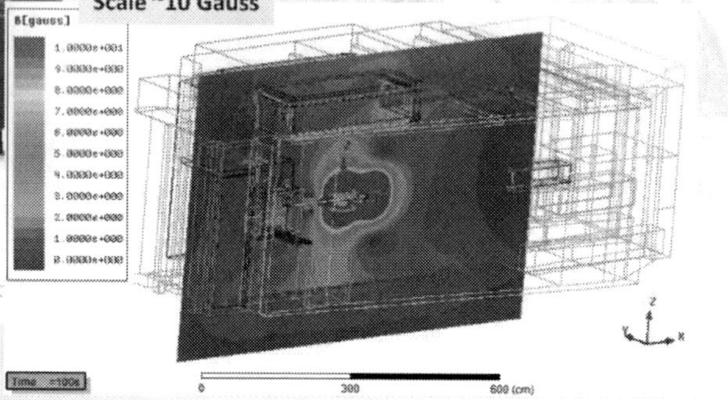


After

31

# Magnetic Field Laboratory

## Beam Line – Emitting Port

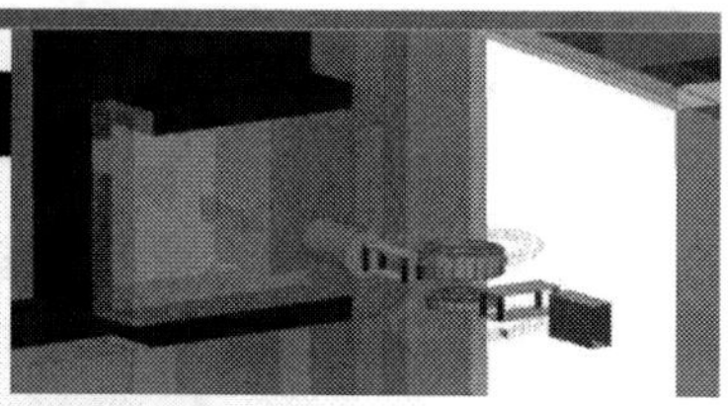

Scale ~10 Gauss

**Before**

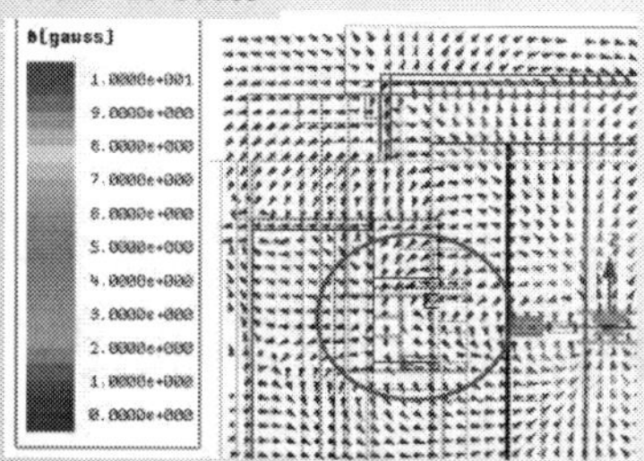


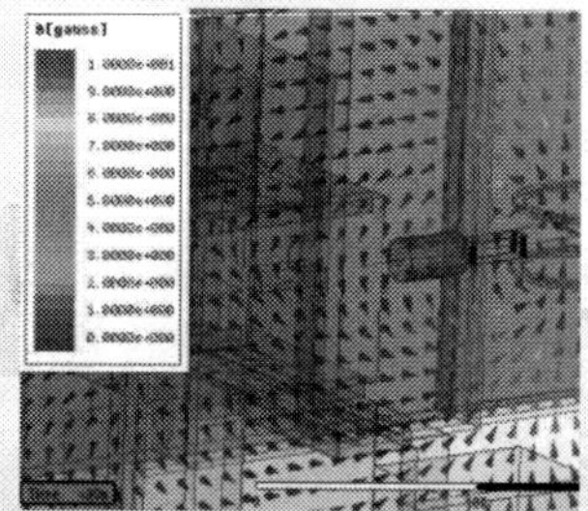

**After**

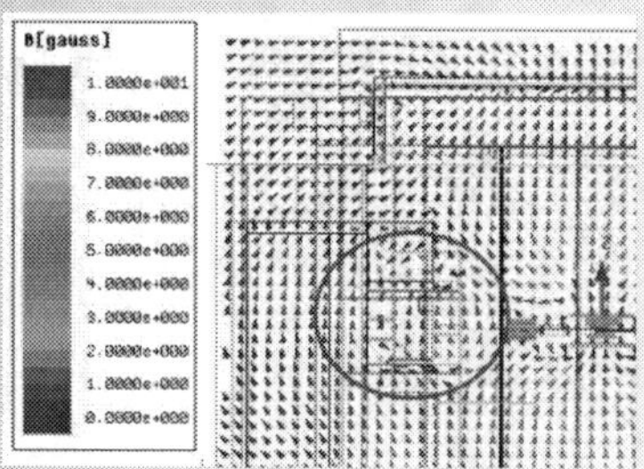


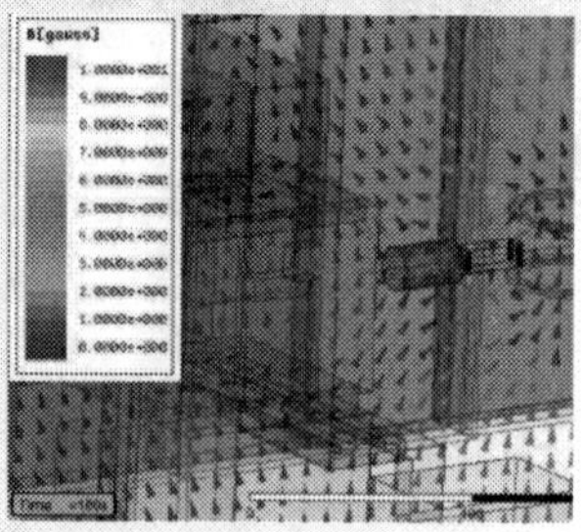

32

2014 ANSYS Electronics Simulation Expo
October 9-10 . Tokyo, Japan

## Beam Line – Emitting Port

Iron structure

**Stainless structure**

33

2014 ANSYS Electronics Simulation Expo
October 9-10 . Tokyo, Japan

## Conclusion

- We succeed to analyze and visualize the influence of outbreak magnetic field, remanence magnetized soft magnetic material by a Superconducting Magnet with ANSYS Maxwell's Vector Hysteresis Modeling.
- Magnetic devices on beam line would be magnetized, but not affect.
- Beam emitting port might influence the beam control by remanence -> Need to change design to non-magnetic material e.g. stainless
- Contribute to investigate the prior validation for the large-scale experimental device under construction.

34

# Comprehensive Solution for Nissan LEAF Motor Performance Prediction

**Dr David Staton**

Motor Design Ltd., Ellesmere, Shropshire, UK
www.motor-design.com

motordesign LIMITED

## Motor Design Ltd (MDL)

- MDL formed in 1998 to:
  - Develop electric motor design software
  - Provide motor design consulting services and motor design training

motordesign LIMITED

2

# Nissan LEAF Motor Simulation

- ANSYS – MDL simulation on LEAF motor
  - proof the synergy of combined solution
  - validates the accuracy prediction of performance criteria on efficiency/loss maps and drive cycle transient thermal analysis
  - quickly evaluate different cooling profiles
  - study to adjust the distribution of coolant using different channel configurations

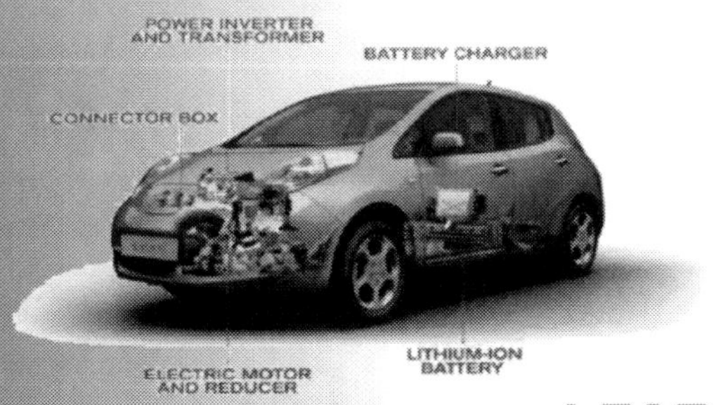


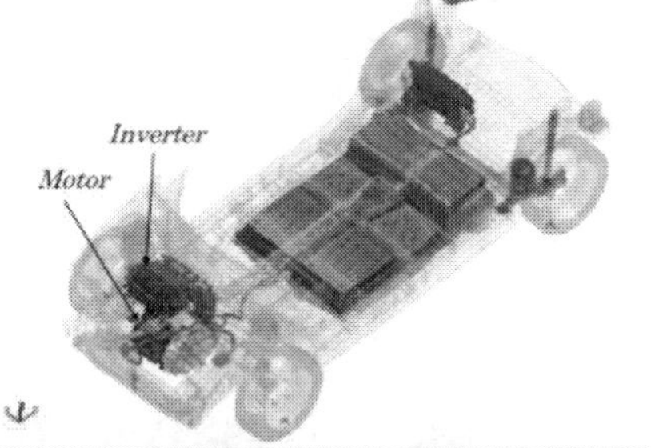




# Electromagnetic Models for LEAF Motor

- Various complexities of electromagnetic model have been created in Maxwell & Motor-CAD EMag

# Maxwell Models for LEAF Motor

- Maxwell 2D & 3D electromagnetic models have been constructed to calculate open circuit and on load to calculate performance criteria such as back emf, torque-speed characteristic, torque ripple, cogging, losses, etc.
  - The 2D model was constructed by two ERASMUS students from Casino University in Italy that were on placement at MDL
  - Their work formed a tutorial from traction motor modelling using Maxwell

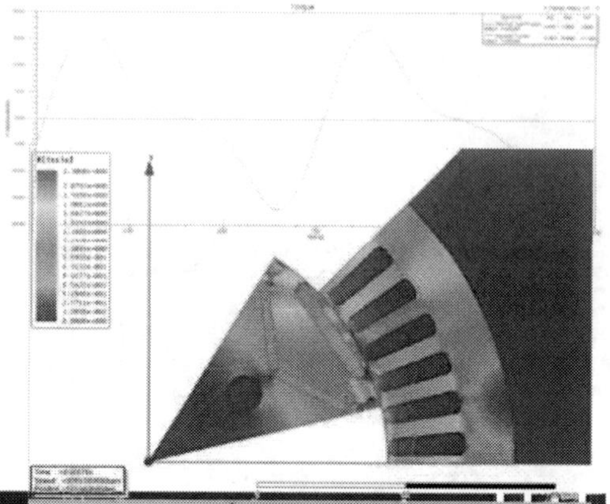

motordesign LIMITED

5

# Maxwell Models for LEAF Motor

**Axial split PM topology to reduce eddy effects**

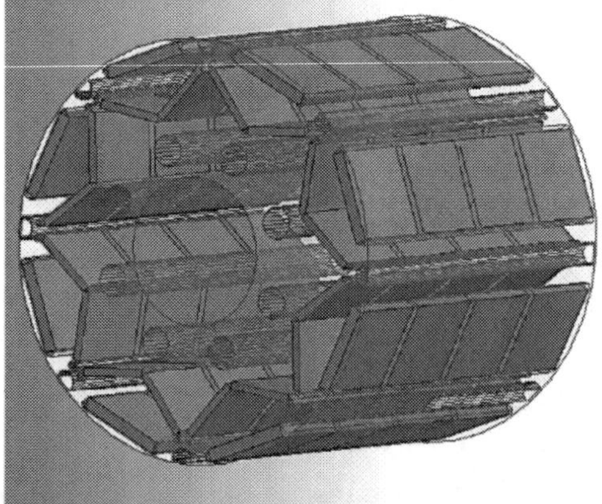

**Field distribution at rated load condition**

motordesign LIMITED

6

# Motor-CAD EMag Model for LEAF Motor

- 2D FEA using parametric geometric model used to give a very fast estimate of the electromagnetic performance
  - Not include exact geometric details of flux barriers/magnet slots
  - More detailed final analysis possible using DFX import of actual geometry
  - The simplified geometry gives relatively accurate estimates of average torque and losses but less accurate prediction of torque ripple

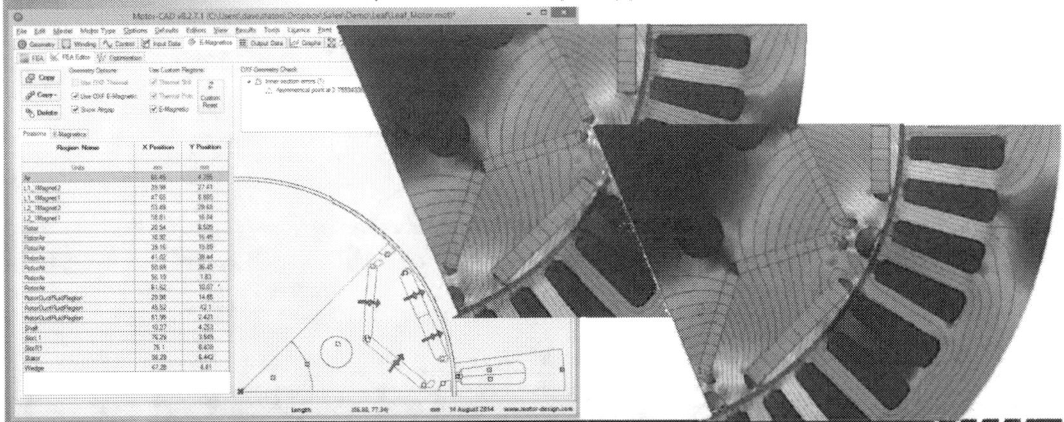



# Electromagnetic Model Validation

- Some assumptions have been made due to lack of exact data on the motor construction and materials used
- We can use Motor-LAB with links to either Maxwell or Motor-CAD EMag to calculate the losses over the full torque/speed
- Motor-LAB also automatically calculates the optimum phase advance for either maximum torque/amp or maximum efficiency control
  - In this case also apply a 80kW max. power limit as in the test procedure
- Optimum use of a minimum number of FEA calculations to model saturation and losses gives a very fast calculation
  - Map calculated within a few minutes

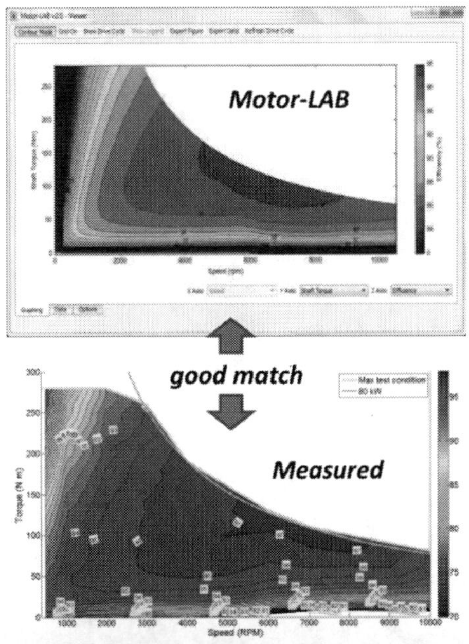

## Motor-CAD Therm Model for LEAF Motor

- ➢ Thermal component such as the water jacket also defined
- ➢ Validated as match measured thermal transient at 50, 60, 70 & 80kW

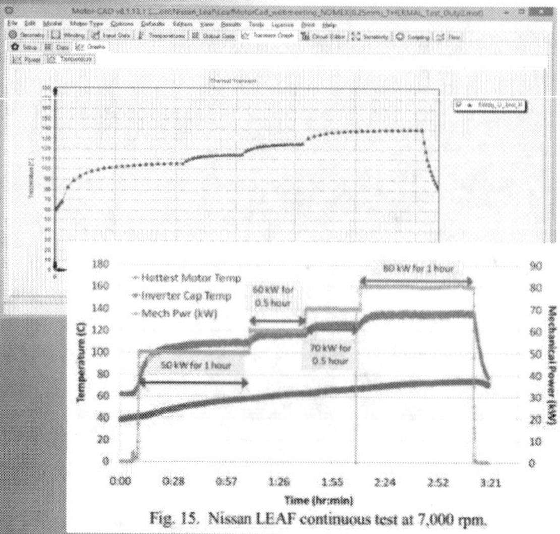


Fig. 15. Nissan LEAF continuous test at 7,000 rpm.

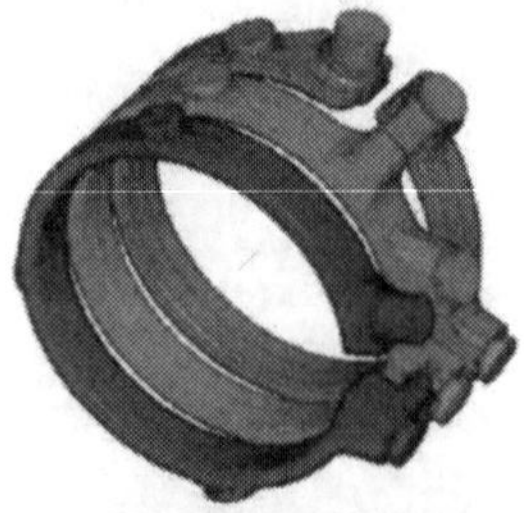

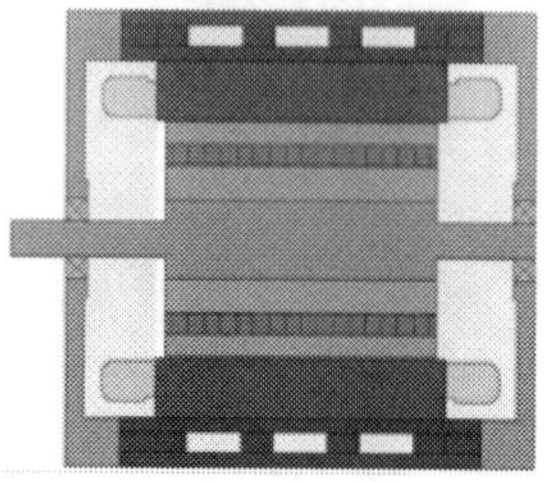

motor**design** LIMITED

9

## 50kW, 60kW, 70kW and 80kW Duty Cycle

- ➢ 50, 60, 70 & 80kW thermal transient set up in Motor-LAB
- ➢ Loss data then transferred to Motor-CAD Therm model

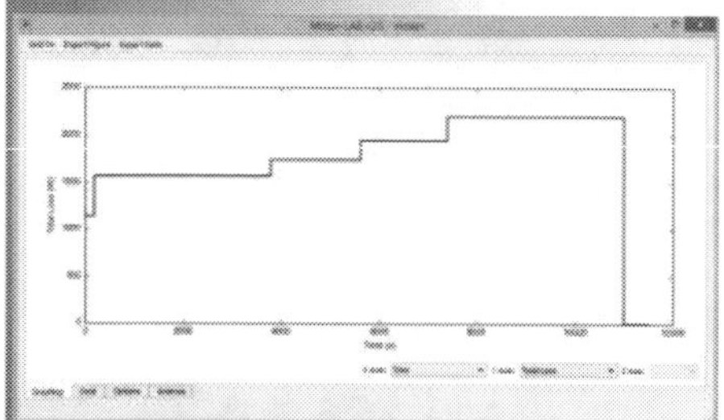

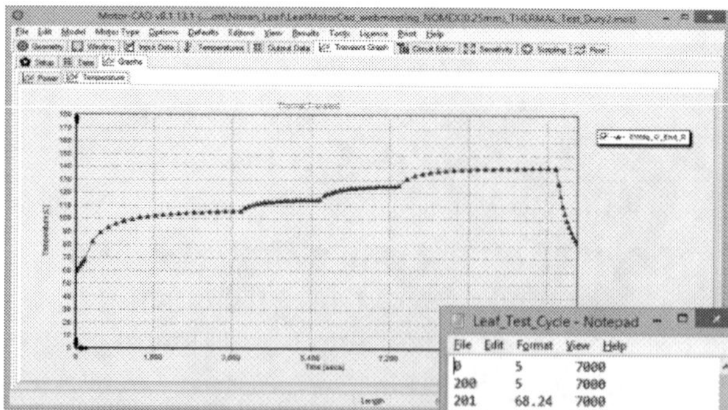

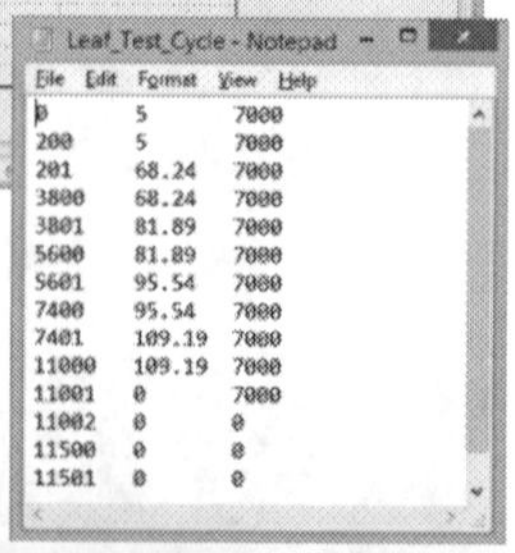


| Period | Elapsed Time | Time | Points | Stator Copper | Stator Back Iron | Stator Tooth | Magnet | Rotor Back Iron | Embedded Magnet Pole | Friction Front Bearing | Friction Rear Bearing | Windage | Windage Exterior Fan | Speed [Start] | Speed [End] | Fault |
|---|---|---|---|---|---|---|---|---|---|---|---|---|---|---|---|---|
| Units | secs | secs |  | Watts | Watts | Watts | Watts | Watts | Watts | Watts | Watts | Watts | Watts | rpm | rpm |  |
| 1 | 200 | 200 | 5 | 41.52 | 583.5 | 366 | 0.4861 | 46.52 | 0 | 0 | 0 | 0 | 0 | 7000 | 7000 | ☐ |
| 2 | 201 | 1 | 1 | 41.52 | 583.5 | 366 | 0.4861 | 46.52 | 0 | 0 | 0 | 0 | 0 | 7000 | 7000 | ☐ |
| 3 | 3800 | 3599 | 20 | 404.5 | 625.4 | 385.7 | 1.924 | 40.16 | 0 | 0 | 0 | 0 | 0 | 7000 | 7000 | ☐ |
| 4 | 3801 | 1 | 1 | 404.5 | 625.4 | 385.7 | 1.924 | 40.16 | 0 | 0 | 0 | 0 | 0 | 7000 | 7000 | ☐ |
| 5 | 5600 | 1799 | 20 | 554.1 | 644.7 | 395.9 | 2.476 | 38.5 | 0 | 0 | 0 | 0 | 0 | 7000 | 7000 | ☐ |

motor**design** LIMITED

10

# Motor-CAD Therm Model for LEAF Motor

- cuboidal elements used to accurately model the winding conductive heat transfer

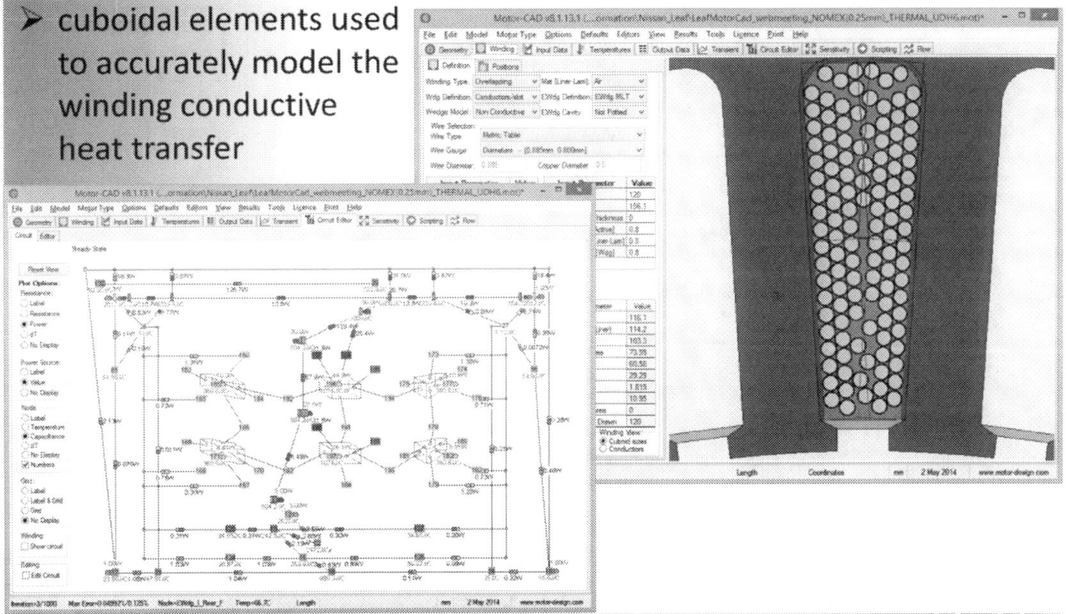

motordesign LIMITED 11

# Calibration of stranded winding models

- By estimating the placement of the conductors in a slot, we can automatically create a thermal FEA model and check if the conduction heat transfer in the slot matches the 3D cuboidal resistance model
- Calibration of effective thermal conductivity of copper, enamel, impregnation and air (no calibration - excellent results below)
- To make the calibration, we remove end effects in the lumped circuit model and then apply them again after calibration.

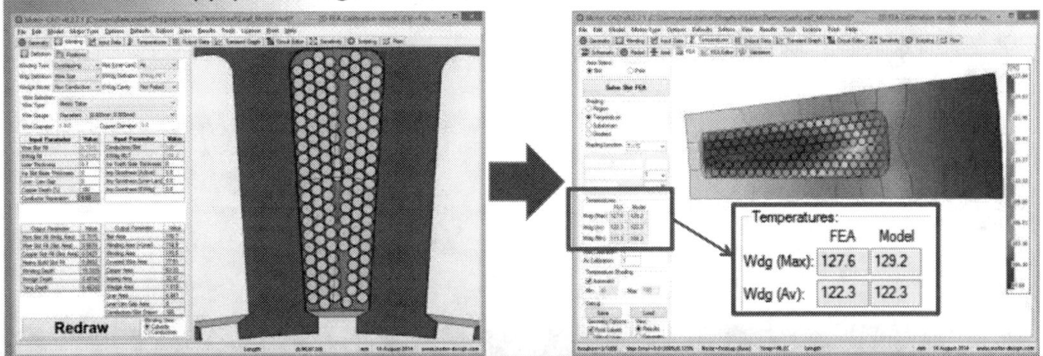


motordesign LIMITED 12

# LEAF Efficiency Map & US06 Drive Cycle

- Using Motor-LAB all the points on US06 drive cycle within efficiency map can be calculated in a few minutes
  - Accounting for saturation, optimum field weakening angle and accurate estimate off losses

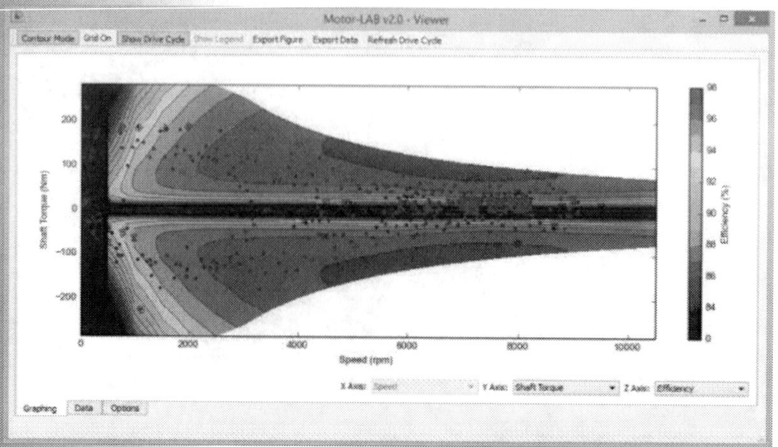


motordesign LIMITED

13

# Efficiency/Loss Maps in Motor-LAB

- Efficiency and loss maps can be plotted

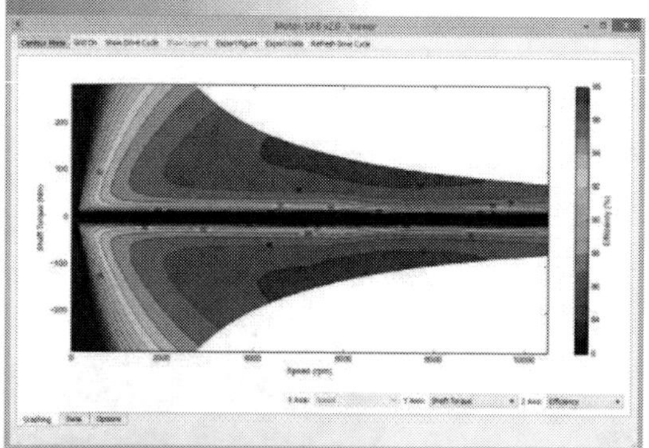

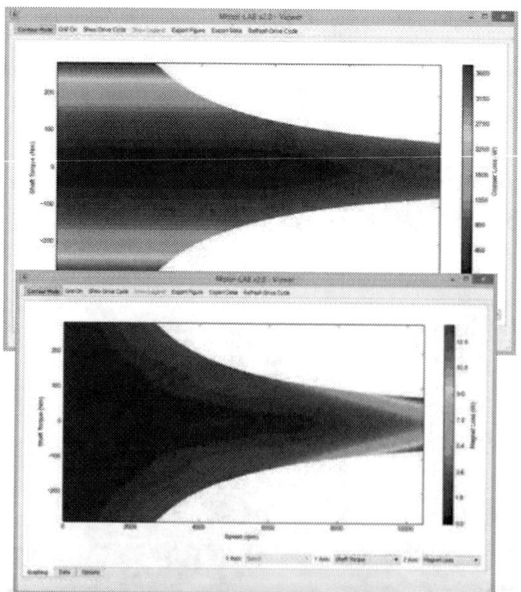

motordesign LIMITED 14

14

# US06 Drive Cycle in Motor-LAB

- ➢ Torque vs time

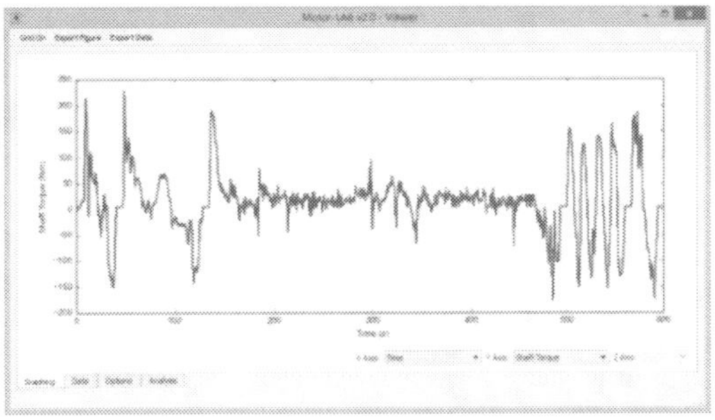

- ➢ Speed vs time

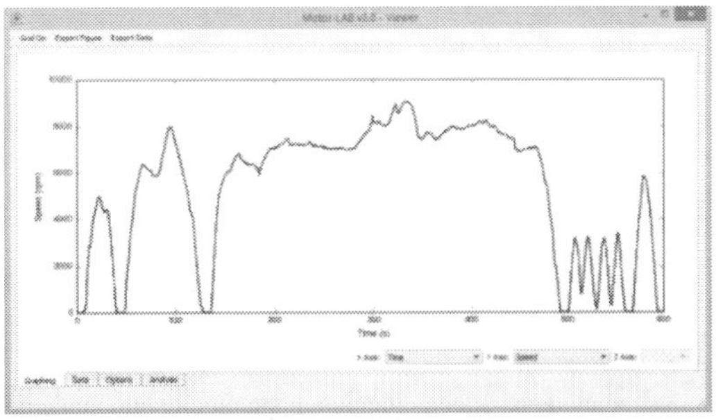

motordesign LIMITED 15

# US06 Drive Cycle in Motor-CAD Therm

- ➢ transient loss data is calculated across the duty cycle in Motor-LAB and transferred to Motor-CAD for thermal analysis

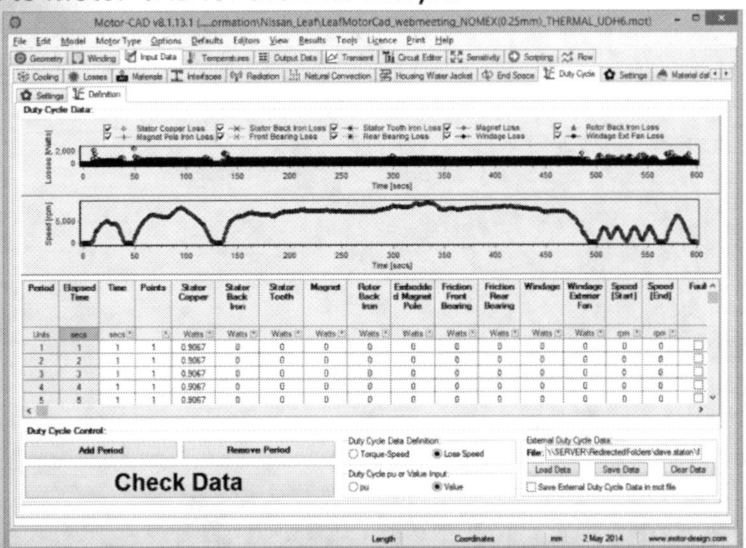


motordesign LIMITED 16

# US06 Thermal Transient in Motor-CAD

- Ten cycles of US06 thermal transient takes less than 8 minutes to compute (Lenovo T440)

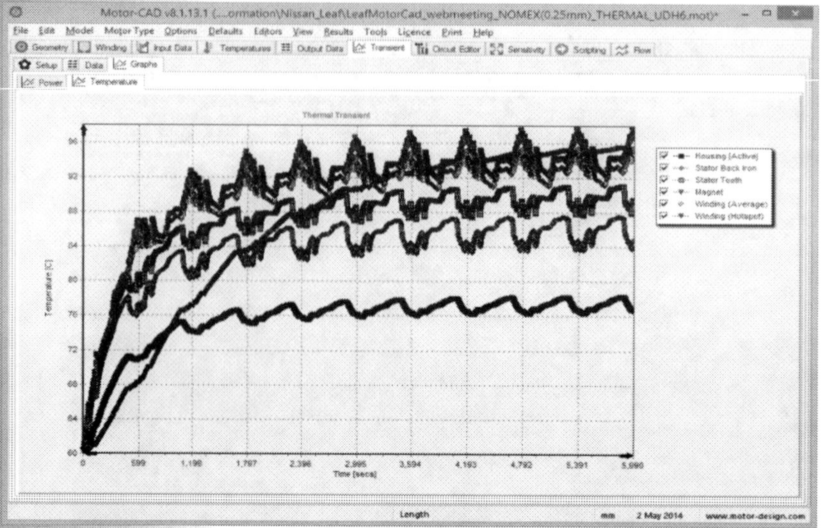


motordesign LIMITED



# More Extensive FEA/CFD Using ANSYS

- Automated links to ANSYS helps speed up creation of 2D/3D FEA and CFD models to make more detailed calculations
  - 3D electromagnetic effects
  - CFD to check/calibrate flow calculations
  - AC loss in winding
  - etc.

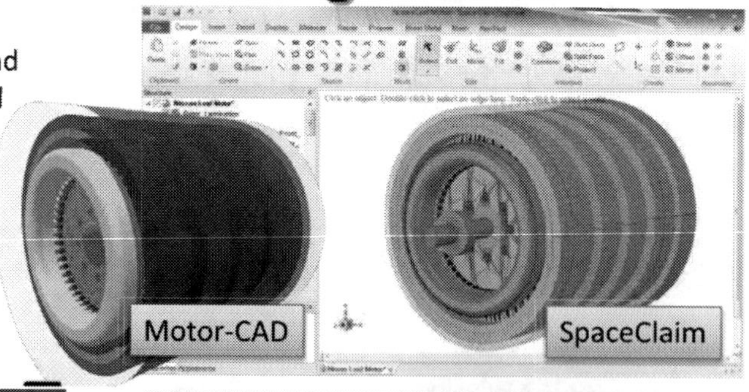


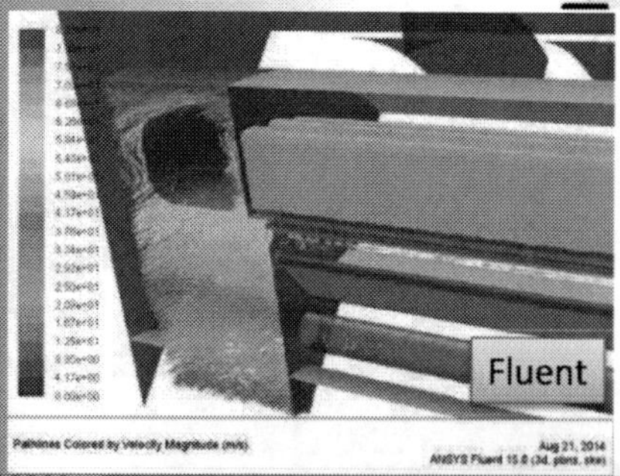


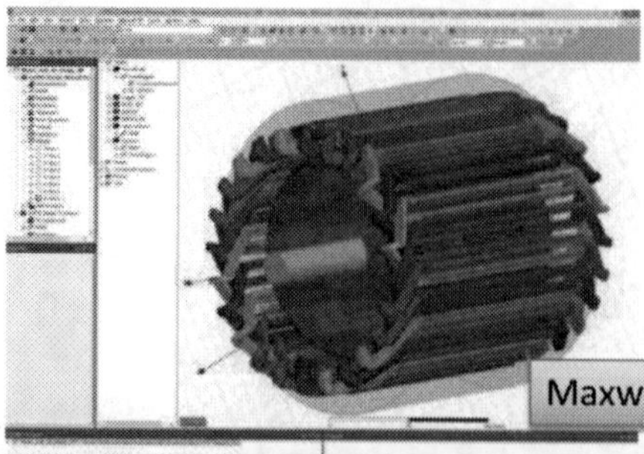


motordesign LIMITED

# Modeling Water Jacket Cooling Channels

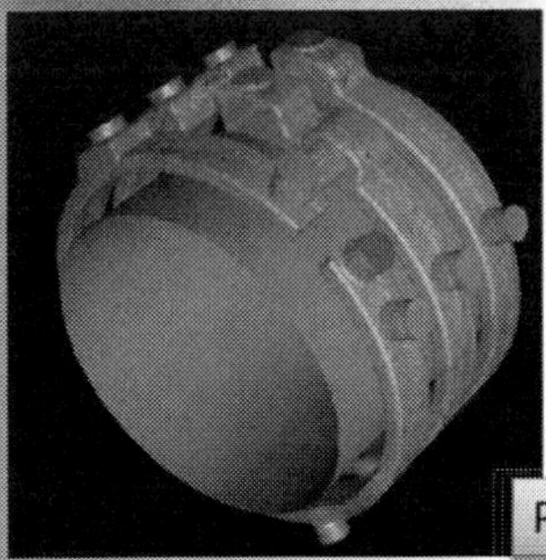

Path Lines

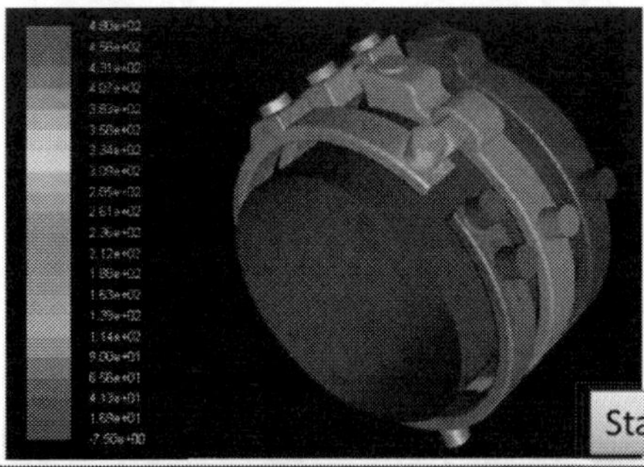

Static Pressure

motordesign LIMITED

19

# Modeling Water Jacket Cooling Channels

Distribution of coolant using different channel configurations

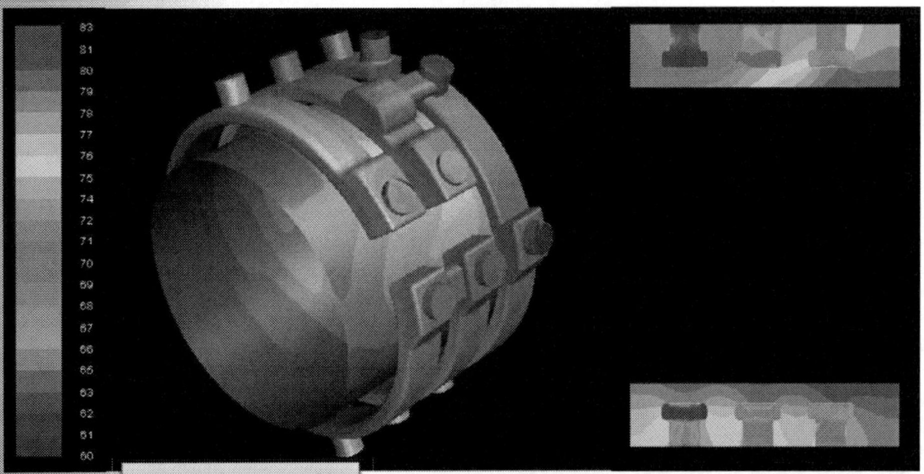


Temperature Profile

motordesign LIMITED

20

## Reduced Node Modelling of LEAF Motor

- useful for full system simulation
- matrix reduction method to calculate a smaller R-C network that gives same transient thermal response
- keep enough nodes to give good indication of temperatures across entire model
- full model of 65 nodes versus 7 nodes in reduced model

**Full Node Model**

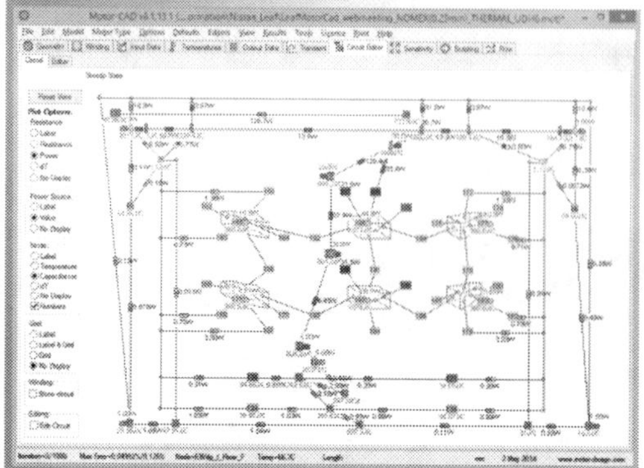

**Reduced Node Model**

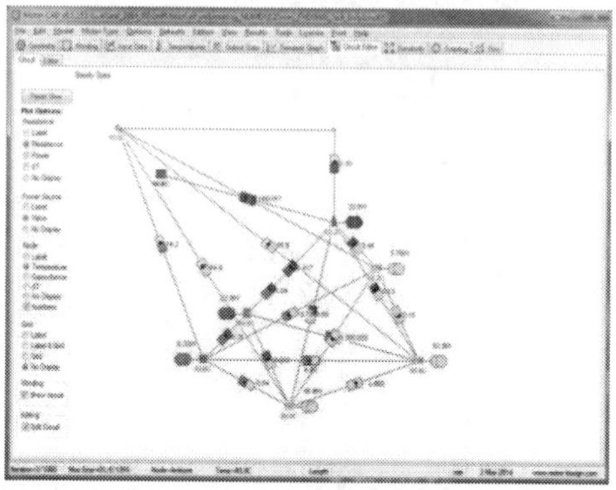

motordesign LIMITED

21

## Reduced Node Modelling of LEAF Motor

- Retains full model accuracy in steady state
- Reduced model thermal transient very good agreement with full model
- Solves in a fraction of time of the full model

| Nodes | % of Total nodes (65) | Steady State accuracy (%) | Transient Time (15 seconds for full) | Transient Accuracy |
|---|---|---|---|---|
| 7 | 10 | 99.75 | < 2 seconds | Excellent |

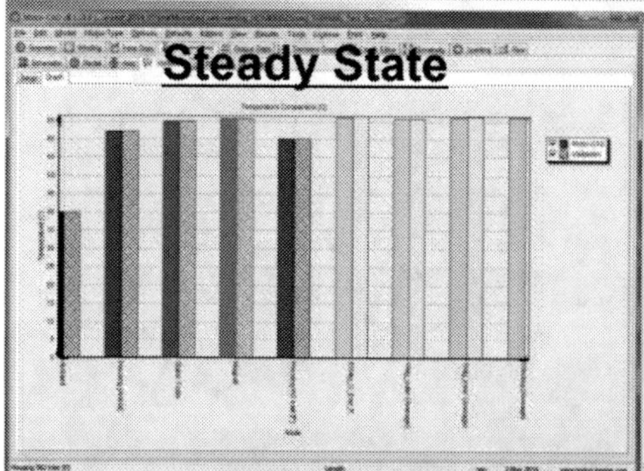


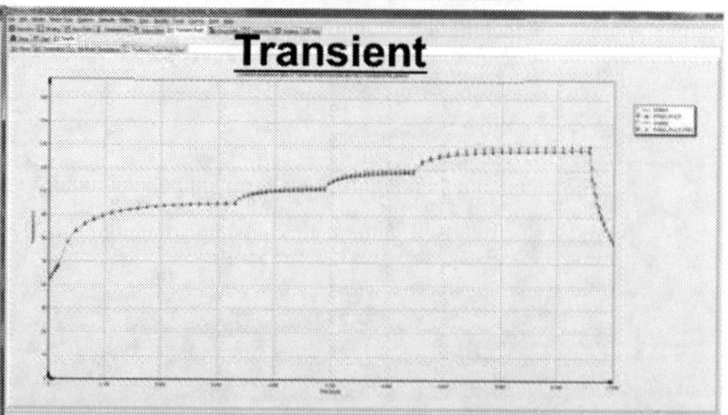


motordesign LIMITED

22

# Comparison of US06 duty cycle

- compare the complex US06 duty cycle with full model against the 7 node model
- the results compare very well

**Full node model**

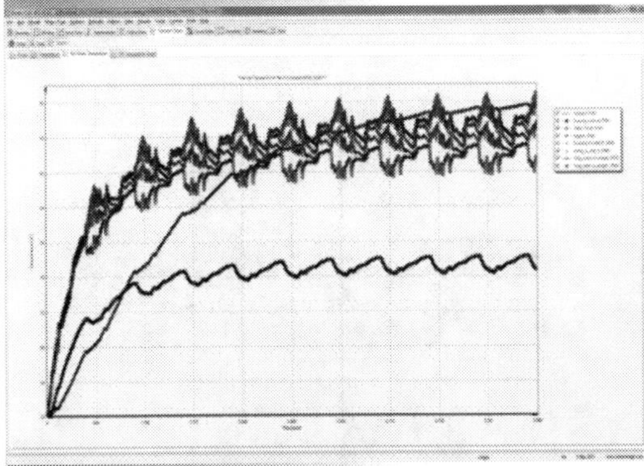

**Reduced node model**

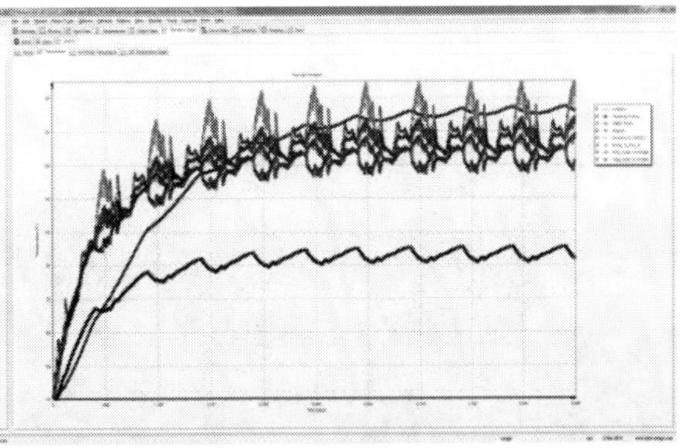



# Summary

- Demonstrated a range of integrated analytical and numerical software tools we use to perform electromagnetic and thermal design of electric motors
  - Used Nissan LEAF motor to demo tools used
- The different tools have different levels of complexity and different calculation and setup speeds
- For complex duty cycle thermal analysis its is best use the faster analytical network solution with the slower numerical tools used to help set up and calibrate the models

**Acknowledgments:**

- **Mike Slack** - **ANSYS CFD Expert, UK**
- **Manoj Nagulapally** - **ANSYS CFD Expert, USA**
- **Eddie Chong** - **MDL CFD Expert, UK**

MOTOR DESIGN LIMITED
Thank you for your attention!
Questions are welcomed
motordesign
LIMITED

## High Performance Real-Time Simulation
ETAS Test System and ANSYS Maxwell

ANSYS | ETAS

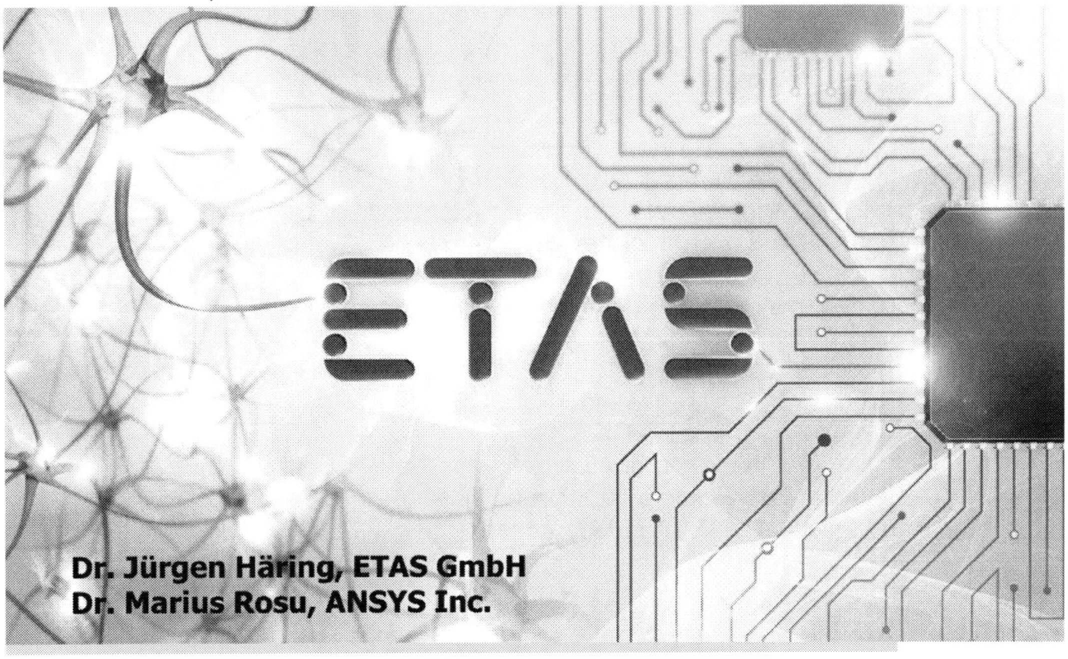

**Dr. Jürgen Häring, ETAS GmbH**
**Dr. Marius Rosu, ANSYS Inc.**

 

DRIVING EMBEDDED EXCELLENCE

## High Performance Real-Time Simulation
Overview

ANSYS | ETAS

| Content | |
|---|---|
| **1** | **Motivation** |
| 2 | E-Motor Modelling with Maxwell |
| 3 | Using Maxwell Parameters with ETAS E-Motor Models |
| 4 | Summary |

 

DRIVING EMBEDDED EXCELLENCE

## High Performance Real-Time Simulation
Design Challenge

**ANSYS Maxwell builds a comprehensive and reliable FE solution for the e-motor model employed in ETAS HIL e-drive platform**

- Modern automotive design architectures rely on complex power electronics, electrical and mechanical systems, which must be designed, tested and integrated concurrently
- These systems must also be engineered within
  - shorter development times
  - specified budget
  - high degree of quality and reliability
- Early design of the e-motor controller is possible even before the real e-motor is available by using a HiL based simulation of the e-motor based on ETAS LABCAR.

 

## High Performance Real-Time Simulation
Overview

| Content | |
|---|---|
| 1 | Motivation |
| **2** | **E-Motor Modelling with Maxwell** |
| 3 | Using Maxwell Parameters with ETAS E-Motor Models |
| 4 | Summary |

 

## High Performance Real-Time Simulation

Maxwell FE e-Motor Model

ANSYS | ETAS

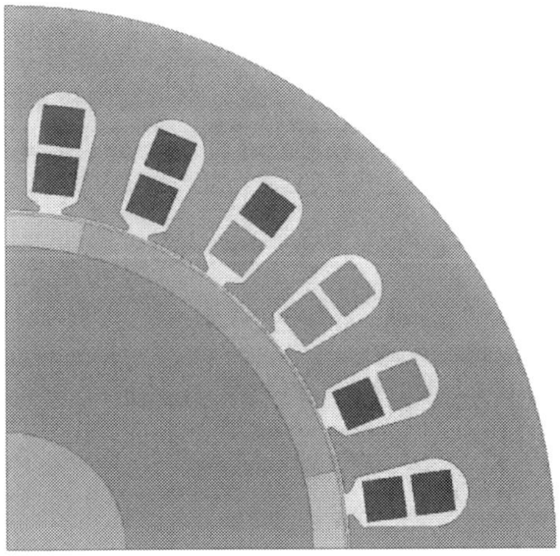

- Convenient Setup in ANSYS Maxwell Transient (2D/3D)
  - Convenient motion setup
  - Winding topology part of design setup
- More Flexible
  - No model creation during parameter sweep
  - Inputs: 2 variations (Id&Iq or Iα&Iβ, Im&$\beta$) for 3 phase (A, B, C) currents and one variation for rotor position
  - Outputs: $\lambda_a, \lambda_b \lambda_c$ or $\lambda_d \lambda_q \lambda_0$ and Torque
- More Capable
  - Recover full operation range using symmetry and periodical conditions
- More Efficient
  - No re-mesh during parametric sweep
  - Much less sweeps without sacrificing accuracy

  DRIVING EMBEDDED EXCELLENCE

ANSYS | ETAS

## Maxwell FE e-Motor Model

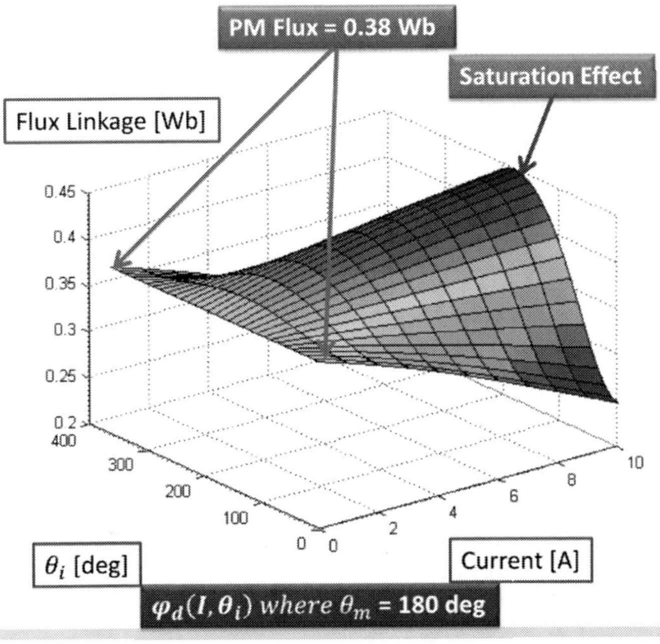


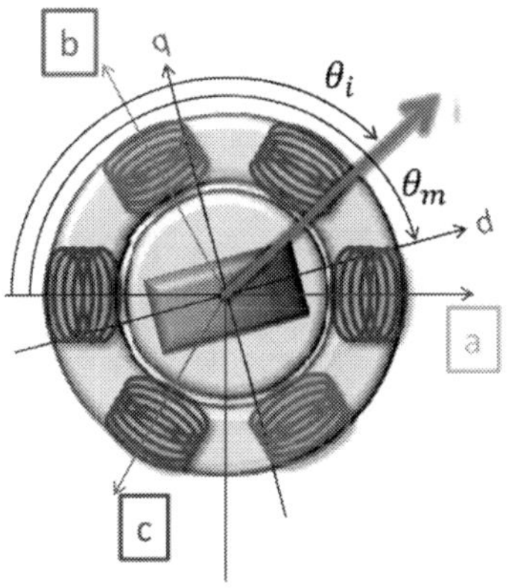


  DRIVING EMBEDDED EXCELLENCE

## Maxwell FE e-Motor Model

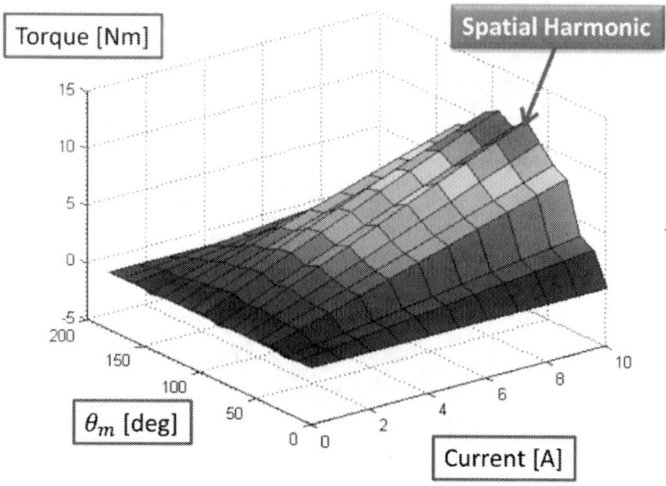


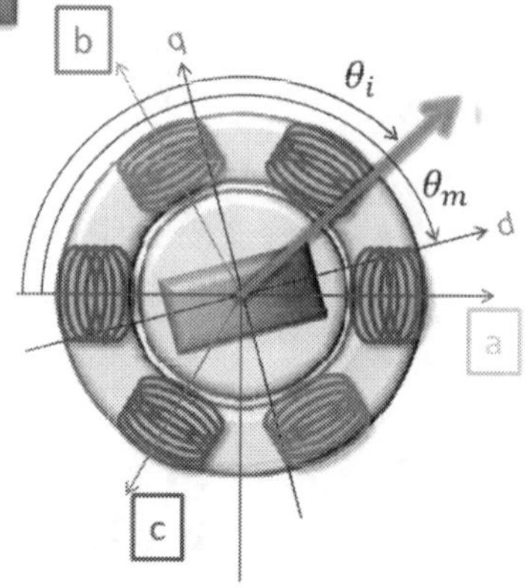


**$T(I, \theta_m)$ where $\theta_i$ = 0 deg**



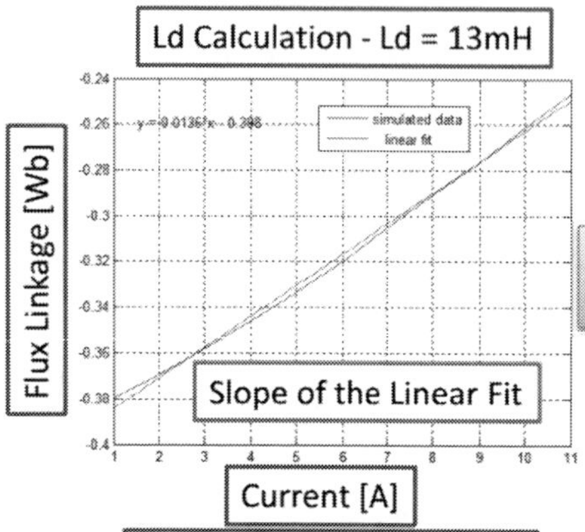


$\theta_i$ = 0 deg
$\theta_m$ = 0 deg

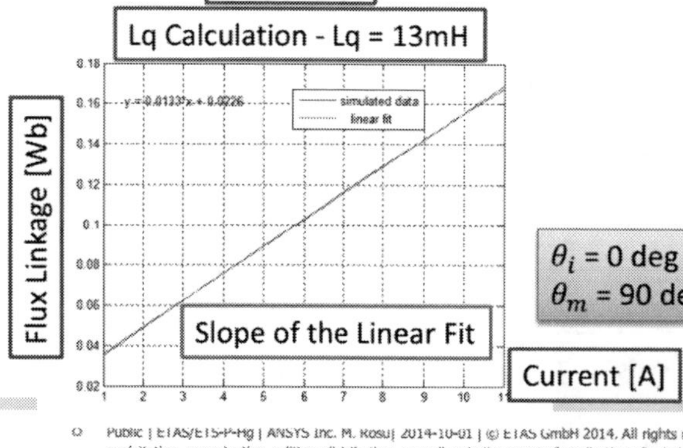


$\theta_i$ = 0 deg
$\theta_m$ = 90 deg

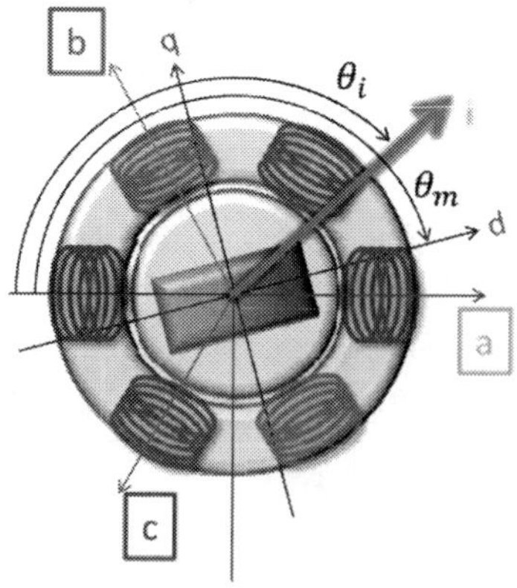


**Parameter Extraction from FEA**

## High Performance Real-Time Simulation
Overview

Content






## High Performance Real-Time Simulation
Hardware-in-the-Loop Architecture

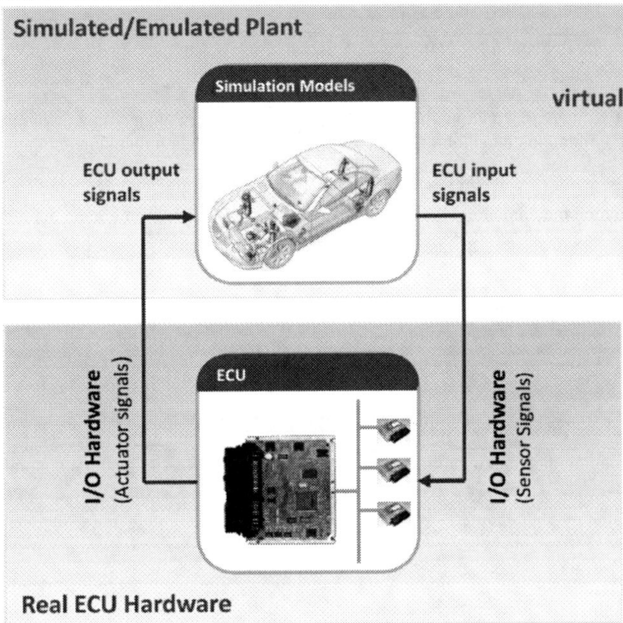


ETAS LABCAR as virtual environment for the ECU

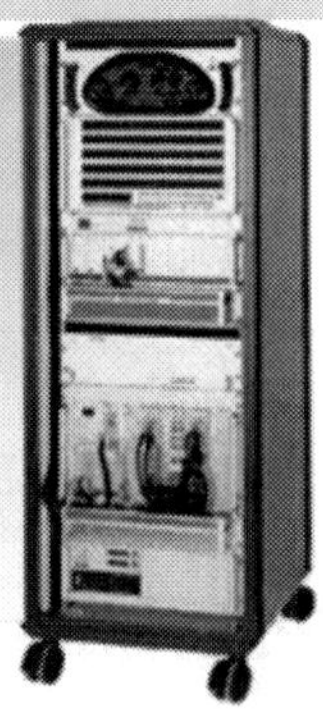

## High Performance Real-Time Simulation
Signal-Level Testing of E-Motor Control Units

**Signal Level Testing: No safety critical voltages and currents**

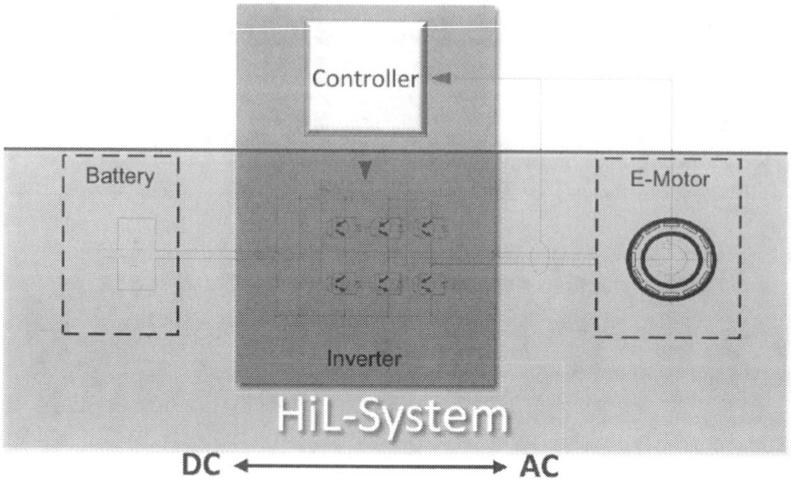


11 Public | ETAS/ETS-P-Hg | ANSYS Inc. M. Rosu| 2014-10-01 | 

DRIVING | EMBEDDED EXCELLENCE

## High Performance Real-Time Simulation
Signal-Level Testing of E-Motor Control

**FPGA based e-motor model required to implement low latencies**

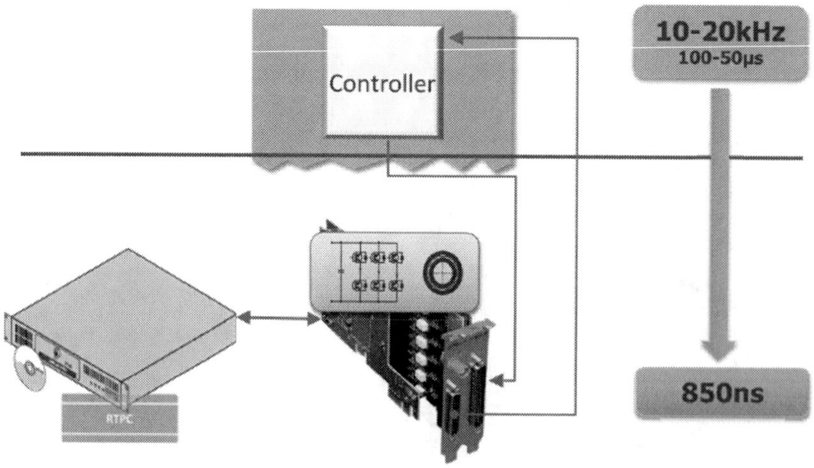


12 Public | ETAS/ETS-P-Hg | ANSYS Inc. M. Rosu| 2014-10-01 | 

DRIVING | EMBEDDED EXCELLENCE

## High Performance Real-Time Simulation
Signal-Level Testing of E-Motor Control

**Closed Loop System Demo with Control Unit for PMSM**

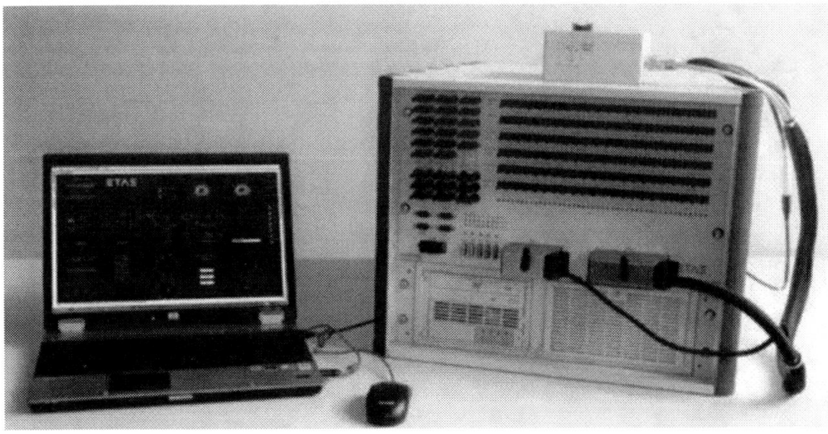



## High Performance Real-Time Simulation
LABCAR and Maxwell Used in a Seamless Workflow

**ETAS PMSM Model Using Parameters from Maxwell E-Motor Design**

- Parameters from E-Motor design in Maxwell are extracted to run ETAS PMSM simulation model on ETAS LABCAR
- Figures show measurements from a closed loop setup (controller & model)

**E-Motor Torque**
(red: desired, blue: actual)

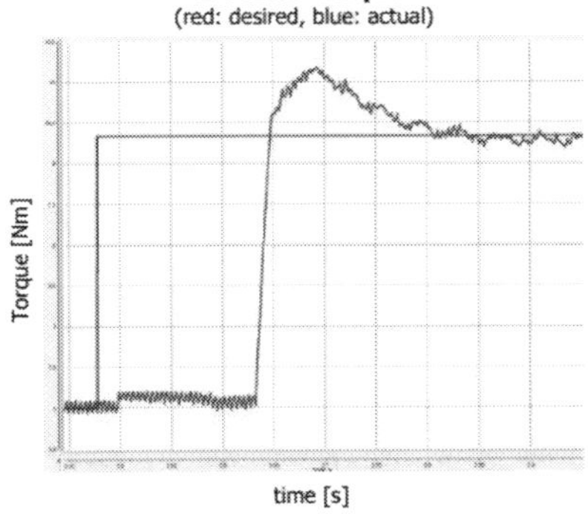


**E-Motor Currents**
(three phases)

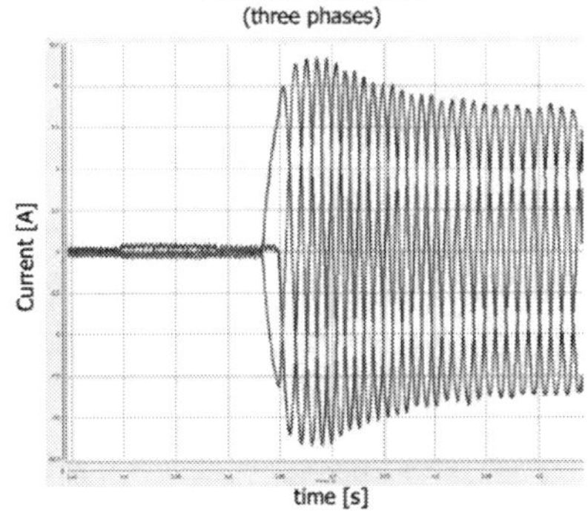

## High Performance Real-Time Simulation
Overview

| Content | |
|---|---|
| 1 | Motivation |
| 2 | E-Motor Modelling with Maxwell |
| 3 | Using Maxwell Parameters with ETAS E-Motor Models |
| **4** | **Summary** |

 



## High Performace Real-Time Simulation
Maxwell e-Motor Solution Advantage

**Summary**

- We show the seamless workflow to re-use parameters of a Maxwell e-motor design in ETAS LABCAR HiL systems running a real time model
- Maxwell model extraction provides with a dynamic and flexible environment enabling the designers to customize the e-motor parameters maintaining the required level of motor model accuracy;
- Maxwell solution enabled by Distributed Solve option expedites the e-motor extraction work employing multi-core technology;
- Maxwell solution enables automotive companies to successfully conduct numerous HIL tests on the motor drive controller reducing the costs related to the prototyping and verification processes, increasing the robustness of the entire drive system

 

2014 Automotive Simulation World Congress

# High-frequency, High-power Magnetic Component Design with Maxwell 3D From Geometry Creation to Component Optimization

Dr. Jenna Pollock
Tesla Motors

---



## Objective

- **Design magnetic components for automotive power conversion systems that are:**
  - Light weight
  - Low cost
  - Reliable
  - Easy to build
- **Meet electrical specs**
  - Inductances, Resistances, Capacitances
  - Winding Loss, Core Loss
- **Meet mechanical specs**
  - Volume
  - Weight
  - Mounting

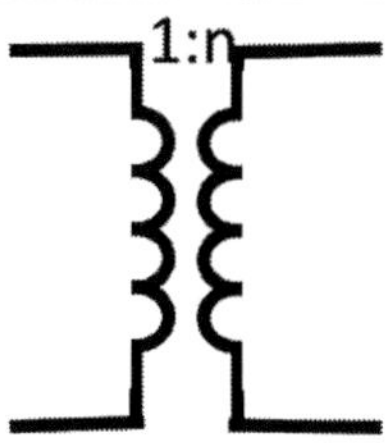

## Simulation-driven Design of an Isolation Transformer with Maxwell 3D

- Challenges
  - Flux distribution and saturation
  - Leakage fields and inductance
  - Minimize losses
- Optimize winding designs
  - Scripting winding geometry
    - Makes iterating complex windings simple
  - Foil designs
    - Find lowest loss design – determine foil thickness
    - Understand current distribution in foil
  - Minimize Leakage Inductance
  - Litz-wire design
    - Determine litz-wire standing
- Determine fields and losses in near by conductors
  - Housing and heat sinks

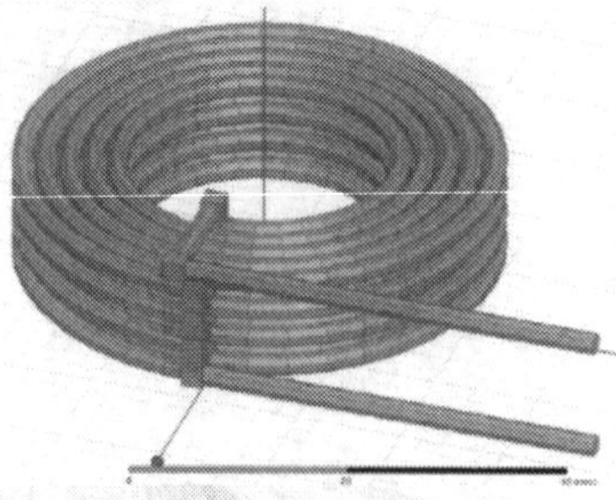

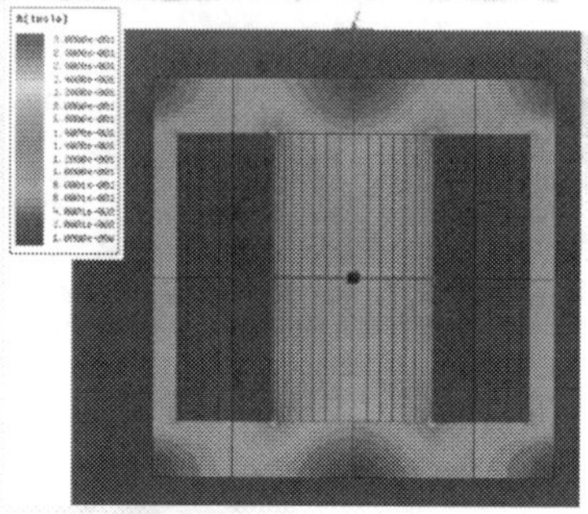



2014 Automotive Simulation World Congress
October 9-10 . Tokyo, Japan

## Transformer Electrical Parameters

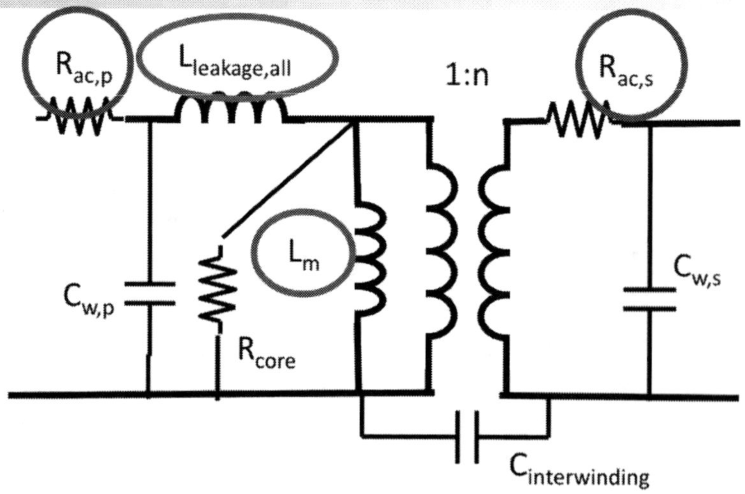

2014 Automotive Simulation World Congress
October 9-10 . Tokyo, Japan

## Transformer Design Example:

| Parameter | |
|---|---|
| Power level | 3 kW |
| Switching Frequency | 120kHz |
| Magnetizing inductance | 1.3 mH |
| Leakage inductance | 4.2 uH |
| Maximum power loss | 1% |
| Turns Ratio | 8:1 |

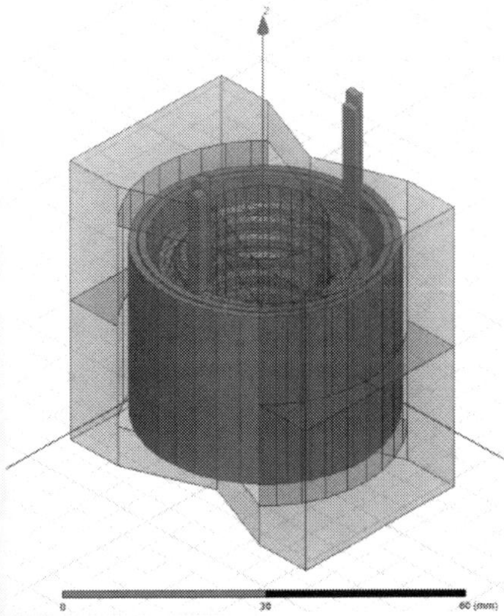

Monday, October 06, 2014 2014 Automotive Simulation World Congress 5

2014 Automotive Simulation World Congress
October 9-10 . Tokyo, Japan

## Preliminary Transformer Design

- Trade off core and windings losses to find an optimal design for given electrical specifications
  - Use FEA to support analytical methods
  - FEA can fine tune design
    - Specific electrical parameters
    - Specific size requirements
- A free transformer design resource:
  - http://ecee.colorado.edu/copec/book/slides/slidedir.html

# First Pass Design

## Use existing analytical methods

- Core Loss:
  - Steinmetz equation
  - Design primary voltage waveform

$$P_c = \mathrm{k_c} f^{\alpha} B^{\beta} \mathrm{V}$$

- Winding loss:
  - Dc resistance
  - Ratio of ac to dc resistance
    - To be determined in coming slides





# Analytical Winding Loss Solutions

| Frequency Range | Loss Solution | |
|---|---|---|
| | Rectangular | Cylindrical |
| Diameter less than skin depth | $P_{loss} = \left(1 + \frac{5p^2 - 1}{45}\left(\frac{h}{\delta}\right)^4\right) R_{dc} I_{ac,rms}^2$ | $P = \frac{\pi\omega^2 \ell (2r)^4 B^2}{128\rho_{cu}} + P_{dc}$ |
| Any Diameter | $P_l = \frac{1}{2}\frac{b\rho}{\delta}\left[(H_a - H_b)^2 F\left(\frac{h}{\delta}\right) + 2H_a H_b G\left(\frac{h}{\delta}\right)\right]$ | $P_l = F(\gamma) I^2 + G(\lambda) H_e^2$ <br> $\gamma = \frac{d}{\delta\sqrt{2}}$ |

# Measured Inductance and Resistance

Leakage Inductance

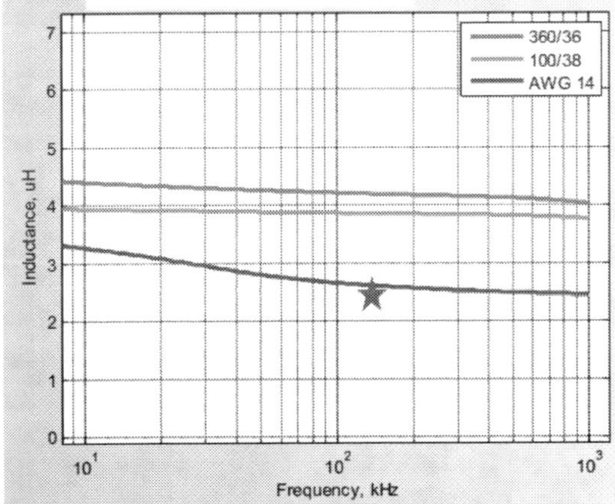


Winding Resistance

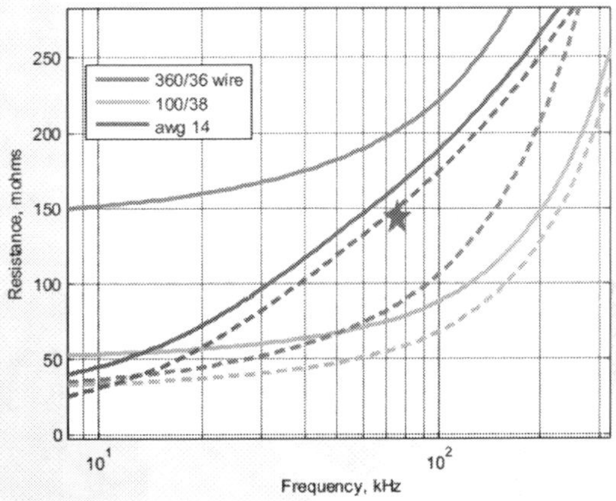


Maxwell frequency sweep data to be added.



2014 Automotive Simulation World Congress
October 9-10 . Tokyo, Japan

# Determine Leakage and Magnetizing

- Mesh size =1.1 million tets

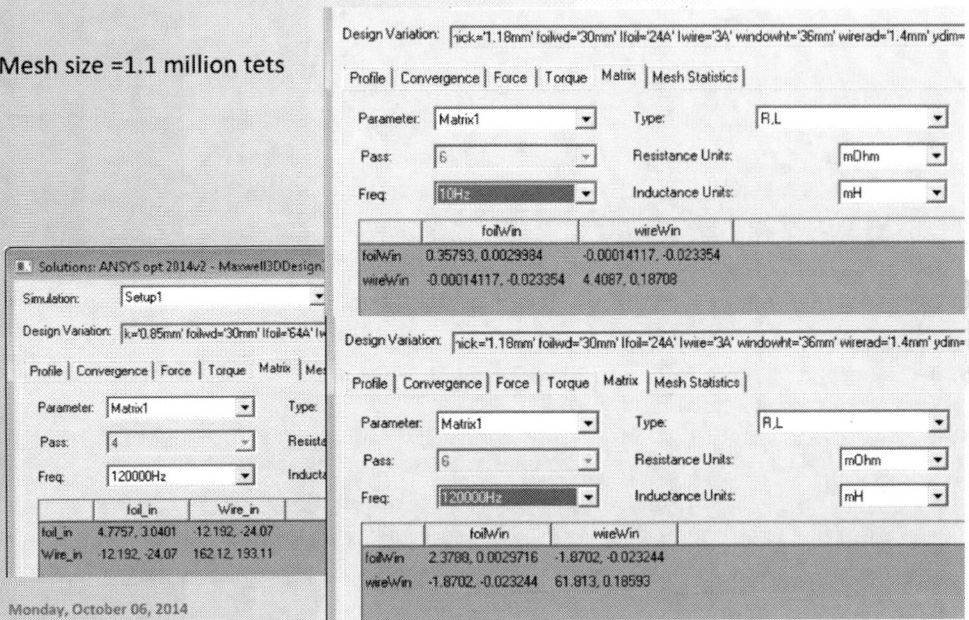

2014 Automotive Simulation World Congress
October 9-10 . Tokyo, Japan

## Leakage Inductance Calculation

- Use total energy from convergence tab
  - E = 9.51 uJ
  - $I_p$ = 8 A – peak current
- Find inductance from: $E = \frac{1}{2} L \frac{I_p}{\sqrt{2}}^2$

$$L_{leak} = \frac{2*2*E}{{I_p}^2}$$

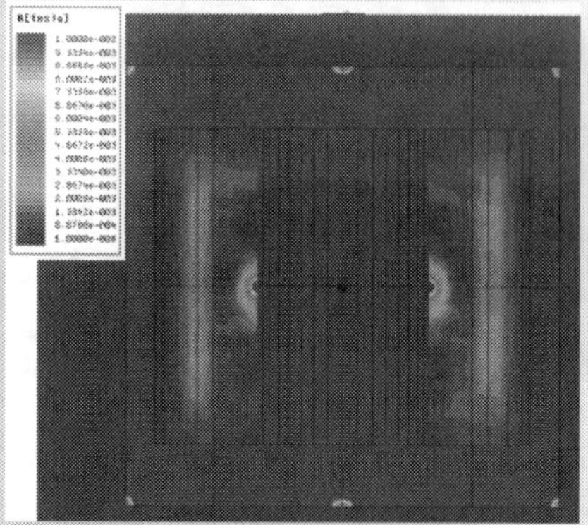

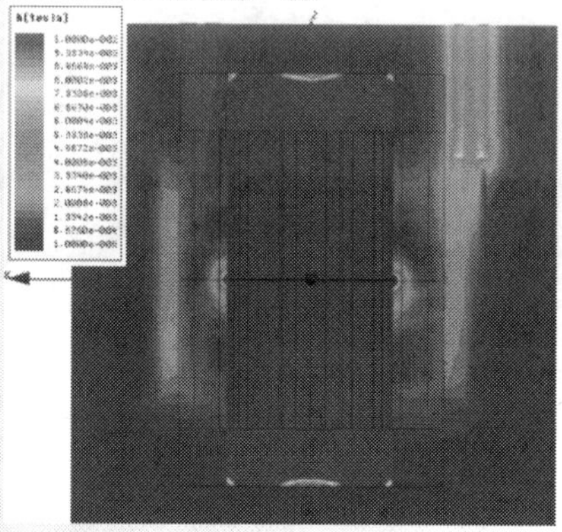



2014 Automotive Simulation World Congress
October 9-10 . Tokyo, Japan

## Core Saturation Simulation

- Plot of inductance vs current
- Field animation of saturation effects

2014 Automotive Simulation World Congress
October 9-10 . Tokyo, Japan

## Foil Winding Design

- Need Maxwell 3D to determine the ac resistance
  - Often ac resistance is to low to measure
  - Analytical methods neglect edge effects, etc
- Use Optimetrics to find foil thickness with lowest ac resistance
  - This means the cost is the "ac resistance"
  - Use eddy current solver
  - Define mesh on foil winding to reflect the skin depth

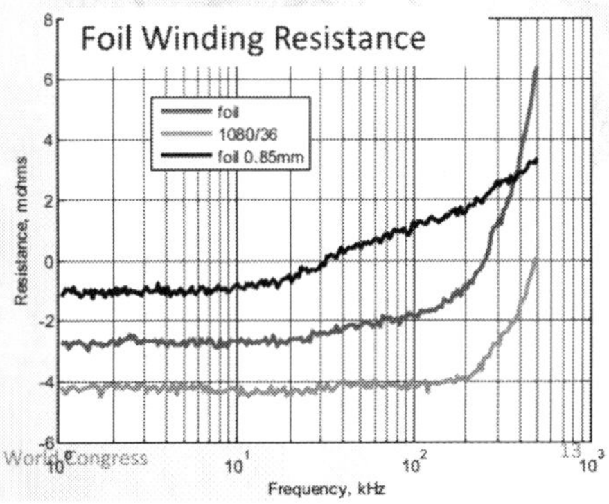


Monday, October 06, 2014 2014 Automotive Simulation World Congress 13

2014 Automotive Simulation World Congress
October 9-10 . Tokyo, Japan

## Optimetrics Results

NEED RDC

- Results – minimize ac resistance
  - Maxwell determines the response surface
  - Updates the design parameters with optimized values

| Evaluation | Foil thickness mm | DC Resistance, mohms | AC Resistance, mohms | Volume, mm3 Maxwell 3D | Weight, g | Cost, US dollar – CU prices |
|---|---|---|---|---|---|---|
| 1 | 0.84 | | 3.02 | 5693.5 | 51 | 0.341 |
| 2 | 0.85 | | | 5767.4 | 51.7 | 0.345 |
| 3 | 1.0 | | 2.84 | 6886. | 61.7 | 0.412 |
| 4 | 1.12 | | 2.72 | 7803.5 | 69.9 | 0.467 |
| 5 | 1.16 | | 2.71 | 8113.4 | 72.7 | 0.486 |
| 6 | 1.28 | | 2.65 | 9056.4 | 81.1 | 0.542 |
| 7 | 1.32 | | 2.63 | 9375.0 | 84.0 | 0.561 |
| 8 | 1.44 | | 2.60 | 10343.9 | 92.7 | 0.619 |
| 9 | 1.48 | | 2.59 | 10671.2 | 95.6 | 0.639 |

## Designing Litz-wire Windings

- **Difficult to pick strand size and number of strands**

$$P = \frac{\pi \omega^2 \ell (2r)^4 B^2}{128 \rho_{cu}} + P_{dc}$$

  – Valid where wire diameter is small compared to skin depth

$$\delta = \sqrt{\frac{\rho}{\pi \mu f}} \qquad d << \delta$$

- Need the average value of magnitude of the flux density over a winding region:

Inductor: $\left\langle |B|^2 \right\rangle$ Transformer: $\begin{bmatrix} \left\langle |\vec{B}_1|^2 \right\rangle & \vec{B}_1 \cdot \vec{B}_2 \\ \vec{B}_2 \cdot \vec{B}_1 & \left\langle |\vec{B}_2|^2 \right\rangle \end{bmatrix}$



2014 Automotive Simulation World Congress
October 9-10 . Tokyo, Japan

## Litz-wire Design Software

http://engineering.dartmouth.edu/inductor/index.shtml

**Dartmouth Magnetic Component and Power Electronics Research**
Transformers and Inductors for Electronics Applications

- Home
- Research Topics
- Publications
- People
- Free Software
- Information for Designers
- Links
- Sponsors

Thayer School of Engineering
Dartmouth College

### Software for Magnetics Design

Here we present some design software you can download or run on-line.

#### LitzOpt

The LitzOpt program optimizes litz wire stranding design to minimize the cost or power loss, for transformers or inductors.

Run LitzOpt online for the simplest user interface.

You can download the LitzOpt program the for use with MATLAB software. It has a primitive user interface, but is more flexible than the web version. Learn more on the LitzOpt page.

#### ShapeOpt

The ShapeOpt program optimizes the shape of an inductor winding to minimize winding losses, including adjusting the number of strands in a litz wire winding.

The ShapeOpt page has more information, or you can proceed directly to run ShapeOpt online, or download the MATLAB program, shapeopt.m.

#### Core Loss Calculator for Nonsinusoidal Waveforms

The Core Loss Calculator program accurately calculates core loss for any piecewise-linear waveform, even for cores with minor loops.

# User Supplied Magnetic Field Data

- Use magnetostatic simulations

Litz

◯ 1-D Internal Field Simulation (more information)

◯ User-supplied Magnetic Field Data (more information)

Both windings conduct with equal and opposite unit N*I

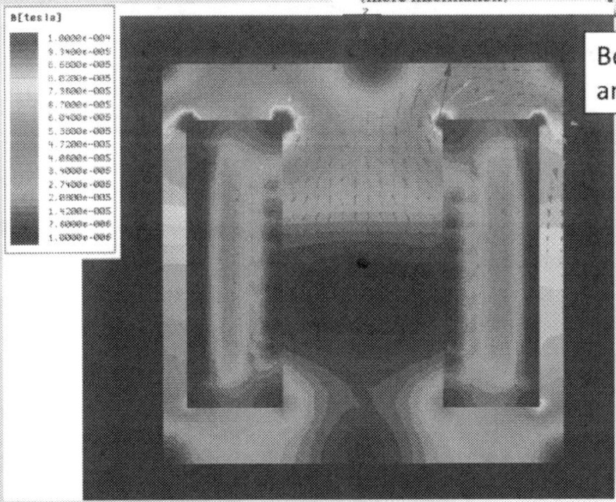

2014 Automotive Simulation World Congress
October 9-10 . Tokyo, Japan

# Component Design Parameters

| | | |
|---|---|---|
| Choose a standard core size or select "user defined" to specify a different size: | PQ50/50 | |
| Breadth of the core window (mm): | 36.1 | mm |
| Breadth of the bobbin window (mm): | 30.4 | mm |

Core gaps:
◯ Center leg gapped ◯ No legs gapped ◉ All legs gapped

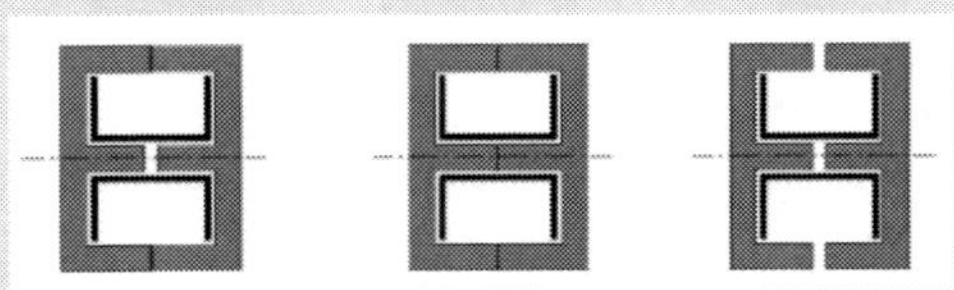

**Current Sinusoidal Waveform Data**

| | | | |
|---|---|---|---|
| Frequency: | 100000 | | Hz |
| | Winding 1: | Winding 2: | |
| Amplitude: | 8 | 64 | A |
| Phase angle: | 0 | 0 | degrees |

**Current Waveforms**

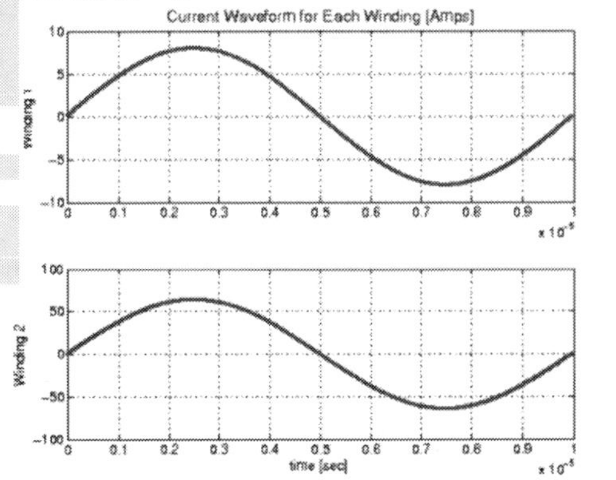

2014 Automotive Simulation World Congress
October 9-10 . Tokyo, Japan

## Magnetostatic Field Calculations

| Variation | i1 | i2 | magB2_foil: None | magB2_wire: None |
|---|---|---|---|---|
| 1 | 0 | 1 | 4.768e-015 | 2.654e-014 |
| 2 | 1 | 0 | 3.159e-016 | 1.724e-014 |
| 3 | 1 | 1 | 5.372e-015 | 5.136e-015 |

**Magnetitic Field Integrals (i.e. from Finite Element Model)**

| | | |
|---|---|---|
| Integral of $B^2$ over winding area with 1 amp in winding 1: | Field integral in winding 1: 3.159e-16<br>Field integral in winding 2: 1.724e-14 | T |
| Integral of $B^2$ over winding area with 1 amp in windings 1 and 2: | Field integral in winding 1: 5.372e-15<br>Field integral in winding 2: 5.136e-15 | T |
| Integral of $B^2$ over winding area with 1 amp in winding 2: | Field integral in winding 1: 4.768e-15<br>Field integral in winding 2: 2.654e-14 | T |



2014 Automotive Simulation World Congress
October 9-10 . Tokyo, Japan

## Winding Design Proposals

- Total winding loss = 2.5 W
  - Now tradeoff
    - Cost
    - Loss
    - Temperature rise
    - Weight
  - To pick the optimal stranding for design specifications

**Buildable Design Table**

| Design Number | Gauge (W1) | Gauge (W2) | Relative Cost | Loss in Watts | NumStrands (W1) | NumStrands (W2) |
|---|---|---|---|---|---|---|
| d1 | 28 | 28 | 0.0114 | 20.318 | 1 | 10 |
| d2 | 30 | 30 | 0.0195 | 12.051 | 3 | 25 |
| d3 | 32 | 32 | 0.0302 | 8.0421 | 7 | 62 |
| d4 | 34 | 34 | 0.0494 | 5.1775 | 18 | 155 |
| d5 | 36 | 36 | 0.0798 | 3.4582 | 44 | 383 |
| d6 | 38 | 38 | 0.133 | 2.3388 | 108 | 948 |
| d7 | 40 | 40 | 0.18 | 1.9746 | 268 | 1466 |
| d8 | 42 | 42 | 0.274 | 1.7679 | 630 | 2267 |
| d9 | 44 | 44 | 0.395 | 1.7354 | 974 | 3506 |
| d10 | 46 | 46 | 0.778 | 1.7338 | 1506 | 5423 |
| d11 | 48 | 48 | 2.12 | 1.7519 | 2329 | 8386 |

**Original Design**

Relative cost of original design = 0.39787
Loss of original design (watts) = 2.5102
Wire gauge in original design = 36 36
Number of strands in original design = 360 1080

# Compare Cost and Weight

| Windings design | Weight, g measured* | Volume, mm3 Maxwell 3D | Rdc, mohms | Rac ,at 120 kHz |
|---|---|---|---|---|
| N=2: 5 mil foil | 12.63 | | 1.8 | |
| N=2: 0.85 mm foil | 61.23 | | 0.224 | |
| N=2: 1080/36 | 37.94 | | 0.310 | |
| N=16: 360/36 | 76.65 | | 5.03 | |
| N=16: awg 14 | 25.52 | | 8.65 | |
| N=16: 100/38 | 11.13 | | 23.19 | |

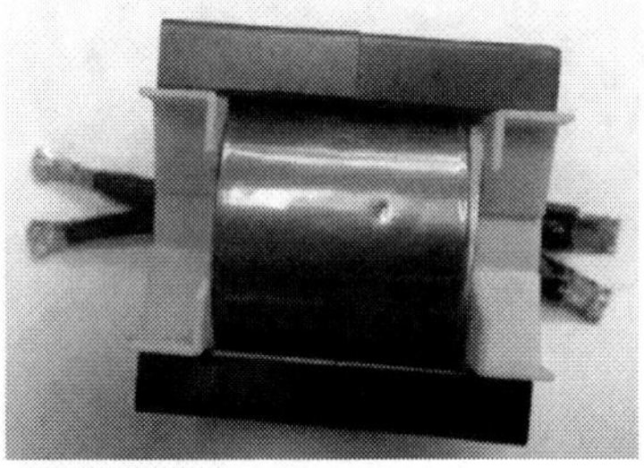



2014 Automotive Simulation World Congress
October 9-10 . Tokyo, Japan

# Integrate Results into Thermal Simulations

- Simulation running now

2014 Automotive Simulation World Congress
October 9-10 . Tokyo, Japan

## Conclusions

- Maxwell 3D powerful tool for power magnetic component design
- Provides insight into field patterns and current distributions at high frequency
- Optimetrics package increase ability to design intelligently with FEA
- Parametric extremely useful for sweeping variables



2014 Automotive Simulation World Congress
October 9-10 . Tokyo, Japan

## Thanks and Questions?

- **Specials thanks to:**
  - Will Shultz, Pavani Gottipati, ANSYS
  - Lev Fedoseyev, Tesla Motors
  - My team, Tesla Motors
  - Thorben Schobre, Intern, Tesla Motors
  - Dr. Charles Sullivan, Dartmouth Magnetic Component and Power Electronics Research Group

2014 Automotive Simulation World Congress
October 9-10 . Tokyo, Japan

# References


- J. Pollock, T. Abdallah and C. R. Sullivan, "Easy-To-Use CAD Tools for Litz-Wire Winding Optimization", IEEE Applied Power Electronics Conference, Feb. 2003, pp. 1157 -1163.
- Litzopt: Dartmouth Magnetic Components and Power Electronics http://engineering.dartmouth.edu/inductor/index.shtml

# MEMS MULTIPHYSICS SIMULATIONS IN ANSYS WORKBENCH

**Metin Ozen, Ph.D., ASME Fellow**

**OZEN ENGINEERING, INC.**

www.ozeninc.com

OZEN ENGINEERING

ANSYS

## WHAT DO WE DO?

- Ozen Engineering, Inc. helps solve challenging and **multidisciplinary engineering problems** with industry leading computational simulation technologies
- We provide advanced
  - **Multi-Physics FEA**
  - **Computational Fluid Dynamics (CFD) simulations**

OZEN ENGINEERING

## THE CAE TOOLS

- ANSYS CHANNEL PARTNER, ANSYS Distributor in California
- Provide sales, marketing, training, technical support and consulting services in California for **ANSYS** software products

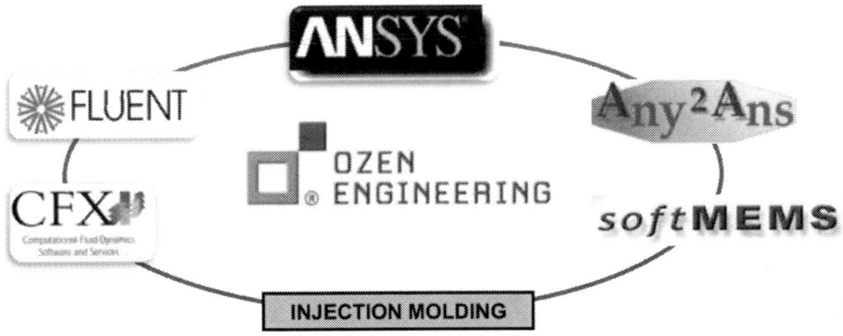


OZEN ENGINEERING

## INDUSTRY SPECIFIC EXPERTISE - SEMICONDUCTOR

Example of analysis we can perform:

- Multi-physics simulations of Semiconductor chambers
- Ball Grid Array Solder Joint Reliability Optimization
- Thermal-Stress
- Seismic vibration of chamber design

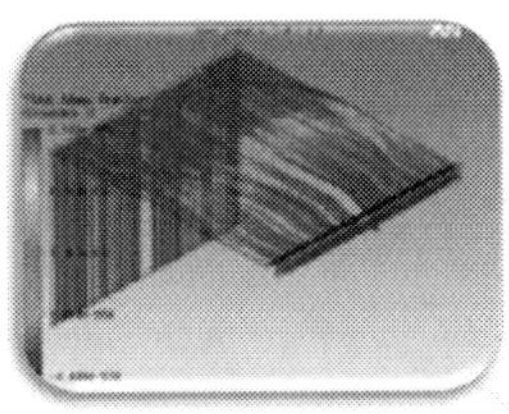

Semiconductor Chambers

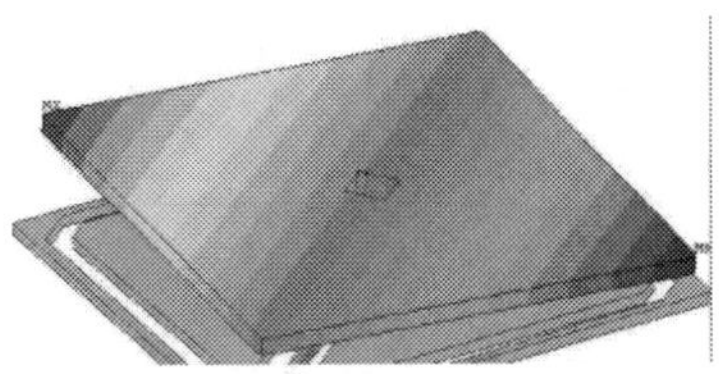

MEMS

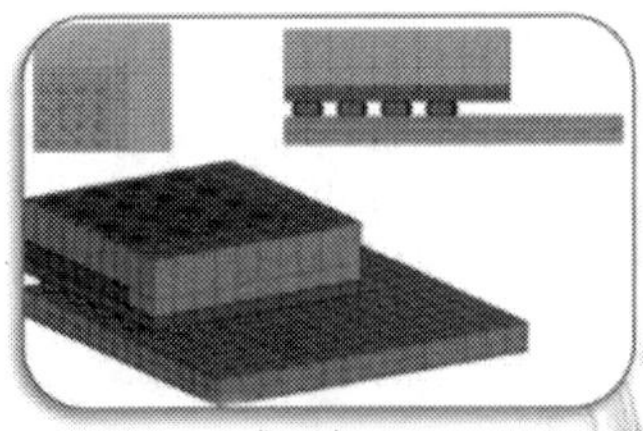

Ball Grid Arrays

OZEN ENGINEERING

## INDUSTRY SPECIFIC EXPERTISE – SOLAR INDUSTRY

Example of analysis we can perform:

- Multi-physics simulations of solar panel and support
- Electrical, thermal, mechanical and structural analysis
- Solar panel design optimization
- Modal analysis
- Virtual Prototyping

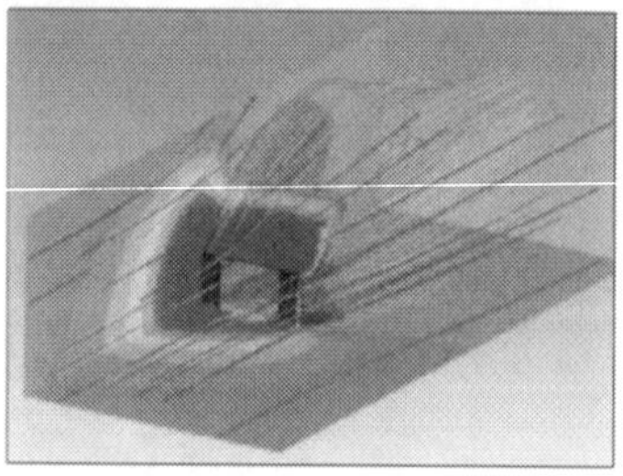

Solar Panel Fluid Structural Analysis

Solar Panel Model

Example of case studies:

- Maximize the solar flux through a surface
- Structural optimization of the pole mount supports of a solar panel in a wind load case study
- Hail Impact on a solar panel

OZEN ENGINEERING

## INDUSTRY SPECIFIC EXPERTISE – DESIGN OPTIMIZATION

**Capabilities:**

- Parametric Exploration
- Mono and Multi-Objective Design Optimization (MDO)
- Process Integration
- Sensitivity Analysis
- Robust Design
- Decision Making Criteria and Tools

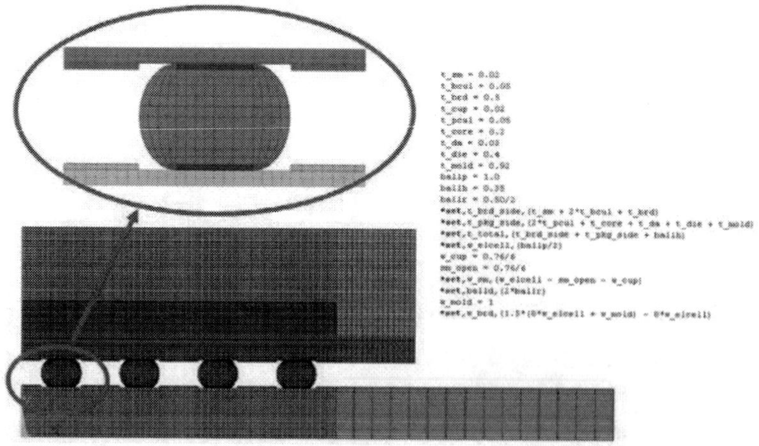

BGA OPTIMIZATION

OZEN ENGINEERING

## INDUSTRY SPECIFIC EXPERTISE – ELECTRONICS

Example of analysis we can perform:

- BGA Solder Joint Reliability
- Theta Jc Thermal Characterization
- Thermal-Stress
- Fracture Mechanics & Fatigue
- Board & System Level CHT

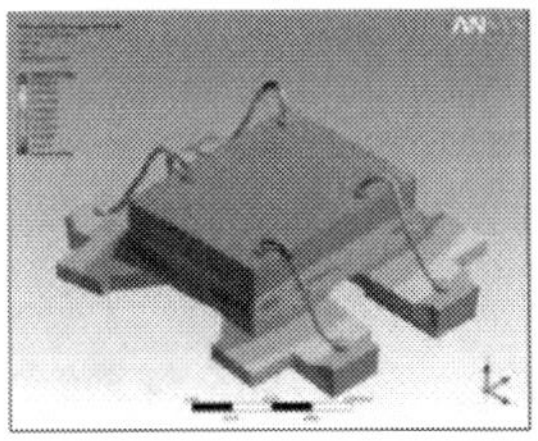

Chip Level Simulation

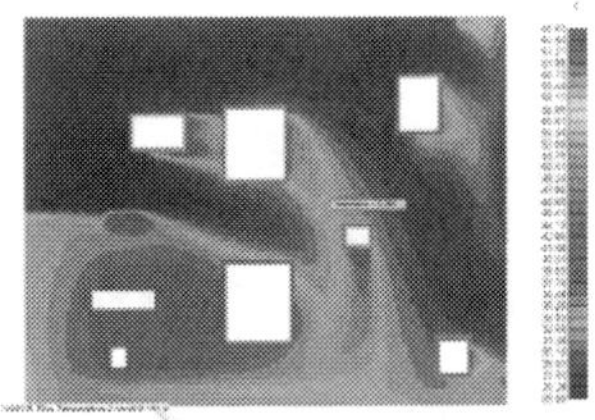

Board Level Simulation

System Level Simulation

OZEN ENGINEERING

## INDUSTRY SPECIFIC EXPERTISE – CONSUMER PRODUCTS

**Example of analysis** we can perform:

- Drop test
- Impact analysis
- Injection Molding
- Failure Analysis
- Reliability Simulation
- Fatigue Analysis

**Example of case studies:**

- Drop test for cell phones
- Car crash

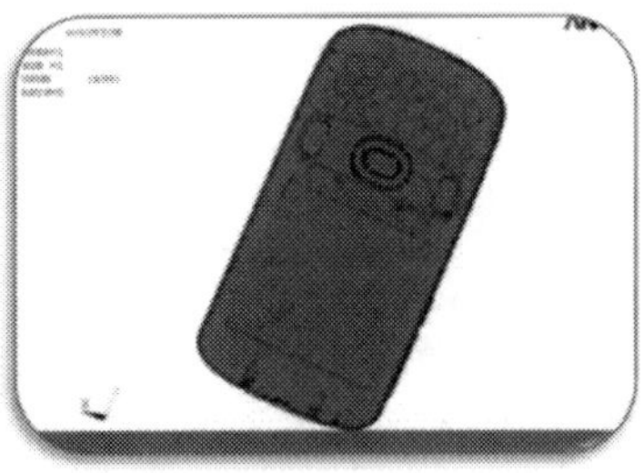

Smart Phone Drop Test

OZEN ENGINEERING

# INDUSTRY SPECIFIC EXPERTISE – BIOMEDICAL INDUSTRY

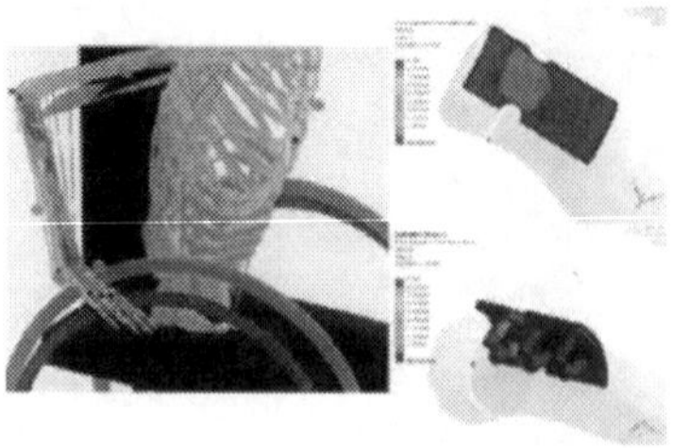

Capabilities:

- Simulating how the human body performs when interacting with the environment
- Model the body, but also the objects it interfaces with
- Optimization of movement patterns
- Implant virtual prototyping

- Analysis of working movements and postures, scale results to population or subject anthropometric data
- Virtually assessing the exertion requirements of a new product or process
- Perform computational assessments and quantitatively investigate ergonomic consequences related to changes in design parameters.

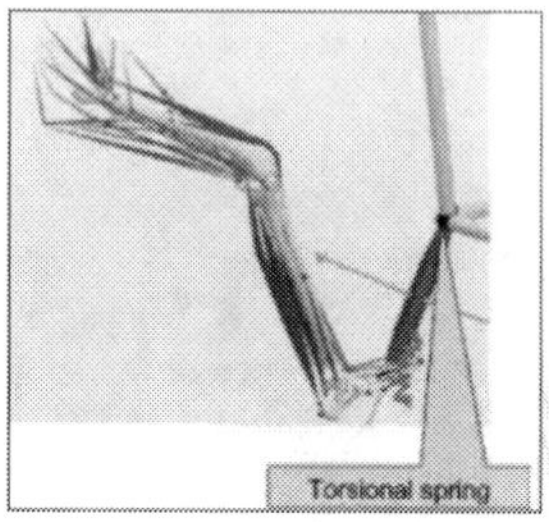


OZEN ENGINEERING

# INDUSTRY SPECIFIC EXPERTISE – INJECTION MOLDING

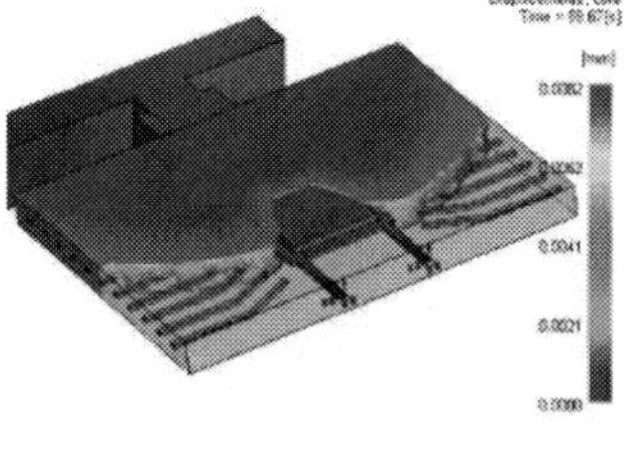

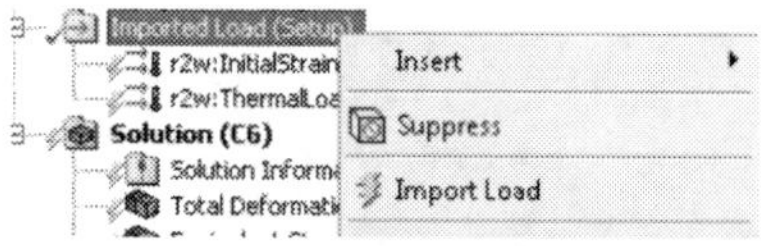


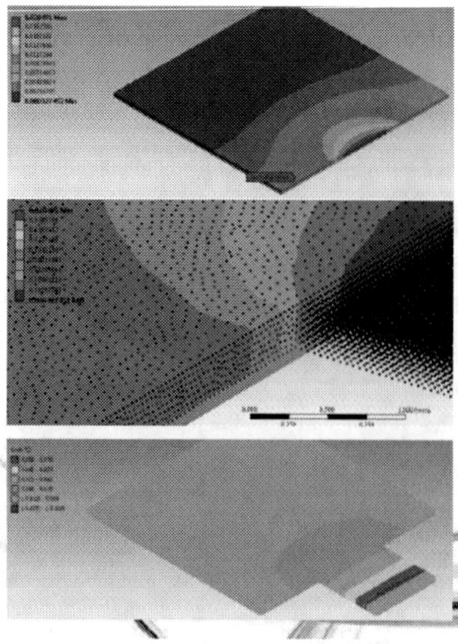

OZEN ENGINEERING

## COUPLED FIELD MEMS SIMULATIONS

FIELD COUPLING:

- Structural Thermal
- Piezoresistive
- Electroelastic
- Piezoelectric
- Thermal-Electric
- Structural Thermoelectric
- Thermal-Piezoelectric
- Structural-Diffusion
- Thermal-Diffusion
- Structural-Thermal-Diffusion
- Structural-Electric
- Acoustic-Structural
- Electromagnetic
- Electromigration
- ...

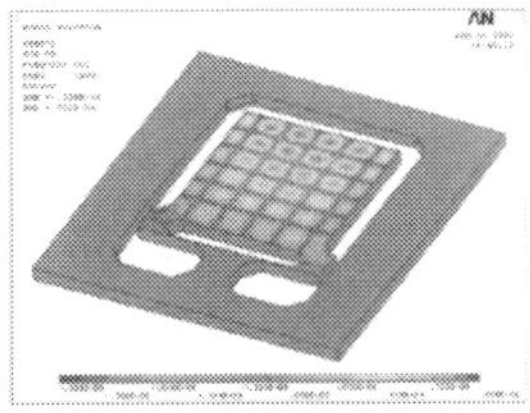

ANSYS has extensive multiphysics coupling capability as can be seen from the list to the left.

OZEN ENGINEERING

## COUPLED FIELD MEMS SIMULATIONS

ADVANTAGES OF DIRECT COUPLED FIELD ELEMENTS:

- Allows for solutions to problems otherwise not possible with usual finite elements.
- Simplifies modeling of coupled-field problems by permitting one element type to be used in a single analysis pass.

OZEN ENGINEERING

## COUPLED FIELD MEMS SIMULATIONS

DISADVANTAGES OF DIRECT COUPLED FIELD ELEMENTS:

- Increases problem size (unless a segregated solver is used).
- Inefficient matrix reformulation (if a section of a matrix associated with one phenomena is reformed, the entire matrix will be reformed).
- Larger storage requirements.

OZEN ENGINEERING

## COUPLED FIELD MEMS SIMULATIONS

COUPLING METHODS:

1. Strong (also matrix, simultaneous, or full) coupling - where the matrix equation is of the form:

$$\begin{bmatrix} [K_{11}] & [K_{12}] \\ [K_{21}] & [K_{22}] \end{bmatrix} \begin{Bmatrix} \{X_1\} \\ \{X_2\} \end{Bmatrix} = \begin{Bmatrix} \{F_1\} \\ \{F_2\} \end{Bmatrix}$$

and the coupled effect is accounted for by the presence of the off-diagonal submatrices $[K_{12}]$ and $[K_{21}]$. This method provides for a coupled response in the solution after one iteration.

2. Weak (also load vector or sequential) coupling - where the coupling in the matrix equation is shown in the most general form:

$$\begin{bmatrix} [K_{11}(\{X_1\},\{X_2\})] & [0] \\ [0] & [K_{22}(\{X_1\},\{X_2\})] \end{bmatrix} \begin{Bmatrix} \{X_1\} \\ \{X_2\} \end{Bmatrix} = \begin{Bmatrix} \{F_1(\{X_1\},\{X_2\})\} \\ \{F_2(\{X_1\},\{X_2\})\} \end{Bmatrix}$$

and the coupled effect is accounted for in the dependency of $[K_{11}]$ and $\{F_1\}$ on $\{X_2\}$ as well as $[K_{22}]$ and $\{F_2\}$ on $\{X_1\}$. At least two iterations are required to achieve a coupled response.

OZEN ENGINEERING

# COUPLED FIELD MEMS SIMULATIONS

| Analysis Category | Coupling Method Used | Example Applications |
|---|---|---|
| Thermal-Structural Analysis | S, W | High temperature turbine |
| Magneto-Structural Analysis (Vector Potential) | W | Solenoid, high energy magnets (MRI) |
| Magneto-Structural Analysis (Scalar Potential) | | |
| Electromagnetic Analysis | S | Current fed massive conductors |
| Electro-Thermo-Structural Analysis | W | Electro-thermal MEMS actuators |
| Electro-Magneto-Thermo-Structural Analysis | W | Direct current electromechanical devices in general |
| Electro-Magneto-Thermal Analysis | | |
| Piezoelectric Analysis | S | Transducers, resonators |
| Electroelastic Analysis | W | Dielectric elastomers |
| Thermo-Piezoelectric Analysis | S, W | Sensors and actuators for smart structures |
| Piezoresistive Analysis | W | Pressure and force sensors |
| Thermo-Pressure Analysis | S, W | Piping networks |
| Acoustic-Structural Analysis | S, W | Acoustics |
| Thermo-Electric Analysis | S, W | High temperature electronics, Peltier coolers, thermoelectric generators |
| Magnetic-Thermal Analysis | W | Direct current transients: power interrupts, surge protection |
| Circuit-Magnetic Analysis | S | Circuit-fed solenoids, transformers, and motors |
| Structural-Diffusion Analysis | S, W | Hygroscopic swelling of polymers in electronics packaging |
| Thermal-Diffusion Analysis | W | Temperature-dependent moisture migration |
| Structural-Thermal-Diffusion Analysis | S, W | Sodium expansion in aluminum reduction cells |

OZEN ENGINEERING

# BASIC GEOMETRIC NONLINEARITIES

- Many MEMS devices undergo large rotation
  - Switches
  - Micro mirrors
  - Gyroscopes
- Switches are usually constrained at both ends, and undergo lateral deflection
- Both of these situations introduce geometric nonlinear effects into the finite element simulation. Without proper treatment, solution results can have significant error.

OZEN ENGINEERING

## BASIC GEOMETRIC NONLINEARITIES

- Consider two phenomena associated with geometric nonlinearities:

1. If an element's *orientation* changes (rotation), the transformation of its local stiffness into global components will change.

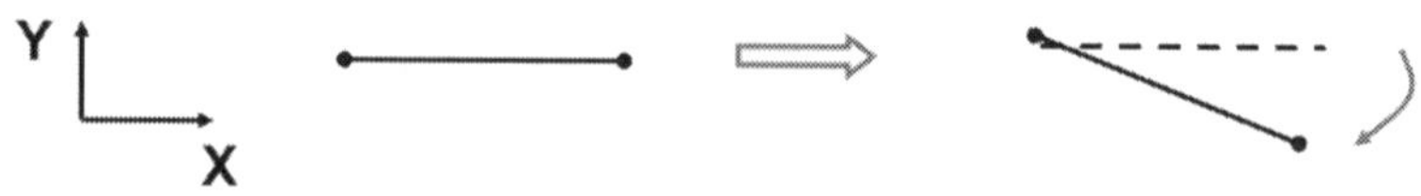


- This is referred to as "LARGE DEFLECTION"

OZEN ENGINEERING

## BASIC GEOMETRIC NONLINEARITIES

2. If an element's strains produce a significant in-plane stress state (*membrane stresses*), the out-of-plane stiffness can be significantly affected.

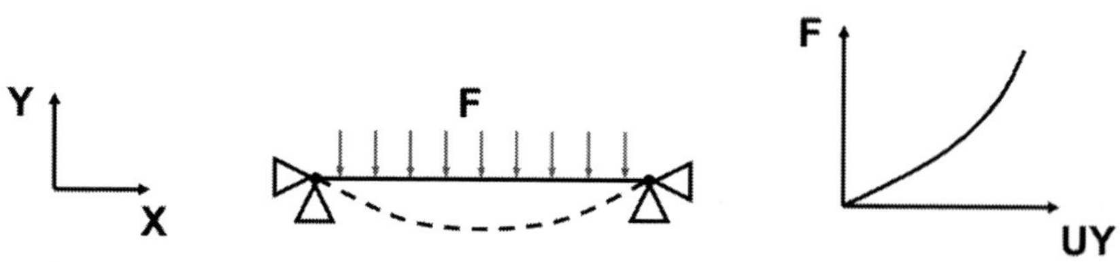


As the vertical deflection increases (UY), significant membrane stresses (SX) lead to a stiffening response.

This is referred to as LARGE DEFLECTION WITH STRESS STIFFENING.

Many MEMS devices exhibit large deflection and stress stiffening.

OZEN ENGINEERING

## INITIAL STRESS

What is Initial Stress?

- Fabrication processes for MEMS structures often leave significant residual stresses in the device
- Residual stresses can significantly effect the performance characteristics of the device
  - Pull-in voltage
  - Eigen frequencies
  - Deflection
- An initial stress state can be prescribed for selected finite elements to simulate residual stresses

OZEN ENGINEERING

## SIMULATION OF MEMS DEVICES

- ANSYS/Mechanical APDL
- ANSYS Workbench
- ACT & ANSYS Workbench = MEMSBENCH!
  - o Note: "Memsbench" is not an ANSYS, Inc. designated label.

- Application Customization Toolkit

OZEN ENGINEERING

## ELECTROSTATIC-STRUCTURAL

- Performance Characterization of a Micromirror
- Micromirror actuation and "pull-in":
  - It is known (analytically) that flat-plate capacitors are such that pull-in occurs when electrostatic gap is reduced by 1/3
  - For torsional actuation, it can be shown analytically that pull-in occurs when micromirror gap at edge is reduced by ~44% ("full-size" micromirror electrode)
  - Further increases in voltage can result in "catastrophic" pull-in: Entire micromirror structure collapses onto stop(s)
  - Note: "Stiction" (important with MEMS devices) is *not* included here

OZEN ENGINEERING

## ELECTROSTATIC-STRUCTURAL

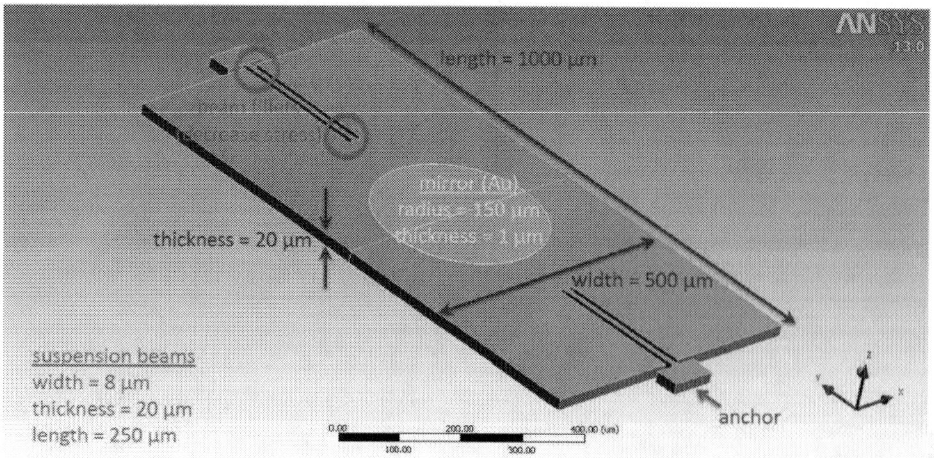


Dimensions of the geometry to be analyzed

OZEN ENGINEERING

## ELECTROSTATIC-STRUCTURAL

- Electrostatic gap between the micromirror bottom (grounded) and the actuation electrode is 3 µm.
- Voltage is varied from 0 to 50 volts, then back down from 50 volts to 0. (Only ~12 volts would be required for likely actuation requirements, but an increased range is considered to investigate overall physical performance.)
- Nonlinear, coupled, electromechanical transducer elements (TRANS126) are used for coupling of the electrostatic and structural domains. A contact stop offset of 0.1 µm from the actuation electrode is assumed.
- Displacement of the bottom edge of one side of the micromirror and the top edge of the other side are plotted vs. applied voltage.
- Various effects are observed, including hysteresis and electrostatic pull-in, collapse, and release. The various physical phenomena are described in more detail on subsequent slides.
- Nonlinear, path-dependent solutions are obtained.

OZEN ENGINEERING

## ELECTROSTATIC-STRUCTURAL

- TRANS126 EMT Elements

- Allows for direct (rather than sequential) coupling of the electrostatic and structural domains.
- Has built-in contact functionality to stop contact between an electrode and the opposing ground plane.

OZEN ENGINEERING

## ELECTROSTATIC-STRUCTURAL

- TRANS126 elements are generated using the EMTGEN macro.
- Requires MAPDL commands; entered by inserting a Command Snippet in the Static Structural Analysis.
- Create a Named Selection for the component of nodes corresponding to the mirror electrode.

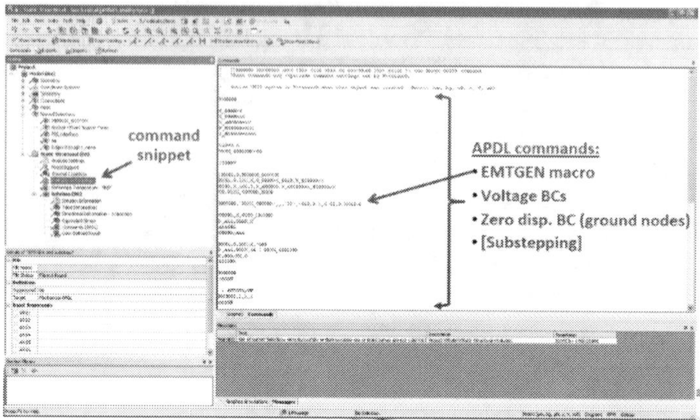


OZEN ENGINEERING

## ELECTROSTATIC-STRUCTURAL

- Hysteresis

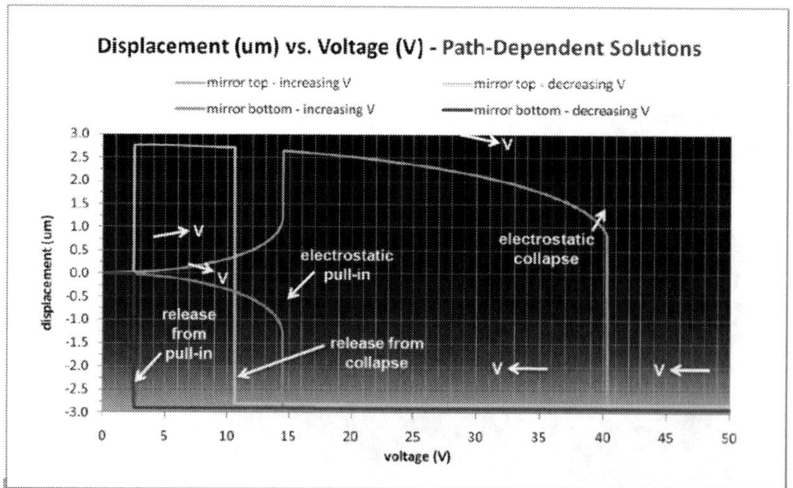


Explanation of the displacement hysteresis is explained in the next page.

OZEN ENGINEERING

## ELECTROSTATIC-STRUCTURAL

- Increasing voltage
  - $0 < V \le 14.44$: Stable equilibrium states of rotational displacement.
  - ~14.46 V:Electrostatic pull-in.
  - $14.48 \le V \le 40.25$:Pulled-in state with micromirror bottom edge contacting stop (torsional flexure of suspension beams).
  - ~40.31 V:Electrostatic collapse.
  - $40.38 \le V \le 50$:Collapsed state with entire micromirror bottom contacting stop (vertical flexure of suspension beams).
- Decreasing voltage
  - $50 \ge V \ge 10.60$:Structure remains in collapsed state.
  - ~10.58 V:Release from electrostatic collapse state to pull-in state.
  - $10.55 \ge V \ge 2.45$:Structure remains in pulled-in state.
  - ~2.42 V:Release from electrostatic pull-in state.
  - $2.40 \ge V > 0$:Stable equilibrium states of rotational displacement.

OZEN ENGINEERING

## ELECTROSTATIC-STRUCTURAL

- Pull-in/collapse and corresponding release states occur at (very) different voltages.
- Example of *path-dependent solutions* –at a given voltage there can be multiple stable solutions. *Very* difficult nonlinear mathematical problem, but ANSYS handles it very well.
- Note that if the problem is solved using only one loadstep with an applied voltage of 20 V the solution obtained corresponds to an electrostatic collapse state, not a pull-in state -> substepping and/or multiple load steps are useful, if not necessary.

OZEN ENGINEERING

## ELECTROSTATIC-STRUCTURAL

At 14.4375 V, structure is in equilibrium state of rotational (torsional) displacement.

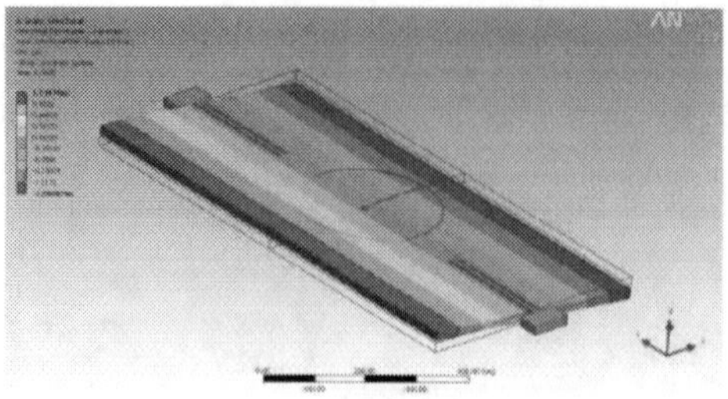

At 14.475 V, structure undergoes electrostatic pull-in.

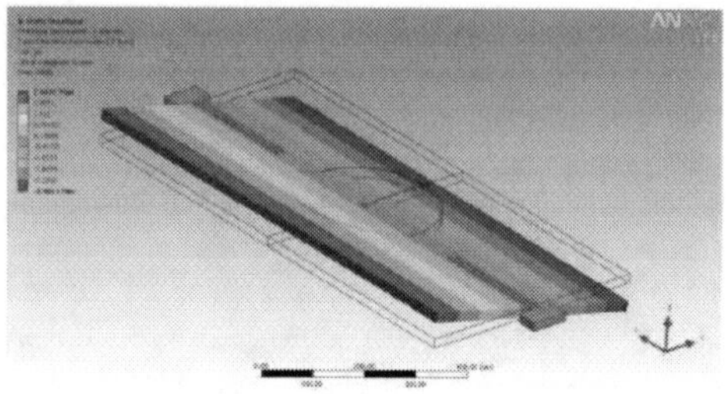

OZEN ENGINEERING

## ELECTROSTATIC-STRUCTURAL

At 40.25 V, structure is still in pulled-in state.

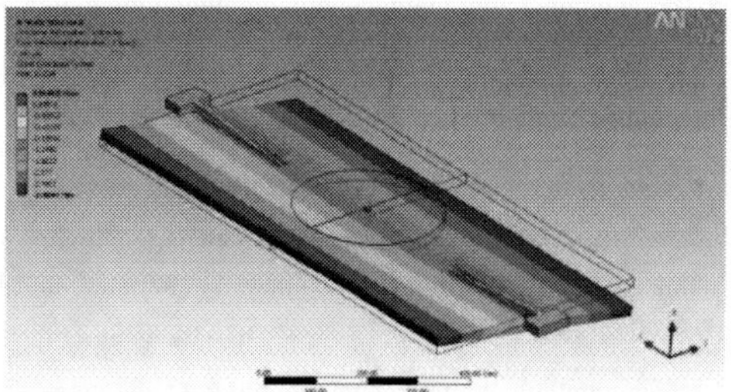

At 40.38 V, electrostatic collapse occurs.

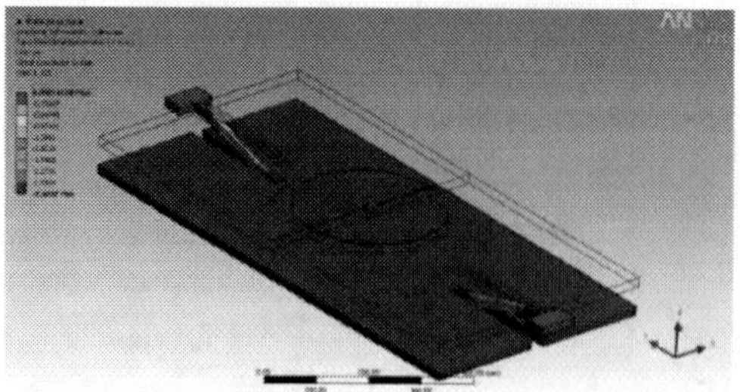

OZEN ENGINEERING

## ELECTROSTATIC-STRUCTURAL

At 10.60 V, structure is still in collapsed state.

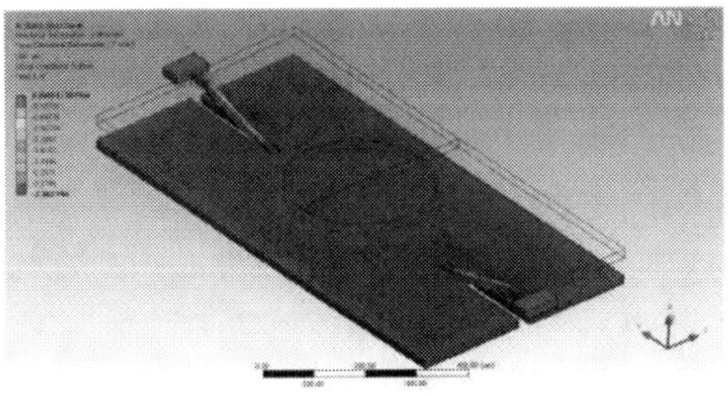

At 10.55 V, structure is released to pull-in state.

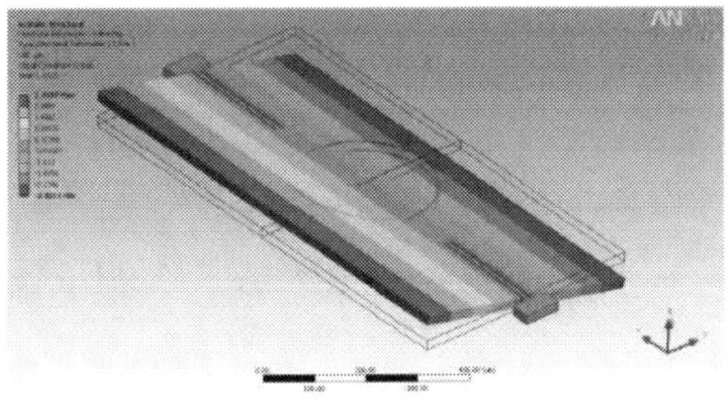

OZEN ENGINEERING

## ELECTROSTATIC-STRUCTURAL

At 2.45 V, structure is still in pull-in state.

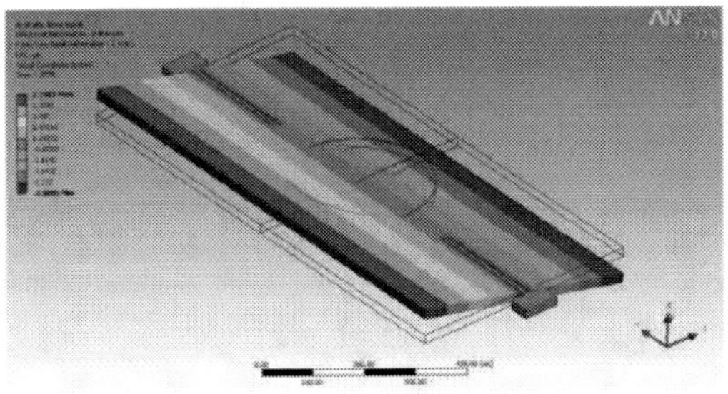

At 2.40 V, structure is released to equilibrium state of rotational displacement.

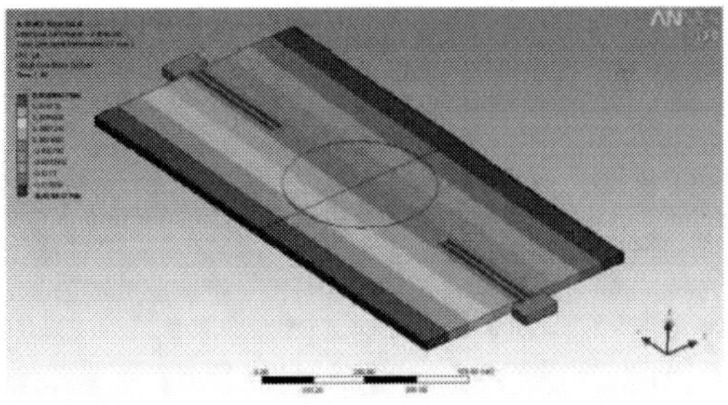

OZEN ENGINEERING

## TRANSIENT ELECTROSTATIC-STRUCTURAL-FLUIDIC

- TRANS126 elements are used for transient analyses
- Two-way fluid-structure interaction (FSI) analysis allows for determination of air damping
- In reality, all structures exhibit damping in some form
- Transient structural analyses can include damping
  - An inverse problem is solved: Determine what parameters for transient structural damping result in comparable results obtained for air damping via FSI

OZEN ENGINEERING

## TRANSIENT ELECTROSTATIC-STRUCTURAL-FLUIDIC

- Response time for MEMS structures is often of high importance.
- Particularly true for micromirrors.
- Here use a different baseline geometry: Suspension beams have *length* x *width* x *thickness* of 200 x 14 x 20 µm, rather than 250 x 8 x 20 µm, i.e. considerably increased stiffness → higher resonant frequencies and faster response times.

- Modal results:

| 250 x 8 x 20 um | | |
|---|---|---|
| mode | f (kHz) | 1/f (usec) |
| 1 | 8.3 | 120.5 |
| 2 | 14.4 | 69.4 |
| 3 | 31.6 | 31.6 |
| 4 | 177.9 | 5.6 |
| 5 | 265.5 | 3.8 |
| 6 | 489.6 | 2.0 |

→

| 200 x 14 x 20 um | | |
|---|---|---|
| mode | f (kHz) | 1/f (usec) |
| 1 | 18.6 | 53.7 |
| 2 | 48.1 | 20.8 |
| 3 | 53.6 | 18.7 |
| 4 | 195.6 | 5.1 |
| 5 | 334.1 | 3.0 |
| 6 | 572.9 | 1.7 |

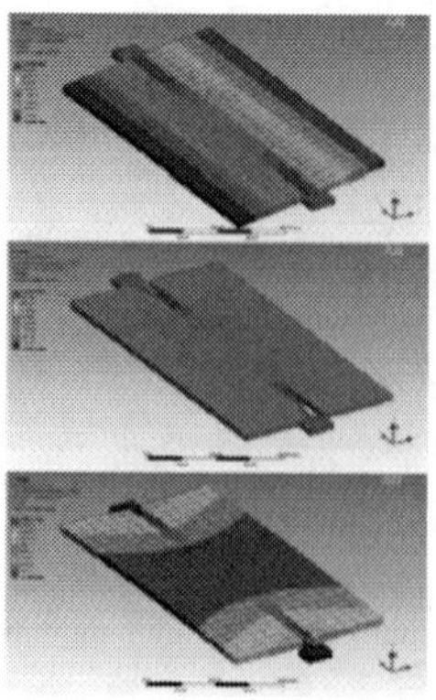

1st mode: rotation about y-axis – beam torsion

2nd mode: horizontal/in-plane – beam y-flexure

3rd mode: vertical/out-of-plane – beam z-flexure

- First three modes are the most important.
- Large separation between 3rd and higher modes.

OZEN ENGINEERING

## TRANSIENT ELECTROSTATIC-STRUCTURAL-FLUIDIC

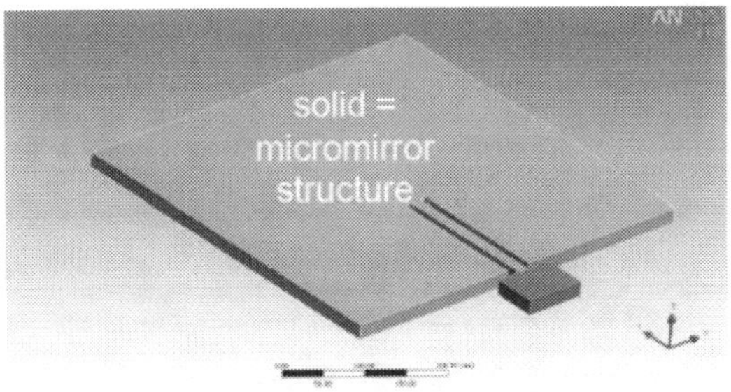


- In solid model (left), tapering (fillets) on suspension beam is removed for easier meshing.
- Air gap beneath structure = electrostatic gap = 3 μm.
- Air gap to each side is 100 μm (overly large, yes).
- Air gap above structure is 12 μm.

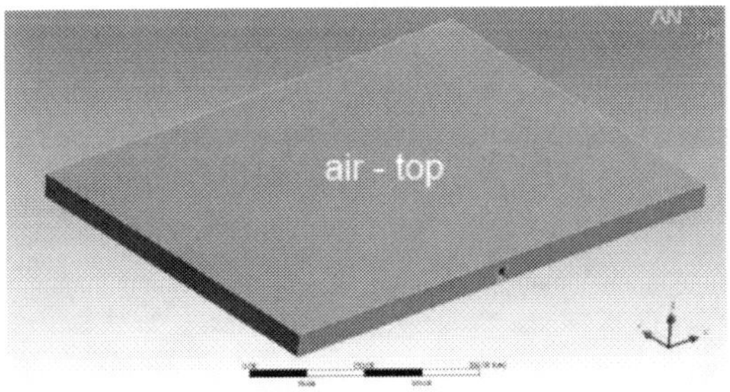


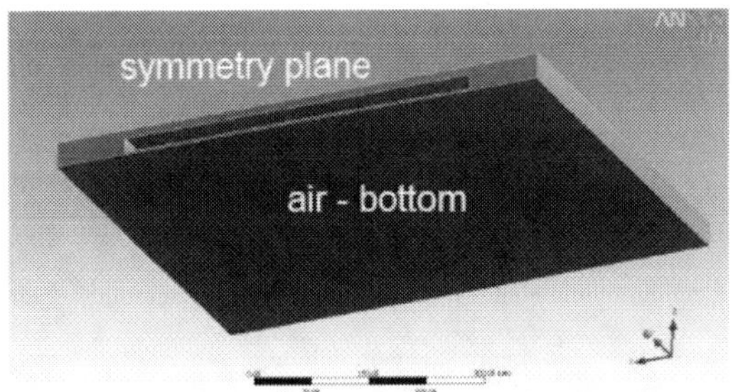


OZEN ENGINEERING

## TRANSIENT ELECTROSTATIC-STRUCTURAL-FLUIDIC

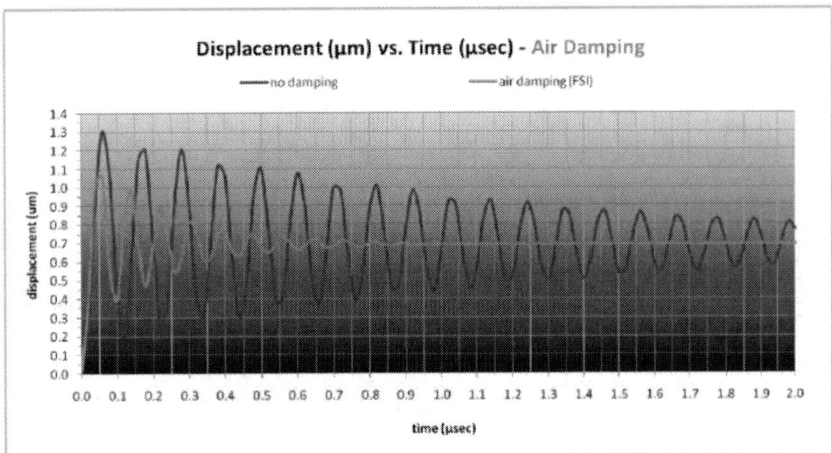


OZEN ENGINEERING

## ELECTROMAGNETIC-THERMAL-STRUCTURAL

- Electromagnetic MEMS devices are not very common but it is worth looking into its performance.

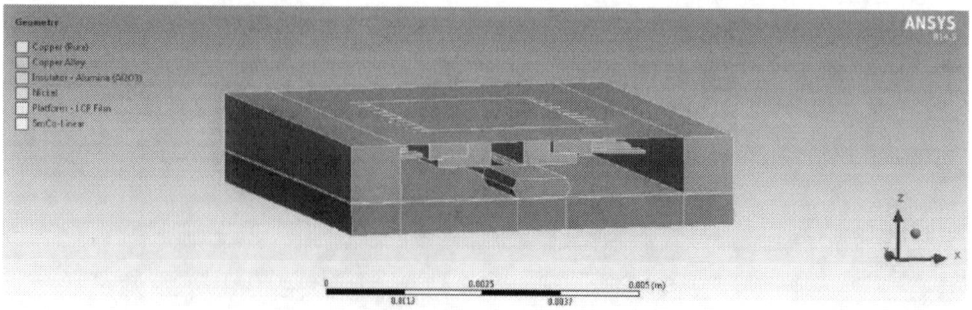


OZEN ENGINEERING

## ELECTROMAGNETIC-THERMAL-STRUCTURAL

- Thermal rise due to current

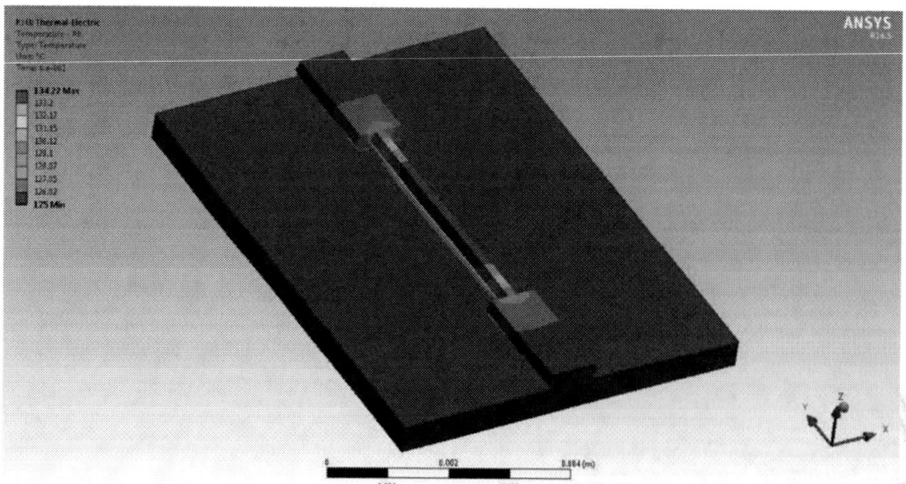


OZEN ENGINEERING

## ELECTROMAGNETIC-THERMAL-STRUCTURAL

- EM Force on the magnets as a function of position:

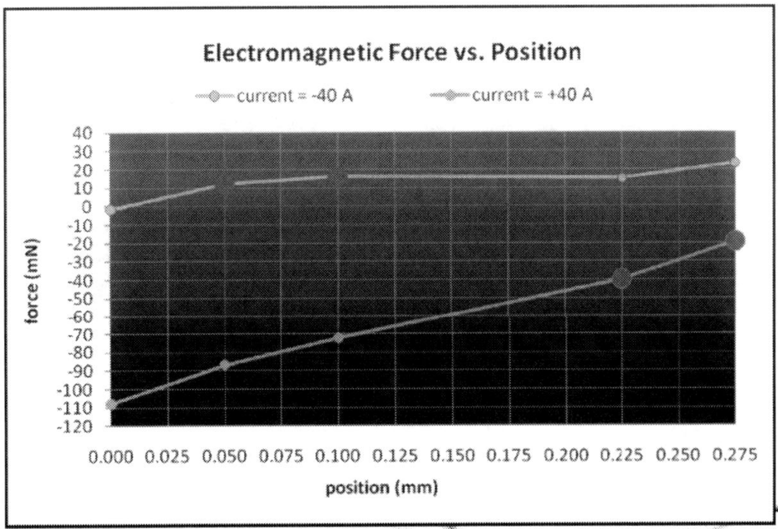


OZEN ENGINEERING

## ACT + WORKBENCH

- ACT exposes existing MAPDL acoustics features in Mechanical without any command objects:
  - Define MEMS related multiphysics elements
  - Add specific MEMS material properties
  - Apply MEMS related boundary conditions
  - ...

OZEN ENGINEERING

## Developed By Ozen: EKILL/EALIVE ACT

- Developed by Ozen Engineering Inc
- Works on bodies & contacts
  - Bodies: 'killed' bodies have elements deactivated and set to a low stiffness material until 'alived' in a deformed, stress free state
  - Contacts: Contacts can be turned on & off throughout analysis

Details of "Element Death"

| Definition | |
|---|---|
| Type | Contact |
| Kill Element Step | 1 |
| Alive Element Step | 3 |
| Contact | Frictionless - Solid To Solid <28> |

OZEN ENGINEERING

## Developed By Ozen: ELECTROMIGRATION ACT

- Developed by Ozen Engineering Inc.
- Electromigration calculations based on a thermal-electric to static structural analysis sequence
  - Automatically sets up and runs a 3rd analysis to resolve total Atomic Diffusion Gradient
- Next steps in extension development:
  - Delete elements as indicated by Atomic Diffusion Gradient & rerun cycle
  - Implement units (custom units will be required)

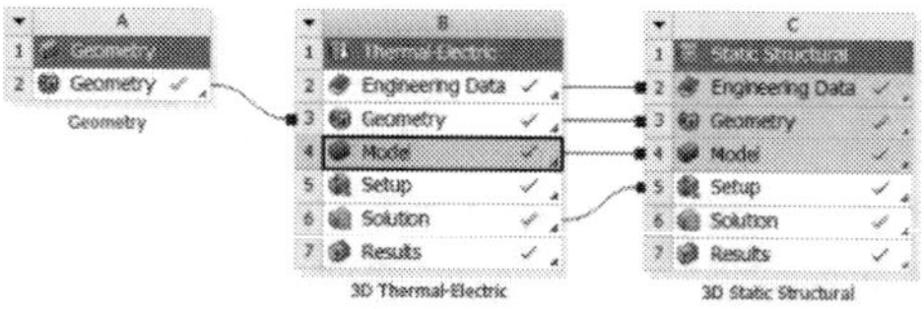


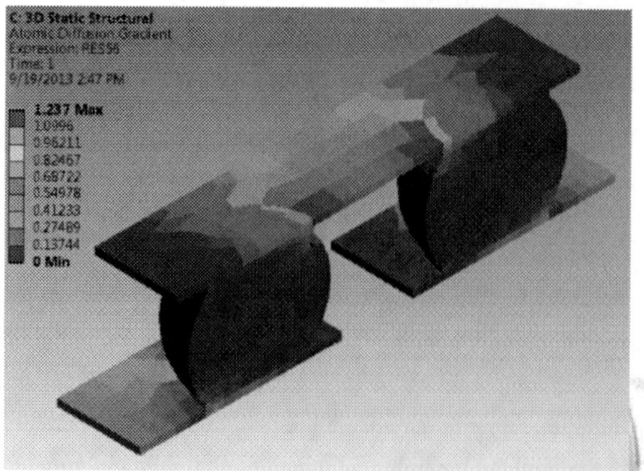


Details of "Electromigration Tool"

| | |
|---|---|
| **Main** | |
| Scoping Method | Named Selection |
| Named Selection | MAIN |
| **Inlet** | |
| Scoping Method | Named Selection |
| Named Selection | INLET |
| **Outlet** | |
| Scoping Method | Named Selection |
| Named Selection | OUTLET |
| **Exterior** | |
| Scoping Method | Named Selection |
| Named Selection | EXTERIOR |
| **Definition** | |
| Thermo-Electric Analysis | Steady-State Thermal-Electric Conduction |
| Structural Analysis | Static Structural |
| dx (discretization) | 1E-05 |
| z* | -33 |
| kT (Current) | 4.14E-21 |
| kT (Temp) | 0.026 |
| kT (Stress) | 4.14E-21 |
| e | 1.6E-19 |
| Q* | 0.0094 |
| Omega | 2.48E-29 |

OZEN ENGINEERING

## SUMMARY

- ANSYS is an excellent Multiphysics simulation tool for MEMS applications.
- Although MEMS simulations started in ANSYS Mechanical APDL, they can now be run in ANSYS Workbench.
- ANSYS Workbench and ACT (Application Customization Toolkit) enables the best environment for creating customized MEMS simulations.

OZEN ENGINEERING

# THANK YOU FOR YOUR ATTENTION!

**FOR FURTHER INFORMATION, PLEASE CONTACT:**


**OZEN ENGINEERING, INC.**
1210 E. ARQUES AVE. SUITE: 207
SUNNYVALE, CA 94085 USA
+1 (408) 732-4665
[illegible]@ozeninc.com
[illegible]ozeninc.com


OZEN ENGINEERING

TOYOTA

2014.10.10

2014 Automotive Simulation World Congress

# Measurement and Prediction Technology of Cooling Capability for Hybrid Drivetrain Components

Tadashi Yamada
TOYOTA MOTOR CORPORATION

---

## Contents

1, Introduction
2, Cooling mechanism of the Hybrid drivetrain
3, Measurement technology
4, Prediction technology(CFD)
5, Improvement of cooling capability (Case study)
6, Summary

# 1, Introduction

### Design of the HV unit

Realize the aim performance by control a flow of the oil, and reduction the rebellion

### Cooling performance

Important in securing the fuel economy of the vehicle, function and durability of the unit

Difficult to achieve an aim in severe heat environment in the engine room

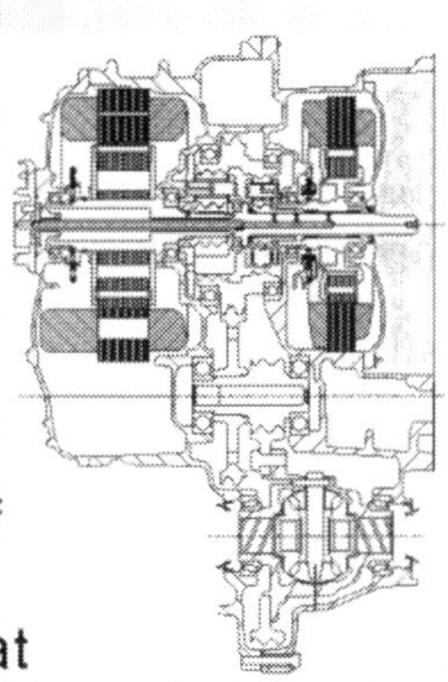

**Aim performance**

*Cooling*

*Lubrication*

*Power transmission / control*

**Rebellion**

*Stirring loss*

*Oil leak*

*Vibration / Noise*

*Substance transportation*

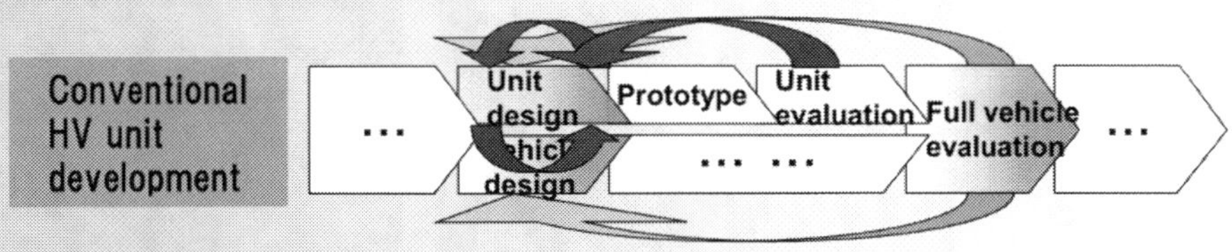


Repeat a practical vehicle evaluation (Wind Tunnel)

Needs to front loading

# 2, Cooling mechanism of the hybrid drive unit

### Heat condition of HV unit

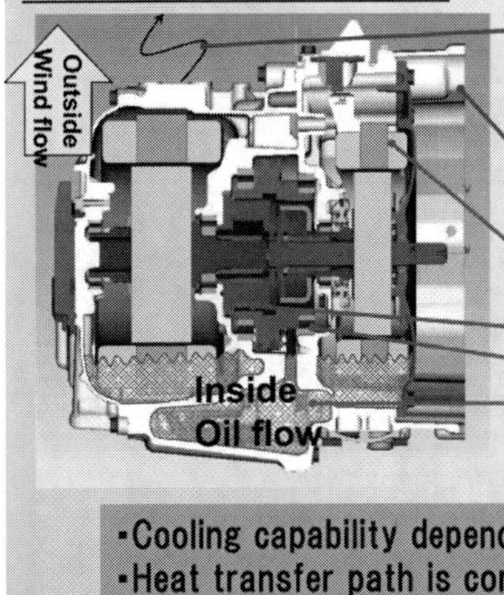


**Heat transfer**

*①To outside air*

*②To water jacket*

*③To oil cooler*

*④To neighbor parts*

**Heat generation**

*①Motor generator (iron , copper loss)*

*②Gear engage*

*③Bearing friction*

*④Oil stirring*

### Heat transfer path

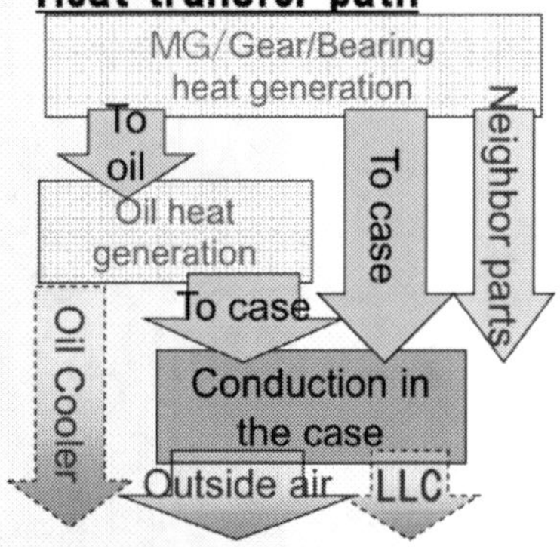


- Cooling capability depends on inside oil flow and outside wind flow
- Heat transfer path is complicated, and difficult to grasp it

### Target technology

【Measurement】 Enable the measurement in a unit simple substance

【Prediction】 Enable the prediction of temperature and heat flux

# 3, Measurement technology (Thermal VRS bench)

**Measurement evaluation technology that simulated a practical vehicle running condition**

**The VRS bench which realizes the drive of the unit simple substance**
⇒Simulate practical vehicle driving condition
- *Measure torque loss*
- *Measure electrical power loss*

**+**

**Thermoregulation wind tunnel**
⇒Simulate practical vehicle driving condition
- *Control and measure outside air cooling and water cooling*

**+**

**Developed multi points measurement technology of unit surface temperature and heat flux**

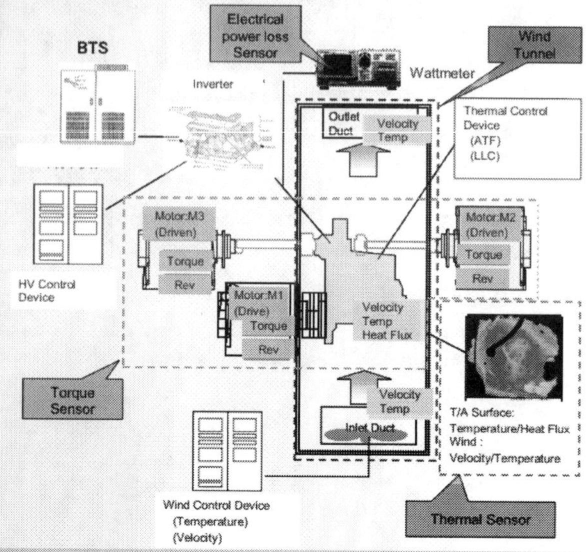


Thermal VRS bench enabled :
Measurement evaluation of each part heat flux and temperature in the environment that simulated the practical vehicle (Without needing practical vehicle)

# 4, Prediction technology (CFD)

**Temperature prediction : Inside oil flow, outside airflow and coupled solid analysis**

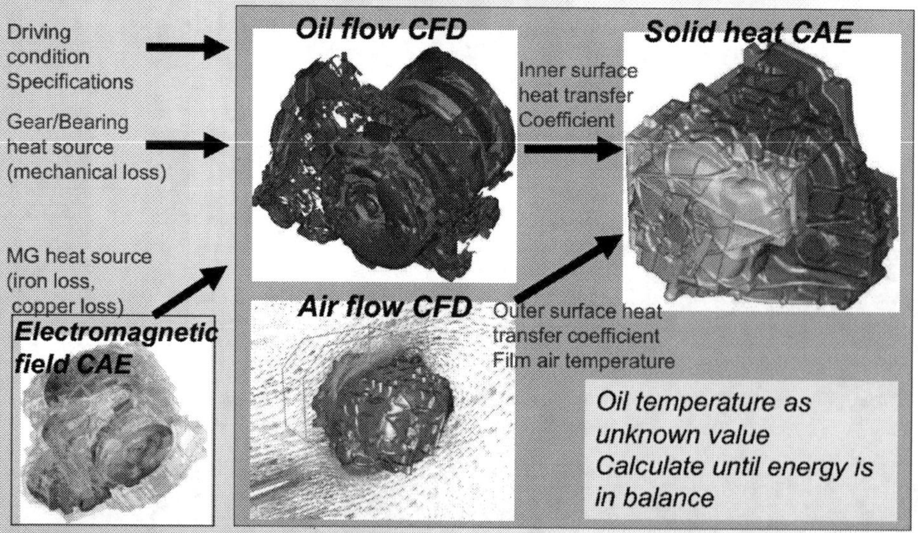


To calculate in a short term :
At first calculate inside oil flow and outside air flow separately
Finally perform in coupled heat calculation

## 4, Prediction technology (CFD)

### Approach to Inside oil flow

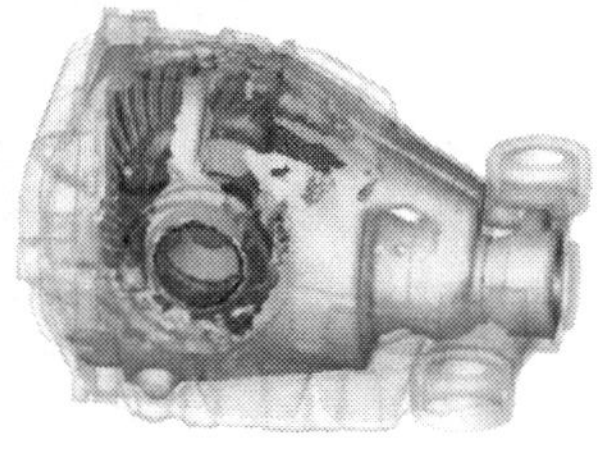

[Required function]

- Multi phase flow of oil and the air. (free surface)
- Moving/rotation objects such as gear (moving boundary)
- Calculate time averaged heat transfer coefficient

[Realization method in the CFD]

- Free surface is considered by VOF (Volume Of Fluid) method.
- Combination of Moving mesh and Grid interface (Really rotate gear).
- By integration of the user defined function, the heat transfer coefficient that was averaged by time is calculated from the transient flow.
- The method that become basic of this technology had been already developed (Reported in 2011JSAE, 2013ANSYS user meeting)

## 4, Prediction technology (CFD)

### Problem of the inside oil flow CFD

Large scale and complex shape
Large number of assy parts
Space volumes increase largely
⇒ A scale is big, and an application of the CFD is difficult

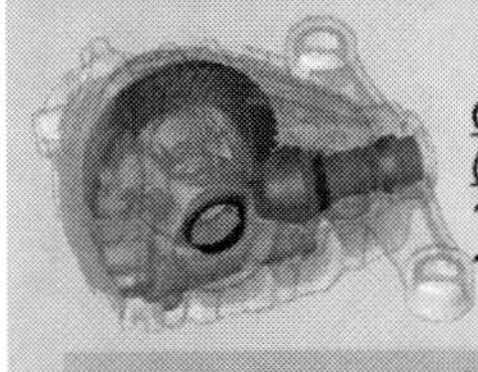

Conventional model (differential)
**1,300,000** fluid cells
**16** parallel CPUs

HV unit model
**5,400,000** fluid cells
**256** parallel CPUs

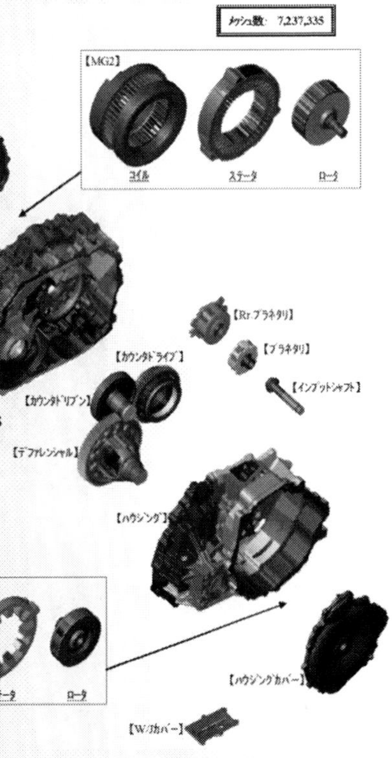


Enabled analysis by:

- Simplification of the part which does not contribute to heat.
- Development of the modeling technique and standardization.
- increasing computer resources and solver licenses

# 4, Prediction technology (CFD)

## Inside oil flow CFD

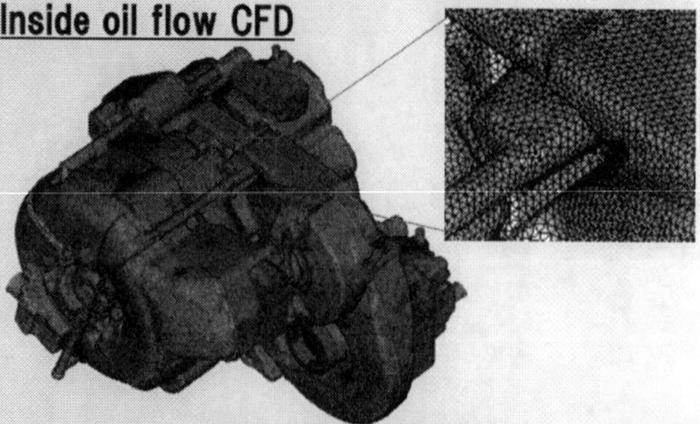

Calculation model(mesh)

Explicit

Implicit

Oil distribution

Calculation condition of inside oil flow

| | |
|---|---|
| Solver | ANSYS FLUENT Ver13(Transient) |
| Turbulence Model | RNG k-$\varepsilon$ |
| Wall Function | Enhanced Wall Treatment |
| Multi Phase Method | **VOF Compressive (Implicit)** |
| Number of Element | 5,362,144 |
| Vehicle Speed | 35、100、170、180(km/h) |

# 4, Prediction technology (CFD)

## Result of inside oil flow CFD

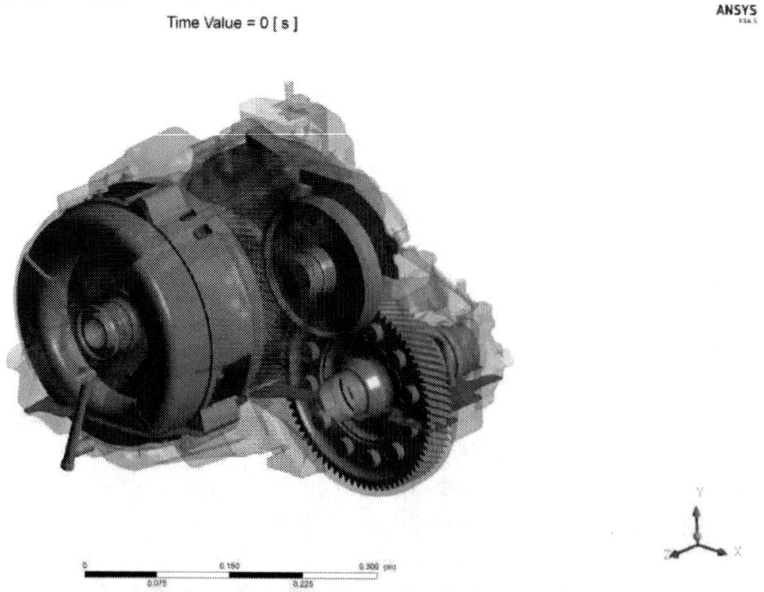


CFD result : Oil distribution (100km/h)

## 4, Prediction technology (CFD)

### Validation of inside oil flow CFD

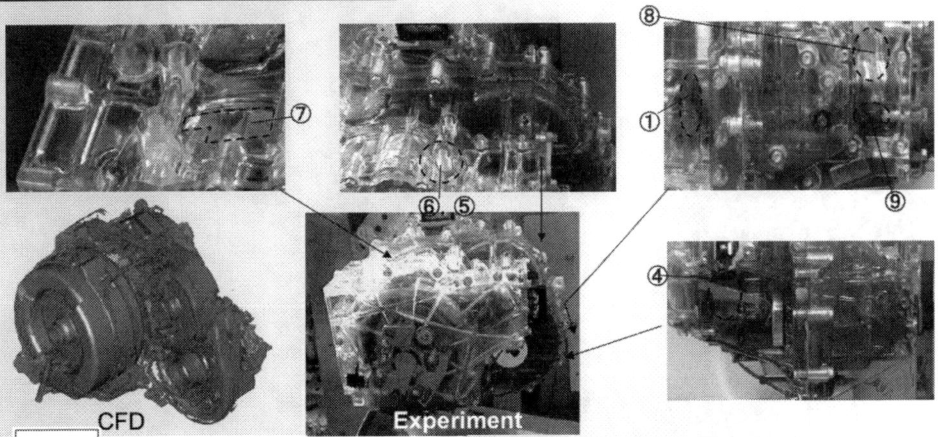


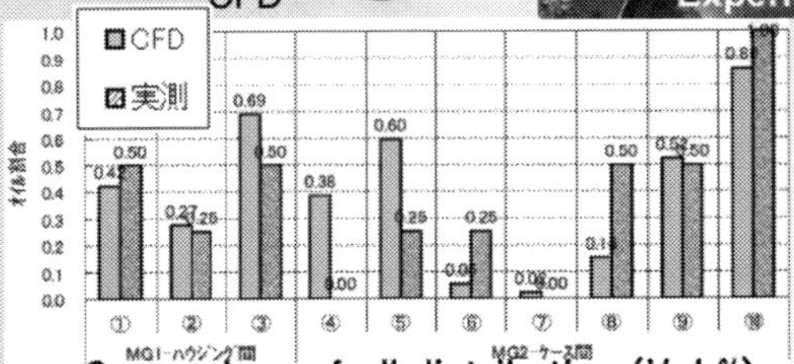


Comparison of oil distribution (Vol %)

CFD expresses oil movement and distribution of experiment exactly

↓

Accuracy of CFD is good

## 4, Prediction technology (CFD)

### Progress

Oil distribution
(Transient)

# 4, Prediction technology (CFD)

## Progress

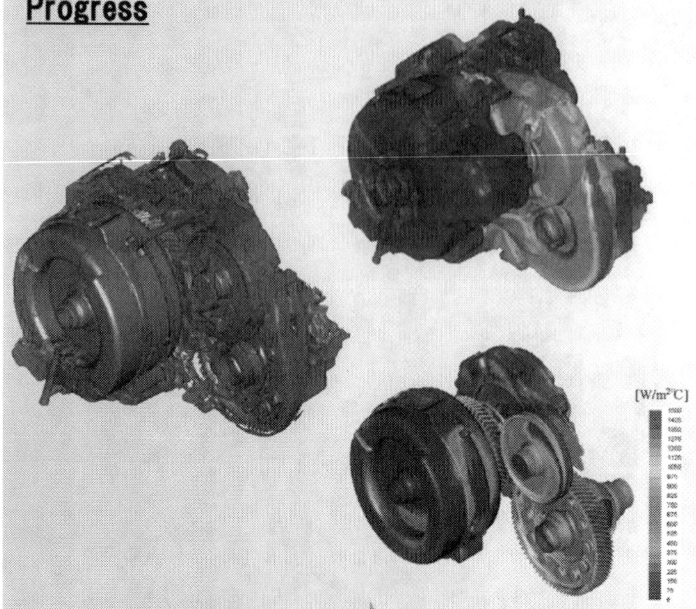


Oil distribution (Transient) ⇒ Heat transfer coefficient (Time averaged)

# 4, Prediction technology (CFD)

## Outside air flow CFD

Calculation condition of outside air flow

| | |
|---|---|
| Solver | ANSYS Fluent R15.0.7 |
| Turbulence Model | Realizable k-epsilon |
| Wall Function | Enhanced Wall Treatment |
| Number of Element | 22,000,000 |
| Wind Velocity | 5、6、10、12、20(m/s) |
| Wind Temp | 40(℃) |

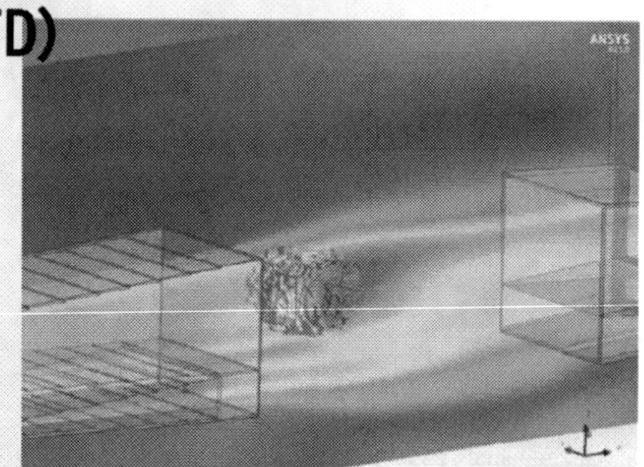

Distribution of wind velocity

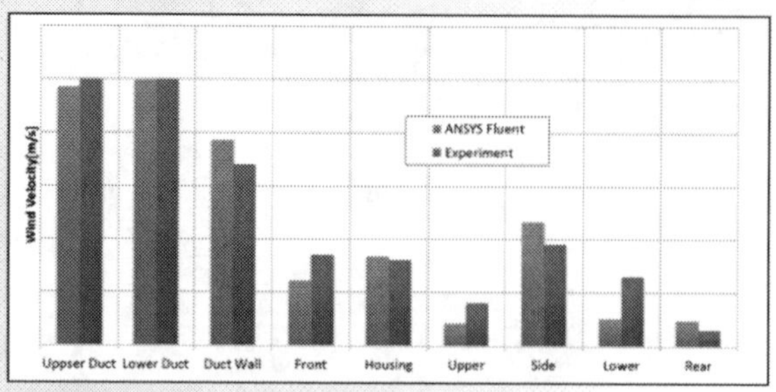


Comparison of wind velocity

Air flow CFD fits measured value well ⇒ Accuracy of CFD is good

# 4, Prediction technology (CFD)

## Progress

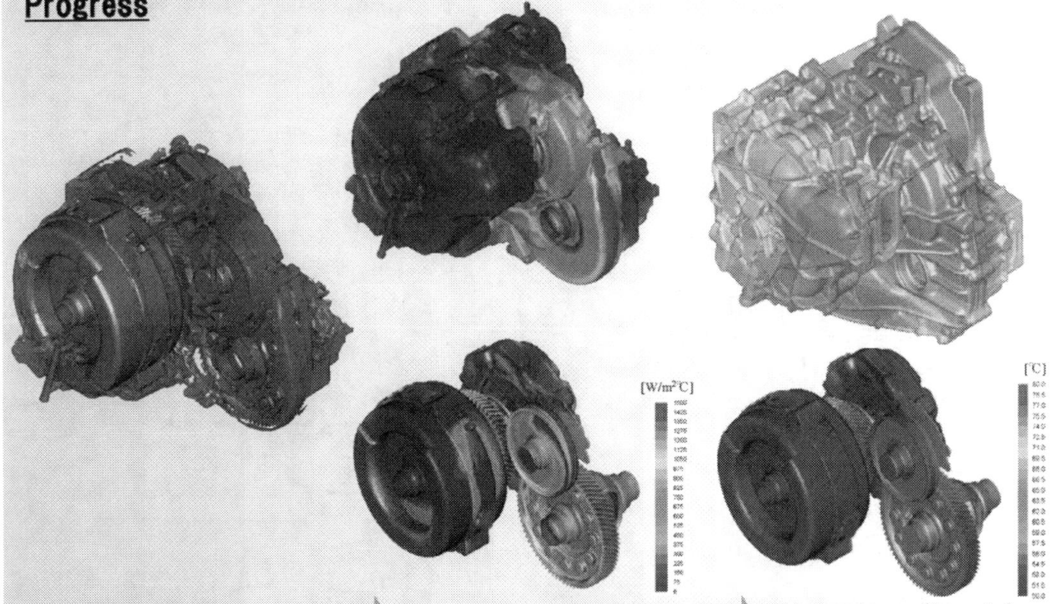


Oil distribution (Transient) ⇒ Heat transfer coefficient (Time averaged) ⇒ Temperature (Steady)

# 4, Prediction technology (CFD)

## Validation of temperature prediction

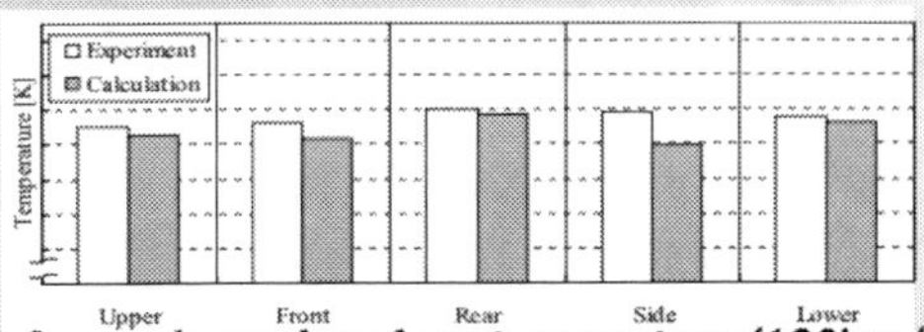


Comparison of surface temperature (180km/h)

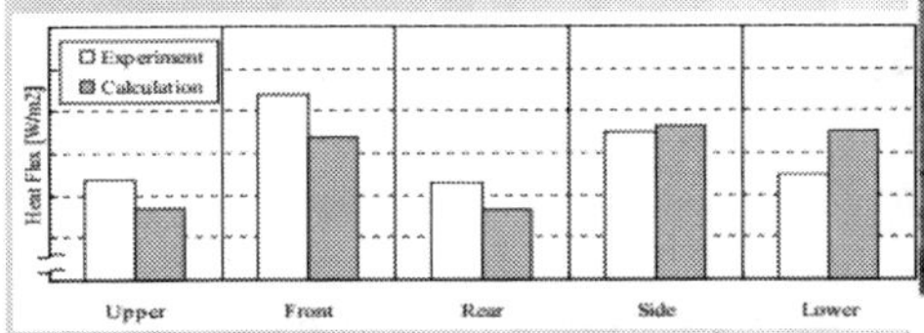


Comparison of surface heat flux (180km/h)

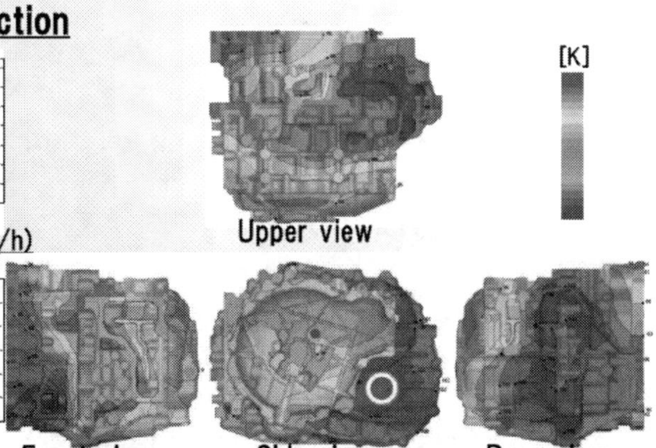


Upper view

Front view Side view Rear view

Distribution of surface temperature

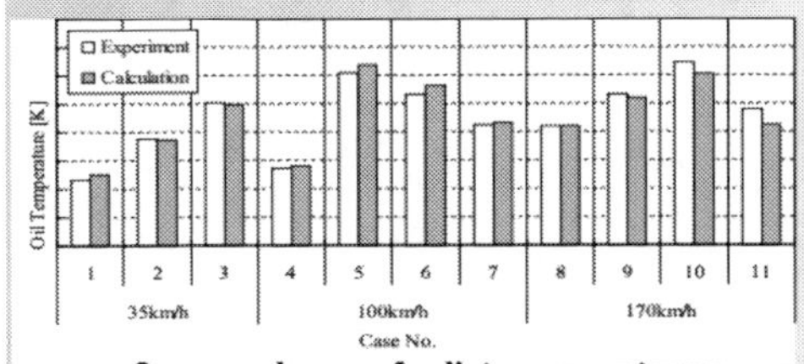


Comparison of oil temperature

Calculation results fit measured value well

⇒ ***It enabled analysis of cooling mechanism under various driving condition of the vehicle***

# 5, Improvement of cooling capability

**Improvement investigation of cooling capability (Case study)**

**Before**

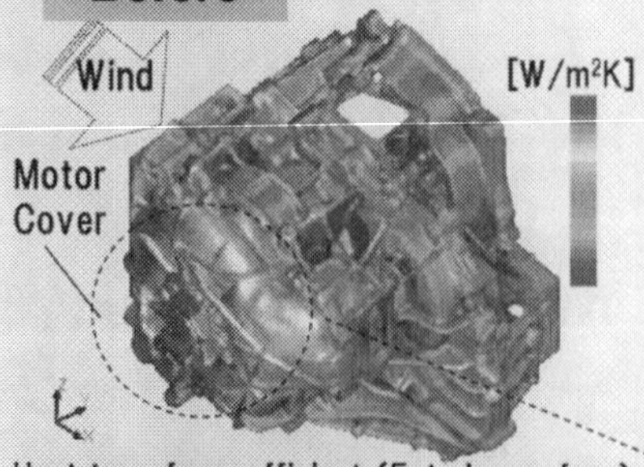


Heat transfer coefficient (Exterior surface)

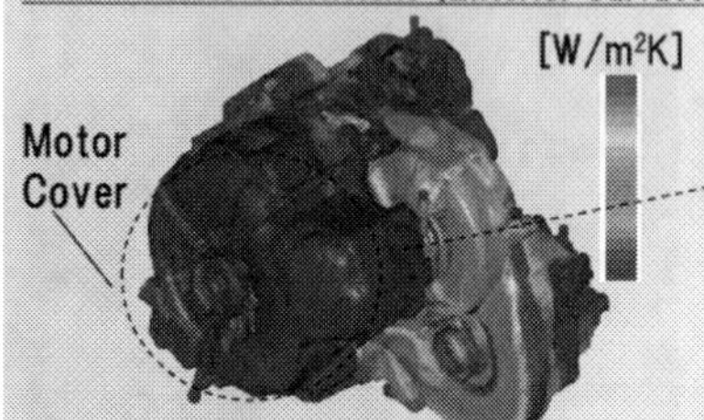


Heat transfer coefficient (Interior surface)

**After**

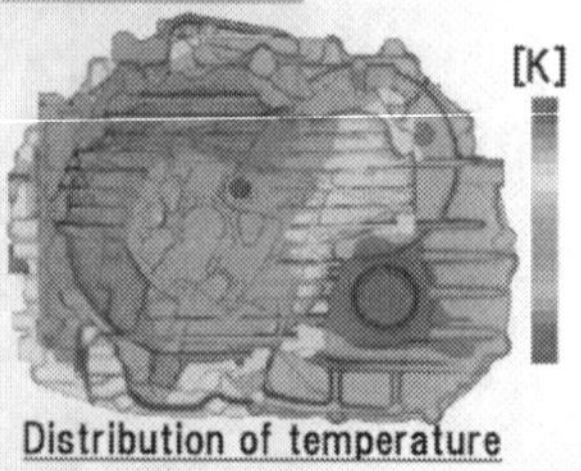


Distribution of temperature

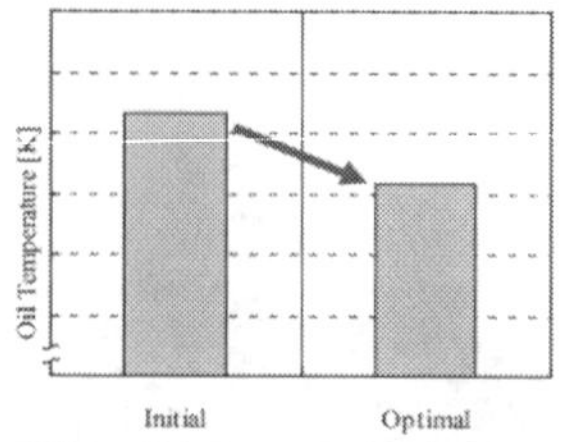


Comparison of oil temperature

Motor Cover:
- Heat transfer coefficient is high in outer surface ⇒ Add fin
- Heat transfer coefficient is low in inner surface ⇒ Supply hot oil

**Made it possible to improve cooling capability that is effective depending on mechanism**

# 6, Summary

**Conclusion**

Thermal VRS bench enabled measurement evaluation of each part heat flux and temperature in the environment that simulated the practical vehicle.

CFD enabled analysis of cooling mechanism under various driving condition of the vehicle.

Prediction technology made it possible to improve cooling capability that is effective depending on mechanism at the stage of design.

## 6, Summary

Next theme

This prediction technology is applied only to a steady running condition (vehicle speed and oil temperature are constant). Continue engineering development toward the prediction of the transient running condition.

Development of the one dimension tool which assumed heat transfer characteristic provided in this development technology. Utilization as a cooling capability simple examination tool at early period of design.

Thank you

2014 Automotive Simulation World Congress

# Evaluation Method for Pressure Loss using Energy Dissipation

Hiroshi Okumura,Keiji Komura,
Denso Techno Corporation

DENSO

デンソーテクノ株式会社



---

2

DENSO
デンソーテクノ株式会社

| | |
|---|---|
| Established | 1984/4/3 |
| Head office | Obu,Aichi,Japan |
| Technical centers | KariyaTC, AnjoTC, AnjoTC2, FukuokaTC |
| Group company | DENSO TECHNO PHILIPPINES |
| Business | Powertrains, Vehicle control, Information communication, Air-conditionin, Body control, Production technology, Technical information systems, HEMS, CAE |
| Capital | 180 million yen |
| Sales | 46, 500 million yen (2013) |
| No. of employees | 2, 466 (April,2014) |

http://www.densotechno.co.jp/

3

# DENSO

デンソーテクノ株式会社

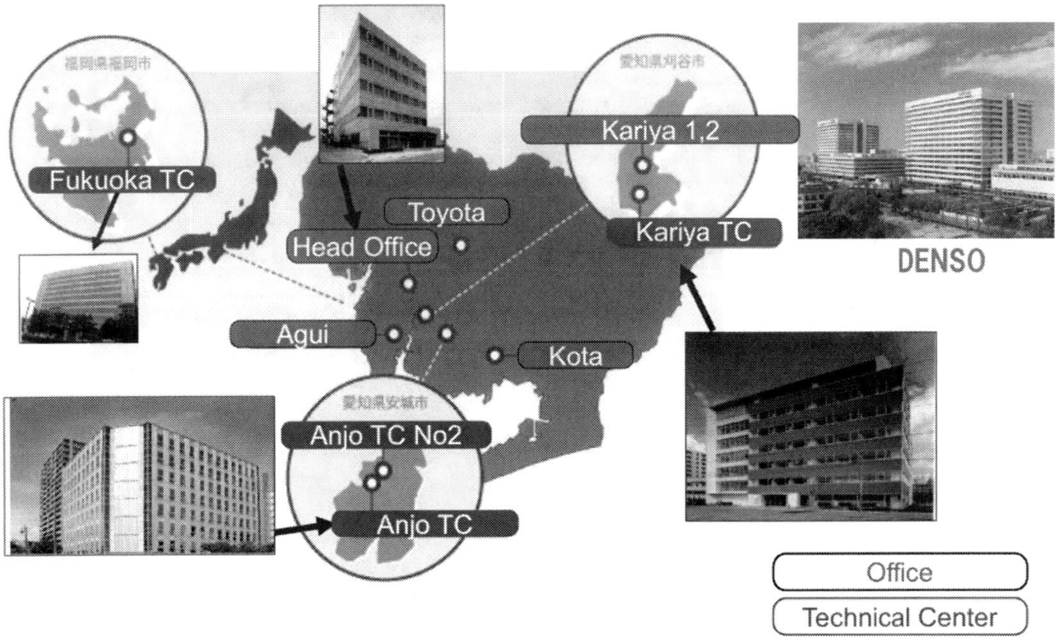


4

# Finding the loss using CFD

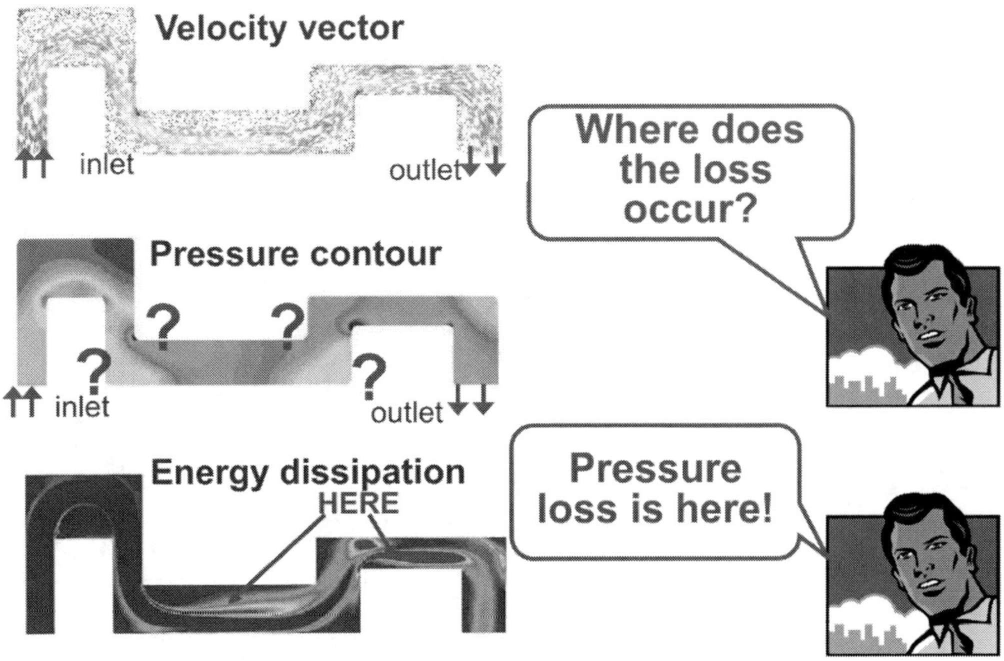

## Energy dissipation

5

**Two energy dissipations** (cause) ⇒ Pressure loss (effect)

**Thermal dissipation**

$$\nu\left(\frac{\partial \overline{u_i}}{\partial x_j}+\frac{\partial \overline{u_j}}{\partial x_i}\right)\frac{\partial \overline{u_i}}{\partial x_j} \quad [\mathrm{W/m^3}]$$

$\nu$ : molecular viscosity

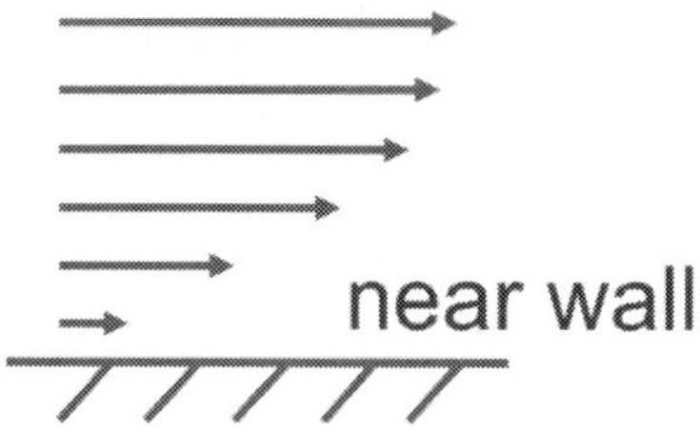


**Turbulent dissipation**

$$-\overline{u_i'u_j'}\frac{\partial \overline{u_i}}{\partial x_j} \quad [\mathrm{W/m^3}]$$

$-\overline{u_i'u_j'}$ : turbulent stress

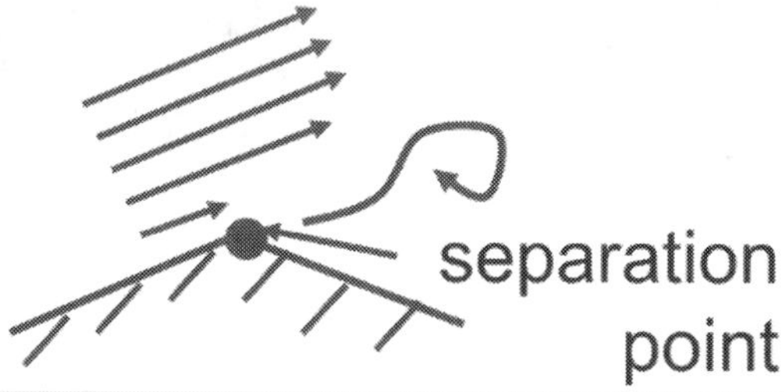


## Design process using CFD

6

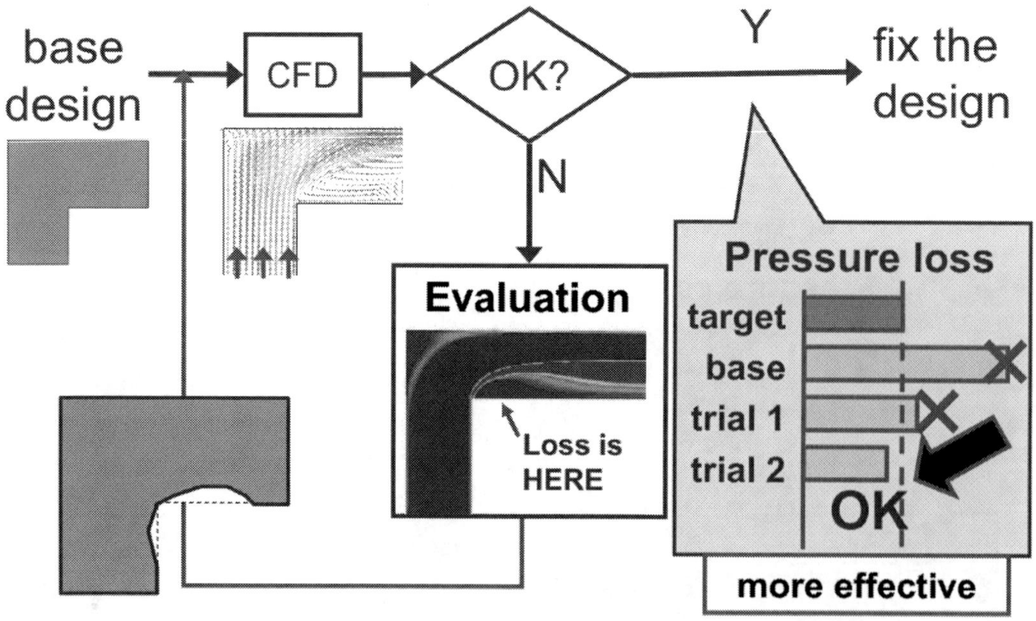

## Product duct application

7

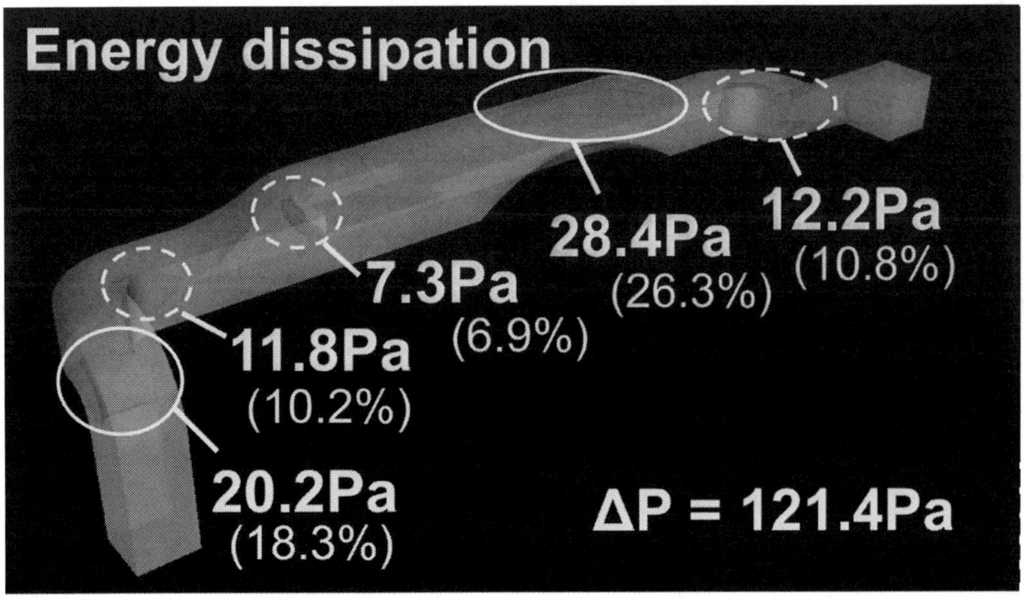


## ANSYS-JAPAN Supported

8

DENSO
デンソーテクノ株式会社

ANSYS
アンシス・ジャパン株式会社

**ANSYS Fluent**

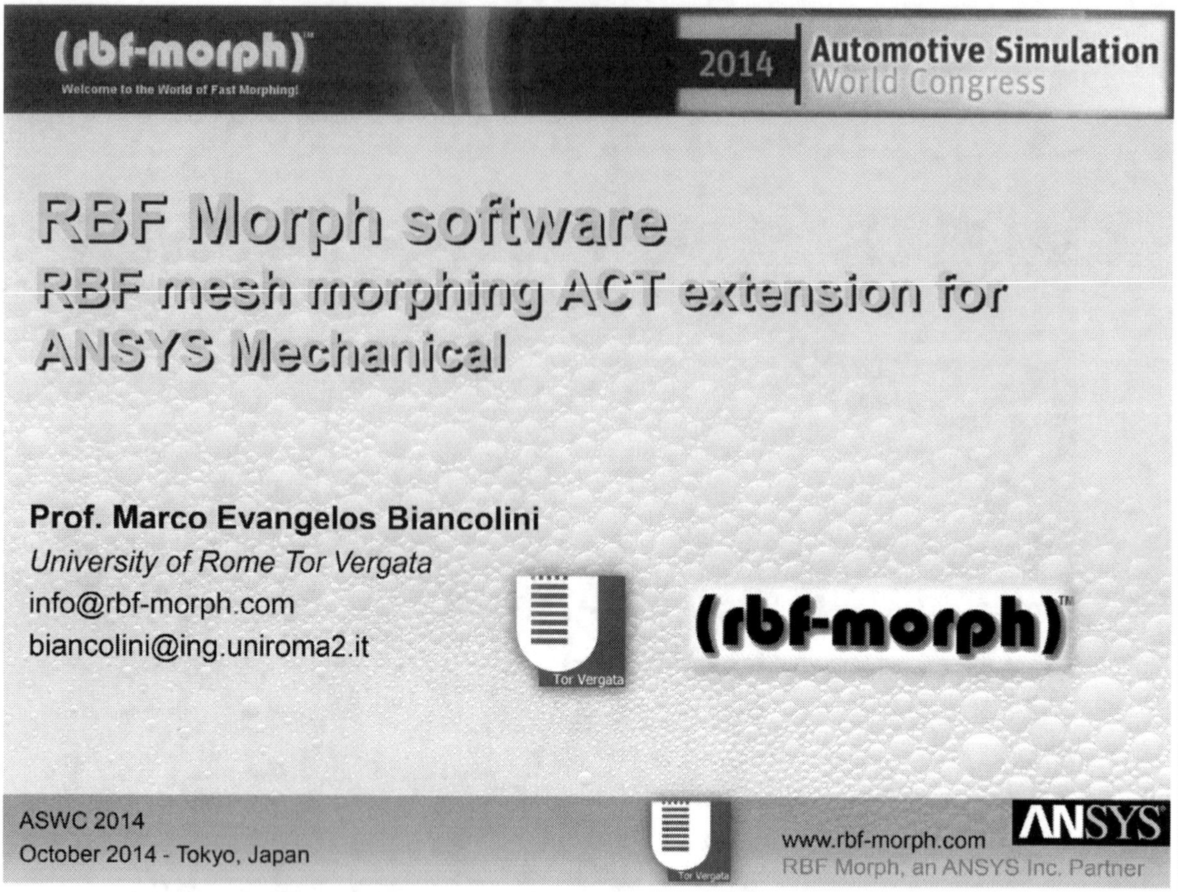
(rbf-morph)™
Welcome to the World of Fast Morphing!
2014 Automotive Simulation World Congress
RBF Morph software
RBF mesh morphing ACT extension for ANSYS Mechanical
Prof. Marco Evangelos Biancolini
University of Rome Tor Vergata
info@rbf-morph.com
biancolini@ing.uniroma2.it
Tor Vergata
(rbf-morph)™
ASWC 2014
October 2014 - Tokyo, Japan
www.rbf-morph.com
ANSYS
RBF Morph, an ANSYS Inc. Partner

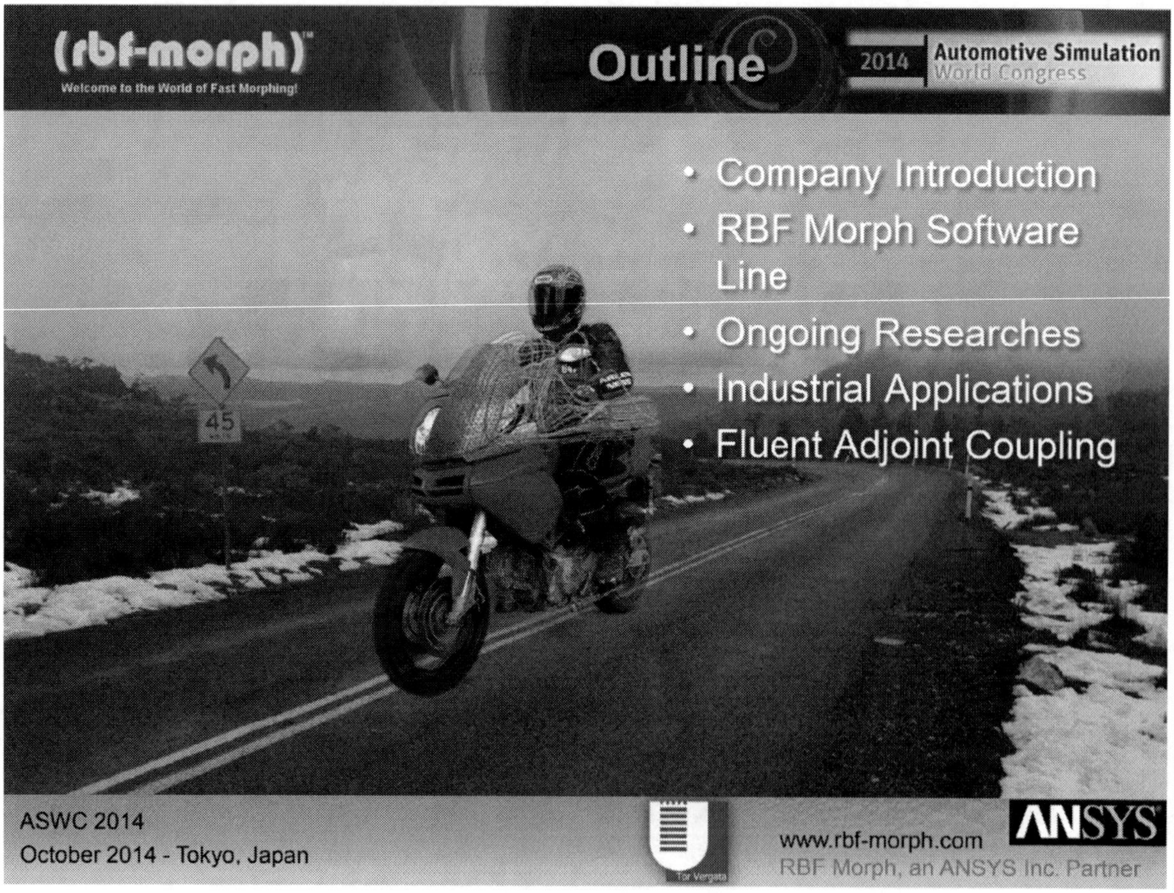
(rbf-morph)™
Welcome to the World of Fast Morphing!
Outline
2014 Automotive Simulation World Congress
• Company Introduction
• RBF Morph Software Line
• Ongoing Researches
• Industrial Applications
• Fluent Adjoint Coupling
45
ASWC 2014
October 2014 - Tokyo, Japan
www.rbf-morph.com
ANSYS
RBF Morph, an ANSYS Inc. Partner

(rbf-morph)™ Welcome to the World of Fast Morphing!

## Company Introduction

Automotive Simulation World Congress 2014

RBF Morph is a pioneer and world-leading provider of numerical morphing techniques and solutions conceived to efficiently handle shape optimization studies concerning most challenging industrial applications. We are an independent software-house and vendor. Our main product is **RBF Morph™**, that is a unique morpher that combines a very accurate control of the geometrical parameters with an extremely fast mesh smoothing properly designed to be integrated in advanced computational optimization procedures.

The **RBF Morph** tool is currently available in the market mainly as add-on of the CFD commercial code ANSYS® Fluent®.

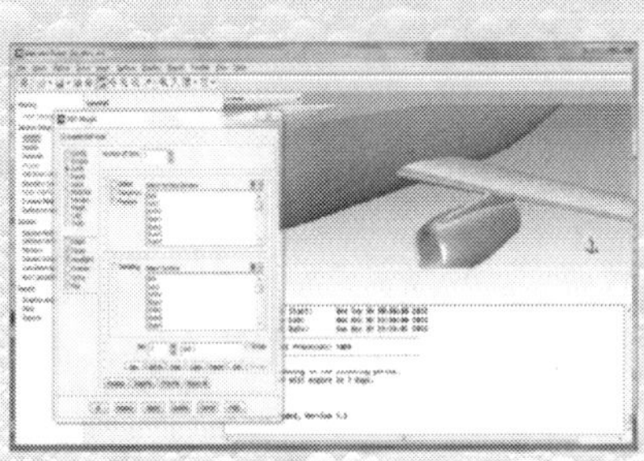

ASWC 2014
October 2014 - Tokyo, Japan

www.rbf-morph.com
RBF Morph, an ANSYS Inc. Partner

(rbf-morph)™ Welcome to the World of Fast Morphing!

## Company Introduction

Automotive Simulation World Congress 2014

The **RBF Morph** tool had its inception in 2008 as on-demand solution for a Formula 1 top team. The need was a novel technology able to change the shape of large CFD numerical models as fast as possible. The final result had been so good that the technology was packaged in a commercial software product and launched onto the market.

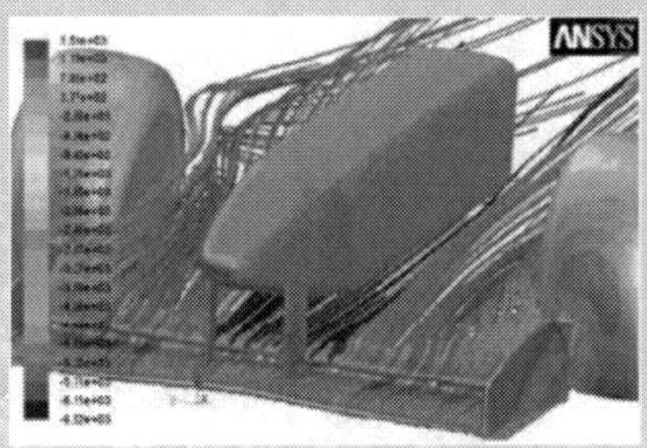

At present, Dr. Marco Evangelos Biancolini is the unique owner of the **RBF Morph** technology and, as Director, avails himself of the collaboration of several experts for the deliver of products and services.

ASWC 2014
October 2014 - Tokyo, Japan

www.rbf-morph.com
RBF Morph, an ANSYS Inc. Partner

(rbf-morph)™ Welcome to the World of Fast Morphing!

## Company Introduction

2014 Automotive Simulation World Congress

- Morphing-based numerical tools and services
- RBF Morph Milestones
  - ✓ 2008: tool implementation for Formula 1 top team consultancy activity
  - ✓ 2009: founded in Italy
  - ✓ 2009: Software Partner of ANSYS
  - ✓ 2009: at EASC **RBF Morph** won the *Most Advanced Approach Award Most Innovative Approach using Simulation Methods*
  - ✓ 2011: strategic partnership with Tor Vergata University (Rome)
  - ✓ 2012: OEM partner of ANSYS
  - ✓ 2013: beneficiary of an FP7 AAT Project RBF4AERO
  - ✓ 2013: at ASWC **RBF Morph** awarded for the *Best use of HPC*
  - ✓ 2013: Partner of Enginsoft
  - ✓ 2014: beneficiary of FP7 Project RIBES
  - ✓ 2014: beneficiary of FP7 Fortissimo

ANSYS

ENGIN SOFT

easco EUROPEAN AUTOMOTIVE SIMULATION CONFERENCE

2013 Automotive Simulation World Congress

ASWC 2014
October 2014 - Tokyo, Japan

Tor Vergata

www.rbf-morph.com ANSYS
RBF Morph, an ANSYS Inc. Partner

(rbf-morph)™ Welcome to the World of Fast Morphing!

## Knowledge Galaxy

2014 Automotive Simulation World Congress

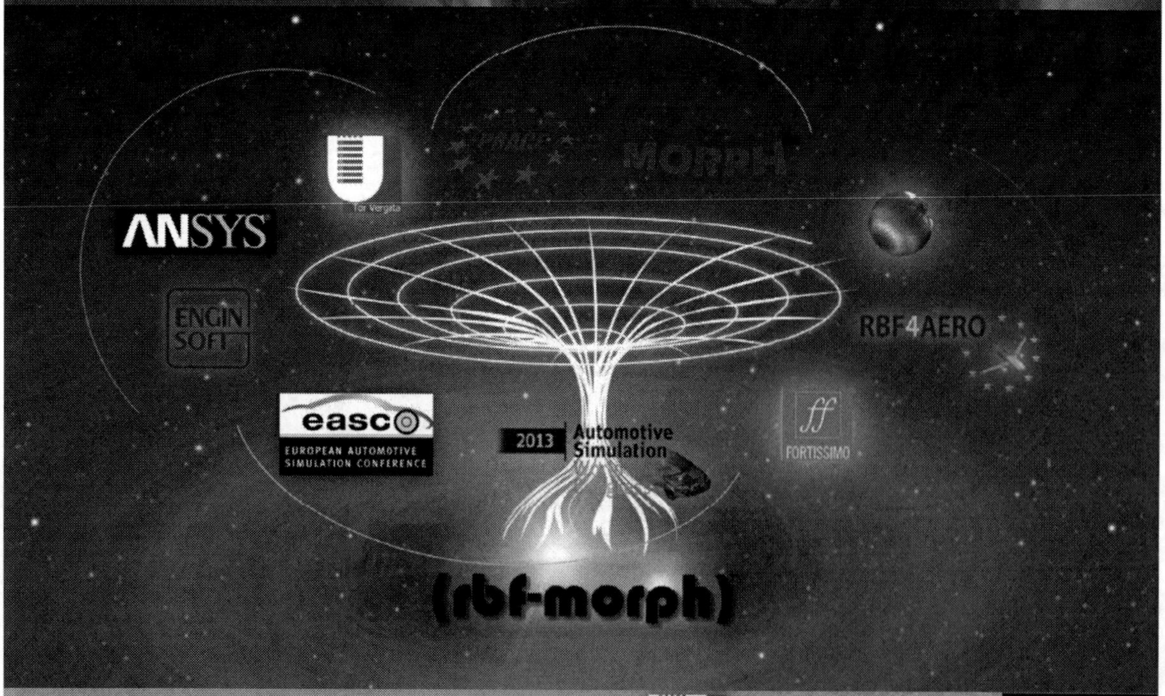


ASWC 2014
October 2014 - Tokyo, Japan

Tor Vergata

www.rbf-morph.com ANSYS
RBF Morph, an ANSYS Inc. Partner

(rbf-morph)™
Welcome to the World of Fast Morphing!

2014 Automotive Simulation World Congress

# RBF Morph software line

ASWC 2014
October 2014 - Tokyo, Japan

Tor Vergata

www.rbf-morph.com
RBF Morph, an ANSYS Inc. Partner

ANSYS

---

(rbf-morph)™
Welcome to the World of Fast Morphing!

# Mesh Morphing with RBF

2014 Automotive Simulation World Congress

- A system of **Radial Basis Functions** is used to **fit** a **solution** for the mesh movement/**Morphing**, from a list of **source points** and their **displacements**.
- The RBF problem definition does not depend on the mesh
- Radial Basis Function interpolation is used to derive the displacement in **any location** in the space, each component of the displacement is interpolated:

$$\begin{cases} v_x = s_x(\mathbf{x}) = \sum_{i=1}^{N} \gamma_i^x \phi\left(\|\mathbf{x} - \mathbf{x}_{k_i}\|\right) + \beta_1^x + \beta_2^x x + \beta_3^x y + \beta_4^x z \\ v_y = s_y(\mathbf{x}) = \sum_{i=1}^{N} \gamma_i^y \phi\left(\|\mathbf{x} - \mathbf{x}_{k_i}\|\right) + \beta_1^y + \beta_2^y x + \beta_3^y y + \beta_4^y z \\ v_z = s_z(\mathbf{x}) = \sum_{i=1}^{N} \gamma_i^z \phi\left(\|\mathbf{x} - \mathbf{x}_{k_i}\|\right) + \beta_1^z + \beta_2^z x + \beta_3^z y + \beta_4^z z \end{cases}$$

- RBF are recognized as on of the **best mathematical tool** for mesh morphing. The main issue is about performances required for the solution of **large dataset**.

ASWC 2014
October 2014 - Tokyo, Japan

Tor Vergata

www.rbf-morph.com
RBF Morph, an ANSYS Inc. Partner

ANSYS

(rbf-morph)™ Welcome to the World of Fast Morphing!

## RBF Morph software line

Automotive Simulation World Congress

- HPC RBF **general purposes** library (state of the art algorithms, parallel, GPU). This is the numerical kernel of our software. **Millions** of RBF centers can be fitted in a short time.
- Awarded mesh morphing software available as an add-on for **ANSYS Fluent** CFD solver
- **Stand alone** morphing software + smoothing commands for different mesh formats
- ANSYS Mechanical **ACT module** (first beta version already working since June 2014)

ASWC 2014
October 2014 - Tokyo, Japan

Tor Vergata

www.rbf-morph.com
RBF Morph, an ANSYS Inc. Partner

ANSYS

(rbf-morph)™ Welcome to the World of Fast Morphing!

## Fluent add-on

2014 Automotive Simulation World Congress

- **Add on** fully integrated within **Fluent** (GUI, TUI & solving stage), **Workbench** and **Adjoint Solver** NEW!
- **Mesh-independent** RBF fit used for surface mesh morphing and volume mesh smoothing
- **Parallel** calculation allows to morph **large size** models (many millions of cells) in a short time
- Management of **every kind of mesh** element type (tetrahedral, hexahedral, polyhedral, etc.)
- Support of the **CAD re-design** of the morphed surfaces
- **Multi fit** makes the Fluent case truly parametric (only 1 mesh is stored)
- **Precision**: exact nodal movement and exact feature preservation (**RBF** are better than **FFD**)

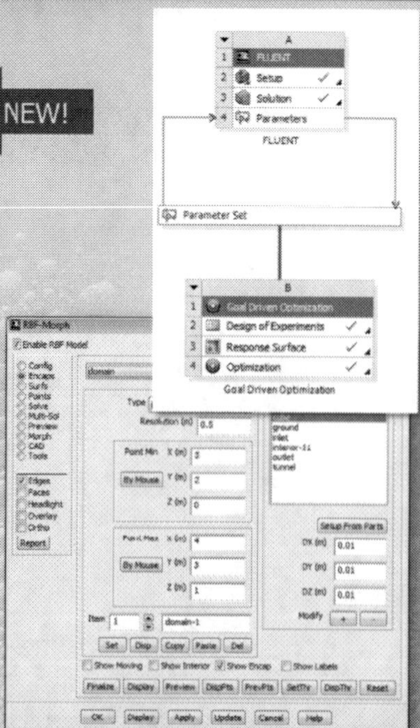


ASWC 2014
October 2014 - Tokyo, Japan

Tor Vergata

www.rbf-morph.com
RBF Morph, an ANSYS Inc. Partner

ANSYS

## (rbf-morph)™ ACT module for Mechanical

Welcome to the World of Fast Morphing!

Automotive Simulation World Congress

- Deeply integrated in ANSYS Mechanical: same look & feel, same interaction logic
- Nested in the usual Mechanical tree as an added object, shares its scoping tools for **geometrical** and **mesh elements** selections
- Written in **python** and **xml**, uses external RBF Morph core libraries
- Child **hierarchical logic** for complex morphings (two steps, three steps, …, n steps setups)

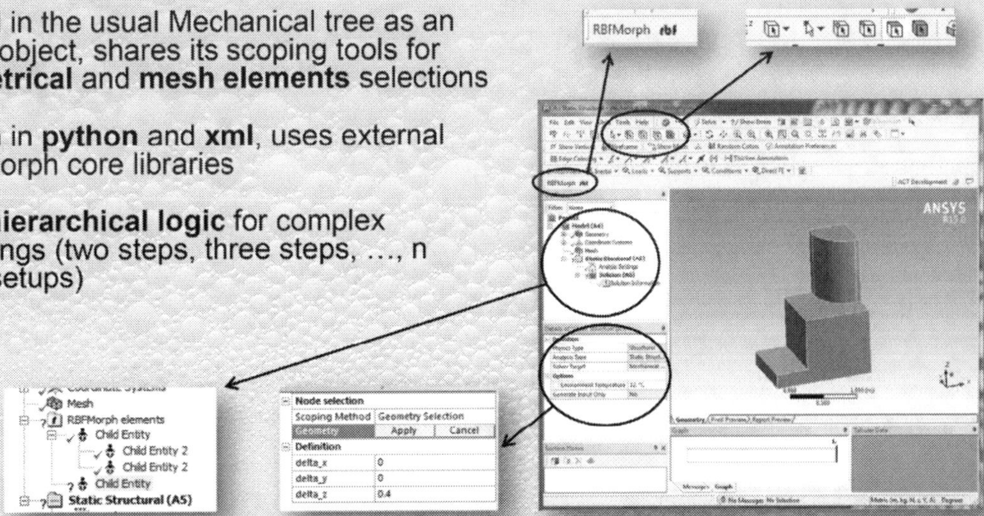


ASWC 2014
October 2014 - Tokyo, Japan

www.rbf-morph.com
RBF Morph, an ANSYS Inc. Partner

ANSYS

## (rbf-morph)™ RBF Morph Stand Alone

Welcome to the World of Fast Morphing!

Automotive Simulation World Congress

- RBF solutions are fully compatible and **exchangeable** between add-on and standalone versions
- Support for STL and CGNS file formats. Selected morphed surfaces can be exported in STL format and **back to CAD** is possible via STEP files
- **Add-on-like** interface
- **Solver independent** process currently supports many mesh formats
- Functions **scriptable** via tcl
- Global supported bi-harmonic functions and $C^0$, $C^2$, $C^4$ compact supported functions available

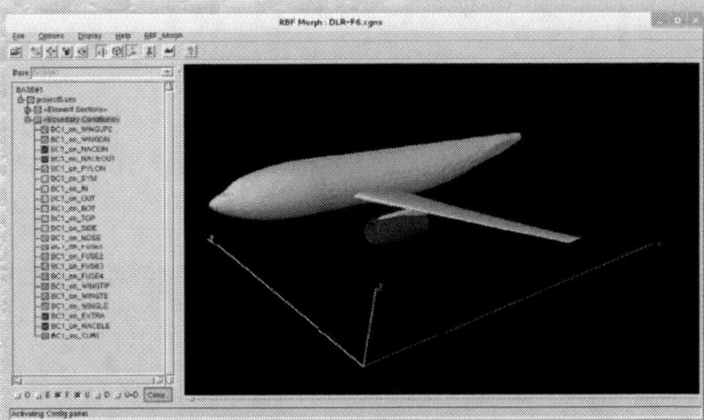


ASWC 2014
October 2014 - Tokyo, Japan

www.rbf-morph.com
RBF Morph, an ANSYS Inc. Partner

ANSYS

(rbf-morph)™ Welcome to the World of Fast Morphing!

2014 Automotive Simulation World Congress

# Ongoing RBF Morph Researches

ASWC 2014
October 2014 - Tokyo, Japan

Tor Vergata

www.rbf-morph.com
RBF Morph, an ANSYS Inc. Partner

ANSYS

(rbf-morph)™ Welcome to the World of Fast Morphing!

# Ongoing RBF research

2014 Automotive Simulation World Congress

- RBF Morph and Adjoint coupling: Adjoint sculpting, Adjoint preview, Augmented DOE
- **STL** targeting, **CAD** controlled surfaces
- **Mesh to CAD** features
- Mapping of **magnetic** and **pressure** loads
- Interpolation of **hemodynamic** flow fields acquired *in vivo*
- Strain and **stress calculation** (experimental data, coarse FEM, isostatic lines)

ASWC 2014
October 2014 - Tokyo, Japan

Tor Vergata

www.rbf-morph.com
RBF Morph, an ANSYS Inc. Partner

ANSYS

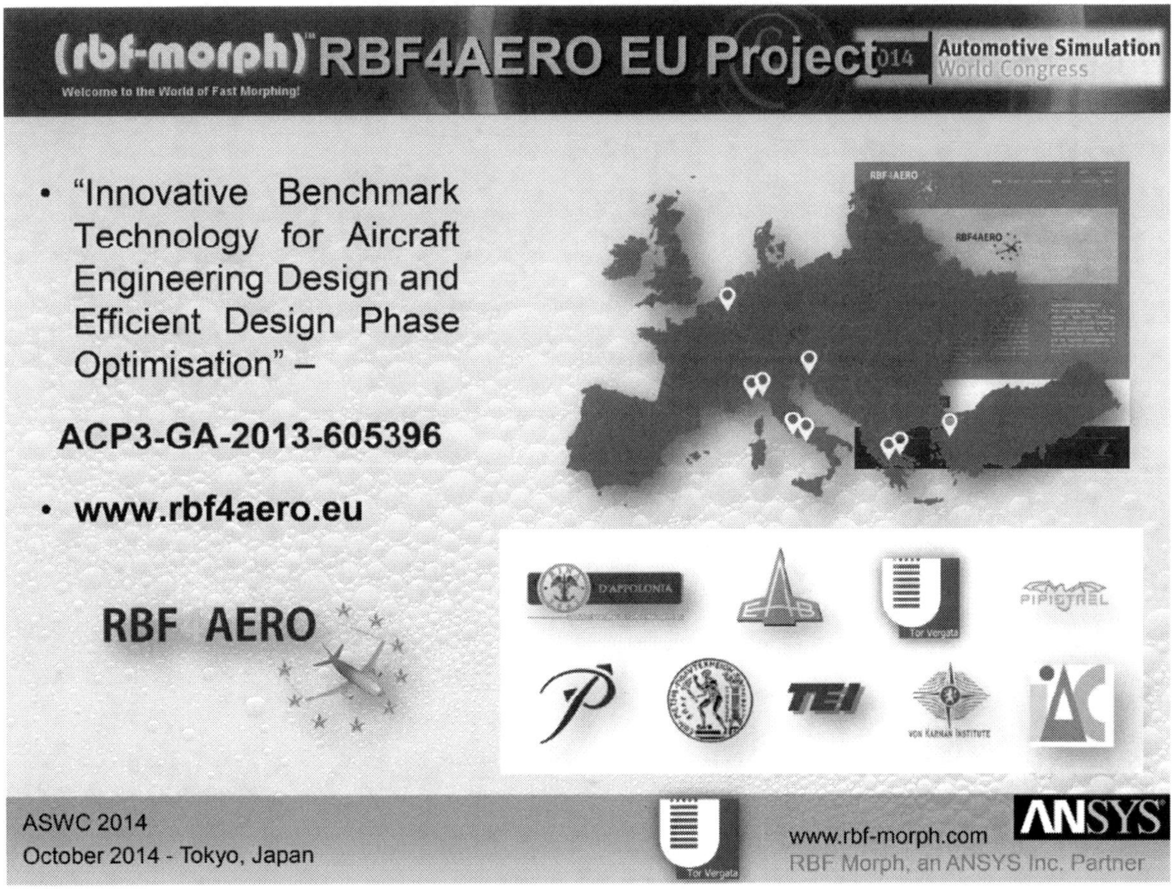
(rbf-morph)™ RBF4AERO EU Project
Welcome to the World of Fast Morphing!
2014 Automotive Simulation World Congress
• "Innovative Benchmark Technology for Aircraft Engineering Design and Efficient Design Phase Optimisation" –
ACP3-GA-2013-605396
• www.rbf4aero.eu
RBF AERO
D'APPOLONIA
PIPISTREL
Tor Vergata
TEI
VON KARMAN INSTITUTE
ASWC 2014
October 2014 - Tokyo, Japan
www.rbf-morph.com
RBF Morph, an ANSYS Inc. Partner
ANSYS

(rbf-morph)™ RIBES EU Project
Welcome to the World of Fast Morphing!
2014 Automotive Simulation World Congress
• Radial basis functions at fluid Interface Boundaries to Envelope flow results for advanced Structural analysis
JTI-CS-2013-GRA-01-052
RIBES
CLEANSKY
Tor Vergata
ASWC 2014
October 2014 - Tokyo, Japan
www.rbf-morph.com
RBF Morph, an ANSYS Inc. Partner
ANSYS

(rbf-morph)™ Welcome to the World of Fast Morphing!

# Fortissimo EU Project

2014 Automotive Simulation World Congress

- **F**actories **O**f the Future **R**esources, **T**echnology, **I**nfrastructure and **S**ervices for **SI**mulation and **MO**delling
- Approved experiment: "Virtual Automatic Rapid Prototyping Based on Fast Morphing on HPC Platforms"

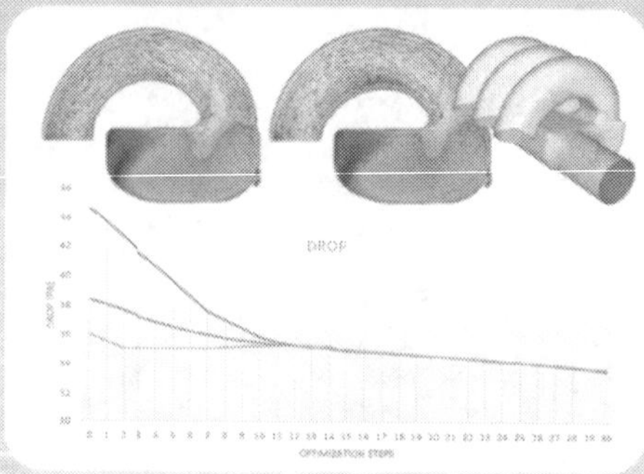

FORTISSIMO

HSL Tor Vergata CINECA SCAI SuperComputing Applications and Innovation

(rbf-morph)™ Welcome to the World of Fast Morphing!

2014 Automotive Simulation World Congress

# RBF mesh morphing ACT extension for ANSYS Mechanical

ASWC 2014
October 2014 - Tokyo, Japan
Tor Vergata
www.rbf-morph.com
ANSYS
RBF Morph, an ANSYS Inc. Partner

(rbf-morph)™ Welcome to the World of Fast Morphing!

## Why mesh morphing?

2014 Automotive Simulation World Congress

- It allows to have parametric shape mesh that preserves the original topology. **Remeshing noise** is avoided.
- It allows to update the shape of a **validated FEM model** without rebuilding a new mesh.
- New shapes can be investigated even if the underlying **CAD geometry** is missing.
- The mesh can be updated to measured shapes (i.e. accounting for **manufacturing tolerances**)
- It's usually faster than remeshing.

ASWC 2014
October 2014 - Tokyo, Japan

Tor Vergata

www.rbf-morph.com
RBF Morph, an ANSYS Inc. Partner

ANSYS

(rbf-morph)™ Welcome to the World of Fast Morphing!

## Why RBF morphing?

2014 Automotive Simulation World Congress

- RBF Morph is a best in class product crafted to deal with challenging CFD application (**huge meshes**)
- 7 years of experience on industrial applications of Radial Basis Functions (RBF)
- RBF are **recognized** as one of the best mesh morphing tool available in the industrial and scientific community
- A new vision (we have started the new project from scratch) to put in the hands of **ANSYS Mechanical** Users the fastest and easiest mesh morphing tool
- Satisfy the needs of **many users** asking for such a kind of technology available in ANSYS Mechanical

ASWC 2014
October 2014 - Tokyo, Japan

Tor Vergata

www.rbf-morph.com
RBF Morph, an ANSYS Inc. Partner

ANSYS

## (rbf-morph)™ ACT module specs

Welcome to the World of Fast Morphing!

2014 Automotive Simulation World Congress

- Fully **embedded** in ANSYS Mechanical with same "look & feel"
- Integration in the tree and founded on the working principles of **ANSYS Workbench**
- Based on the ACT (Application Customization Toolkit) **extension** concept
- Geometrical and mesh scoping to set-up the mesh morphing problem
- Parametric (coming soon! In ANSYS 16 ACT modules parameters will be exposed natively)
- RBF **fast solver** (including parallel and GPU support)

ASWC 2014
October 2014 - Tokyo, Japan

Tor Vergata

www.rbf-morph.com
RBF Morph, an ANSYS Inc. Partner

ANSYS

## (rbf-morph)™ ACT module specs

Welcome to the World of Fast Morphing!

2014 Automotive Simulation World Congress

- Easy to use, flexible and **expressive**.
- Powered by multi-step RBF technology (which effectiveness has been extensively proven in RBF Morph)
- RBF **fitting and mesh morphing** happen as a unique **HPC process** at each shape update. This allows the maximum flexibility with respect to parameterization
- RBF component set-up data stored and persistent with the ANSYS Mechanical project.
- Advanced **interaction with DM** to enable CAD resynchronization after morphing (coming soon)

ASWC 2014
October 2014 - Tokyo, Japan

Tor Vergata

www.rbf-morph.com
RBF Morph, an ANSYS Inc. Partner

ANSYS

(rbf-morph)™ Welcome to the World of Fast Morphing!

# Current prototype

2014 Automotive Simulation World Congress

- **Hierarchical** multi-step RBF approach
- Each morphing **target** can be controlled using the superposition of a constant translation (other modifiers will be introduced) and an RBF field generated by its **sources** (if any)
- **Source points** are extracted using the motion of all the **children**
- If the child is a leaf its movement is known otherwise **it becomes a target** and has to be solved descending the tree
- Sounds complicated? Let's explain it with an example!

ASWC 2014
October 2014 - Tokyo, Japan

www.rbf-morph.com
ANSYS
RBF Morph, an ANSYS Inc. Partner

(rbf-morph)™ Welcome to the World of Fast Morphing!

# Cube example

2014 Automotive Simulation World Congress

- A simple **cube** is loaded on the top and clamped to the bottom
- **Structural mechanics** tell us that this structure can be optimized adopting a tapered profile
- A parabola is obtained if we look for constant **bending stress** beam
- The theory is well known in gear design as the **Lewis formula** used for the stress assessment of a teeth

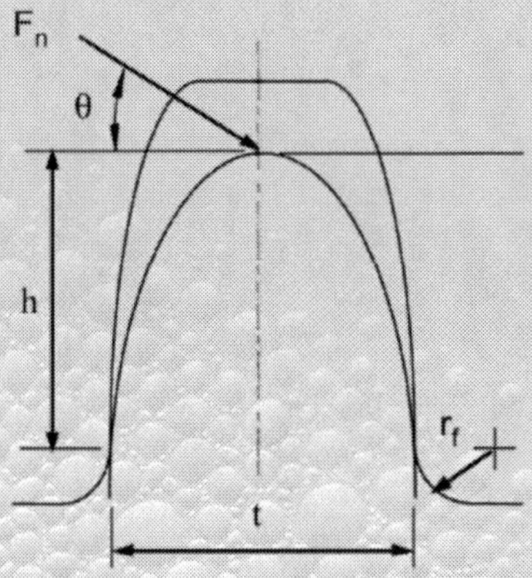


(b)

ASWC 2014
October 2014 - Tokyo, Japan

www.rbf-morph.com
ANSYS
RBF Morph, an ANSYS Inc. Partner

# (rbf-morph)™ Cube example

Welcome to the World of Fast Morphing!

2014 Automotive Simulation World Congress

- Boundary conditions

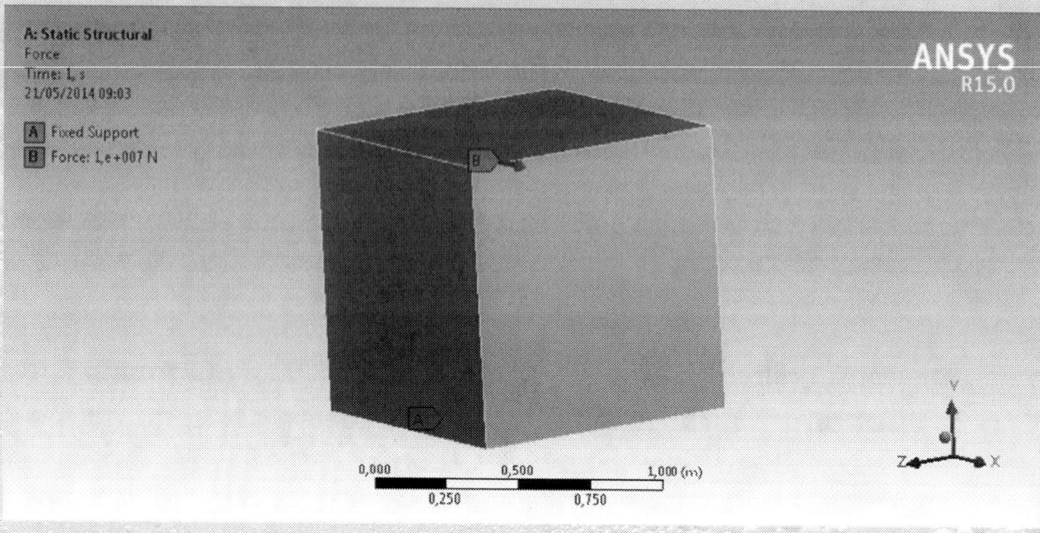




# (rbf-morph)™ Cube example

Welcome to the World of Fast Morphing!

2014 Automotive Simulation World Congress

- Baseline solution

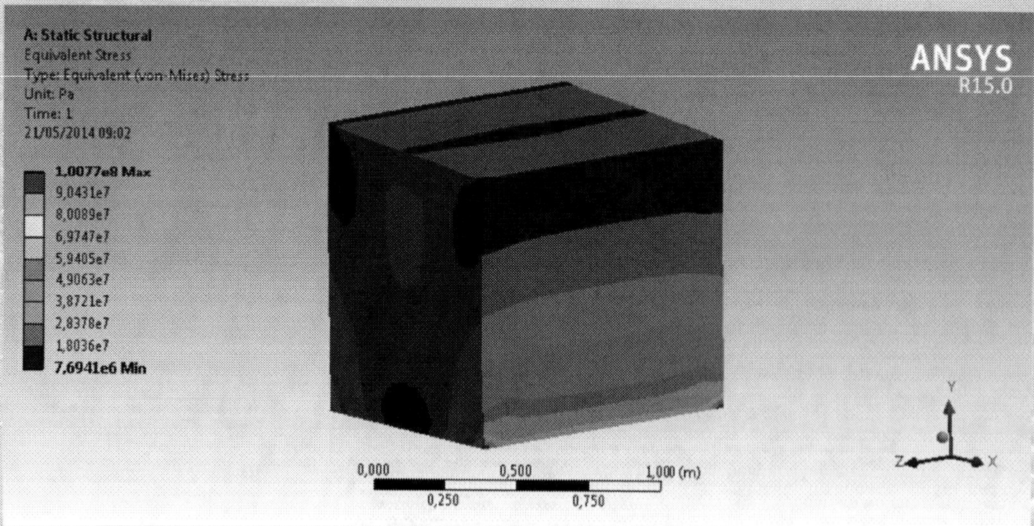

- First morphing set-up!

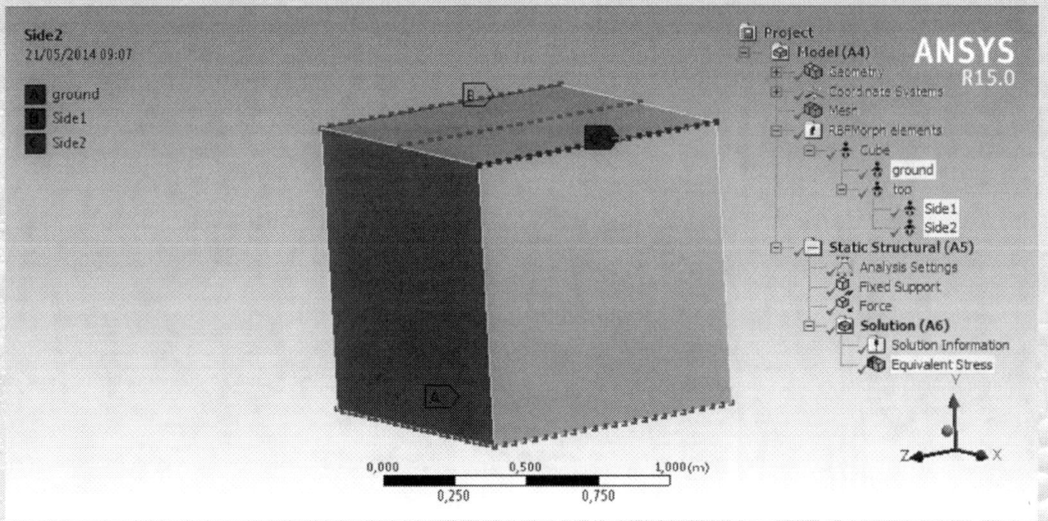




(rbf-morph) Cube example
Welcome to the World of Fast Morphing!
2014 Automotive Simulation World Congress

- The first level set **Cube** (volume scoping) define the mesh to be morphed. Controlled by ground and top.
- At second level (surface scoping) we have two sets: **ground** (fixed and leaf of the tree) and **top** (controlled by Side1 and Side2).
- At third level (edge scoping) we have **Side1** (fixed) and **Side2** (moving in negative X direction).

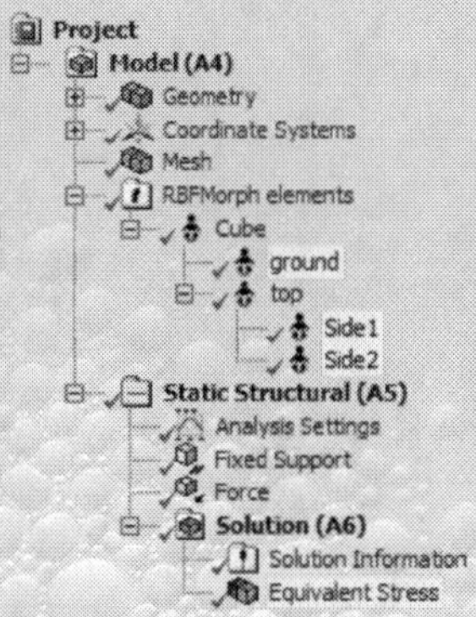


ASWC 2014
October 2014 - Tokyo, Japan
www.rbf-morph.com
RBF Morph, an ANSYS Inc. Partner

## (rbf-morph)™ Cube example

Welcome to the World of Fast Morphing!

2014 Automotive Simulation World Congress

- Morphed solution #1

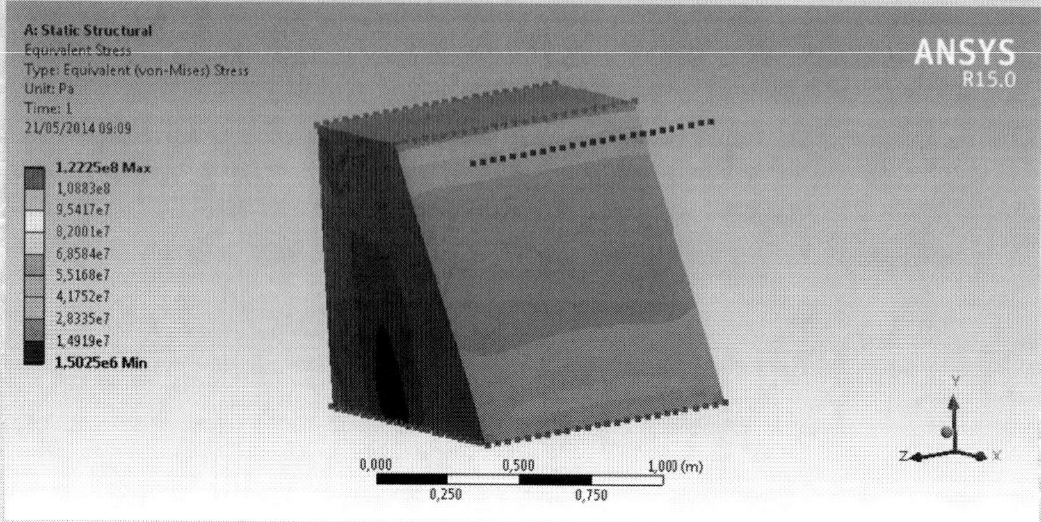


ASWC 2014
October 2014 - Tokyo, Japan

www.rbf-morph.com
RBF Morph, an ANSYS Inc. Partner

## (rbf-morph)™ Cube example

Welcome to the World of Fast Morphing!

2014 Automotive Simulation World Congress

- Second morphing set-up!

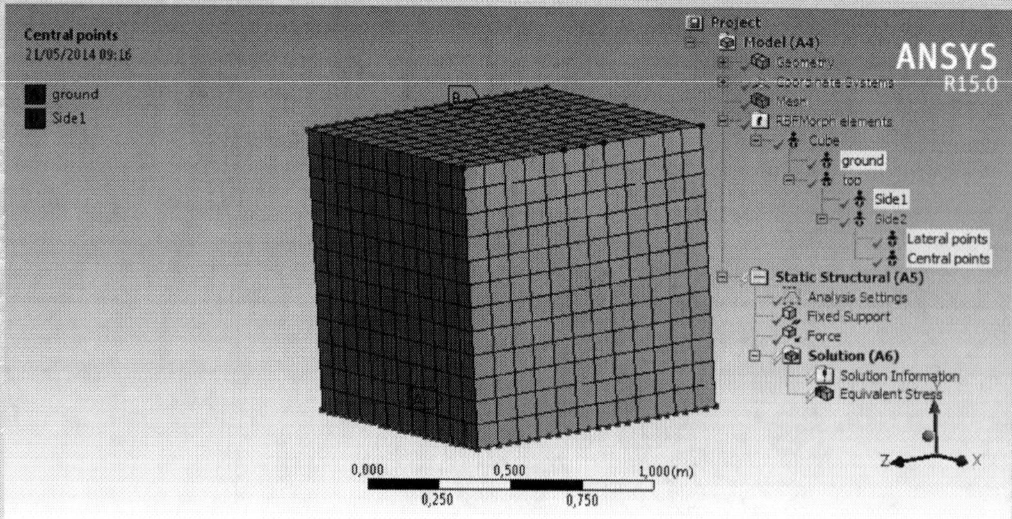


ASWC 2014
October 2014 - Tokyo, Japan

www.rbf-morph.com
RBF Morph, an ANSYS Inc. Partner

## (rbf-morph)™ Cube example

Welcome to the World of Fast Morphing!

2014 Automotive Simulation World Congress

- The previous set-up has been refined and now we have a fourth level (scoping points and nodes) to control the shape of **Side2** that is controlled by Lateral Points and Central Points
- Individual movements are imposed in the negative X direction to **Lateral Points** and **Central Points**
- At this level it would be good to control curvature changing the **order** of RBF function (coming soon!)

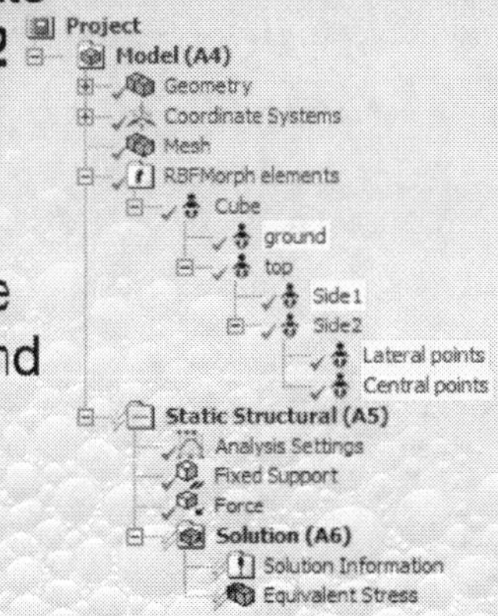


ASWC 2014
October 2014 - Tokyo, Japan

www.rbf-morph.com
RBF Morph, an ANSYS Inc. Partner

## (rbf-morph)™ Cube example

Welcome to the World of Fast Morphing!

2014 Automotive Simulation World Congress

- Morphed solution #2

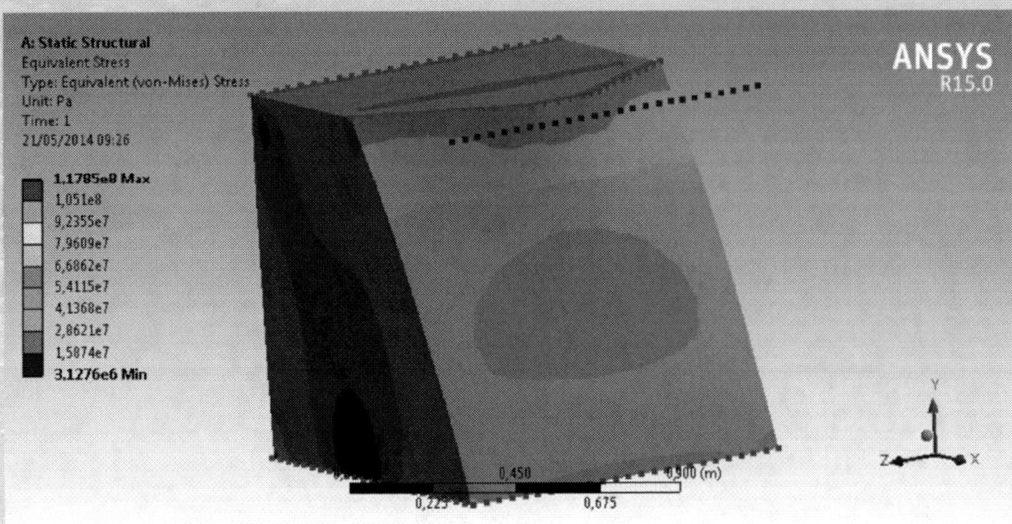


ASWC 2014
October 2014 - Tokyo, Japan

www.rbf-morph.com
RBF Morph, an ANSYS Inc. Partner

# Cube example



- The set-up #1 and #2 are explained in detail.
- The set-up can be further enriched to fulfill the desired shape.
- Some example follows…



(rbf-morph)™ Welcome to the World of Fast Morphing!

# Cube example

2014 Automotive Simulation World Congress

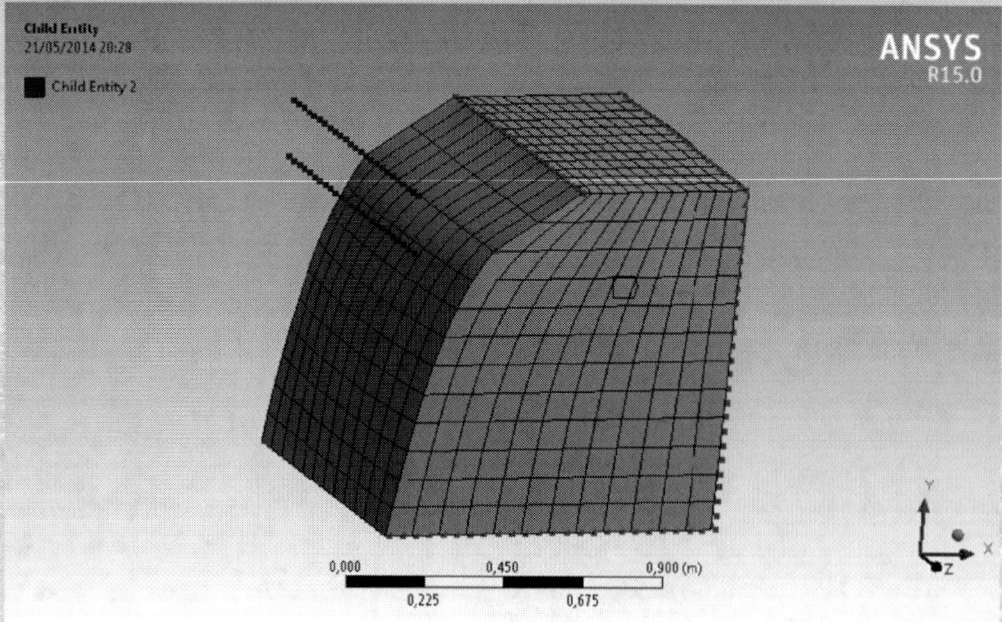


ASWC 2014
October 2014 - Tokyo, Japan
Tor Vergata
www.rbf-morph.com
ANSYS
RBF Morph, an ANSYS Inc. Partner

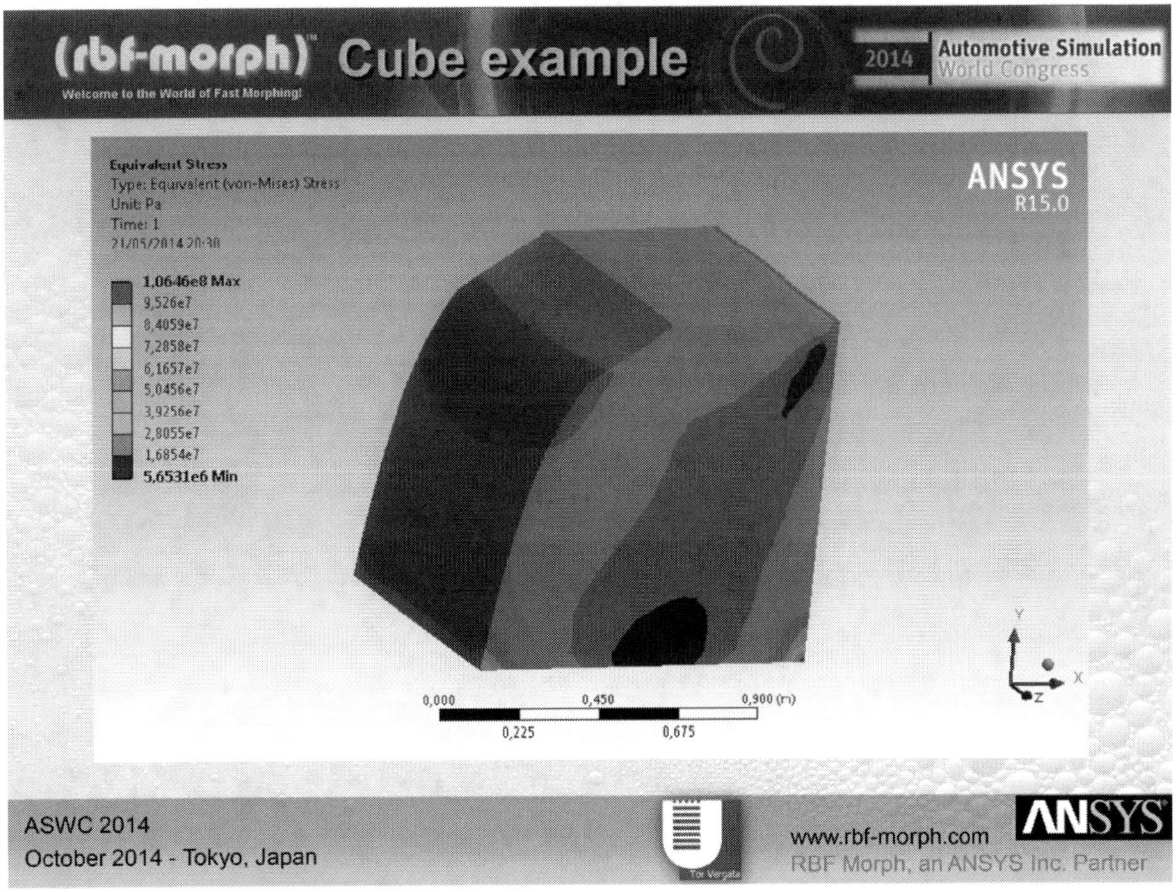
(rbf-morph)™ Cube example
Welcome to the World of Fast Morphing!
2014 Automotive Simulation World Congress
Equivalent Stress
Type: Equivalent (von-Mises) Stress
Unit: Pa
Time: 1
21/05/2014 20:30
1,0646e8 Max
9,526e7
8,4059e7
7,2858e7
6,1657e7
5,0456e7
3,9256e7
2,8055e7
1,6854e7
5,6531e6 Min
ANSYS
R15.0
0,000
0,450
0,900 (m)
0,225
0,675
Y
X
Z
ASWC 2014
October 2014 - Tokyo, Japan
Tor Vergata
www.rbf-morph.com
RBF Morph, an ANSYS Inc. Partner
ANSYS

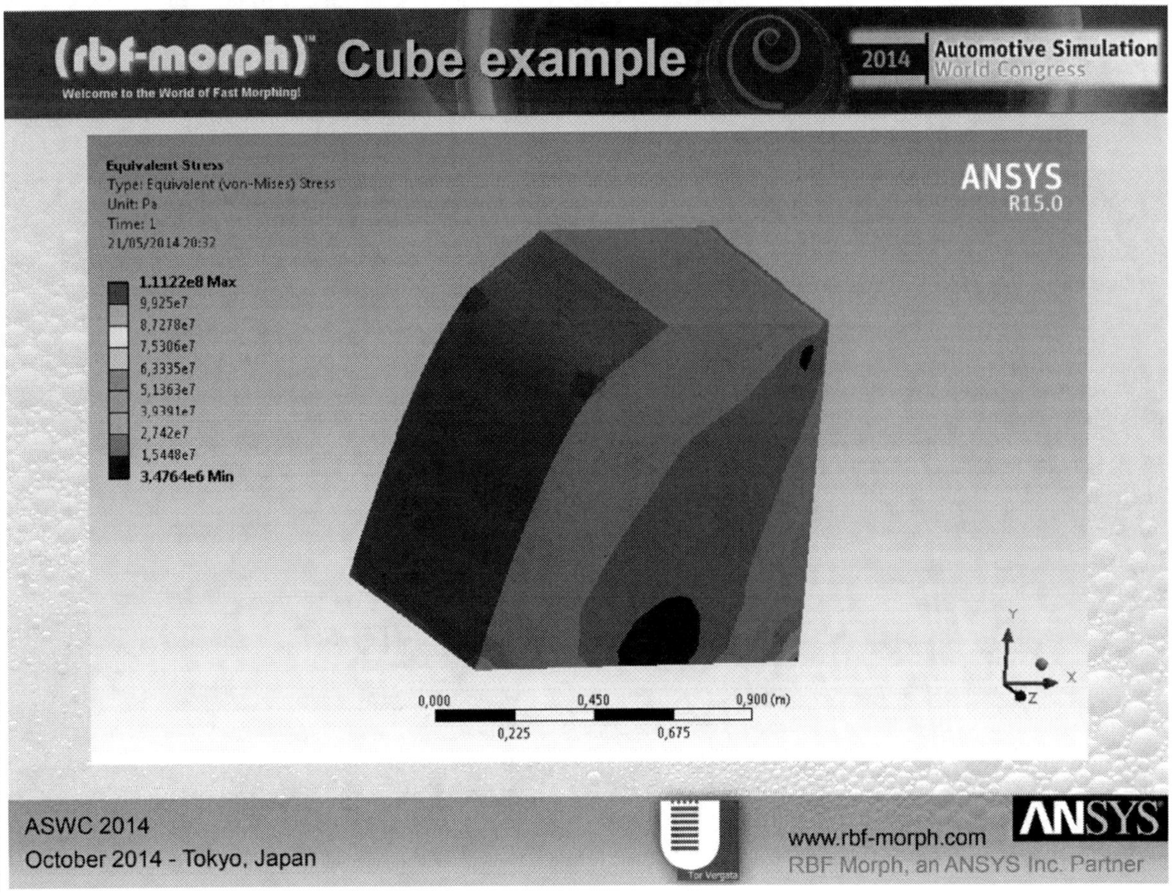
(rbf-morph)™ Cube example
Welcome to the World of Fast Morphing!
2014 Automotive Simulation World Congress
Equivalent Stress
Type: Equivalent (von-Mises) Stress
Unit: Pa
Time: 1
21/05/2014 20:32
1,1122e8 Max
9,925e7
8,7278e7
7,5306e7
6,3335e7
5,1363e7
3,9391e7
2,742e7
1,5448e7
3,4764e6 Min
ANSYS
R15.0
0,000
0,450
0,900 (m)
0,225
0,675
Y
X
Z
ASWC 2014
October 2014 - Tokyo, Japan
Tor Vergata
www.rbf-morph.com
RBF Morph, an ANSYS Inc. Partner
ANSYS

## (rbf-morph)™ Cube example

Welcome to the World of Fast Morphing!

2014 Automotive Simulation World Congress

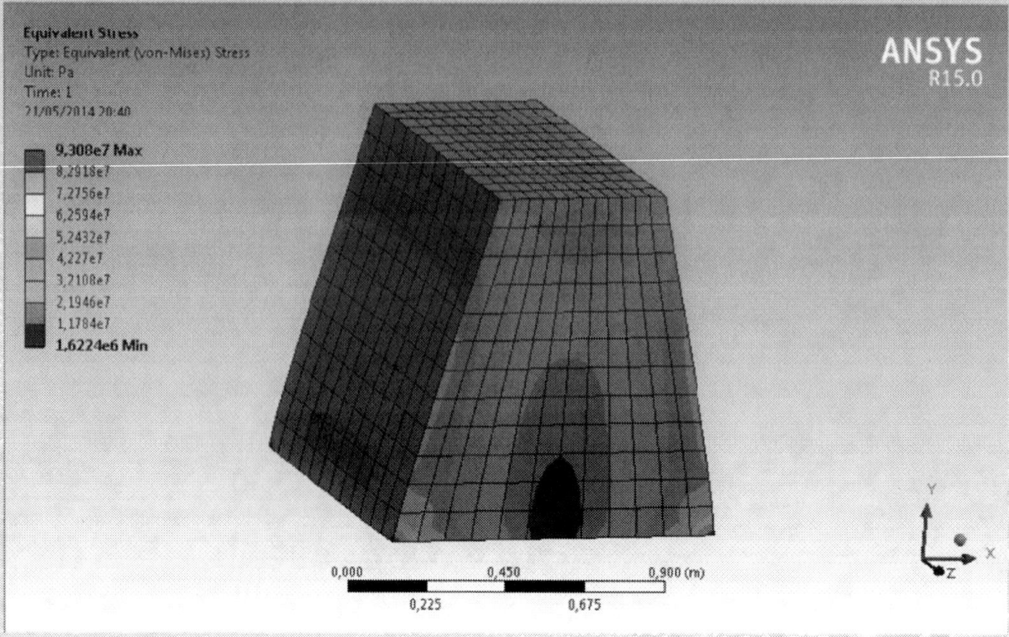


ASWC 2014
October 2014 - Tokyo, Japan

www.rbf-morph.com
RBF Morph, an ANSYS Inc. Partner

## (rbf-morph)™ Cube example

Welcome to the World of Fast Morphing!

2014 Automotive Simulation World Congress

- Mass reduction 36% Stress reduction 12%

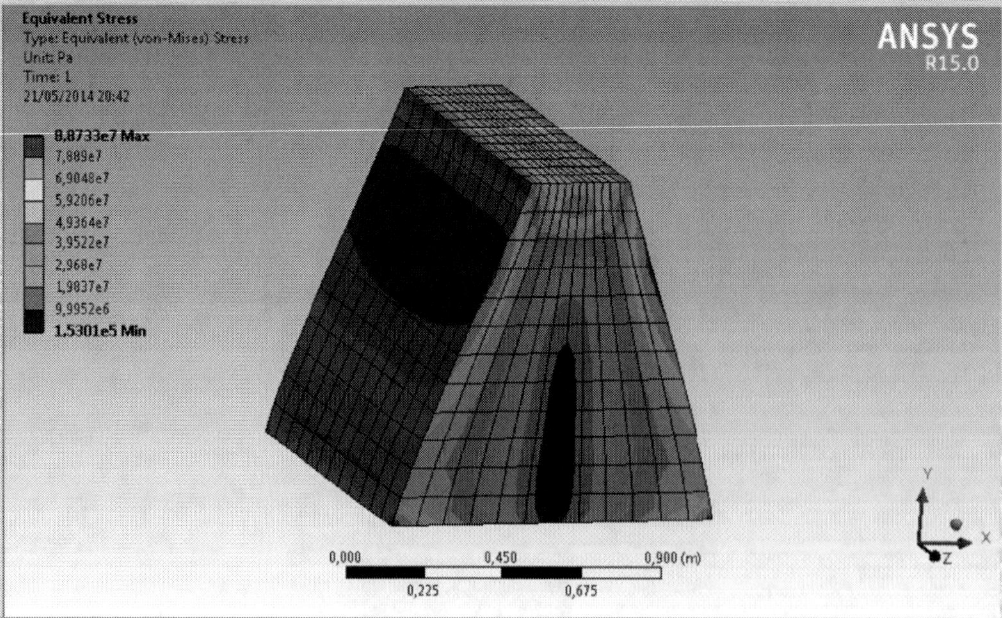


ASWC 2014
October 2014 - Tokyo, Japan

www.rbf-morph.com
RBF Morph, an ANSYS Inc. Partner

(rbf-morph)™ Welcome to the World of Fast Morphing!

# Press basement

2014 Automotive Simulation World Congress

SACMI

- This is a simple mesh morphing **feasibility study** for a geometry relevant for SACMI
- Part studied is a quarter of the basement of a press
- Area to be optimized is the one highlighted in **green**
- **Simple boundary** conditions are used to stress the part (red fixed, yellow symmetry, blue loaded)

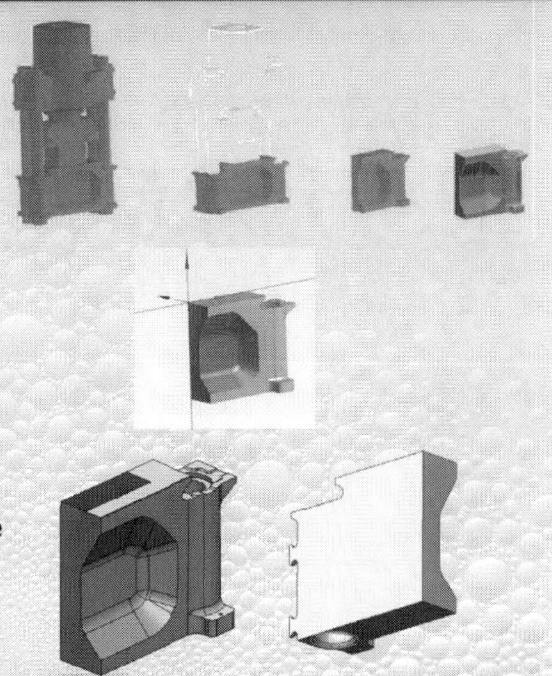

ASWC 2014
October 2014 - Tokyo, Japan

SACMI Tor Vergata

www.rbf-morph.com
RBF Morph, an ANSYS Inc. Partner

ANSYS

(rbf-morph)™ Welcome to the World of Fast Morphing!

# Press basement

2014 Automotive Simulation World Congress

- Mesh morphing has been used to reduce the stress concentration acting on the shape of the fillet.

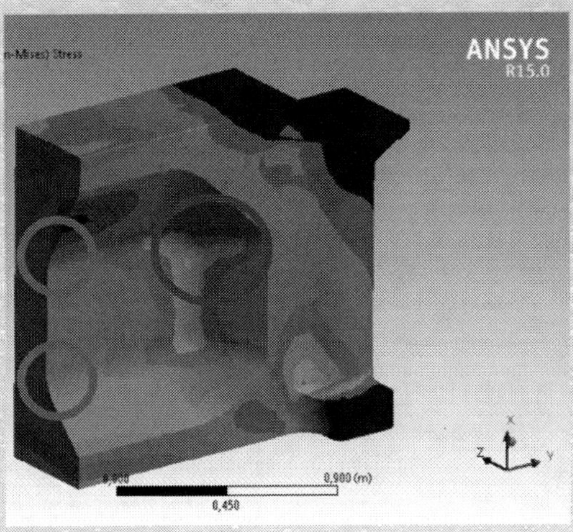


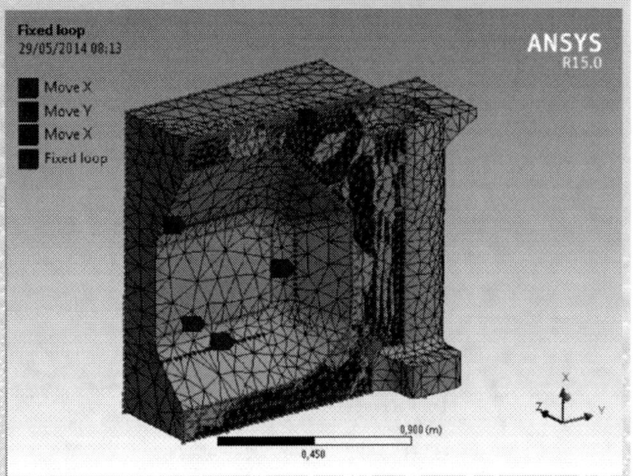


ASWC 2014
October 2014 - Tokyo, Japan

SACMI Tor Vergata

www.rbf-morph.com
RBF Morph, an ANSYS Inc. Partner

ANSYS

## (rbf-morph)™ Press basement

Welcome to the World of Fast Morphing!

2014 Automotive Simulation World Congress

- As the fillet is smoothed a stress redistribution is observed.

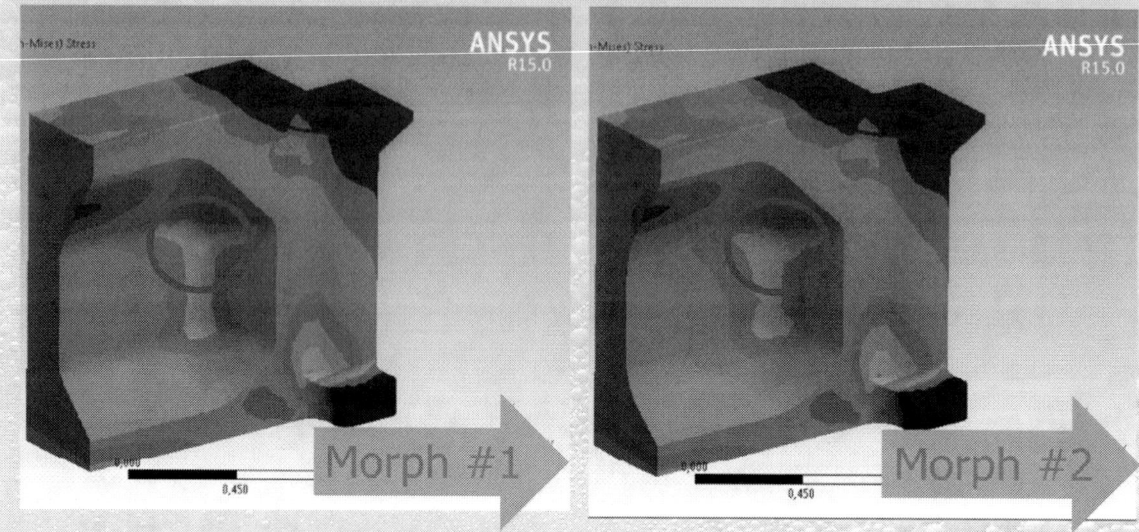


ASWC 2014
October 2014 - Tokyo, Japan

SACMI

Tor Vergata

www.rbf-morph.com
RBF Morph, an ANSYS Inc. Partner

ANSYS

## (rbf-morph)™ Press basement

Welcome to the World of Fast Morphing!

2014 Automotive Simulation World Congress

- Notice that high mesh deformation is properly accommodated thanks to RBF mesh morphing

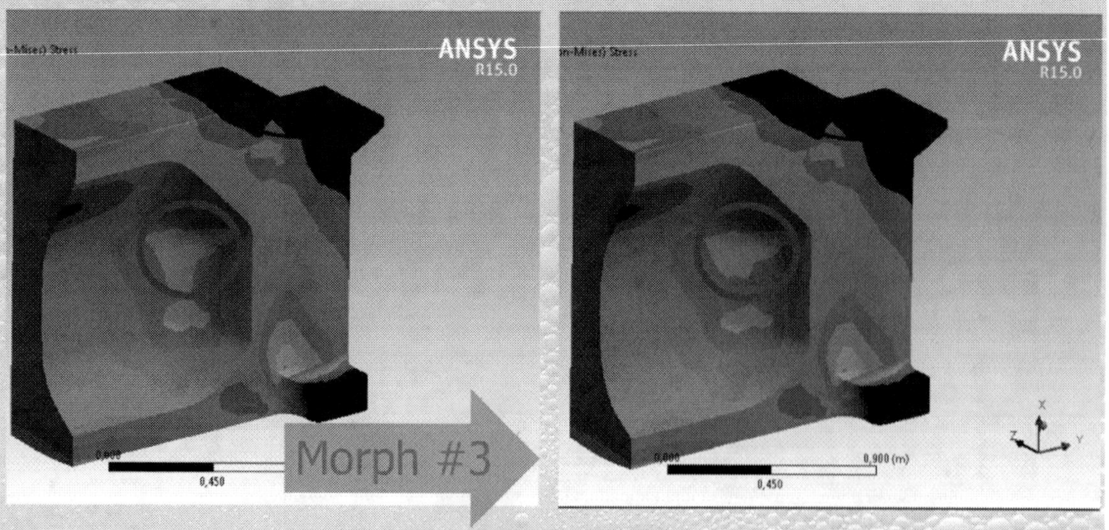


ASWC 2014
October 2014 - Tokyo, Japan

SACMI

Tor Vergata

www.rbf-morph.com
RBF Morph, an ANSYS Inc. Partner

ANSYS

(rbf-morph)™ Welcome to the World of Fast Morphing!

# Press basement

2014 Automotive Simulation World Congress

- Original vs. Optimised (13% reduction of stress peak)

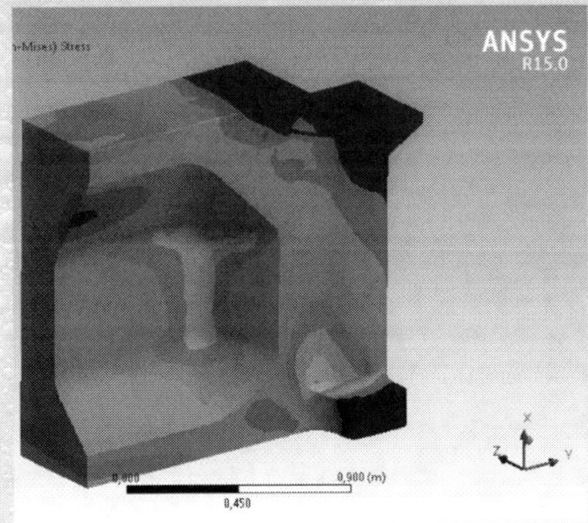


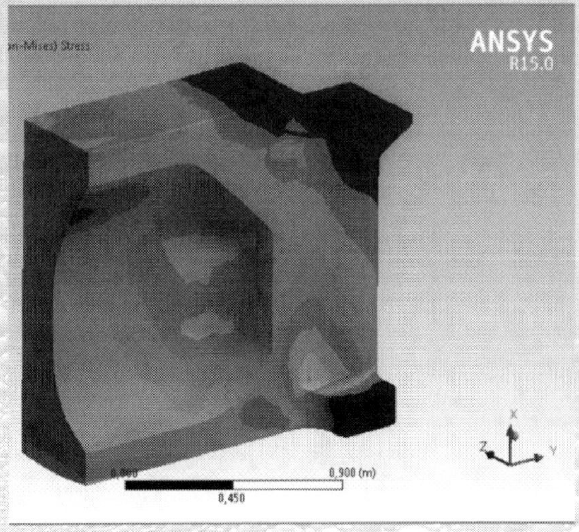


ASWC 2014
October 2014 - Tokyo, Japan

SACMI

www.rbf-morph.com
RBF Morph, an ANSYS Inc. Partner

ANSYS

(rbf-morph)™ Welcome to the World of Fast Morphing!

# Conclusions

2014 Automotive Simulation World Congress

- A novel mesh morphing tool has been implemented in **ANSYS Mechanical** using ACT extension technology
- **Radial Basis Functions** are used for multistep set-up
- The new software benefits of past experience on RBF Morph **ANSYS Fluent** add-on (mainly CFD)
- Nevertheless we have restarted all the developments **from scratch** i.e. reusing **knowledge** and **ideas** (no cross compatibility between tools)
- Basic capability of the first software prototype are demonstrated on a simple FEM mesh and on an **industrial FEM model**
- **How can we do better? Please do not hesitate to tell us your needs!**

ASWC 2014
October 2014 - Tokyo, Japan

www.rbf-morph.com
RBF Morph, an ANSYS Inc. Partner

ANSYS

(rbf-morph)™
Welcome to the World of Fast Morphing!
2014
Automotive Simulation
World Congress
ご清聴ありがとうございました!
Prof. Marco Evangelos Biancolini
E-mail: info@rbf-morph.com
Web: www.rbf-morph.com
Goo.gl/1svYd
linkedin.com/company/rbf-morph
Twitter.com/RBFMorph
youtube.com/user/RbfMorph
ASWC 2014
October 2014 - Tokyo, Japan
Tor Vergata
www.rbf-morph.com
RBF Morph, an ANSYS Inc. Partner
ANSYS

2014 Automotive Simulation World Congress

# CAEBAT Battery Thermal Management Project by General Motors, ANSYS and ESim

Presenter: Sandeep Sovani, Ph.D., ANSYS USA

Based On Work of:
Taeyoung Han, Ph.D., General Motors
Lewis Collins and Dimitrios Tselepidakis, Ph.D., ANSYS Inc
Prof. Ralph White, ESim and University of South Carolina
Gi-Heon Kim, Ph.D., National Renewable Energy Lab, USA

## Battery is the Key to Vehicle Electrification



- 2012 DoE *EV Everywhere* Grand Challenge goal: make plug-in electric vehicles as affordable and convenient as today's gasoline-powered vehicles by 2022

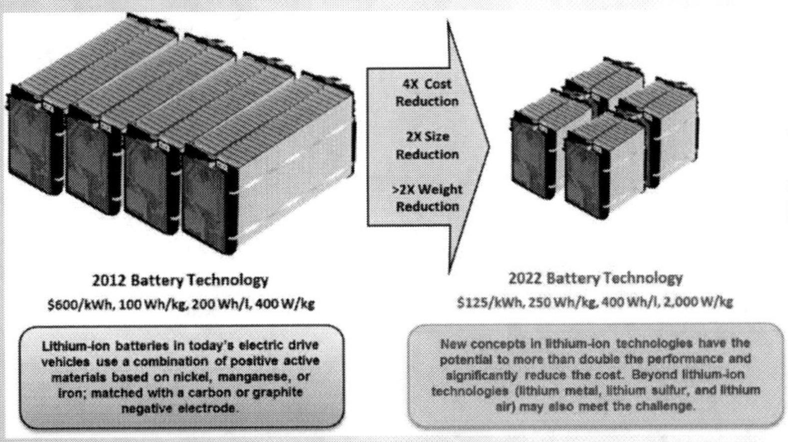


Image credit: "FY 2013 Annual Progress Report", Energy Storage R&D, Vehicle Technologies Office, U.S. Department of Energy

## Engineering Challenges in Battery Development

2014 Automotive Simulation World Congress
October 9-10 . Tokyo, Japan

- Cost
- Performance (power and energy density)
- Durability and service life (in disparate environments and drive cycles)
- Safety (tolerance to abusive conditions)

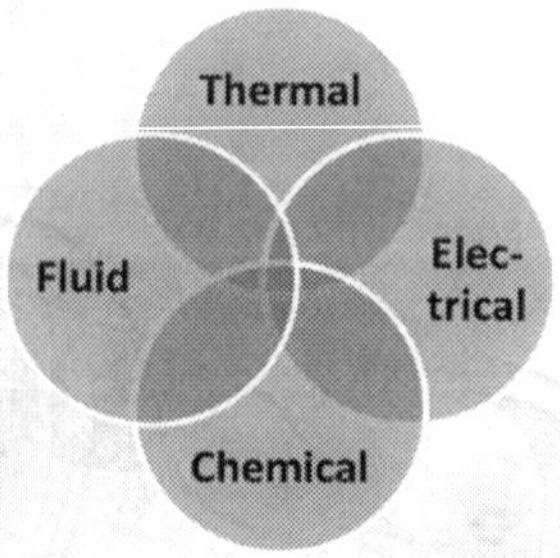


- Complex multi-scale, multi-physics system
- Rapidly evolving materials and design concepts
- Existing software tools not "tuned" for batteries

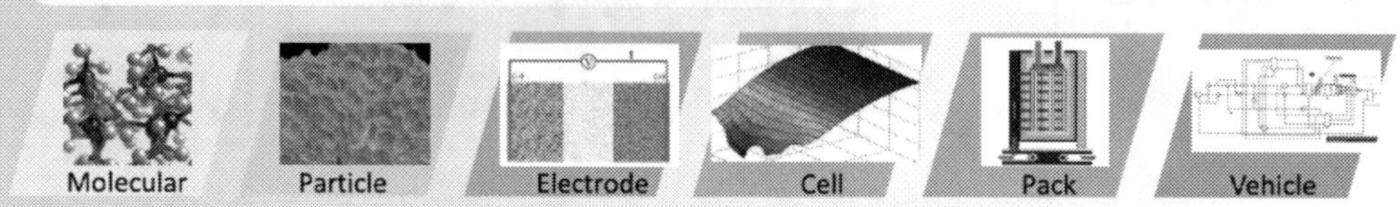


## Central Importance of Thermal Management

2014 Automotive Simulation World Congress
October 9-10 . Tokyo, Japan

- Performance and life sensitive to cumulative thermal history
- Minimize intra-cell and inter-cell temperature variations

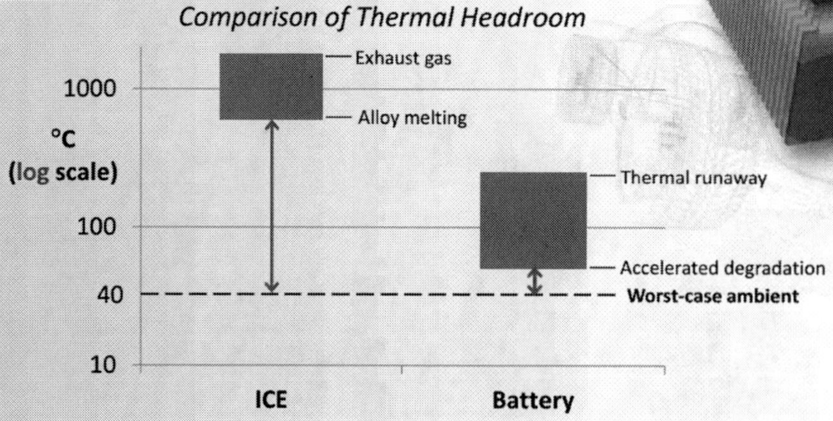

## CAEBAT Project

- Sponsor: U.S. Dept of Energy (DoE) Vehicle Technologies Office
  - High-risk/high-reward R&D to advance transportation technologies
- Computer Aided Engineering of Electric Drive Vehicle BATteries
  - Improved simulation is vital to accelerate design innovation
- Goal: develop and validate more accurate, efficient, modular, flexible, accessible, battery-oriented software tools
- Collaborative team project 2011-14

NREL NATIONAL RENEWABLE ENERGY LABORATORY | GM | ANSYS | ESim

2014 Automotive Simulation World Congress
October 9-10 . Tokyo, Japan

## Project Objectives

Faster design cycles and optimize batteries (cells and packs) for improved performance, safety, life, and low cost.

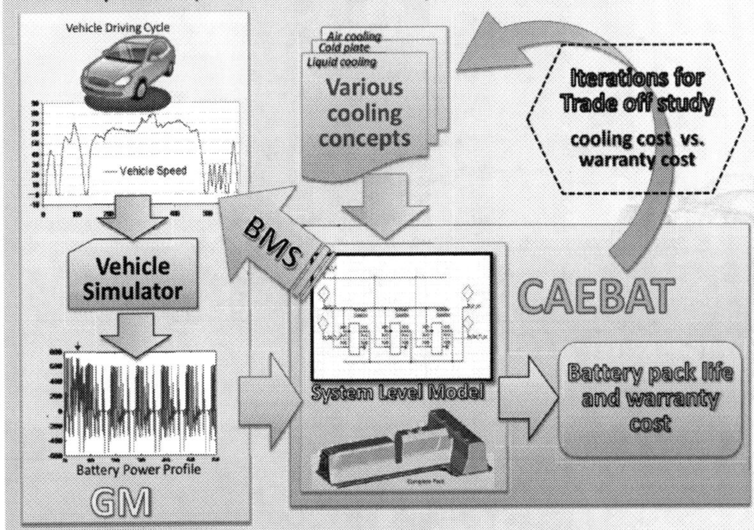


- Address multi-scale, multi-physics interactions
- Provide flexibilities
- Expandable framework
- Validate models

**Ability to provide trade off studies between various cooling concepts and the battery pack life.**

6

## Project Technical Approach



- Blend field (CFD) simulation, system simulation, novel scale-coupling and reduced-order methods

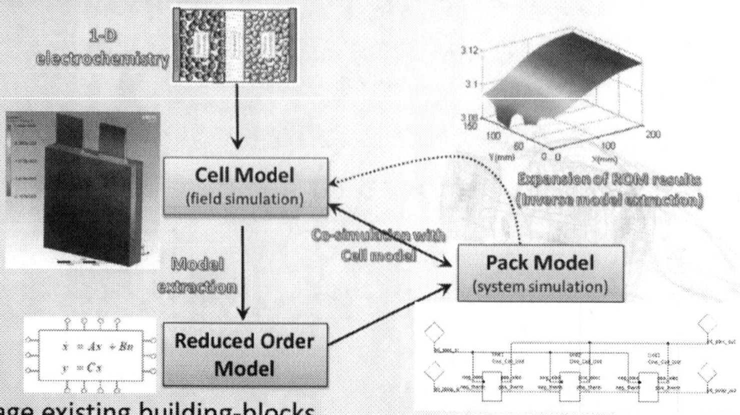


- Leverage existing building-blocks
  - ANSYS Workbench, Fluent, Simplorer
  - Interoperability via Open Architecture Software (OAS)

## Project Technical Approach (continued)

2014 Automotive Simulation World Congress
October 9-10 . Tokyo, Japan

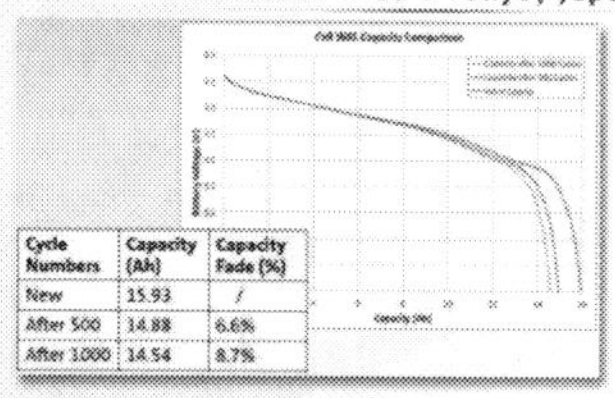


- Address gaps in methods and models (NREL, ESim)

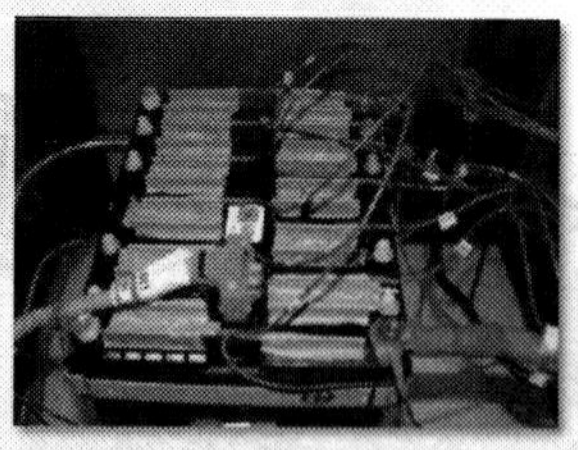

- Experimental validation (GM)

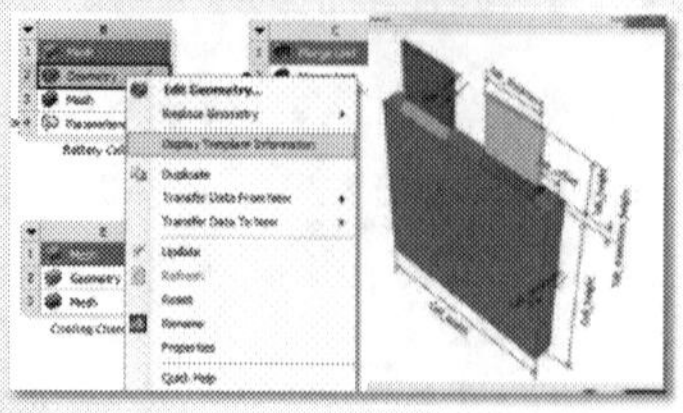

- Sustainable deployment to industry (ANSYS)

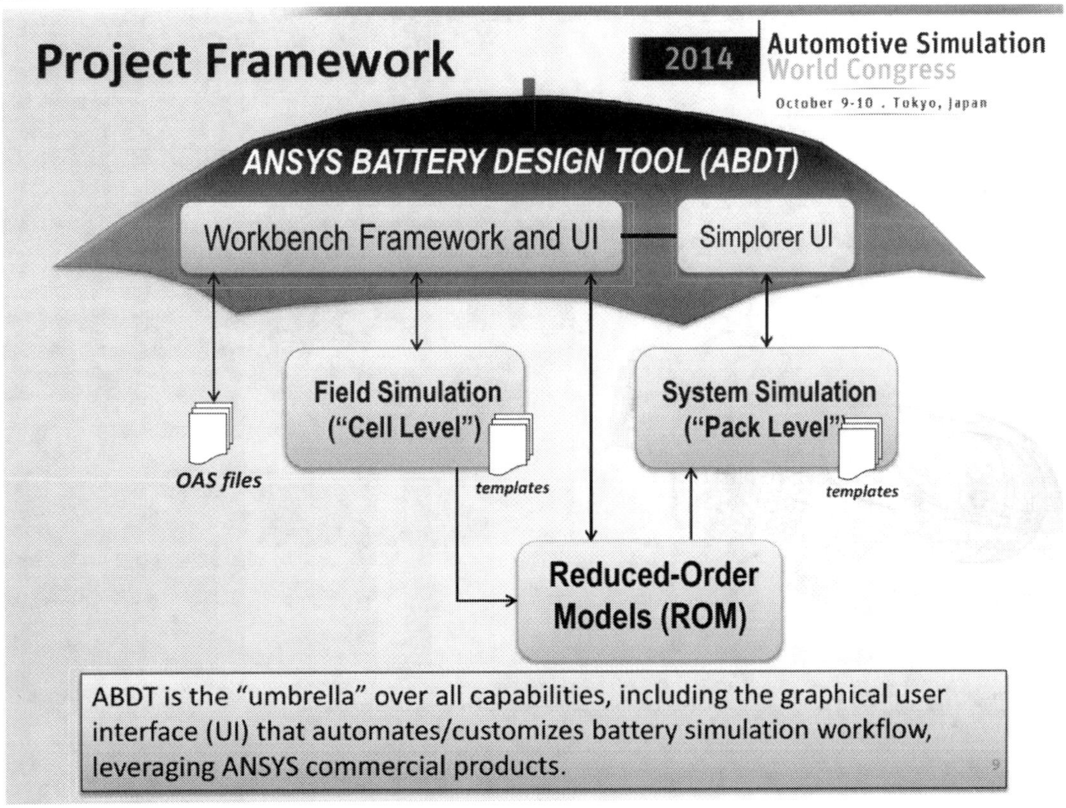
Project Framework
2014 Automotive Simulation World Congress
October 9-10 . Tokyo, Japan
ANSYS BATTERY DESIGN TOOL (ABDT)
Workbench Framework and UI
Simplorer UI
Field Simulation ("Cell Level")
System Simulation ("Pack Level")
OAS files
templates
templates
Reduced-Order Models (ROM)
ABDT is the "umbrella" over all capabilities, including the graphical user interface (UI) that automates/customizes battery simulation workflow, leveraging ANSYS commercial products.
9

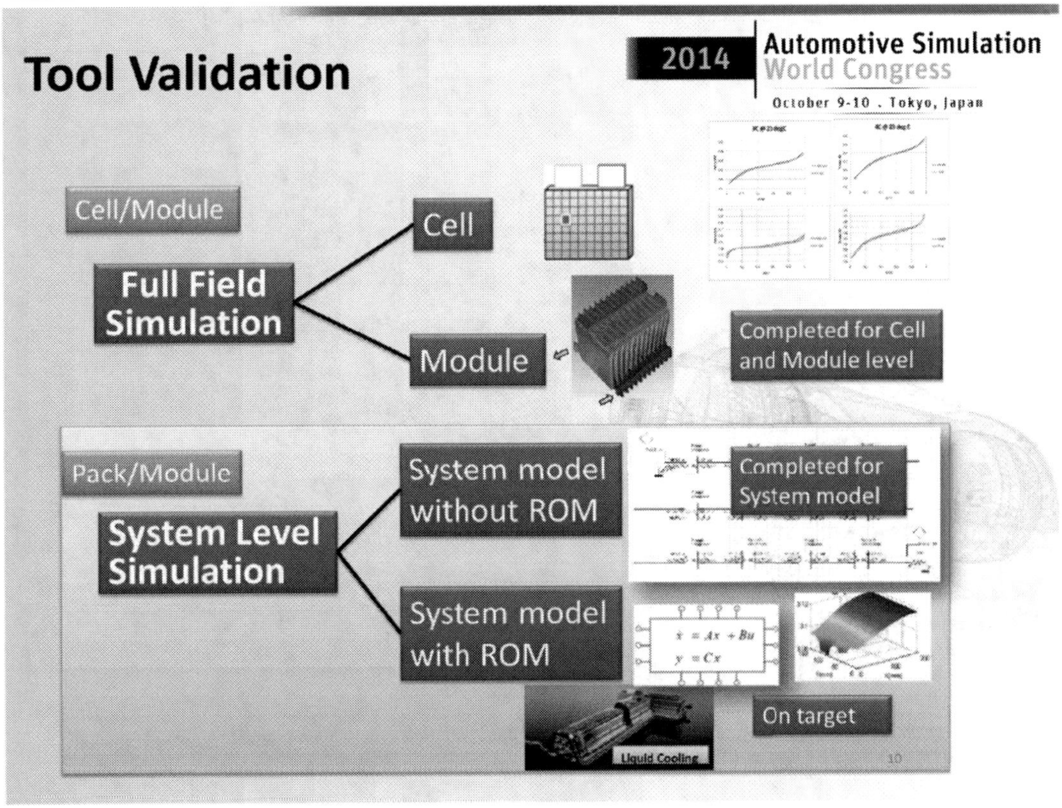
Tool Validation
2014 Automotive Simulation World Congress
October 9-10 . Tokyo, Japan
Cell/Module
Full Field Simulation
Cell
Module
Completed for Cell and Module level
Pack/Module
System Level Simulation
System model without ROM
Completed for System model
System model with ROM
$\dot{x} = Ax + Bu$
$y = Cx$
On target
Liquid Cooling
10

# Validation of Full Field Simulation

2014 Automotive Simulation World Congress
October 9-10 . Tokyo, Japan

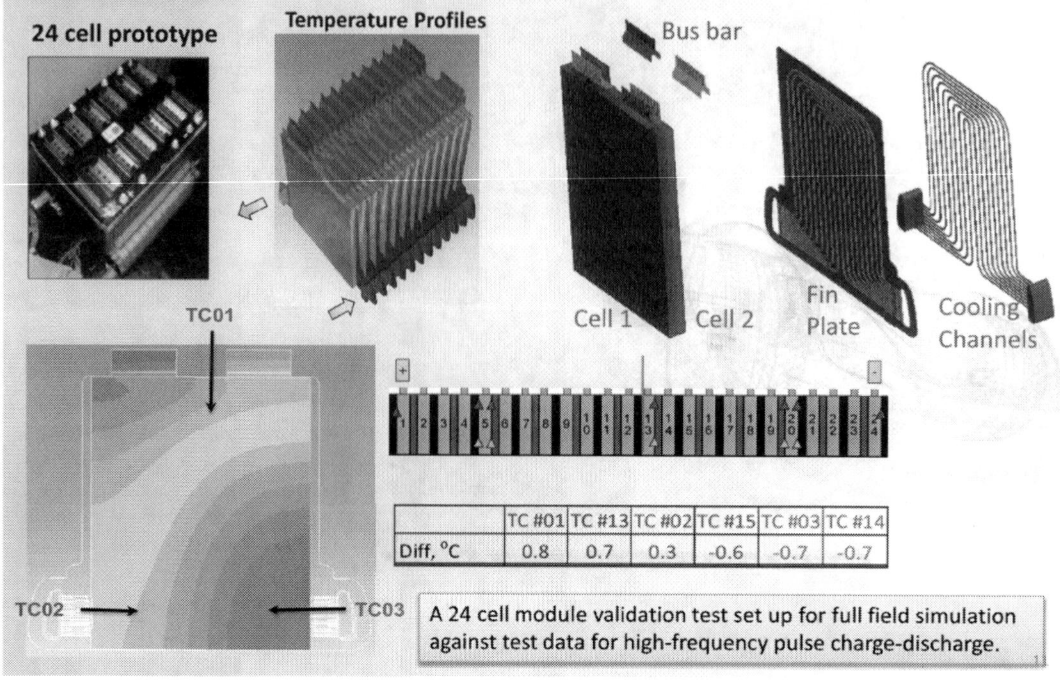


| | TC #01 | TC #13 | TC #02 | TC #15 | TC #03 | TC #14 |
|---|---|---|---|---|---|---|
| Diff, °C | 0.8 | 0.7 | 0.3 | -0.6 | -0.7 | -0.7 |

A 24 cell module validation test set up for full field simulation against test data for high-frequency pulse charge-discharge.

# Validation for 24 Cell Module

2014 Automotive Simulation World Congress
October 9-10 . Tokyo, Japan

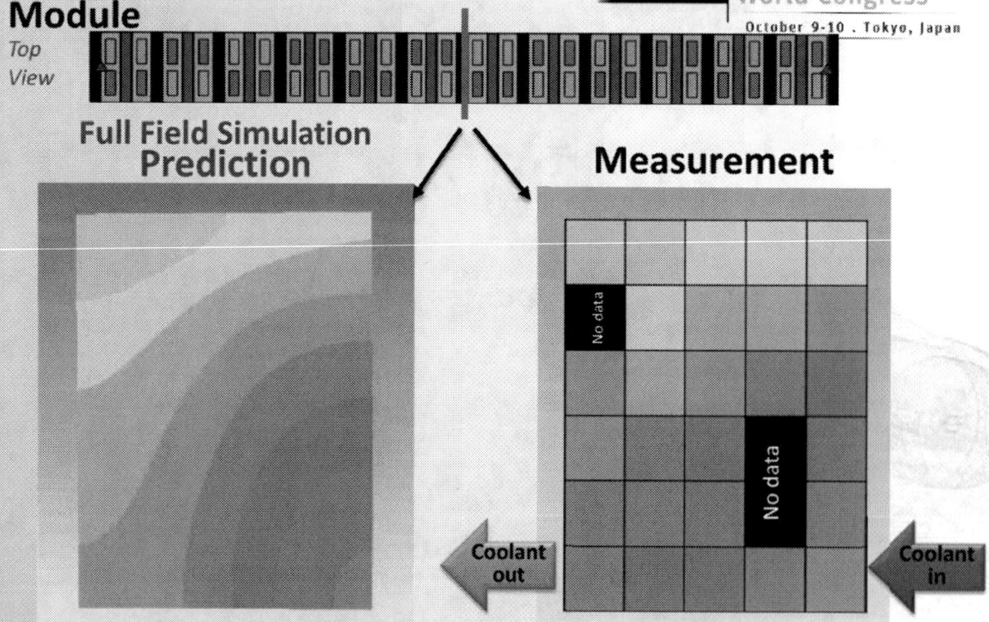


Maximum difference between the prediction and the measurement is within 1 ºC.

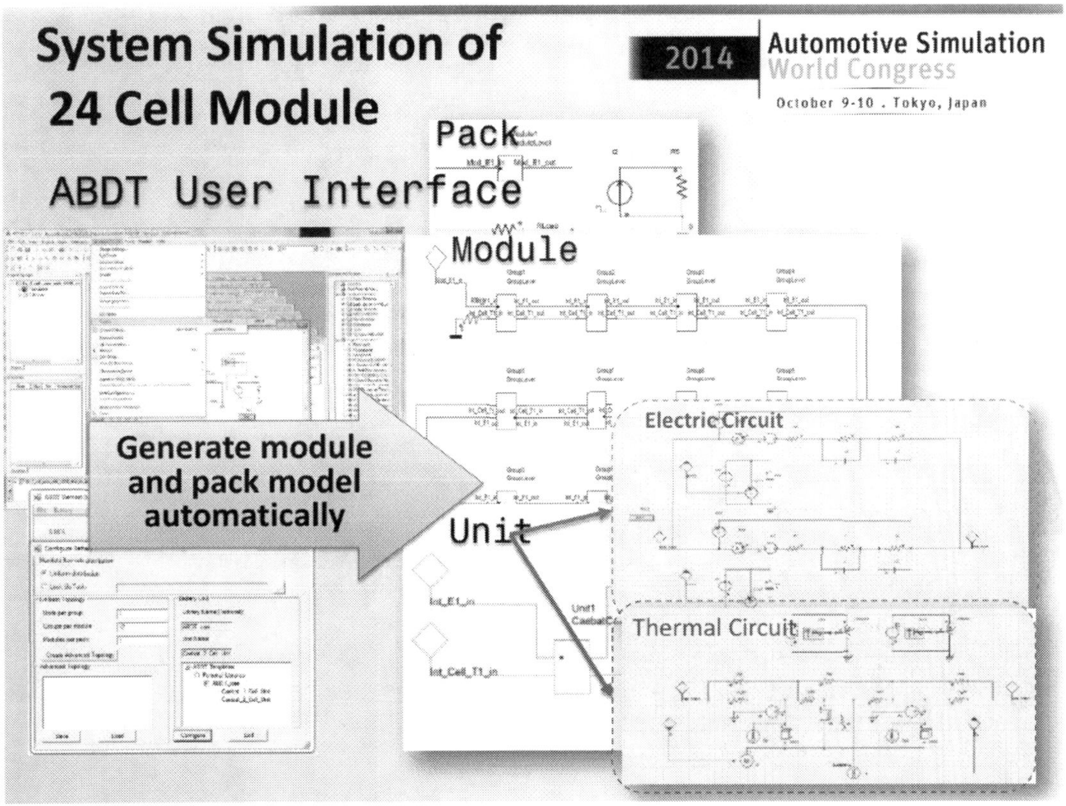
System Simulation of
24 Cell Module
2014
Automotive Simulation
World Congress
October 9-10 . Tokyo, Japan
ABDT User Interface
Pack
Module
Generate module
and pack model
automatically
Unit
Electric Circuit
Thermal Circuit

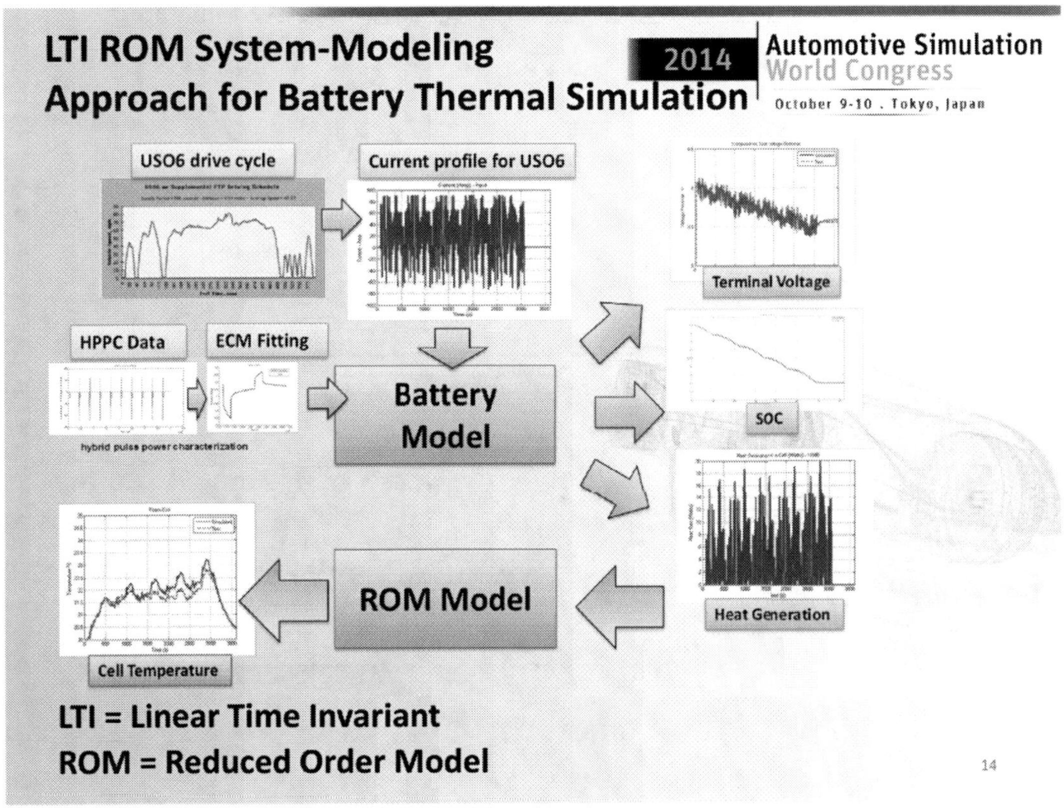
LTI ROM System-Modeling
Approach for Battery Thermal Simulation
2014
Automotive Simulation
World Congress
October 9-10 . Tokyo, Japan
US06 drive cycle
Current profile for US06
Terminal Voltage
HPPC Data
ECM Fitting
hybrid pulse power characterization
Battery
Model
SOC
Heat Generation
ROM Model
Cell Temperature
LTI = Linear Time Invariant
ROM = Reduced Order Model
14

## Validation of System Simulation

2014 Automotive Simulation World Congress
October 9-10 . Tokyo, Japan

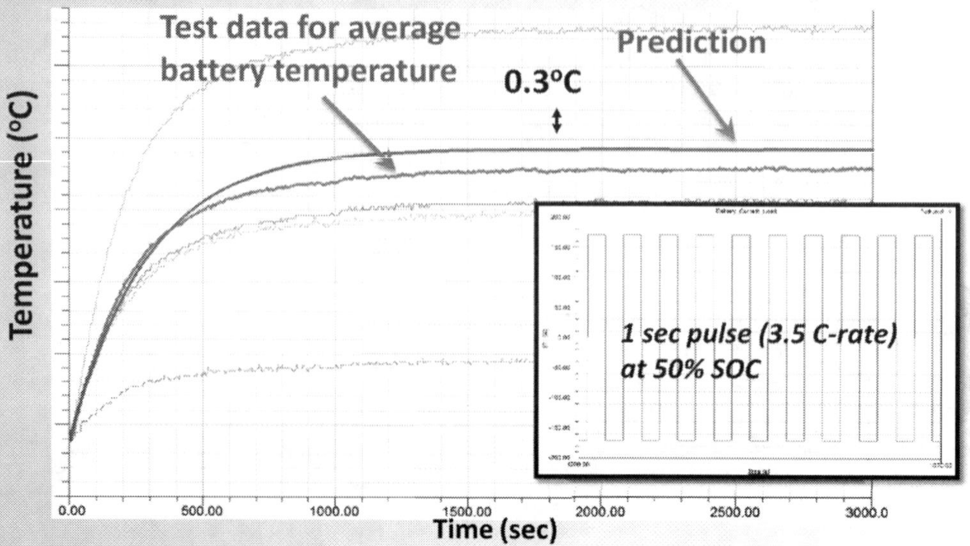


CPU runtime for system simulation is less than 1 minute (Dell Z800 PC)
Compared to the full field simulation that takes 4 ~ 5 days with 64 processor HPC

## Validation of System Simulation for USO6 Drive Cycle


2014 Automotive Simulation World Congress
October 9-10 . Tokyo, Japan


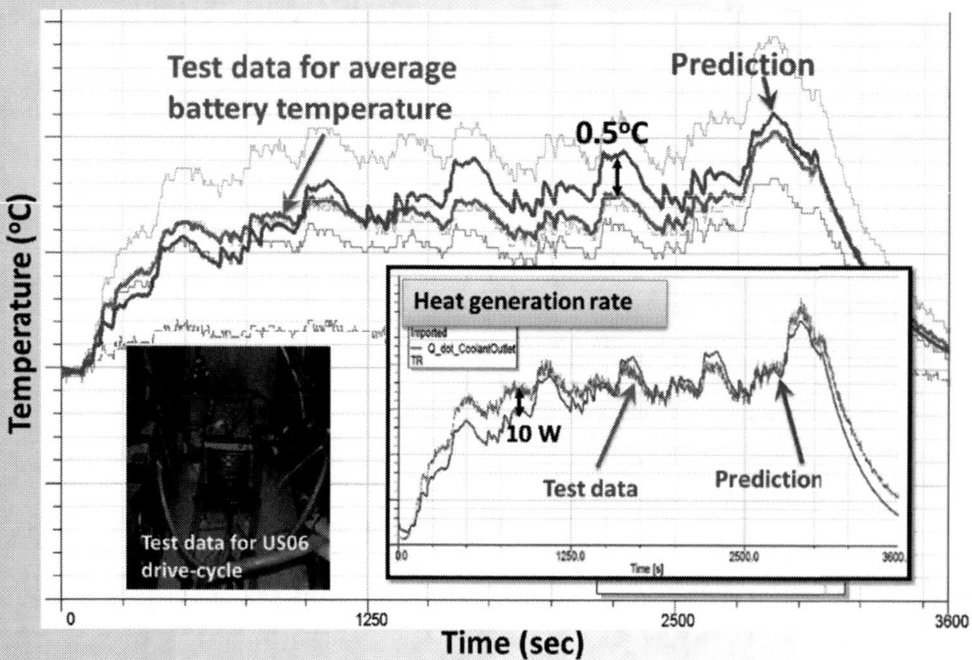

# Comparison Between Simulation and Test Data for US06 Drive-Cycle



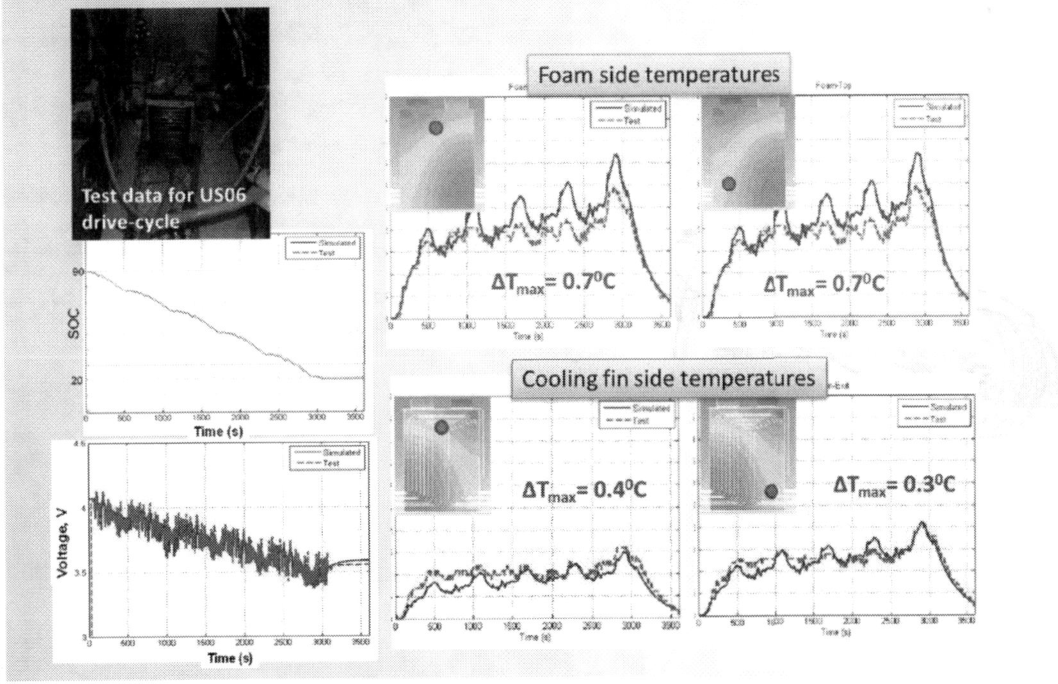


# Summary: Sustainable Software Tool - ADBT

2014 Automotive Simulation World Congress
October 9-10 . Tokyo, Japan

**Support of DOE CAEBAT**

*The automotive industry requires CAE design tools that include the following capabilities.*

- **Modular:**
  - —Integrate physics and chemistry in a computationally efficient manner.
- **Provide Flexibilities:**
  - —Provide a platform to enable various simulation strategies.
- **Provide Expandable Framework:**
  - —Enable future users to easily add new physics of interest.
  - —OAS-compatible.
- **Validated :**
  - —Ensure model predictions agree with experimental data by performing carefully designed experiments.
- **Easy to use**

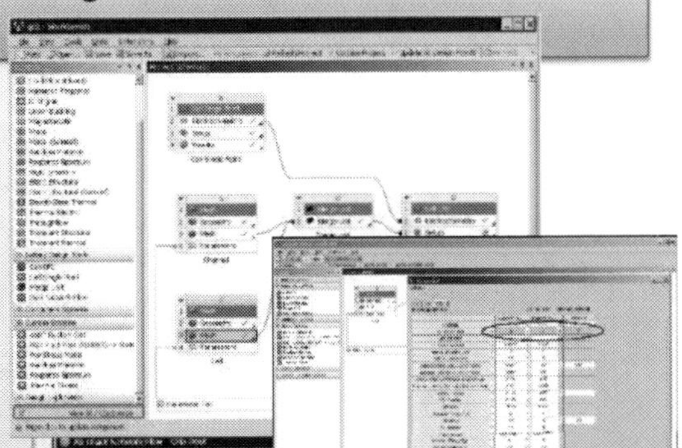

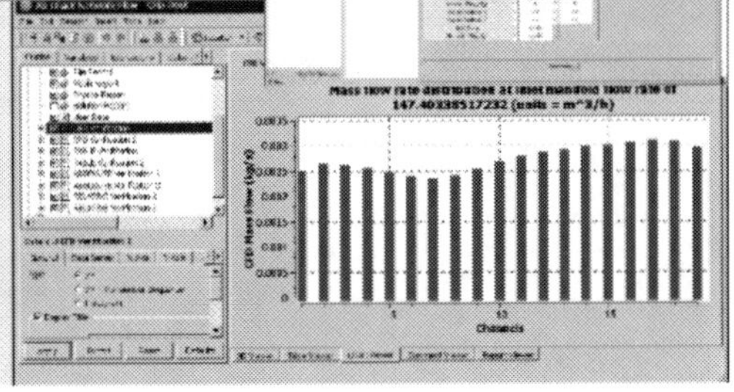

2014 Automotive Simulation World Congress
October 9-10 . Tokyo, Japan

# Thank you!

ありがとう

Contact:
Sandeep.Sovani@ansys.com

2014 Automotive Simulation World Congress

# Process Integration and Comprehensive Solution Paradigms: Engine Combustion, Conjugate Heat Transfer, Durability and Fatigue

Laz Foley, ANSYS Inc.

## Outline

- **Simulation Trends**
- **CAE in Powertrain Development**
- **Solution Paradigm Challenges and Requirements**
- **A Comprehensive Solution Practice Example**
  - Turbocharger Design
- **Conjugate Heat Transfer Work Flows**
  - Cylinder Head Design
- **Co-Simulation**
  - 1D-3D Coupling
- **Non-Linear FEA**
  - Gasket Optimization
- **CAE Toolkit Integration, Automation and Customization**
  - Solution Scenario: Comprehensive Combustion/CHT Vision

## Simulation Trends

| Old Paradigms | New Paradigms |
|---|---|
| • Single Component Simulation | • Multi-Domain Simulation (Component and System) |
| • Single Physics | • Multi-Physics Simulation (flow, thermal, structural, electronic, systems) |
| • Few Design Points | • Extensive Design Exploration |





## CAE in Powertrain Development

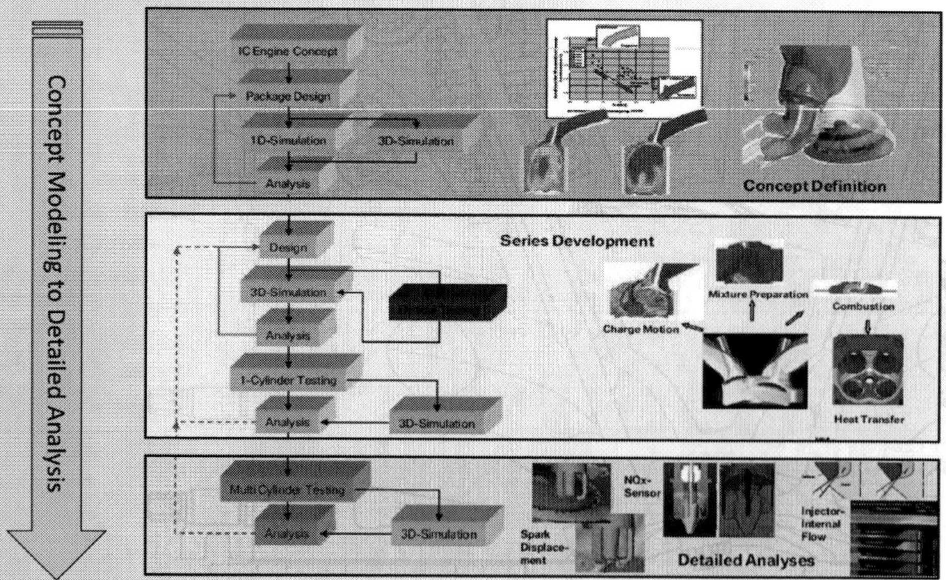


Courtesy: BMW AG

# Solution Paradigm Challenges and Requirements

- Interoperability across physics, scales and user personas
- Workgroup productivity via HPC usage
- Design exploration with embedded capabilities
- Integrated multi-physics and co-simulation with variety of tools in simulation toolkit
- Multi-Disciplinary Optimization (MDO) and Topology Optimization (TO) for all physics disciplines: Structural, Fluids, FSI
- Integrating FMEA work groups with CAE and driving CAT-CAE analyses including simulation as a calibration tool (MiL, SiL, HiL etc.)
- Cloud computing infrastructure, knowledge management, simulation data management/re-use

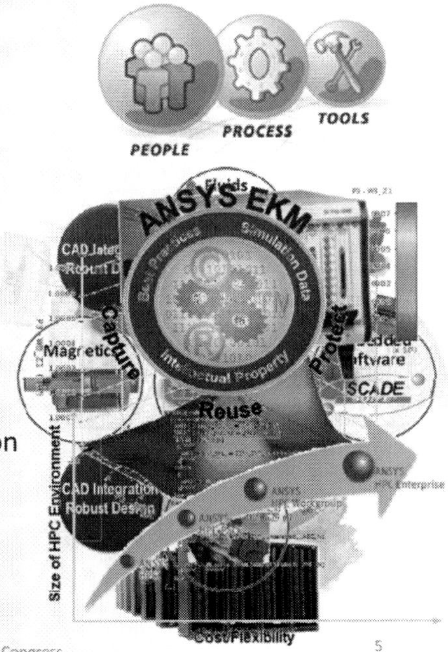






# Conceptual Design Via Parameterization and Optimization

- Parametric CAD Model
  - Bi-directional connectivity with Workbench to enable CAD to PPT "push-button" process
  - Automated and persistent work flows
  - Easily extended to design exploration/optimization
- Non-Parametric CAD Model
  - Topology Optimization via mesh morphing (Fluent, RBF-Morph) and/or adjoint methods

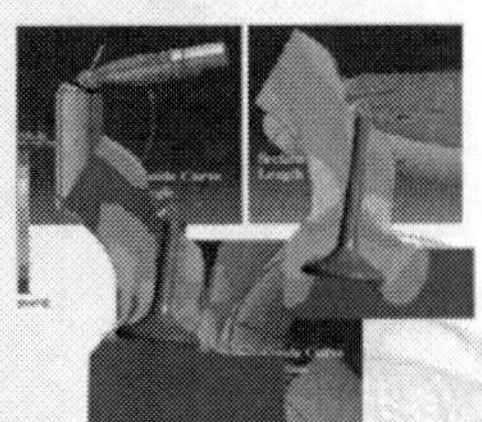

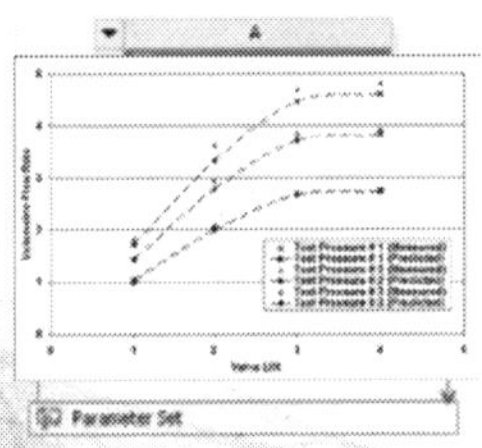


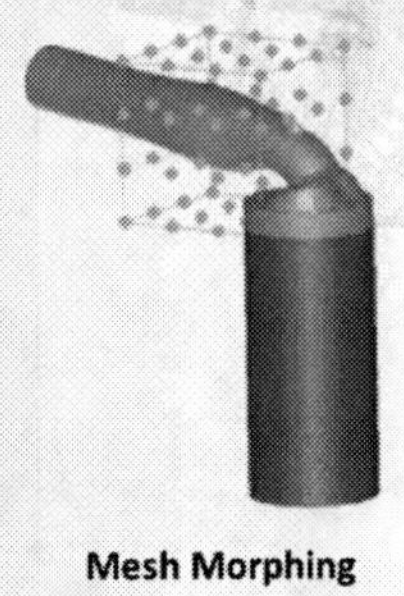

Mesh Morphing Control Volume

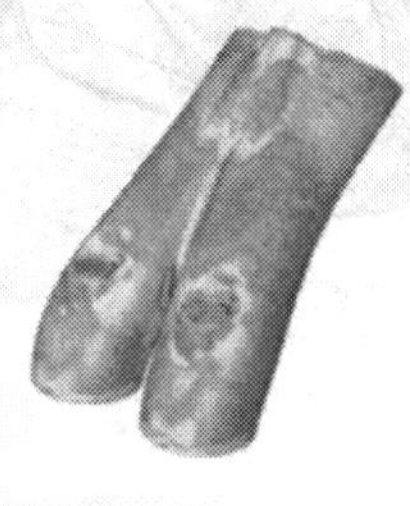

Adjoint Shape Sensitivity Contours

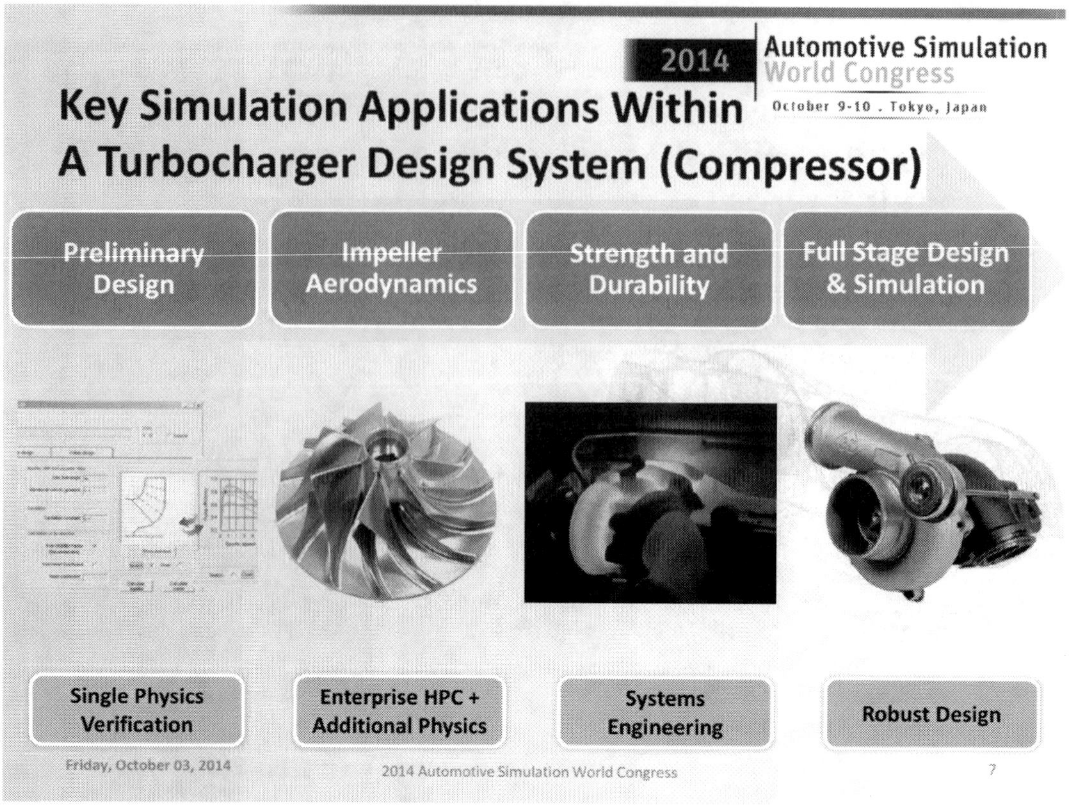
2014
Automotive Simulation
World Congress
October 9-10 . Tokyo, Japan
Key Simulation Applications Within
A Turbocharger Design System (Compressor)
Preliminary Design
Impeller Aerodynamics
Strength and Durability
Full Stage Design & Simulation
Single Physics Verification
Enterprise HPC + Additional Physics
Systems Engineering
Robust Design
Friday, October 03, 2014
2014 Automotive Simulation World Congress
7

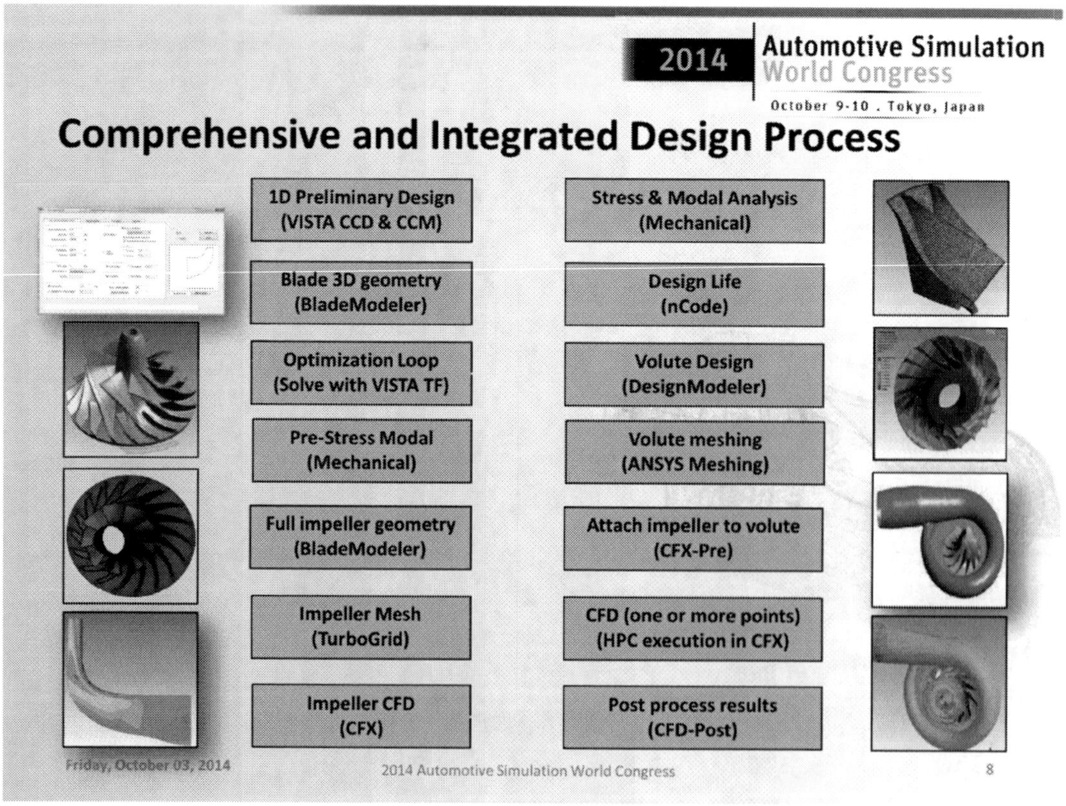
2014
Automotive Simulation
World Congress
October 9-10 . Tokyo, Japan
Comprehensive and Integrated Design Process
1D Preliminary Design (VISTA CCD & CCM)
Blade 3D geometry (BladeModeler)
Optimization Loop (Solve with VISTA TF)
Pre-Stress Modal (Mechanical)
Full impeller geometry (BladeModeler)
Impeller Mesh (TurboGrid)
Impeller CFD (CFX)
Stress & Modal Analysis (Mechanical)
Design Life (nCode)
Volute Design (DesignModeler)
Volute meshing (ANSYS Meshing)
Attach impeller to volute (CFX-Pre)
CFD (one or more points) (HPC execution in CFX)
Post process results (CFD-Post)
Friday, October 03, 2014
2014 Automotive Simulation World Congress
8

2014 Automotive Simulation World Congress
October 9-10 . Tokyo, Japan

## Turbocharger Durability and Fatigue

- Fluid-Structure Interaction (FSI) including Conjugate Heat Transfer (CHT)
- Blade Flutter Analysis
  - Specify blade oscillation via FEM modal analysis
  - Calculate aero-damping via CFD
- Forced Response Analysis
  - Blade excitation forces via CFD
  - Calculate deformation and stress via FEM or ROM
- Life Predictions

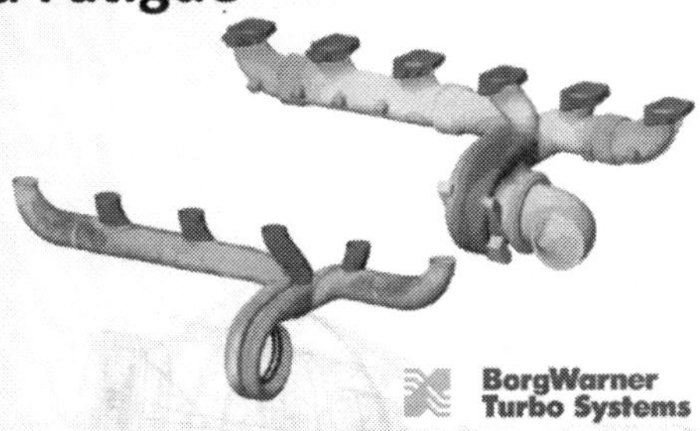

BorgWarner Turbo Systems

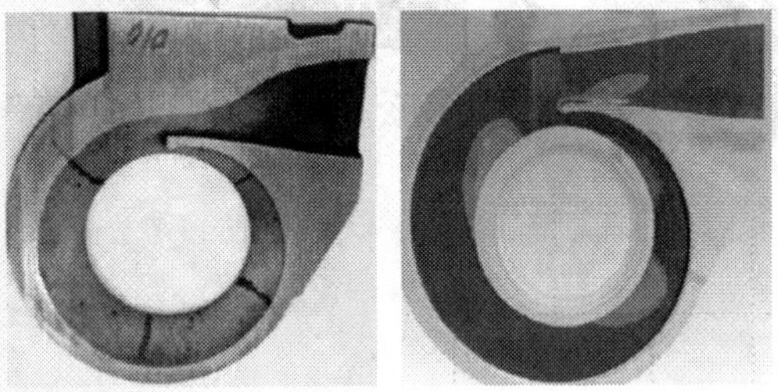

VOITH

Friday, October 03, 2014 2014 Automotive Simulation World Congress 9

2014 Automotive Simulation World Congress
October 9-10 . Tokyo, Japan

## Conjugate Heat Transfer (CHT) For Combustion System Design

- Variety of cross-physics and data exchange approaches used
  - Use of empirical heat data in a zonal combustion face map
  - Measured or predicted heat release profile imposed as energy source in gas exchange CFD model
  - Iterative combustion to non-combustion approach
- Challenges include
  - Different user-personas (Designer, CFD/FEM analyst), multiple simulation tools (1D/2D/3D), multiple functional groups (CFD, FEA, System)

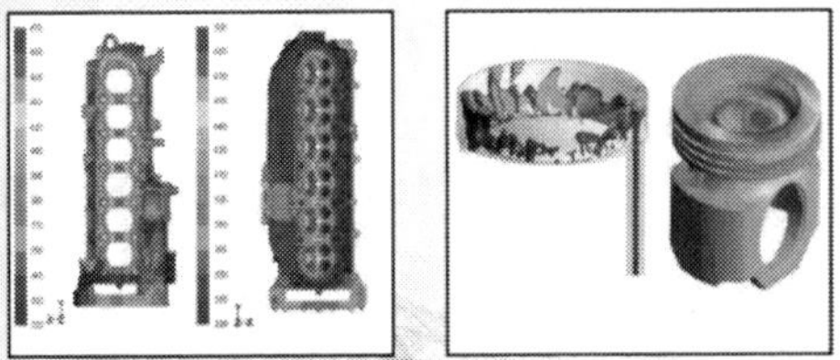

**Courtesy: Mercury Marine**

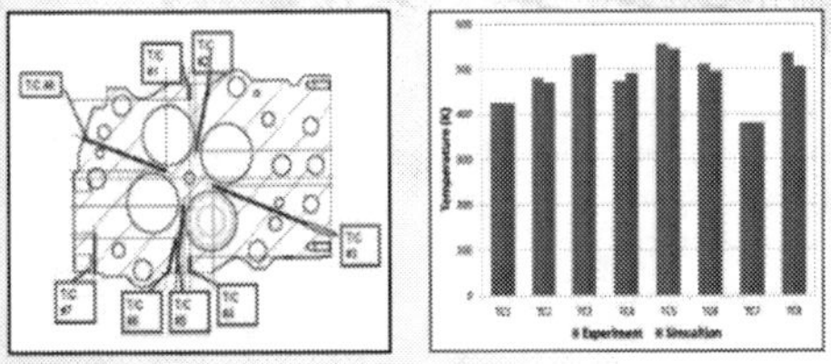

**Cylinder Head Temperature Predictions**
**Courtesy: Cummins**

Friday, October 03, 2014 2014 Automotive Simulation World Congress 10

## CAD to Fatigue for Cylinder Head Design: Process Overview

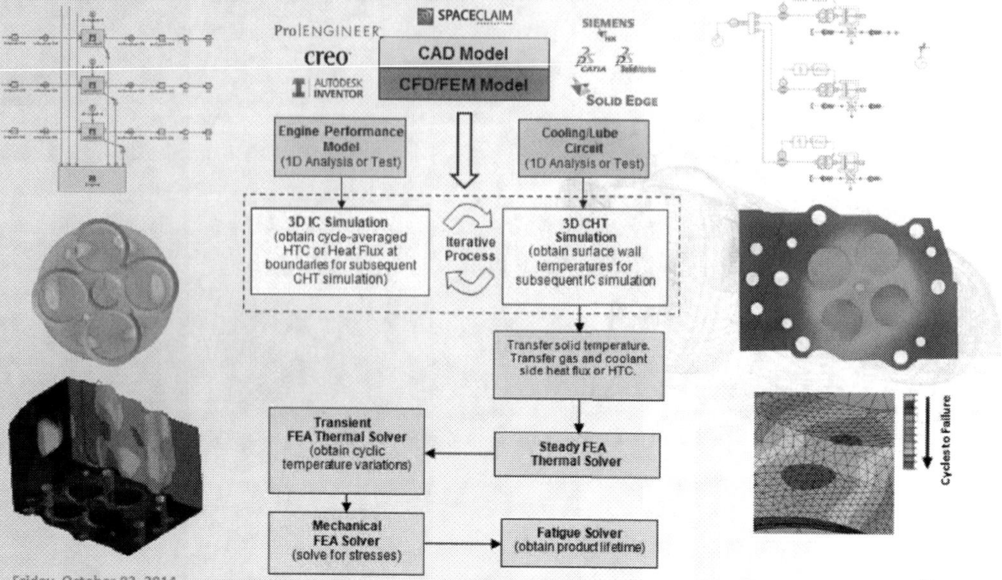




2014 Automotive Simulation World Congress
October 9-10 . Tokyo, Japan

## CAD to Fatigue for Cylinder Head Design: Key Requirements/Deliverables

- Parametric CAD for Design Points and Optimization
- Predictive Combustion (CFD)
- Conjugate Heat Transfer Simulation (CFD)
  - Head, Liners, Valves, Seats, Guides etc.
  - Thermal Resistance/Contact
  - Boiling
- Thermal/Stress/Fatigue Simulation (FEA)
  - Non-Linear Materials/Contact, Gaskets, Bolt Pre-Tension, Threading, Life

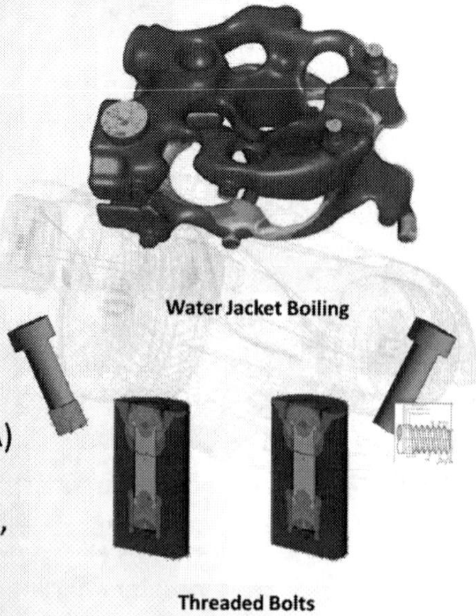

## Co-Simulation: 1D/3D Coupling

- Typical Coupling Applications
  - Intake/Exhaust Manifolds, AIS, Filters, Mufflers, EGR/PCV, Catalysts, Cylinders
- Dynamic Coupling with 1D Simulation Tools
  - GT-Suite, WAVE, Virtual Engines for engine simulation
- Solution Benefits
  - Capture of complete system level interactions, accurate CFD boundary conditions etc.
  - Insights into cylinder-to-cylinder performance, variations etc.
  - One coupled engine cycle with ~1M CFD cells < 3h on 64 CPU's (large model)

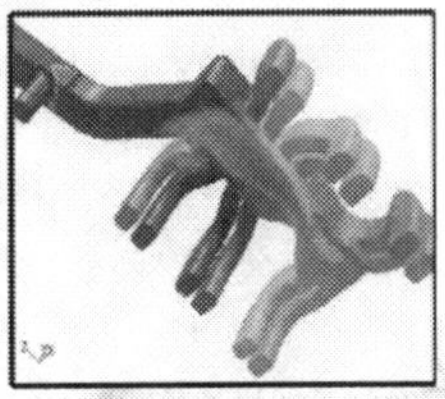

Diesel Intake Manifold (EGR)
Courtesy: Cummins

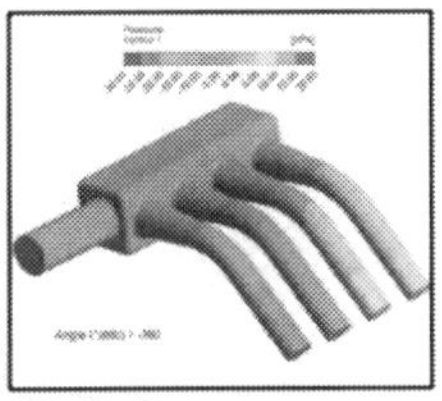

Gasoline
Intake Manifold

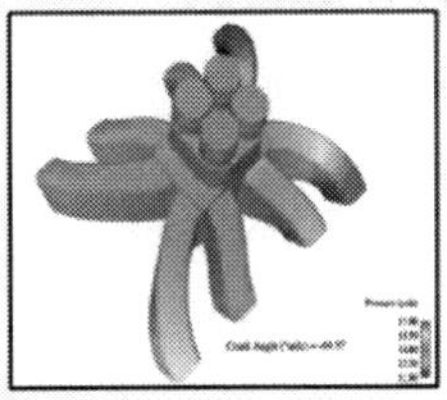

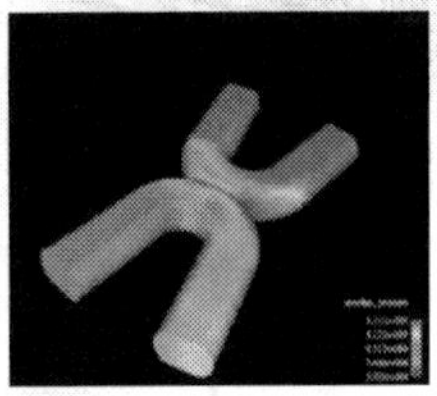

Motorsports: Intake Manifold & Exhaust Cross-Pipe





## Non-Linear FEA: Gasket Optimization

- Durability and Fatigue prediction challenges include
  - Highly non-linear material
  - Contact status modeling
- Solution Benefits
  - Rich suite of hyper-elastic models
  - Special gasket elements
  - Curve fitting module to determine material input from test data

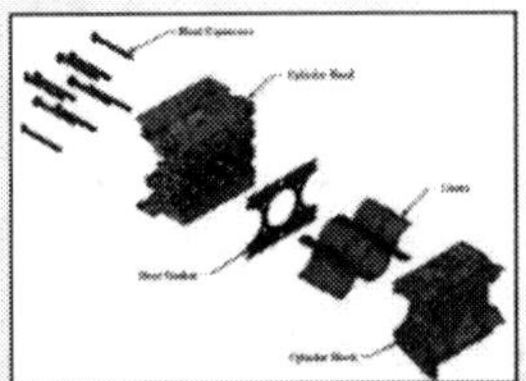

Cylinder Head Gasket Model
Courtesy: Cummins

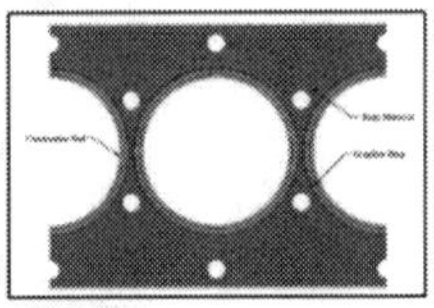

Gasket Components

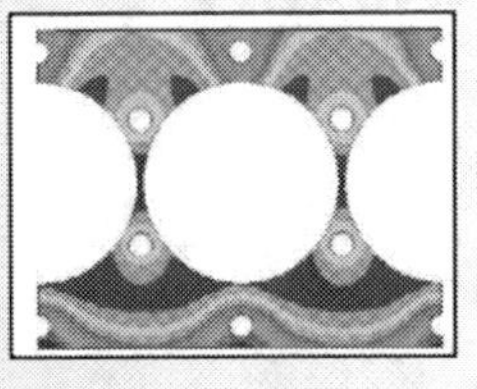

Soft Gasket

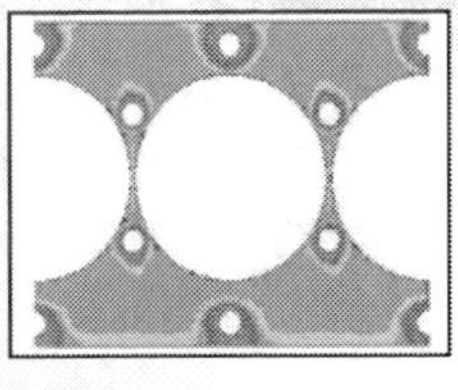

Stiff Gasket

Compressive Stress Contours

2014 Automotive Simulation World Congress
October 9-10 . Tokyo, Japan

## Steady State vs. Transient Loading and Fatigue

- Steady state loads do not always produce the highest stresses
- Cyclic stresses can cause failure at loads lower than material tensile/yield strength
- Transient loading behavior critical for fatigue/life predictions

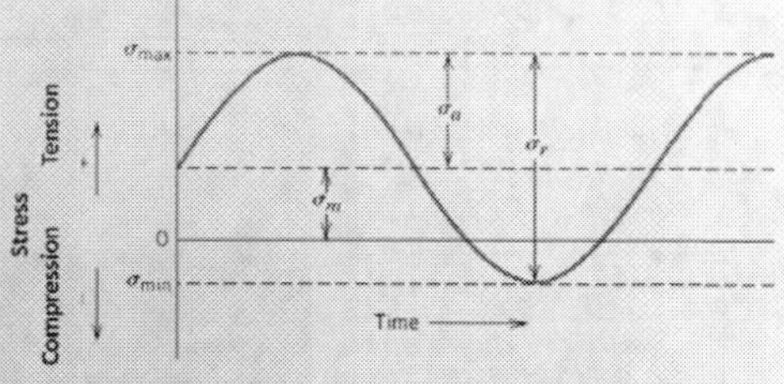


Mean stress: $\sigma_m = (\sigma_{max} + \sigma_{min}) / 2$
Range of stress: $\sigma_r = (\sigma_{max} - \sigma_{min})$
Stress amplitude: $\sigma_a = \sigma_r/2 = (\sigma_{max} - \sigma_{min}) / 2$
Stress ratio: $R = \sigma_{min} / \sigma_{max}$

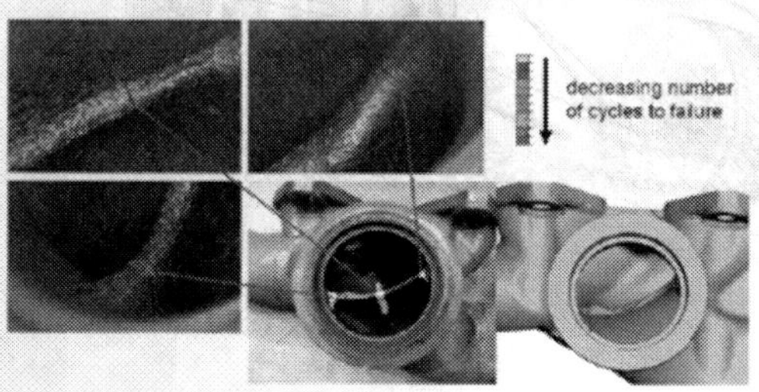


Fraunhofer IWM



2014 Automotive Simulation World Congress
October 9-10 . Tokyo, Japan

## ANSYS Customization Suite

- Two-Tier Customization Suite
  - SDK: Workflow and data integration of 3rd party software
  - ACT: Custom interfaces for WB/Mechanical and Beyond
- SDK already used by partners such as nCode, fe-safe and many others
- ACT has been used to customize interfaces and build vertical applications

nCode DesignLife

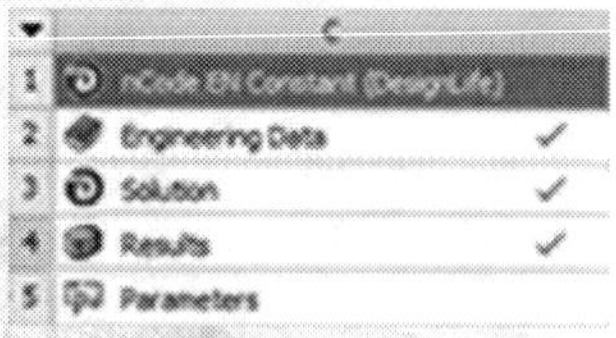


fe-safe

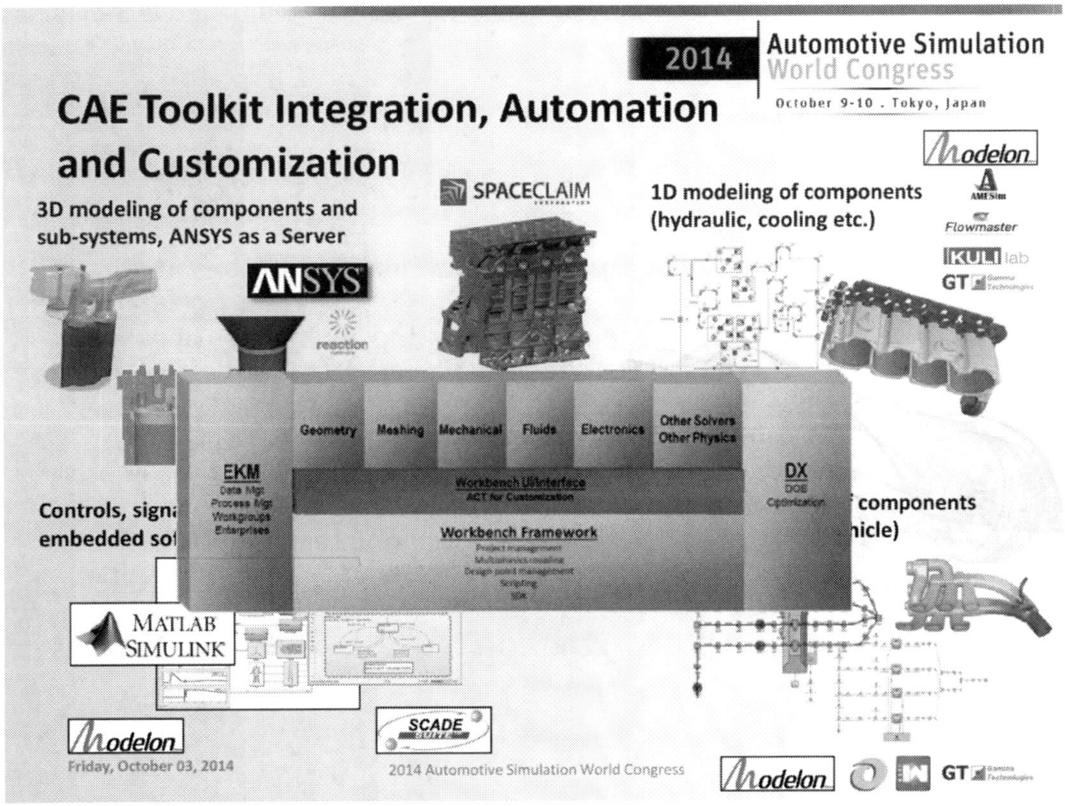
2014
Automotive Simulation
World Congress
October 9-10 . Tokyo, Japan
CAE Toolkit Integration, Automation and Customization
3D modeling of components and sub-systems, ANSYS as a Server
ANSYS
reaction
SPACECLAIM
1D modeling of components (hydraulic, cooling etc.)
Modelon
Flowmaster
KULI lab
GT
Geometry
Meshing
Mechanical
Fluids
Electronics
Other Solvers
Other Physics
EKM
Data Mgt
Process Mgt
Workgroups
Enterprises
Workbench UI/Interface
ACT for Customization
Workbench Framework
DX
DOE
Optimization
Controls, signa
embedded so
components
hicle)
MATLAB
SIMULINK
Modelon
SCADE
Friday, October 03, 2014
2014 Automotive Simulation World Congress

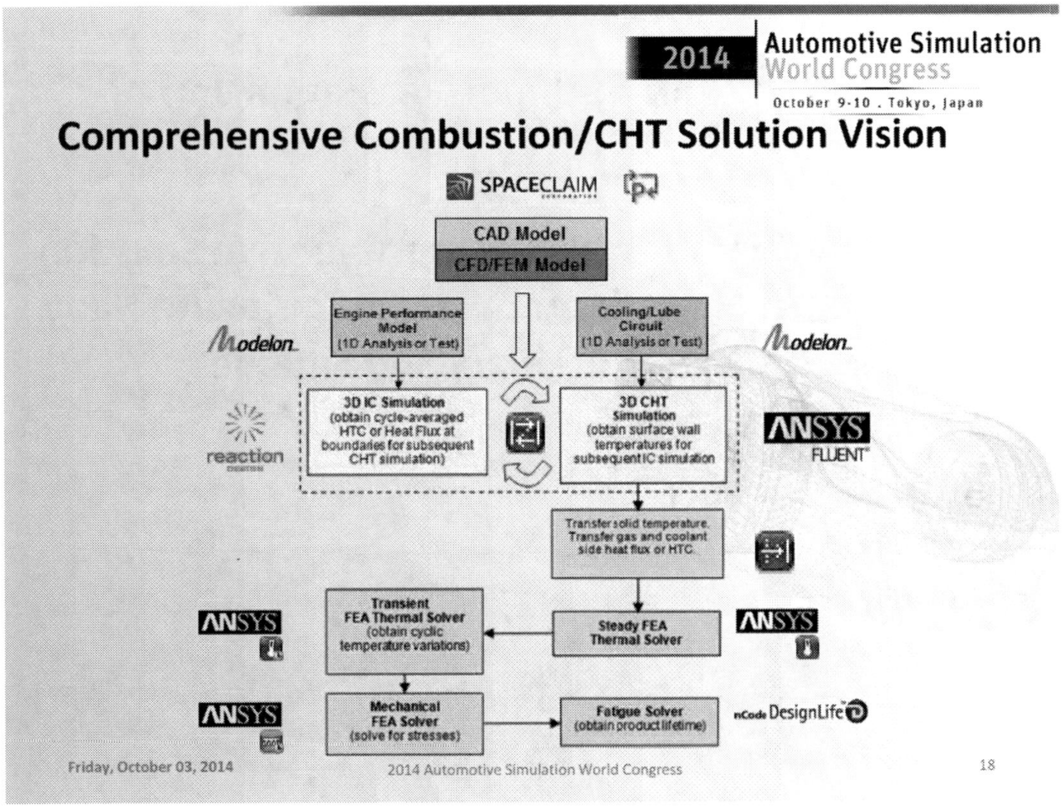
2014
Automotive Simulation
World Congress
October 9-10 . Tokyo, Japan
Comprehensive Combustion/CHT Solution Vision
SPACECLAIM
CAD Model
CFD/FEM Model
Engine Performance Model (1D Analysis or Test)
Cooling/Lube Circuit (1D Analysis or Test)
Modelon
Modelon
3D IC Simulation (obtain cycle-averaged HTC or Heat Flux at boundaries for subsequent CHT simulation)
3D CHT Simulation (obtain surface wall temperatures for subsequent IC simulation)
reaction
ANSYS
FLUENT
Transfer solid temperature. Transfer gas and coolant side heat flux or HTC.
Transient FEA Thermal Solver (obtain cyclic temperature variations)
Steady FEA Thermal Solver
ANSYS
ANSYS
Mechanical FEA Solver (solve for stresses)
Fatigue Solver (obtain product lifetime)
ANSYS
nCode DesignLife
Friday, October 03, 2014
2014 Automotive Simulation World Congress
18

2014 Automotive Simulation World Congress
October 9-10 . Tokyo, Japan

## Summary

- Some key take ways from this comprehensive solution paradigm
  - Engine design is much more than developing a combustion system
  - Depth and diversity within all physics and at all stages of a design process is necessary (combustion, fluid, thermal, stress, durability, fatigue)
  - Tools and processes that cater to different user personas and disciplines within a design process is important (Designer, Analyst, Systems)
  - A collaborative and common platform for integrated CAE is a requirement given product design demands (engine calibration, packaging, controls development, system optimization etc.)
- ANSYS is committed to the continuous advancement of its integrated solution work flows and processes

Friday, October 03, 2014 2014 Automotive Simulation World Congress 19

2014 Automotive Simulation World Congress

# Process Integration and Comprehensive Solution Paradigms: Engine Combustion, Conjugate Heat Transfer, Durability and Fatigue

Laz Foley, ANSYS Inc.

2014 Automotive Simulation World Congress

# ANSYS Customization Solutions

Padmesh Mandloi
ANSYS Inc.

## Outline

- Introduction
- Advantages of ANSYS Workbench Environment
- Examples of Customization
- Next Steps
- Summary

2014 Automotive Simulation World Congress
October 9-10 . Tokyo, Japan

# Introduction

**Table I: Top Five Pressures on Engineering Executives**

| Pressures | All Respondents |
|---|---|
| Project schedules for engineering products are shortening | 60% |
| Cost of raw goods is increasing (e.g. steel, aluminum, etc.) | 33% |
| Development budgets for product engineering are shrinking | 32% |
| Decreasing target product price-points driving down product cost targets | 28% |
| Market or customer requirements demand increasingly "smarter" products | 22% |

Source: Aberdeen Group, June 2008

**Business pressures continue to drive engineering organizations to accomplish more with less**



2014 Automotive Simulation World Congress
October 9-10 . Tokyo, Japan

# ANSYS Workbench

- Software EASIER to USE and more INDUSTRY SPECIFIC
  - Custom workflow, vertical applications
  - Implement customer "best practices" into ANSYS unified GUI in order to fit and enhance the customer's process

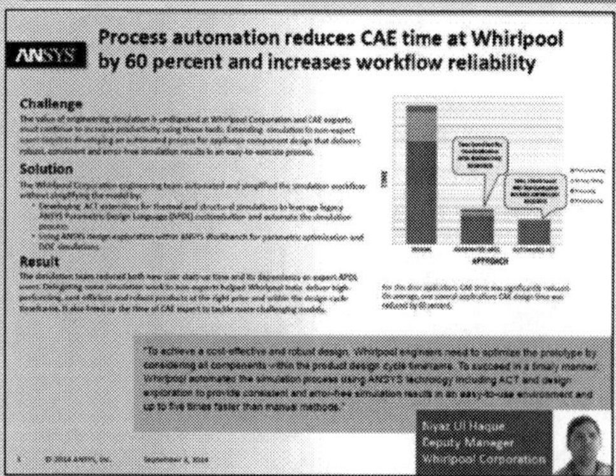


**"When ANSYS developed ACT, it was a giant step for industries. This amazing tool can be an instrument of knowledge diffusion of internal know-how. For WEG, it resulted in an instantaneous increase of knowledge."**

***Cassiano A. Cezario***
***R&D Center Simulation Leader***
***WEG Electric Motors***

## ANSYS Workbench Journaling and Scripting

– "Basic ingredient" to many customizations, from simple to complex

WB Solution from Excel Application

| Inputs: | Dp1 | Dp2 |
|---|---|---|
| catalysator length (mm)x10 | 20 | 40 |
| inlet mass flow rate (kg/s) | 0.001 | 0.001 |
| Dynamic viscosity (Pa s) | 1.79E-05 | 1.79E-05 |
| 1/Permeability Constant (m^2) | 2.00E-08 | 2.00E-08 |
| Density (kg/m^3) | 1.225 | 1.225 |
| CrossSectional Area(m^2) | 2.41E-02 | 2.41E-02 |

Simulate

Provide the Parameters for the Design Point study

Both the DPs are solved, Project is saved.

Pressure drop values from those analysis are updated in Excel

| Results : | ANSYS FLUENT Solution | | Analytical Solution | | Relative Error ,% | |
|---|---|---|---|---|---|---|
| | Dp1 | Dp2 | Dp1 | Dp2 | Dp1 | Dp2 |
| Pressure drop, Pa | 6.068 | 12.13 | 6.07 | 12.14 | 0.0 | 0.1 |





## Wizards and Templates

- Streamlined UI to guide user through complete simulation process
- Adopt the best practices

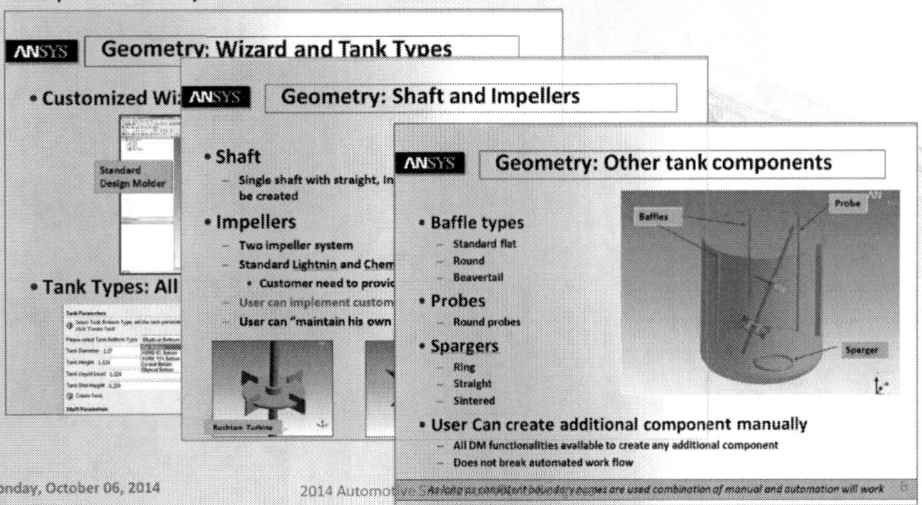

2014 Automotive Simulation World Congress
October 9-10 . Tokyo, Japan

# FlowModeler (Wizard in Meshing)

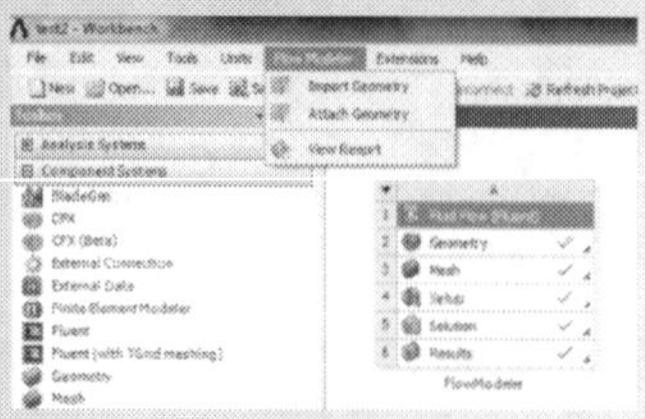

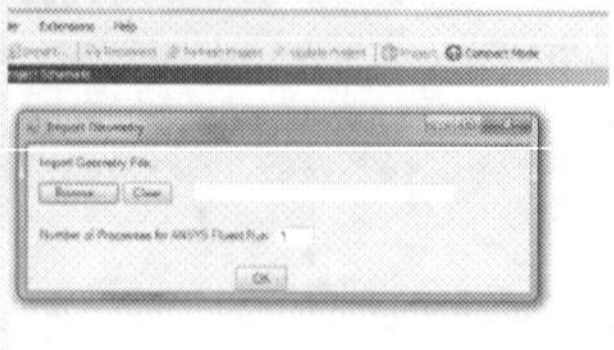

Custom import panel

Custom process inputs

Custom setup inputs

Custom report

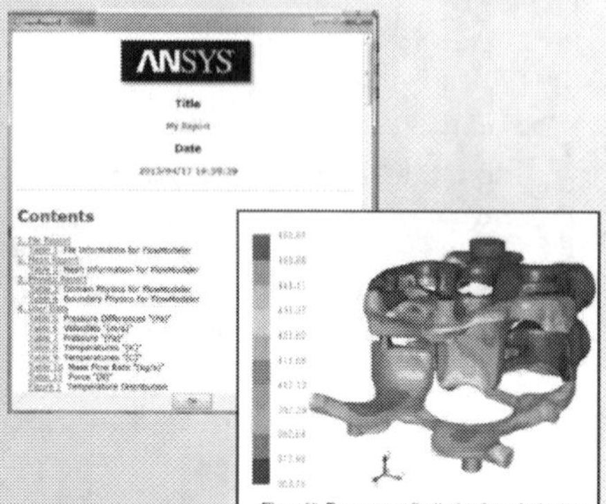


*Figure 10. Temperature distribution from single phase*

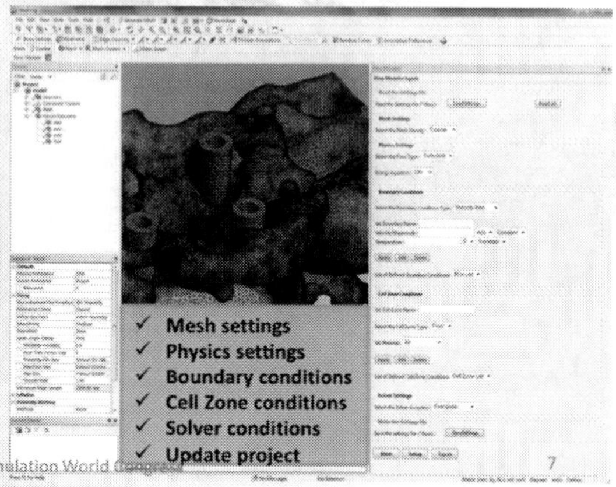




2014 Automotive Simulation World Congress
October 9-10 . Tokyo, Japan

# Exposing CMS in ANSYS Meshing (ACT)

- Solving larger FE models still remains a challenge...
- Approach of using "Superelements" will provide efficient solution for large assemblies without losing accuracy.
- Extension helps user to define/organize the Superelements, export/reuse in different analyses

| | Full finite element model | Model with 3 superelements |
|---|---|---|
| Nodes in the model | 379375 | 392 (in use pass) |
| Modal Analysis time (sec) | 145 | 256 (Gen: 250; Use: 6) |
| Harmonic Analysis time (sec) | 2134 | 43 (Gen: 0 **; Use: 43) |
| Total time (sec) | 2279 | 299 |
| Total Deformation Plot | 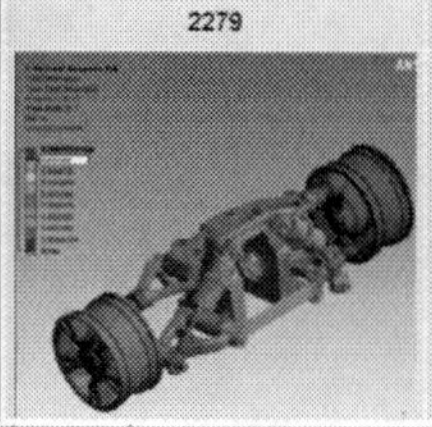 | 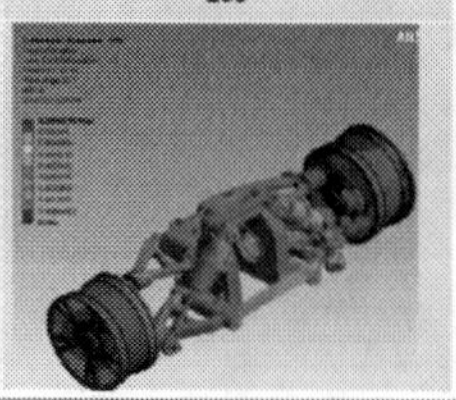 |

# Mesh Independent Spot Welds

Typically, a car body contains about 5000 spot welds joining sheets of different thicknesses.

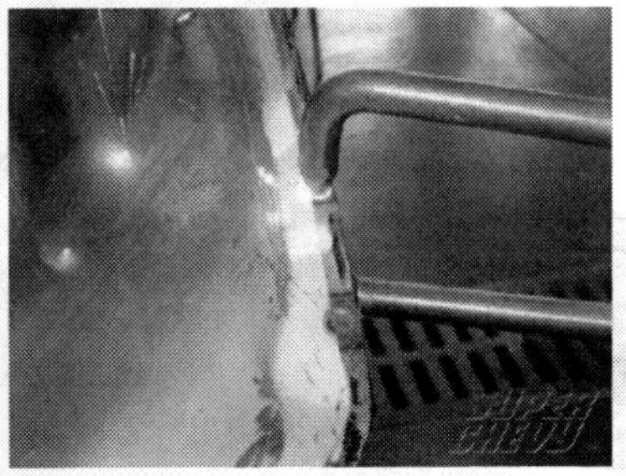

- Perhaps the most common application of spot welding is in the automobile manufacturing industry, where it is used almost universally to weld the sheet metal to form a car.
- The extension exposes the feature to Mechanical for creating spot welds using locations



2014 Automotive Simulation World Congress
October 9-10 . Tokyo, Japan

# Mesh Independent Spot Welds

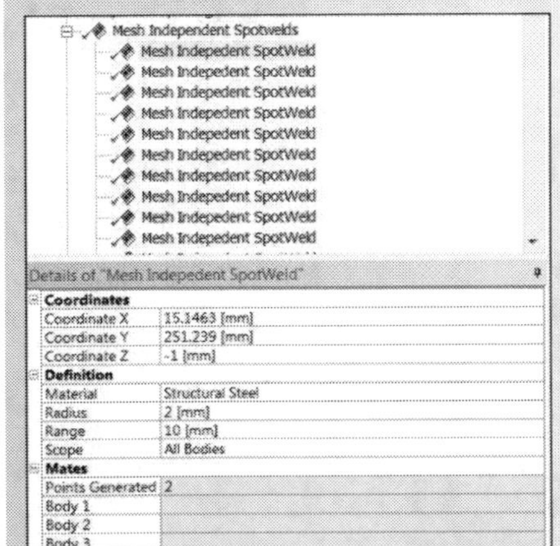


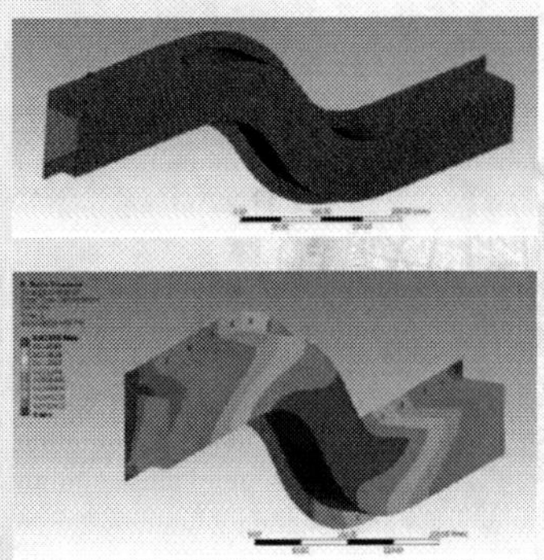

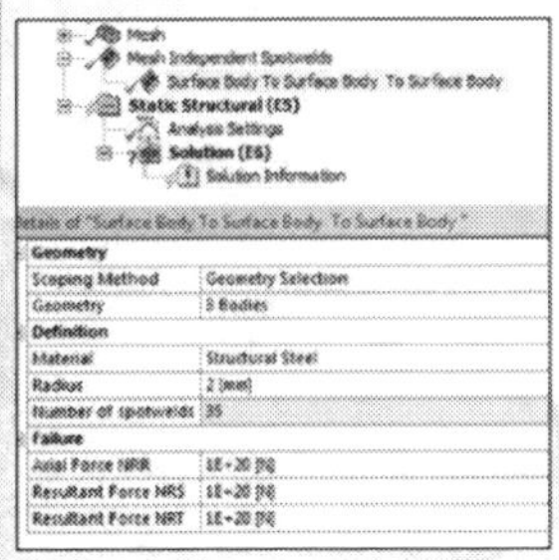


**Individual spotwelds or Multi spotwelds Connection can be created by projection on the geometry using point coordinate files**

# Mesh Independent Spot Welds

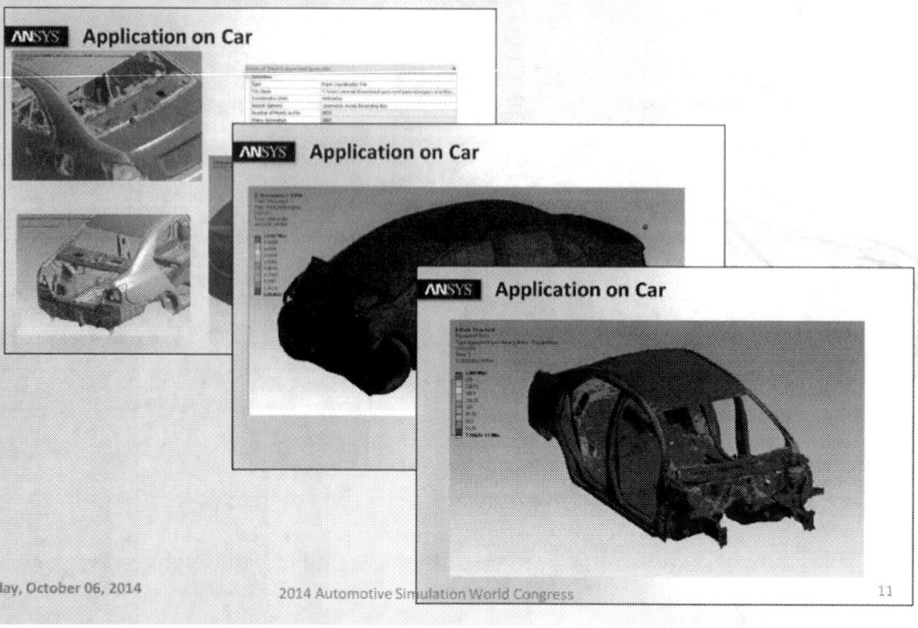




2014 Automotive Simulation World Congress
October 9-10 . Tokyo, Japan

# Create Bolt Imprints in DM (ACT)

- Typically, a car body contains about 100 bolts joining sheets of with other components
- Finding the stress in the these joins is most critical part of the mechanical simulations
- Creating imprints on the surfaces is a repetitive task for designers/CAE experts

# Create Bolt Imprints in DM (ACT)

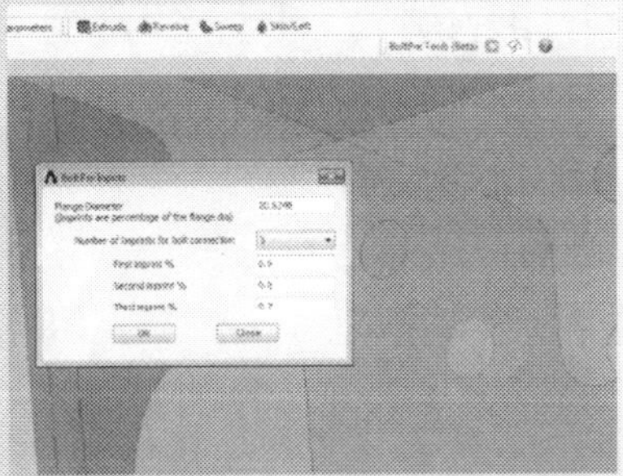

- Select edges and set the bolt imprint percentages
- Operations are automated to create the imprints using custom and standard features of DesignModeler
- Repeat the operations to create imprints of bolts on sheet bodies (surface bodies)

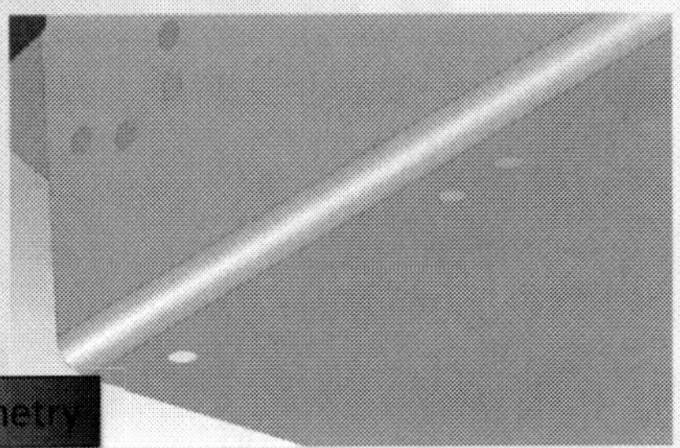

Initial Geometry

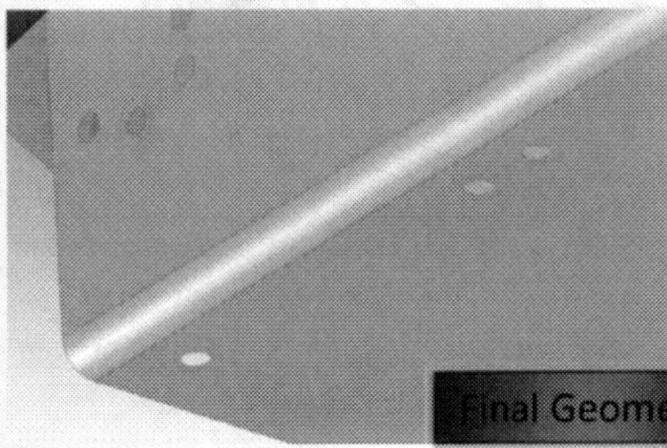

Final Geometry



2014 Automotive Simulation World Congress
October 9-10 . Tokyo, Japan

# HVAC Extension

- Complete setup in DM using ACT objects
- Smart and Rapid object creation
- Room, Person, Blocks, Fans, Openings and Vents
- Support for CAD Import and definition of bodies
- Visual representation in DM similar to Mechanical
- Full automation in DM. (No need to visit Meshing, Fluent and Post)

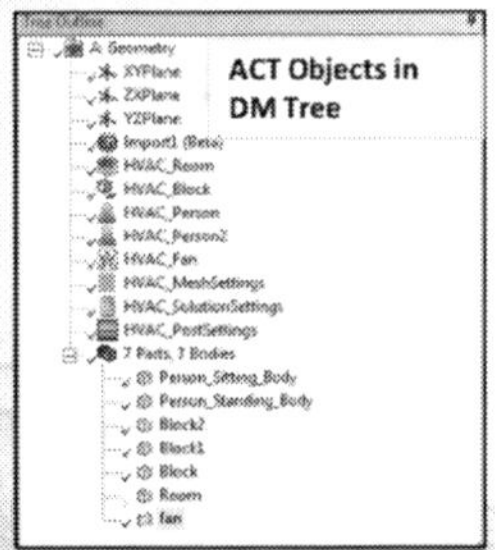


Rapid Model Creation
(Model Created Under 5 minutes)

Extension Manager

ACT Tool Bar in DM

Room Block Fan Person Mesh Post & Solution

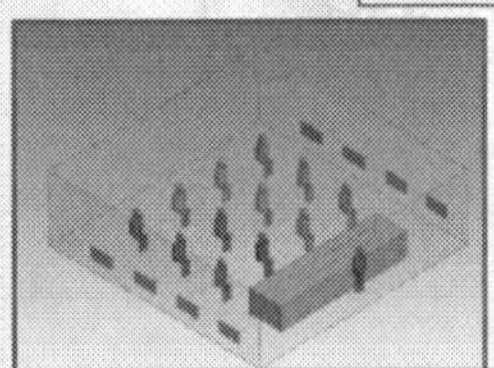

Automated Report and Images

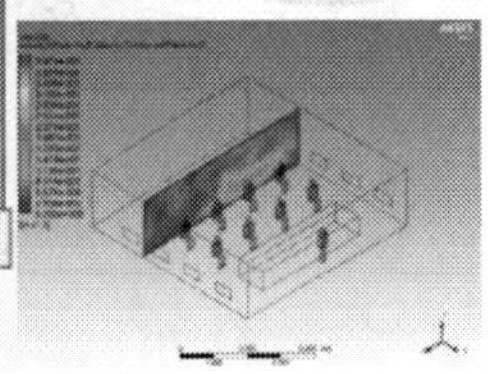

2014 Automotive Simulation World Congress
October 9-10 . Tokyo, Japan

## Next Steps

- Resources available for Customization
- Customer portal
- Training
- Technical Support
- Consulting Services

Monday, October 06, 2014 2014 Automotive Simulation World Congress 15

2014 Automotive Simulation World Congress
October 9-10 . Tokyo, Japan

## Summary

- We will continue to be challenged to accomplish "more with less"
- Engineering simulation is a key enabler
- The ANSYS Workbench environment can be customized in a variety of ways to:
- Automate repetitive tasks
- Create simplified UI's
- Expose advanced features
- Capture and deploy best practices
- Let's discuss how ANSYS can help you take advantage of customization

Monday, October 06, 2014 2014 Automotive Simulation World Congress 16

## Acknowledgements

- Sudesh Palase, ANSYS India
- Rajiv Rath, ANSYS India
- Simon Mendy, ANSYS France

# Leveraging SGI's UV™ and VizServer® Platform to Enable Multiple CAE Users to Run ANSYS® Software with Remote Visualization in a Data Center

Tony DeVarco, Director Manufacturing Solutions

# Customers Challenges and Needs

**CHALLENGES:**

- Maintaining multiple geographically distributed CAE engineering groups.
- Engineering visualization of large models requires high cost workstations.
- Remote workstations can't interact with the compute resources and data in the data center.
- Remote collaboration requires transfer of large data sets.

**NEEDS:**

- Consolidate hardware, applications, and data into centralized data centers.
- Replace workstations with low-cost remote clients.
- Enable remote collaboration.

# Typical HPC Scenario

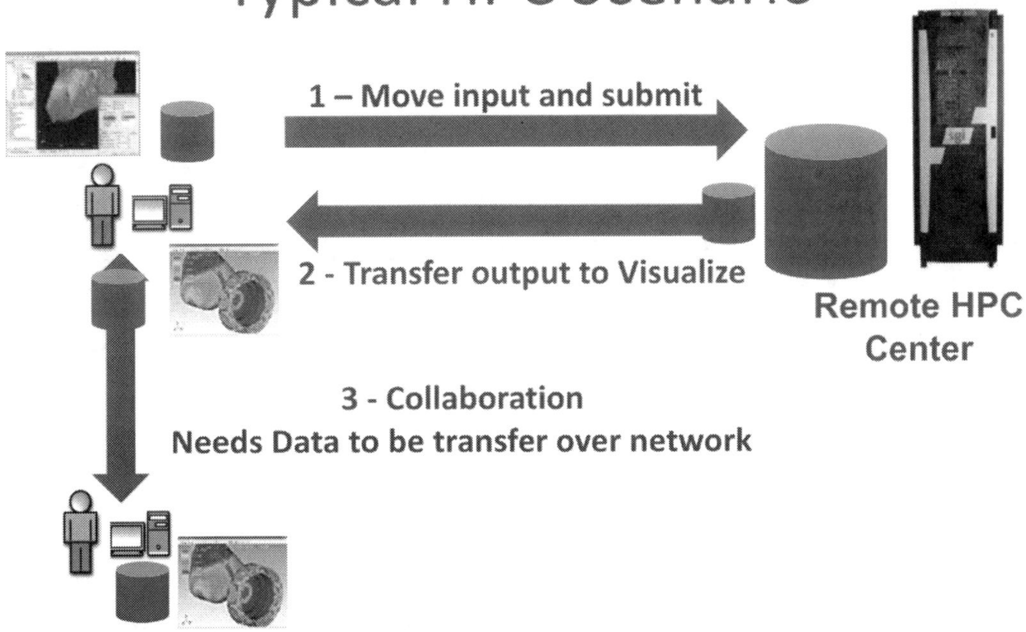


# Problems With This Model

- NETWORK
  Network overload leading to poor performance and response times all round
- COST
  Expensive, dedicated workstations (GPU, memory, ...) with short lifecycle
- IT MANAGEMENT
  Support, update and replace tens to tens of thousands of workstations
- WORKSTATION SIZING
  Workstations have to be sized for the largest expected models
- SECURITY
  Moving sensitive data around (in/out organization) is always risky
- WORKFORCE WORKING METHODS
  Current models do not support a diverse mobile workforce

# SGI® VizServer® with NICE Desktop Cloud Visualization™ (DCV) and EnginFrame™ Views Software

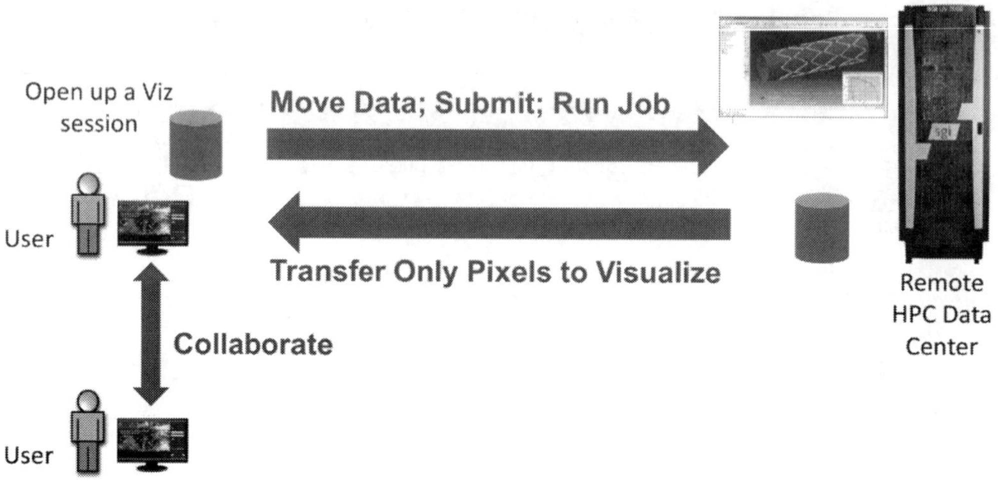


# What the Solution Solves

**NETWORK**
Network is no longer a bottleneck and data loads faster

**COST**
Centralized & Shared servers are less expensive to buy & manage

**IT MANAGEMENT**
Support, update and replacement are more efficient & do not affect users

**WORKSTATION SIZING**
Resources are dynamically sized based on users needs

**SECURITY**
Sensitive data remain within protected data center with full access control

**WORKFORCE**
Users can virtually connect & collaborate from anywhere with any client

# CAE Enterprise Solution Example

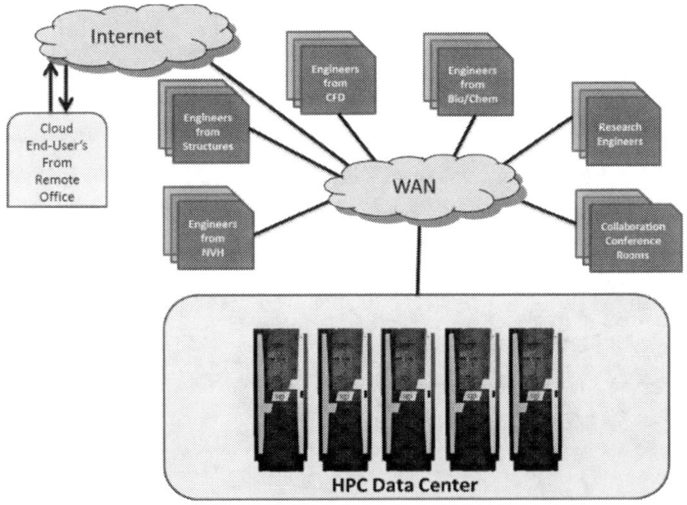


# SGI® VizServer® with DCV and EnginFrame Views Software Stack

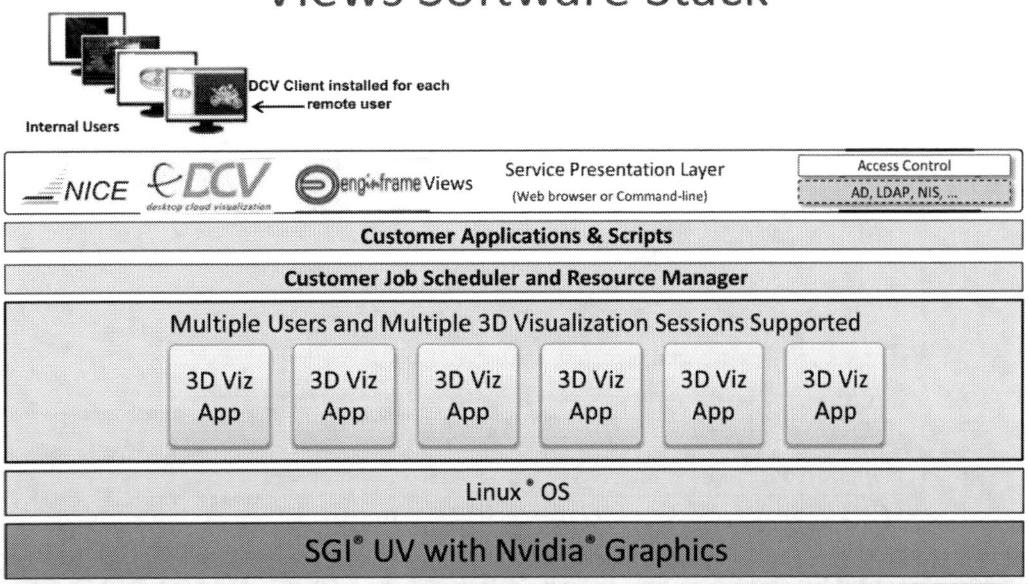

# Solutions for CAE-Compute, Viz & Storage

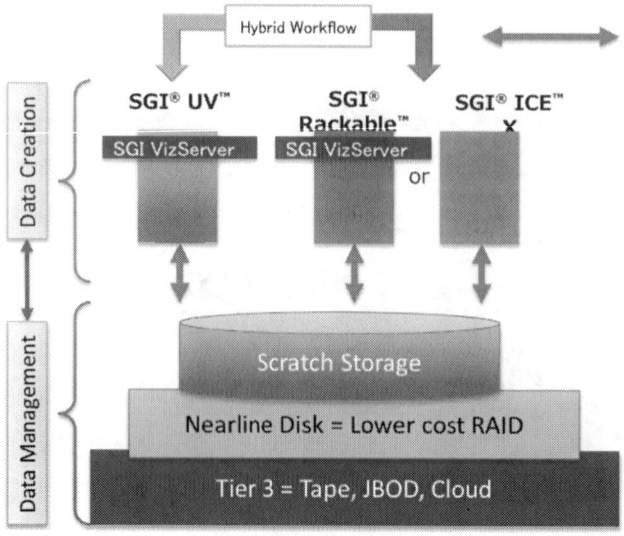


**Typical CAE Workflow**

- CAD model creation
- Mesh-generation
- Model decomposition
- Running solvers
- Viewing results
- Adjust and repeat

**Data Management Options**

1. **Scratch Storage**
   - High Performance-disk
2. **Nearline Storage**
   - Lower performance/ cost disk
3. **Persistent Storage**
   - Virtualized Storage Tiers

SGI Proprietary & Confidential

# Characterization of ANSYS Applications

Typical HPC Behavior of a Single Job for Various Applications

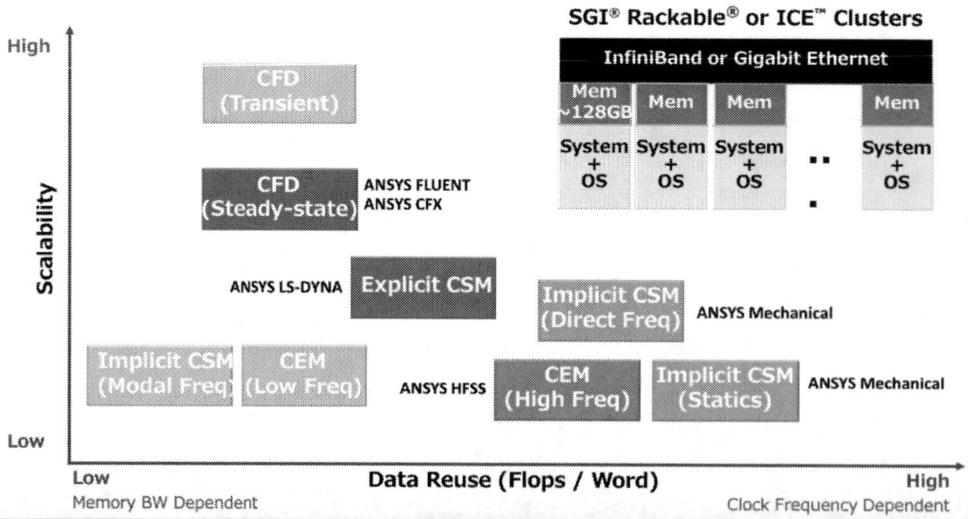

# Characterization of ANSYS Applications

Typical HPC Behavior of a Single Job for Various Applications

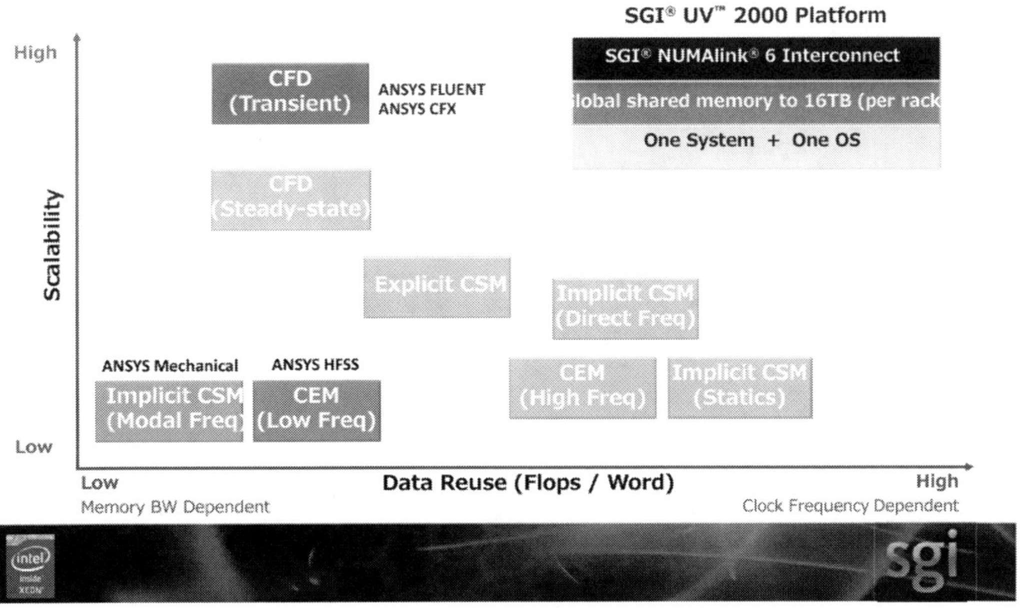


# Characterization of ANSYS Applications

SGI UV can provide the environment for both

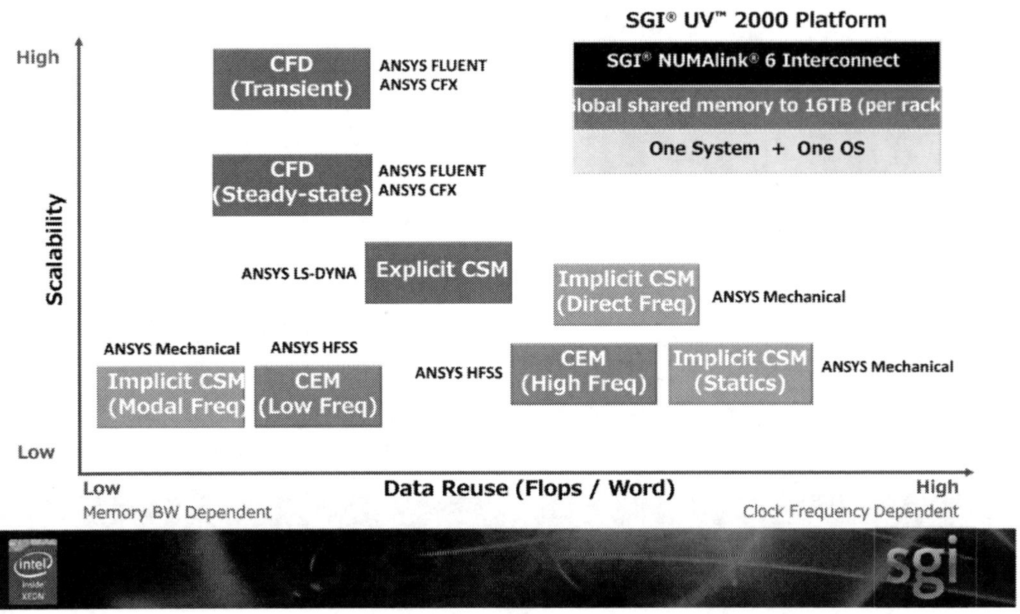

## CAE Workflow Using SGI VizServer

- CAD model creation
- Mesh-generation
- Model decomposition
- Running solvers
- Viewing results
- Adjust and repeat

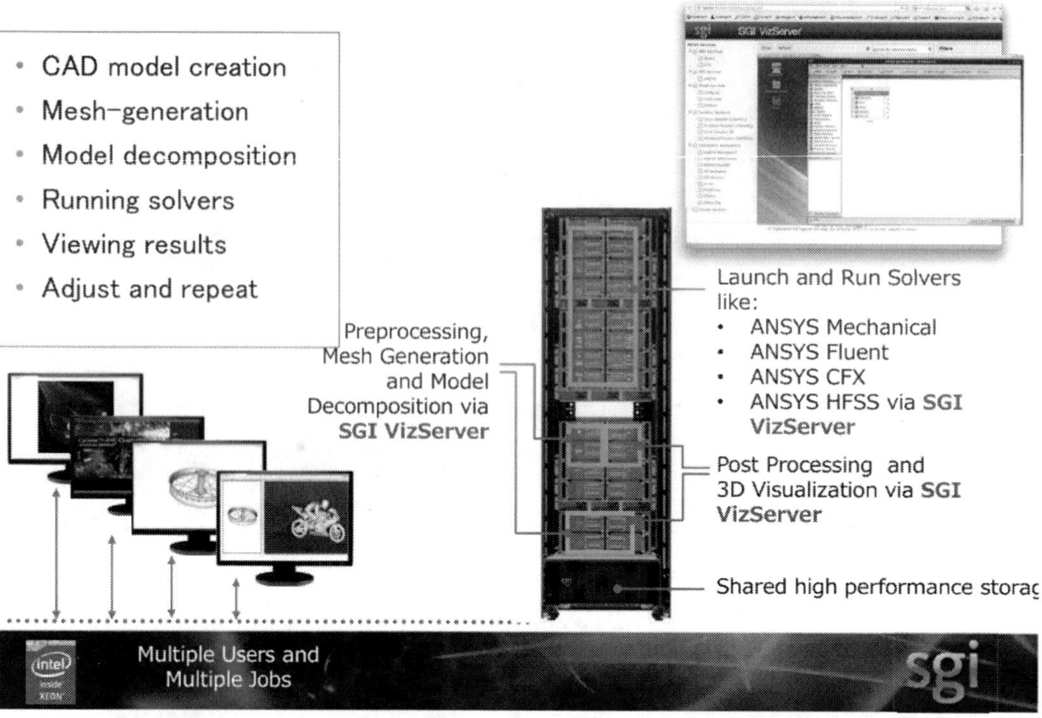


## Demo Environment for Automotive Customer

SGI Japan office (TOKYO) <-> US. SGI (MN)

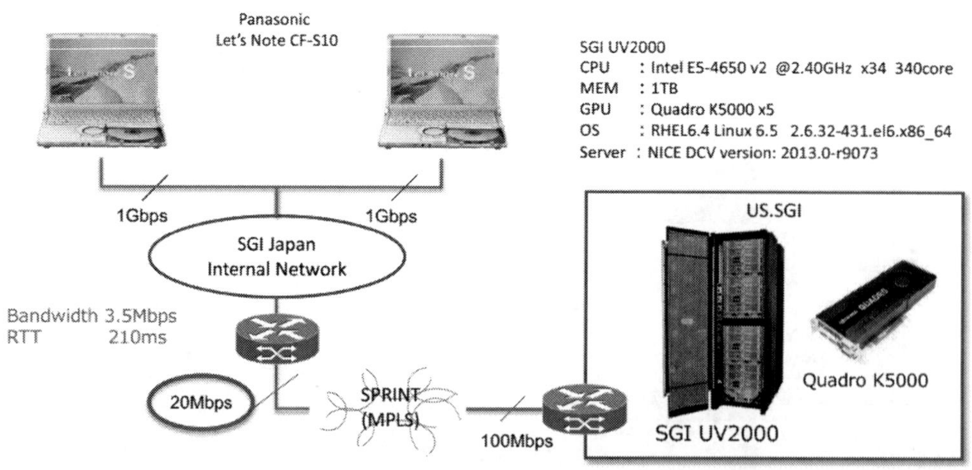

# Remote Visualization Session Management Tools

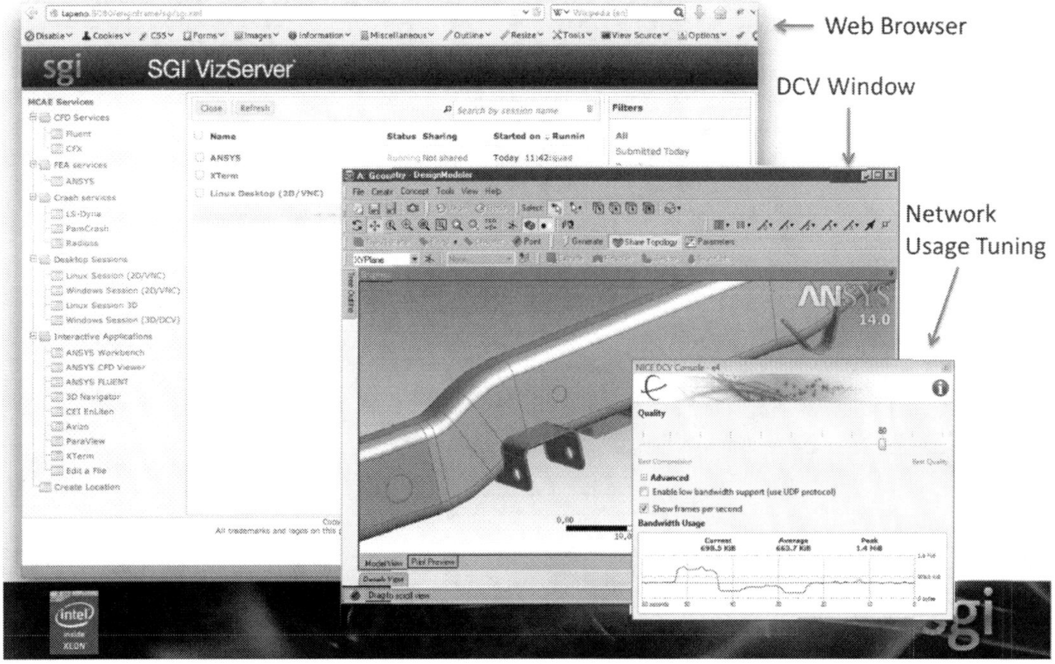


# Multiple Sessions Example

Measuring memory consumption with multiple users run multiple applications at the same time

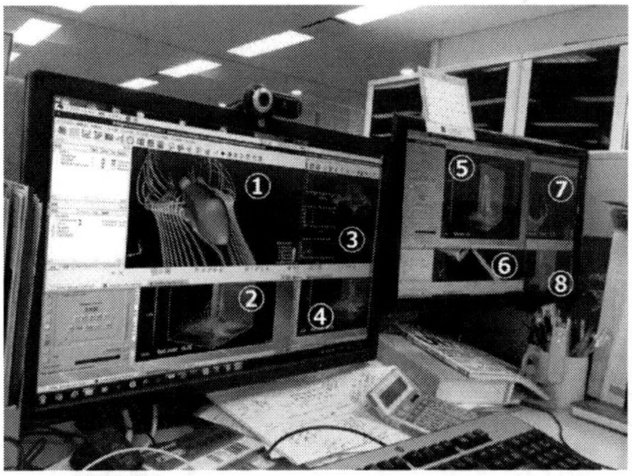


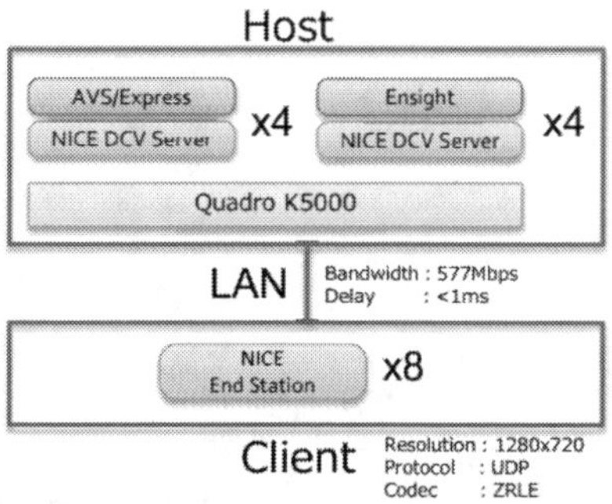


| Memory consumption [MByte] | | | | |
|---|---|---|---|---|
| sessions | AVS/Express | Ensight | DCV Server | Sub total |
| 1 | | 320 | 45 | 365 |
| 2 | 1,100 | | 36 | 1,136 |
| 3 | 1,000 | | 36 | 1,036 |
| 4 | | 306 | 48 | 354 |
| 5 | | 317 | 45 | 362 |
| 6 | 565 | | 36 | 601 |
| 7 | | 273 | 48 | 321 |
| 8 | 1,000 | | 37 | 1,037 |
| Sub total | 3,665 | 1,216 | 331 | 5,212 |
| Ave. | 916 | 304 | 41 | 652 |

[ Result ]
Memory consumption of DCV Server per 1 session
SystemMemory : 36~48MByte
GraphicsMemory : 25MByte

* Depending on application and model data size

# Built-in collaboration

# Collaboration Example

Multiple users and any device can share same desktop

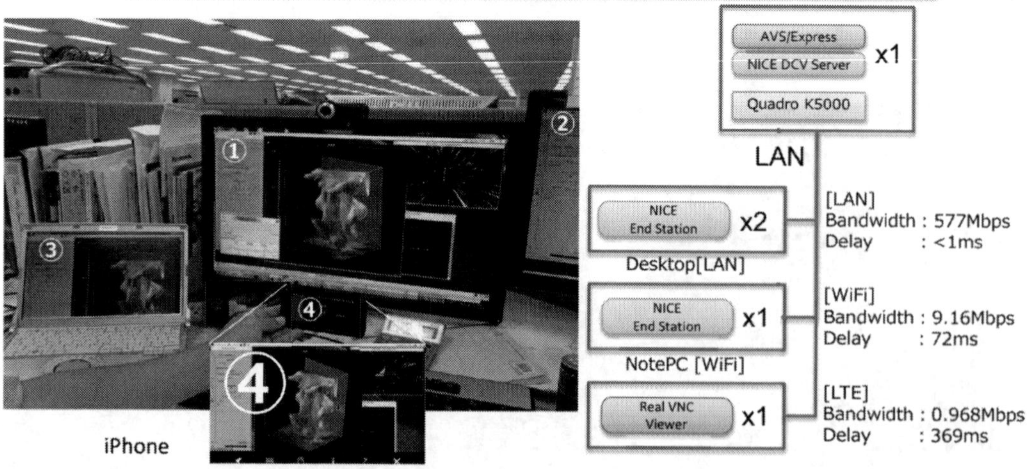

# Questions

Contact Information:
Tony DeVarco
tdevarco@sgi.com
C: 1-510-364-0408

## An example of aerodynamics development with CFD for low speed racing car (Formula SAE)

2014.10.10
Tetsuya Fujimoto
M2, Mechanical engineering, Sophia University

# Table of Contents

# Table of Contents

- What is the "Formula SAE"
- Special feature of Formula SAE racing car
- Historical aerodynamics developments at Sophia Racing
- Latest instance

What is the "Formula SAE" -1

# History of Formula SAE

- Formula SAE® is a student design competition organized by SAE International (formerly Society of Automotive Engineers).
- Students are requested to develop a small Formula-style race car with a fictional manufacturing company
- The car is to be evaluated for its potential as a production item
- Each student team designs, builds and tests a prototype based on a series of rules.
- The prototype race car is judged in a number of different events.

(SAE INTERNATIONAL™ HP : http://students.sae.org/cds/formulaseries/about.htm)

What is the "Formula SAE" -2

## Formula SAE participants

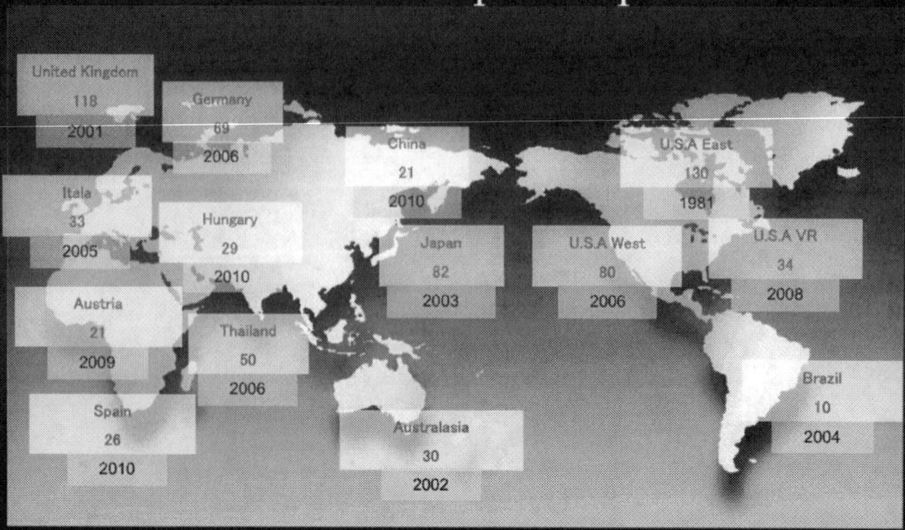


Participants distribution (Country, number of teams, first year)

1981
6 CV teams
→
2013
503 CV teams
64 EV teams

# Table of Contents

## Special features of formula SAE racing car

- The engine should be 4-stroke and smaller than 610cc.

- In order to limit the power capability from the engine, a single circular restrictor must be placed in the intake system between the throttle and the engine and all engine airflow must pass through the restrictor.

- The vehicle must be open-wheeled and open-cockpit (a formula style body) with four (4) wheels that are not in a straight line.

- In plain view, no part of any aerodynamic device, wing, under tray or splitter can be further forward than 762 mm (30 inches) forward of the fronts of the front tires, no further rearward than 305 mm (12 inches) rearward of the rear of the rear tires and no wider than the outside of the front tires or rear tires measured at the height of the hubs, whichever is wider.

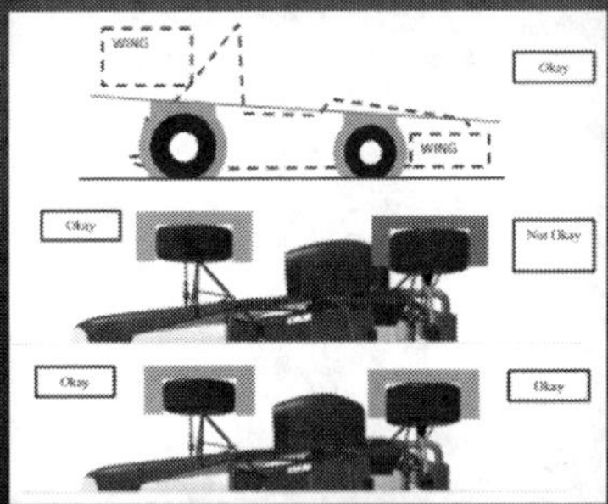


- No power device may be used to move or remove air from under the vehicle except fans designed exclusively for cooling. Power ground effects are prohibited.

(From 2013 FSAE Rules)

What is the "Formula SAE" -3

## Formula SAE events

There are two kind of events that are

| | |
|---|---|
| Dynamic events<br>675pts total | To judge vehicle dynamic potential through the several events<br>(Acceleration, Skidpad, Autocross, Endurance, Fuel consumption) |
| Static events<br>325pts total | To judge vehicle marketing potential through the several events<br>(Presentation, Cost, Design) |

# Special features of formula SAE racing car -2

**D7.2 Autocross Course Specifications & Speeds**

D7.2.1 The following standard specifications will suggest the maximum speeds that will be encountered on the course. **Average speeds should be 40 km/hr (25 mph) to 48 km/hr (30 mph).**

**Constant Turns:** 23 m (75 feet) to 45 m (148 feet) diameter.

(From 2013 FSAE Rules)

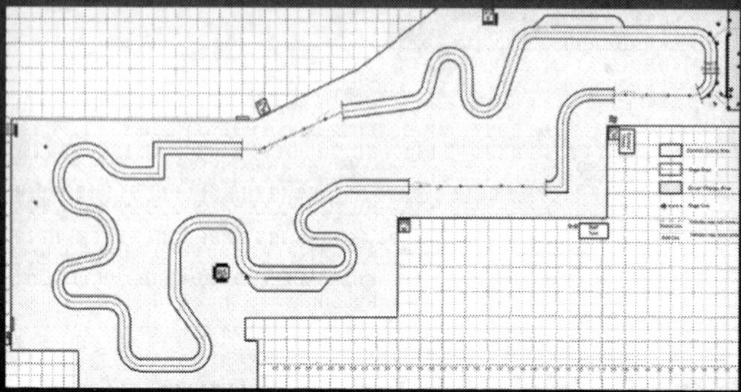

➡ Formula style racing car specialized for that slow speed is required

# Table of Contents

- What is the "Formula SAE"
- Special feature of Formula SAE racing car
- Historical aerodynamics developments at Sophia Racing
- Latest instance

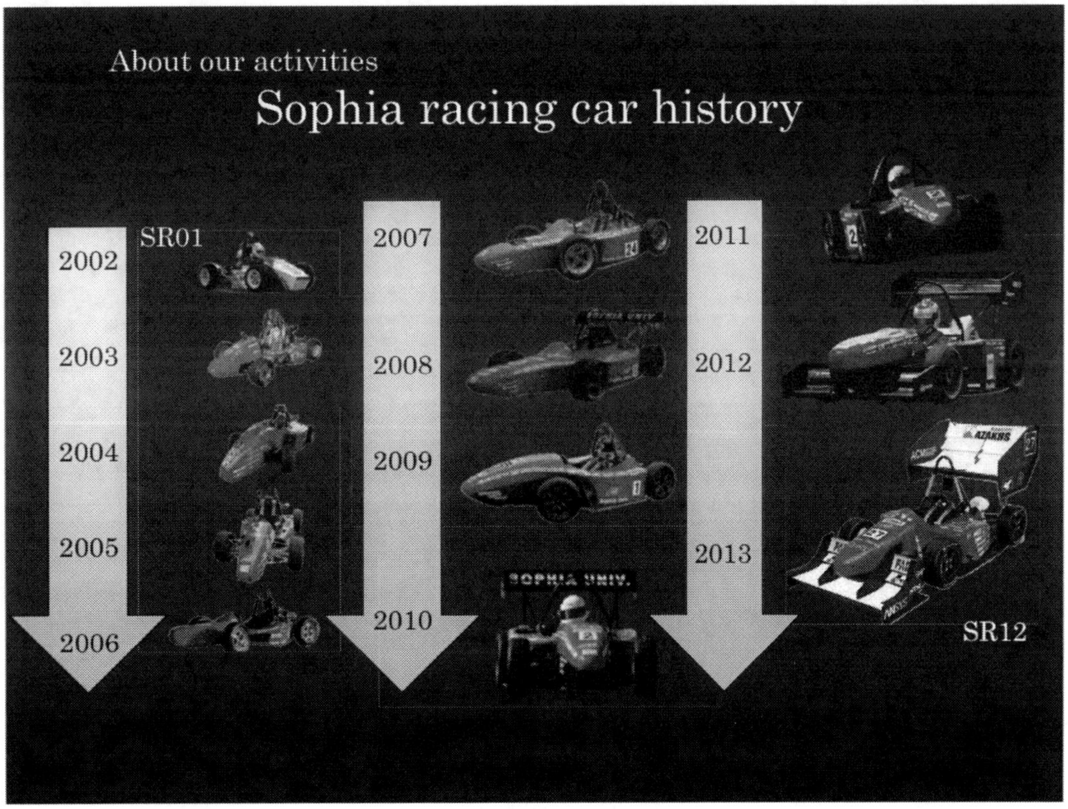
About our activities
Sophia racing car history
2002
SR01
2003
2004
2005
2006
2007
2008
2009
2010
SOPHIA UNIV.
2011
2012
2013
SR12

Wind tunnel testing at Monashmotorsport
Many teams doesn't afford testing with wind tunnel
→CFD is the most popular tool for Formula SAE

# SR12 vehicle specification (2013)

Table : SR12 vehicle specification

| | |
|---|---|
| Car weigh | 185kg (W/O wings)<br>210kg (W/ wings) ⊿25kg |
| Wheelbase | 1560mm |
| Track width | 1200mm |
| Bodywork | Carbon fiber monocoque |
| Engine | YAMAHA WR450F |
| Maximum power | 53HP |
| Downforce | 1200N at 60km/h<br>31500N at 300km/h |

Figure : SR12

Table : typical F1 vehicle specification

| | |
|---|---|
| Car weigh | 600~800kg |
| Wheelbase | ~3300mm |
| Track width | <1800mm |
| Bodywork | Carbon fiber monocoque |
| Maximum power | 700~800HP |
| Downforce | 225N at 60km/h<br>5500N at 300km/h |

Figure : Formula one car image

About our activities

# Sophia racing car history

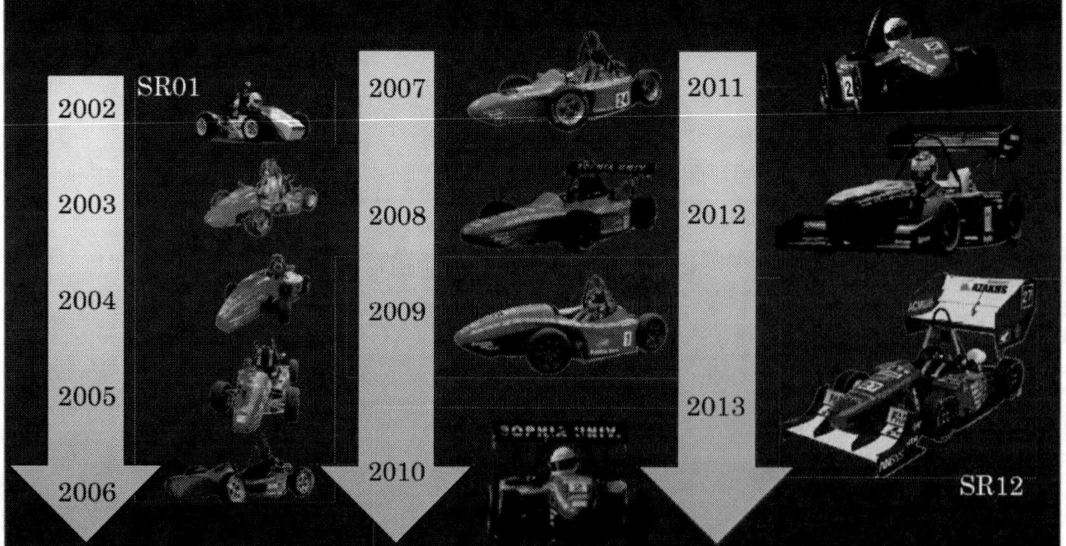


It is generally said "diffuser is effective"
Is it true even for Formula SAE cars?

## SR12 compared with typical formula one car from top

Narrower track width (75%)
Shorter wheelbase (45%)
Slower air speed (25%)
=Less inertia, Low Re
→Less downforce under the car

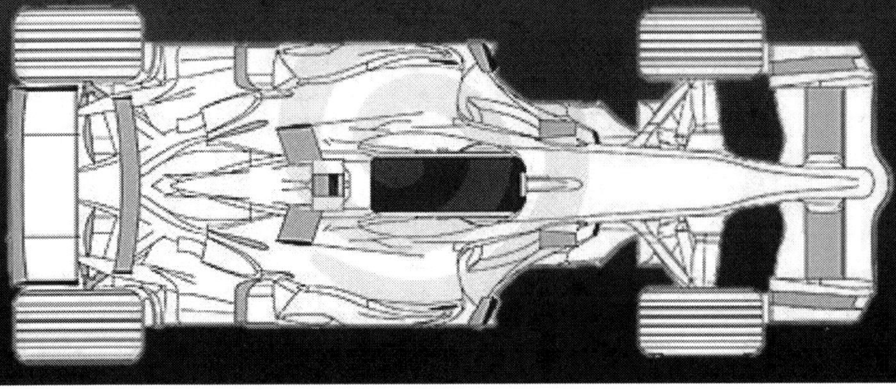

## How does aerodynamics on FSAE car work

2012

2013

| Section | Item | 2012 ～ 2013 | Unit | points delta | |
|---|---|---|---|---|---|
| PT | Engine torque | -36 | % | -35 | 8.2 |
| | Fuel consumption | -0.3 | L | 6 | |
| | Weight | -30 | kg | 37.2 | |
| AD | Downforce | 600 | N at 60km/h | 107 | 67 |
| | Dragforce | 250 | N at 60km/h | -40 | |
| | DRS | -200 | N at 60km/h | 11 | |
| | Center of gravity height | 20 | mm | -11 | |

Downforce / Dragforce / Mass (at60km/h)

Front wing 400N / 100N / 8.5kg
Rear wing 600N / 270N / 10kg
Diffuser 200N (Δ300N) / 10N / 6.5kg

・Mass of diffuser should be taken into account.
・Downforce from diffuser can easily be changed by the vehicle state.

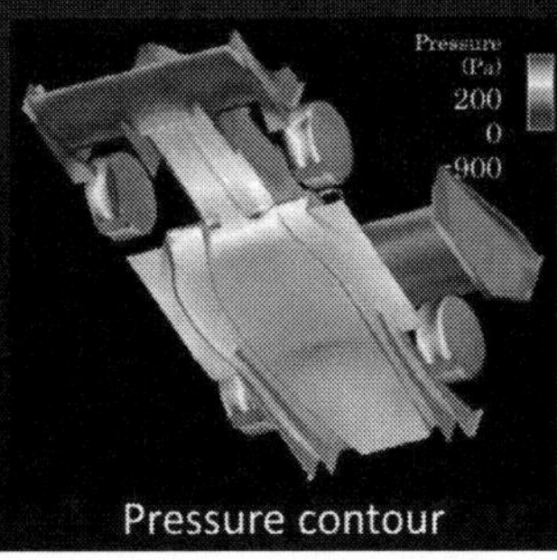


Pressure contour

# CFD 計算条件

| Software | ANSYS ICEM CFD/FLUENT |
|---|---|
| Number of cells | 2500～3300万 |
| Mesh type | Tetra/prism |
| Viscous model | k-ω |
| Calculation type | 10h/model |

Grid example of SR12 analysis

# Wings

## Unsprung wing mount

Both wings are mounted on unsprung position

- Uncontact to the ground
- Constant ground height
- Wide range of suspension setup without regard to wing ground height
- Front wing with bigger downforce had been designed, then moved to rear wing.

Cut view of SR12 indicating pitch center

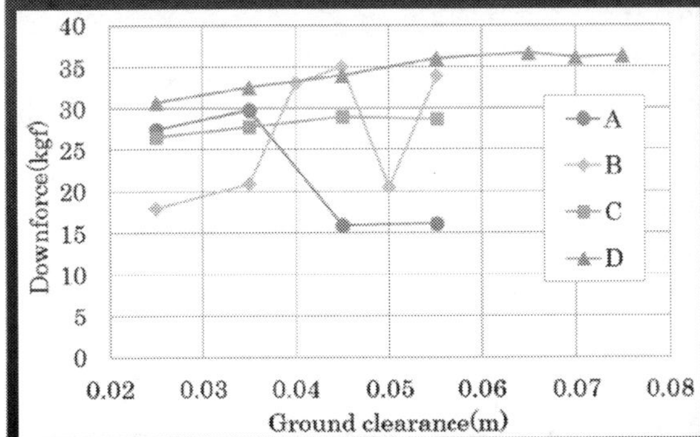


Ground height vs. Downforce

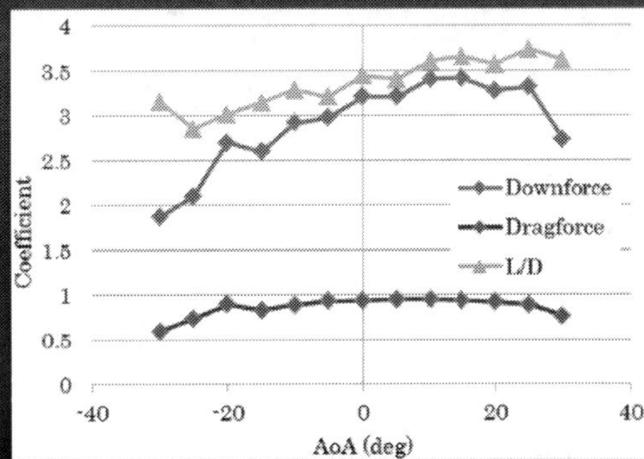


AoA vs. Coefficients

## Front wing

Benzing based front wing (2013-2014)

- Front wing had been modified manually.
- Designing them with Ajdoint solver is next challenge.

Sophia Racing

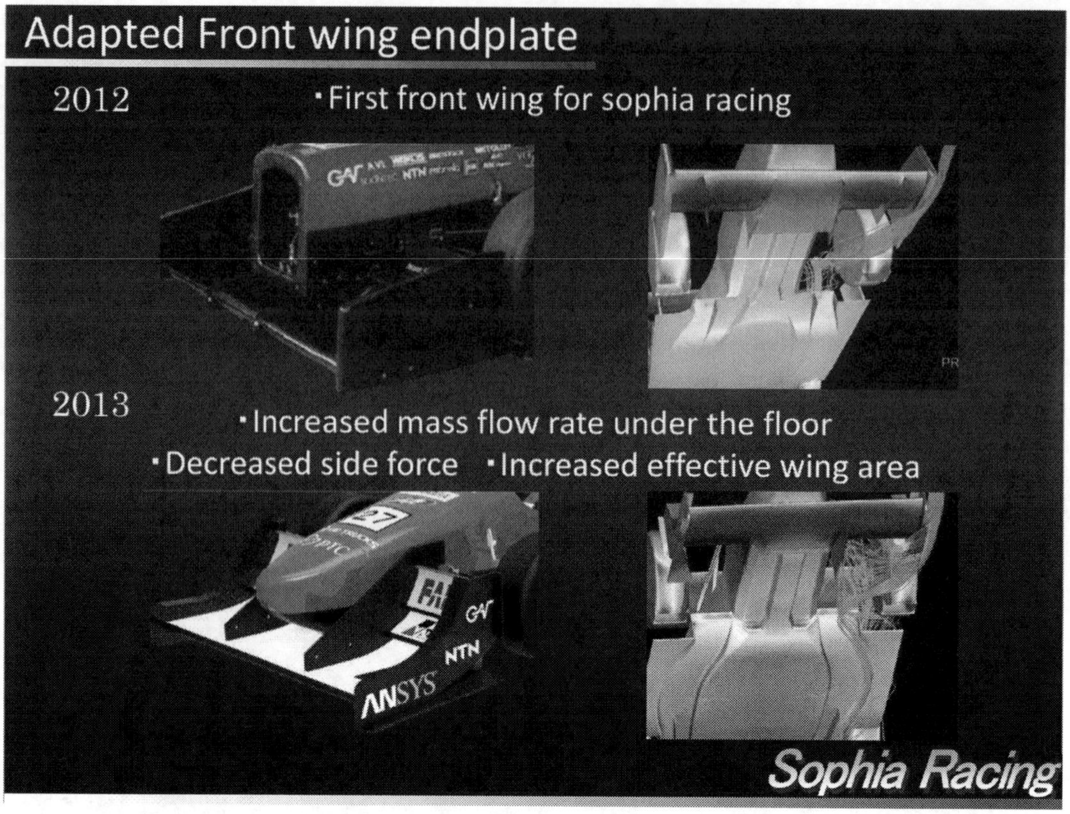
Adapted Front wing endplate
2012
・First front wing for sophia racing
2013
・Increased mass flow rate under the floor
・Decreased side force ・Increased effective wing area
Sophia Racing

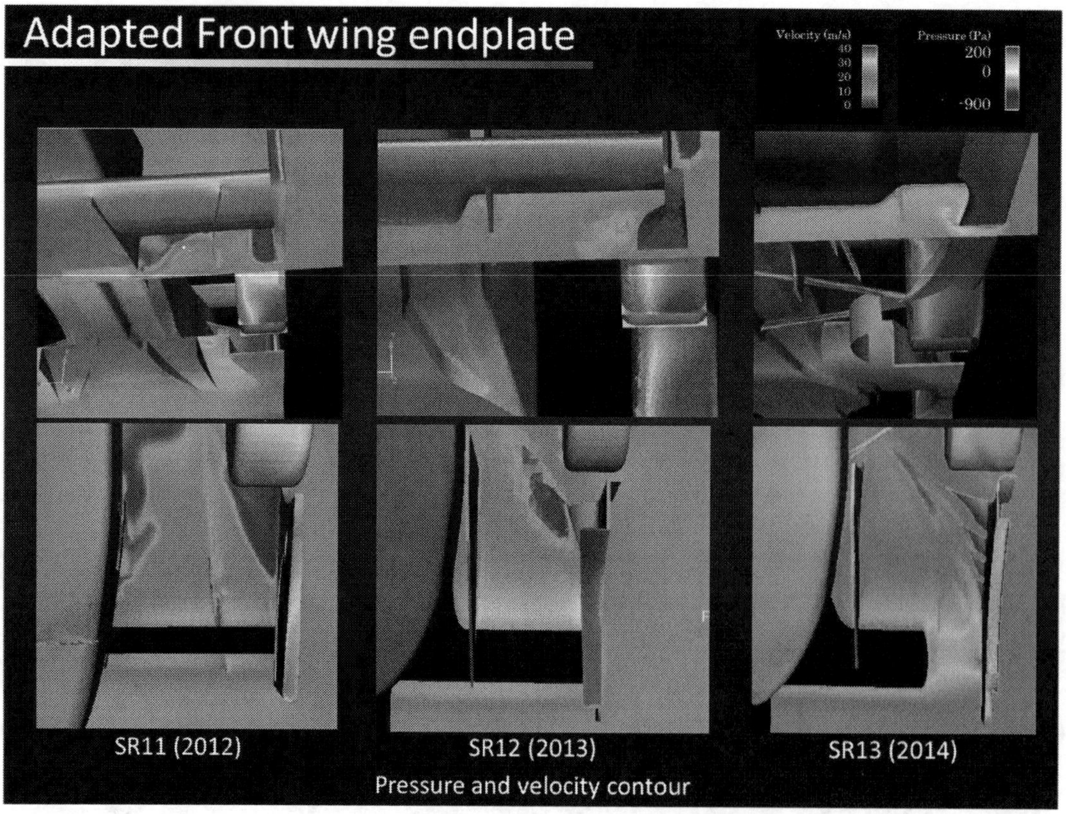
Adapted Front wing endplate
Velocity (m/s)
40
30
20
10
0
Pressure (Pa)
200
0
-900
SR11 (2012)
SR12 (2013)
SR13 (2014)
Pressure and velocity contour

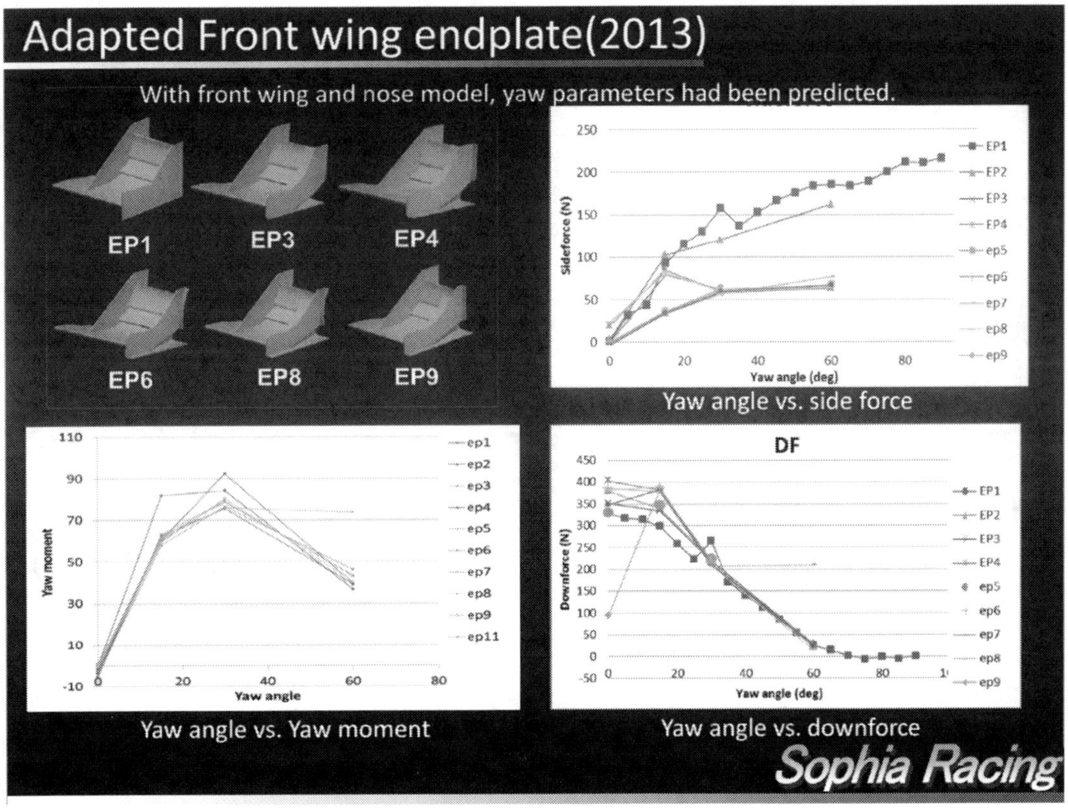
Adapted Front wing endplate(2013)
With front wing and nose model, yaw parameters had been predicted.
EP1
EP3
EP4
EP6
EP8
EP9
Sideforce (N)
Yaw angle (deg)
EP1
EP2
EP3
EP4
ep5
ep6
ep7
ep8
ep9
Yaw angle vs. side force
Yaw moment
Yaw angle
ep1
ep2
ep3
ep4
ep5
ep6
ep7
ep8
ep9
ep11
Yaw angle vs. Yaw moment
DF
Downforce (N)
Yaw angle (deg)
Yaw angle vs. downforce
Sophia Racing

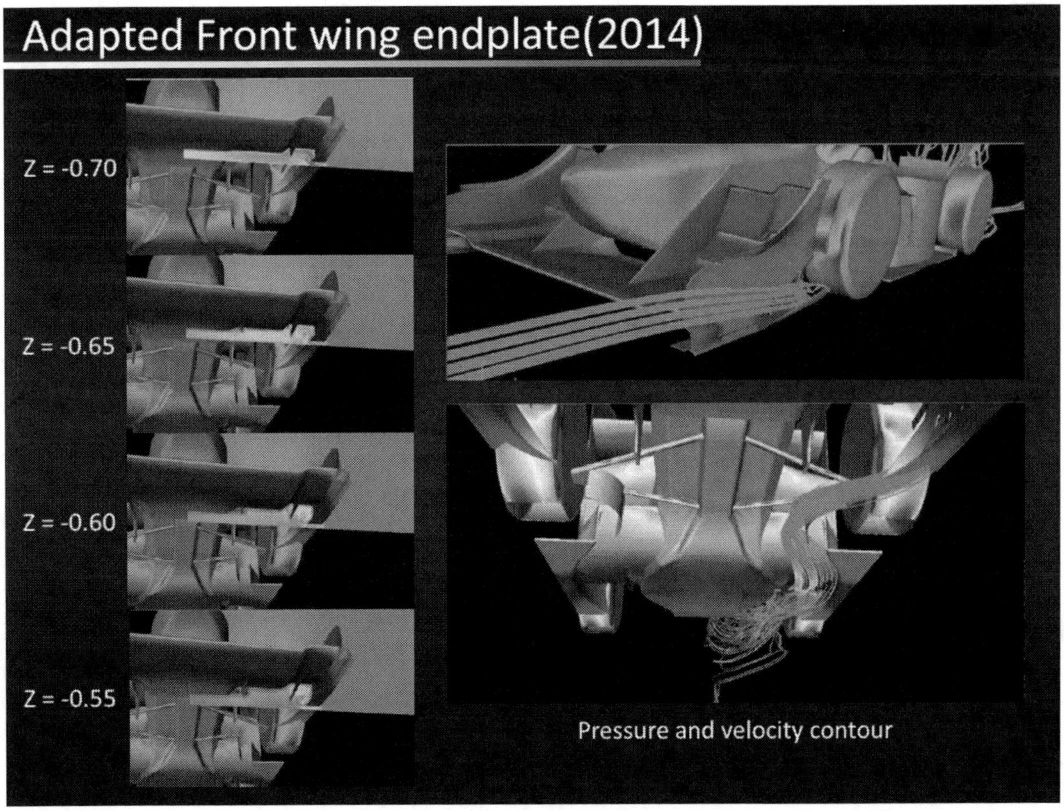
Adapted Front wing endplate(2014)
Z = -0.70
Z = -0.65
Z = -0.60
Z = -0.55
Pressure and velocity contour

## Rear wing with DRS

View of closed (Left) and activated DRS (Right)

DRS closed
Doenforce 620N
Dragforce 270N

Drag -200N
-75%
+15pts

DRS open
Downforce 250N
Dragforce 70N

Unsteady Retouch / Detouch analysis is required for the next stepto advance DRS effectiveness

## Rear wing

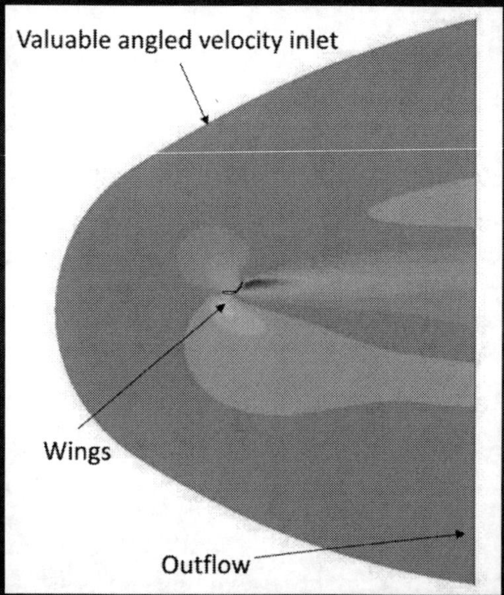


・Rear wing had been modified automatically with adjoint solver and jouanal based on 2013 2D model.

・Valuable angled velocity inlet with air speed of 60km/h and k-ω turbulent viscosity model had been applied.

Sophia Racing

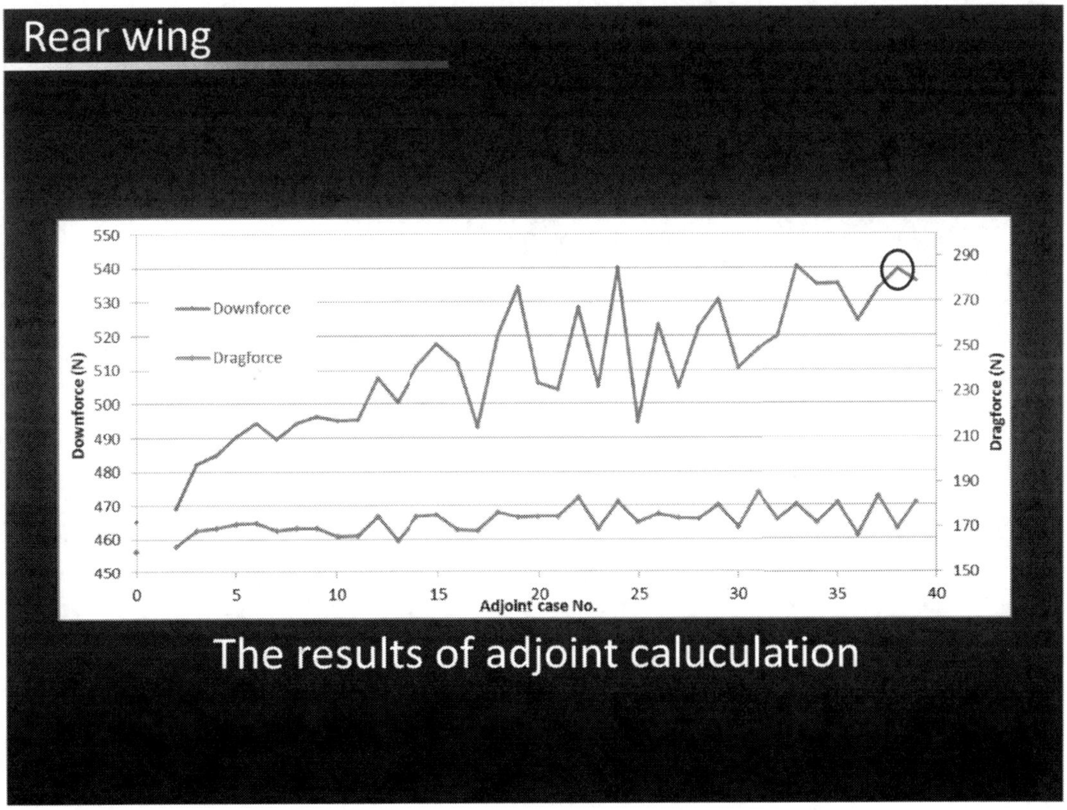
Rear wing
Downforce
Dragforce
Downforce (N)
Dragforce (N)
Adjoint case No.
550
540
530
520
510
500
490
480
470
460
450
290
270
250
230
210
190
170
150
0
5
10
15
20
25
30
35
40
The results of adjoint caluculation

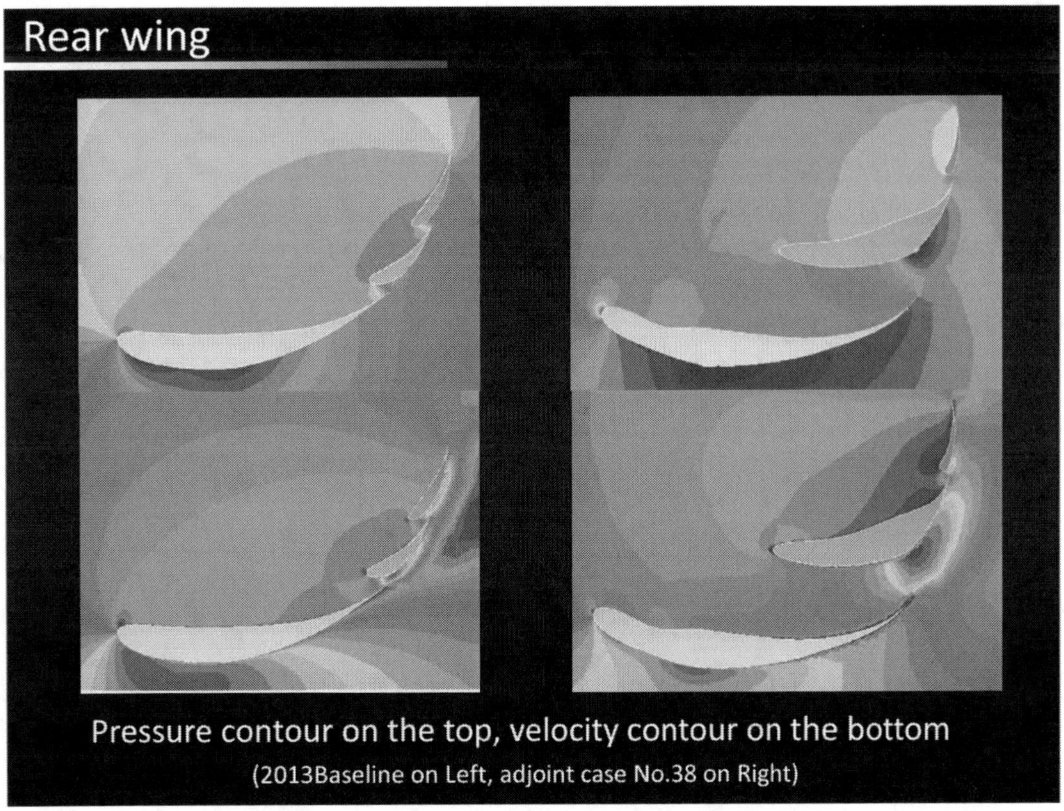
Rear wing
Pressure contour on the top, velocity contour on the bottom
(2013Baseline on Left, adjoint case No.38 on Right)

## Rear wing

Based on 2D analysis of No.38 profile,
some manual modification had been done for only main plane with regard to DRS function.

2013Baseline on the Left, 2014 Adapted (No.18) on the Right.

Pressure contour on the top, velocity contour at the bottom.

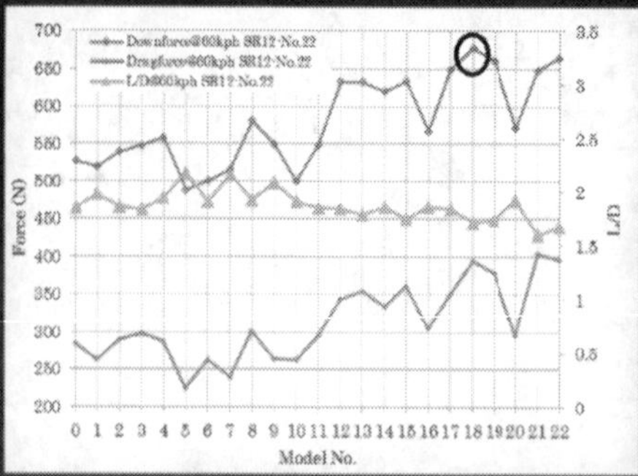


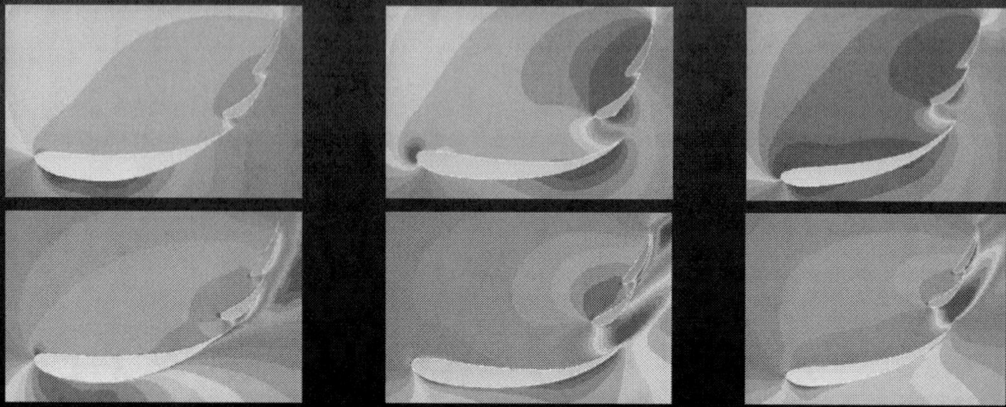

2013 | 2014 No.18 | 2014 Final

## Rear wing

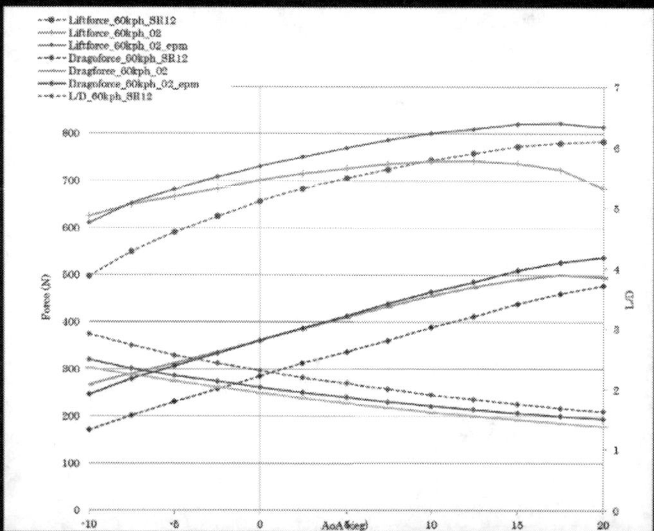


AoA vs. Force

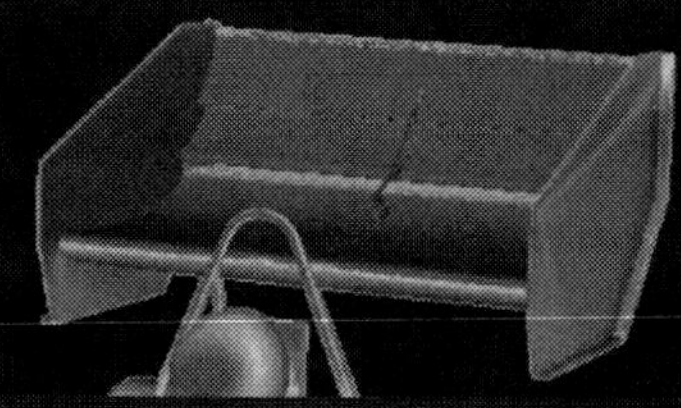

Rear wing of SR12

Rear wing of SR13

Based on manual adapted profile, 3d analysis had been done.
For the new profile, endplates had been changed and downforce increased for the wide range of AoA.

## Rear wing

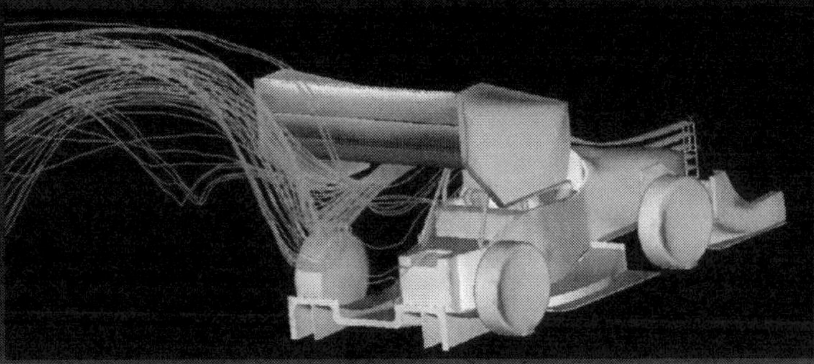

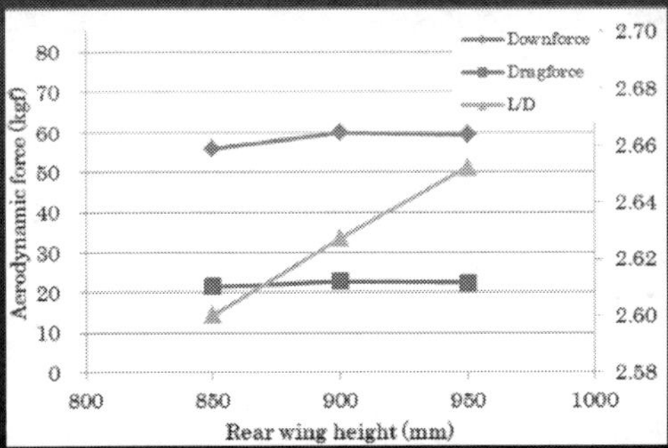


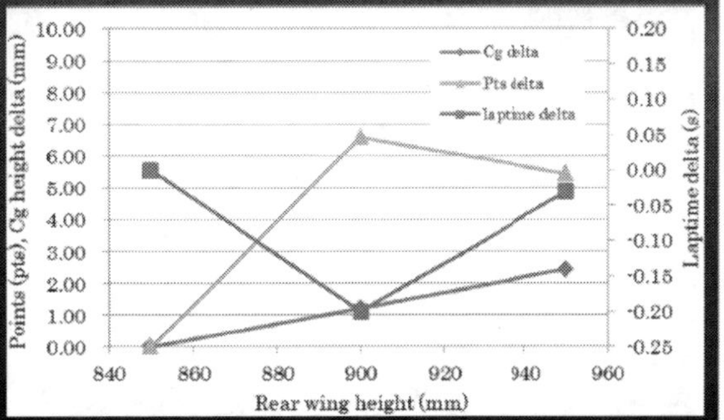


Rear wing height had been decided with regard to center of gravity, drag force, downforce, down force distribution with maximum predicted point gain.

## Use of side force

### High speed corner

$V_{trans} + \omega_{small}$

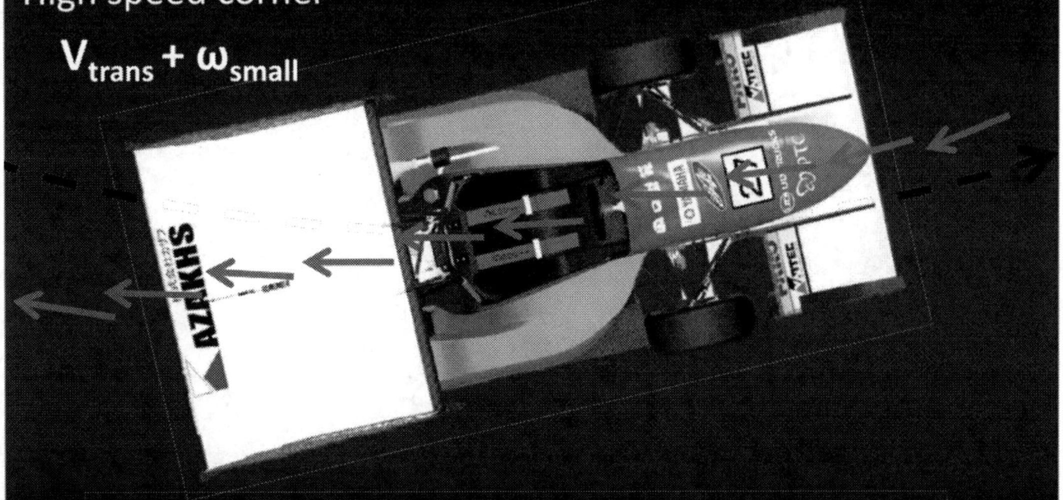

Total velocity around the car at high speed corner

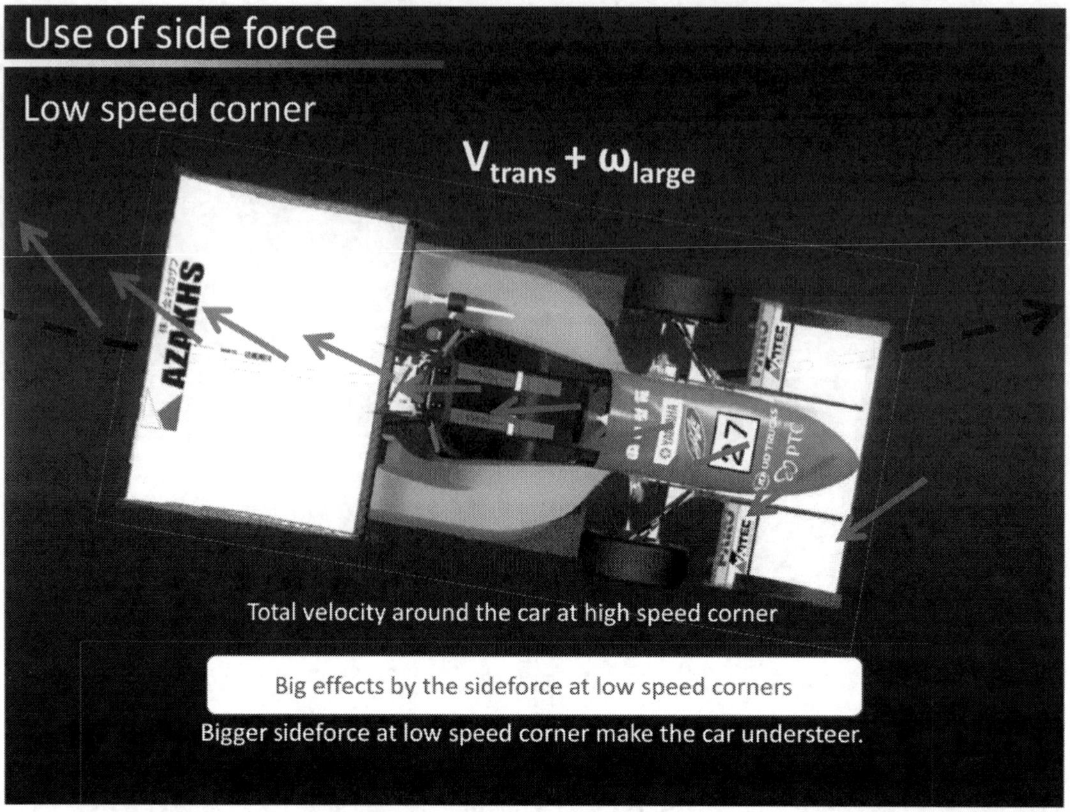
Use of side force
Low speed corner
V trans + ω large
Total velocity around the car at high speed corner
Big effects by the sideforce at low speed corners
Bigger sideforce at low speed corner make the car understeer.

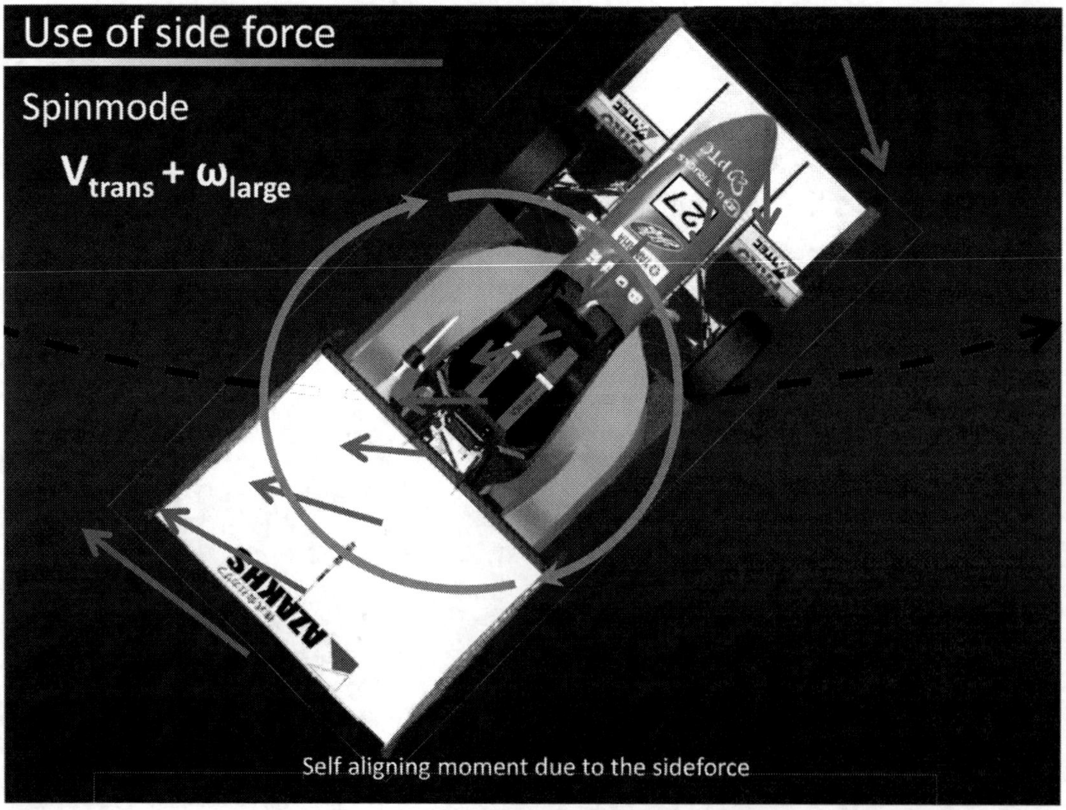
Use of side force
Spinmode
V trans + ω large
Self aligning moment due to the sideforce

## Diffuser

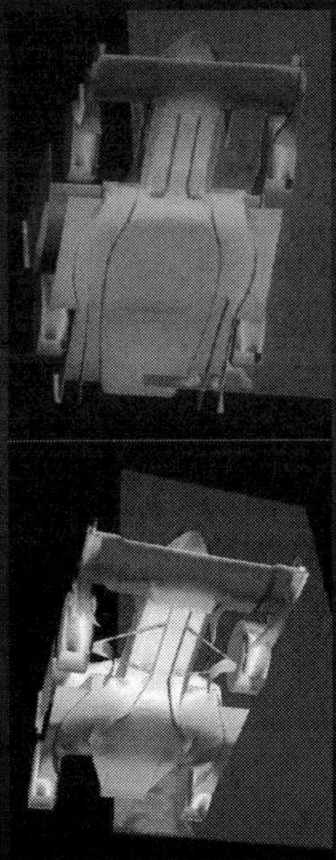

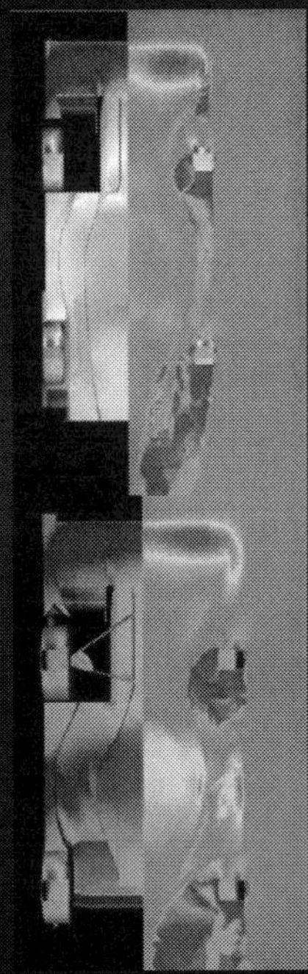

- Larger A/R : effective
- Shorter length : less sensitive to pitch
- Turning flow around the endplate helps generating more downforce

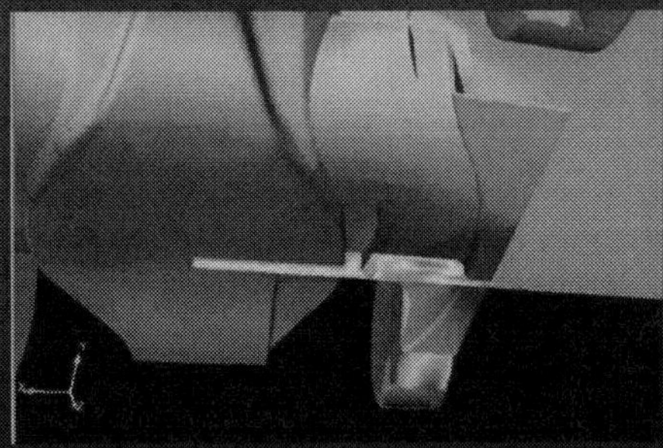

Splitter under the floor prevents flow separation and generates vortex

Pressure and TKE contour

Pressure and velocity contour

## Fuel tank sloshing analysis

| Software | ANSYS ICEM CFD/FLUENT |
|---|---|
| Number of cells | 0.43 million |
| Cell type | Tetra |
| Multi phase model | Volume of Fluid |
| Viscous model | Laminar |
| Time step | $1.0 \times 10^{-3}$ s |

Video Presentation in Original e-media

Fuel sloshing analysis

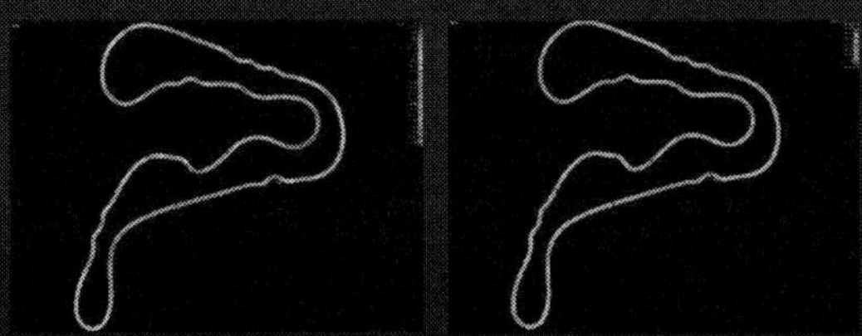

Lateral (Left) and longitudinal (right) G force

## Intake analysis

| Software | ANSYS ICEM CFD/FLUENT |
|---|---|
| Number of cells | 1.5 million |
| Cell type | Tetra/prism |
| Viscous model | Spalart-Allmalas |
| Time step | $1.5 \times 10^{-4}$ s |
| Calculation cycle | 60 (cycle) |

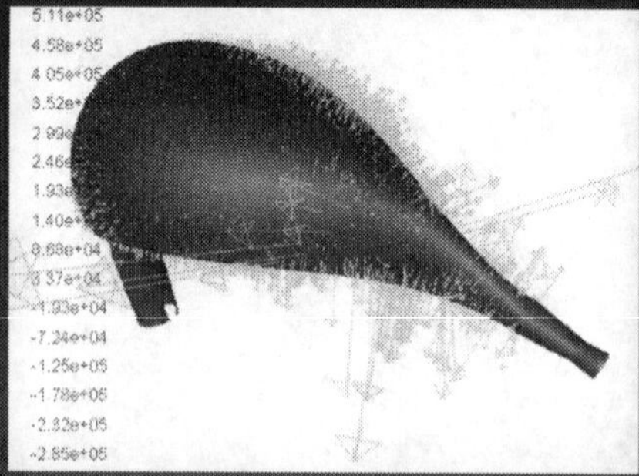


Adjoint analysis result

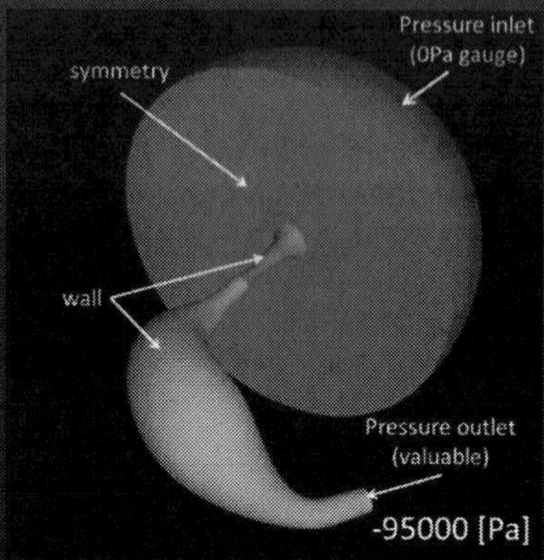


Model view

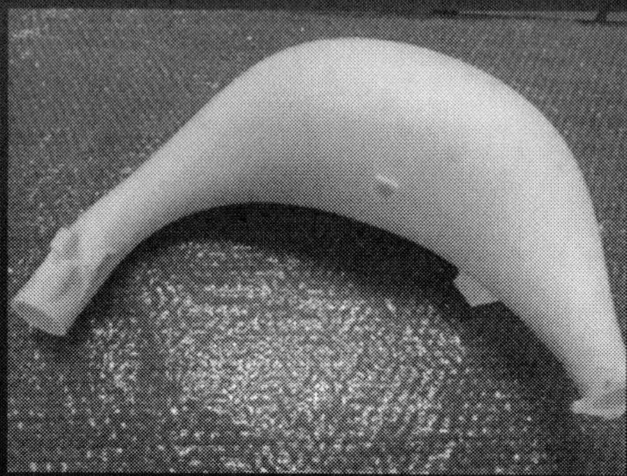

Intake plenum manufactured with rapid prototype

## Future work

・Front wing will be shorten, cut in front of the front tires and mounted more rigidly for load test.

・Rear wing will be shorten, narrowed and lowered.

Wings are more likely to F1 style in 2015.
"Area (Volume) effective" aero package rather than "drag effective" is strongly required in the future.

ANSYS adjoint solver is assumed to assist our new challenge for the future work.

Only for the presentation

2014 Automotive Simulation World Congress

# Rain Water Management

Padmesh Mandloi
ANSYS Inc.



## Contents

- Motivation
- Design Challenges in Rain Water Management
- Modeling Requirements
- Examples
- Summary

## Motivation

- Air is not the only medium encountered by a vehicle on the road
- Vehicles must be designed to handle the sometimes harsh natural conditions found on the road
- Water, snow, ice, dirt, rocks, and other particles of debris are kicked up off the road by tires or wind, or are present due to weather conditions
- Prediction of worst case scenarios and assessment of safety and comfort

2014 Automotive Simulation World Congress
October 9-10 . Tokyo, Japan

## RWM Design Challenges

**Aesthetics & Contamination**
- Vehicle soiling
- Contamination of sideview mirror, windshield and windows

**Safety**
- Defogging
- Defrosting
- Water entering HVAC inlet
- Water entering brake liners
- Visibility

**Reliability and Durability**
- Impact of water on underbody components due to wading

## Modeling Requirements

**Aerodynamics**

- Steady simulation with RANS models
- Transient simulation with URANS and LES models

**Multiphase Flows**

- Free-surface multiphase models like VOF
- Discrete-phase-model (DPM)

**Wall Film**

- Lagrangian wall film model
- Eulerian wall film model

**Phase Change**

- Solidification and melting model
- Evapration and condensation model





## VOF Model Improvements

- Comparison of schemes : Pure advection of shapes

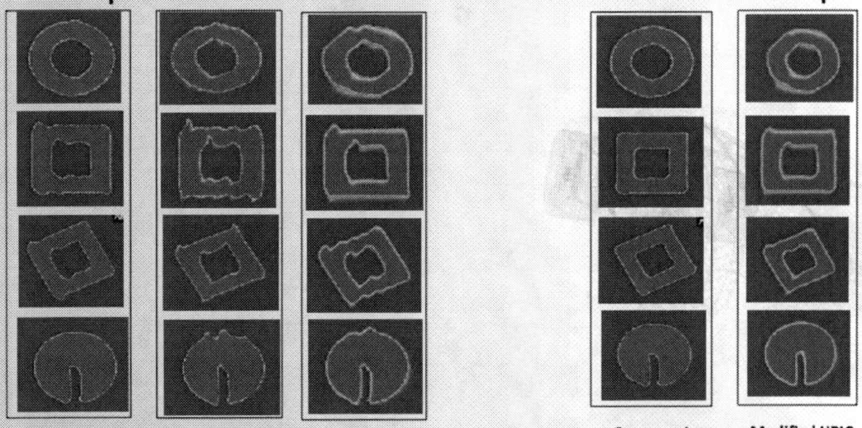

Compressive CICSAM Modified HRIC

**VOF Explicit, First order time**

Compressive Modified HRIC

**VOF Implicit, Second order time**

## Modeling Wall-Film

**Wall-Film**

| Volume of Fluid (VOF) | DPM Wall Film | Eulerian Wall Film |
| --- | --- | --- |
| + Direct modeling<br>+ Complex exchange surface<br>- Meshing cost<br>-At least 10 cells for film | + 3Eqn's sets<br>+ DPM particles to model the wall film<br>-No particles no film<br>-Only unsteadiness<br>-Particle tracking is huge | +3Eqn's sets<br>+ free analysis type<br>+ No mesh restrictions<br>+ Few parameters to adjust<br>-Only 3D simulations |





## Eulerian Wall Film Model

- Solves for film mass, momentum, heat transfer
- EWF can be coupled with Eulerian-Lagrangian(DPM) and Eulerian-Eulerian multiphase frame work.

**Assumption:**

- EWF model assumes that film always flows parallel to the surface so normal component of film velocity is zero.
- The film is assumed to have:
  - A parabolic velocity profile &
  - A bilinear temperature profile

  across its depth.

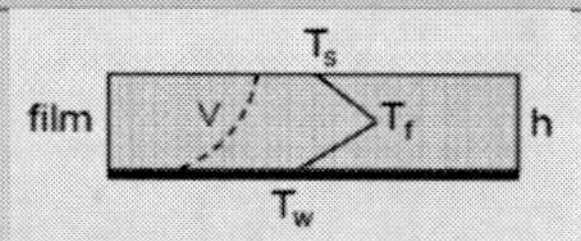


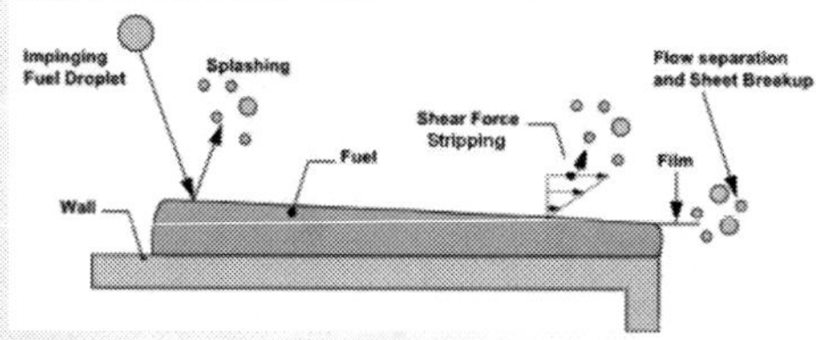


| | Physics |
| --- | --- |
| 1 | Liquid Droplet Collection |
| 2 | Film Transport |
| 3 | Particle Splashing, Stripping and Separation |
| 4 | Film Heat Transfer with gas and wall |
| 5 | Viscous & Kinetic Heating |
| 6 | Evaporation & Condensation |
| 7 | Melting & Solidification |
| 8 | Sublimation |

## Water Wading

- Integrity of underbody functional parts, like bumper cover, engine under-tray, plastic sill cover when vehicle passes through water of different depth at different speeds

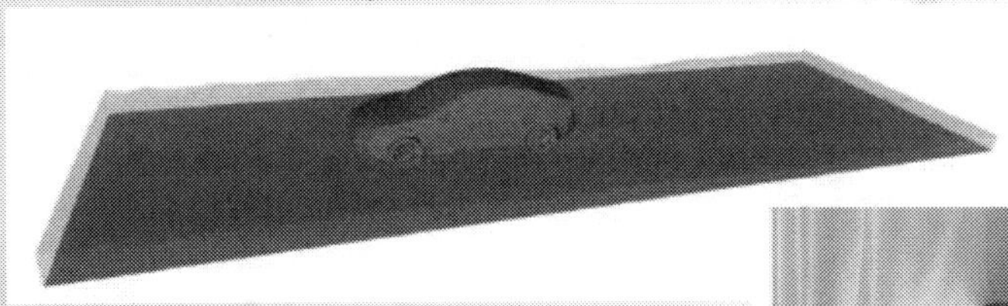

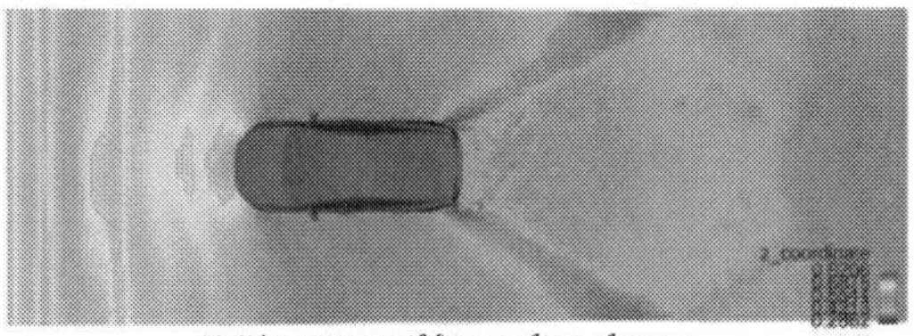

*Numerical Computation of Vehicle Wading – A Study using the VOF Method*, Master's Thesis in Fluid Dynamics, Shayan Rahat Varnaousfaderani, Chalmers University of Technology, Sweden, 2013

Height contour of free surface of water





## Water Wading

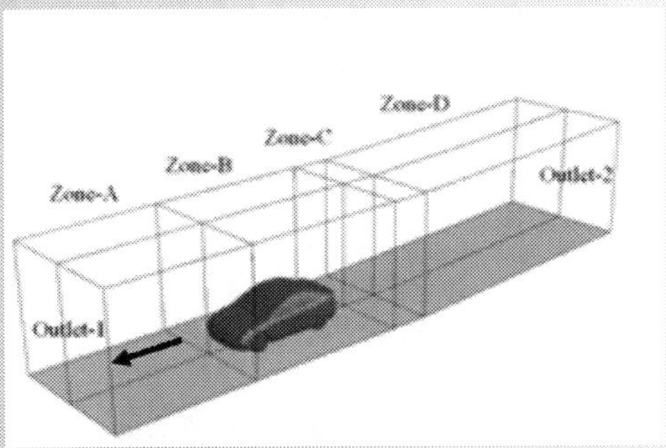

- Dynamic Mesh Modeling
  - Layering
  - Mesh Replacement
- Multiphase Modeling
  - VOF Compressible (implicit/explicit?) scheme
  - Other details?

## Tire Water Splash

- Understanding of splash pattern from tire is an important precurser to effective vehicle soiling analysis
- Splash pattern is usually a function of vehicle speed (or tire RPM), thickness of water pool, details of tire tread and aerodynamics field around the vehicle

http://www.flickr.com/photos/route9autos/3050629286/



2014 Automotive Simulation World Congress
October 9-10 . Tokyo, Japan

## Tire Water Splash

**Vehicle Velocity- 60 kmph**

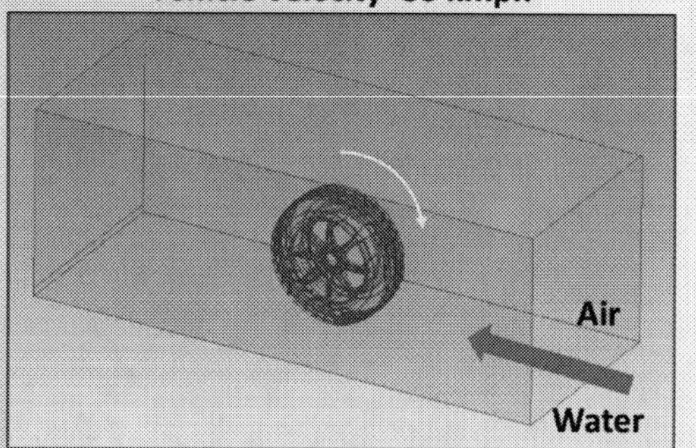


**Water depth- 65mm**

- Tire surface modeled as rotating wall
- Translational motion to air, water and road
- A thin cell zone around tire specified with constant angular speed

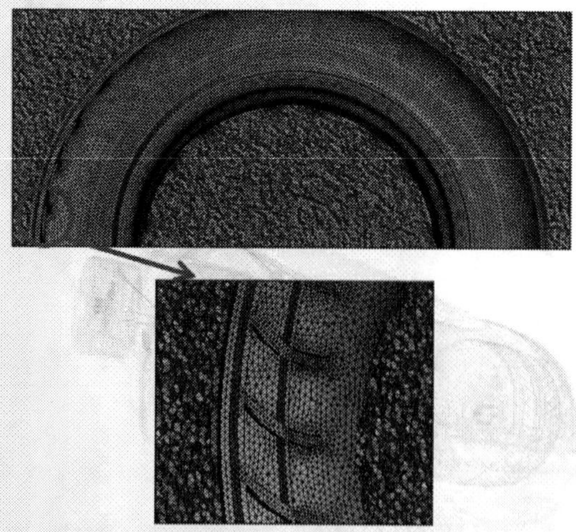

- High fidelity mesh
- Mesh Count = 34 million tetrahedral cells
- Mesh size = ~5 mm near tire, ~20 mm rest of the domain

## Tire Water Splash

- Multiphase Volume of Fluid ( VOF) model to track the air-water interface
- SST K-Omega Turbulence Model
- Air-Water surface tension and wall adhesion are included in the model

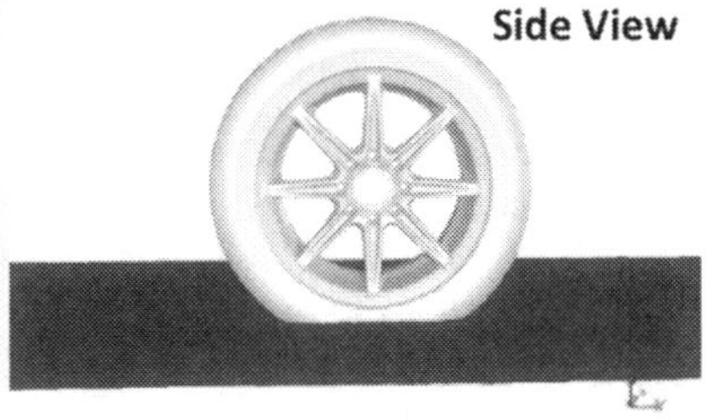

## Vehicle Soiling

- Dirt accumulation due tire splashing
- Vehicle aerodynamics and car shape decide soiling pattern
- Dirt particles can be modeled using Discrete Phase Model (DPM)

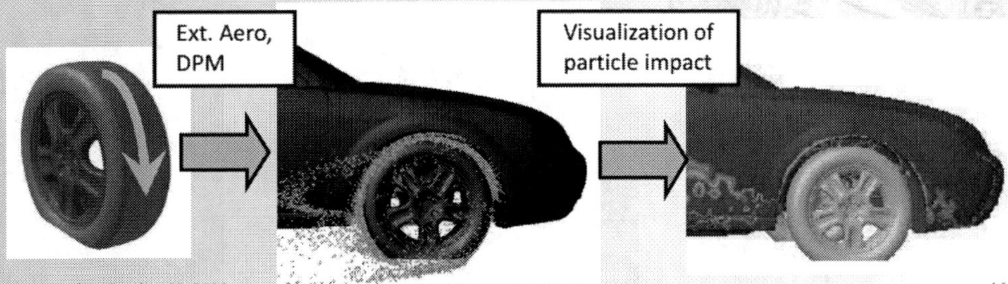

## Vehicle Soiling

- Solve vehicle aerodynamics
- Freeze flowfield and define DPM injections using "File-Injection"
- Turbulent Dispersion: Discrete Random Walk Model with 4 tries
- Enable Gravity
- Turn off flow, turbulence and energy equations and do one iteration or do particle tracks
- Use UDF to post-process particle accumulation

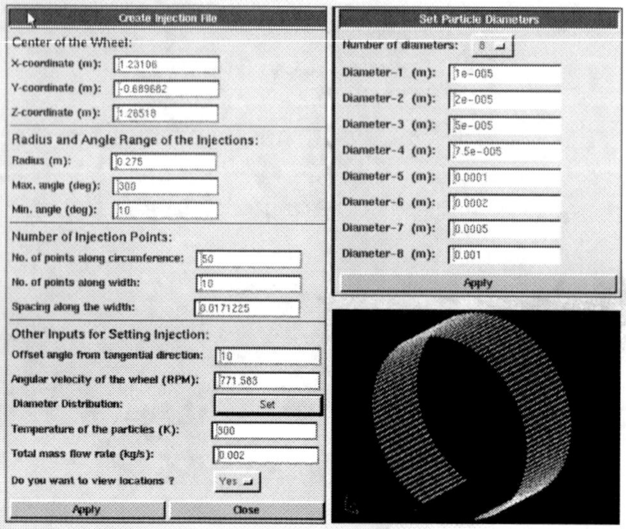




2014 Automotive Simulation World Congress
October 9-10 . Tokyo, Japan

## Vehicle Soiling

| Sum of Facet Values User Memory 8 | |
|---|---|
| car-prsm-ext--headlight-left | 0 |
| car-prsm-ext--headlight-right | 0 |
| car-prsm-ext--light-rear-left | 5 |
| car-prsm-ext--light-rear-right | 1 |
| car-prsm-ext--mirror-left | 0 |
| car-prsm-ext--mirror-right | 0 |
| car-prsm-ext--sidewindows-left | 0 |
| car-prsm-ext--sidewindows-right | 0 |
| car-prsm-ext--trunk | 22 |
| car-prsm-ext--window-rear | 6 |
| car-prsm-ext--windscreen | 0 |
| car-prsm-ext-body | 2723 |
| car-prsm-ext-bumper | 0 |
| car-prsm-ext-front-grille | 0 |
| car-prsm-ext-rear | 58 |
| car-prsm-ext-top-grille | 0 |
| u-axis-front | 77 |
| u-axis-rear | 897 |
| u-exhaust-pipe | 57 |
| u-spare-wheel | 33 |
| u-steering-knuckle-left | 59 |
| u-steering-knuckle-right | 97 |
| u-tank | 4 |
| u-underbody | 23582 |
| Net | 27621 |

History of particle accumulation on different surfaces of vehicle

# Water Ingestion in HVAC Plenum

- Water sliding through the windshield enters the HVAC plenum
- Design consideration:
  - The wiper motor as well as HVAC inlet should not submerge under heavy flowrates of water
- Modeling Approach:
  - VOF Multiphase
  - Compressive Implicit Scheme

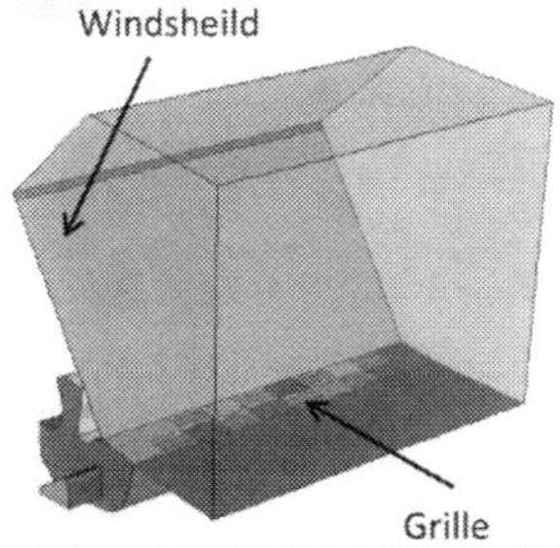


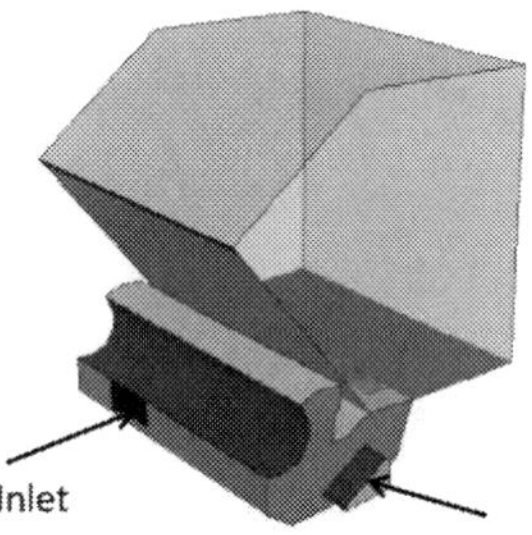




2014 Automotive Simulation World Congress
October 9-10 . Tokyo, Japan

# Water Ingestion in HVAC Plenum

- Results?

2014 Automotive Simulation World Congress
October 9-10 . Tokyo, Japan

## Rain Water Interaction with Side View Mirror

- Rainwater collects on the side mirrors of cars.
- It may be dirty and become redistributed where it may interfere with visibility
- Eulerian Wall Film model coupled with DPM model
- Custom injection created upstream of the mirror

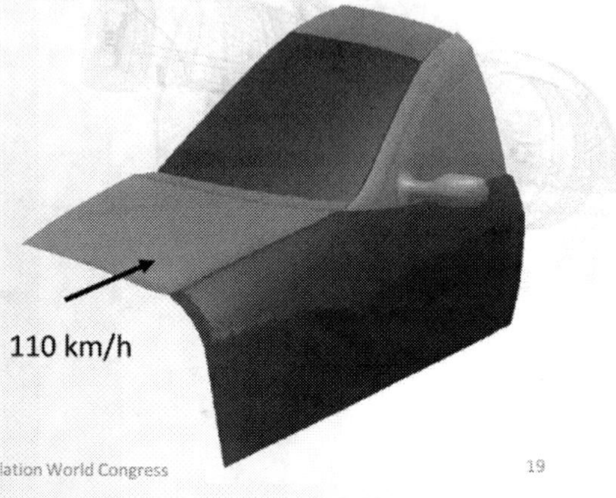




2014 Automotive Simulation World Congress
October 9-10 . Tokyo, Japan

## Rain Water Interaction with Side View Mirror

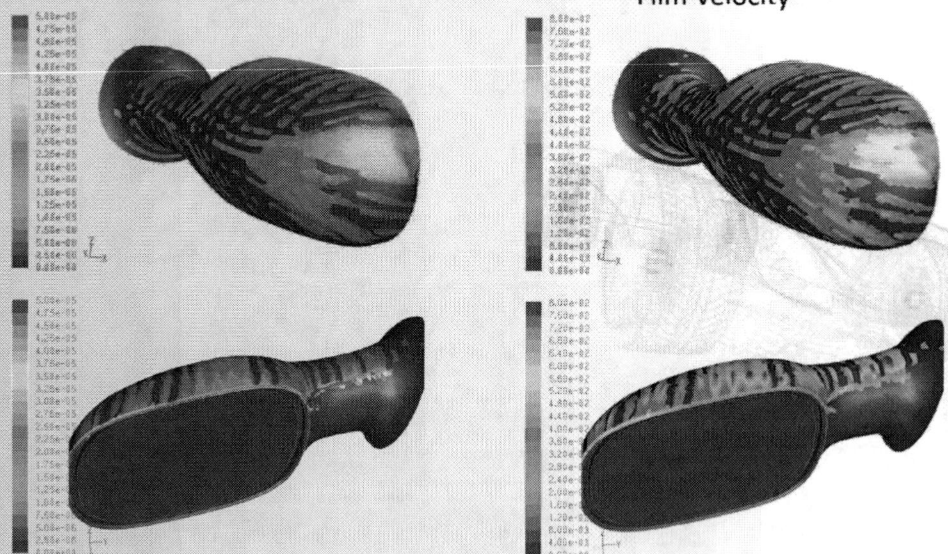

2014 Automotive Simulation World Congress
October 9-10 . Tokyo, Japan

## Rain Water Interaction with Side View Mirror

Film Thickness

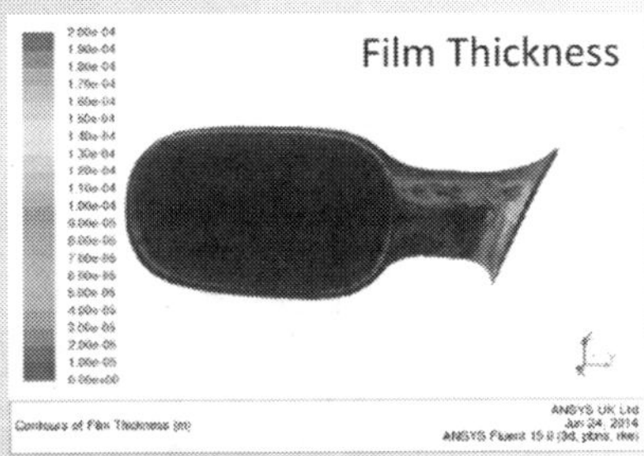

Film Thickness

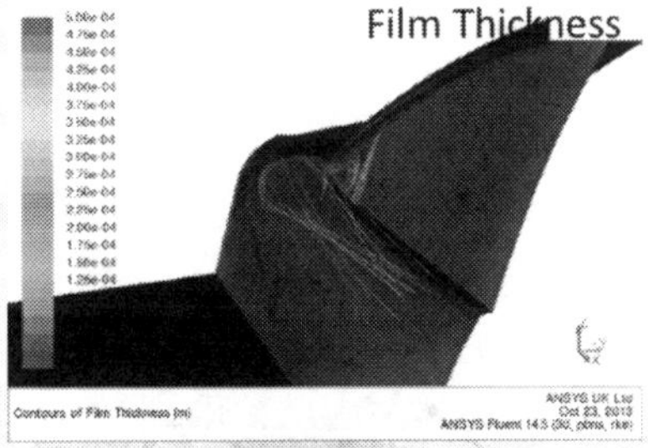

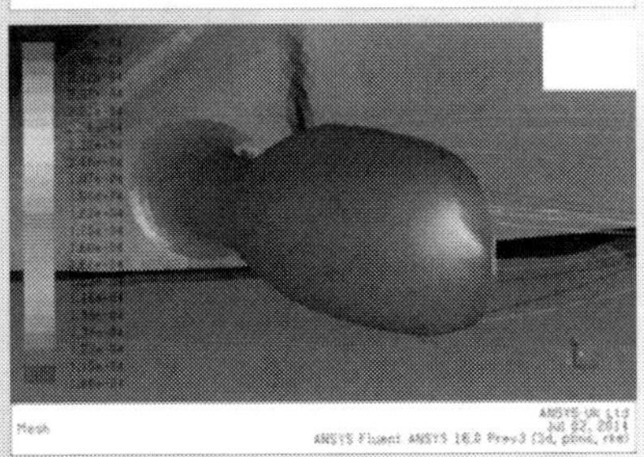

Particle Re-release

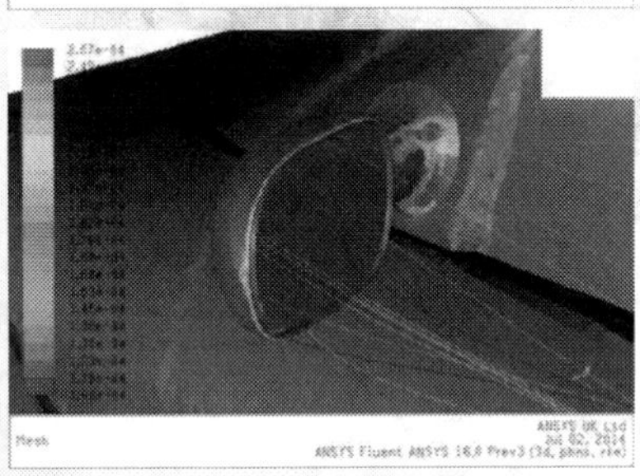

Particle Re-release



2014 Automotive Simulation World Congress
October 9-10 . Tokyo, Japan

## Rain Water Interaction with Side View Mirror

- In regions of the film where the shear forces are high and the depth is above a local critical value, droplet streams will be automatically released into the flow.
- Release rate is governed by local Weber number

Droplet Re-Release

2014 Automotive Simulation World Congress
October 9-10 . Tokyo, Japan

# Rain Water Interaction with a Vehicle

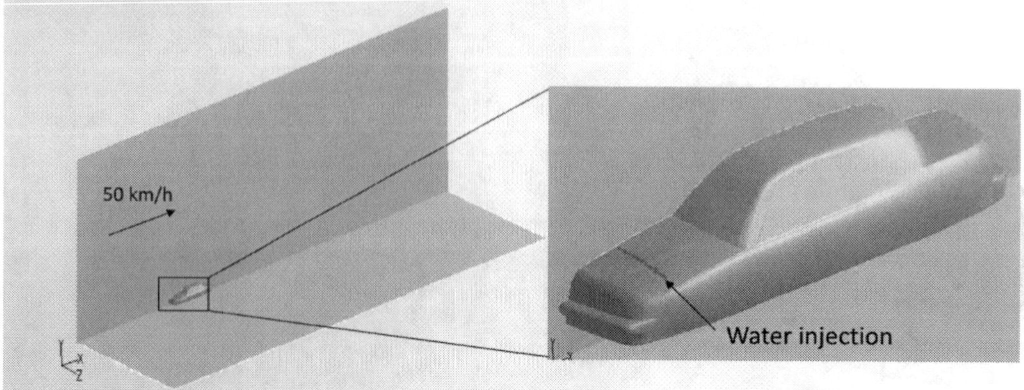




2014 Automotive Simulation World Congress
October 9-10 . Tokyo, Japan

# Rain Water Interaction with a Vehicle

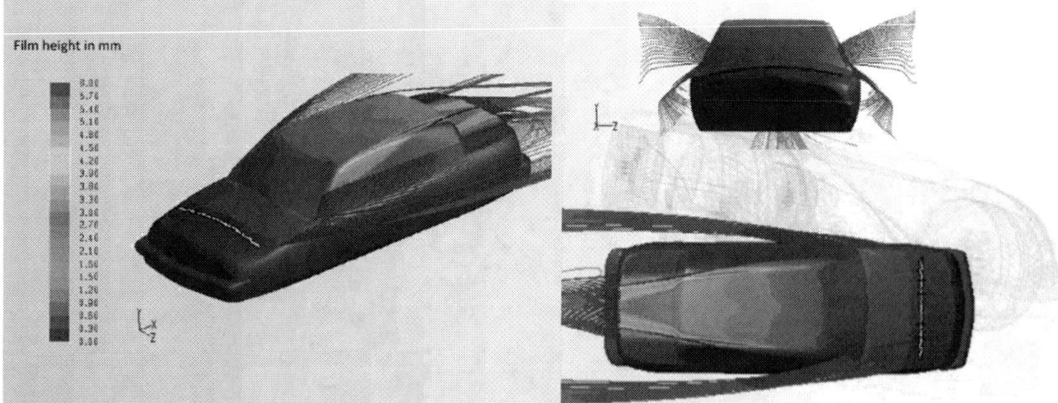

## Interaction of Cleaning Fluid with Windshield

Velocity 0[m/s]

Velocity 40[m/s]

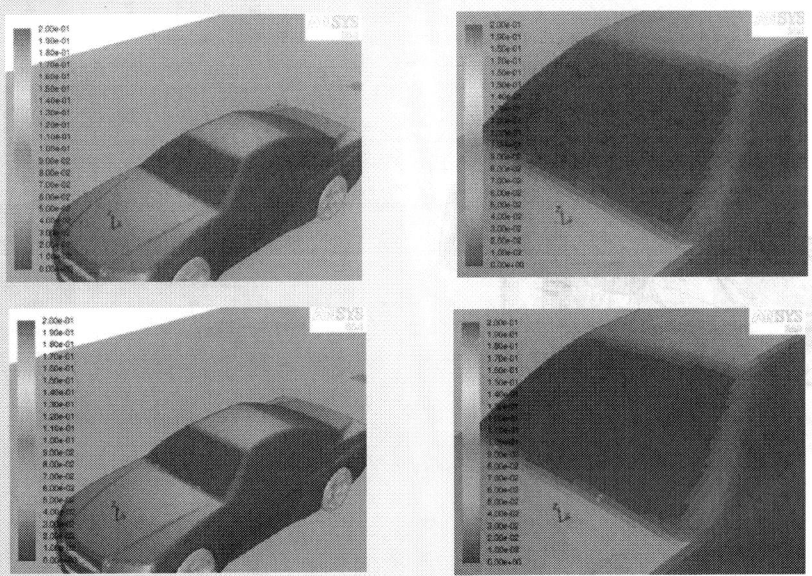




2014 Automotive Simulation World Congress
October 9-10 . Tokyo, Japan

## Summary

- ANSYS has strong capabilities in modeling Automotive Rain Water Management
- Eulerian and lagrangian multiphase models along with eulerian and lagrangian wall film models and phase change models are available to simulate different RWM applications
- Best practice strategies for all RWM applications exist

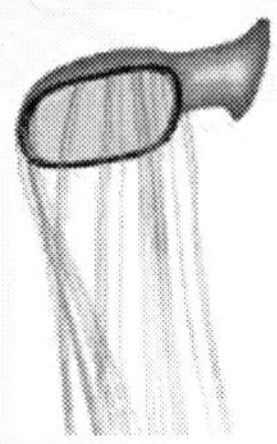

2014 Automotive Simulation World Congress
October 9-10 . Tokyo, Japan


## Acknowledgements

- Lubeena R, ANSYS India
- Eijijro Suzuki, ANSYS Japan
- Oliver Zuehlke, ANSYS Germany
- Torbjorn Virdung, ANSYS Sweden
- Adam Anderson, ANSYS UK

2014 Automotive Simulation World Congress

# ANSA-TGrid: A Common Platform for Automotive CFD Preprocessing

Xingshi Wang, PhD, ANSYS Inc.
Mohammad "Peyman" Davoudabadi, PhD, ANSYS Inc.

## Overview

- ❑Motivation
- ❑Advantages
- ❑Use Case I: External Aero Dynamic Simulation
- ❑Use Case II: Vehicle Thermal Management Simulation

# Motivation
# Conventional vs. Integrated Approach



To create a leakage free, conformal mesh and ensure physical conduction paths, all the overlapping, intersecting, and gap regions have to be fixed and connected.

**Topo Level: Manual, Time consuming**

- Traditional approach requires users to manually remove gaps, resolve overlaps and intersections for *ALL* components in the model.
- In the conventional approach, user has to somehow manually create cap and contact faces to patch leakages and preserve conduction paths.

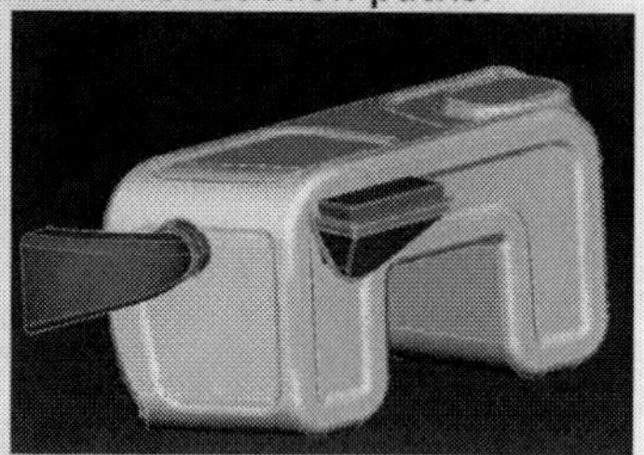

**Mesh Level: Semi-automated, More efficient**

- Leverage wrapping technology as much as possible with minimum manual efforts. Wrap surface is leakage free.
- Invest manual efforts only to create high fidelity surface mesh and preserve conduction path between different materials.
- Resolve connection on the mesh level is more efficient and flexible.





# ANSA-TGrid Interface

2014 Automotive Simulation World Congress

## Unique Merits of ANSA-TGrid Process

- ❑ Highly automated; maximizes the outcome of engineering time
- ❑ Extensive control of mesh quality suitable to all automotive simulations
- ❑ Parallel execution
- ❑ Flexible and heterogeneous inputs (CAD, existing mesh, previous analys
  - From components to assembly
  - Throughout the design cycle
  - Analysis to analysis
  - Mesh tech. to meshing technology

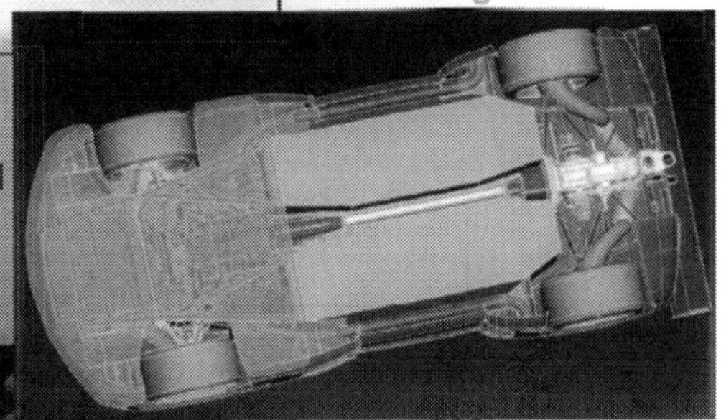

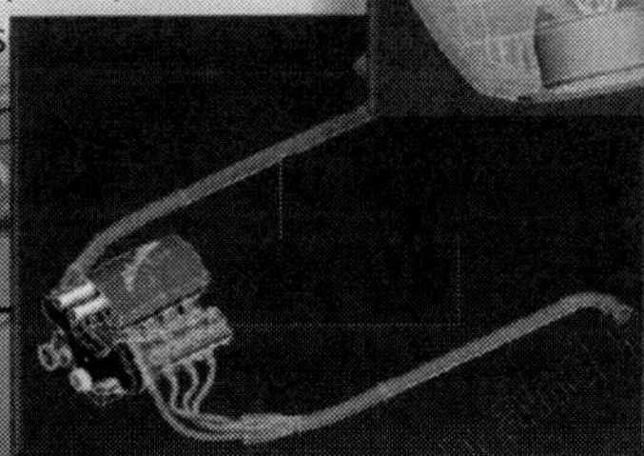

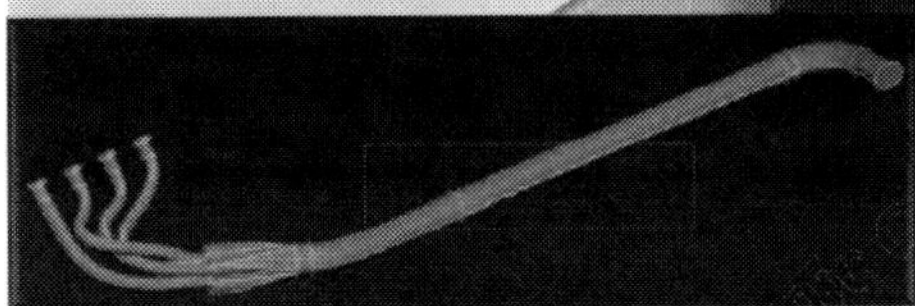

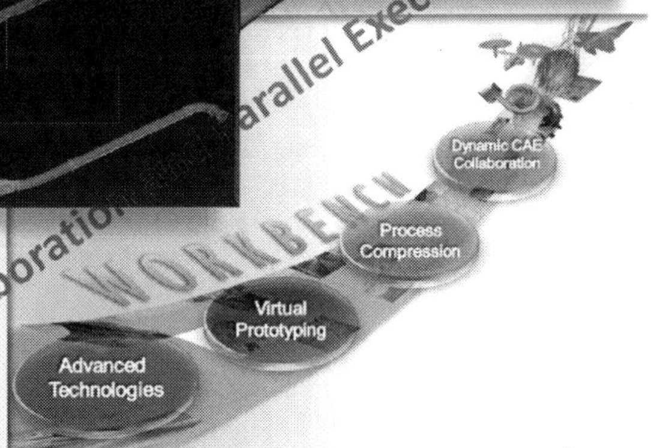




2014 Automotive Simulation World Congress
October 9-10 . Tokyo, Japan

## Use Case I
## External Aero Dynamics Simulation

❑ Objective

- Provide accurate prediction for Drag and Lift coefficients.
- Evaluate design changes for optimization to reduce drag and provide enough down force at the same time.

❑ Challenges for CFD Pre-processing

- Capture features accurately especially on styling surfaces
- Generate boundary layers to achieve $y^+ \sim 1$ on styling surfaces
- Huge cell counts for typical applications
- Fast turn around
- Go through large design changes rapidly

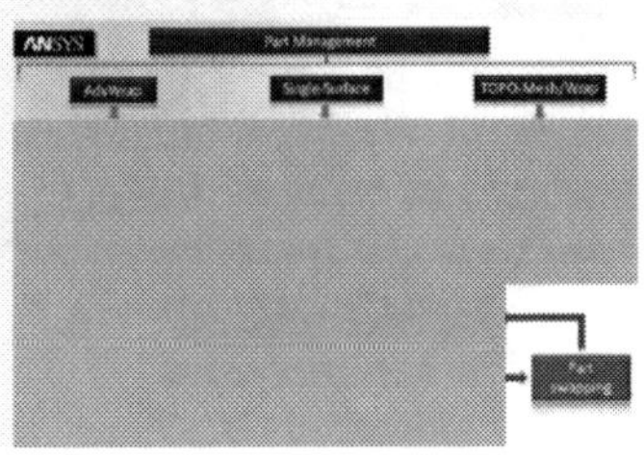

2014 Automotive Simulation World Congress
October 9-10 . Tokyo, Japan

## Preprocessing Strategy for External Aero Dynamics

- ❑ Pristine surface mesh on styling surfaces
  - Perfect feature capturing
  - Full control on elements size and quality
  - Fine resolution to resolve boundary layer
- ❑ AdvWrap drive, underhood/underbody and in-cabin components
  - Walk over geometry defects
  - Ensure leakage free surface mesh
  - Preserve important features
  - Reduce engineering time by automation

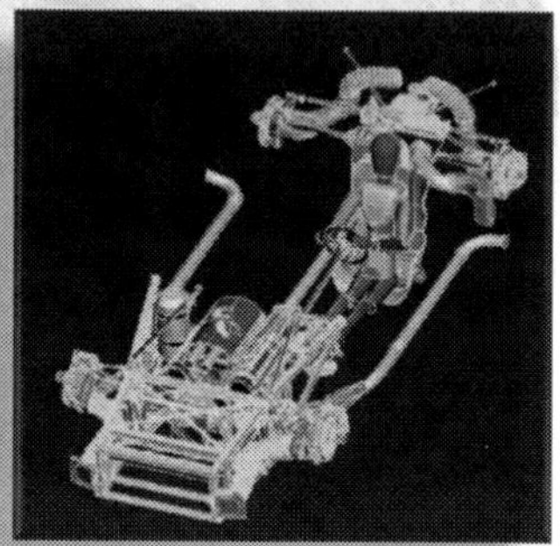

Monday, October 06, 2014 2014 Automotive Simulation World Congress

2014 Automotive Simulation World Congress
October 9-10 . Tokyo, Japan

## AdvWrap Group

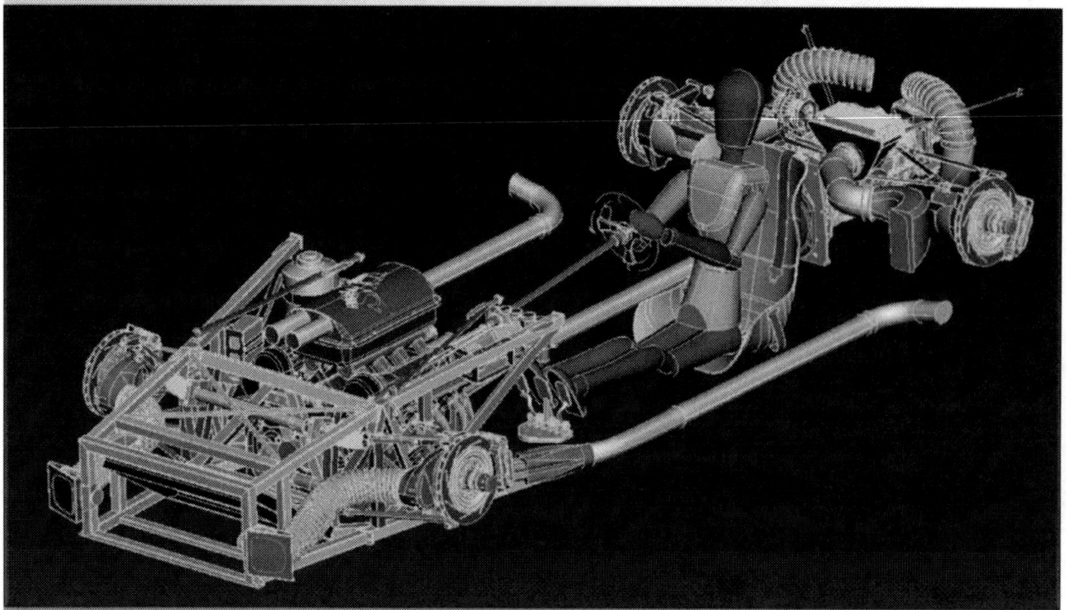

Monday, October 06, 2014 2014 Automotive Simulation World Congress 8

2014 Automotive Simulation World Congress
October 9-10 . Tokyo, Japan

# AdvWrap Results

2014 Automotive Simulation World Congress
October 9-10 . Tokyo, Japan

# Single-Surface and TOPO-Mesh Group

## Single-Surface and TOPO-Mesh Results

2014 Automotive Simulation World Congress
October 9-10 . Tokyo, Japan

## Local Connect

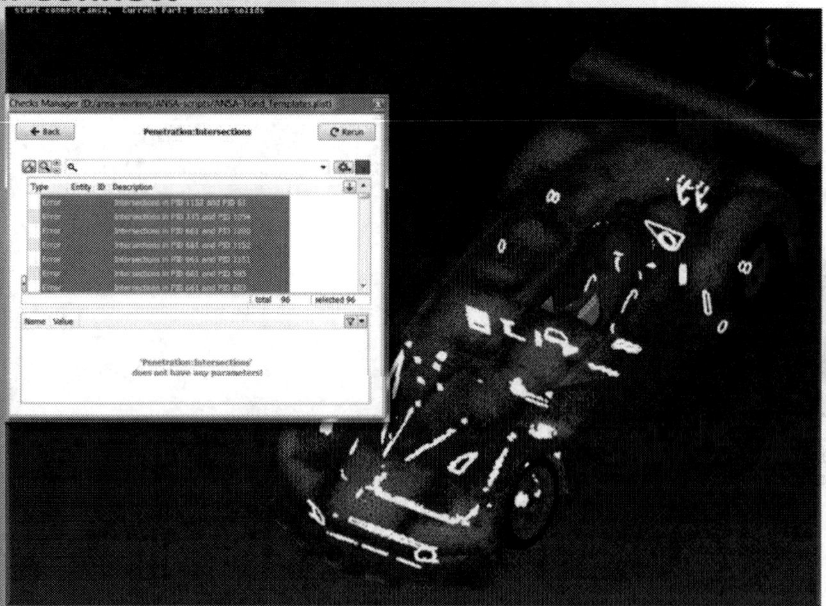

2014
Automotive Simulation
World Congress
October 9-10 . Tokyo, Japan
Surface Mesh Ready for Volume Mesh
start-connect-finish-reset-prism.ansa, Current Part: incabin-solids
Monday, October 06, 2014
13

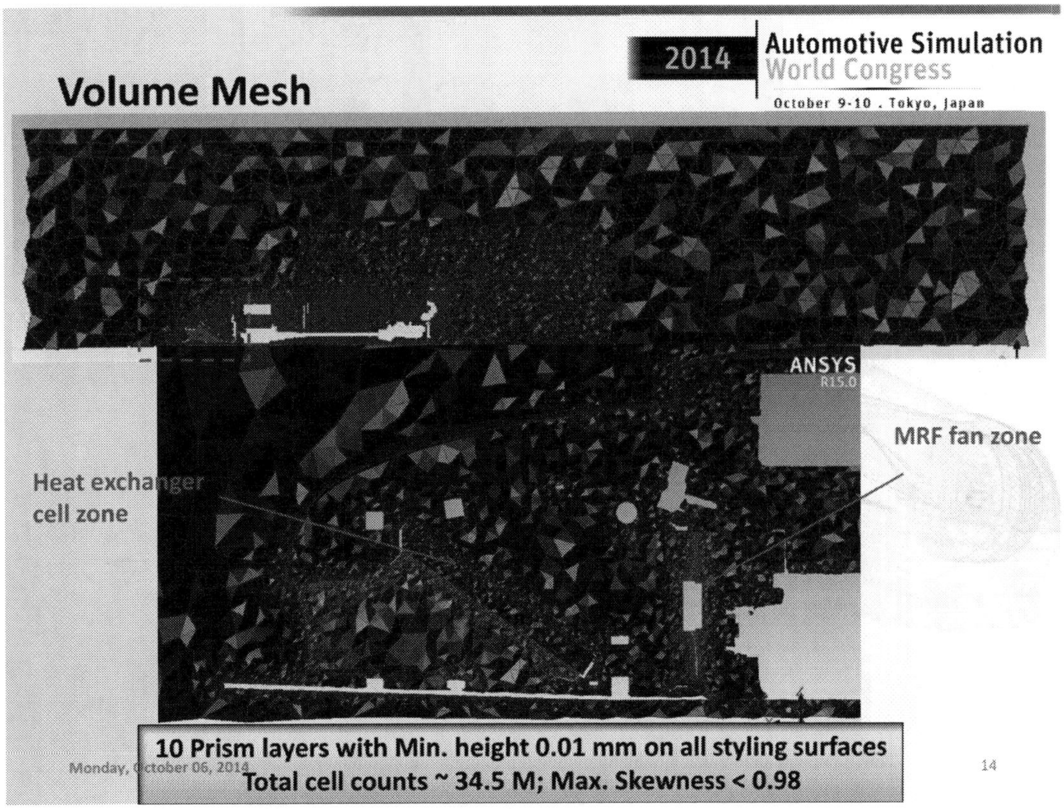
2014
Automotive Simulation
World Congress
October 9-10 . Tokyo, Japan
Volume Mesh
ANSYS
R15.0
MRF fan zone
Heat exchanger cell zone
10 Prism layers with Min. height 0.01 mm on all styling surfaces
Total cell counts ~ 34.5 M; Max. Skewness < 0.98
Monday, October 06, 2014
14

2014 Automotive Simulation World Congress
October 9-10 . Tokyo, Japan

## Simulation Results

2014 Automotive Simulation World Congress
October 9-10 . Tokyo, Japan

## Turnaround

| Subprocess | Man Time (Hour) | CPU Time (Hour) |
|---|---|---|
| Part Management | 1 | N/A |
| AdvWrap | 0.5 | 3 |
| TOPO-Mesh | 16 | N/A |
| Prism Layer Validation | 4 | 4 |
| Connect | 6 | N/A |
| Surface Mesh Improvement | 2 | N/A |
| Volume Mesh | 0.5 | 1.5 |
| **Total** | **30** | **8.5** |

2014 Automotive Simulation World Congress
October 9-10 . Tokyo, Japan

# Use Case II
# Vehicle Thermal Management Simulation

## ❑ Objective

- Model full vehicle to predict flow and heat transfer through cooling module.
- Provide accurate temperature prediction for critical components under difference road conditions to improve vehicle quality and reliability.

## ❑ Challenges for CFD Pre-processing

- Complex and dirty geometry
- Separate fluid cell zones are required to model secondary fluids/flows
- Separate solid cell zones are required to obtain accurate temperature prediction in Conjugate Heat Transfer (CHT) simulation
- Complex case setup
- Fast turn around
- Go through large design changes rapidly

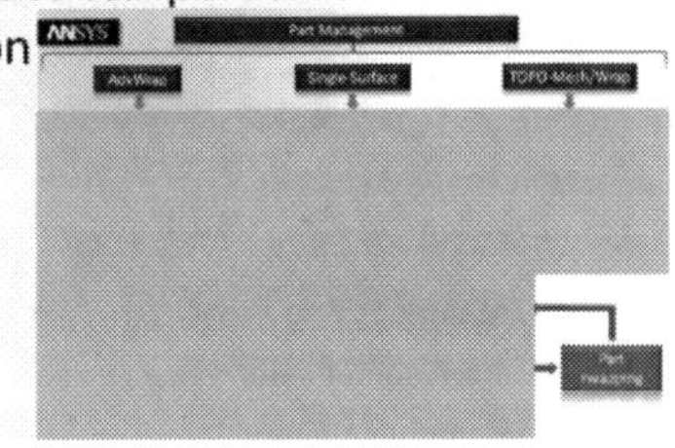



2014 Automotive Simulation World Congress
October 9-10 . Tokyo, Japan

# Preprocessing Strategy for Vehicle Thermal Management

## ❑ Individual TOPO-Mesh/wrap on components critical to thermal analysis

- High quality for thermal analysis
- Ensure leakage free surface mesh
- Obtain separate cell zones for secondary flows
- Obtain separate cell zones for different materials
- Conduction paths persevered by local connect

## ❑ Single-Surface thin solids

- Leverage shell conduction in FLUENT
- Reduce cell count and improve mesh quality

## ❑ AdvWrap components not critical to thermal analysis

- Walk over geometry defects
- Ensure leakage free surface mesh
- Preserve important features
- Reduce engineering time by automation

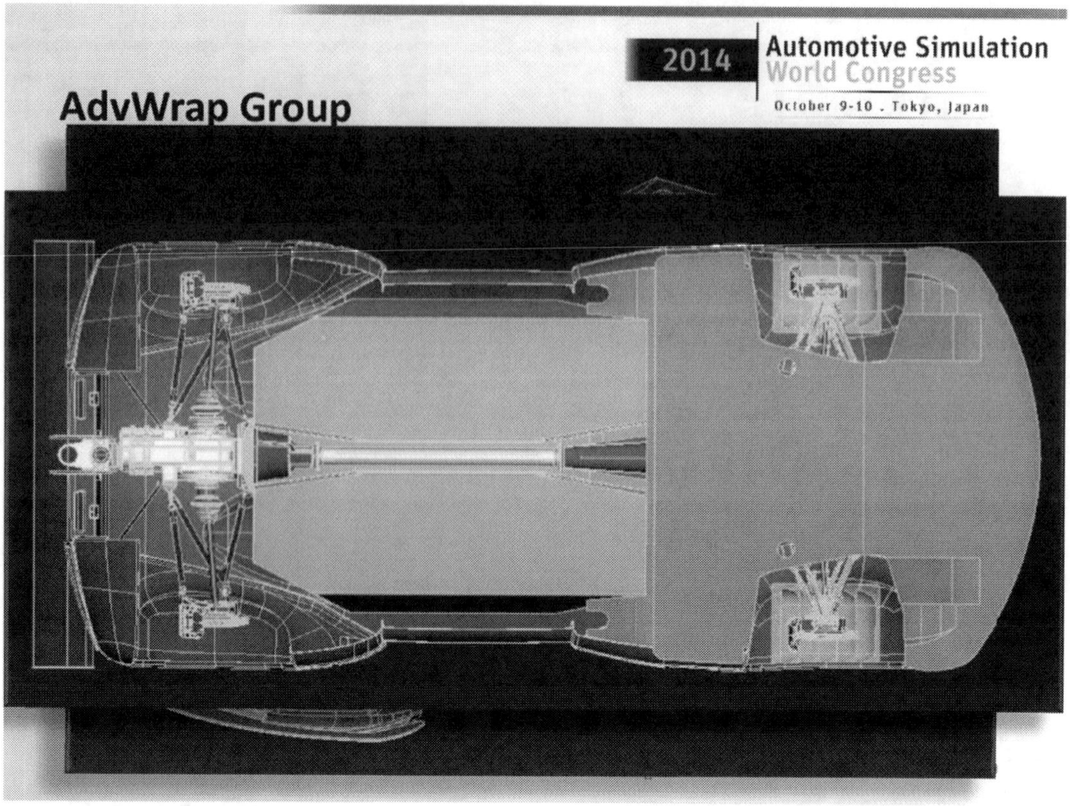
AdvWrap Group
2014
Automotive Simulation
World Congress
October 9-10 . Tokyo, Japan

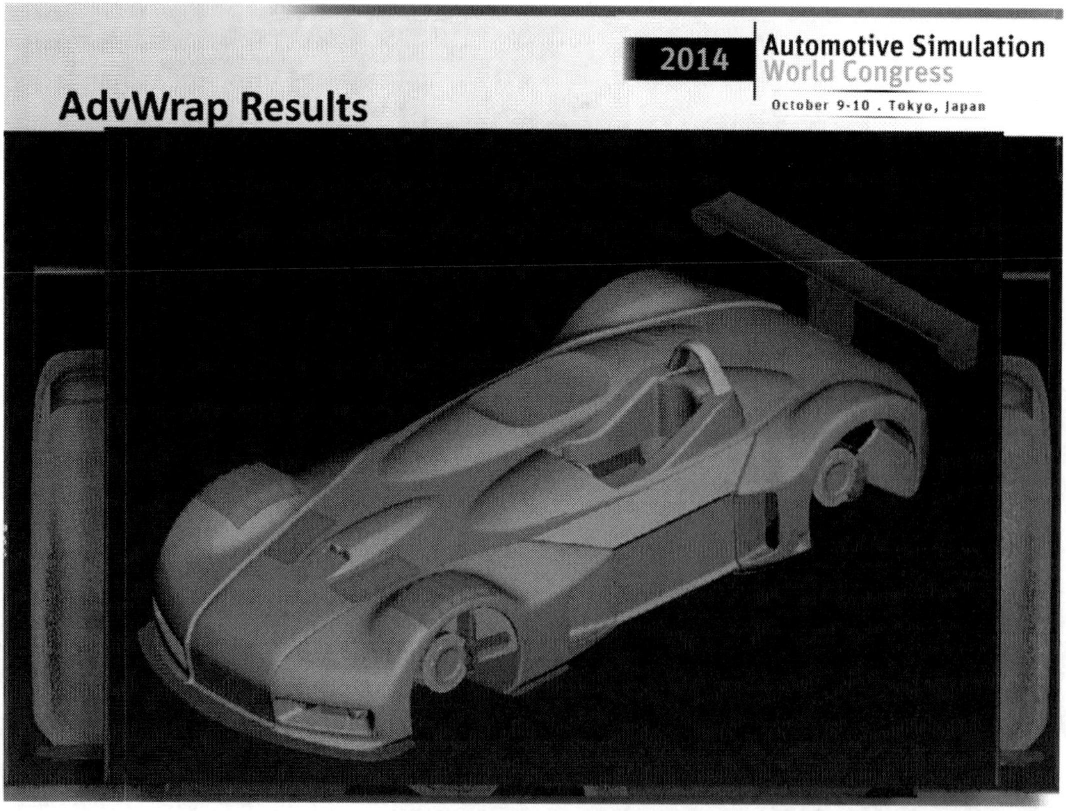
AdvWrap Results
2014
Automotive Simulation
World Congress
October 9-10 . Tokyo, Japan

2014 Automotive Simulation World Congress
October 9-10 . Tokyo, Japan

## Single-Surface Group and Results

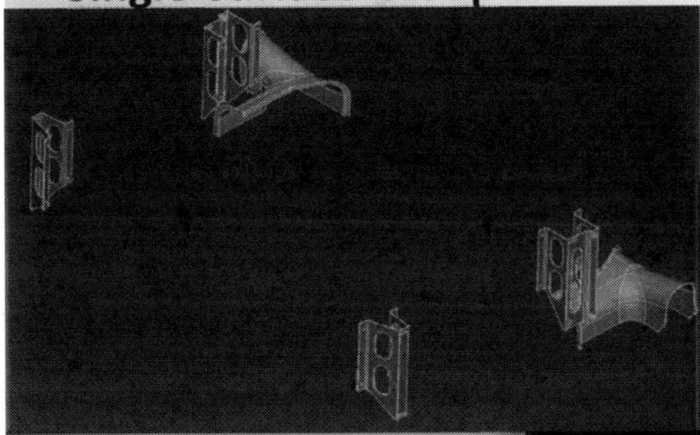

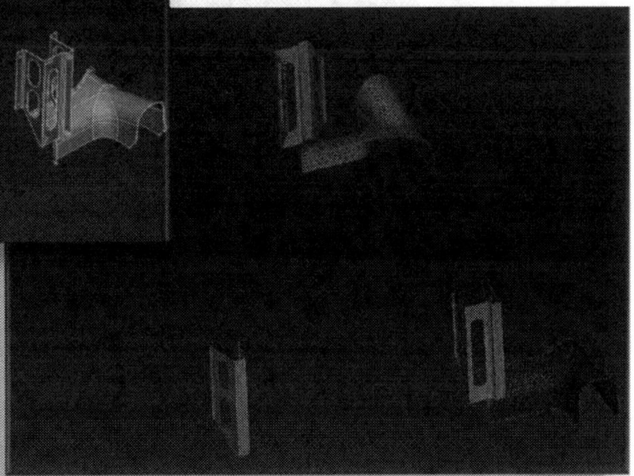



2014 Automotive Simulation World Congress
October 9-10 . Tokyo, Japan

## TOPO-Mesh/Wrap Group

2014 Automotive Simulation World Congress
October 9-10 . Tokyo, Japan

# TOPO-Mesh/Wrap Results

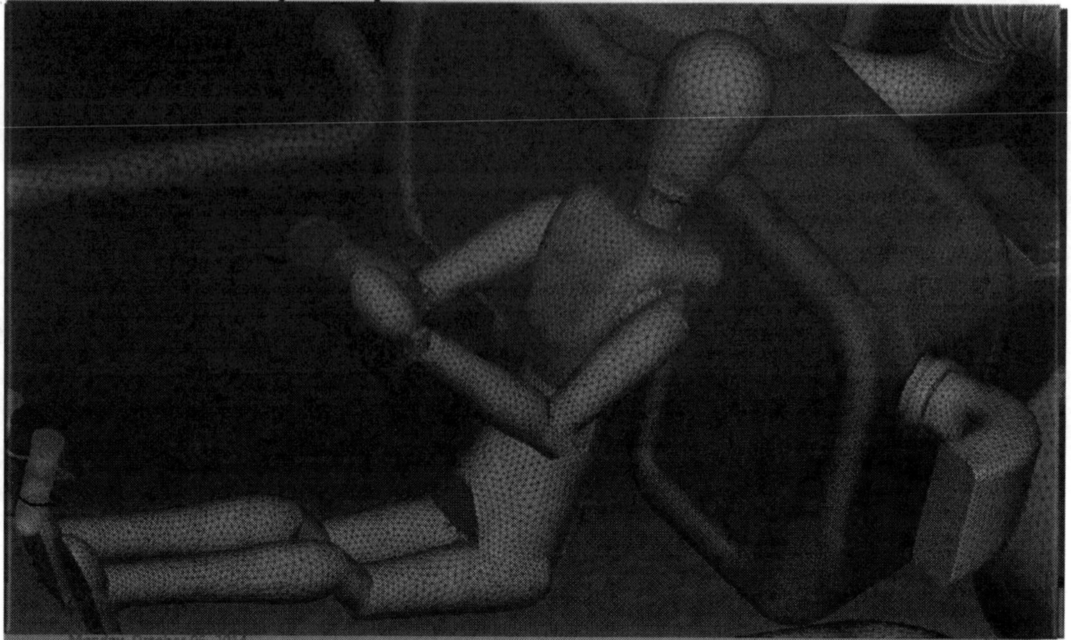



2014 Automotive Simulation World Congress
October 9-10 . Tokyo, Japan

# Local Connect

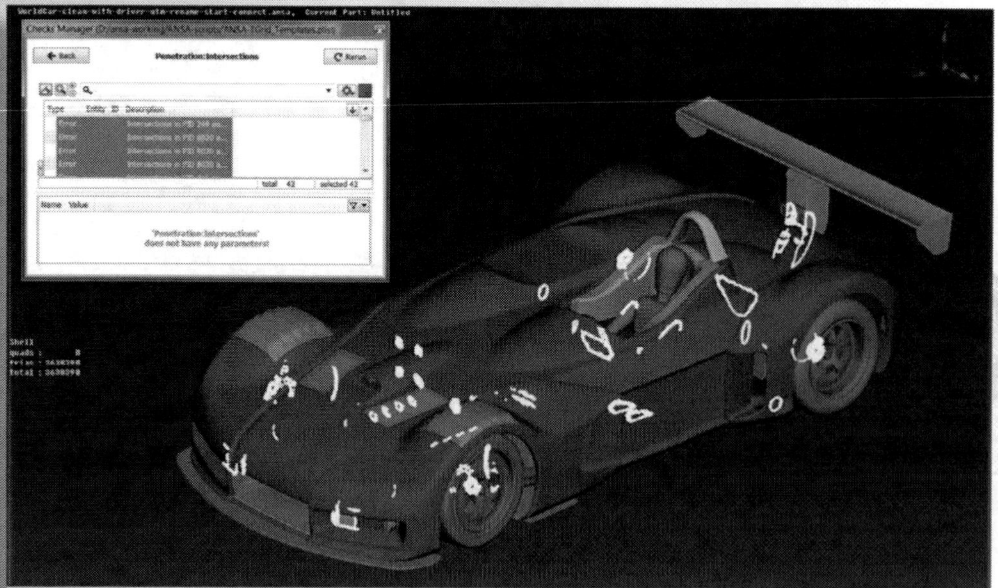

2014 Automotive Simulation World Congress
October 9-10 . Tokyo, Japan

## Surface Mesh Ready for Volume Mesh

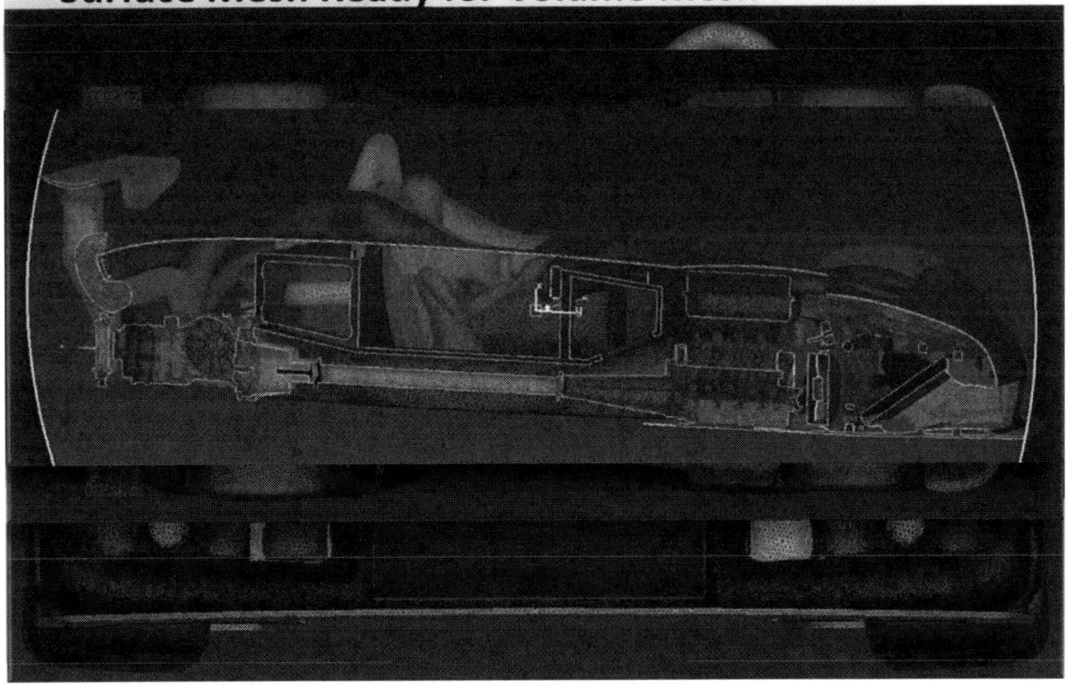

2014 Automotive Simulation World Congress

## Volume Mesh

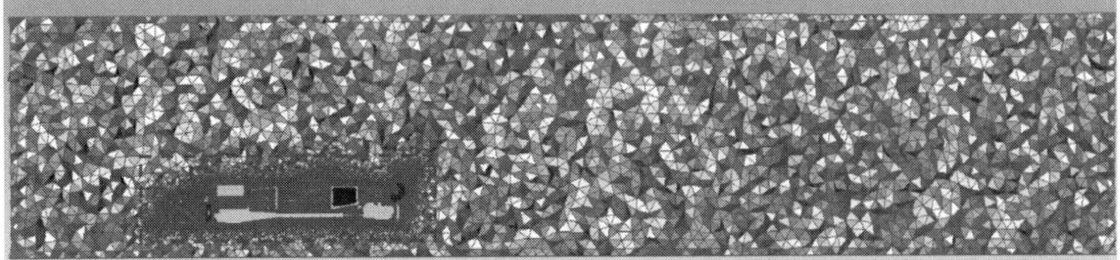

10 Fluid cell zones, 30.3M tet., Max. Skewness < 0.96
29 Solid cell zones, 3.05M tet., Max. Skewness < 0.95

Monday, October 06, 2014

## Simulation Results

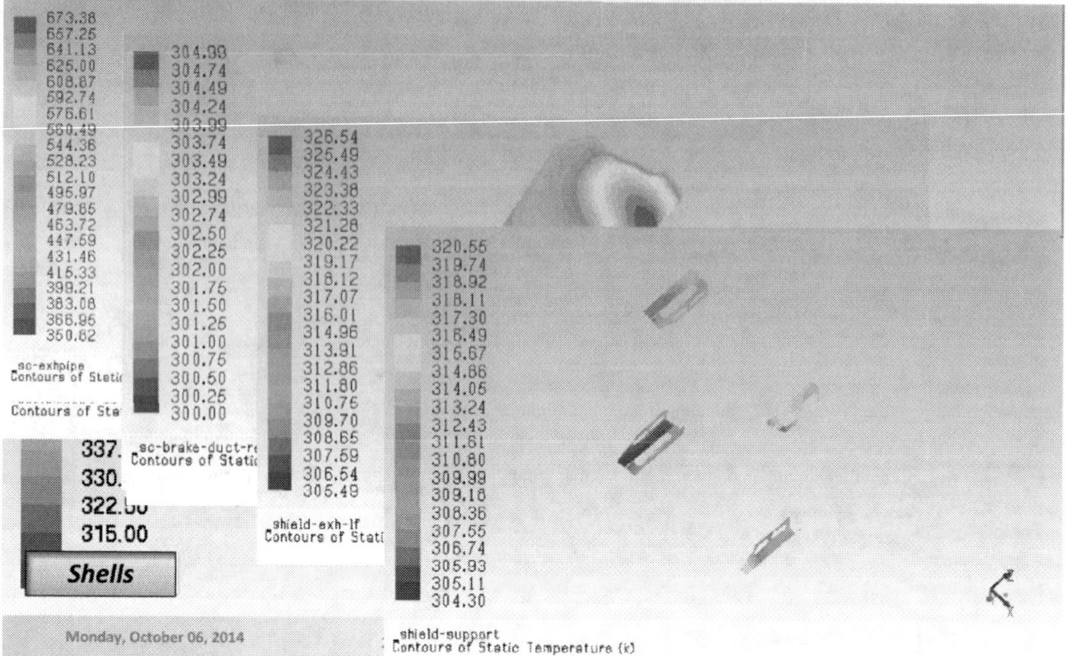




2014 Automotive Simulation World Congress
October 9-10 . Tokyo, Japan

## Turnaround

| Subprocess | Man Time (Hour) | CPU Time (Hour) |
|---|---|---|
| Part Management | 1 | N/A |
| AdvWrap | 0.5 | 4 |
| TOPO-Mesh/Wrap | 4 | N/A |
| Prism Layer Validation | 0.5 | N/A |
| Connect | 5 | N/A |
| Surface Mesh Improvement | 0.5 | N/A |
| Volume Mesh | 0.5 | 1.5 |
| **Total** | **12** | **5.5** |

# Typical Industrial Productivity Gains

| | Subprocess | Method | No. of days |
|---|---|---|---|
| 1 | Translate CAD to ANSA | | |
| 2 | Upper body including styled surfaces (external aero) | Topo-Mesh | 5 |
| 3 | Surface mesh generation (external aero) | Topo-Mesh | 2 |
| 4 | Brakes | Topo-Mesh | 2 |
| 5 | Part Management (for naming and wrap) | Includes fixing large leakages/holes | 3 |
| 6 | Main wrap of UH and UB components | WrapAdv.; HX+Fan | 1 |
| 7 | Solids surface mesh | Surface-mesh or Wrap | 3 |
| 8 | Connect | ANSA-TGrid | 3 |
| 9 | 7 Design changes | | 1 |
| | **Total** | | **20** |

- ❑ Full vehicle with all details (~10K components)
- ❑ One common model built for external aero, front-end cooling, brake and thermal simulations
- ❑ CAD to Solution time reduced by 3X with no sacrifice of accuracy
- ❑ Maximizing efficiency across multiple teams

Subprocesses 2-4 and 5-7, executed in parallel.



2014 Automotive Simulation World Congress
October 9-10 . Tokyo, Japan

# Conclusion

Semi-automated, streamlined workflow that employs the best of both ANSA and TGrid

- ❑ Increased throughput (w.r.t. man hrs and user comfort); Reusability; Scalability
- ❑ Flexibility and control over mesh quality and accuracy
- ❑ Enables dynamic collaboration and parallel execution
- ❑ PLM/PDM connectivity/portability
- ❑ Allows for higher level of automation

And can be readily integrated in the OEMs' processes

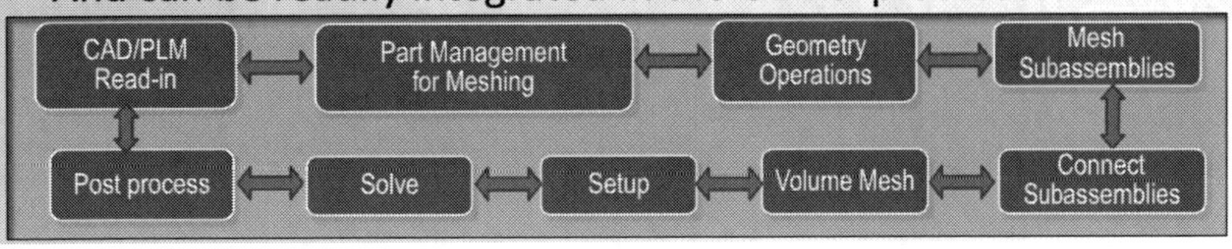

2014 ANSYS Automotive Simulation World Congress
ANSYS Electronics Simulation Expo Oct. 2014

AIST FREA FUKUSHIMA RENEWABLE ENERGY INSTITUTE, AIST

# CFD Simulations with Chemical Kinetics for Advanced Engine Combustion

National Institute of Advanced Industrial Science and Technology (AIST)
Fukushima Renewable Energy institute, AIST (FREA)

Taku TSUJIMURA



FREA

## Overview

- ☐ Motivation for using CFD simulation of engine combustion
- ☐ Policies on using CFD codes and models
- ☐ Applications for dual fuel engine combustion (Natural Gas/Diesel)
- ☐ Applications for HCCI engine combustion
- ☐ Summaries

1

# Overview

- Motivation for using CFD simulation of engine combustion
- Policies on using CFD codes and models
- Applications for dual fuel engine combustion (Natural Gas/Diesel)
- Applications for HCCI engine combustion
- Summaries

2

# Motivation for using CFD simulation of engine combustion

- Fundamental works on chemical and physical phenomena have been progressing to find "truths of nature"
- Physical and chemical models for a CFD simulations are becoming more detailed and complicated
- A CFD code should have become a strong tool to analyze unknown phenomena even for development of products without experiments

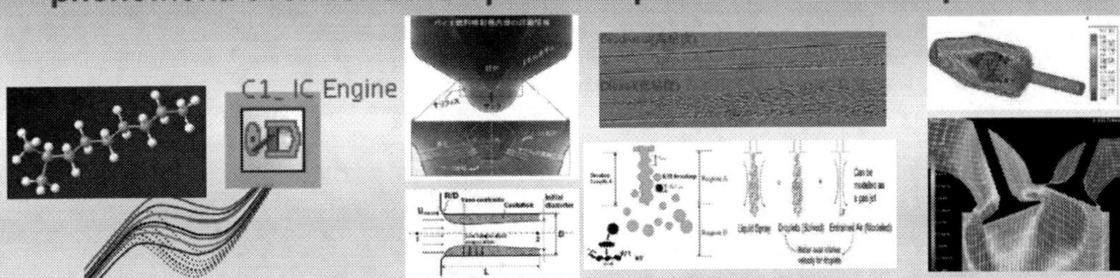


Detailed Chemical Kinetic Models
Fundamental Experiments

→

2D thru 3D Dimensional Combustion
Simulations in Practical Engine Systems

3

# Overview

- ☐ Motivation for using CFD simulation of engine combustion
- ☐ Policies on using CFD codes and models
- ☐ Applications for dual fuel engine combustion (Natural Gas/Diesel)
- ☐ Applications for HCCI engine combustion
- ☐ Summaries

4

# Policies on using CFD codes and models

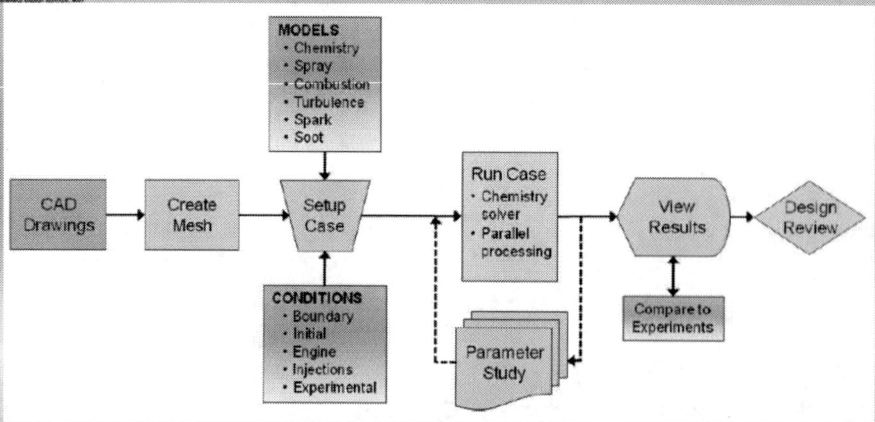


- Installs lots of physical and chemical models

➔ Allows us reproduce various situations

- Requires lots of inputs for engine conditions, models, etc.

➔ Gives us furiously to think

5

## Policies on using CFD codes and models

- UNDERSTAND essentials of physical and chemical models
- KNOW what is known or unknown in experiments
- DO NOT CHANGE parameters just to meet experimental results

ENGINE GEOMETRY
- Easy to reproduce "precise engine geometry" by loading STL files
- How small or large is adequate mesh size?

WALL BOUNDARY
- Nonuniform distribution which depends on engine conditions
- Difficult to measure temperatures around walls

INITIAL CONDITIONS
- Difficult to know initial gas conditions (temp., composition, flow)

CHEMICAL KINETIC MODEL
- Must use reduced mechanisms to achieve reasonable cost

FUEL PROPERTY Unknown supercritical properties etc.

6

## Overview

- ☐ Motivation for using CFD simulation of engine combustion
- ☐ Policies on using CFD codes and models
- ☐ **Applications for dual fuel engine combustion (Natural Gas/Diesel)**
- ☐ Applications for HCCI engine combustion
- ☐ Summaries

7

## Schematic Diagram of Natural Gas/Diesel Dual Fuel Engine

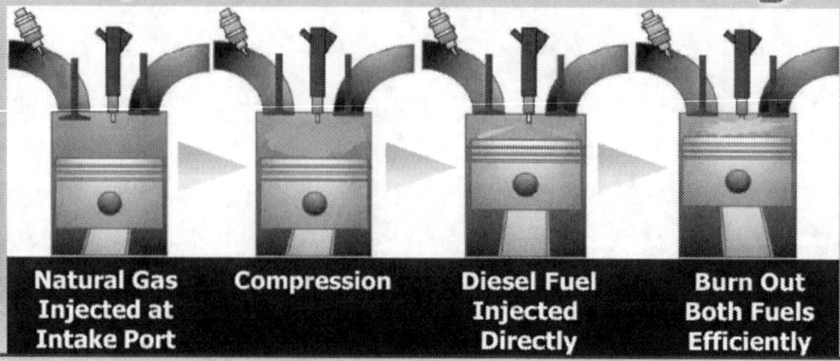


**Milestones**

- Dual fuel combustion has been fascinating engine developers
- Dual fuel combustion can allow us to use various fuels
- The state-of-the-art combustion with various combination of fuels

**Objectives**

- Advanced diesel injection technologies promise enhancing degrees of freedom for controlling dual fuel combustion.
- Deeply understanding dual fuel combustion with advance diesel injection technologies by experiments and CFD simulations

8

## Natural Gas/Diesel Dual Fuel Engine Tests in Experiments

| | | |
|---|---|---|
| Engine type | | 4-stroke single-cylinder diesel engine |
| Bore x stroke | mm | 92 x 96 |
| Displacement | $cm^3$ | 638 |
| Compression ratio | | 15.1 |
| Combustion chamber | | Dish (Cavity $\phi$: 56 mm) |
| Injection type | | CNG: Port, Diesel: Direct |
| Injection equipment | | Diesel: Common rail system @50MPa |
| Nozzle type (Injection angle) | | $\phi$0.085mm x 12 holes ($\beta$: 130 deg.) |
| Intake gas temp. | deg.C | 25±1 |
| Lube. oil and coolant temp. | deg.C | 70±5, 80±2 |

9

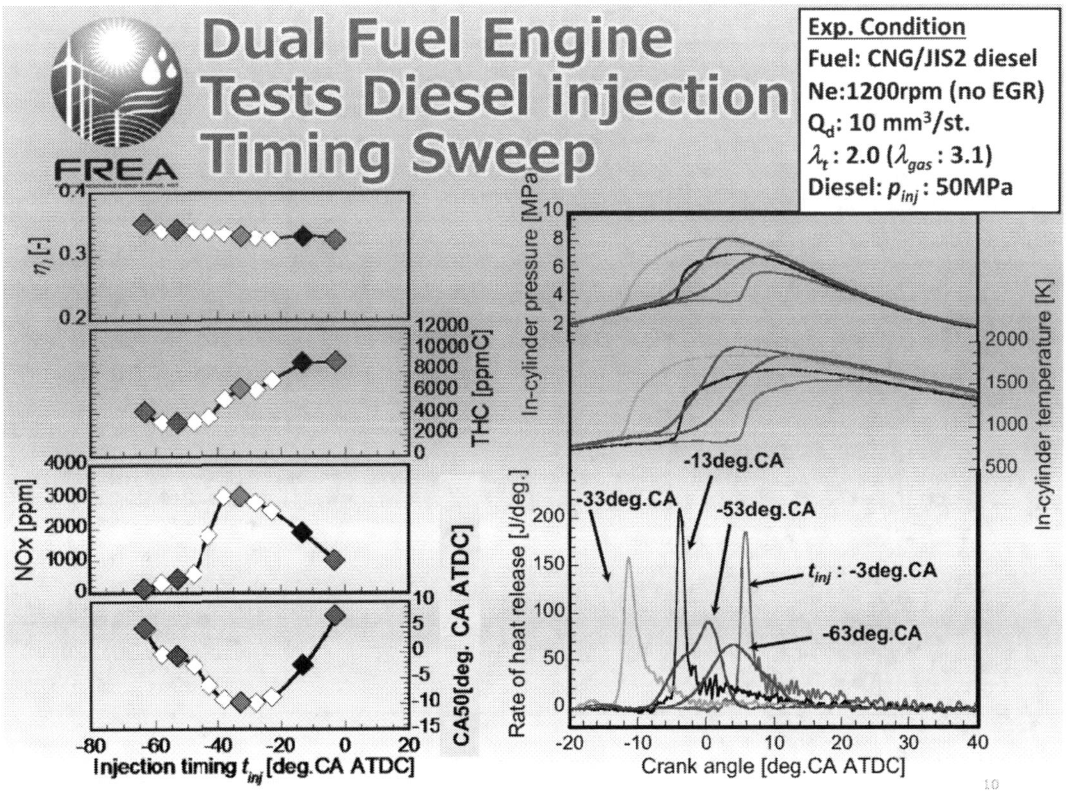
Dual Fuel Engine Tests Diesel injection Timing Sweep
FREA
Exp. Condition
Fuel: CNG/JIS2 diesel
Ne:1200rpm (no EGR)
Q_d: 10 mm³/st.
λ_t: 2.0 (λ_gas: 3.1)
Diesel: p_inj: 50MPa
η_i [-]
THC [ppmC]
NOx [ppm]
CA50[deg. CA ATDC]
Injection timing t_inj [deg.CA ATDC]
In-cylinder pressure [MPa]
In-cylinder temperature [K]
Rate of heat release [J/deg.]
-13deg.CA
-33deg.CA
-53deg.CA
t_inj: -3deg.CA
-63deg.CA
Crank angle [deg.CA ATDC]
10

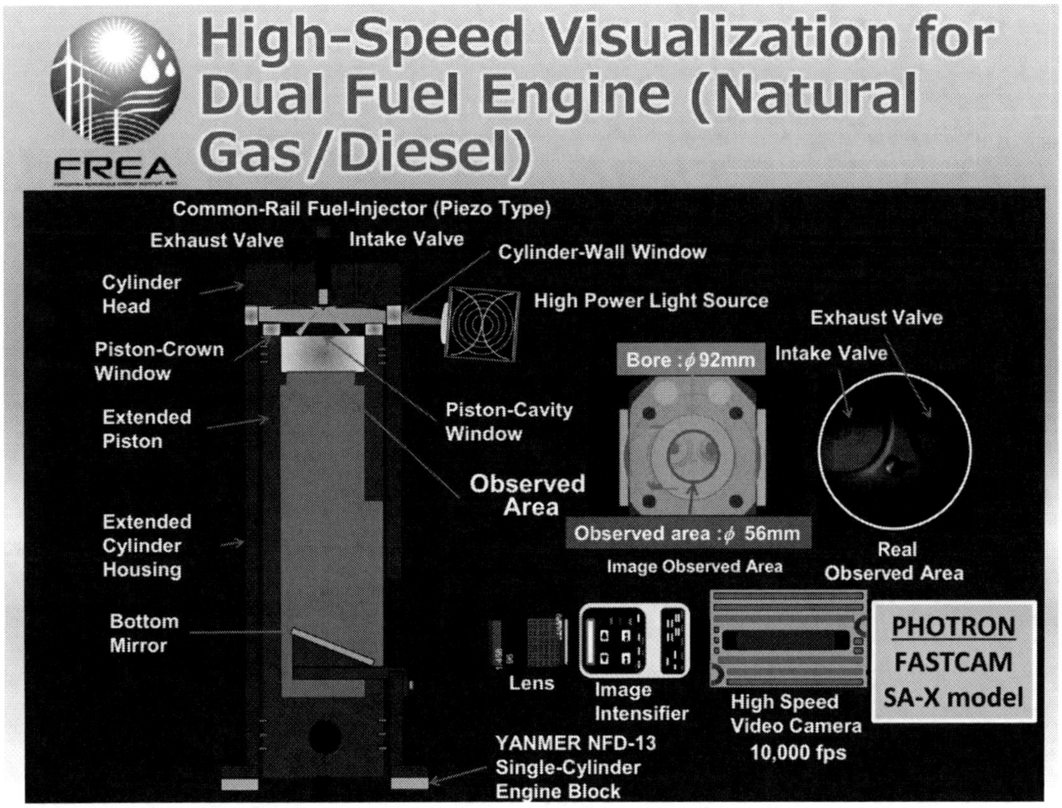
High-Speed Visualization for Dual Fuel Engine (Natural Gas/Diesel)
FREA
Common-Rail Fuel-Injector (Piezo Type)
Exhaust Valve
Intake Valve
Cylinder-Wall Window
Cylinder Head
High Power Light Source
Piston-Crown Window
Extended Piston
Piston-Cavity Window
Observed Area
Extended Cylinder Housing
Bottom Mirror
Bore : φ 92mm
Observed area : φ 56mm
Image Observed Area
Exhaust Valve
Intake Valve
Real Observed Area
Lens
Image Intensifier
High Speed Video Camera 10,000 fps
PHOTRON FASTCAM SA-X model
YANMER NFD-13 Single-Cylinder Engine Block

## High-Speed Visualizations for (Top) Diesel Injected around TDC (Bottom) Diesel Injected 60 deg.CA before TDC

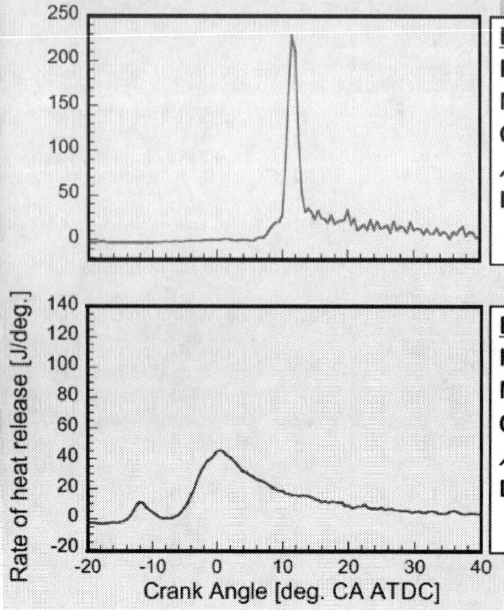


**Exp. Condition**
Fuel: CNG/Solvent
Ne:1200rpm (no EGR)
$Q_d$: 10 mm³/st.
$\lambda_t$: 2.0 ($\lambda_{gas}$: 3.1)
Diesel: $p_{inj}$: 50MPa
$t_{inj}$: TDC

**Exp. Condition**
Fuel: CNG/Solvent
Ne:1200rpm (no EGR)
$Q_d$: 10 mm³/st.
$\lambda_t$: 2.0 ($\lambda_{gas}$: 3.1)
Diesel: $p_{inj}$: 50MPa
$t_{inj}$: -60deg.CA ATDC

12

## Implementing CFD Simulation Computational grid and Procedure for CFD

| | | | |
|---|---|---|---|
| Squish | mm | 0.9275 | |
| Bore | mm | 92.0 | ← 48 div. |
| Stroke | mm | 96.0 | ← 34 div. |
| Cavity depth | mm | 15 | ←15 div. |
| Cavity diameter | mm | 56 | ←28 div. |
| Crevice width | mm | 0.05 | ← 2 div. |

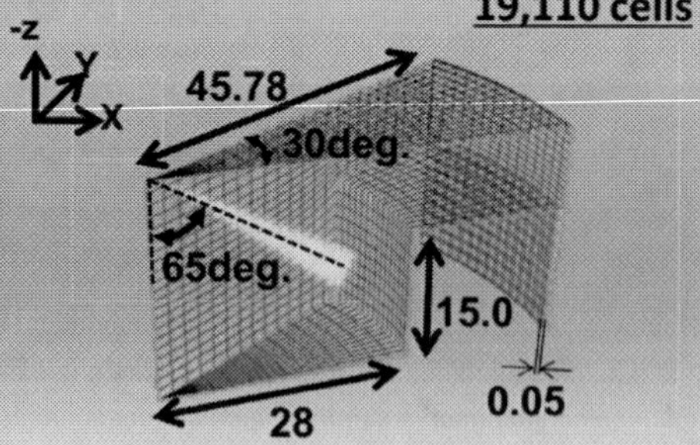


| | | | |
|---|---|---|---|
| No. of nozzle | | 1 | |
| Diameter nozzle | mm | 0.085 | |
| Location (X, Y, Z) | mm | 4.0, 1.0, 2.3 | |
| r | mm | 0.0 | |
| θ | deg | 30.0 | ←15 div. |
| Spray angle | deg | 65.0 | |
| Fuel | Diesel (chem./phys.) : $n\text{-}C_7H_{16}/n\text{-}C_{14}H_{30}$<br>Natural gas : $CH_4$ | | |

| | |
|---|---|
| Code | FORTÉ based on KIVA3V |
| Droplet/Breakup | DDM/KH-RT |
| Gas penetration | Gas jet |
| Interaction | Turbulence-kinetics interaction |
| Reaction model | RD (35sps/74reac)<br>This study (47sps/87reac) |

13

# One of Key Factors for Simulating Results: Turbulence-Kinetics Interaction

FREA

**Turbulence-Kinetics Interaction Model**

- **Implemented by Kong et al. for HCCI and diesel combustion (e.g. SAE 920512, 940523)**

**--- Turbulence-Kinetics Interaction Model included in FORTÉ ---**

- **The mixing time-scale considers that the combustion chemistry should be partly controlled by the breakup of turbulent eddies due to the imperfect mixing**
- **Assumption: All species move toward their local equilibrium with a time scale of $\tau_{eff}$**

**Production rate species $k$** $\dot{\omega}_{,eff} = \frac{Y_{k,EQ} - Y_k}{\tau_{eff}}$

**Effective time scale** $\tau_{eff} = \tau_{chem} + f\tau_{mix}$

**Turbulent scalar mixing time scale** $\tau_{mix} = C_{tki}\tau_{turb} = C_{tki}\frac{k}{\,}$

**Function denotes an effect of turbulence** $f = \frac{1 - exp[-Y_{products}]}{1 - exp[-1]}$

**Mixing time coefficient** $C_{tki}$

# One of Key Factors for Simulating Results: Turbulence-Kinetics Interaction

FREA

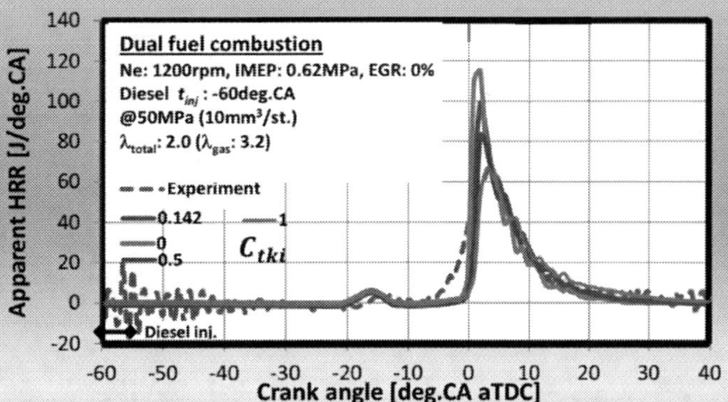


**Case: Early Diesel Injection**

- **$C_{tki}$ was change from 0 to 1, but in term of the slope of hot ignition, little improvement can be seen**
- **In case of well-mixed diesel spray, turbulence-kinetics interaction is supposed to have less effects**

**➔➔ Chemical kinetics must have a greater influence!!!**

15

# Heat Release Rate Profiles Computed with Three Models

FREA

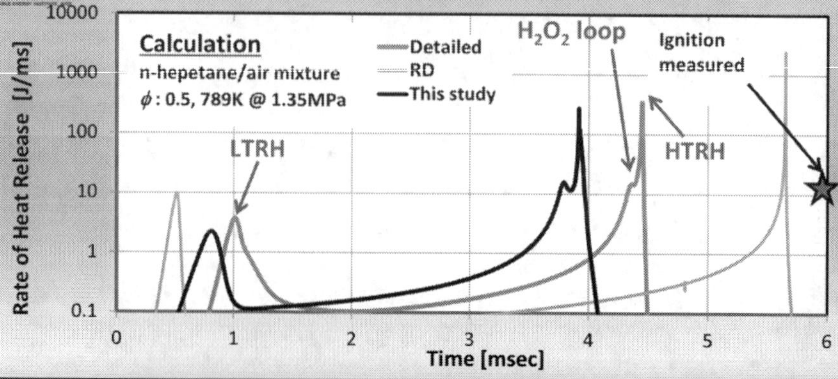


**Comparison of Computed Heat Release Rate Profiles**

- All computed profiles show rapid rises just before steep drop
- RD model can estimate the ignition delay most precisely
- The detailed and current models indicate three-stage heat release
- The second spike should be associated with "so-called" $H_2O_2$ loop

$H_2O_2$ decomposition ↔ $CH_2O+OH \rightarrow HCO+H_2O$

16

# Simulating Dual Fuel Combustion with Early Diesel Injection

FREA

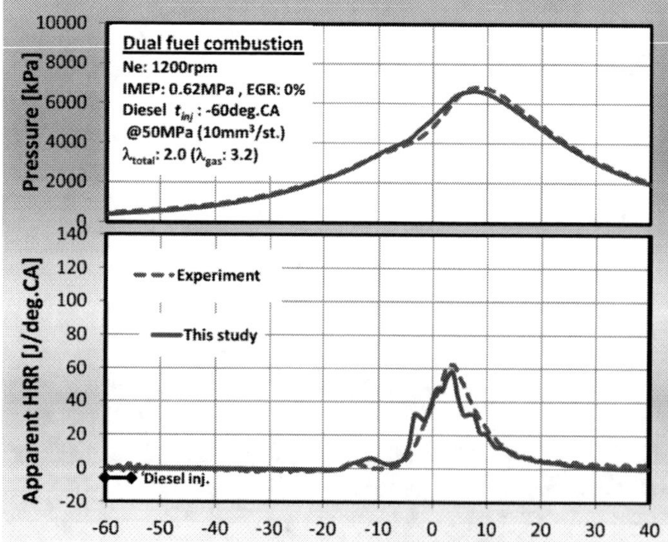


**Validation: The newly developed $n\text{-}C_7H_{16}/CH_4$ reduced model**

- If the same $C_{tki}$ is used, the significant changes can be seen in terms of a slope and peak value of hot ignition
- The current model can reproduce a heat release rate profile as well as onsets of LTHR and HTHR

17

# Simulating Dual Fuel Combustion with Early Diesel Injection

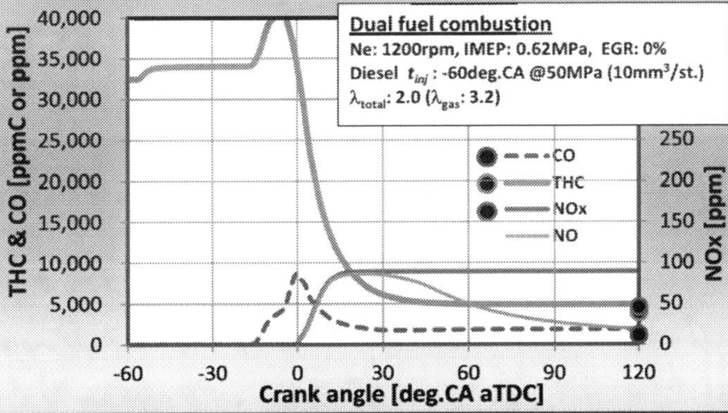


| Measured emission | Unit | Value |
|---|---|---|
| THC | ppmC | 3900 |
| NOx | ppm | 45 |
| CO | ppm | 1130 |

| Simulated emission | Unit | Value |
|---|---|---|
| THC | ppmC | 4890 |
| NOx | ppm | 89 |
| CO | ppm | 1860 |

**Validation: The newly developed n-$C_7H_{16}$/$CH_4$ reduced model**

- It's likely that the current model is capable of predicting THC, CO, and NOx emissions as well as a heat release profile
- But it is required to have more accurate prediction in the future
- It is suggested that a simulation would be conducted with much
- finer grids to get better predictions

# In-Cylinder Distributions

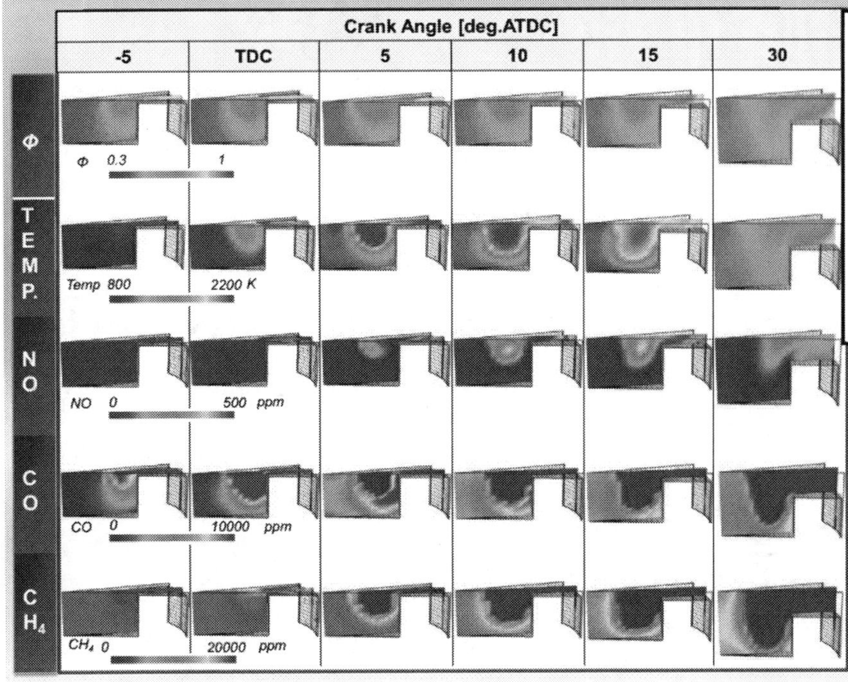


**Cal. Condition**
Fuel: n-heptane/ methane
Ne: 1200rpm
(no EGR)
IMEP: 0.62MPa
$t_{inj}$: -60deg.ATDC @50MPa
$\lambda_{total}$ : 2.0
($\lambda_{gas}$ : 3.2)

19

# Simulating Dual Fuel Combustion for Several Combustion Strategies

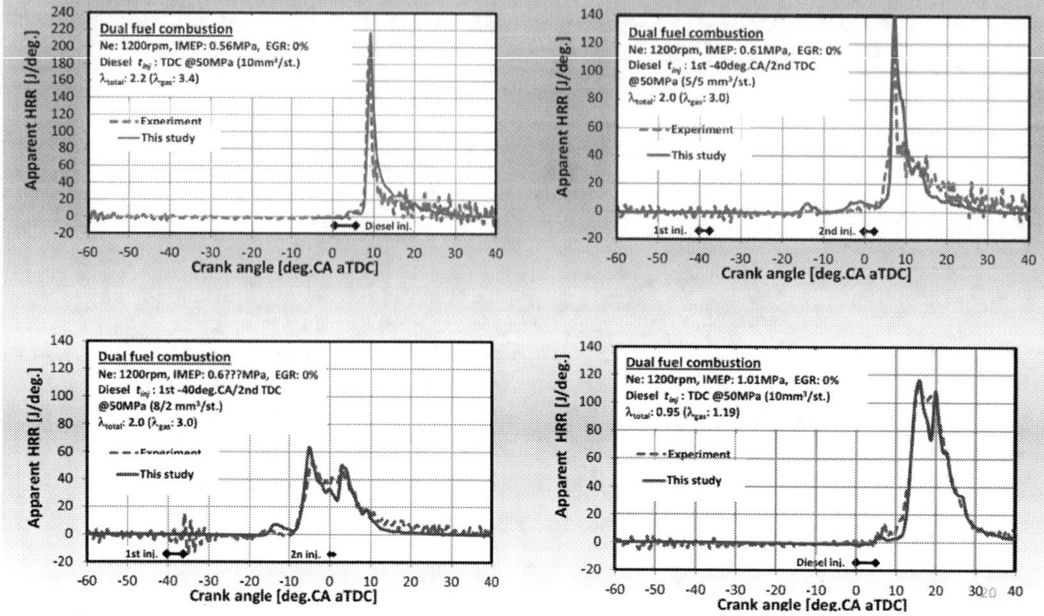


# Simulating Dual Fuel Combustion for Several Combustion Strategies

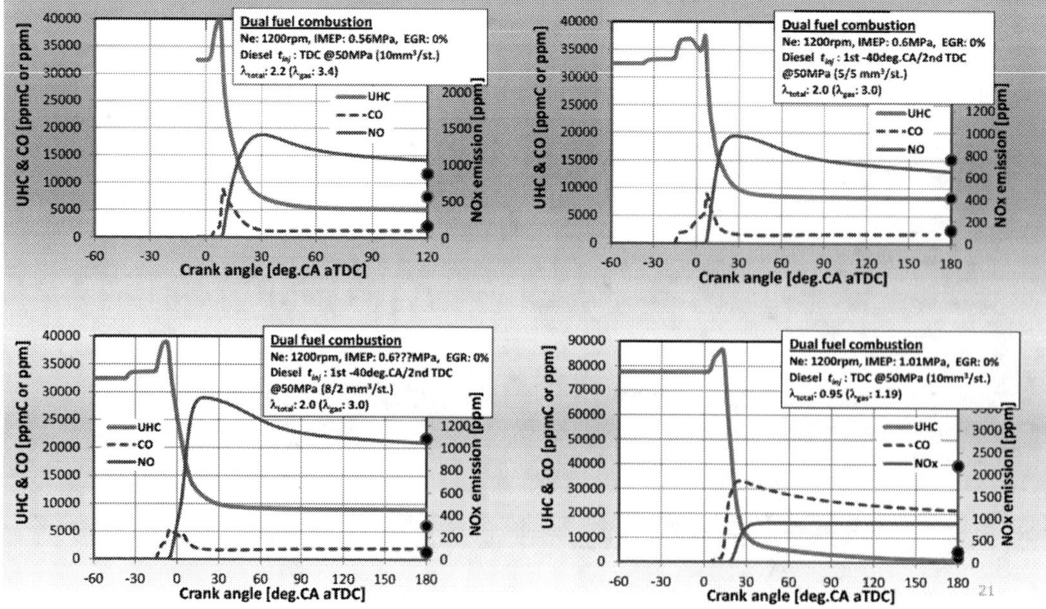

## Overview

FREA

- ☐ Motivation for using CFD simulation of engine combustion
- ☐ Policies on using CFD codes and models
- ☐ Applications for dual fuel engine combustion (Natural Gas/Diesel)
- ☐ **Applications for HCCI engine combustion**
- ☐ Summaries

22

## Applications for HCCI Engine Combustion

FREA

**A Next Generation Bio-Alcohol Fuel:**

- Isopentanol (3-Methyl-1-Butanol or 3 Methylbutane-1-ol) is one of biomass derived alcoholic fuels, like Ethanol

jbei http://www.jbei.org/ Joint BioEnergy Institute

- **One of the targets of JBEI,** alternative to gasoline
- **Sandia National Laboratories** have already experimentally investigated the characteristics of Isopentanol as a fuel for HCCI engine (SAE No. 2010-01-2164)

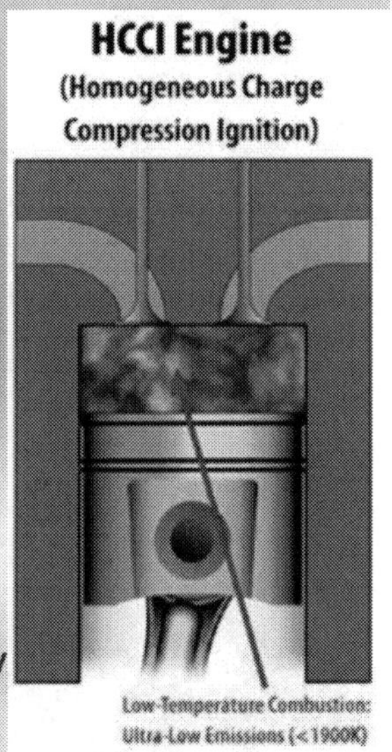


Source: https://www-pls.llnl.gov/?url=science_and_technology-chemistry-combustion

23

# Applications for HCCI Engine Combustion

FREA

**Objectives of HCCI Engine Simulation: Clarify unclear Issues come from the HCCI engine experiments conducted by SNL**

- The chemistry details of ITHR are currently unclear
- Whether the resultant ITHR enhancement is due to the increased pressure and/or decreased temperature is unclear
- How in-cylinder flow and temperature distribution influence heat release rate profiles of HCCI engine combustion is unclear

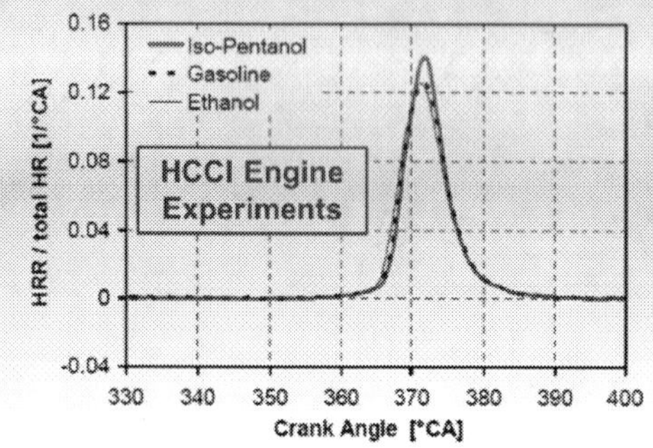


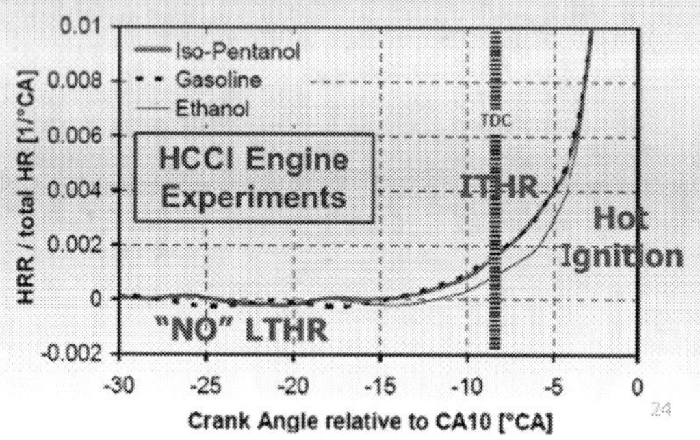


24

# Conditions for HCCI Engine Experiments and Simulations

FREA

**HCCI Engine @CRF SNL**

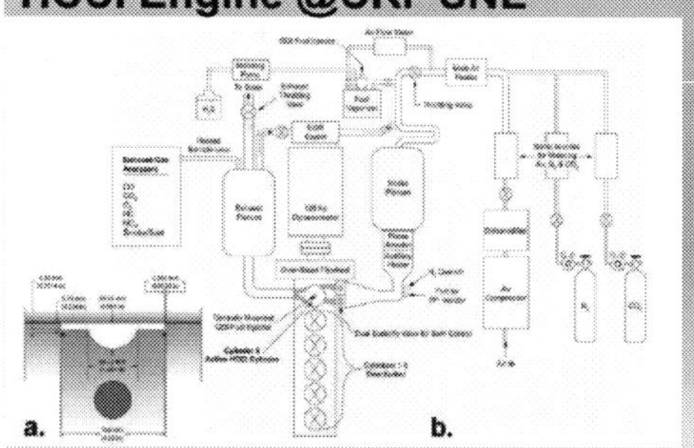


**FORTÉ@AIST, LLNL**

**HCCI Engine Experiments**

Displacement (single-cylinder) ......... 0.981 liters
Geometric Compression Ratio ................. 14:1
Fueling system .......................... Fully Premixed
Engine Speed.....................................1200 rpm
Intake Temperature..........................60 – 189° C
Intake Pressure (abs.) .................100 – 200 kPa
EGR rate ............................................ 0 – 53 %
Coolant Temperature .............................100° C

**HCCI Engine Simulation**

Displacement (single-cylinder) ..........0.981 liters
Geometric Compression Ratio ................. 14:1
Fueling system .......................... Fully Premixed
Engine Speed.....................................1200 rpm
Temperature @BDC..Variable to align with CA10
Intake Pressure (abs.) .................100 – 200 kPa
EGR rate ............................................ 0 – 53 %
Constant Temperature..................200(Piston),
...........................150(Head), 150(Liner) ° C

25

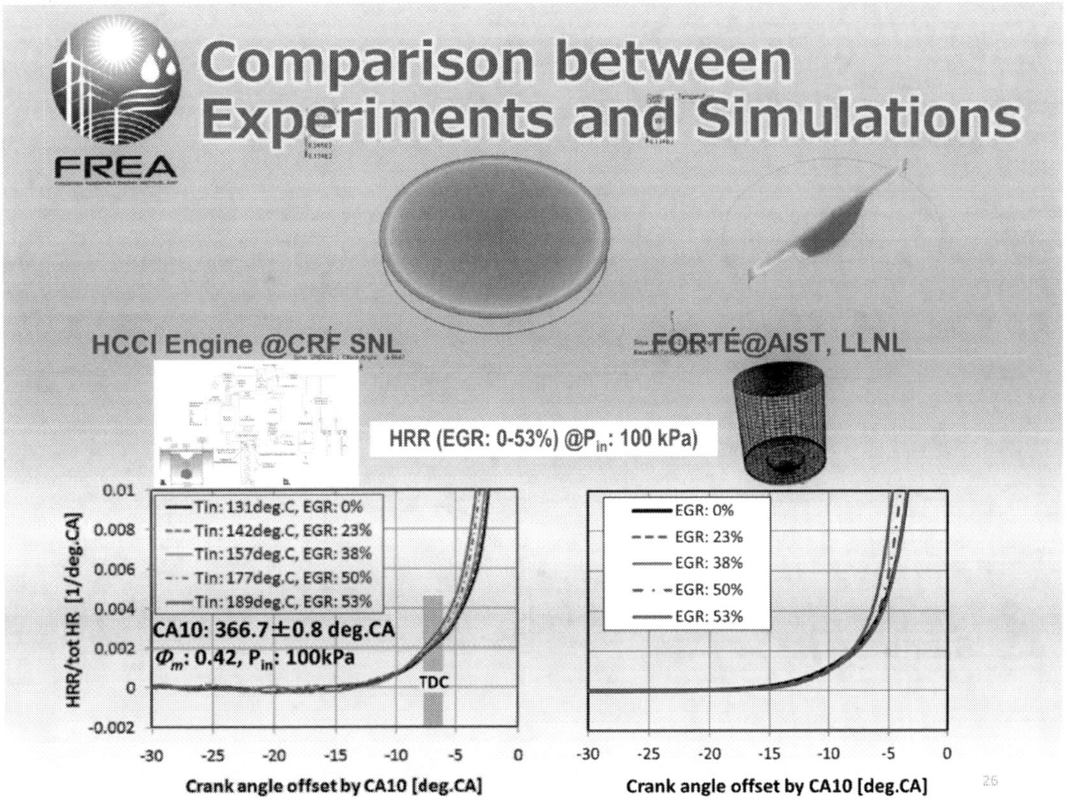

Comparison between Experiments and Simulations
FREA
HCCI Engine @CRF SNL
FORTÉ@AIST, LLNL
HRR (EGR: 0-53%) @Pin: 100 kPa)
Tin: 131deg.C, EGR: 0%
Tin: 142deg.C, EGR: 23%
Tin: 157deg.C, EGR: 38%
Tin: 177deg.C, EGR: 50%
Tin: 189deg.C, EGR: 53%
CA10: 366.7±0.8 deg.CA
Φm: 0.42, Pin: 100kPa
TDC
HRR/tot HR [1/deg.CA]
0.01
0.008
0.006
0.004
0.002
0
-0.002
-30 -25 -20 -15 -10 -5 0
Crank angle offset by CA10 [deg.CA]
EGR: 0%
EGR: 23%
EGR: 38%
EGR: 50%
EGR: 53%
-30 -25 -20 -15 -10 -5 0
Crank angle offset by CA10 [deg.CA]
26


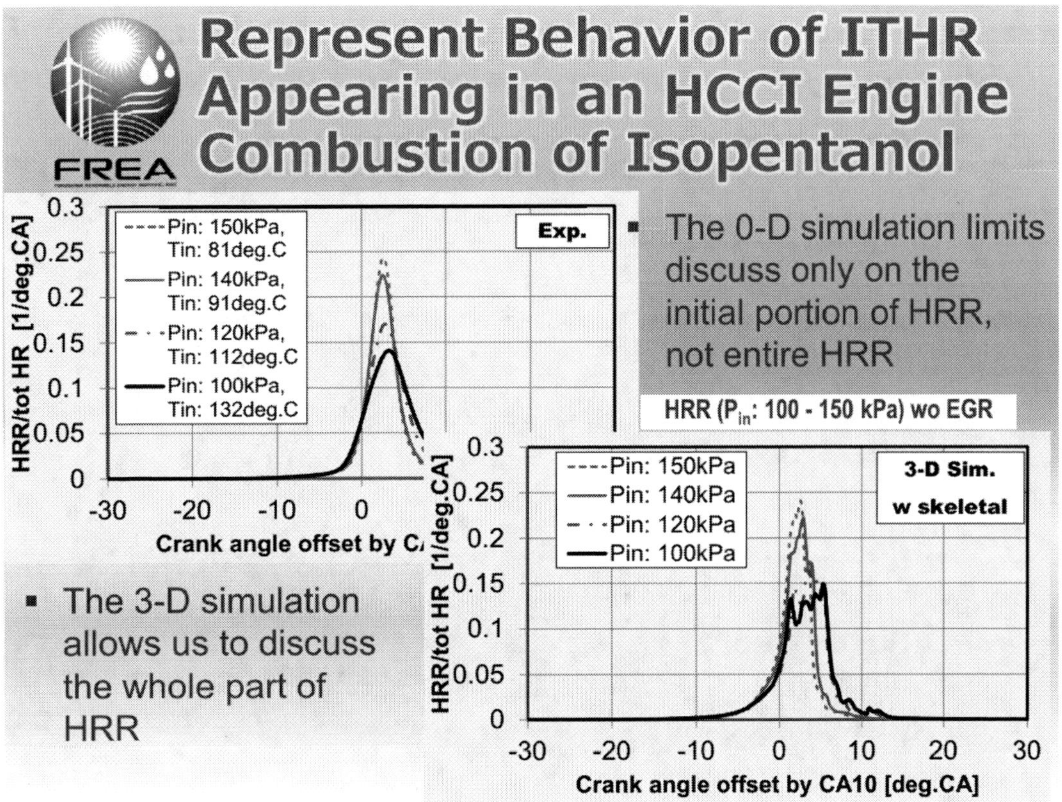

Represent Behavior of ITHR Appearing in an HCCI Engine Combustion of Isopentanol
FREA
Pin: 150kPa, Tin: 81deg.C
Pin: 140kPa, Tin: 91deg.C
Pin: 120kPa, Tin: 112deg.C
Pin: 100kPa, Tin: 132deg.C
Exp.
HRR/tot HR [1/deg.CA]
0.3
0.25
0.2
0.15
0.1
0.05
0
-30 -20 -10 0
Crank angle offset by C
The 0-D simulation limits discuss only on the initial portion of HRR, not entire HRR
HRR (Pin: 100 - 150 kPa) wo EGR
The 3-D simulation allows us to discuss the whole part of HRR
Pin: 150kPa
Pin: 140kPa
Pin: 120kPa
Pin: 100kPa
3-D Sim.
w skeletal
HRR/tot HR [1/deg.CA]
0.3
0.25
0.2
0.15
0.1
0.05
0
-30 -20 -10 0 10 20 30
Crank angle offset by CA10 [deg.CA]

# Overview

FREA

- ☐ Motivation for using CFD simulation of engine combustion
- ☐ Policies on using CFD codes and models
- ☐ Applications for dual fuel engine combustion (Natural Gas/Diesel)
- ☐ Applications for HCCI engine combustion
- ☐ **Summaries**

28

# Summaries

FREA

- CFD simulations with chemical kinetics are conducted to reproduce advanced engine combustion which is more complicated than conventional one
- CFD simulation code becomes a strong tool, but is not good at everything.
- Users should understand essentials of physical and chemical models and what is known or unknown in experiments
- According to simulation results, FORTÉ with originally developed chemical kinetic model works very well
- FORTÉ can reproduce various patterns of natural gas/diesel dual fuel combustion and slight heat release of HCCI combustion

29

FREA

FUKUSHIMA RENEWABLE ENERGY INSTITUTE, AIST

Automotive Simulation World Congress 2014

# Numerical Investigation of Soot Formation in Gasoline Direct Injection Engines

Jun Hashimoto
Yoshito Nouno, Daiki Takahashi, Kimitoshi Tanoue
Oita University, Japan
Yasuo Moriyoshi
Chiba University, Japan

OITA UNIVERSITY   CHIBA UNIVERSITY

Center for Power Source Research for Next-Generation Mobility
http://engine50.tm.chiba-u.jp/mobiltyPSRC/

## PM Emissions from Direct Injection Gasoline Engines

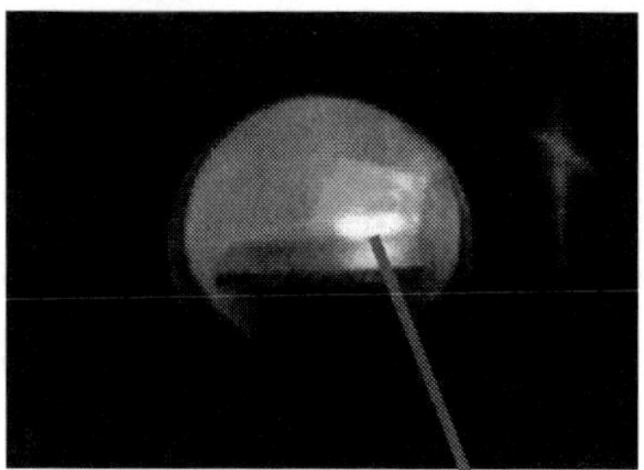

DI Gasoline Engines
PM number limit
$6 \times 10^{11}$[Num./km], Sep. 2014-

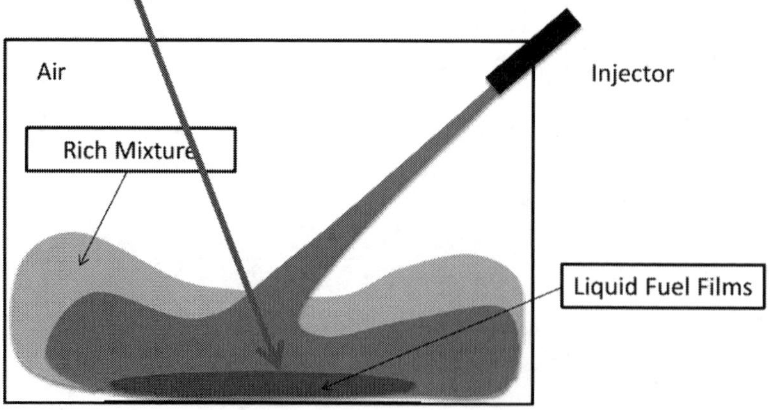

## Objectives and Outline

- Objectives
  - To examine the combustion of the liquid fuel films attached on wall surfaces during the cold startup period by using the numerical analysis
  - To investigate an optimal model to predict particle number emissions
- Outline
  - Computational Method
  - Results and Discussion
    - Combustion characteristics of the liquid fuel films under DI gasoline engine like conditions
    - Prediction of particle number emissions
  - Summary

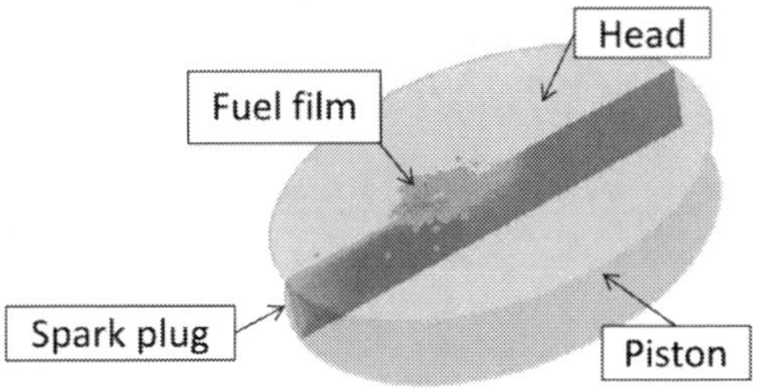


## Computational Model

Rapid Compression and Expansion Machine

| | |
|---|---|
| Bore [mm] | 80.3 |
| Strok [mm] | 95.0 |
| Compression Ratio [-] | 10.5 |
| Displacement [L] | 481.1 |
| Intake Temperature [K] | 300 |
| Intake Pressure [MPa] | 0.101 |
| Engine Speed [rpm] | 900 |
| Ignition Timing [deg. ATDC] | -15 |

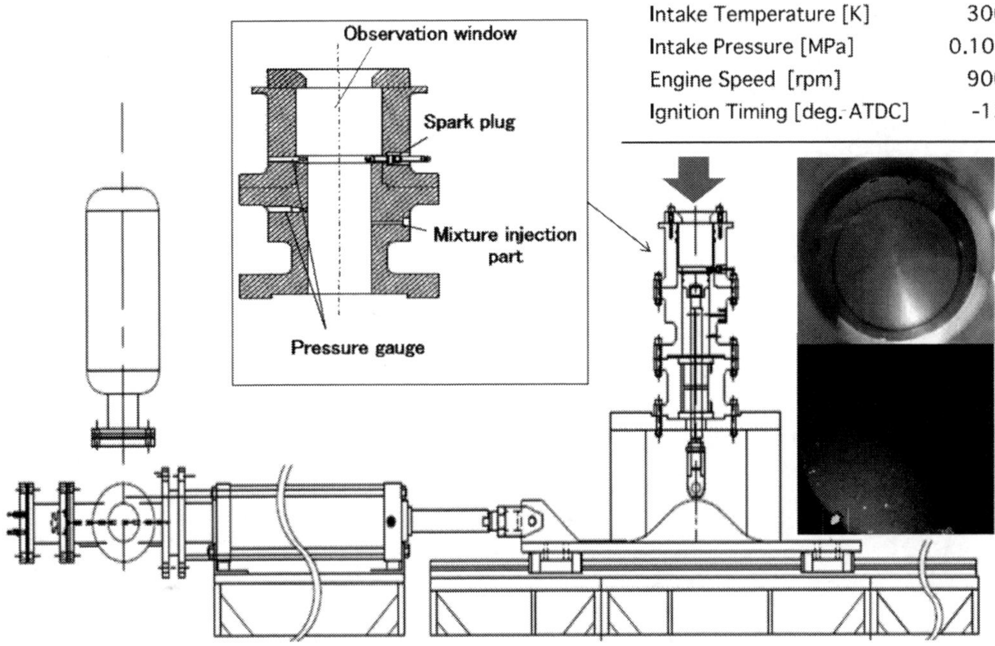

## Computational Model

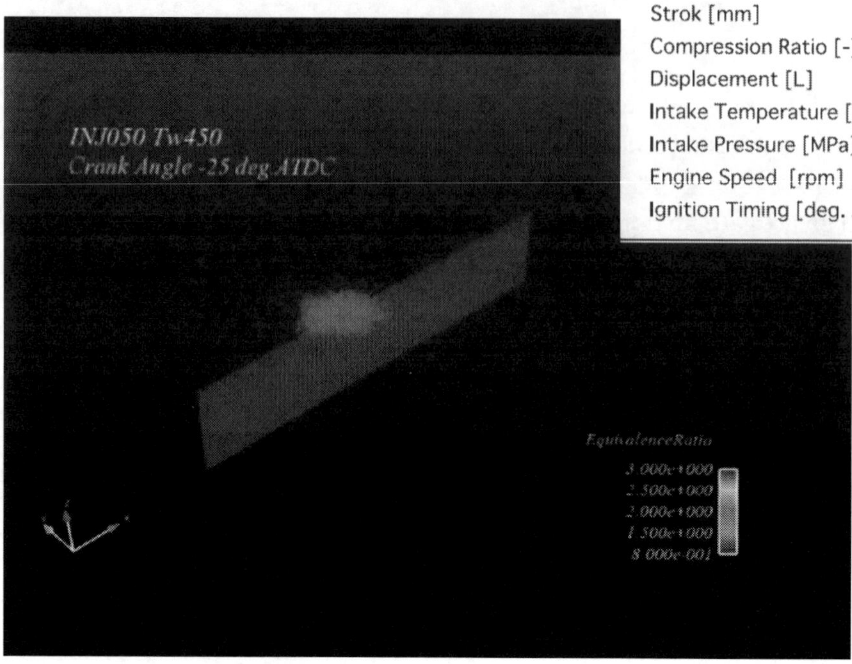


| | |
|---|---|
| Bore [mm] | 80.3 |
| Strok [mm] | 95.0 |
| Compression Ratio [-] | 10.5 |
| Displacement [L] | 481.1 |
| Intake Temperature [K] | 300 |
| Intake Pressure [MPa] | 0.101 |
| Engine Speed [rpm] | 900 |
| Ignition Timing [deg. ATDC] | -15 |

Cell count
110,000 @ TDC
200,000 @ BDC

Ignition Energy
20 J/s * 6deg @ 900 rpm = 20 mJ/st.

## Computational Method

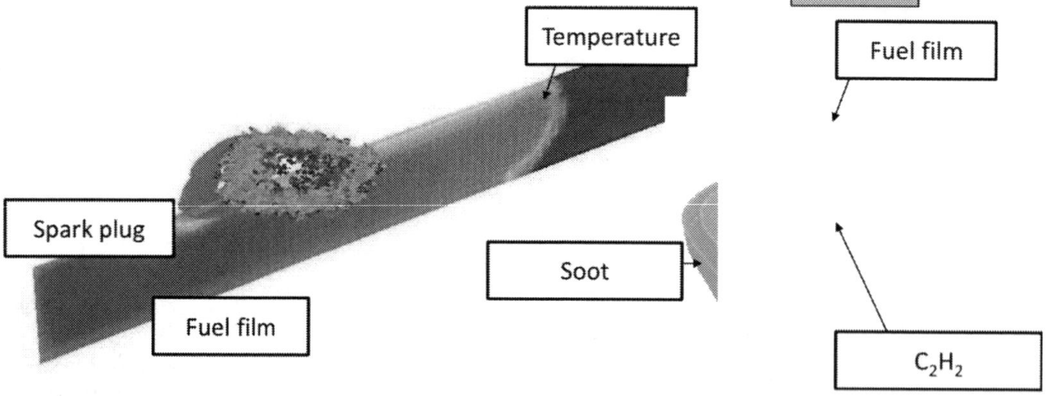


| Fuel | *iso*-Octane | | |
|---|---|---|---|
| | | | 1950 |
| N2 | 0.72026 | 0.72138 | 0.72251 |
| SOI[deg] | | 180 | |
| DOI[deg] | 1.43855 | 2.87711 | 4.31566 |

| Wall Temperature[K] |
|---|
| 350 |
| 400 |
| 450 |
| 500 |

## Computational Method

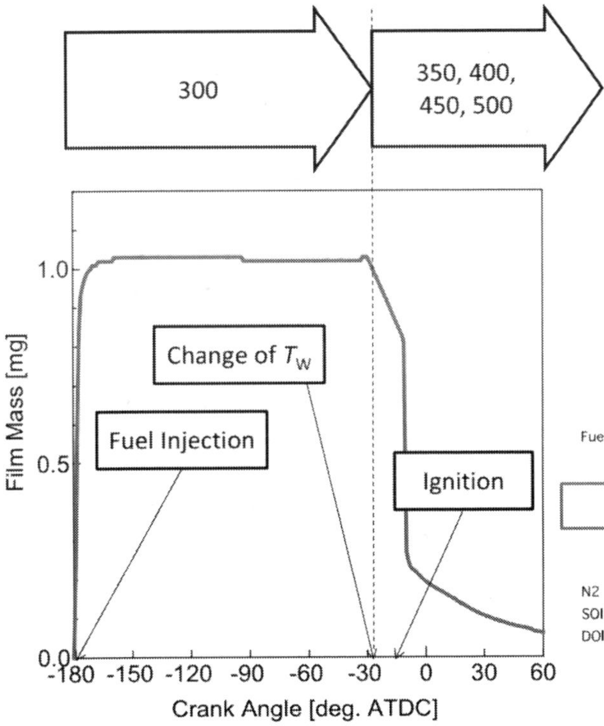


| Wall Temperature[K] |
|---|
| 350 |
| 400 |
| 450 |
| 500 |

| | | | |
|---|---|---|---|
| Fuel | *iso*-Octane | | |

| | | | |
|---|---|---|---|
| | | | 1950 |
| N2 | 0.72026 | 0.72138 | 0.72251 |
| SOI[deg] | | 180 | |
| DOI[deg] | 1.43855 | 2.87711 | 4.31566 |

## Computational Method

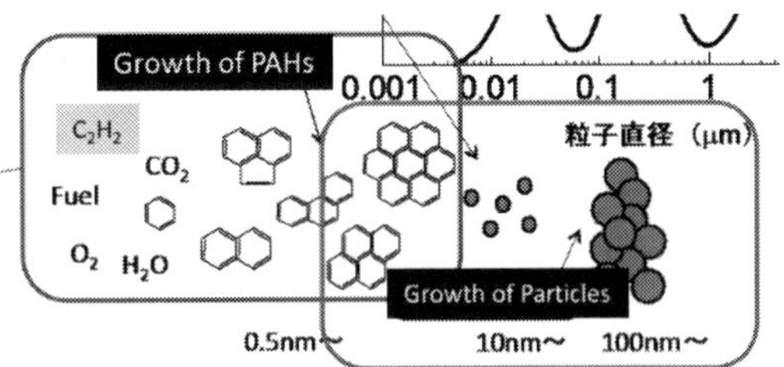


- Gas phase reaction
  - KAUST model, RAJ et al.
  - Gasoline surrogate fuel
    - *iso*-Octane, *n*-Heptane, *n*-Decane, Toluene
  - 231 species and 1350 reactions
  - PAHs growth reactions up to coronene ($A_7$, $C_{24}H_{12}$)
- Soot model
  - Phenomenological model

$$\frac{dM_s}{dt} = \frac{dM_{\mathrm{sf}}}{dt} - \frac{dM_{\mathrm{so}}}{dt}$$

$$\frac{dM_{\mathrm{sf}}}{dt} = K_f M_{\mathrm{pre}} \qquad \frac{dM_{\mathrm{so}}}{dt} = \frac{6MW_c}{\rho_s D_s} M_s R_{\mathrm{total}}$$

$T_W$=450K, Injected Fuel Ratio=5%

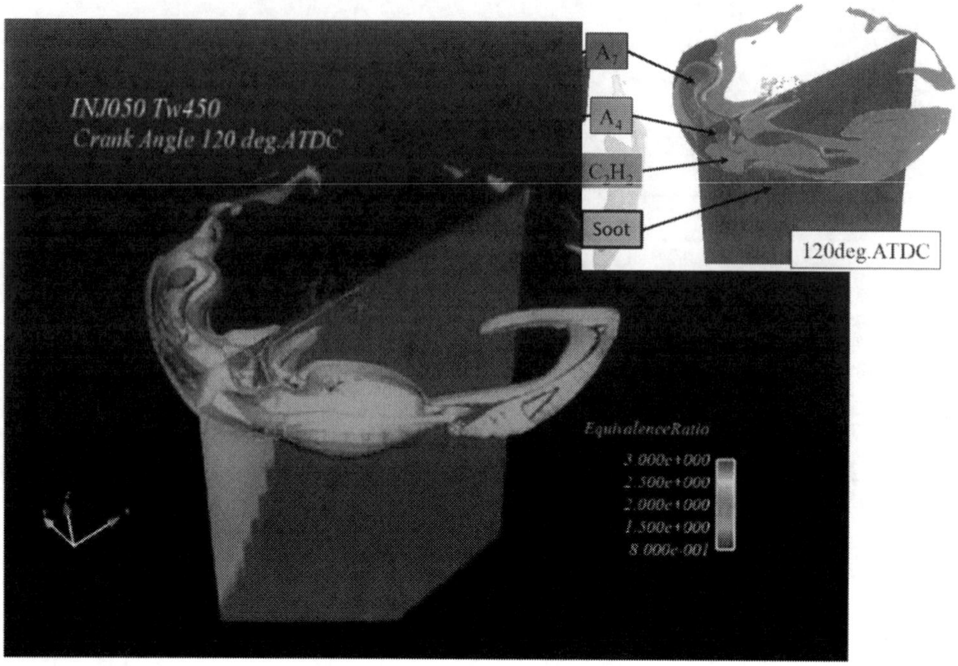


## Crank angle histories of pressure, temperature, CHRR and soot

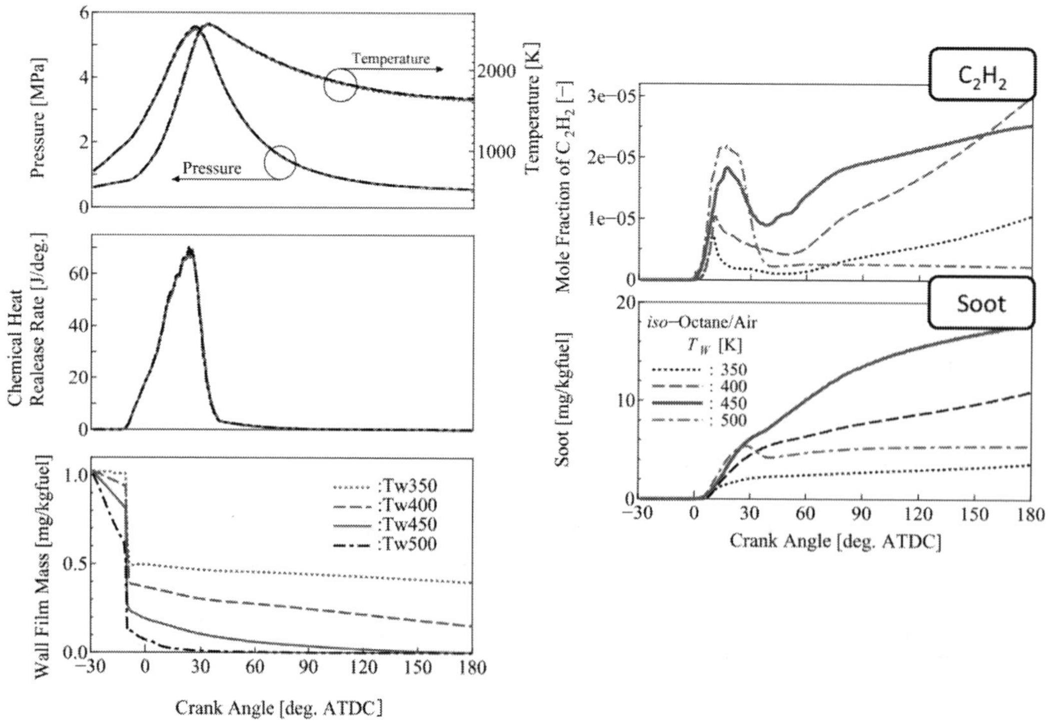

## Emission factors of Unburned Hydrocarbons

ONLY Evaporation characteristics of fuel deposits are changed

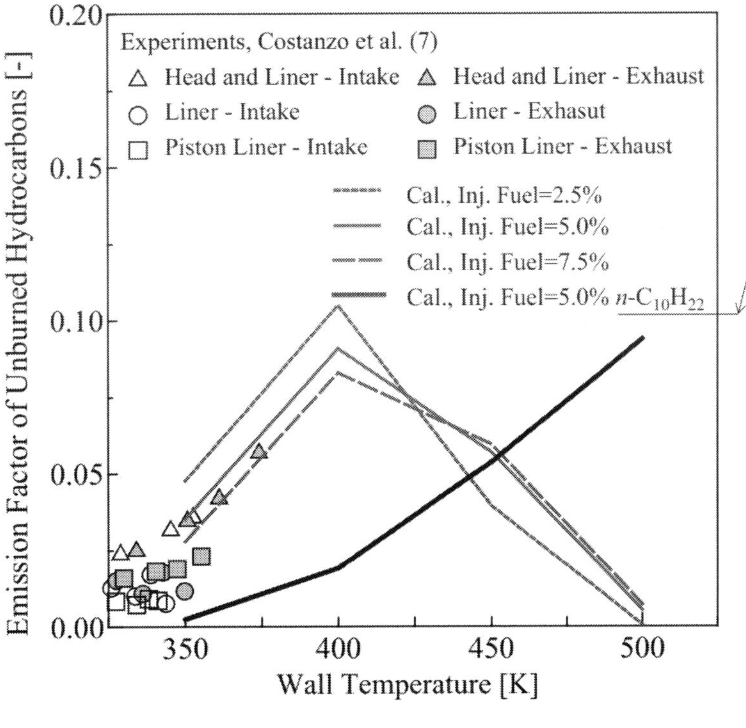


## Residual ratio of the liquid fuel film mass

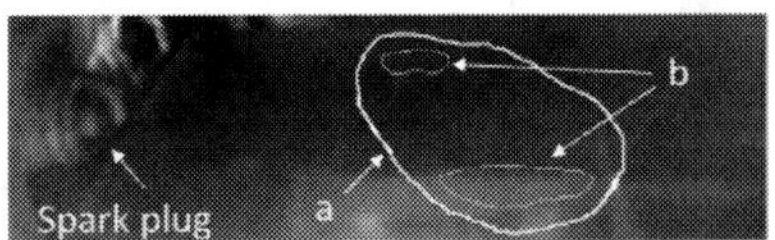


(a: before combustion, b: after combustion)

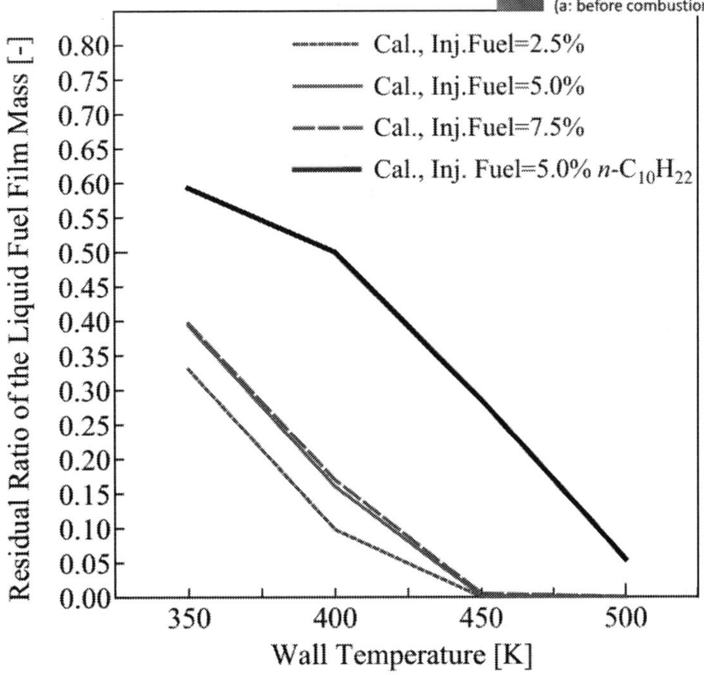

## Emission factor of the soot mass

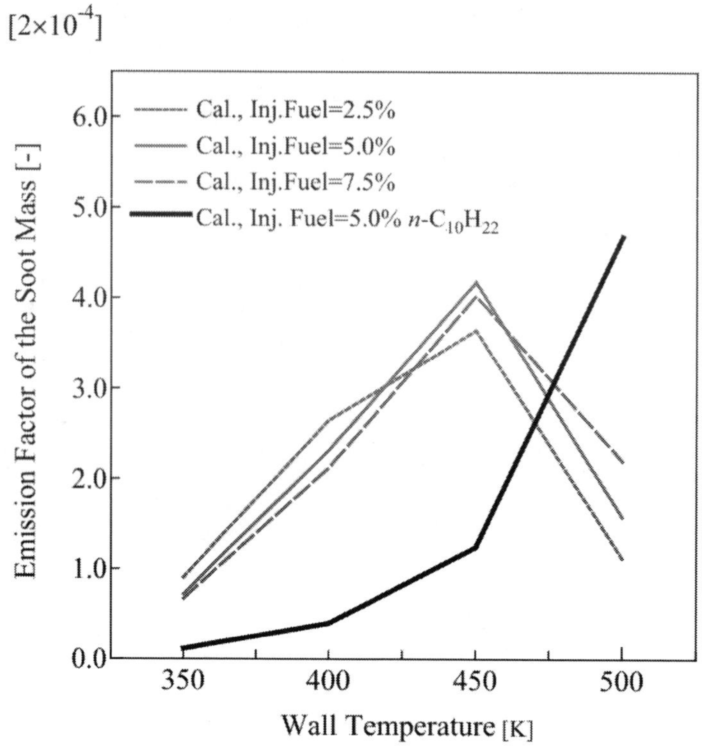


## A simple prediction of soot and UHC mass during cold startup period

- Wall temperature
  - increases in proportion to the time
  - The rate of temperature rising is 150K/600s
- Liquid fuel film mass
  - Amount of the fuel adhering to the head in each cycle is 1mg
  - Total amount of the liquid fuel film mass is sum of the remaining fuel mass and new deposits
- Soot and UHC
  - Calculated from the liquid fuel mass and emission factors

## A simple prediction of soot and UHC mass during CS period

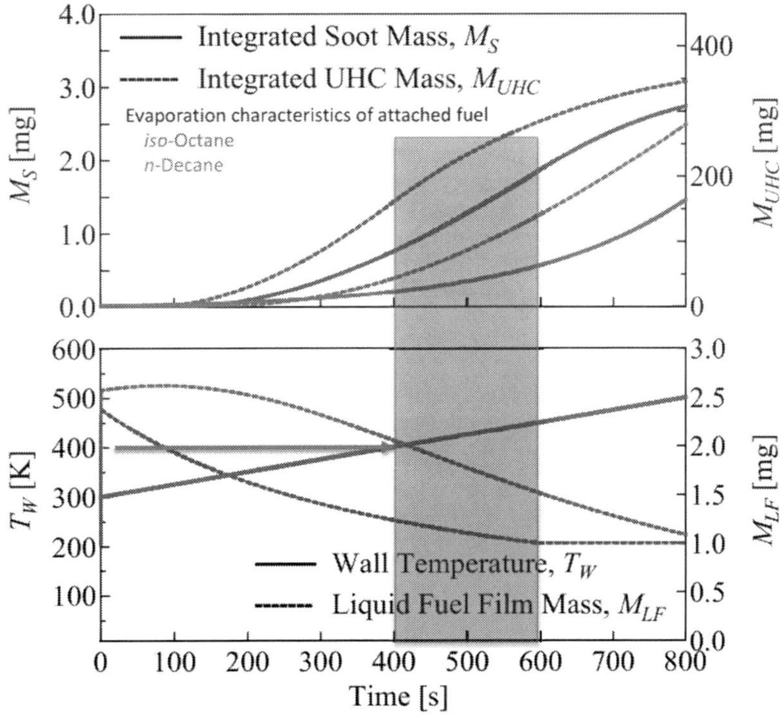


## Crank angle histories of the average mole fraction of precursors

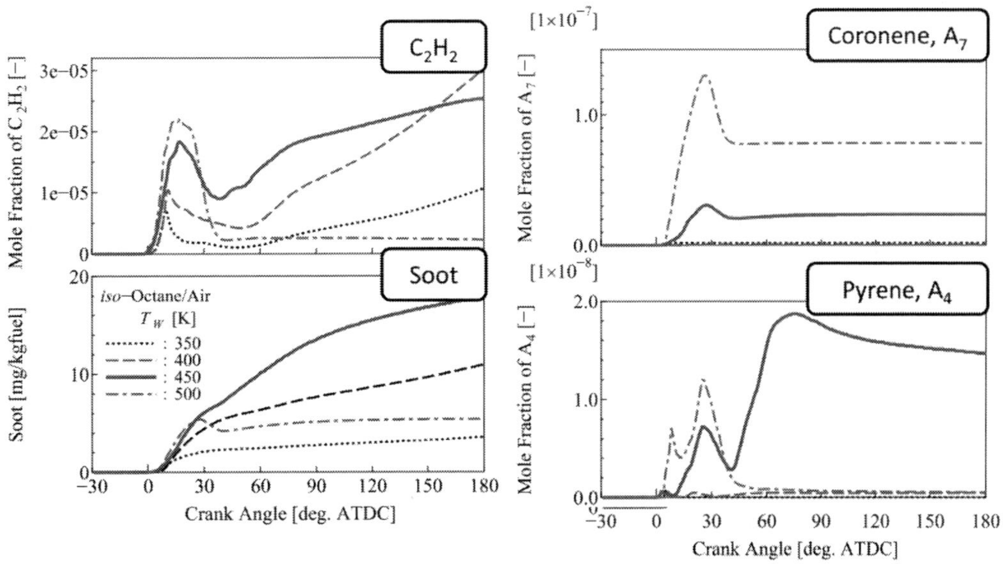

## Precursor of the particulate matter

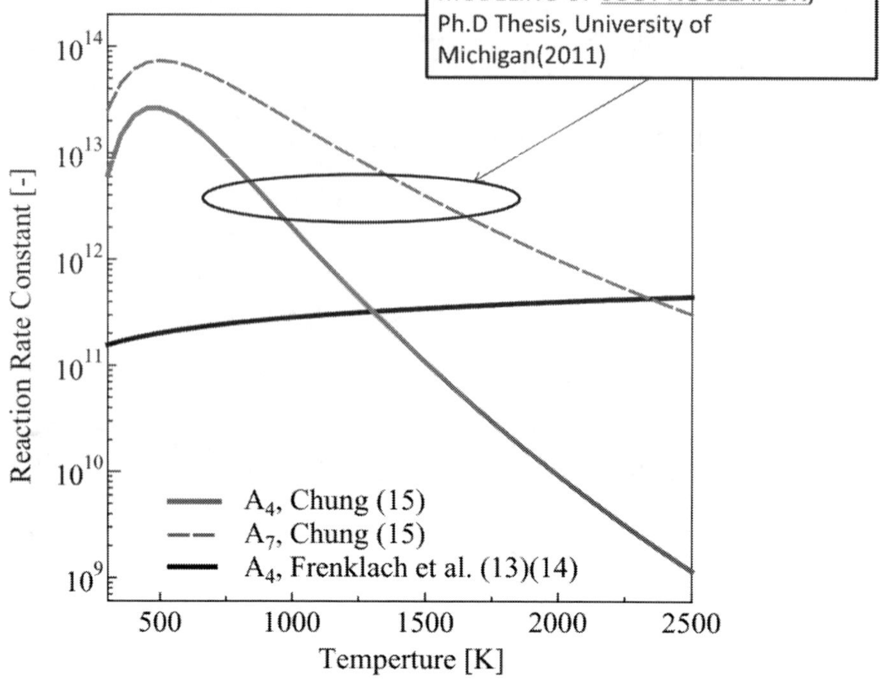


## Surface reactions for Method of Moments

| Reaction | *A* [mole-cm-s] (15) | *n* [-] (15) | *E* [cale/mole] (15) | |
|---|---|---|---|---|
| ! nucleation from PAH | | | | |
| 2A4 => 32C(B) + 20 H(se) + 28.72 open(se) | 1.370E+50 | -11.800 | 11130.0 | Chung (15) |
| 2CORONEN => 48C(B) + 24 H(se) + 39.87 open(se) | 1.360E+35 | -6.785 | 6757.0 | Chung (15) |
| ! HACA growth | | | | |
| H + H(se) = open(se) + H2 | 4.2E12 | 0.0 | 13000.0 | 1/10 |
| open(se) + H => H(se) | 2.0E13 | 0.0 | 0.0 | |
| H(se) + OH = H2O + open(se) | 1.0E9 | 0.734 | 1430.0 | 1/10 |
| open(se) + C2H2 => H(se) + 2C(B) + H | 3.00E+05 | 0.78 | 3800.0 | Murakami et al. (16) |
| ! PAH condensation | | | | |
| A1 + 6H(se) => 6C(B) + 6open(se) + 6H2 | 0.2E-1 | 0.0 | 0.0 | 1/10 |
| A1C2H + 6H(se) => 8C(B) + 6open(se) + 6H2 | 0.21E-1 | 0.0 | 0.0 | 1/10 |
| A2 + 16H(se) => 10C(B) + 16open(se) + 12H2 | 0.1E-1 | 0.0 | 0.0 | 1/10 |
| A2R5 + 16H(se) => 10C(B) + 16open(se) + 11H2 + C2H2 | 0.1E-1 | 0.0 | 0.0 | 1/10 |
| A3 + 20H(se) => 14C(B) + 20open(se) + 15H2 | 0.1E-1 | 0.0 | 0.0 | 1/10 |
| A4 + 20H(se) => 16C(B) + 20open(se) + 15H2 | 0.1E-1 | 0.0 | 0.0 | 1/10 |
| ! Oxidation | | | | |
| OH + open(se) + C(B) => CO + H(se) | 1.8 | 0.0 | 6000. | Neoh |

## Spatial distributions of particle volume fraction and precursors

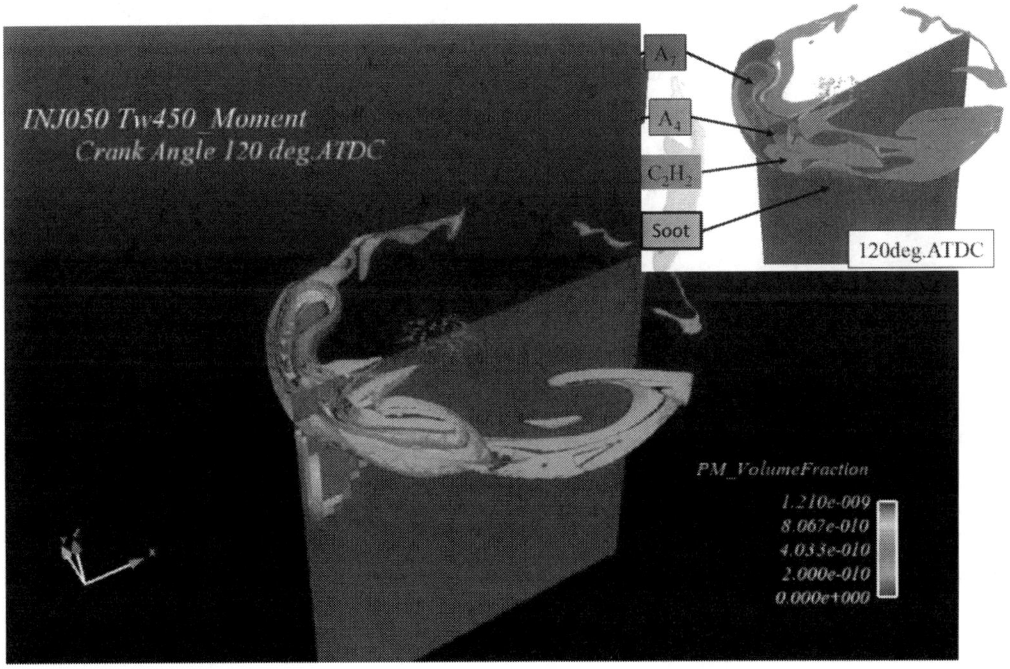


## Prediction of soot by using the method of moments

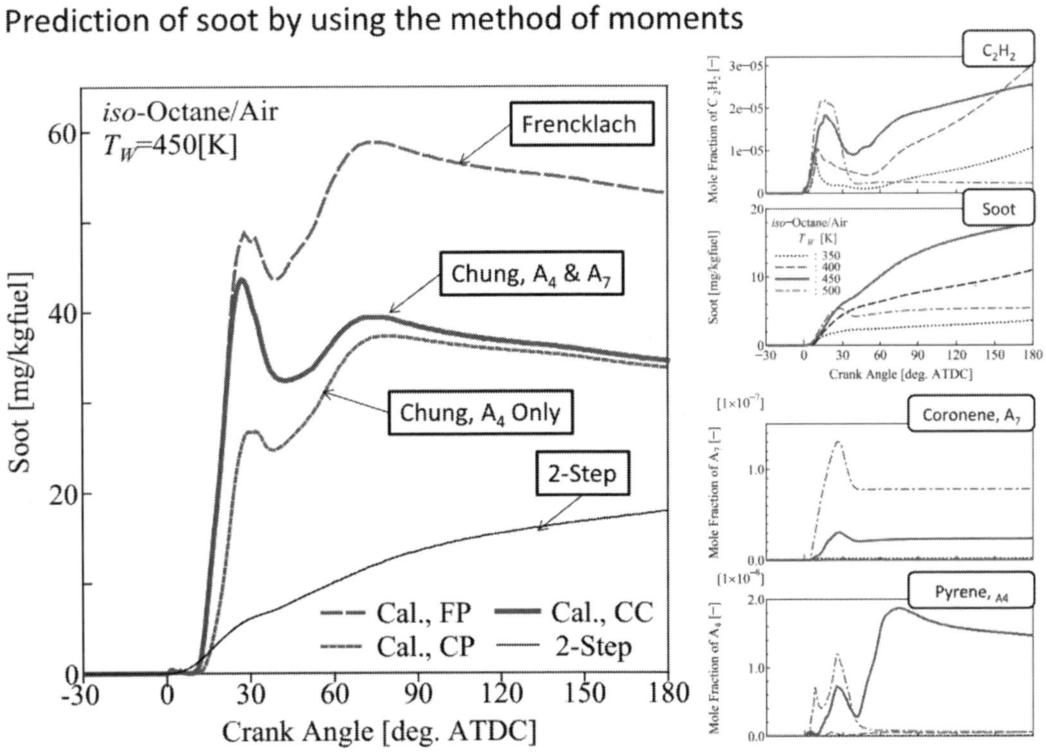

Crank angle histories of the particle number and mean average diameter

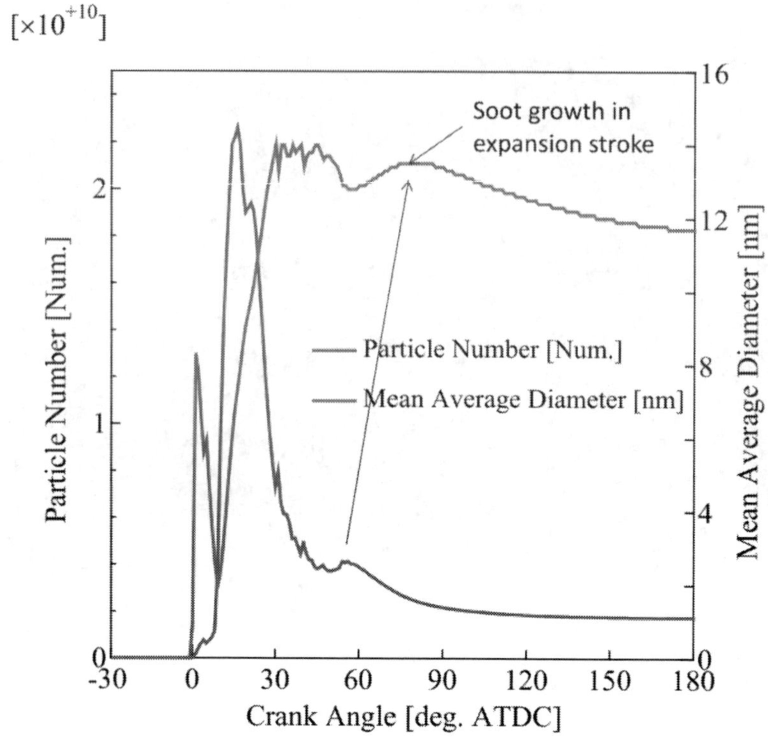


Crank angle histories of particle size distribution

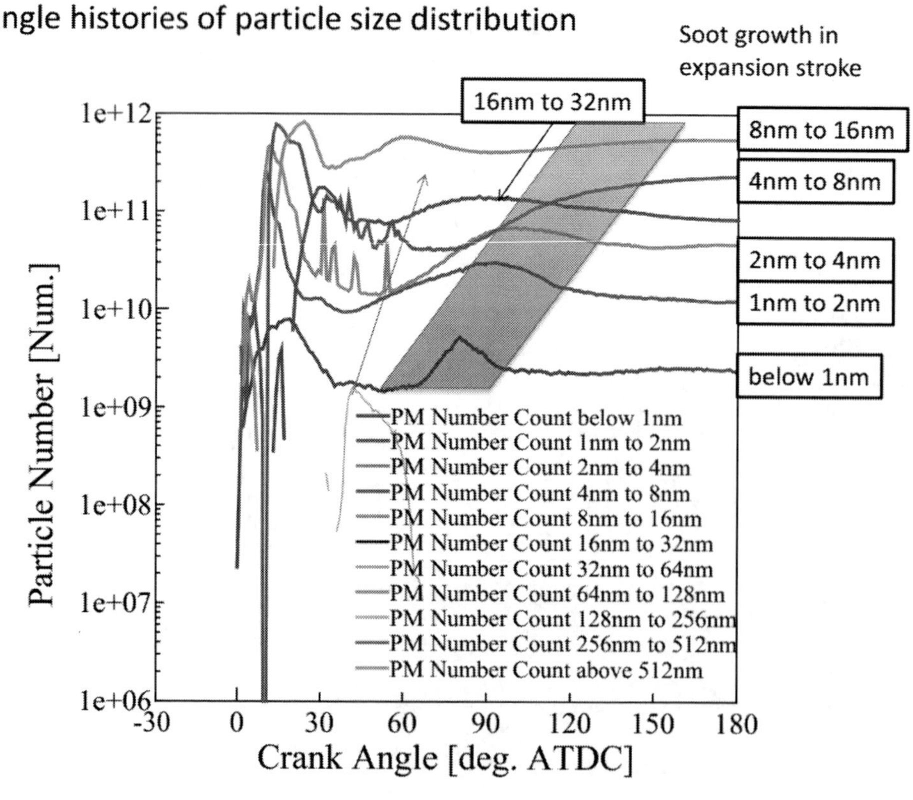

## Summary

- liquid fuel films are emitted as one of followings
  - remaining fuel films
  - unburned hydrocarbons
  - precursors ($C_2H_2$, PAHs..)
  - soot
- Emission factors of UHC and soot can be expressed as a function of the wall temperature
- A surface reaction mechanism including pyrene and coronene as precursors was constructed based on the literature
- Particle number emissions generated from combustion of liquid fuel films during cold startup period were estimated by using the method of moments

2014 Automotive Simulation World Congress

# A Study of Autoignition in an IC Engine by using Numerical Calculation and Optical Measurements

Akira Iijima and Tomomi Miyasaka
Nihon University

2014 Automotive Simulation World Congress

## Contensts

1. Analysis of SI-Knocking by using Optical Engine and **FORTÉ**

2. Analysis of HCCI Reactions by using Spectroscopic Measurement and **CHEMKIN**

## Background

2014 Automotive Simulation World Congress

**Fuel consumption reduction in gasoline vehicle**

✓Increasing Compression Ratio
✓Downsizing (i.e. Boosting)
} **Occurrence of Knocking**

Autoignition
- Low temperature reaction
- HCHO (formaldehyde)

➢ **Requires elucidation of abnormal combustion mechanism**

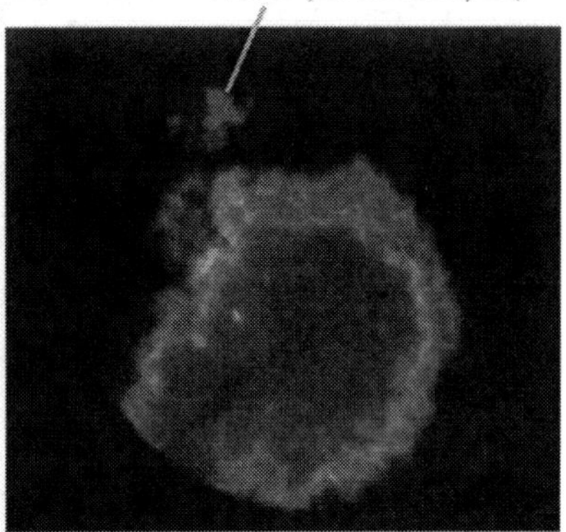

❑ **Approach**

- ☐ Experiment by Optical Engine
- ☐ In-cylinder spectroscopy
- ☐ **Combustion analyses by FORTÉ**



## Optically Accessible Engine

2014 Automotive Simulation World Congress

| Element Resolution | 512 × 128 pixels |
|---|---|
| Frame Rate | 40,000 frames / sec. |

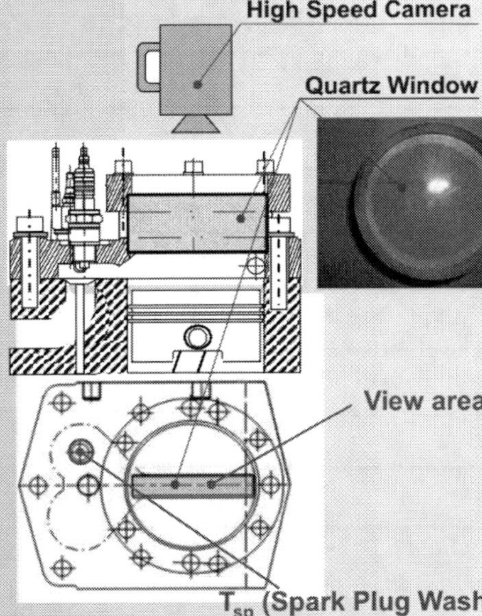


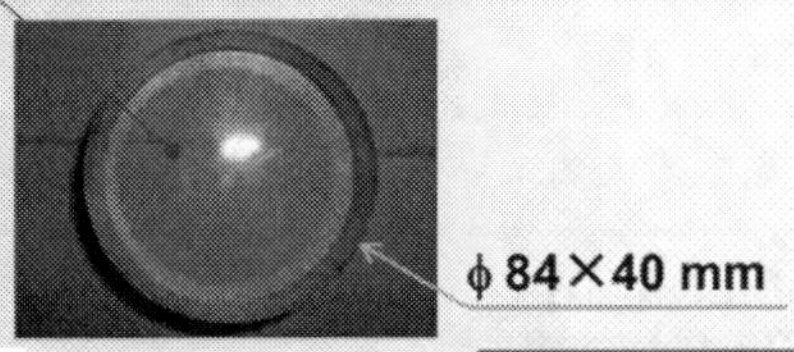

## Spectroscopic Measurement

2014 Automotive Simulation World Congress

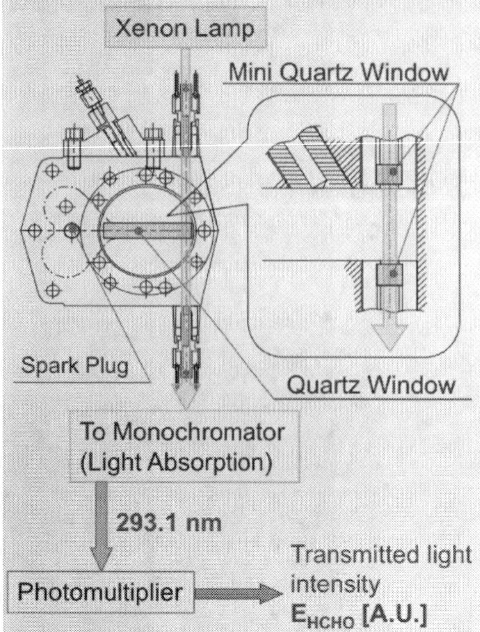


**Absorbance : $A_{HCHO}$**

$$A_{HCHO}\ [-] = \frac{E_{HCHO(AVG)} - E_{HCHO}}{E_{HCHO}}$$

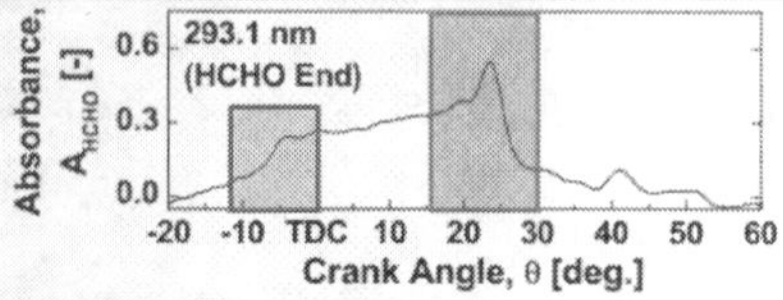


$E_{HCHO(AVG)}$ [A.U.] :
Transmitted light intensity before low-temperature reaction (reference light intensity)

$E_{HCHO}$ [A.U.] :
Transmitted light intensity at any chosen crank angle



2014 Automotive Simulation World Congress

## Engine Specification (Experimental)

| Four-stroke Single Cylinder Air-cooled Gasoline Engine | |
|---|---|
| Bore × Stroke | 85 × 70 mm |
| Displacement | 397 $cm^3$ |
| Compression Ratio | 8 : 1 |
| Valve Arrangement | Side Valves |
| Ignition Timing (I.T.) | - 5 deg. BTDC |
| Boost Pressure | NA : $P_{b(abs)}$ = 98.0±1.0 kPa |
| Test Fuel | PRF 50 |
| Air-Fuel Ratio | Stoichiometric |

## Calculation Tool -FORTÉ-

2014 Automotive Simulation World Congress

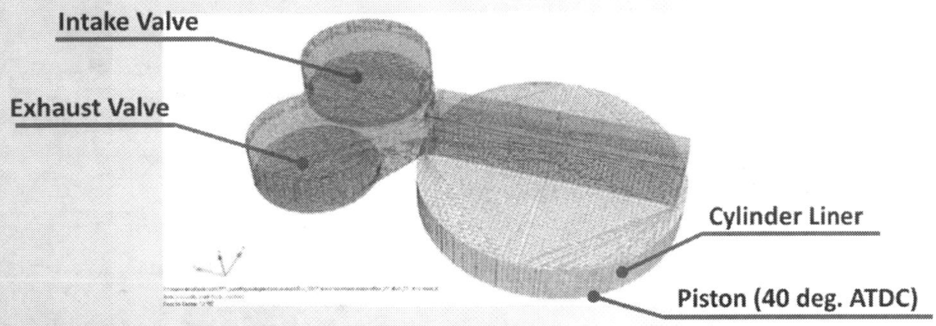


Calculation model

- ✓ Engine CFD with Chemical Reaction
- ✓ Reaction Mechanism for Primary Reference Fuels (PRF)
  - ➤ 178 Species, 1271 Reactions (by Reaction design)
- ✓ Autoignition and Knocking Analysis with Low-temp. Reactions



2014 Automotive Simulation World Congress

## Calculation Conditions

| | |
|---|---|
| Displacement | 397 $cm^3$ |
| Compression Ratio | 8.0 : 1 |
| Ignition Timing (I.T.) | - 5 deg. BTDC |
| Boost Pressure | NA : $P_{b(abs)}$ = 98.0 kPa |
| Test Fuel | PRF 50 |
| Air-Fuel Ratio | Stoichiometric |
| Turbulence Model | RNG k-ε |
| Calculation Range | 40 deg. BTDC ～ 80 deg. ATDC |
| Calculation | 4 Parallel Computation |

# Experimental Result (Knocking)

**Tsp = 475 K $\phi$ = 1.0 N = 1400 rpm**

**Frame Rate: 40,000 frames/sec**

**22 ～ 46 deg.ATDC**

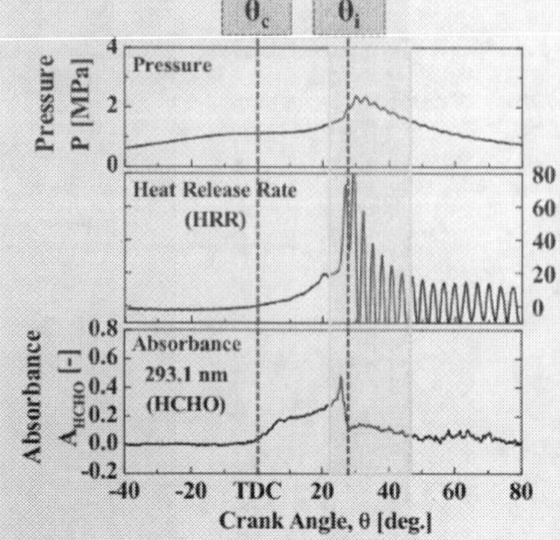


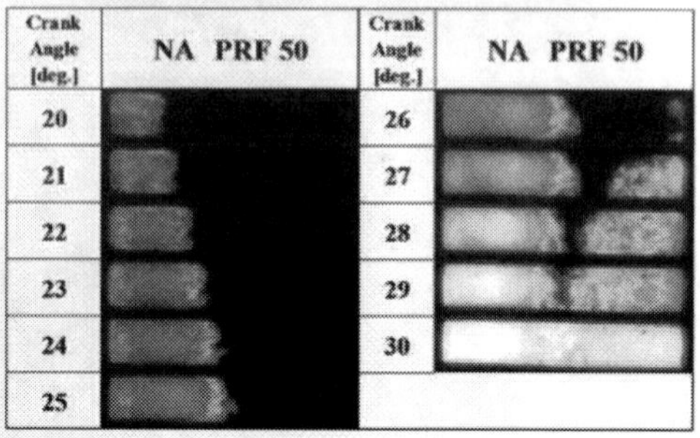


- $\theta_c$: Increase $A_{HCHO}$ → Occurrence of Cool Flame (LTR)
- $\theta_i$ : Decrease $A_{HCHO}$ → Occurrence of Autoignition





# Calculation Results

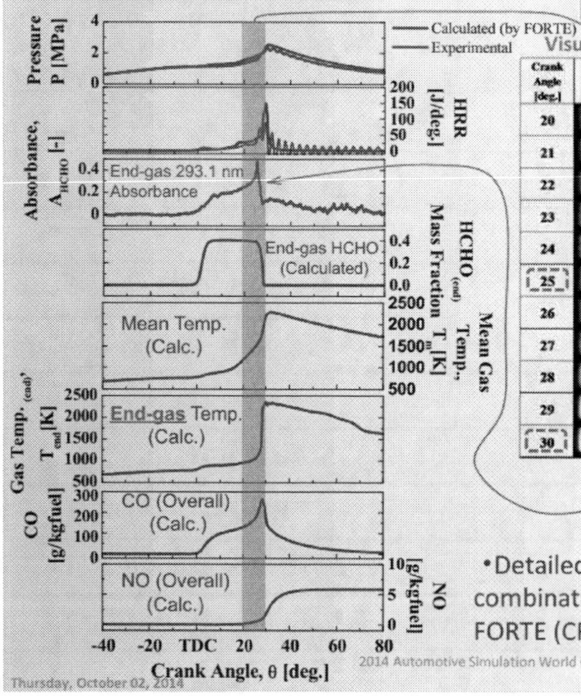


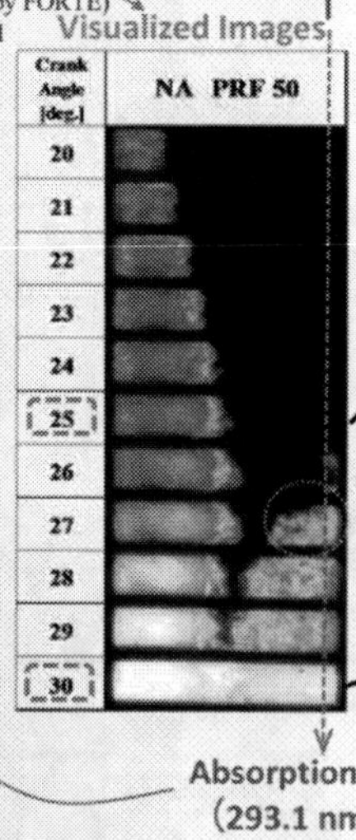


Temp. distribution (Calculated)

25 deg. ATDC

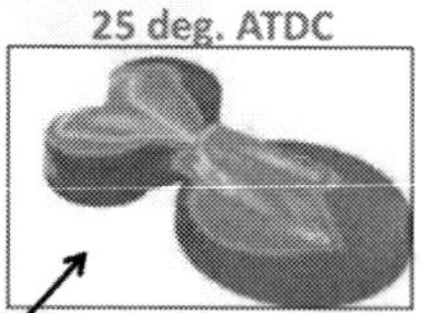

30 deg. ATDC

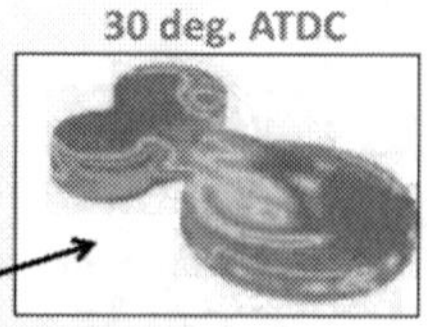

- Detailed analyses are possible by combination use of optical engine and FORTE (CFD)

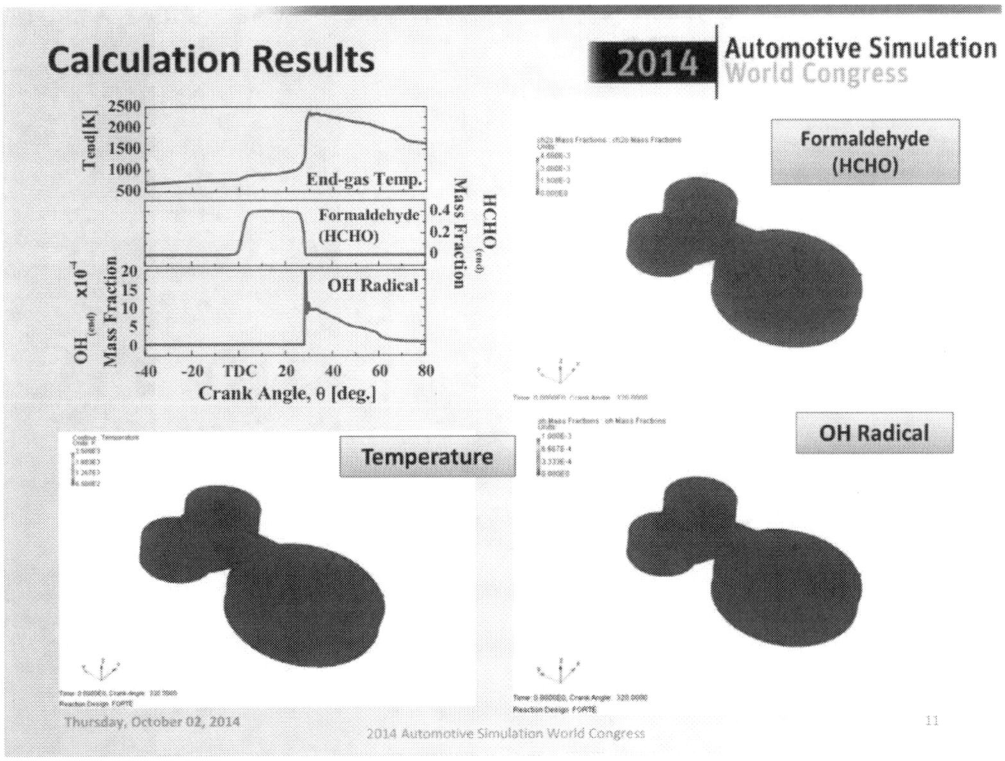
Calculation Results
2014 Automotive Simulation World Congress
Tend[K]
End-gas Temp.
Formaldehyde (HCHO)
HCHO(end) Mass Fraction
OH Radical
OH(end) Mass Fraction x10⁻⁴
Crank Angle, θ [deg.]
Formaldehyde (HCHO)
Temperature
OH Radical
Thursday, October 02, 2014
2014 Automotive Simulation World Congress
11

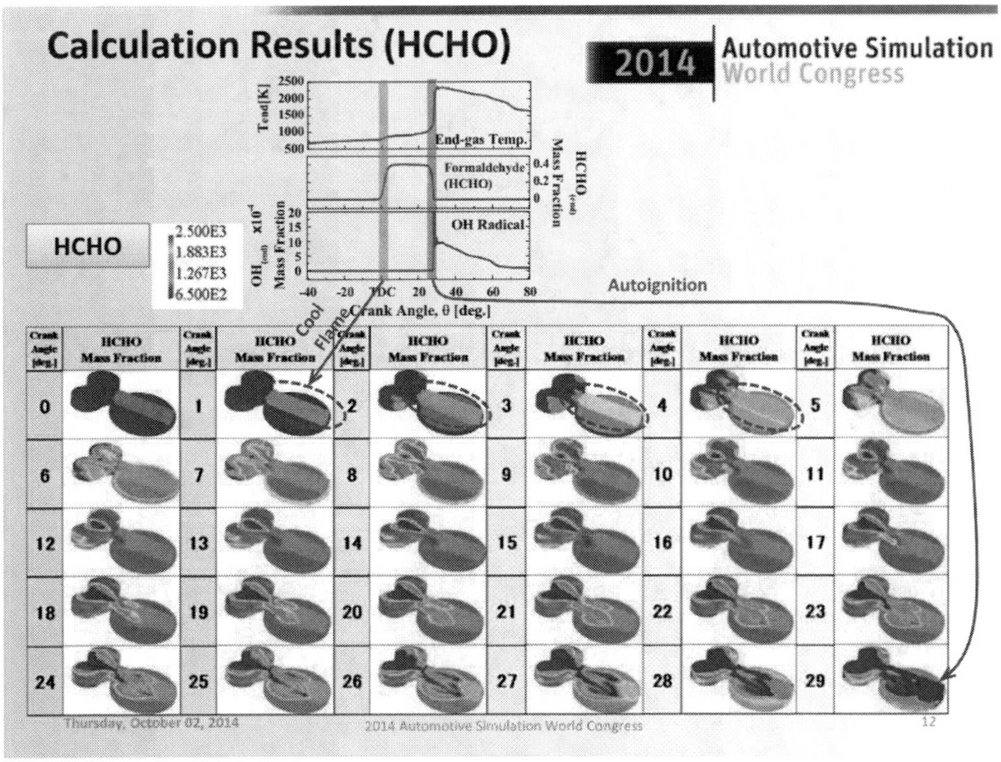
Calculation Results (HCHO)
2014 Automotive Simulation World Congress
Tend[K]
End-gas Temp.
Formaldehyde (HCHO)
HCHO(end) Mass Fraction
OH Radical
OH(end) Mass Fraction x10⁻⁴
Crank Angle, θ [deg.]
HCHO
2.500E3
1.883E3
1.267E3
6.500E2
Cool Flame
Autoignition
Crank Angle [deg.]
HCHO Mass Fraction
0
1
2
3
4
5
6
7
8
9
10
11
12
13
14
15
16
17
18
19
20
21
22
23
24
25
26
27
28
29
Thursday, October 02, 2014
2014 Automotive Simulation World Congress
12

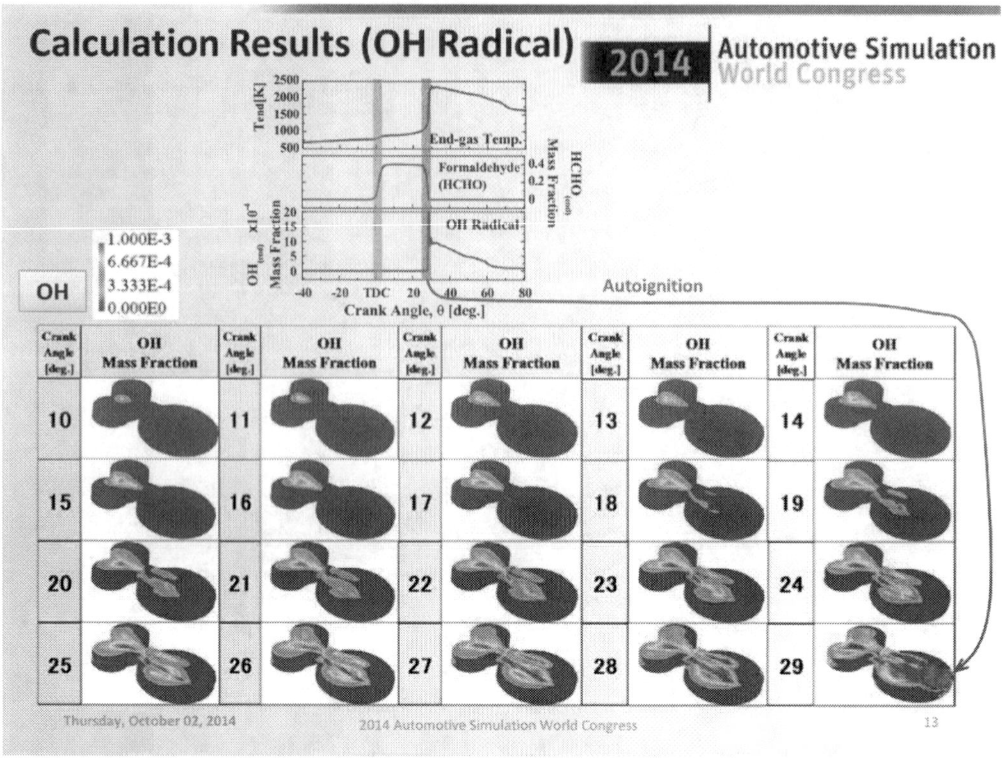
Calculation Results (OH Radical)
2014
Automotive Simulation
World Congress
Tend[K]
End-gas Temp.
Formaldehyde
(HCHO)
HCHO(end) Mass Fraction
OH Radical
OH(end) Mass Fraction x10^-4
Crank Angle, θ [deg.]
TDC
Autoignition
OH
1.000E-3
6.667E-4
3.333E-4
0.000E0
Crank Angle [deg.]
OH Mass Fraction
10 11 12 13 14
15 16 17 18 19
20 21 22 23 24
25 26 27 28 29
Thursday, October 02, 2014
2014 Automotive Simulation World Congress
13

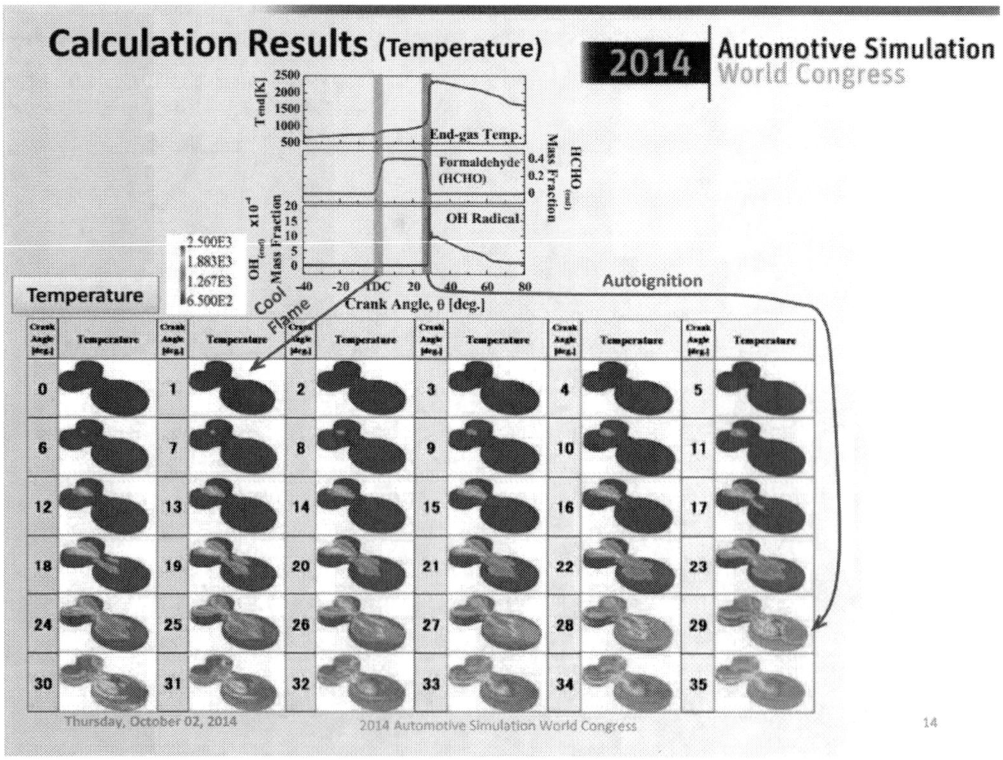
Calculation Results (Temperature)
2014
Automotive Simulation
World Congress
Tend[K]
End-gas Temp.
Formaldehyde
(HCHO)
HCHO(end) Mass Fraction
OH Radical
OH(end) Mass Fraction x10^-4
Crank Angle, θ [deg.]
TDC
Cool Flame
Autoignition
Temperature
2.500E3
1.883E3
1.267E3
6.500E2
Crank Angle [deg.]
Temperature
0 1 2 3 4 5
6 7 8 9 10 11
12 13 14 15 16 17
18 19 20 21 22 23
24 25 26 27 28 29
30 31 32 33 34 35
Thursday, October 02, 2014
2014 Automotive Simulation World Congress
14

2014 Automotive Simulation World Congress

## Conclusions (Part 1)

- Behavior of the low-temperature reaction and autoignition during SI knocking were investigated by using in-cylinder high speed photography, light absorption spectroscopy and CFD analysis by FORTÉ
- Formaldehyde (HCHO) was generated in end-gas by occurrence of cool flame reaction.
- Formaldehyde was rapidly consumed with occurrence of autognition, at the same time, OH radical was rapidly generated.
- Measured HCHO absorption behavior and calculated HCHO mole fraction were consistent. Therefore, combustion analysis by numerical calculation is valid for elucidation of the knocking mechanism in SI engines.



2014 Automotive Simulation World Congress

## Contensts

1. Analysis of SI-Knocking by using Optical Engine and **FORTÉ**

2. Analysis of HCCI Reactions by using Spectroscopic Measurement and **CHEMKIN**

## Why HCCI ?



**HCCI : Homogeneous Charge Compression Ignition**

✓ **Premixture is Introduced into the Cylinder** (like a spark-ignition engine) **and Autoignited by Compression** (like a diesel engine)

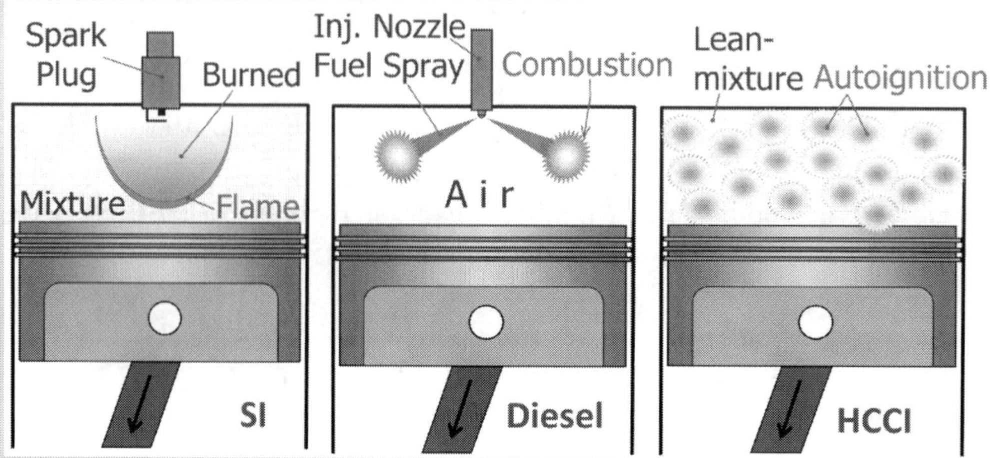




## Combustion Images

2014 Automotive Simulation World Congress

**Visualized Images of SI, SI-Knock and HCCI Combustion**

**1000 rpm**

**SI** **SI-Knock** **HCCI**

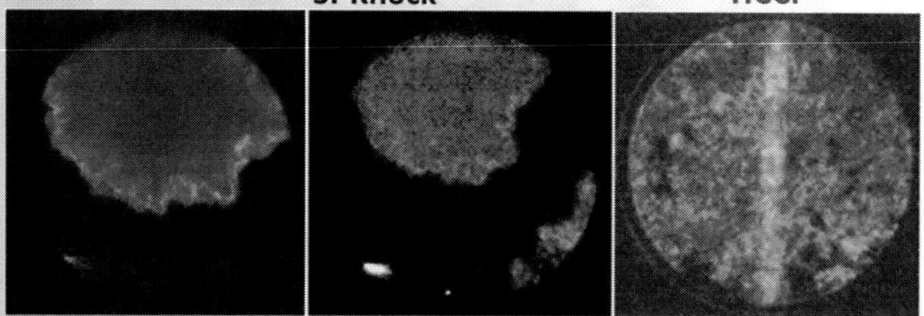

✓ **SI takes place flame propagation**

✓ **In SI-knock, Unburned end-gas autoignited rapidly**

✓ **HCCI combustion initiates at multiple sites by autoignition**

## Background

**2014 Automotive Simulation World Congress**

- **Challenges for HCCI**
  - ✓ Expansion of the Operation Region
    i.e. Ignition Timing Control, **Combustion Rate Control…**
  - ➢ **Necessary to moderate combustion in high load**

**Several approaches**

- ✓ Weak fuel stratified
- ✓ Thermal stratified
- ✓ **Supercharging**
- ✓ **Fuel chemical effect**

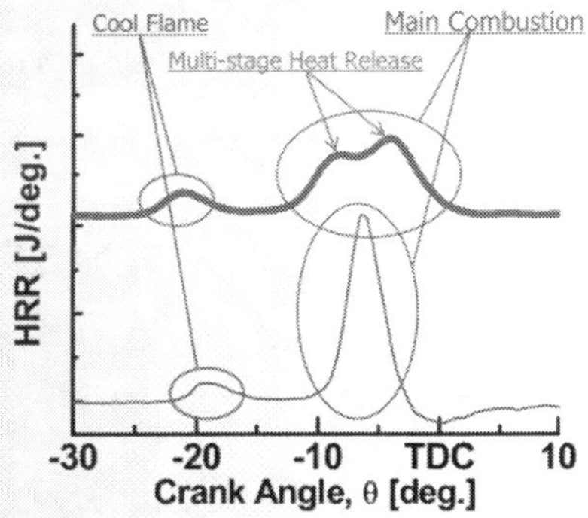


**Purpose**

**Examine the effects of**

- ✓ **Intake Pressure**
- ✓ **Fuel chemical effect**

**on multi-stage High Temperature Heat Release (HTHR) behavior and rapidity of combustion**



## Spectroscopic Measurement

**2014 Automotive Simulation World Congress**

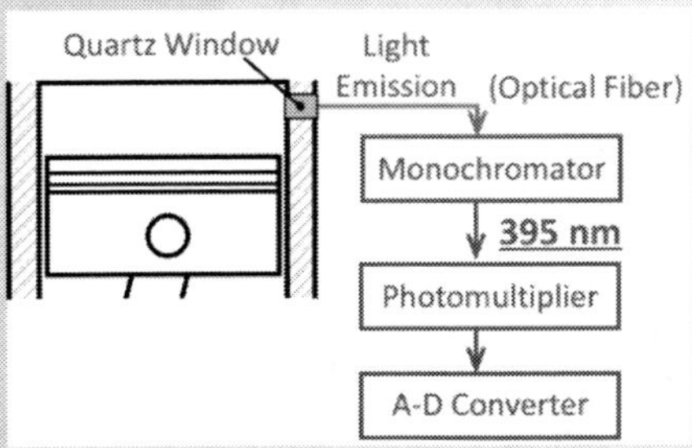


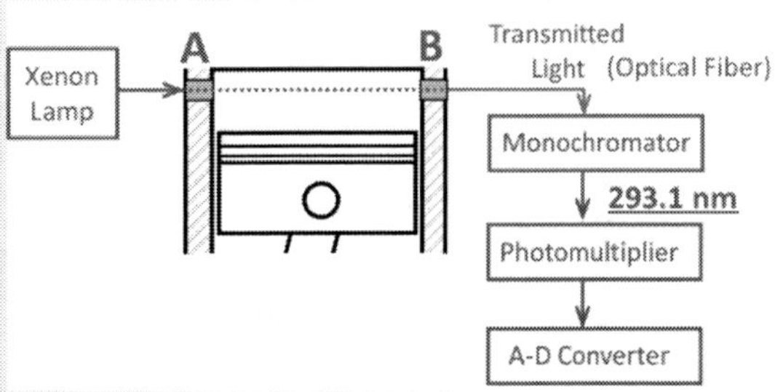


**Why 395 nm ?**

- ✓ Emission Wavelength of **Formaldehyde (HCHO)** in **Cool Flame**
- ✓ Emission Wavelength of **CO-O** Glow in Hot Flame
- ➢ **Cool flame and hot flame behavior can be judged by L.E. measurement**

**Why 293.1 nm ?**

- ✓ Absorption Wavelength of **Formaldehyde (HCHO)** in **Cool Flame to Autoignition**
- ➢ **Cool flame, NTC* and Autoignition behavior can be judged by L.A. measurement**

## Chemical Kinetic Simulation

2014 Automotive Simulation World Congress

**Tool:**

**Calculation Conditions:**

- ✓ Dimensionless (Spatially-homogeneous)
- ✓ Adiabatic Change (No Heat Losses)
- ✓ Volumetric Change by Piston Compression

**Reaction Mechanism:**

- Lawrence Livermore National Laboratory
  - ✓ DME: dimethyl ether

Species: 79 Reactions: 351

Thursday, October 02, 2014 2014 Automotive Simulation World Congress 21

## Two-stage High temp. HRR behavior by using Blended Fuel of DME & $CH_4$ with Supercharging

2014 Automotive Simulation World Congress

$\varepsilon$ = 11:5: 1, N=1400 rpm, $\gamma_{CH4}$=30 %

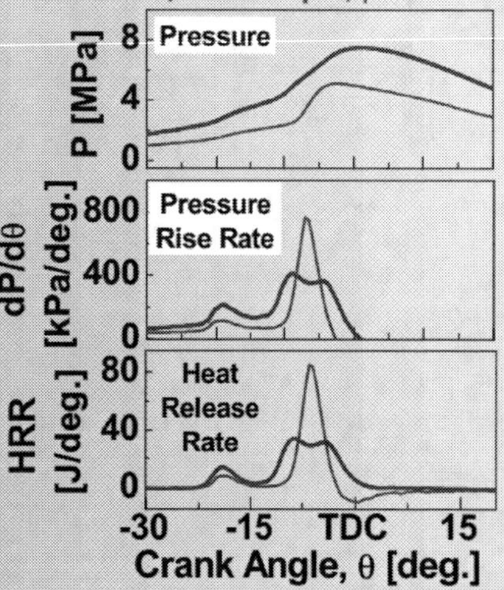


Fuel : $\gamma_{CH4}$ = 30 %
(DME 70%, $CH_4$ 30%)
Red : $P_b$ = 100 kPa
Blue : $P_b$ = 180 kPa

✓ **Two-stage High Temp. Heat Release (HTHR)** occurred at **$P_b$ = 180 kPa**

➢ **Reduce dP/dθ and moderates the combustion**

Thursday, October 02, 2014 2014 Automotive Simulation World Congress 22

## Temp. Dependence on HCCI



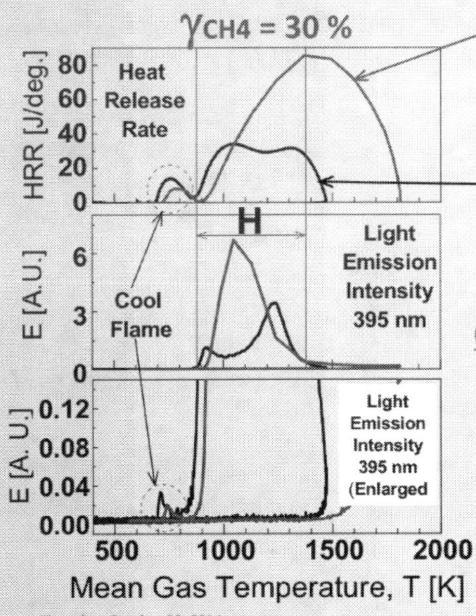


$\gamma_{CH4}$ = 30 %

**Case 1 ($P_b$=100 kPa)**

- ✓ Large HRR and high temp.

➢ **Rapid Single-stage HTHR**

**Case 3 ($P_b$=180 kPa)**

- ✓ Small HRR and low temp.

➢ **Moderate two-stage HTHR**

**Commonality in both cases**

- ✓ Strong light emission occurred same temp. region

➢ **HR reactions occur same temp. in both cases**



## Chemical Kinetic Analysis ( $\gamma_{CH4}$ = 30 % )



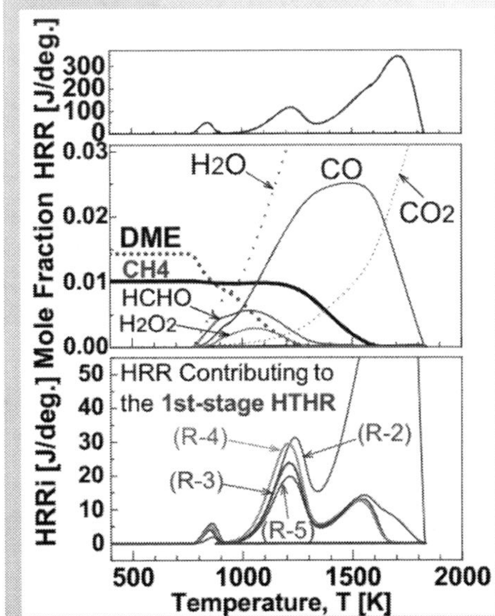


**Cool Flame**

- ✓ Consumption of DME
- ✓ Generation of HCHO and $H_2O_2$

**1st-HTHR**

- ✓ Consumption of DME, HCHO, $H_2O_2$
- ✓ Generation of CO

**2nd-HTHR**

- ✓ Consumption of CO
- ✓ Generation of $CO_2$

➢ Same tendency as experimental spectroscopic data

## Reactions Contributing 1st-stage HTHR

2014 Automotive Simulation World Congress

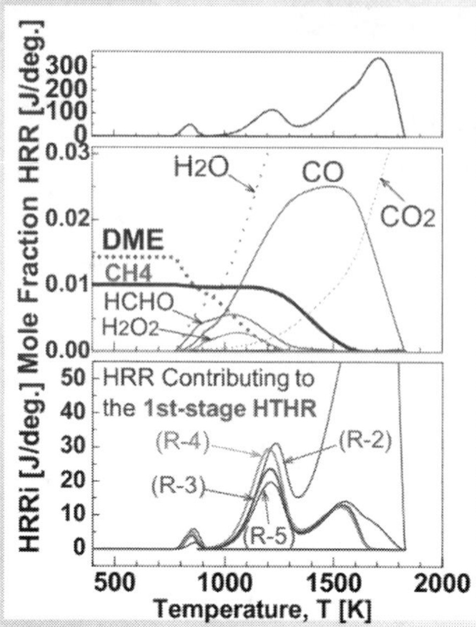


**1st-HTHR**

| | |
|---|---|
| $H + O_2 + M = HO_2 + M$ | (R-2) |
| $HCHO + OH = HCO + H_2O$ | (R-3) |
| $HCO + O_2 = CO + HO_2$ | (R-4) |
| $HO_2 + HO_2 = H_2O_2 + O_2$ | (R-5) |
| $H_2O_2 + M = OH + OH + M$ | (R-6) |

**Overall reaction**

$$2HCHO + O_2 = 2H_2O + 2CO + 473\ \text{kJ}$$

**Ando, H., Sakai, Y. and Kuwahara, K., SAE 2009-01-0948**

➢ 1st-stage HTHR is generation reaction of CO form HCHO derived from DME



## Reactions Contributing 1st-stage HTHR

2014 Automotive Simulation World Congress

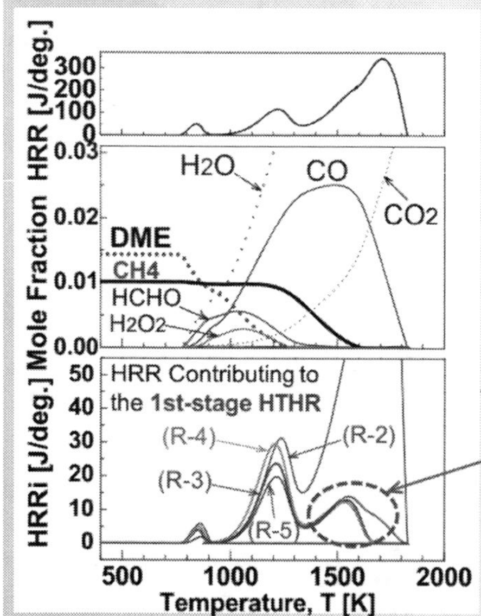


**1st-HTHR**

| | |
|---|---|
| $H + O_2 + M = HO_2 + M$ | (R-2) |
| $HCHO + OH = HCO + H_2O$ | (R-3) |
| $HCO + O_2 = CO + HO_2$ | (R-4) |
| $HO_2 + HO_2 = H_2O_2 + O_2$ | (R-5) |
| $H_2O_2 + M = OH + OH + M$ | (R-6) |

**Role of $CH_4$**

✓ $CH_4$ react to CO (and generate heat) at higher temperature

➢ 1st-stage HTHR decrease with increasing $\gamma_{CH4}$

## Reactions Contributing 2nd-stage HTHR

2014 Automotive Simulation World Congress

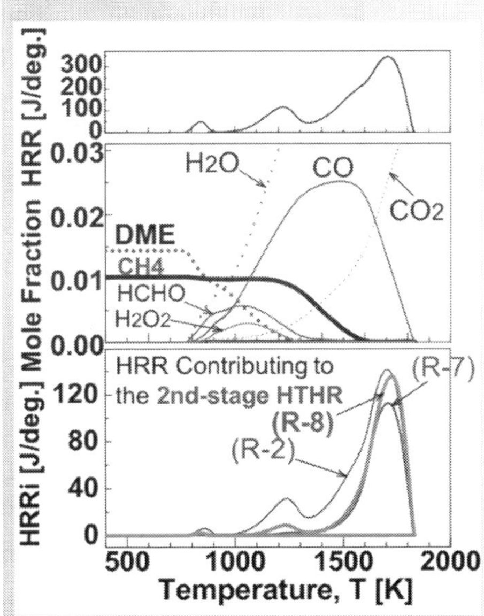


**2nd-HTHR**

$CO + OH = CO_2 + H$ **(R-7)**

$HO_2 + OH = H_2O + O_2$ (R-8)

✓ Generation reactions of final products ($CO_2$ and $H_2O$)



## Conclusions (Part 2)

2014 Automotive Simulation World Congress

This study examined the effects of intake pressure and fuel blend on 2-stage HTHR behavior of an HCCI engine. Light emission and absorption spectroscopy and chemical kinetic simulation were used to investigate the reaction behavior in detail.

- Use of DME + $CH_4$ blended fuel together with a suitable intake air flow rate corresponding to the fuelling rate gave rise to 2-stage HTHR, which had the effect of moderating combustion.
- 1st-stage HTHR was one that produced CO from HCHO. The results also indicated that the main reactions of the second stage of the hot flame heat release were ones that produced the final products of $CO_2$ and $H_2O$ from CO and $HO_2$.
- DME was consumed by the cool flame and in the 1st-stage HTHR, whereas $CH_4$ was mainly consumed in the just prior to second stage. That explains why the heat release of the hot flame occurred into two distinct stages.

# Concept for Combustion Chamber Design with High Compression Ratio

Munemasa Hashimoto Hideaki Osada Tetsuya Murayama Noboru Uchida

New A.C.E Institute Co., Ltd.

2

## Contents

1. Introduction and test engine.
2. Investigation into process of emissions deterioration with higher compression ratio and study of alternative cavity shapes in high load.
3. Application of alternative cavity shape in low load.
4. Experimental result and conclusion.

## Contents

1. Introduction and test engine.
2. Investigation into process of emissions deterioration with higher compression ratio and study of alternative cavity shapes in high load.
3. Application of alternative cavity shape in low load.
4. Experimental result and conclusion.



## Concept of high compression ratio

- **Passenger cars:**
  - **Low compression ratio**
    - Limitation of peak firing pressure
    - Reduction of exhaust emissions and friction loss
- **Heavy duty vehicles:**
  - **High compression ratio**
    - Optimized injection timing for fuel consumption improvement with higher allowable cylinder pressure
    - Reduction of friction loss by down speeding and higher BMEP operation

**In this study, potential and problem were investigated, by using piston with higher compression ratio in order to improve fuel consumption in heavy duty diesel engine.**

# Test engine and operating mode

Engine : heavy duty diesel equipped with 2-stage boosting system.
Target mode : Japanese HDV emissions & fuel consumption test cycle (JE05).

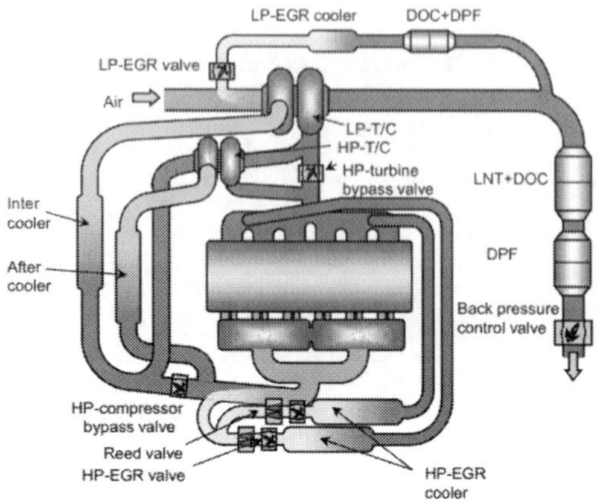


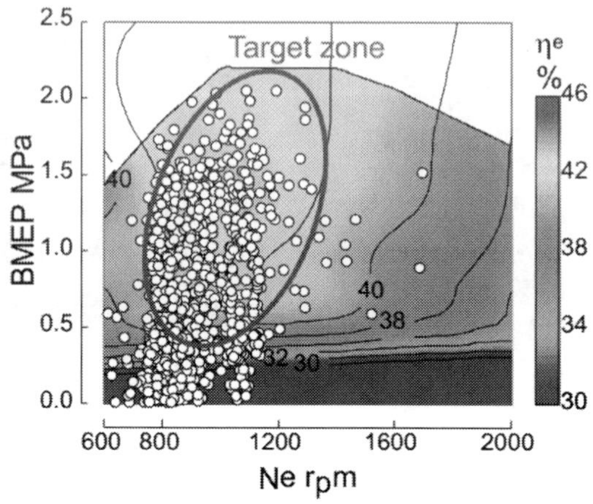




# Items for improvement in fuel consumption

These items were selected in order to improve fuel consumption.

**2-stage boosting;**

High boosting decreases combustion temperature in any speed region

**Piston with higher compression ratio;**

Improvement in thermal efficiency

**Retarded IVC;**

Decrease of effective compression ratio and suppressing of peak firing pressure

## Experimental results

### Experimental Results

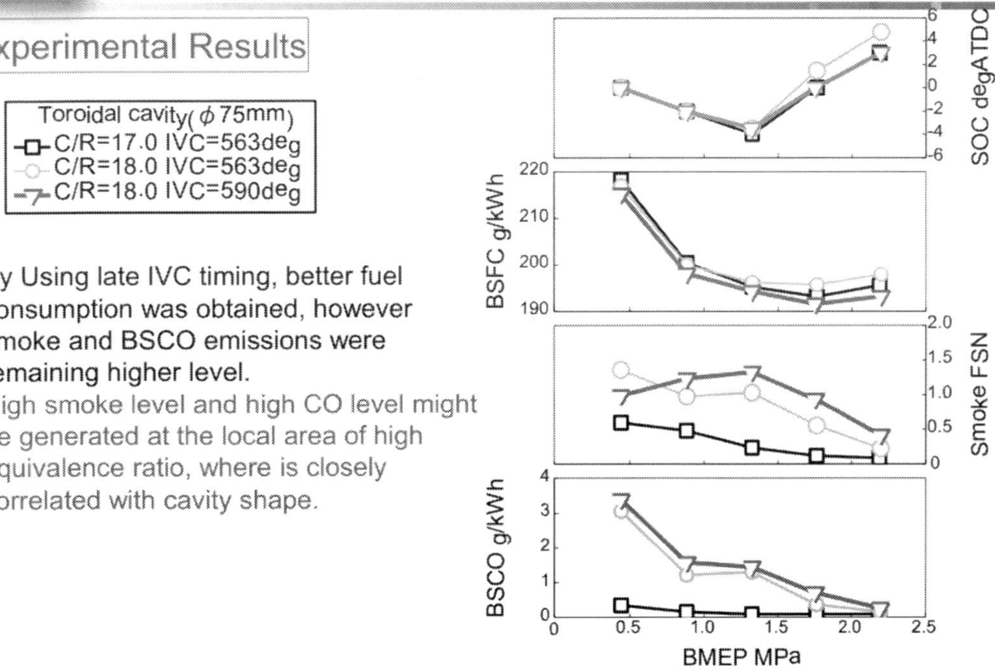


By Using late IVC timing, better fuel consumption was obtained, however smoke and BSCO emissions were remaining higher level.
High smoke level and high CO level might be generated at the local area of high equivalence ratio, where is closely correlated with cavity shape.



## Contents

1. Introduction and test engine.
2. **Investigation into process of emissions deterioration with higher compression ratio and study of alternative cavity shapes in high load.**
3. Application of alternative cavity shape in low load.
4. Experimental result and conclusion.

## Calculated Condition in high load with higher compression ratio



As a first, Toroidal shape was selected. Cavity height was reduced to achieve higher compression ratio.

| Condition | | High Load |
|---|---|---|
| Ne | | 1200rpm |
| q (BMEP) | | 211mm³/st (1.65MPa) |
| Swirl Ratio Ks | | 1.58(PWA) |
| Initial Condition | Pcyl | 240kPa |
| | Tcyl | 391K |
| | Lambda | 1.29 |
| | EGR | 23.5% |
| SOI | | -3deg ATDC |
| Nozzle | | 8-ϕ0.173mm - 155deg |

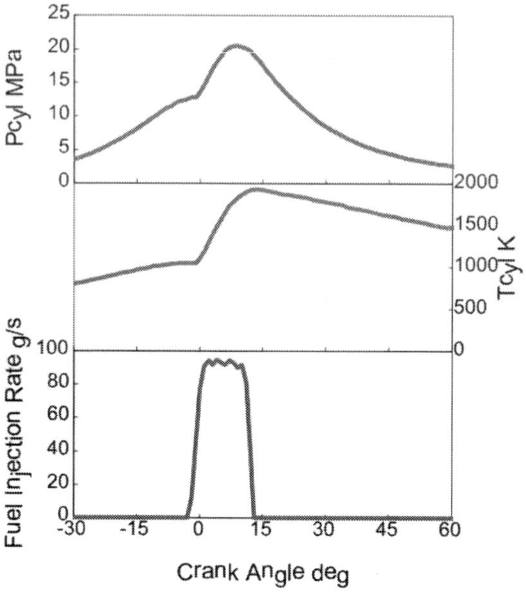


## Local area of high equivalence ratio in cavity



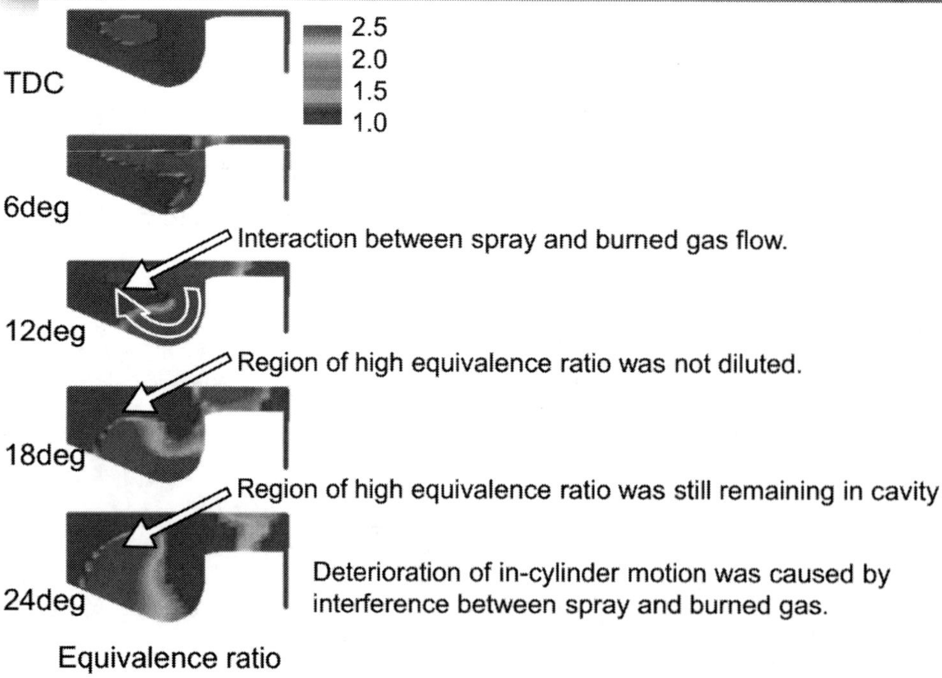


Deterioration of in-cylinder motion was caused by interference between spray and burned gas.

11

## Alternative cavity design

### Baseline cavity

Utilize of spray impingement in order to increase in in-cylinder motion.

### Alternative cavity

Decrease of interaction between spray and burned gas flow.

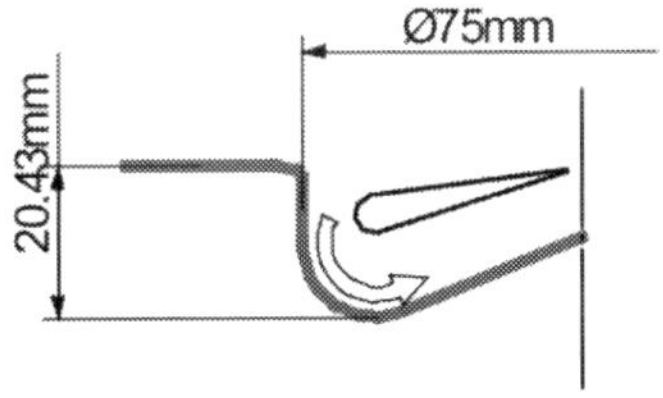


Step1: Toroidal (Φ75mm)

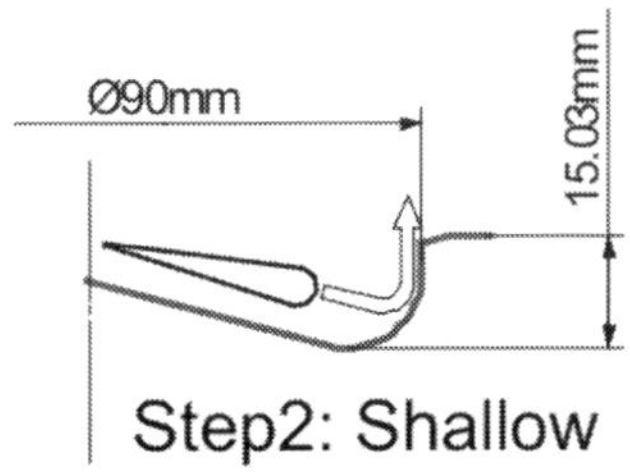


Step2: Shallow (Φ90mm)

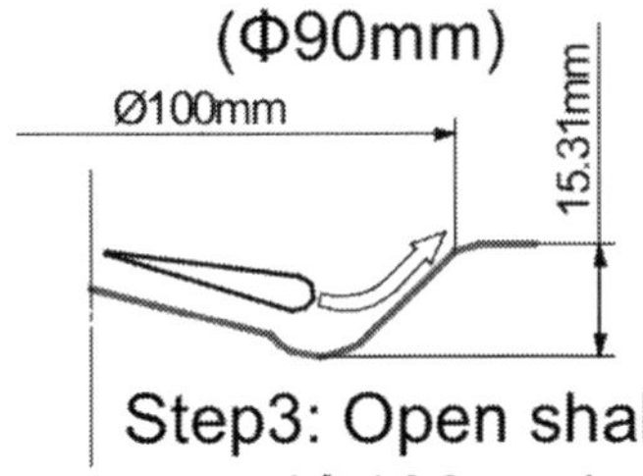


Step3: Open shallow (Φ100mm)

12

## Local area of high equivalence ratio in cavity

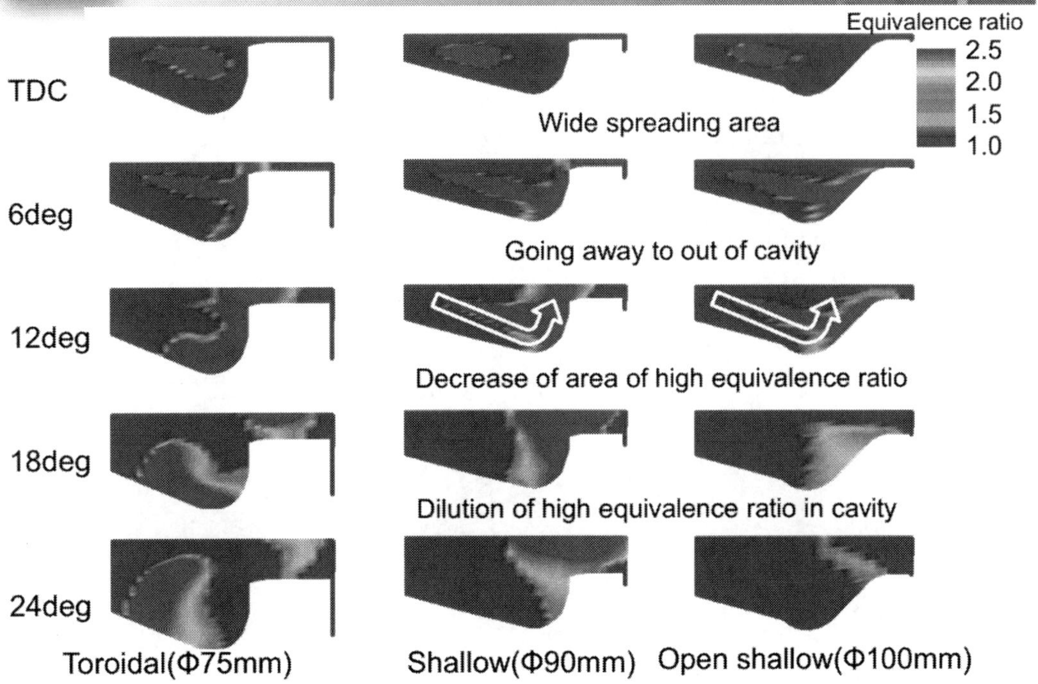

## Comparison of CO mass fraction in the cavity at high load

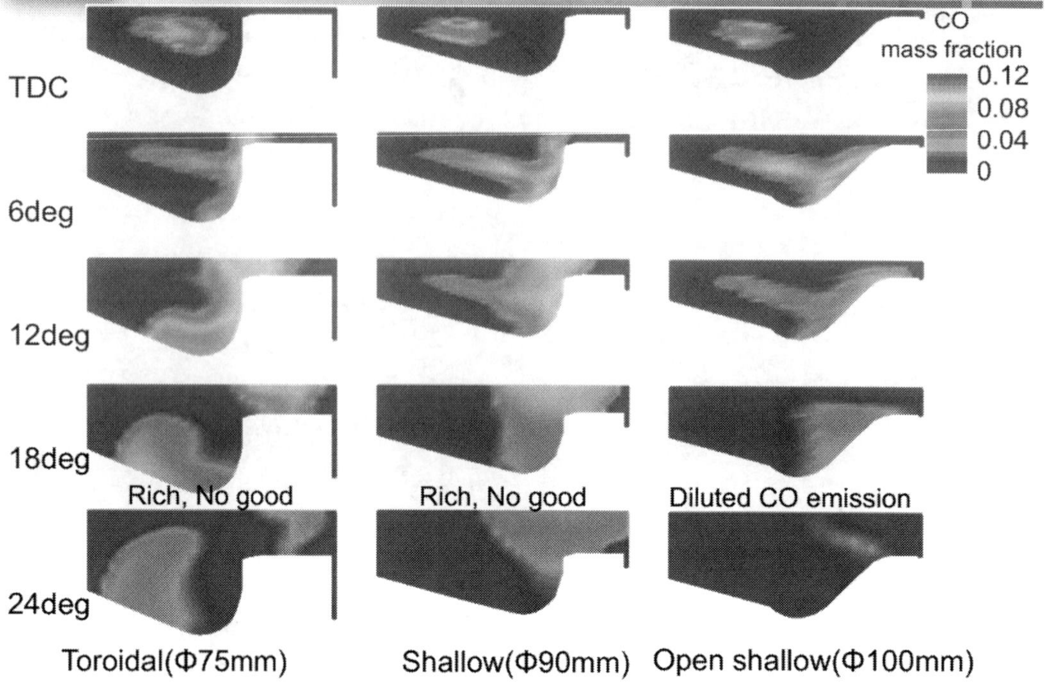




## Comparison of in-cylinder pressure

In early period, rate of heat release was almost same.
But in latter period, rate of heat release in open shallow was higher than that in toroidal.
As a result, cumulative heat release, averaged pressure and temperature in the cylinder were increased by open shallow type cavity.

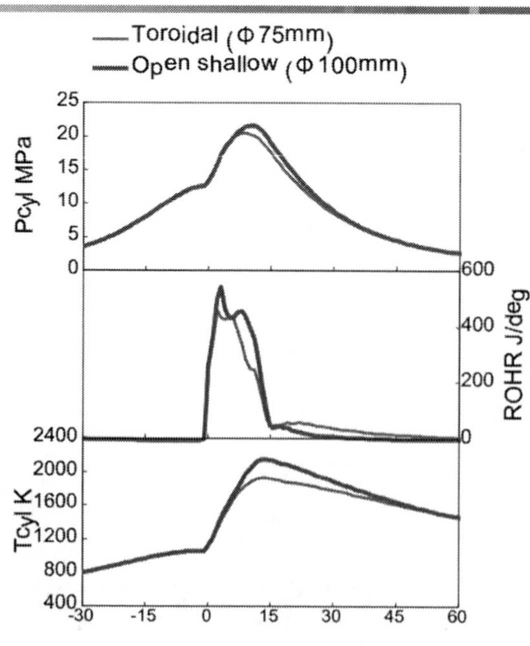

## Contents

1. Introduction and test engine.
2. Investigation into process of emissions deterioration with higher compression ratio and study of alternative cavity shapes in high load.
3. Application of alternative cavity shape in low load.
4. Experimental result and conclusion.



## Comparison of in-cylinder pressure at low load condition

| Condition | | Low Load |
|---|---|---|
| Ne | | 1200rpm |
| q (BMEP) | | $55mm^3$/st (0.44MPa) |
| Swirl Ratio Ks | | 1.58(PWA) |
| Initial Condition | Pcyl | 203kPa |
| | Tcyl | 567K |
| | Lambda | 1.41 |
| | EGR | 34.1% |
| SOI | | -8.5deg ATDC |
| Nozzle | | 8-φ0.173mm - 155deg |

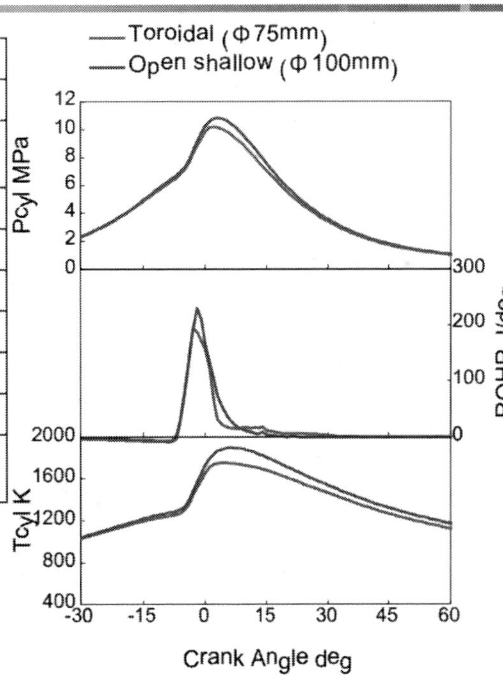

# Suppressing CO emissions deterioration by higher boosting pressure

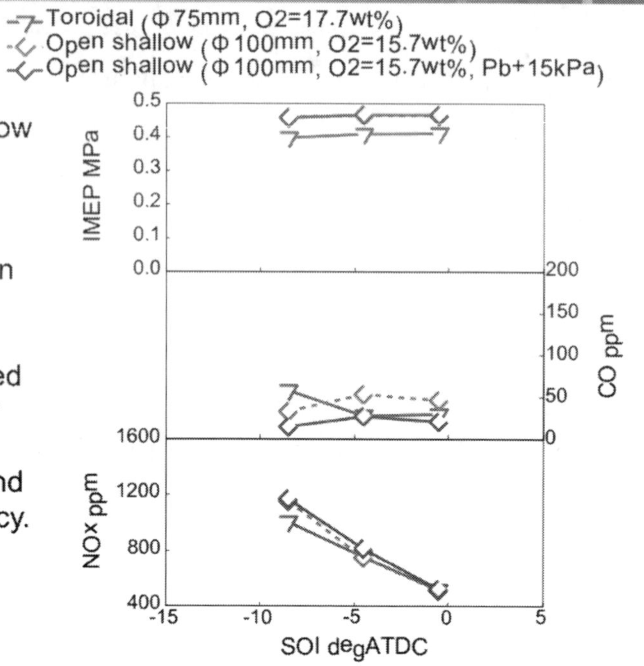


CO emissions in open shallow were slightly higher than those in toroidal, because initial O2 mass fraction in open shallow was lower than that in toroidal.

CO emissions were improved by high boosting, but PMEP might be increased even though it was primary depend on the turbocharger efficiency.



# Comparison of CO mass fraction in the cavity at low load

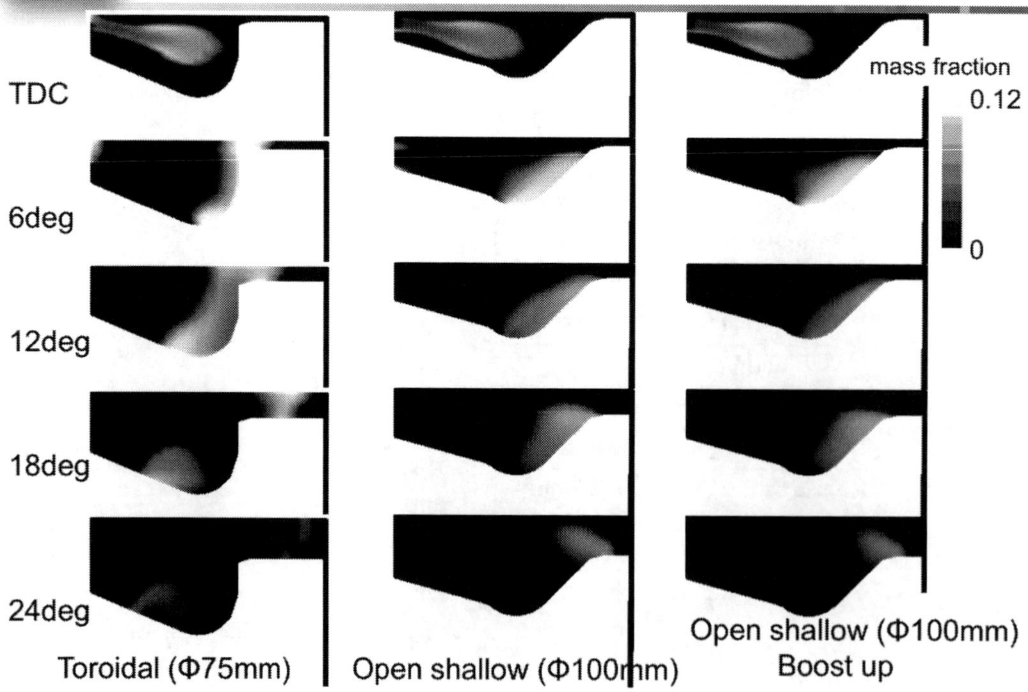

## Contents

1. Introduction and test engine.
2. Investigation into process of emissions deterioration with higher compression ratio and study of alternative cavity shapes in high load.
3. Application of alternative cavity shape in low load.
4. Experimental result and conclusion.



## Comparison of rate of heat release

Experimental Results

Experimental condition
Ne=1200rpm,
BMEP=1.65MPa,
BSNOx=1.0g/kWh,
SOI=-4degATDC

In new shallow dish cavity shape,

- premixed combustion was decreased,
- diffusion combustion was increased,
- duration of combustion was shorter than in toroidal cavity shape.

Better diffusion combustion

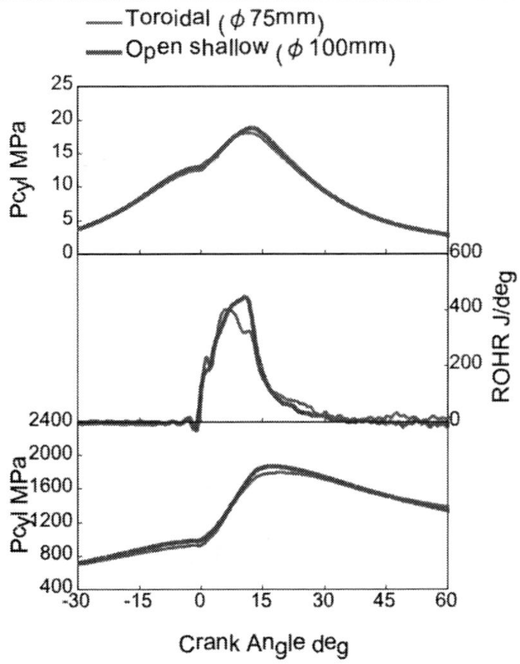

## Experimental Result

21

### Experimental Results

Experimental condition
Ne=1200rpm

New shallow cavity can achieve both fuel consumption and better exhaust emissions, mainly middle-high road region

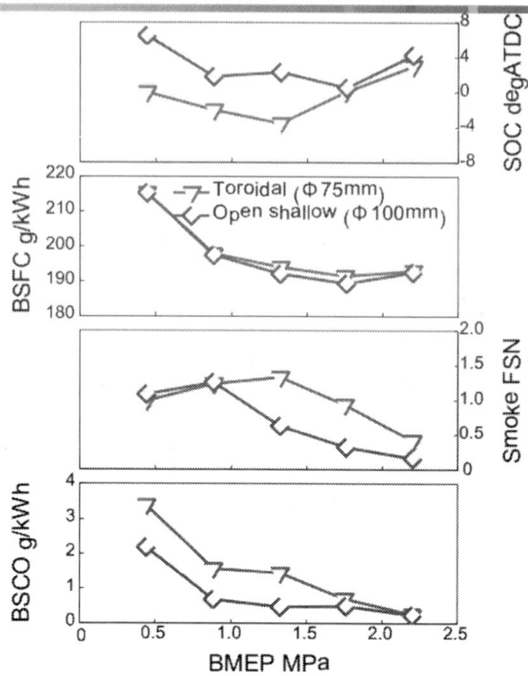


## Conclusion

22

Optimized cavity shape with high compression ratio was investigated by using simulation combined with detailed chemical reaction; **FORTÉ** code, in order to improve fuel consumption and exhaust emissions in a heavy duty diesel engine.

- Optimized cavity shape which improves both combustion and emissions for higher compression ratio was found within a short period.
- Both fuel consumption and exhaust emissions in the experiments were well predicted by the numerical simulation.

END

2014 Automotive Simulation World Congress

# ANSYS Strategy for Internal Combustion Engine Simulations

Ellen Meeks, ANSYS, Inc.



## IC Engine Products at ANSYS

- Workbench – ICE System (WB-ICE)
  - ANSYS Workbench based ICE analysis
- FORTÉ
  - Full 3D CFD analysis for IC engine designers
- Model Fuels Library
  - Best-validated fuel models in the market (~60)
- CHEMKIN PRO
  - Ignition, flame-speed tables, zonal models
- CHEMKIN-PRO REACTION WORKBENCH
  - Fuel analysis and mechanism reduction

## The ANSYS strategy for IC engine simulations is: FORTÉ CFD

- We are investing in the ANSYS FORTÉ solution:
  - Integrating RD Products into ANSYS Workbench and WB-ICE
  - Integrating with FLUENT for conjugate heat transfer (CHT)
  - Developing new capabilities
  - Performing extensive verification and validation tests
  - Continuing development of fuel models & fuel modeling tools
- FLUENT (with WB-ICE) remains a supported option for cold-flow and scale-resolved turbulence simulation

2014 Automotive Simulation World Congress
October 9-10 . Tokyo, Japan

## A brief history of FORTÉ CFD

- Partnered with Profs. Rolf Reitz and Chris Rutland
  - State of the art spray models
  - Engine design expertise
- Reaction Design
  - Combustion kinetics and kinetics solvers
- Model Fuels Consortium experience
  - 7+ year, industry-led effort
  - Comprehensive library of fuel models for engine combustion simulation
- Now part of ANSYS, Inc.!
  - Full suite of analysis tools and integration

## Key advantages of FORTÉ CFD

- Set up tailored to IC Engines
- Automatic mesh generation
  - On-the-fly, dynamic mesh motion
- Well validated fuel-combustion kinetics with *fast* solver
  - Capture fuel effects
  - Predict ignition, emissions
  - Predict flame-front location and knocking in SI engines
- State-of-the-art spray and spray-wall models
  - Capture physics accurately and robustly

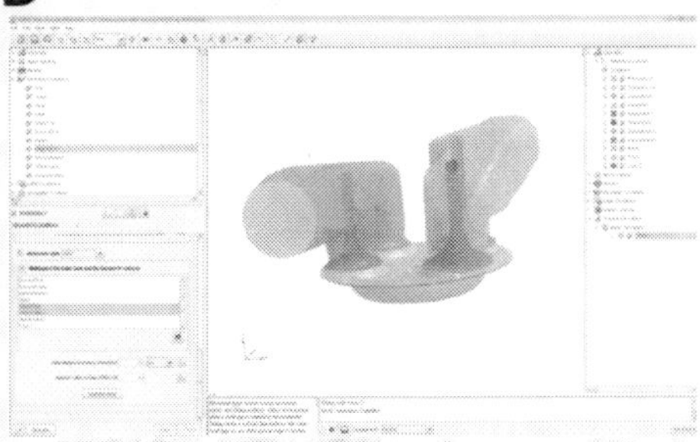

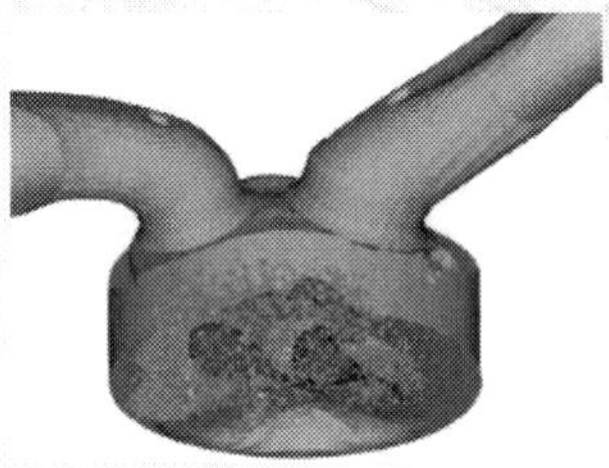

2014 Automotive Simulation World Congress
October 9-10 . Tokyo, Japan

## FORTÉ provides auto-generated, Cartesian immersed-boundary mesh

- Dynamic mesh from CAD:
  - CAD STL data read directly into FORTÉ
  - WB-ICE can be used to provide geometry
    - E.g., Fluent *.msh file read in => surface mesh
  - FORTÉ generates the mesh on-the-fly
    - With automated refinement around valve gaps and small features
    - Preview the mesh in the Simulation Interface
- Flexible mesh options:
  - Fully automatic Cartesian meshing
  - Body-fitted sector meshing for quick diesel design studies
    - Sector Mesh Generator included
    - ICEM-CFD can also be used

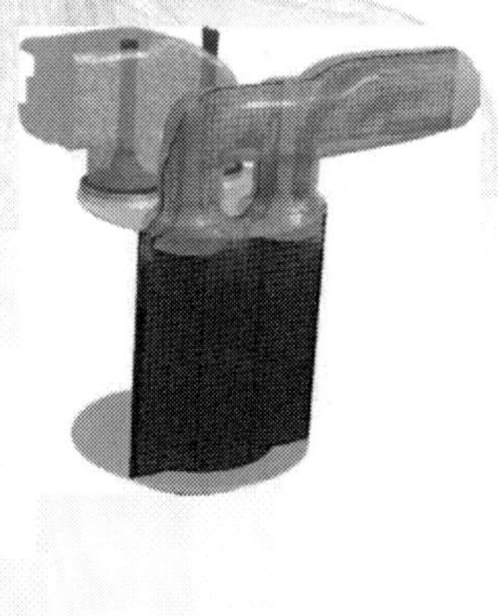

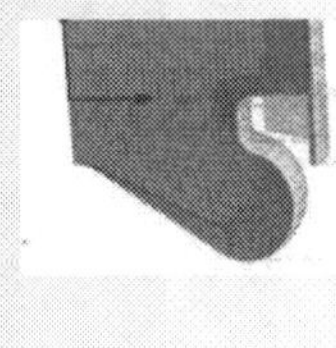

## FORTÉ integration with WB-ICE allows quick geometry preparation from CAD

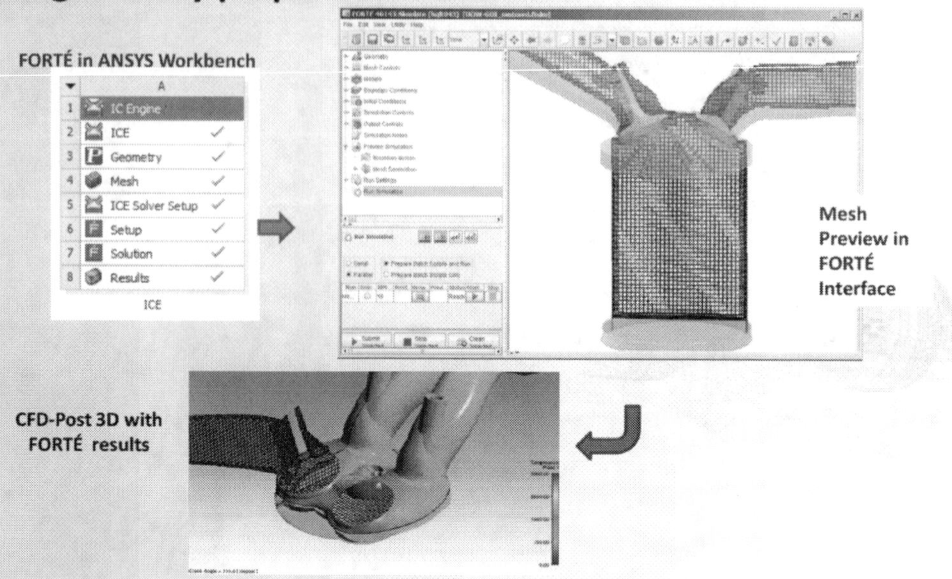


2014 Automotive Simulation World Congress
October 9-10 . Tokyo, Japan

## Advanced solution techniques allow practical turn-around with chemistry

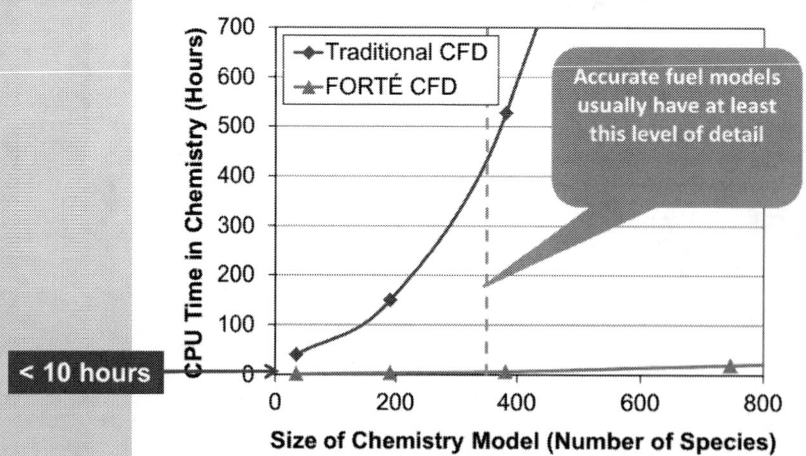


A good chemistry mechanism is key to predicting engine combustion behavior

2014 Automotive Simulation World Congress
October 9-10 . Tokyo, Japan

## Example: Dual-fuel heavy-duty engine

- RCCI Simulation with 470-species for Diesel & Gasoline
- CPU time (16 CPUs): < 4 hours

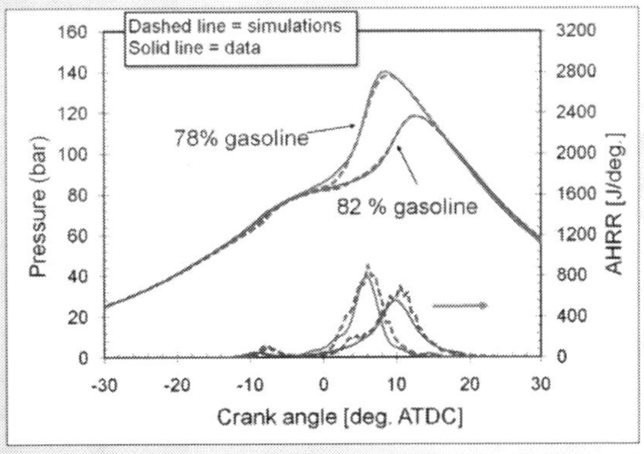


CO and UHC agree with experiment

*From: K. Puduppakkam, et al., SAE 2011 World Congress*

2014 Automotive Simulation World Congress
October 9-10 . Tokyo, Japan

## Spark ignition and flame propagation use state-of-the-art modeling

**The spark ignites a kernel of gas that is very small at first**

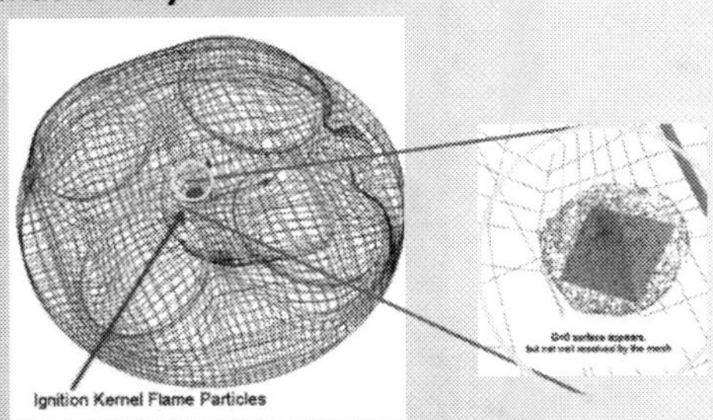


Particles are used to track the initial kernel growth and flame location

**The flame (typically <~0.01 mm thick) propagates through the unburned gas**

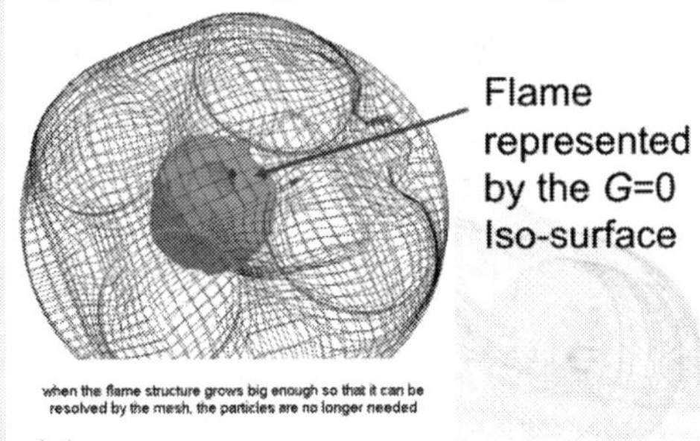


Afterwards, the level-set (G-eq.) method takes over in tracking the flame front

The flame propagation responds to local turbulence conditions and chemistry is solved throughout

***SAE 2003-01-0722, 2003-01-1843, 2006-01-0243, 2007-01-0165***

2014 Automotive Simulation World Congress
October 9-10 . Tokyo, Japan

## FORTÉ spray model captures all the physics with reduced mesh dependence

- Even with a coarse mesh, the spray characteristics are well represented
  - Allows quick exploratory studies as well as detailed
  - Removes need to iteratively test for mesh "convergence"

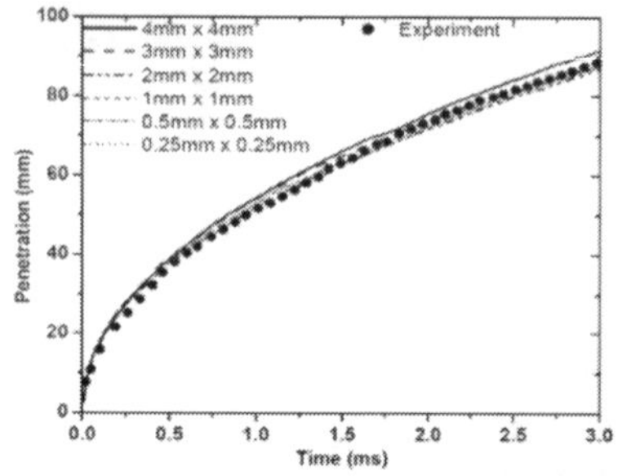


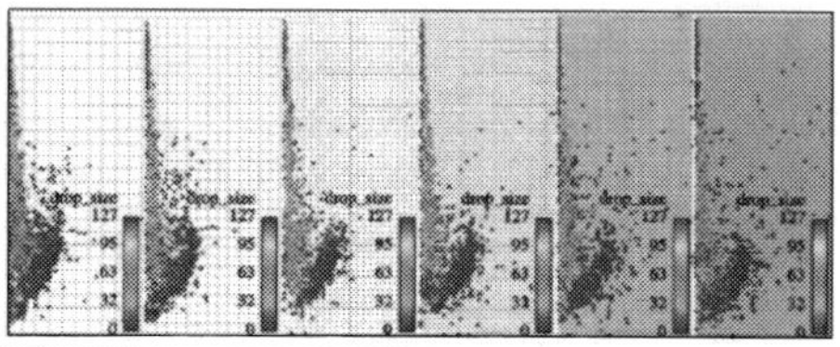

*From N. Abani, A. Munnannur, and **R. D. Reitz**, ILASS 2006*
*(Data from Naber and Siebers, SAE960034)*

2014 Automotive Simulation World Congress
October 9-10 . Tokyo, Japan

## FORTÉ includes advanced spray-wall interactions

- Important to many engine simulations:
  - Cold start of PFI engines
  - Fuel/air mixing in DISI and GDI engines
  - Soot from small-bore diesel and GDI engines
- FORTÉ includes physics for:
  - Droplet stick, rebound
  - Droplet splash
  - Film spread, separation (stripping)

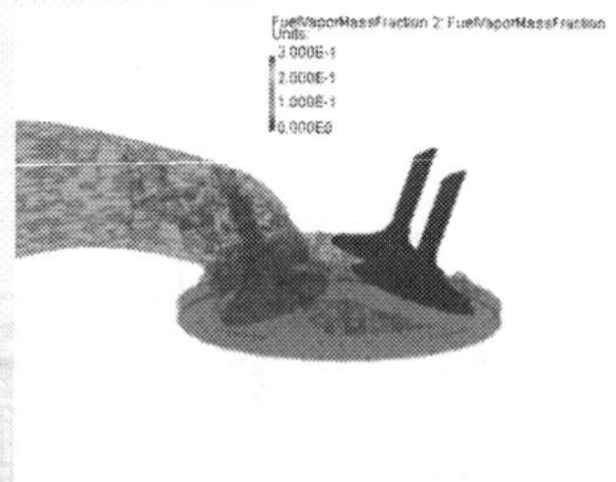

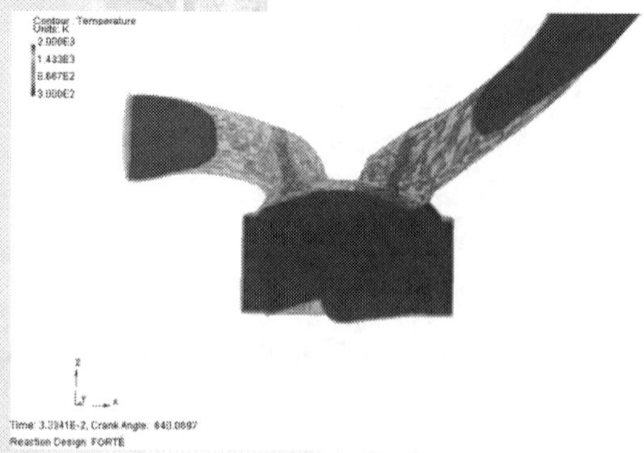

## Work-in-progress for FORTÉ and WB-ICE

- Optional gradient-driven mesh refinement
- Multi-cylinder simulations
- Conjugate Heat Transfer coupling with Fluent
- Full automation & scripting through WB
  - Parameterization of geometry inputs
  - Optimization of geometry as well as other parameters

2014 Automotive Simulation World Congress
October 9-10 . Tokyo, Japan

## Summary of key capabilities in FORTÉ

- **Ease of Use**
  - Automatic mesh generation with moving boundaries
  - Workflow specifically for engine combustion simulations
- **Accuracy**
  - Detailed kinetics tied to Model Fuels Library to capture fuel effects
  - Track flame front location using level-set method, DPIK for SI engines
  - State-of-the-art spray and spray-wall models
  - Capture correct physics on practical mesh
  - Direct prediction of ignition, emissions, particle size
- **Speed**
  - Fast and accurate chemistry solver
  - Parallel (MPI) simulation using automatic domain decomposition
- **Integration with other simulation**
  - Integration with WB-ICE, ANSYS Workbench for fluid-structural interactions

2014 Automotive Simulation World Congress

# Applying FORTÉ to Design Simulations: Importance of the Fuel Model

2014 Automotive Simulation World Congress

October 9-10 . Tokyo, Japan

## Changing engine-design requirements shift the focus to the fuel model

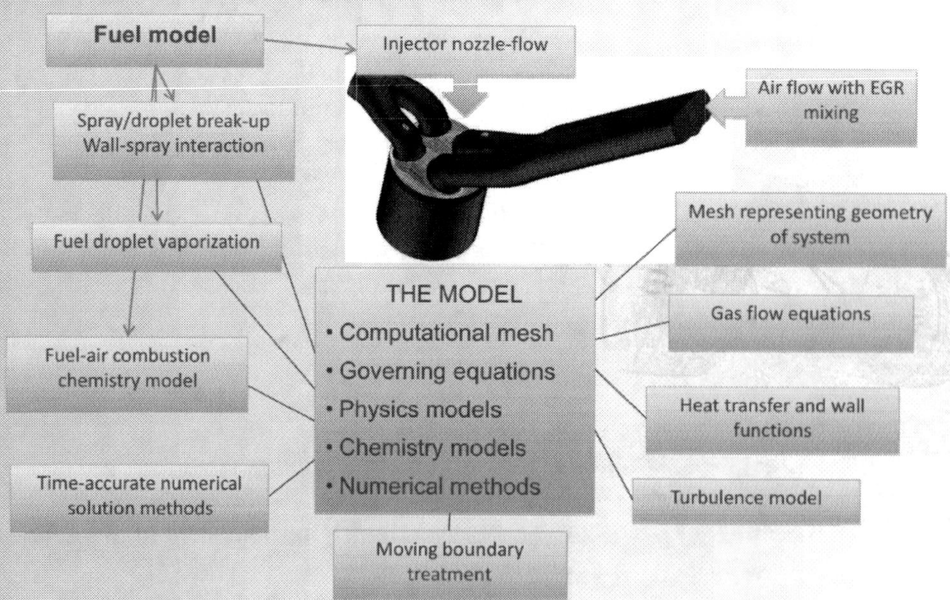

2014 Automotive Simulation World Congress
October 9-10 . Tokyo, Japan

## Model Fuels Library (MFL)

- The MFL enables detailed fuel modeling
  - Represent complex "real" fuels with appropriate model fuels
  - Extensively validated chemistry
- Results from the MF Consortium:
  - 7 years and $7 million invested
  - Involved 20 companies
- The MFL is now continuously updated and maintained through a subscription program
  - Internal expertise adds & improves models
  - Advised by world-wide experts

2014 Automotive Simulation World Congress
October 9-10 . Tokyo, Japan

## A model fuel should match real fuel properties

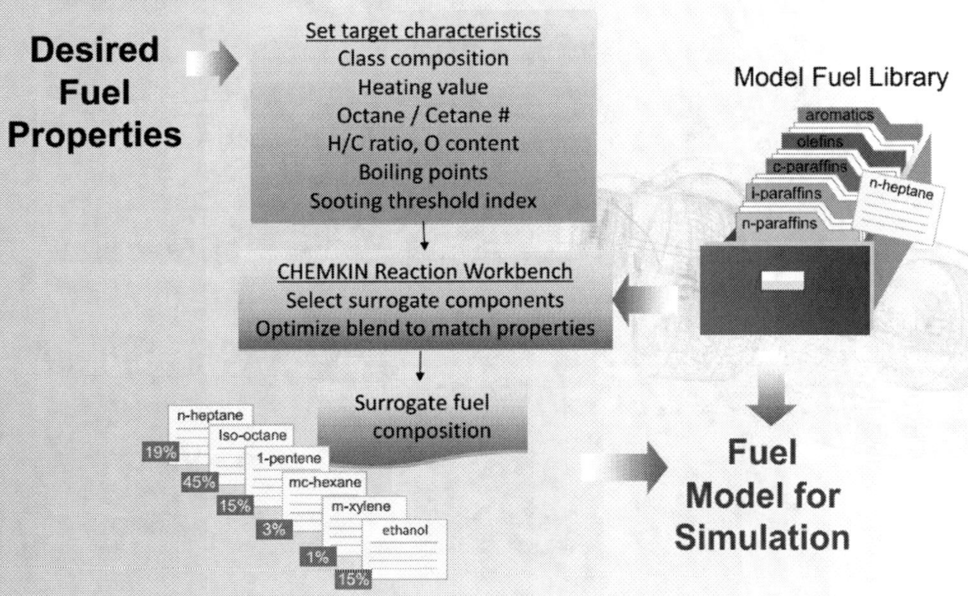

2014 Automotive Simulation World Congress
October 9-10 . Tokyo, Japan

## A good chemistry mechanism is key to describing the fuel combustion behavior

The engine is a chemical plant:

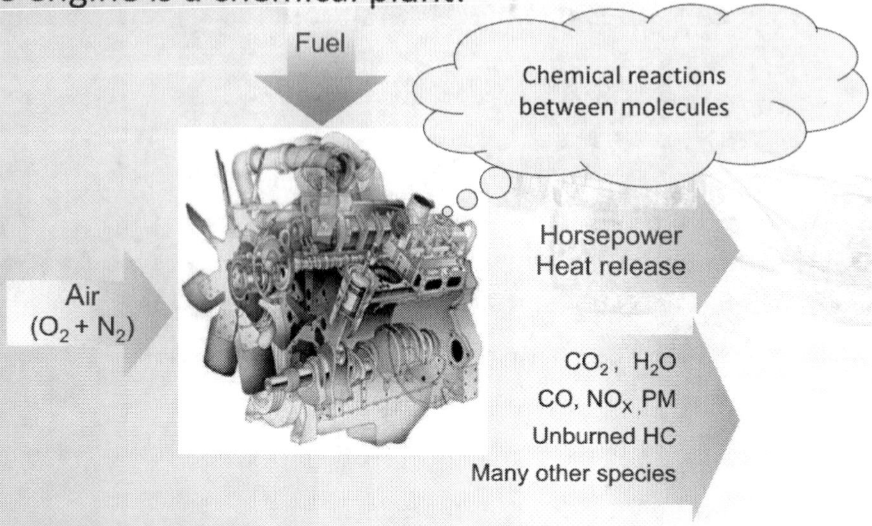


2014 Automotive Simulation World Congress
October 9-10 . Tokyo, Japan

## The Model Fuel Library is a large set of validated fuel-component mechanisms

**~ 60 Validated Fuel Models**

- Biodiesel Components
  - Diesel and Jet-Fuel Surrogates
    - Gasoline Surrogates
      - Gaseous Fuels

\+

Detailed Particle-based and Pseudo-gas Soot Models

- Common core chemistry
- Consistent rules for reaction rates
- Validated fuel-blending behavior

2014 Automotive Simulation World Congress
October 9-10 . Tokyo, Japan

## With a good fuel model, we match emissions in high-performance Diesel engines

- Detailed (7-component fuel model), 500 gas-phase species, multi-step soot model
- **Combustion and emissions trends captured with no change to model or model approach**

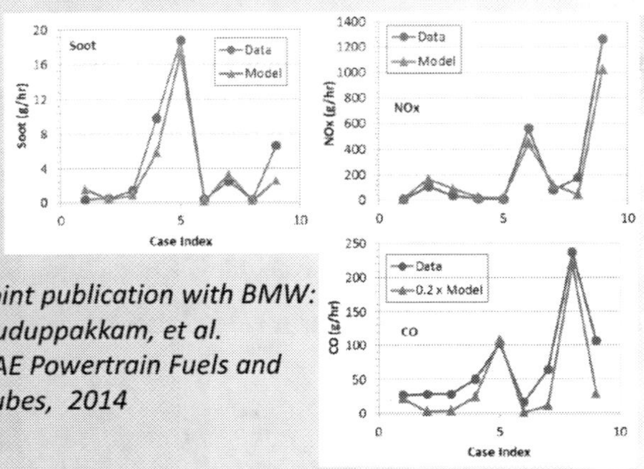


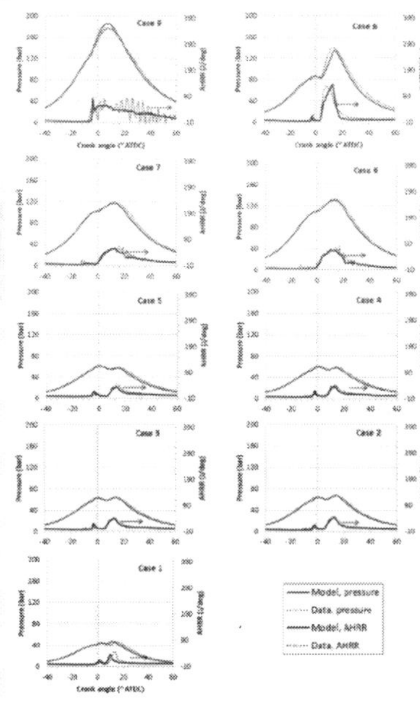


*Joint publication with BMW: Puduppakkam, et al. SAE Powertrain Fuels and Lubes, 2014*

2014 Automotive Simulation World Congress
October 9-10 . Tokyo, Japan

## Detailed kinetics also enable prediction of soot in GDI engines

*SAE 2014-01-1135, Naik, et al..*

7-component gasoline fuel model
230 species mechanism
Full cycle simulations using RNG k-eps

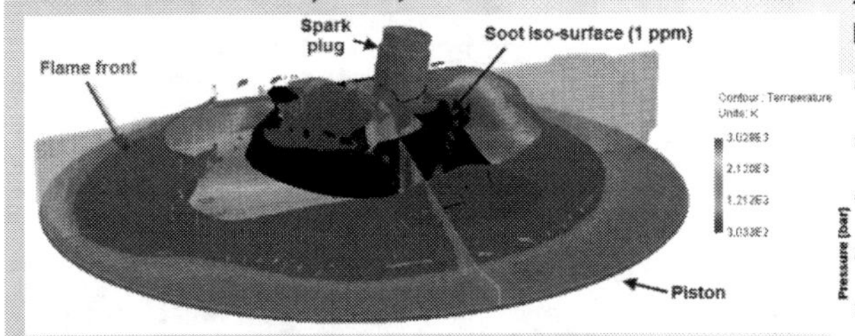


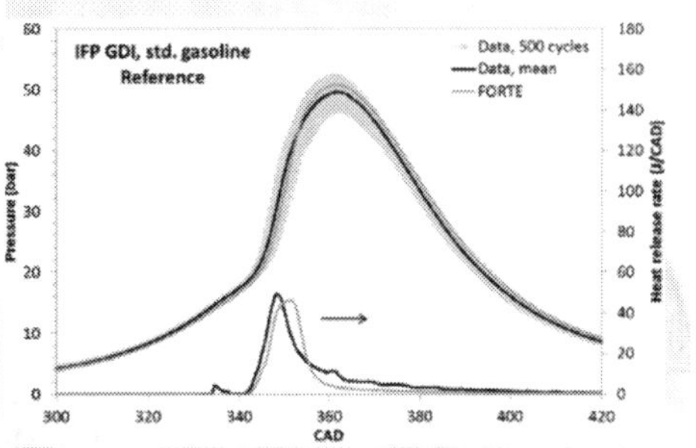


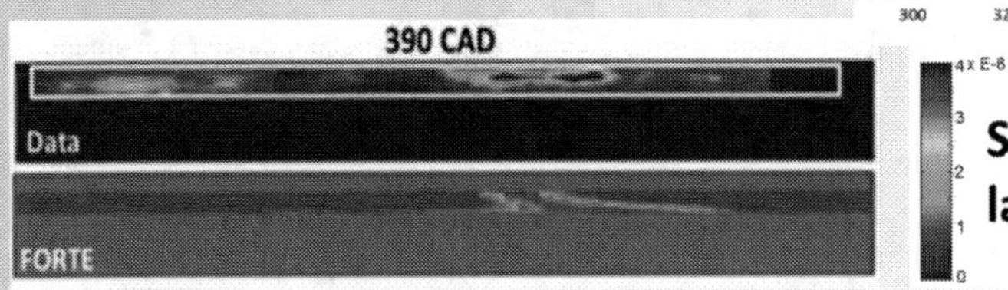


**Soot in laser plane**

*Data from IFPEN, Pires da Cruz et al.*

## FORTÉ can predict knocking tendency

- Detailed chemistry predicts end-gas auto-ignition
- Pressure sensor monitors, just like the test stand

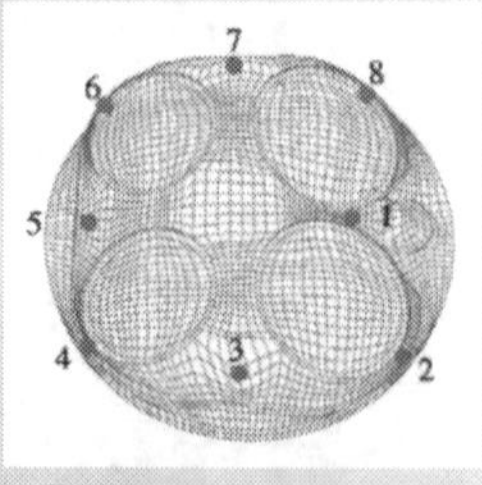


**Pressure sensors in FORTÉ simulation match experimental setup**

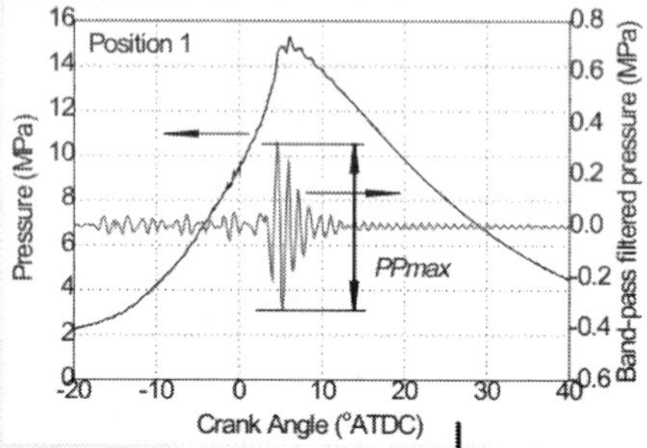


**FORTÉ calculates Knock Intensity**

- Allows exploration of trade-offs
  - Fuel effects, spark timings, higher compression ratios, boost, etc.

*SAE 2007-0Cal1-0165*

2014 Automotive Simulation World Congress
October 9-10 . Tokyo, Japan

## Knocking can also be visualized in FORTÉ

- White/Black grid line intersections shows flame front location
- Contour plot shows temperature distribution in the engine

<u>No knocking observed</u> <u>Significant knocking observed</u>

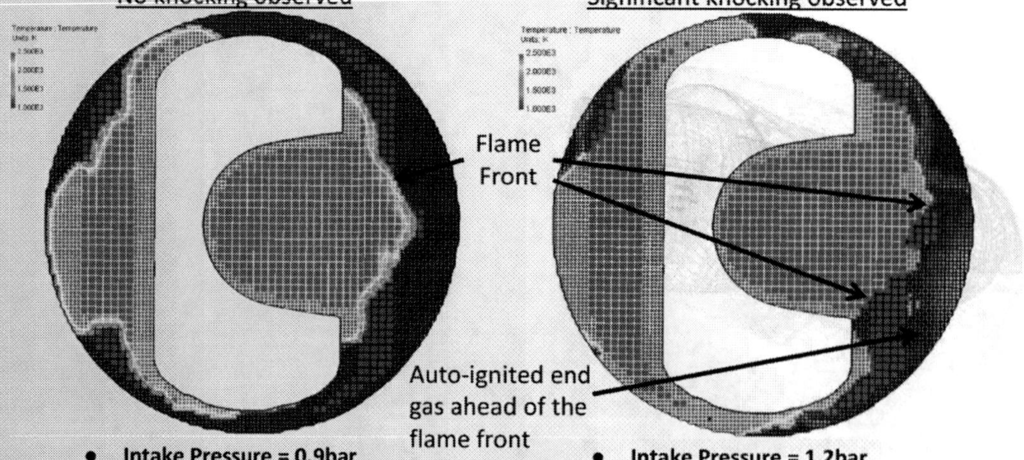


- **Intake Pressure = 0.9bar**
- **Spark Timing = -32CA ATDC**

- **Intake Pressure = 1.2bar**
- **Spark Timing = -32CA ATDC**

24

## Better fuel model = better predictions

- Multi-component fuel model captures knock in the engine where a single-component fuel does not

Same engine conditions

The only difference is the fuel model

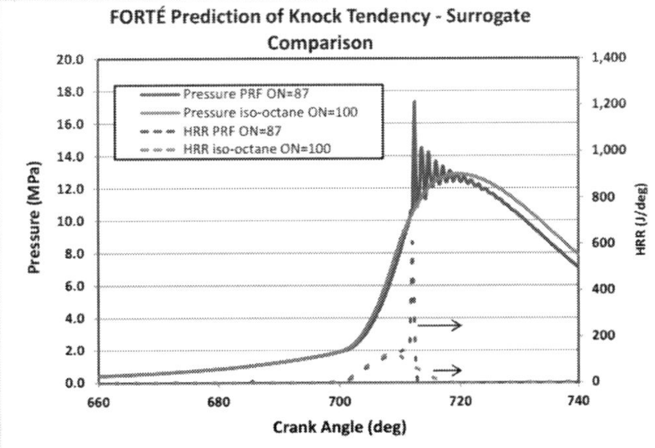




2014 Automotive Simulation World Congress
October 9-10 . Tokyo, Japan

## Overall Summary

- ANSYS is committed to advancing IC Engine Simulation
- FORTÉ is the core of the ANSYS IC engine strategy
  - New capabilities and WB integration well underway
  - Integration with Fluent for CHT
  - Continuous improvement on performance and scalability
  - Validation and verification is an active and continuous process
- The Model Fuels Library is key to accurate combustion, and for soot and knocking predictions
- We are working closely with customers to fine-tune our long-term plan and address critical short-term needs

2014
Automotive Simulation
World Congress
Case study of automotive ESD analysis
Hongkeun Kim / SL Corporation


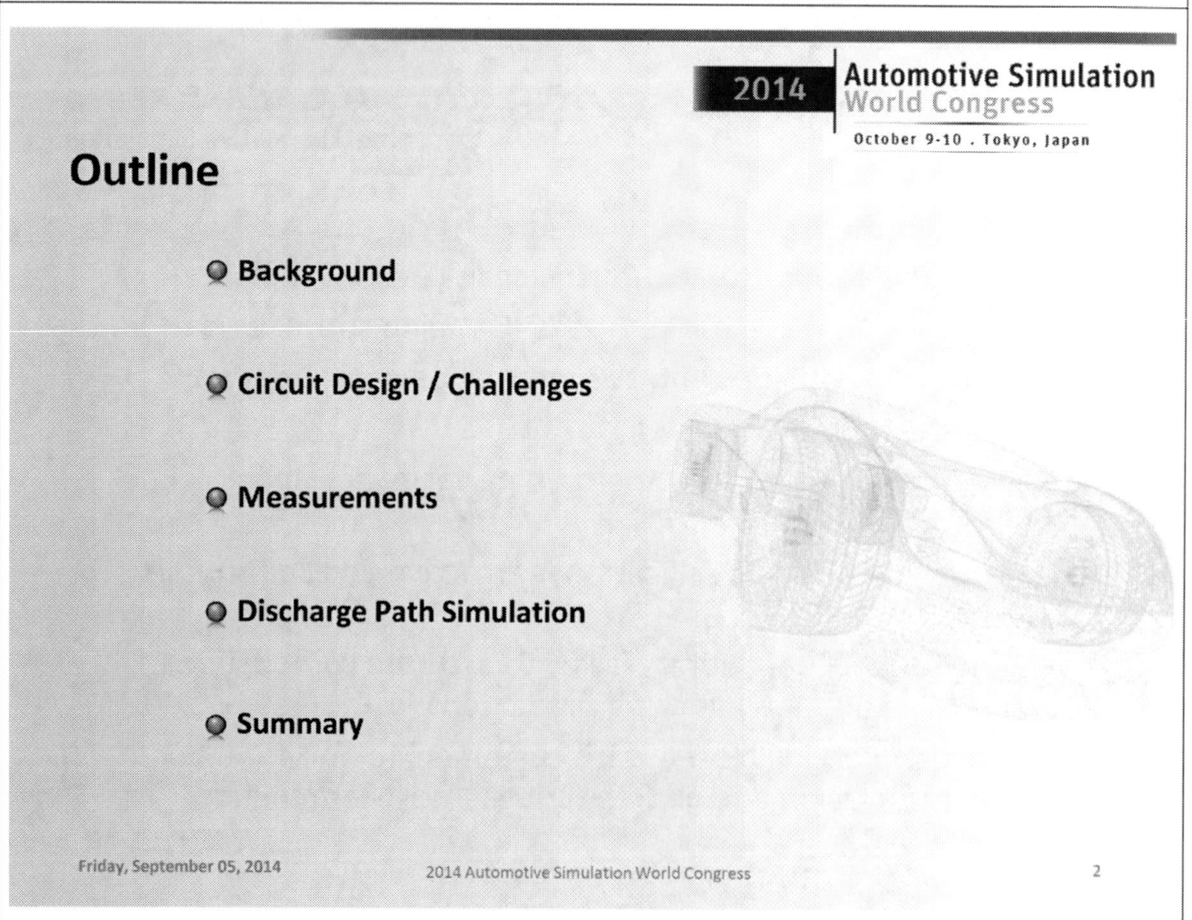

2014
Automotive Simulation
World Congress
October 9-10 . Tokyo, Japan
Outline
Background
Circuit Design / Challenges
Measurements
Discharge Path Simulation
Summary
Friday, September 05, 2014
2014 Automotive Simulation World Congress
2

# Background

**ESD(Electrostatic Discharge)**

- **System of ESD test**

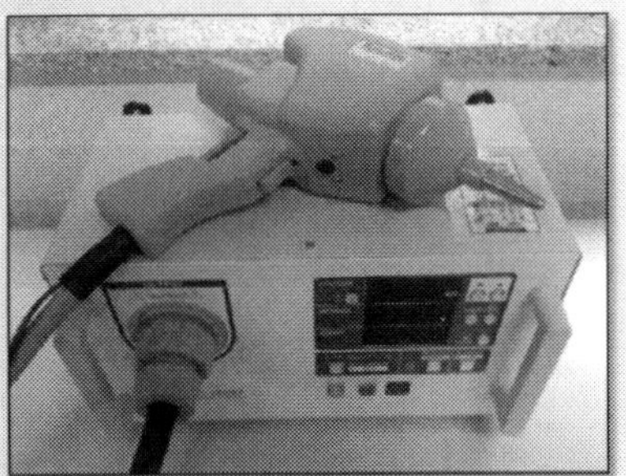

**- Contact Discharge**

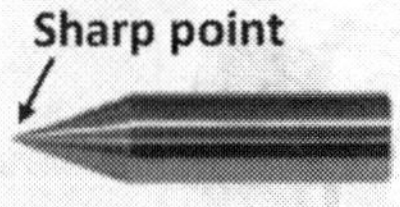


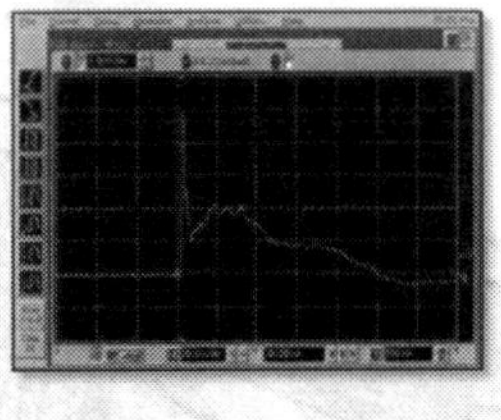

**- Non-contact(Air) Discharge**

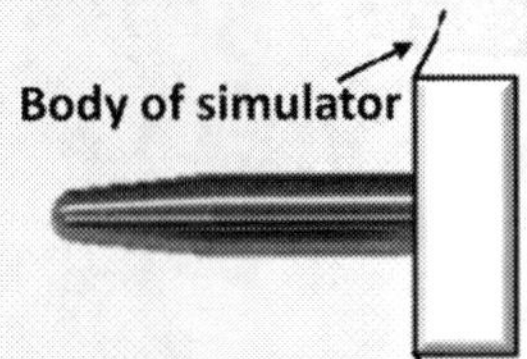


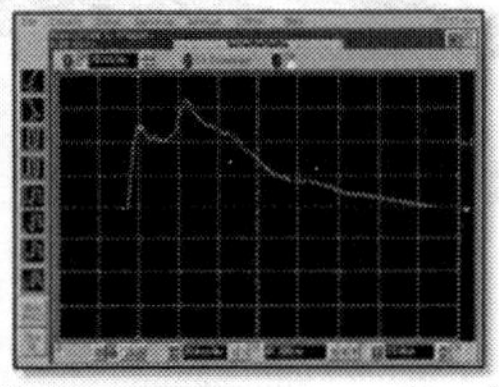

**– Case 1. Voltage, Current characteristics when ESD source was excited**



2014 Automotive Simulation World Congress
October 9-10 . Tokyo, Japan

# ESD Analysis

**Modeling of ESD Gun**

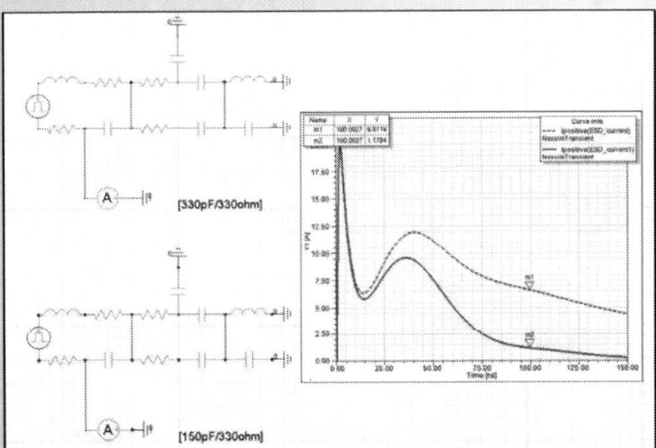


**– For 150pF/330pF 330Ω and 5kV**

- INTERNATIONAL STANDARD ISO10605

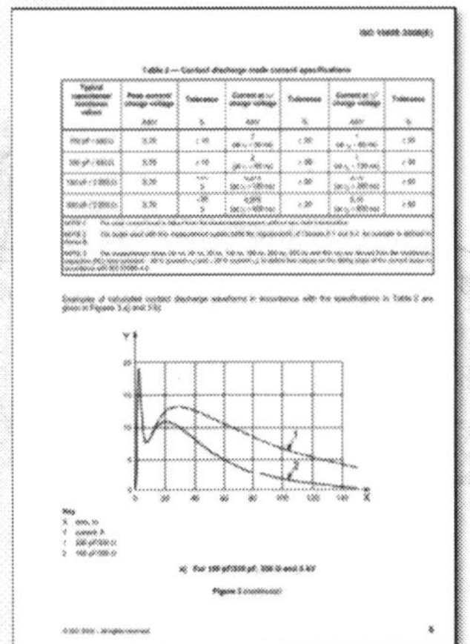

- Byoung-Su Seol, David Pommerenke "A Circuit Model for ESD Performance Analysis of Printed Circuit Boards

# ESD Analysis

## Passive Modeling

CQ1608X7R472K Capacitor Impedance

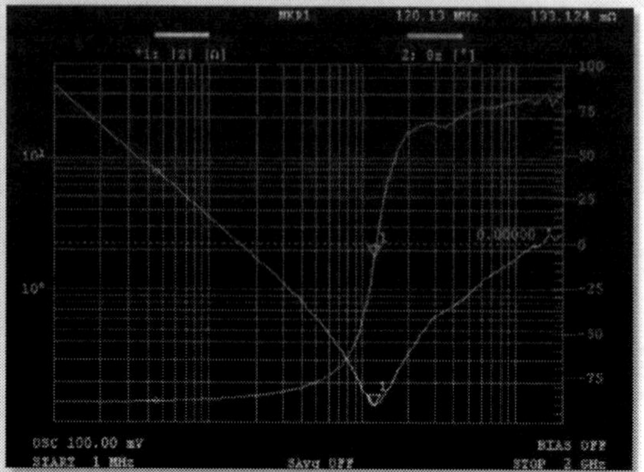

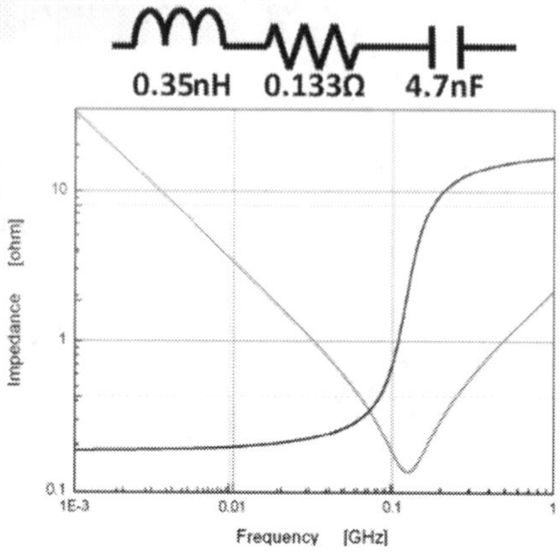


- Model all passive components with impedance analyzer
- To create frequency dependent equivalent network for device

2014 Automotive Simulation World Congress
October 9-10 . Tokyo, Japan

# ESD Analysis

## PCB/Circuit Simulation

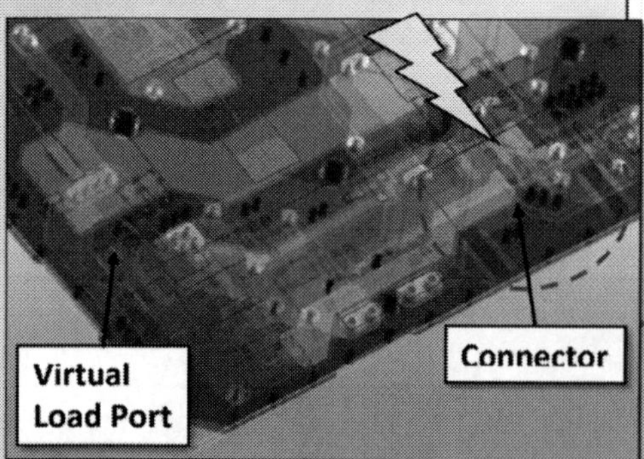


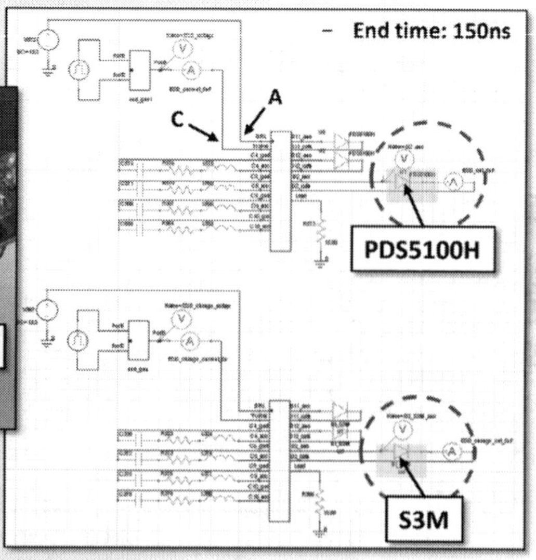

# ESD Analysis

## Voltage Level

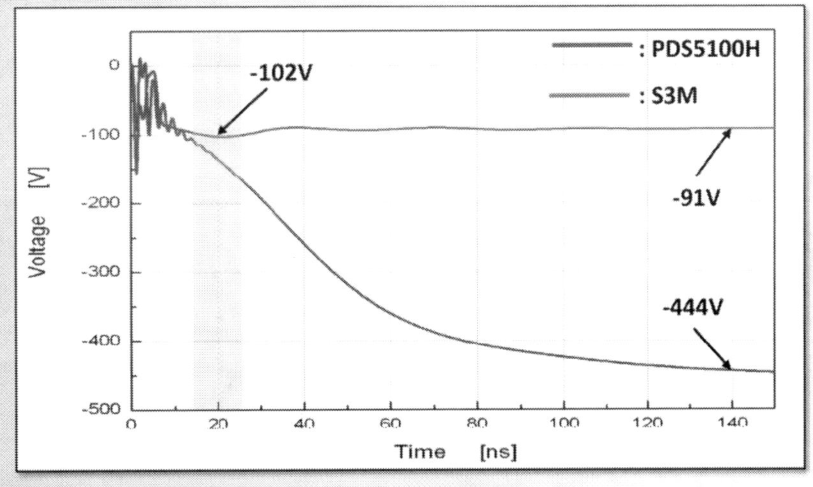


- Simulation result of the voltage PDS5100H, S3M anode

2014 Automotive Simulation World Congress
October 9-10 . Tokyo, Japan

# ESD Analysis

## Voltage Level

Peak Repetitive Reverse Voltage($V_{RRM}$): 100V

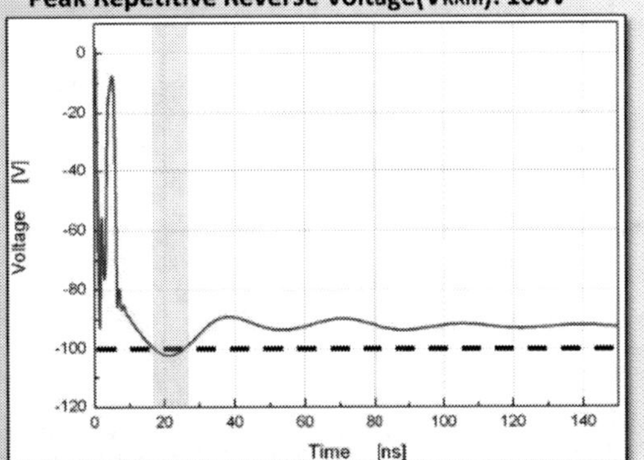


Peak Repetitive Reverse Voltage($V_{RRM}$): 1000V

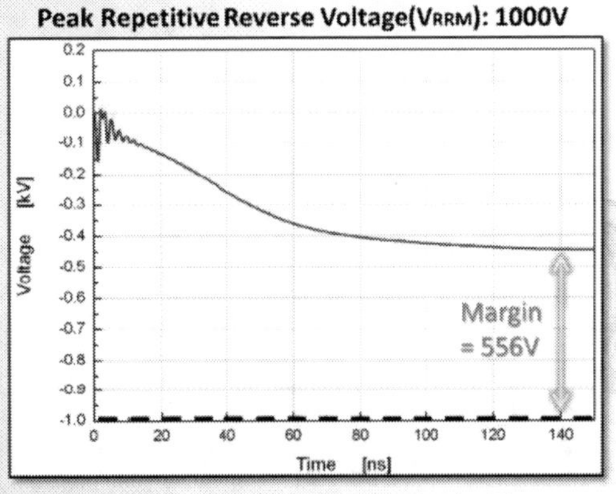


- PDS5100H: There is no margin compared with $V_{RRM}$
- S3M: There is 556V margin compared with $V_{RRM}$

# ESD Analysis

## Current Level

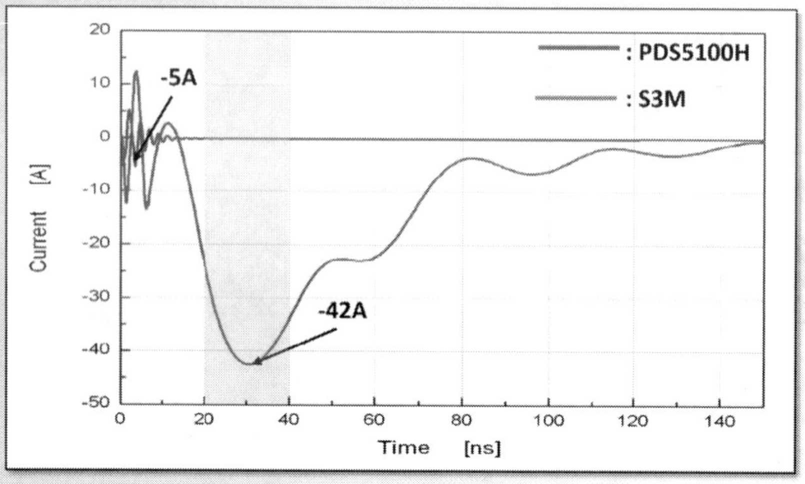


- Simulation result of the current PDS5100H, S3M cathode

2014 Automotive Simulation World Congress
October 9-10 . Tokyo, Japan

# ESD Analysis

## 2.5D PCB Simulation

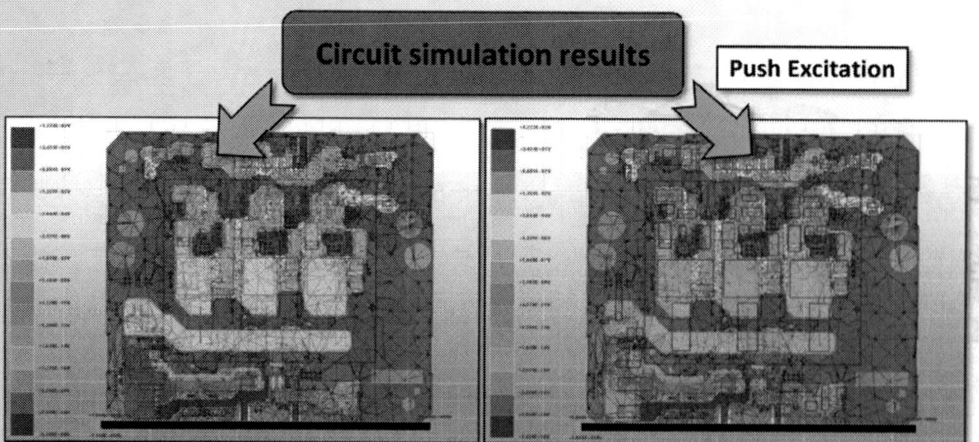


[PDS5100H] [S3M]

# ESD Analysis

### 2.5D PCB Simulation

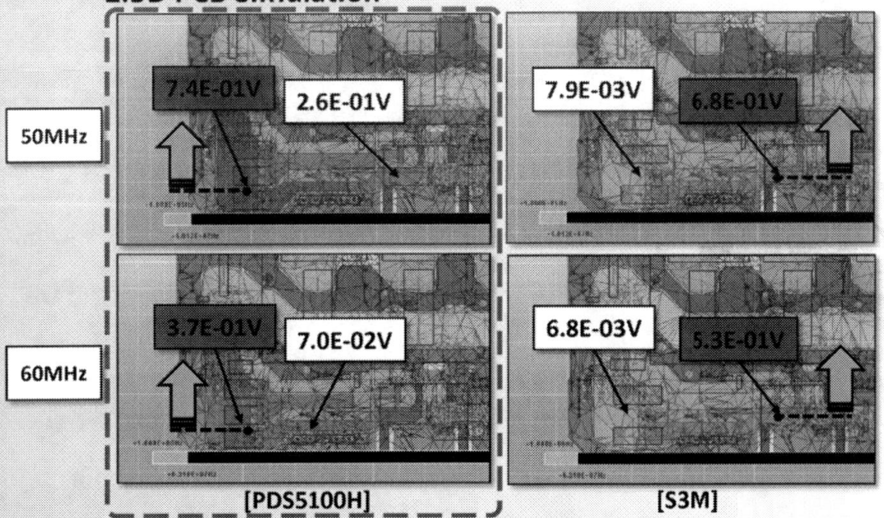


- Reverse Voltage occurred in Schottky Diode.(50, 60, 242, 289MHz)

2014 Automotive Simulation World Congress
October 9-10 . Tokyo, Japan

# Discharge Path Simulation

### PCB and ESD Source Modeling(HFSS)

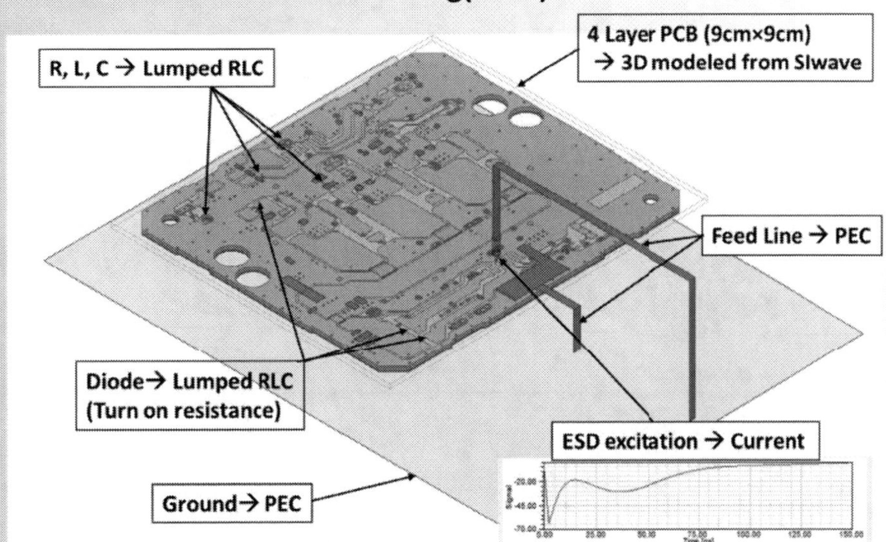

# Discharge Path Simulation

## Analysis Summary and Test Case

- Analyzing Mode : HFSS Transient Solver
- Analyzing Time : ~24 hours with 12 CPU(3GHz)
- Test Case
  1. Without RLC : bare PCB without any component
  2. With RLC : only R/L/C modeled with Lumped RLC
- Analyzing Mode : SIwave Frequency sweep
- Test Case
  1. With RLC





# Discharge Path Simulation

## ESD Propagation

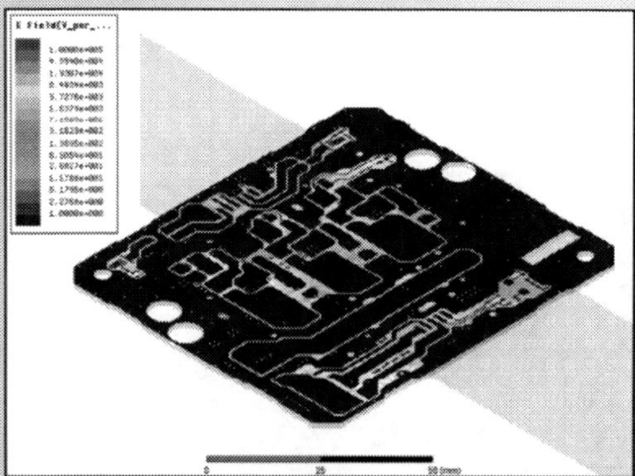

[Top Layer]

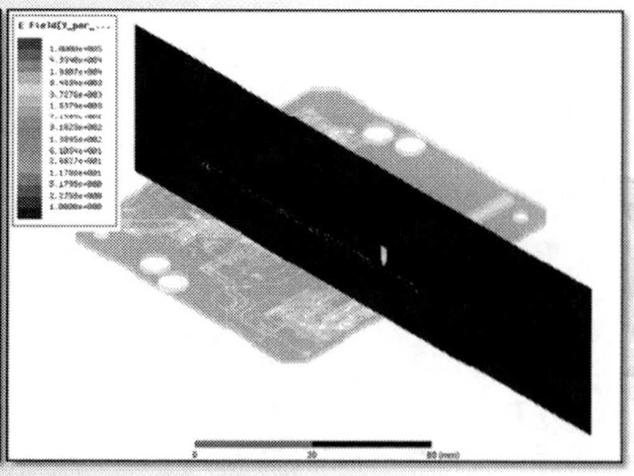

[Cross Section]

2014 Automotive Simulation World Congress
October 9-10 . Tokyo, Japan

# Discharge Path Simulation

1 Layer

4 Layer

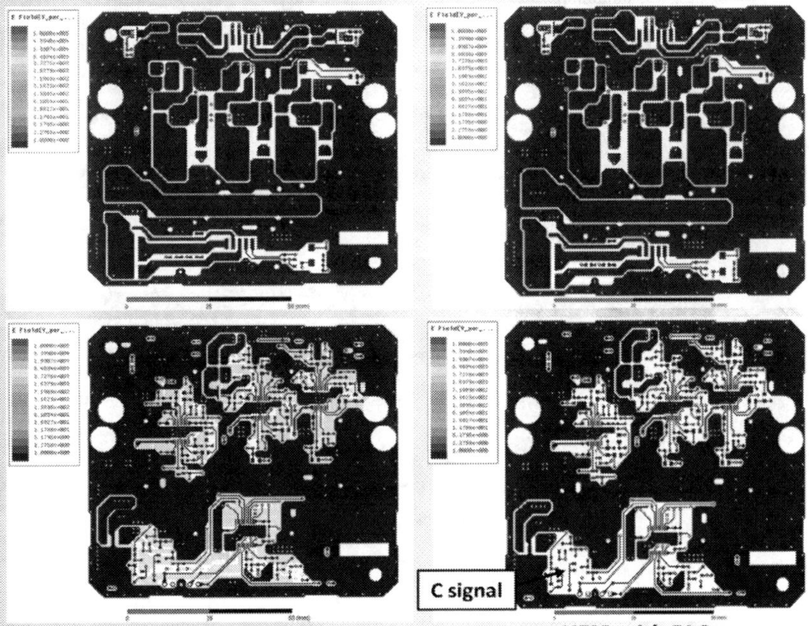


HFSS No RLC

HFSS with RLC



2014 Automotive Simulation World Congress
October 9-10 . Tokyo, Japan

# Discharge Path Simulation

3 Layer

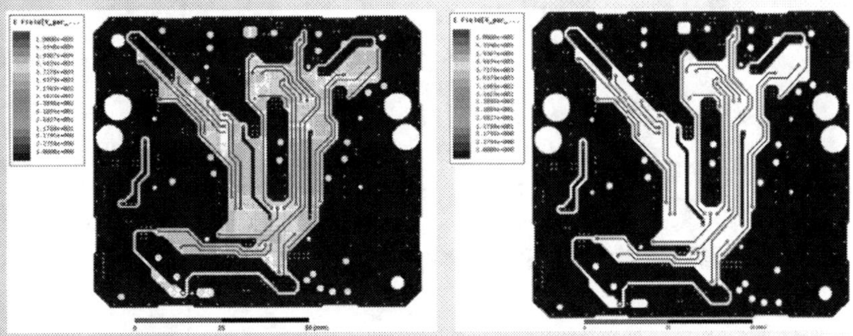

HFSS No RLC

HFSS with RLC

## Discharge Path Simulation

1 Layer

4 Layer

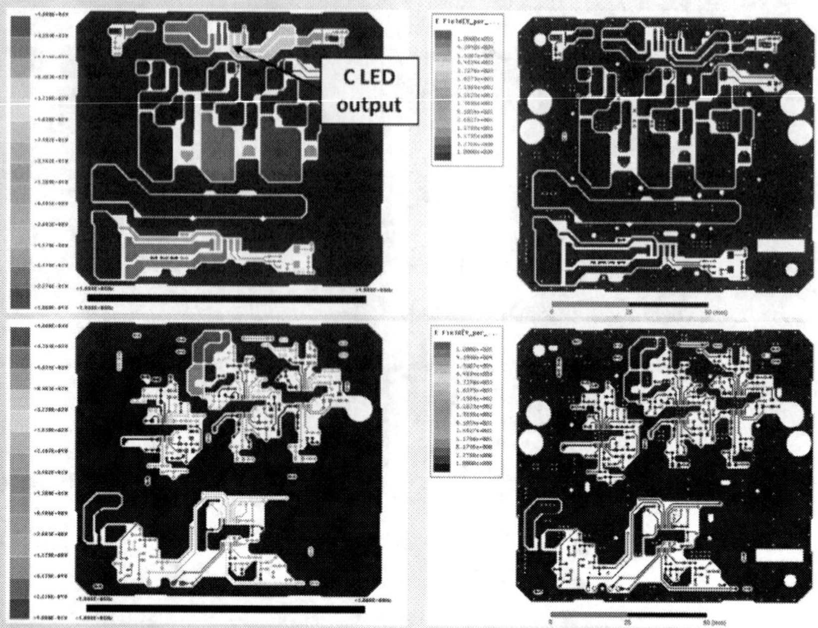


Siwave with RLC

HFSS with RLC





## Background

3 Layer

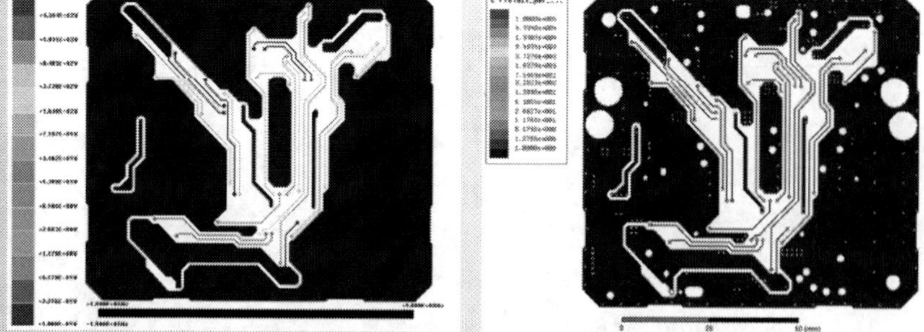

Siwave with RLC

HFSS with RLC

2014 Automotive Simulation World Congress
October 9-10 . Tokyo, Japan

## Summary

- We can analyze ESD voltage, current level using SIwave & Desinger SI
  - Consider actual load placement / Circuit operation
- Also reverse voltage characterization similarly occurred like the actual situation
- Prediction of Discharge path with RLC model
- Future work for HFSS transient(ESD)
  - Add active component characterization
  - Solving Method of big size model

# Simulation of stray magnetic couplings in EMC filters for automotive power electronic devices

EMC: Electromagnetic compatibility

October 10th, 2014

Takashi Masuzawa
DENSO CORPORATION

DENSO 2014 Automotive Simulation World Congress



## Contents



- Background and motivation
- Simplification method of stray magnetic coupling
- Application to EPS-ECU
  (EPS: Electronic Power Steering, ECU: Electronic Control Unit)
- Conclusion

DENSO 2014 Automotive Simulation World Congress

## Difficulty of EMC filter design



Performance can't easily meet expected value.

Filter components are added and redesigned until required performance is reached.

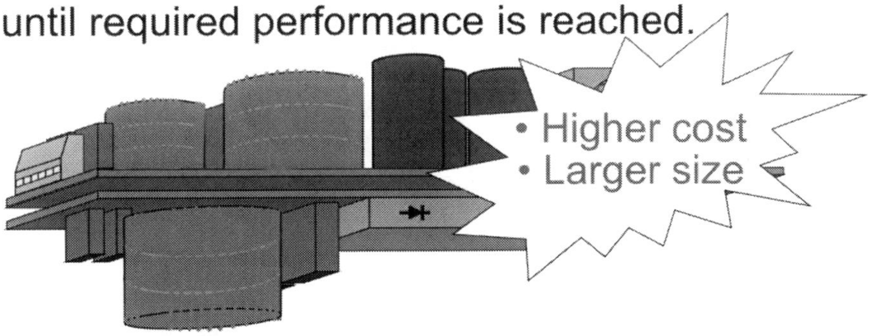


Accurate filter simulation is needed.




## Filter simulation for EMC filter design



**Level 0 : Ideal passive component**

**Level 1.0 : Non-ideal passive component**

**Level 2.0 : 1.0 + stray impedance of PCB track**

**Level 3.0 : 2.0 + stray magnetic coupling**

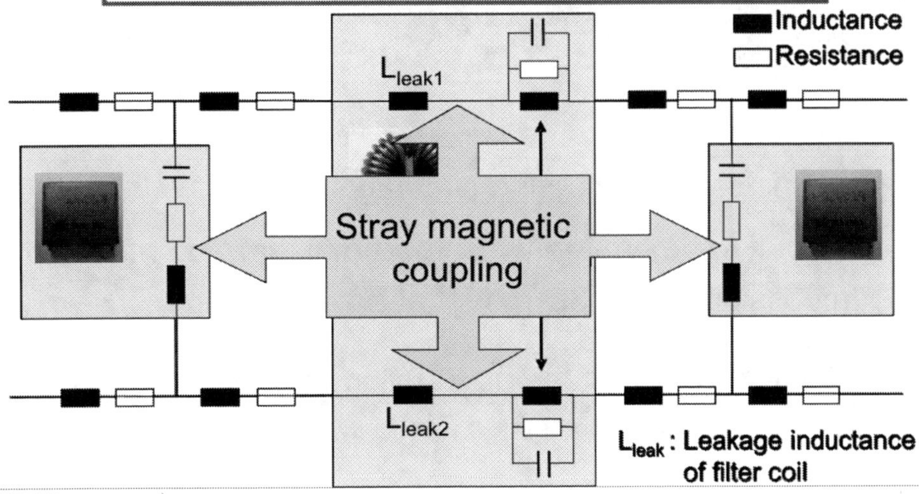

## Complexity to consider stray magnetic coupling



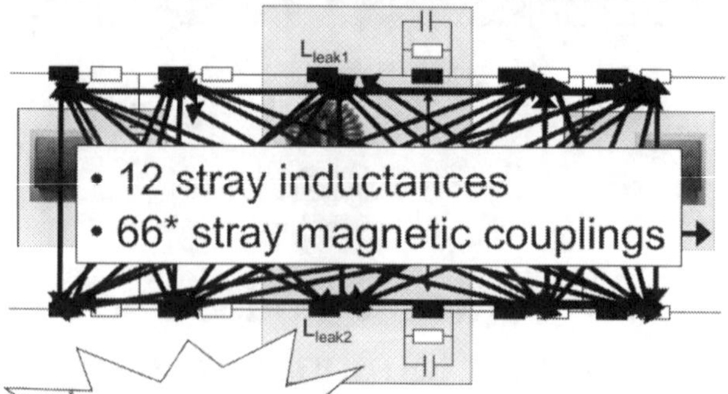


(*)66 = $_{12}C_2$

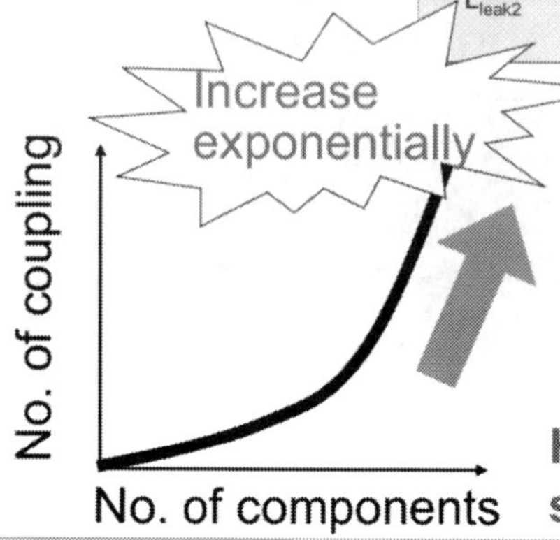

**Hundreds of couplings should be considered !!**





## Motivation for topic of research



**Circuit design (Circuit simulation)** **Structural design (Magnetic simulation)**

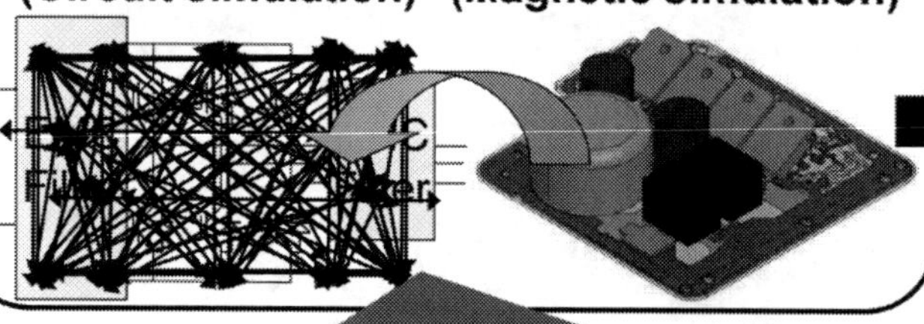

**Performance evaluation**

OK

NG

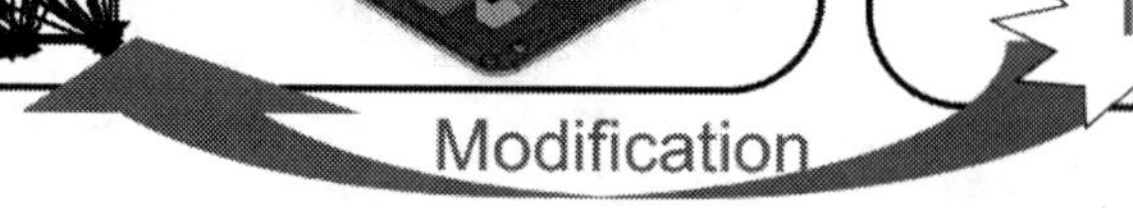


<u>Problems of process using simulation level 3.0</u>

- Difficult to determine which part of filter to be modified
- Impossible to get 3D model at an early stage

Simplification to extract major couplings has to be developed.

(Simulation level 3.5: 2.0 + major stray magnetic coupling)

# Contents



- Background and motivation
- Simplification method of stray magnetic coupling
- Application to EPS-ECU
  (EPS: Electronic Power Steering, ECU: Electronic Control Unit)
- Conclusion




# Mechanism of stray magnetic couplings



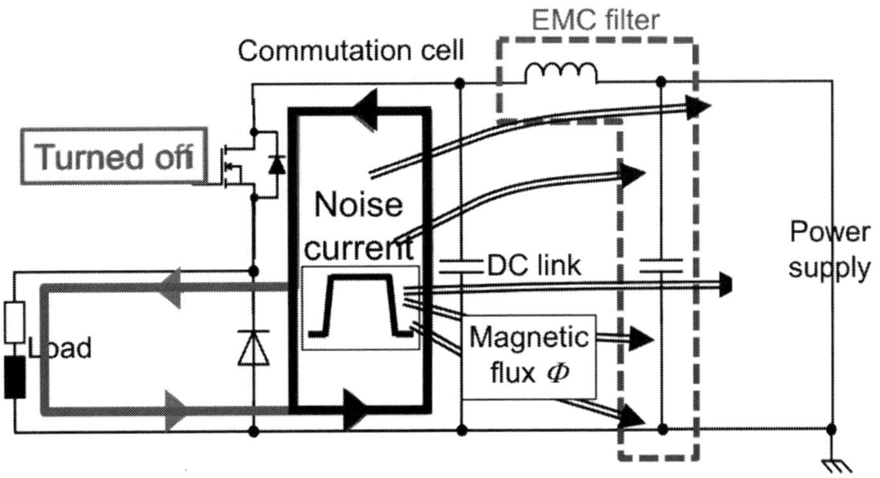


The noise current flowing in the commutation cell generate magnetic flux having significant influences on the EMC filter.

## Influence of stray magnetic couplings



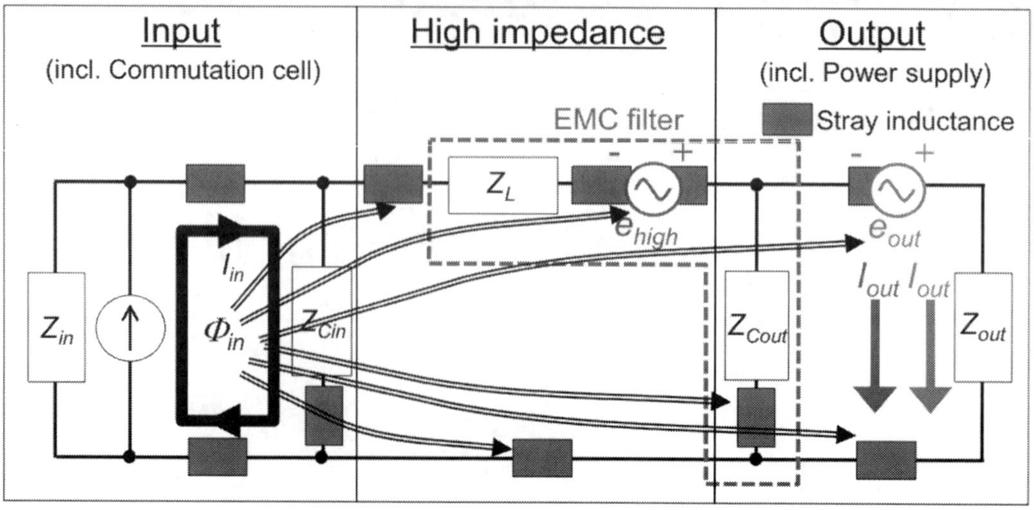


$I_{in}$: Noise current in the input loop, $I_{out}$: Noise current in the output loop,
$\Phi_{in}$: Magnetic flux generated by $I_{in}$, $Z_{in}$: Impedance of input, $Z_{out}$: Impedance of output,
$Z_L$: Impedance of coil, $Z_{Cin}$: Impedance of input capacitor, $Z_{Cout}$: Impedance of output capacitor
$e_{high}$: Induced voltage in the high impedance area, $e_{out}$: Induced voltage in the output loop




## $I_{out}$ caused by $e_{high}$



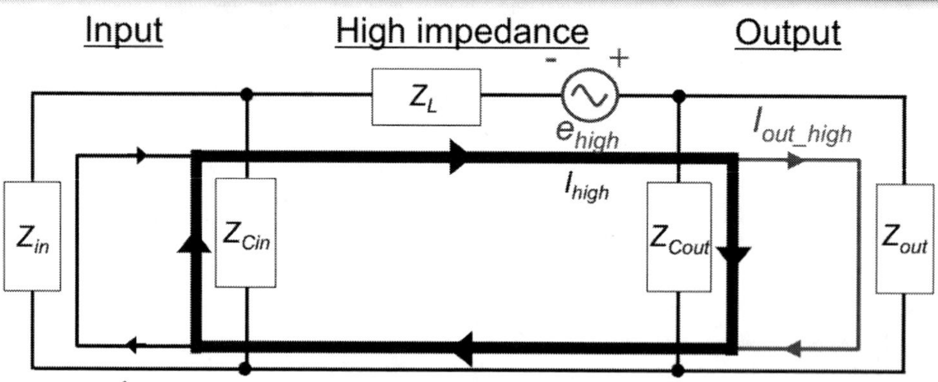


[ $I_{high}$: Noise current in the high impedance area, $I_{out\_high}$ : $I_{out}$ caused by $e_{high}$ ]

$$I_{out\_high} = \frac{Z_{Cout}}{Z_{Cout} + Z_{out}} I_{high}$$

$$= \frac{Z_{Cout}}{Z_{Cout} + Z_{out}} \frac{e_{high}}{Z_L}$$

Since

$$Z_L >> Z_{Cin}, Z_{Cout}$$

$$I_{out\_high} = \frac{Z_{Cout}}{Z_{Cout} + Z_{out}} \frac{e_{high}}{Z_L}$$

## $I_{out}$ caused by $e_{out}$

10 / 20

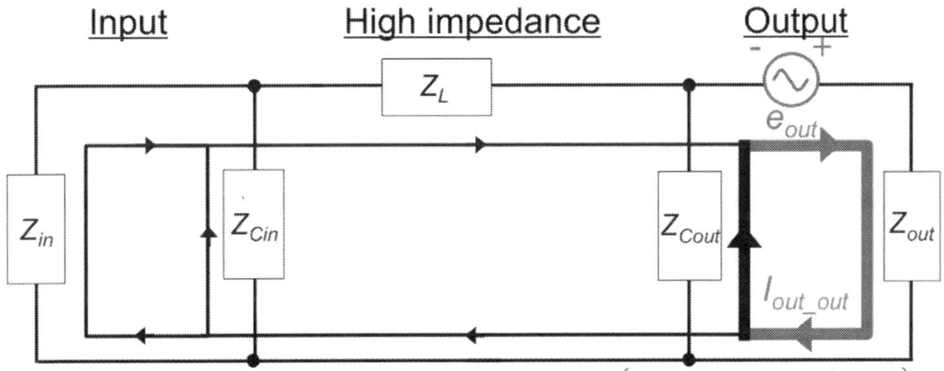


[ $I_{out\_out}$ : $I_{out}$ caused by $e_{out}$ ]

$$I_{out\_out} = \frac{e_{out}}{(Z_{in} // Z_{Cin} + Z_L) // Z_{Cout} + Z_{out}}$$

$$= \frac{e_{out}}{Z_L Z_{Cout} / Z_L + Z_{out}}$$

Since

$$Z_L >> Z_{Cin},\ Z_{Cout}$$

$$\Rightarrow I_{out\_out} = \frac{e_{out}}{Z_{Cout} + Z_{out}}$$





## Extraction of major stray magnetic coupling

11 / 20

$$I_{out\_high} = \frac{Z_{Cout}}{Z_{Cout} + Z_{out}} \frac{e_{high}}{Z_L}$$

$$I_{out\_out} = \frac{e_{out}}{Z_{Cout} + Z_{out}}$$

**Ratio of output current $r_{Iout}$**

$$\left| \frac{I_{out\_high}}{I_{out\_out}} \right| = \left| \frac{Z_{Cout}}{Z_L} \right| \left| \frac{M_{in-L}}{M_{in-out}} \right| << 1$$

$$e_{out} = M_{in-out} \frac{dI_{in}}{dt}$$

$$\because Z_L >> Z_{Cout}$$

$$e_{high} \cong M_{in-L} \frac{dI_{in}}{dt}$$

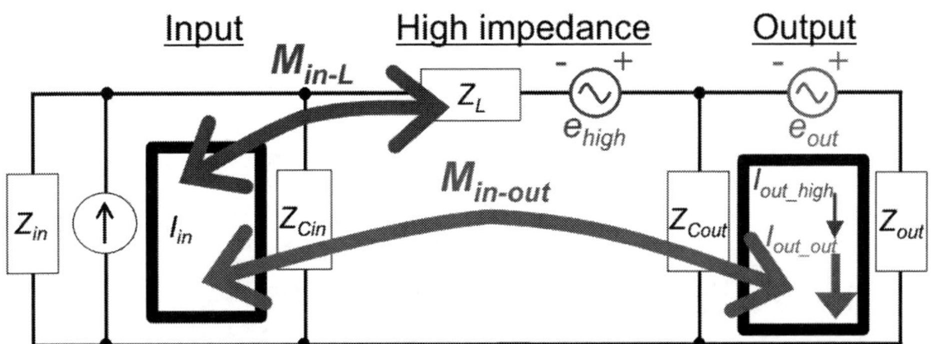


$M_{in-out}$: coupling between input and output loop, $M_{in-L}$: coupling between input loop and filter coil

Major stray magnetic couplings are $M_{in-out}$ and $M_{in-L}$.

# Contents



- Background and motivation
- Simplification method of stray magnetic coupling
- **Application to EPS-ECU**
  (EPS: Electronic Power Steering, ECU: Electronic Control Unit)
- Conclusion




# Tested EMC filter (EPS-ECU)



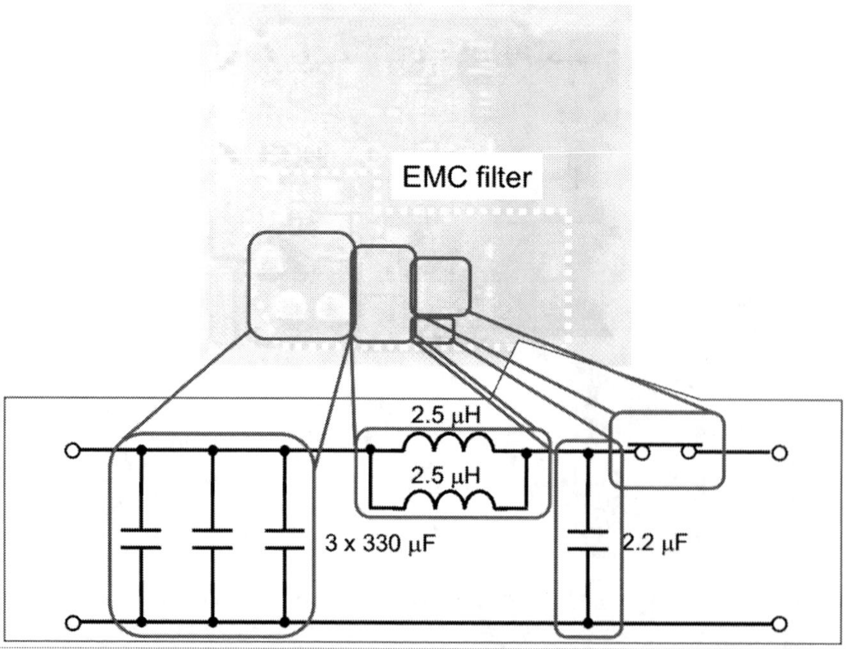

# Simulation result of $M_{in\text{-}L}$ and $M_{in\text{-}out}$



Software: ANSYS Q3D Extractor ANSYS

**Current distribution**

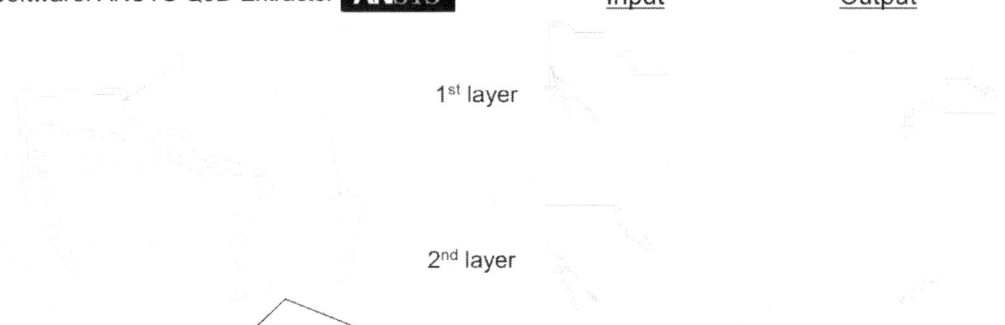


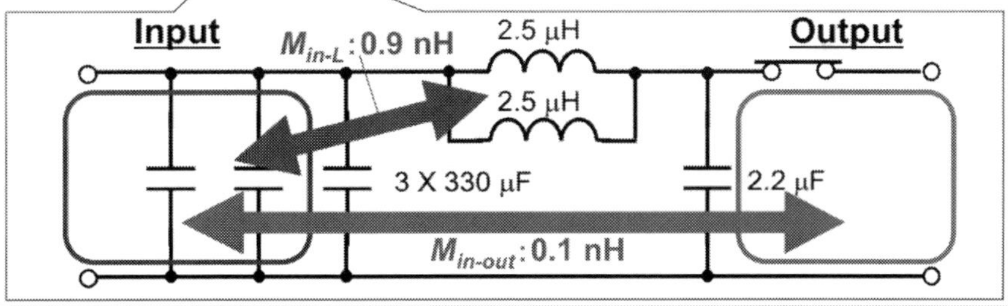






# Extraction of major stray magnetic coupling



**Ratio of output current $r_{Iout}$**

$$\left|\frac{I_{out\_high}}{I_{out\_out}}\right| = \left|\frac{Z_{Cout}}{Z_L}\right| \frac{M_{in-L}}{M_{in-out}} << 1$$

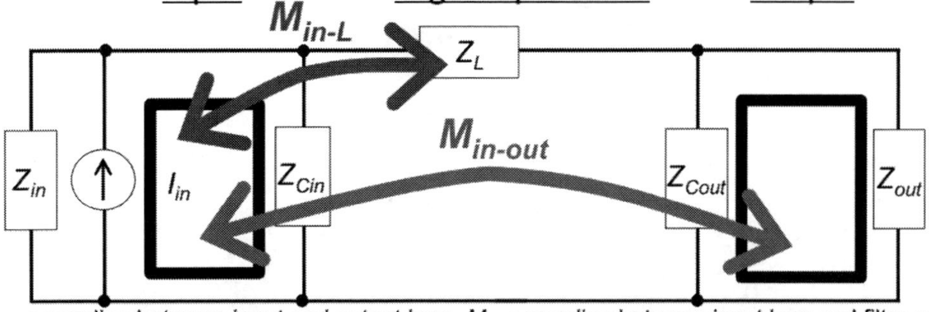


$M_{in\text{-}out}$ : coupling between input and output loop, $M_{in\text{-}L}$: coupling between input loop and filter coil

Major stray magnetic couplings are $M_{in\text{-}out}$ and $M_{in\text{-}L}$.

## Major stray magnetic coupling for tested EMC filter



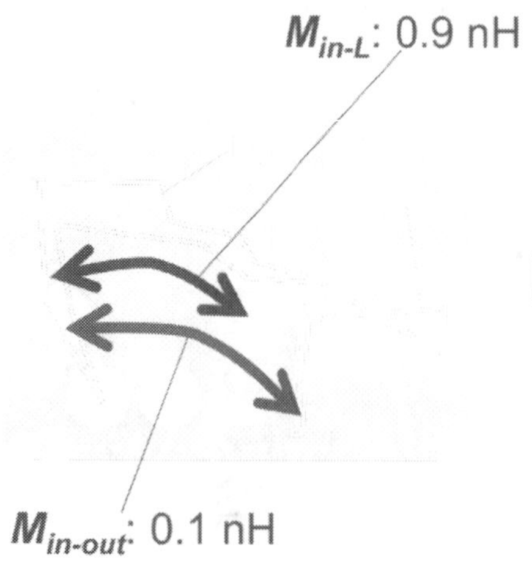


$$r_{Iout} = \left| \frac{I_{out\_high}}{I_{out\_out}} \right| = \left| \frac{Z_{Cout}}{Z_L} \right| \frac{M_{in\_L}}{M_{in\_out}}$$

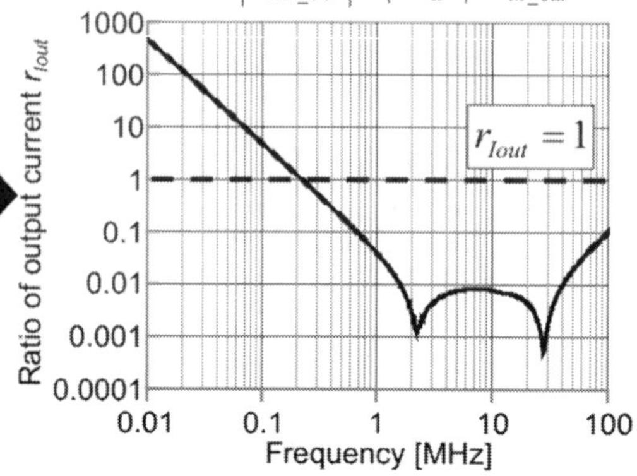


**Major coupling for the tested EMC filter is $M_{in\text{-}out}$.**




## Measurement configuration



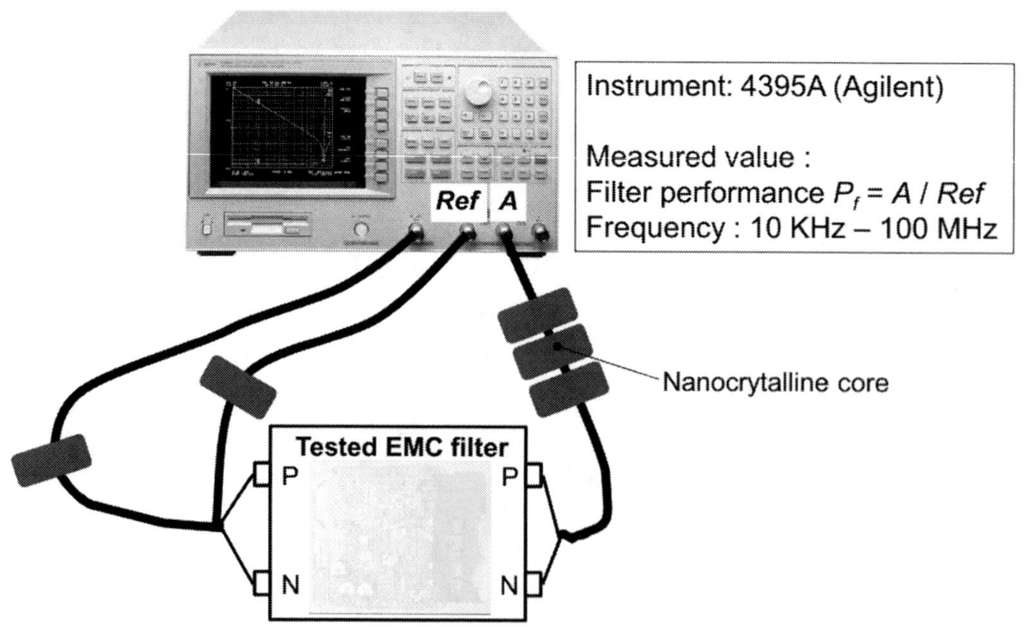

## Circuit simulation model



Simulation level 2.0 : 1.0 + stray impedance of PCB track
Simulation level 3.5 : 2.0 + major stray magnetic coupling

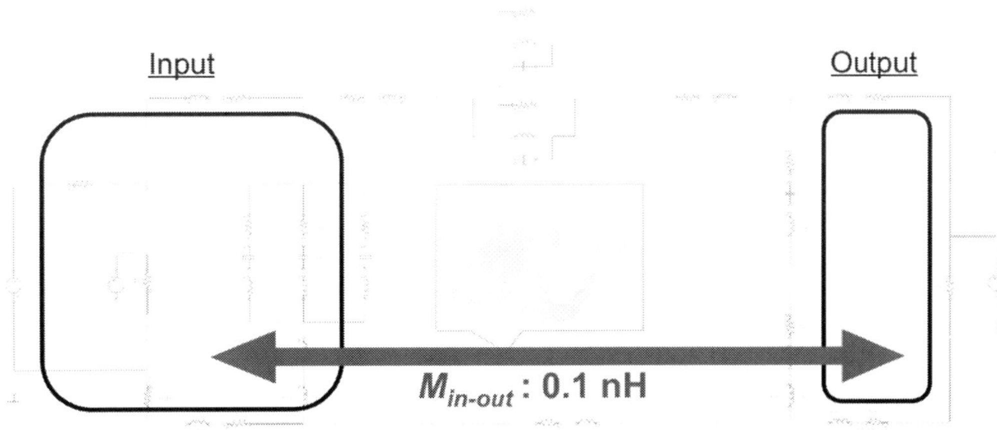


Software: ANSYS Simplorer ANSYS




## Comparison of filter performance $P_f$



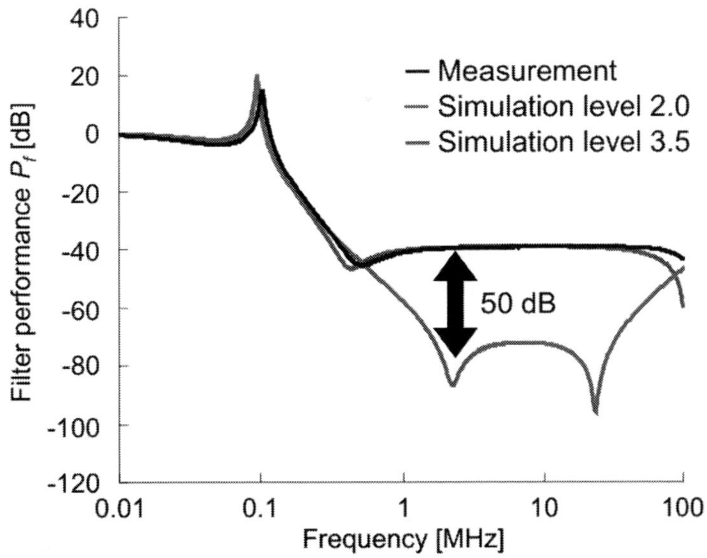


Measurement and simulation level 3.5 show a good agreement.

## Conclusion



- Highly efficient modeling method of stray magnetic couplings was proposed.
- Its effectiveness was verified by comparing between the measurement and the simulation.
- Results of the comparison showed that the proposed modeling method could significantly reduce the number of stray magnetic couplings to be considered for an accurate filter simulation.





DENSO

ANSYS Automotive Simulation World Congress
2D-4

Panasonic

# Applying simulation technologies to the automotive product development

Hiroshi Higashitani

Industrial Business Development Center (IBDC),
Automotive & Industrial Systems Company,
Panasonic Corporation

October 7, 2014

1

## Mission of our organization

Technology / Products / Global & Group

Integrate values of Panasonic

**Total Solution**

Troubles in product design → Solution to problems in circuit design

Next-generation dream → Creation of attractive added values

Customer

Device library
Simulation analysis
Evaluation by measurements

・Proposal of applications & solutions
[ Thermal /Noise /Power /Sensor/ Miniaturization ]

Contribute to enhancement of values of customer's products

2

## Agenda

1. Summary of the noise solution activity
2. Example 1) Radiation noise
3. Example 2) Conduction noise
4. Example 3) Immunity (static electricity)
5. Example 4) Immunity (Radio irradiation)

3

## 1. Summary of the noise solution activity

A product trend and development design issue

**Set trend**

- Multifunction
- Low power operation
- High speed and frequency
- Severe law regulations of the Wireless Telegraphy Act

**development design issue**

- Securing basic performance
- Reliability assurance
- The laws and regulations compliance

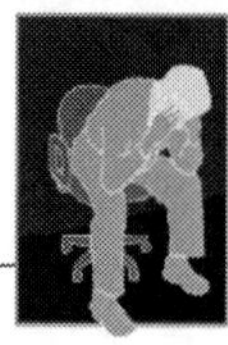

- Complicating noise and thermal issues
- Difficulty for countermeasure

Panasonic provide not only a device, but also "Solution" to solve your problems.

4

# 1. Summary of the noise solution activity

## From noise and thermal measures to design

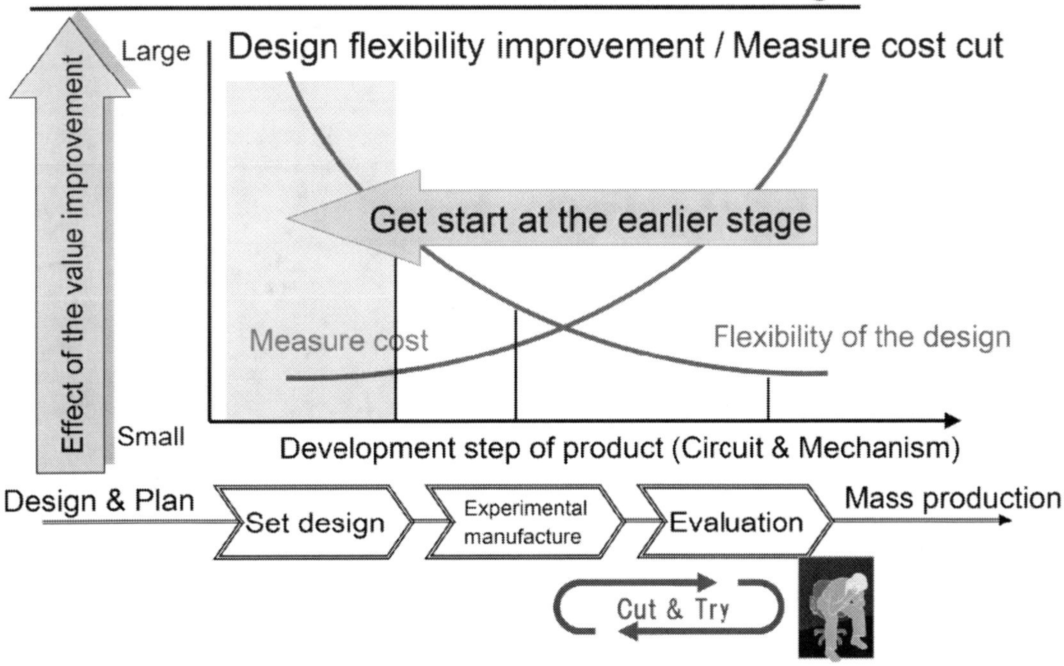


5

# 1. Summary of the noise solution activity

## Drastic measures of noise and thermal issues at the earlier stage

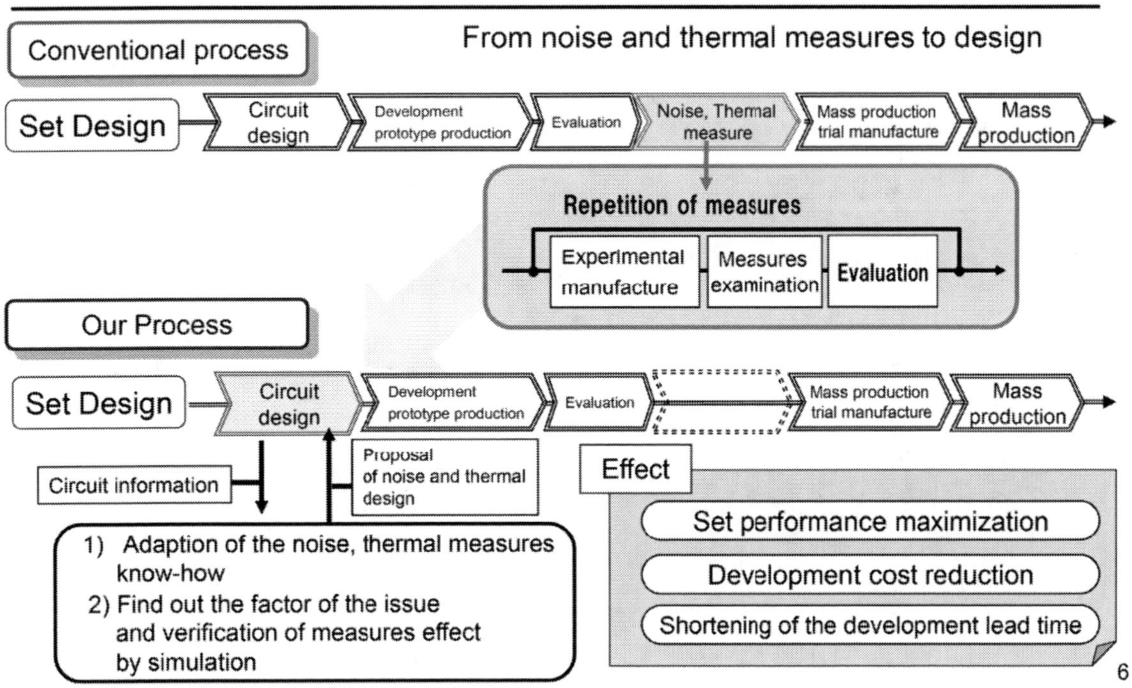


6

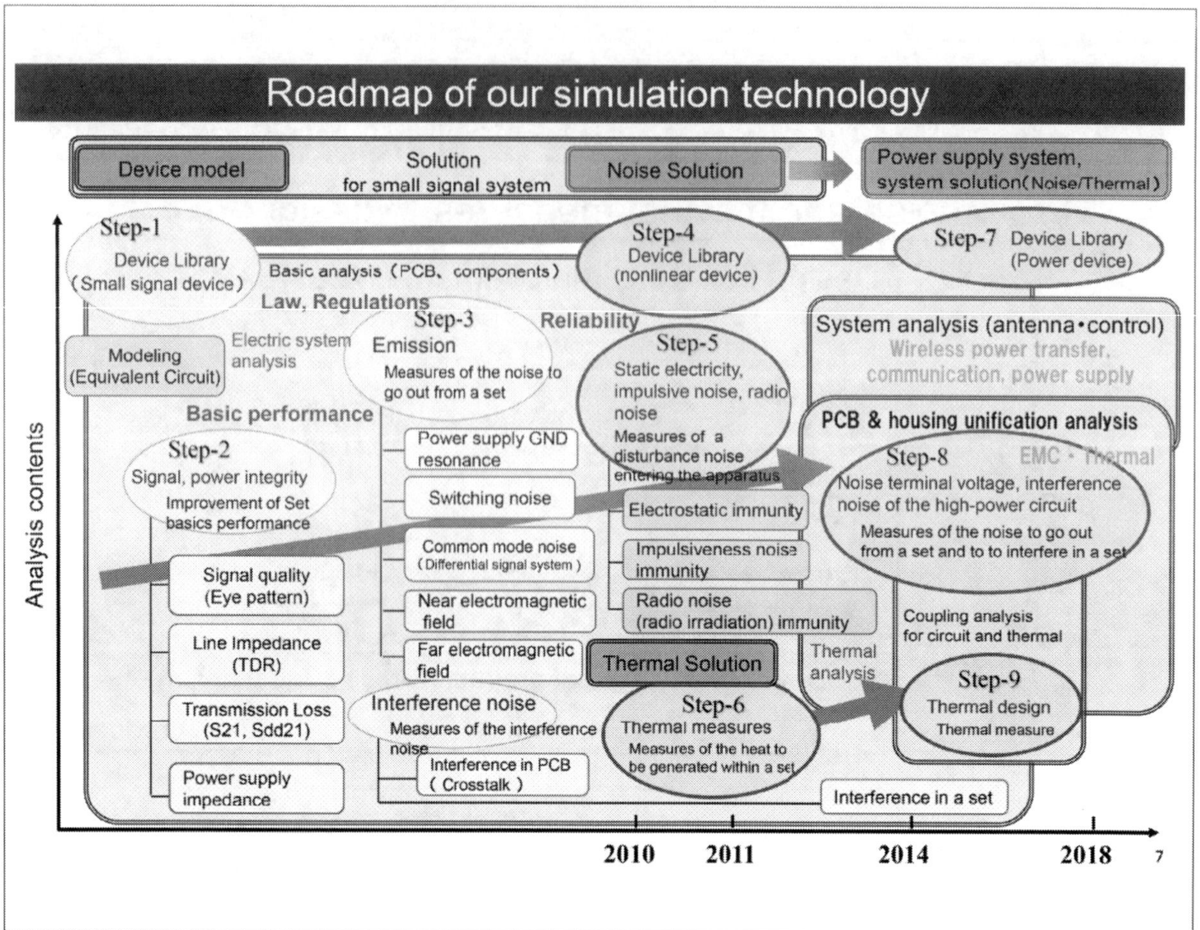
Roadmap of our simulation technology
Device model
Solution for small signal system
Noise Solution
Power supply system, system solution(Noise/Thermal)
Step-1 Device Library (Small signal device)
Basic analysis (PCB, components)
Law, Regulations
Step-4 Device Library (nonlinear device)
Step-7 Device Library (Power device)
Modeling (Equivalent Circuit)
Electric system analysis
Step-3 Emission
Measures of the noise to go out from a set
Reliability
Step-5
Static electricity, impulsive noise, radio noise
Measures of a disturbance noise entering the apparatus
System analysis (antenna•control)
Wireless power transfer, communication, power supply
PCB & housing unification analysis
Basic performance
Step-2 Signal, power integrity
Improvement of Set basics performance
Power supply GND resonance
Switching noise
Common mode noise (Differential signal system)
Near electromagnetic field
Far electromagnetic field
Electrostatic immunity
Impulsiveness noise immunity
Radio noise (radio irradiation) immunity
Step-8 EMC • Thermal
Noise terminal voltage, interference noise of the high-power circuit
Measures of the noise to go out from a set and to to interfere in a set
Coupling analysis for circuit and thermal
Thermal analysis
Signal quality (Eye pattern)
Line Impedance (TDR)
Transmission Loss (S21, Sdd21)
Power supply impedance
Interference noise
Measures of the interference noise
Interference in PCB (Crosstalk)
Thermal Solution
Step-6 Thermal measures
Measures of the heat to be generated within a set
Step-9 Thermal design
Thermal measure
Interference in a set
Analysis contents
2010
2011
2014
2018
7

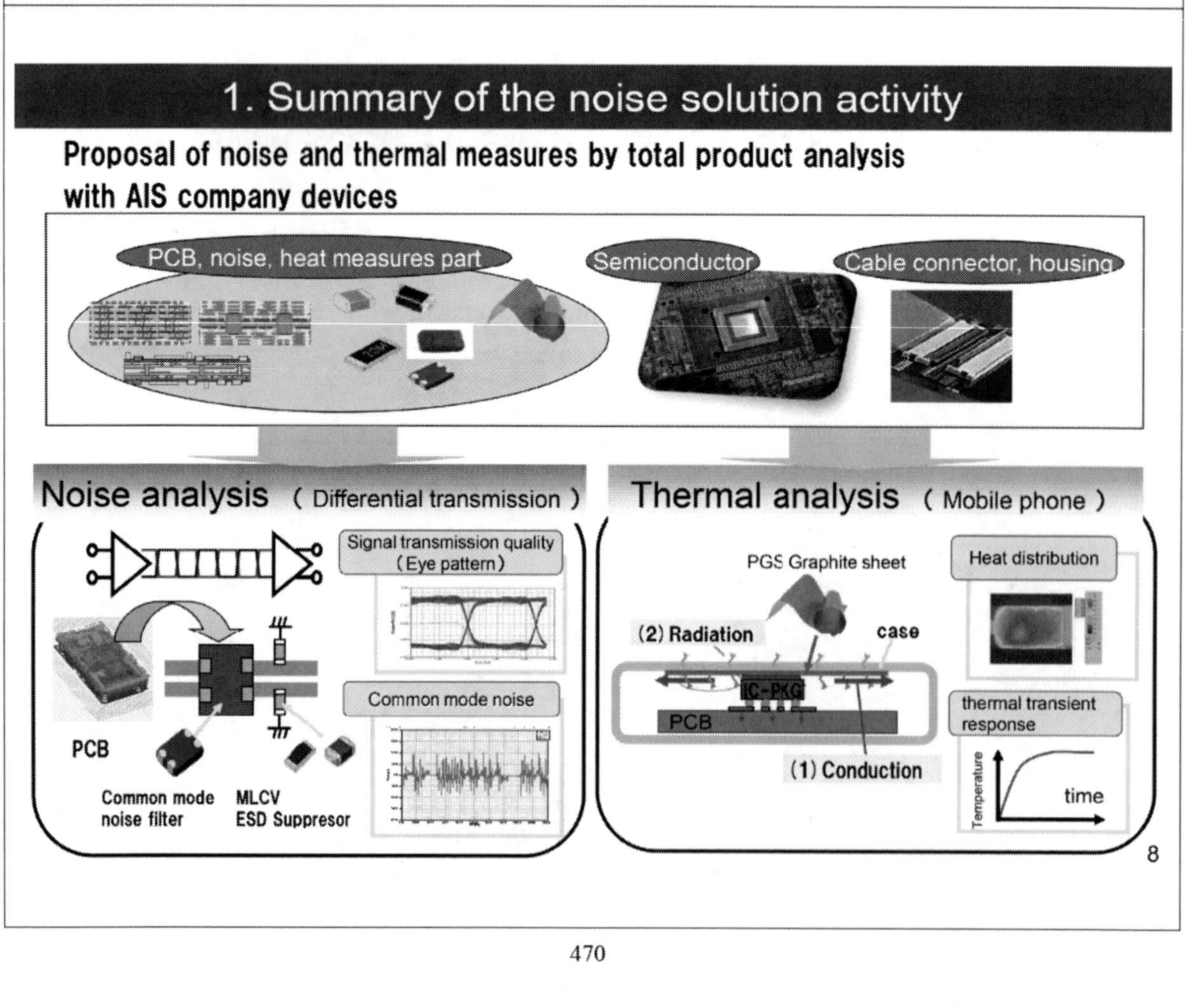
1. Summary of the noise solution activity
Proposal of noise and thermal measures by total product analysis with AIS company devices
PCB, noise, heat measures part
Semiconductor
Cable connector, housing
Noise analysis (Differential transmission)
Signal transmission quality (Eye pattern)
Common mode noise
PCB
Common mode noise filter
MLCV ESD Suppresor
Thermal analysis (Mobile phone)
PGS Graphite sheet
(2) Radiation
case
IC-PKG
PCB
(1) Conduction
Heat distribution
thermal transient response
Temperature
time
8

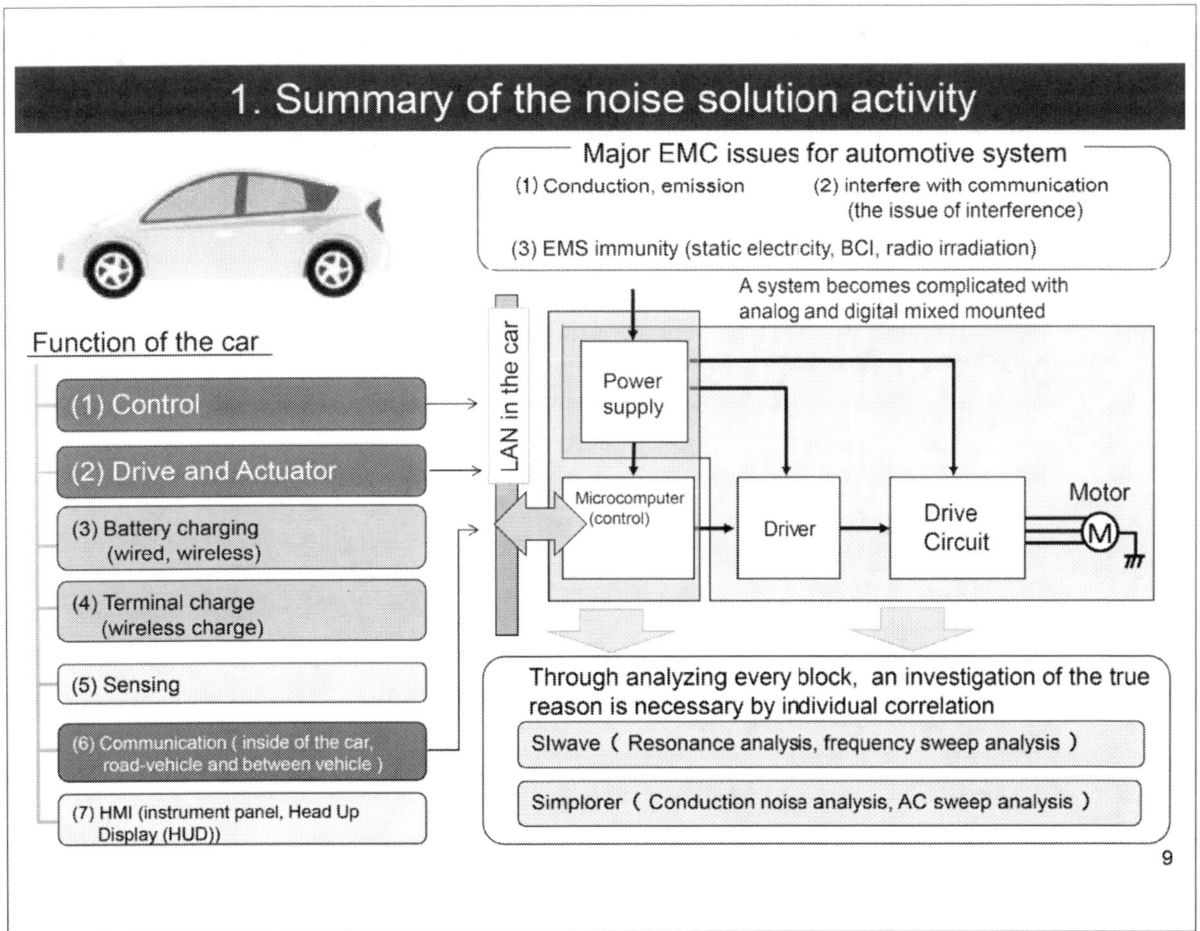
1. Summary of the noise solution activity
Major EMC issues for automotive system
(1) Conduction, emission
(2) interfere with communication (the issue of interference)
(3) EMS immunity (static electricity, BCI, radio irradiation)
A system becomes complicated with analog and digital mixed mounted
Function of the car
(1) Control
(2) Drive and Actuator
(3) Battery charging (wired, wireless)
(4) Terminal charge (wireless charge)
(5) Sensing
(6) Communication ( inside of the car, road-vehicle and between vehicle )
(7) HMI (instrument panel, Head Up Display (HUD))
LAN in the car
Power supply
Microcomputer (control)
Driver
Drive Circuit
Motor
M
Through analyzing every block, an investigation of the true reason is necessary by individual correlation
SIwave ( Resonance analysis, frequency sweep analysis )
Simplorer ( Conduction noise analysis, AC sweep analysis )
9

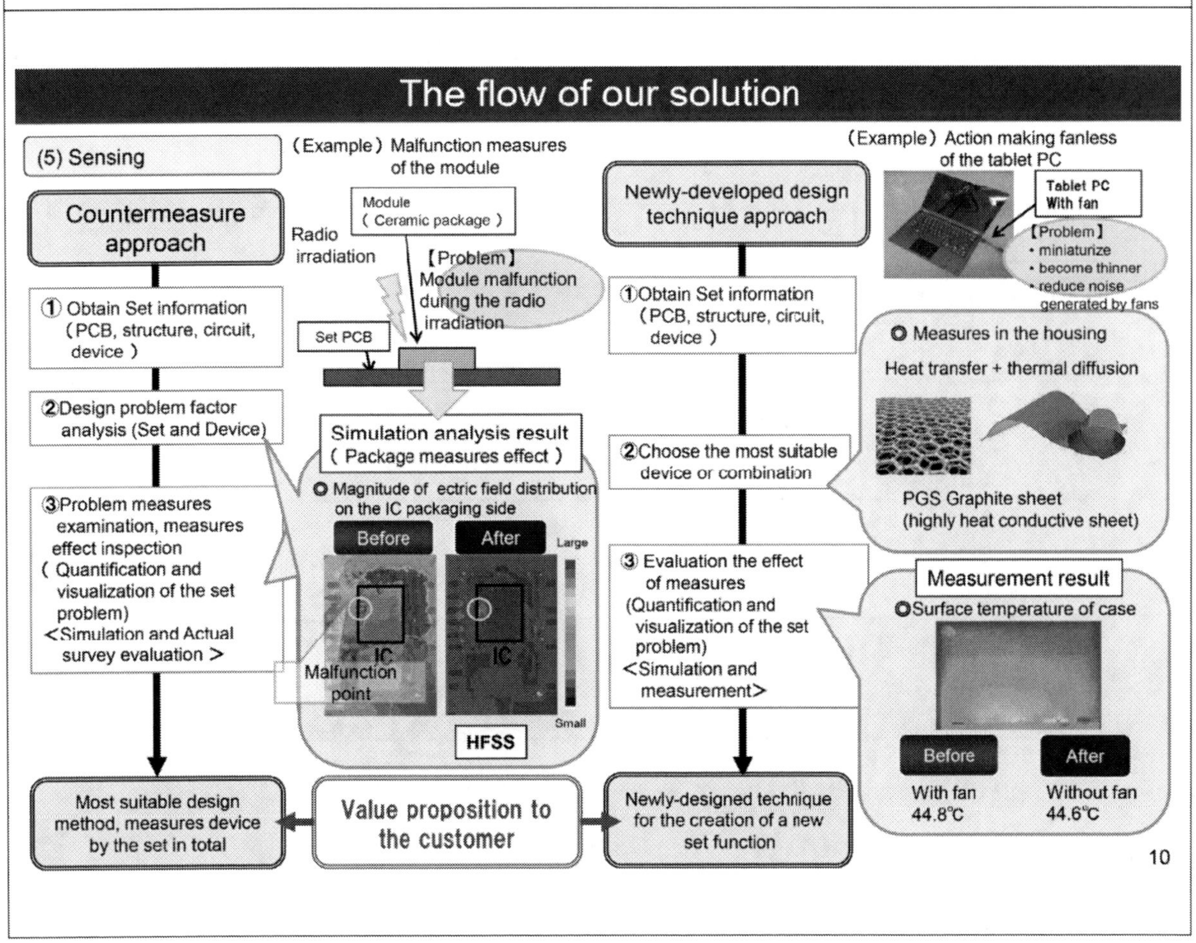
The flow of our solution
(5) Sensing
Countermeasure approach
① Obtain Set information (PCB, structure, circuit, device )
②Design problem factor analysis (Set and Device)
③Problem measures examination, measures effect inspection ( Quantification and visualization of the set problem) <Simulation and Actual survey evaluation >
(Example) Malfunction measures of the module
Module ( Ceramic package )
Radio irradiation
【Problem】 Module malfunction during the radio irradiation
Set PCB
Simulation analysis result ( Package measures effect )
Magnitude of ectric field distribution on the IC packaging side
Before
After
Large
Small
IC
Malfunction point
HFSS
Newly-developed design technique approach
①Obtain Set information (PCB, structure, circuit, device )
②Choose the most suitable device or combination
③ Evaluation the effect of measures (Quantification and visualization of the set problem) <Simulation and measurement>
(Example) Action making fanless of the tablet PC
Tablet PC With fan
【Problem】
• miniaturize
• become thinner
• reduce noise generated by fans
Measures in the housing
Heat transfer + thermal diffusion
PGS Graphite sheet (highly heat conductive sheet)
Measurement result
Surface temperature of case
Before
After
With fan 44.8℃
Without fan 44.6℃
Most suitable design method, measures device by the set in total
Value proposition to the customer
Newly-designed technique for the creation of a new set function
10

# 1. Summary of the noise solution activity

Utilizing models provided by ANSYS Japan to effectively establish our solution

Wireless power transer system (ANSYS Japan)

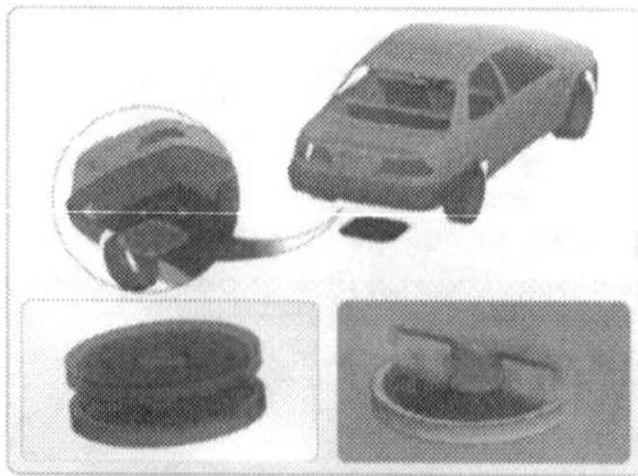

Two or three dimensional electromagnetic field analysis software
**ANSYS Maxwell**

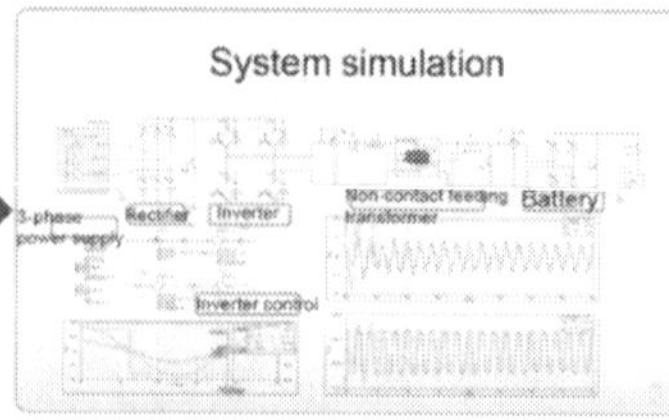


Circuit, system simulator for power electronics
**ANSYS Simplorer**

(3) Battery charging (contact, non-contact)

Method examination

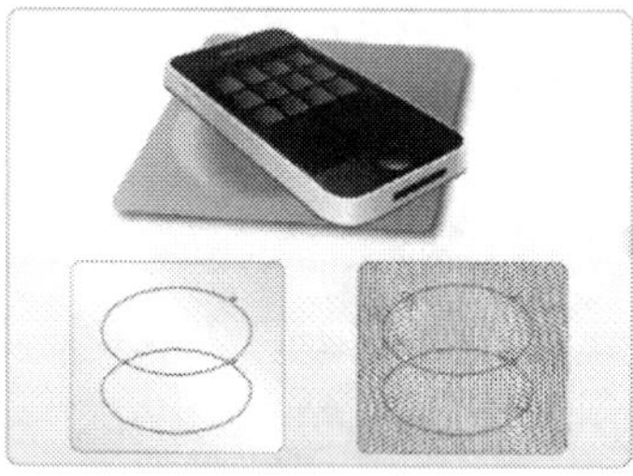

High frequency three dimensionsal electromagnetic field analysis software
**ANSYS HFSS**

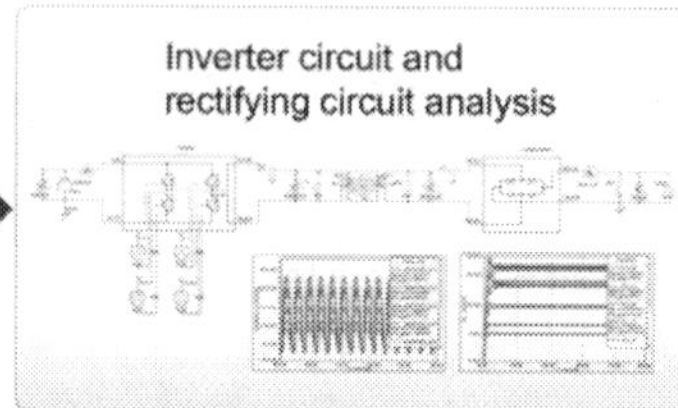


Integrated design environment of electromagnetic field, electronic circuit and system
**ANSYS Designer**

(4) Terminal charge (non-contact charge)

Measures of circuit problem and Optimization of circuit and device

11

# Agenda

1 Summary of the noise solution activity

**2 Example 1) Radiation noise**

3 Example 2) Conduction noise

4 Example 3) Immunity (static electricity)

5 Example 4) Immunity (Radio irradiation)

12

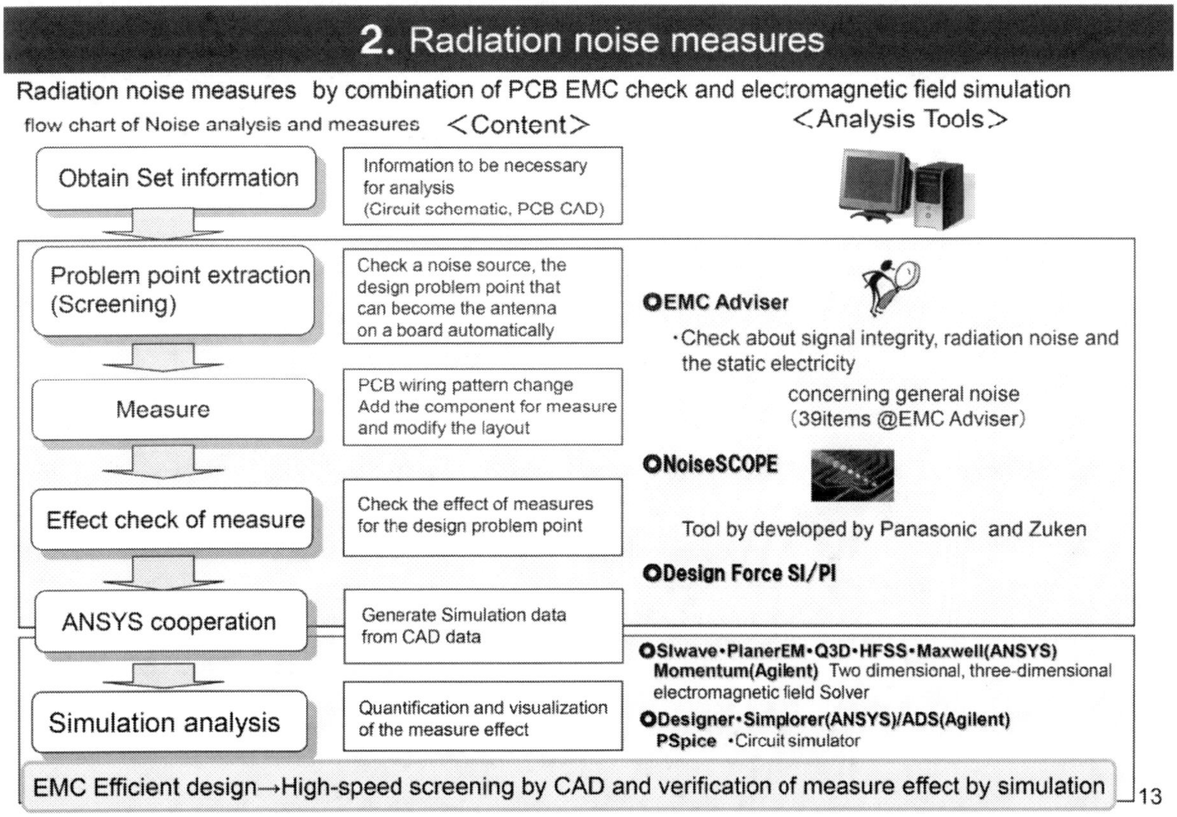
2. Radiation noise measures
Radiation noise measures by combination of PCB EMC check and electromagnetic field simulation
flow chart of Noise analysis and measures
<Content>
<Analysis Tools>
Obtain Set information
Information to be necessary for analysis (Circuit schematic, PCB CAD)
Problem point extraction (Screening)
Check a noise source, the design problem point that can become the antenna on a board automatically
Measure
PCB wiring pattern change Add the component for measure and modify the layout
Effect check of measure
Check the effect of measures for the design problem point
ANSYS cooperation
Generate Simulation data from CAD data
Simulation analysis
Quantification and visualization of the measure effect
OEMC Adviser
·Check about signal integrity, radiation noise and the static electricity
concerning general noise (39items @EMC Adviser)
ONoiseSCOPE
Tool by developed by Panasonic and Zuken
ODesign Force SI/PI
OSIwave·PlanerEM·Q3D·HFSS·Maxwell(ANSYS) Momentum(Agilent) Two dimensional, three-dimensional electromagnetic field Solver
ODesigner·Simplorer(ANSYS)/ADS(Agilent) PSpice ·Circuit simulator
EMC Efficient design→High-speed screening by CAD and verification of measure effect by simulation
13

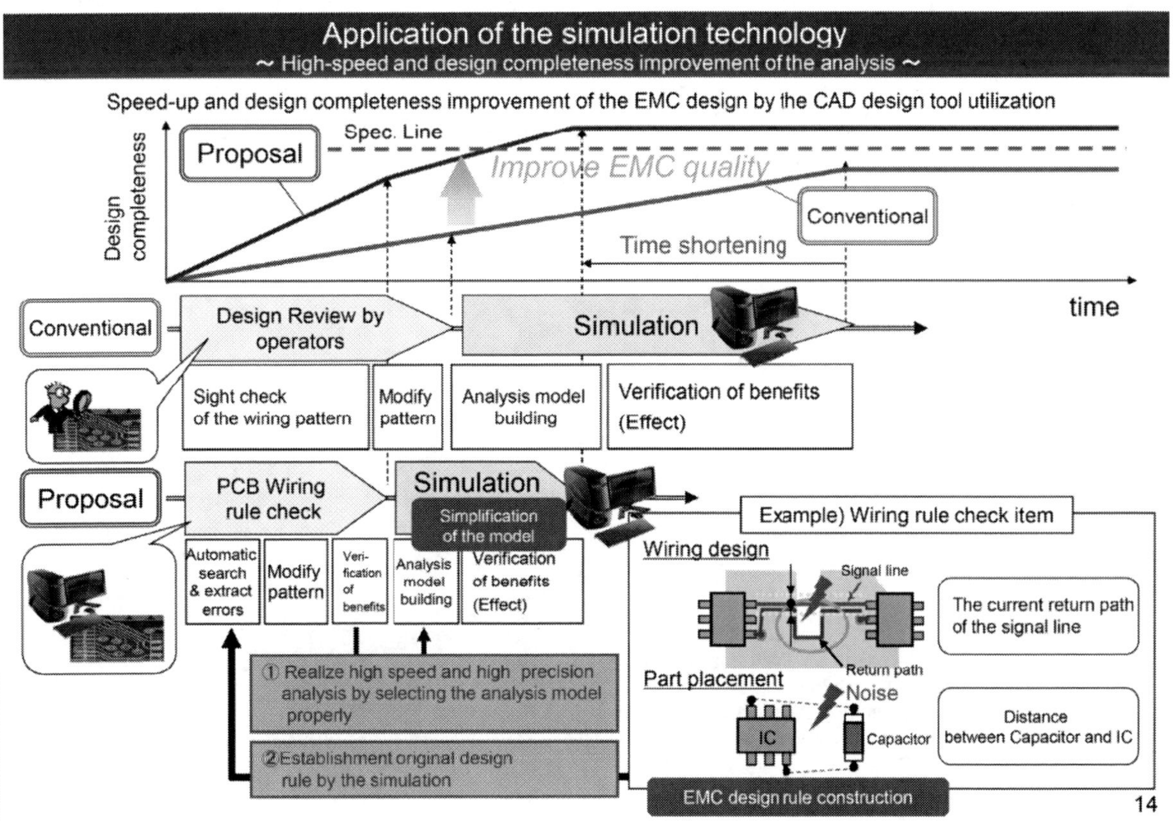
Application of the simulation technology
~ High-speed and design completeness improvement of the analysis ~
Speed-up and design completeness improvement of the EMC design by the CAD design tool utilization
Design completeness
Proposal
Spec. Line
Improve EMC quality
Conventional
Time shortening
time
Conventional
Design Review by operators
Simulation
Sight check of the wiring pattern
Modify pattern
Analysis model building
Verification of benefits (Effect)
Proposal
PCB Wiring rule check
Simulation
Simplification of the model
Automatic search & extract errors
Modify pattern
Verification of benefits
Analysis model building
Verification of benefits (Effect)
① Realize high speed and high precision analysis by selecting the analysis model properly
② Establishment original design rule by the simulation
Example) Wiring rule check item
Wiring design
Signal line
Return path
The current return path of the signal line
Part placement
Noise
IC
Capacitor
Distance between Capacitor and IC
EMC design rule construction
14

## 2−(1). SI analysis &EMI measures of multiple boards

Example of FPC design and measure

～Maximization of noise measure effect by the FPC wiring optimization～

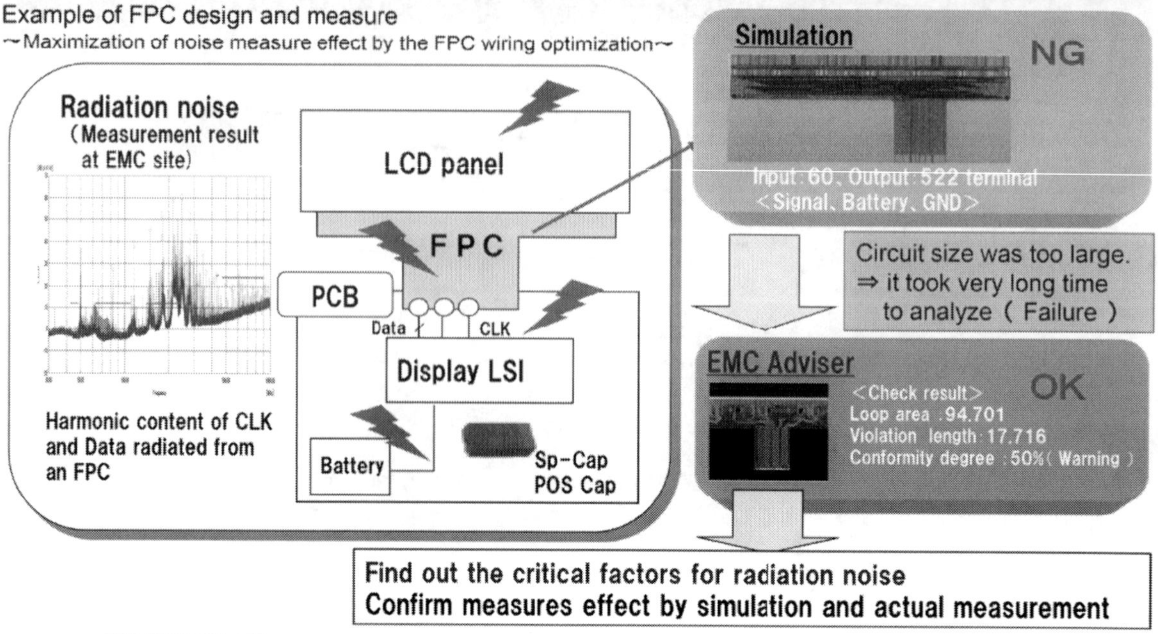


**Confirm an effect of EMC Adviser to analyze radiation noise at FPC.**

15

## 2−(1). SI analysis &EMI measures of multiple boards

EMI design approach by simulation

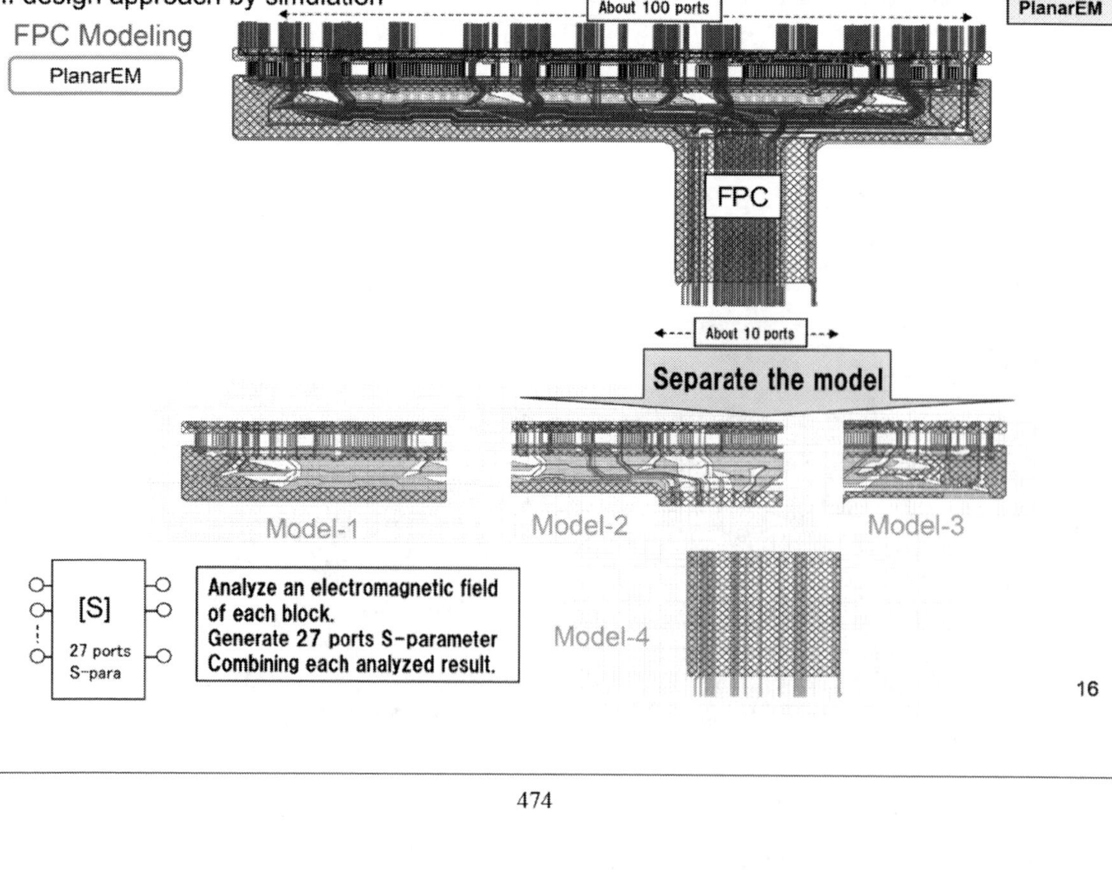


16

# 2－(1). SI analysis &EMI measures of multiple boards

EMI design approach by simulation

ANSYS Designer

Designer/NEXXIM

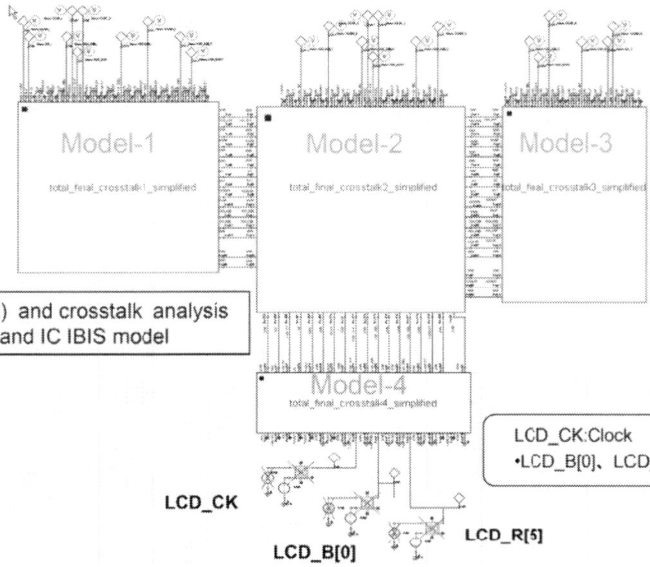


Signal integrity(SI) and crosstalk analysis with S parameter and IC IBIS model

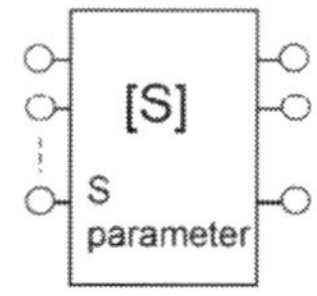


LCD_CK:Clock
•LCD_B[0]、LCD_R[5]:pseudorandom signal

**Even SI analysis takes very long time, it shall be quite difficult to proceed to EMI analysis.**
**（There is the convergent problem of the S parameter, too ）**

17

# 2－(1). SI analysis &EMI measures of multiple boards

EMI design approach by EMC Adviser

Design Force (EMC Adviser)

Summary comparison

Before

| ルール | | 警告 | 注意 | 合格 | 計 | 準拠度 |
|---|---|---|---|---|---|---|
| A- 2 | クローズド・ループ | 0 | 0 | 24 | 24 | 100 |
| A- 3 | オープン・ループ | 0 | 19 | 5 | 24 | 89 |
| A- 4 | インピーダンス整合 | 24 | 0 | 0 | 24 | 26 |
| A- 6 | 配線スタブ長 | 0 | 0 | 24 | 24 | 100 |
| A- 7 | 配線の共振 | 0 | 0 | 24 | 24 | 100 |
| A- 8 | リターン経路 | 6 | 45 | 33 | 84 | 81 |
| E- 1 | 4方向シールディング | 22 | 2 | 0 | 24 | 13 |
| E- 3 | 配線クロストーク | 22 | 0 | 183 | 205 | 90 |
| E- 4 | 電源クロストーク | 3 | 0 | 1 | 4 | 25 |
| E- 9 | 基準プレーン端配線 | 0 | 0 | 0 | 0 | 100 |
| E-10 | 電源ビア配置 | 0 | 0 | 0 | 0 | 100 |
| G- 3 | 面のビア不足 | 40 | 29 | 8 | 77 | 59 |
| H- 3 | 電源GNDループ | 266 | 17 | 50 | 333 | 59 |

After

| ルール | | 警告 | 注意 | 合格 | 計 | 準拠度 |
|---|---|---|---|---|---|---|
| A- 2 | クローズド・ループ | 0 | 0 | 24 | 24 | 100 |
| A- 3 | オープン・ループ | 0 | 20 | 4 | 24 | 85 |
| A- 4 | インピーダンス整合 | 24 | 0 | 0 | 24 | 40 |
| A- 6 | 配線スタブ長 | 0 | 0 | 24 | 24 | 100 |
| A- 7 | 配線の共振 | 0 | 0 | 24 | 24 | 100 |
| A- 8 | リターン経路 | 1 | 44 | 39 | 84 | 88 |
| E- 1 | 4方向シールディング | 24 | 0 | 0 | 24 | 10 |
| E- 3 | 配線クロストーク | 20 | 0 | 184 | 204 | 90 |
| E- 4 | 電源クロストーク | 4 | 0 | 0 | 4 | 0 |
| E- 9 | 基準プレーン端配線 | 0 | 0 | 0 | 0 | 100 |
| E-10 | 電源ビア配置 | 0 | 0 | 0 | 0 | 100 |
| G- 3 | 面のビア不足 | 28 | 25 | 27 | 80 | 66 |
| H- 3 | 電源GNDループ | 315 | 28 | 67 | 410 | 60 |

**3 ～ 5dB radiation noise reduction in the actual evaluation ！！**

**Warning contents decrease about a return path and lack of Via hole about FPC after measures, and a conformity degree improves.**
**The number of the warning does not have the change about the impedance matching either, but the conformity degree improves.**

18

## 2−(1). SI analysis &EMI measures of multiple boards

EMI design approach by EMC Adviser

LCD G3 line

**Before**

<Check result>
Loop area : 94.701
Violation length : 17.716
Conformity degree : 50%(Warning)

Example for comparison of FPC measures effect
(Return path)

**After**

<Check result>
Loop area : 52.085
Violation length : 17.137
Conformity degree : 78%(Attention)

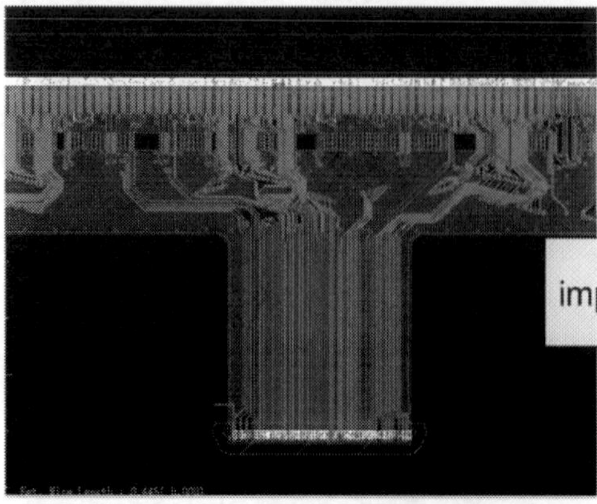

improvement

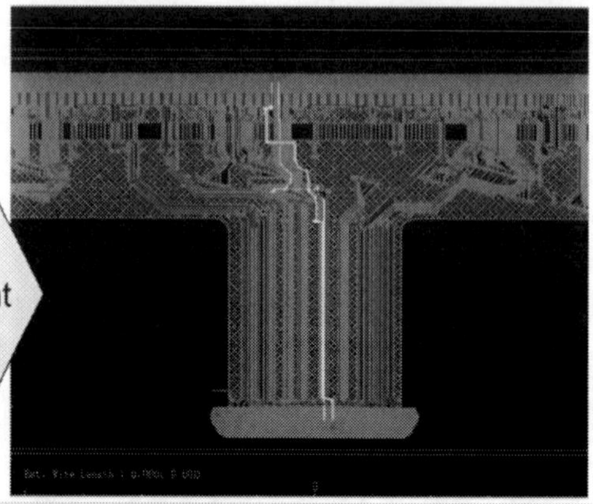

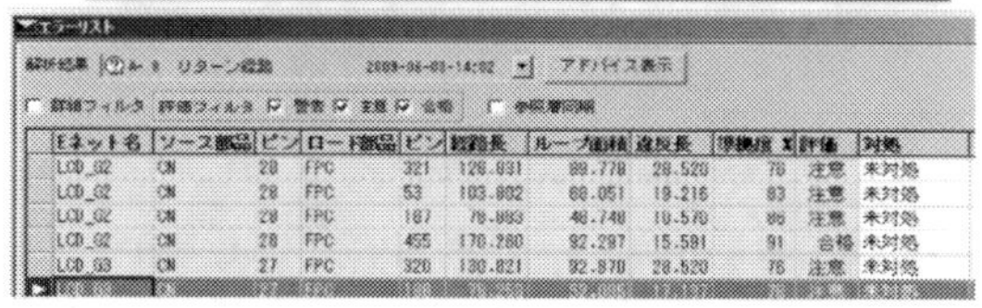

19

## 2−(1). SI analysis &EMI measures of multiple boards

E net select and ANF output function

SIwave

**Design Force**

Choose only E net concerned and export ANF

It does not take time of the net editing on SIwave

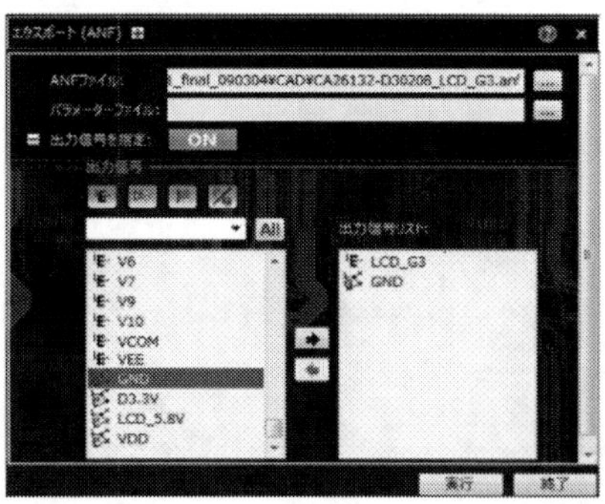


**SIwave**

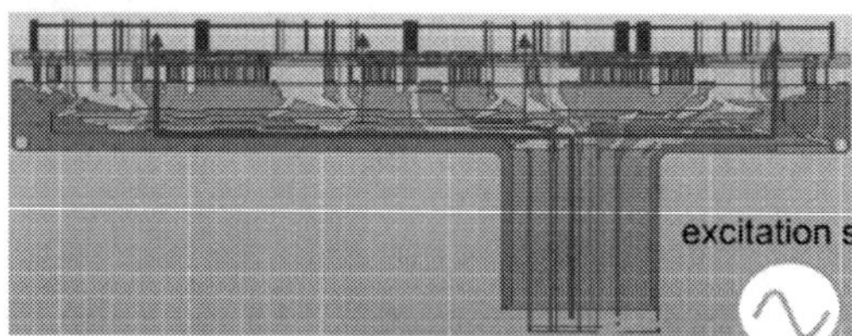

excitation source

**Near-Field analysis (30-100MHz)**

**Before**

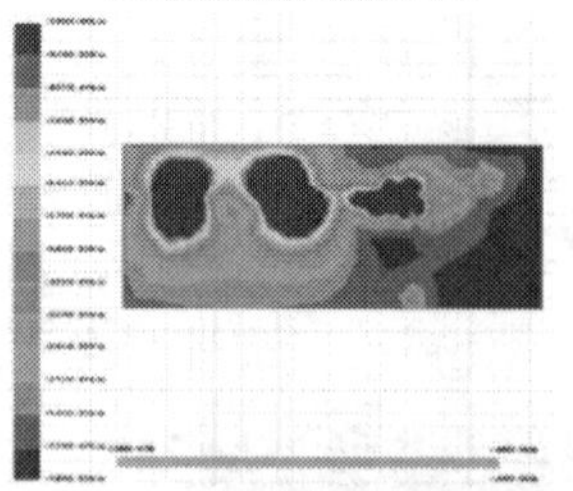

**After**

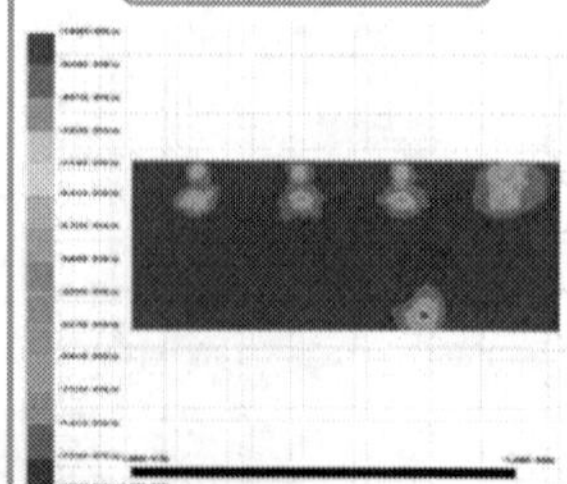

Quantification and visualization of the physical phenomenon

20

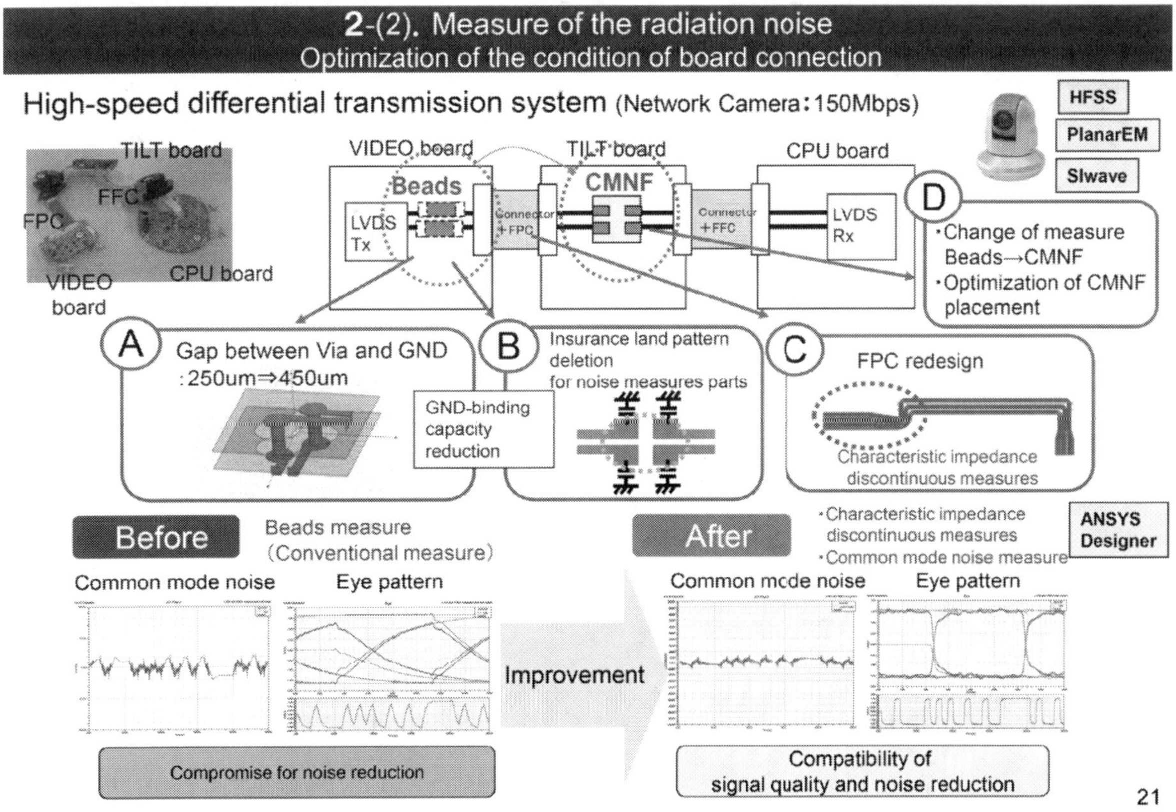
2-(2). Measure of the radiation noise
Optimization of the condition of board connection
High-speed differential transmission system (Network Camera:150Mbps)
HFSS
PlanarEM
SIwave
TILT board
FFC
FPC
VIDEO board
CPU board
VIDEO board
TILT board
CPU board
Beads
CMNF
LVDS Tx
LVDS Rx
Connector +FPC
Connector +FFC
D
·Change of measure Beads→CMNF
·Optimization of CMNF placement
A
Gap between Via and GND :250um⇒450um
GND-binding capacity reduction
B
Insurance land pattern deletion for noise measures parts
C
FPC redesign
Characteristic impedance discontinuous measures
Before
Beads measure (Conventional measure)
After
·Characteristic impedance discontinuous measures
·Common mode noise measure
ANSYS Designer
Common mode noise
Eye pattern
Improvement
Common mode noise
Eye pattern
Compromise for noise reduction
Compatibility of signal quality and noise reduction
21

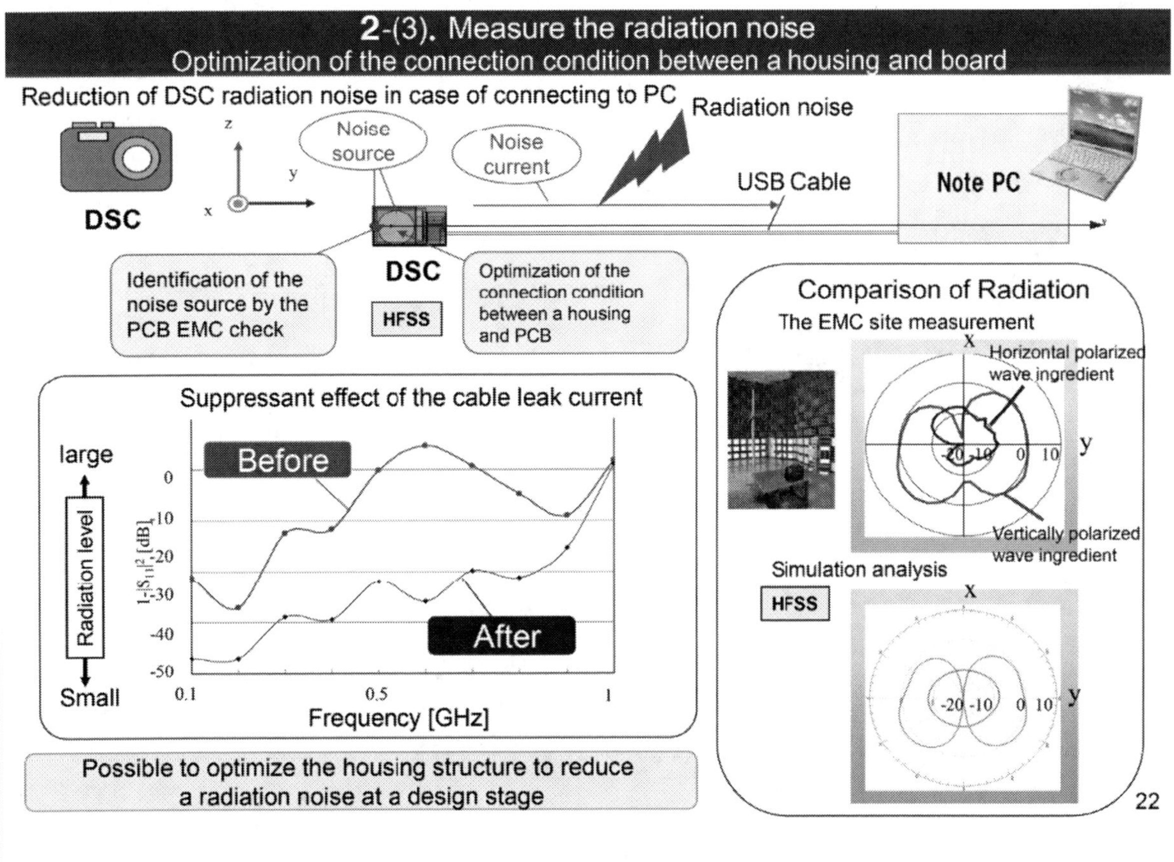
2-(3). Measure the radiation noise
Optimization of the connection condition between a housing and board
Reduction of DSC radiation noise in case of connecting to PC
Radiation noise
DSC
Noise source
Noise current
USB Cable
Note PC
DSC
HFSS
Identification of the noise source by the PCB EMC check
Optimization of the connection condition between a housing and PCB
Suppressant effect of the cable leak current
large
Radiation level
Small
Before
After
1-|S11|2 [dB]
Frequency [GHz]
Comparison of Radiation
The EMC site measurement
Horizontal polarized wave ingredient
Vertically polarized wave ingredient
Simulation analysis
HFSS
Possible to optimize the housing structure to reduce a radiation noise at a design stage
22

# 2-(4). Measure the radiation noise

~ Change of emission properties by structure and the connection condition ~

HFSS

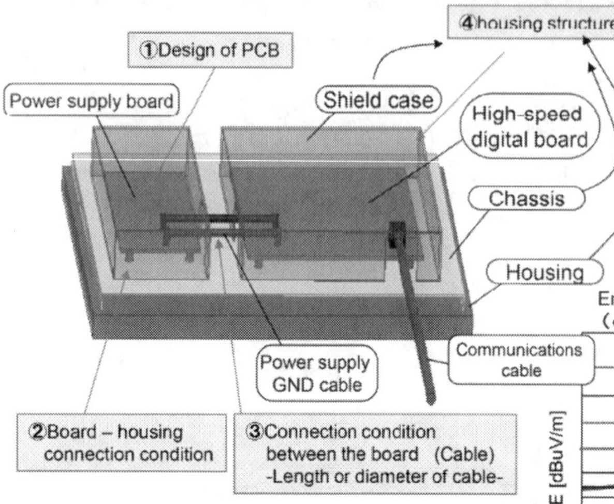


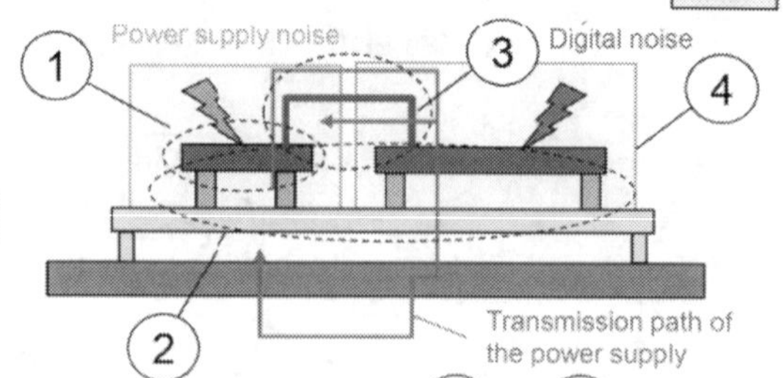


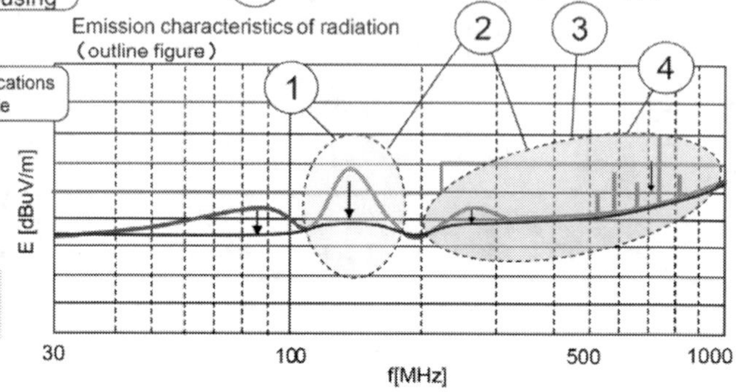


Series and parallel resonance on the transmission path cause radiation noise.

Before the countermeasure of a housing or the board individually, it is importan: to understand the radiation properties based on structure. Able to expect the large improvement of the radiation noise by the optimization of both structure and PCB.

23

# 2−(5). Measure the radiation noise ~ Electric field strength analysis around the coil ~

Coupled analysis of magnetic field and electromagnetic field analysis

Quantification and visualization of the physical phenomenon

AC magnetic field analysis

Low frequency electromagnetic field analysis for the coil winding

AC electromagnetic field analysis

Evaluating electric field strength of the space and emission properties using the source data mapped from the low frequency electromagnetic field analysis

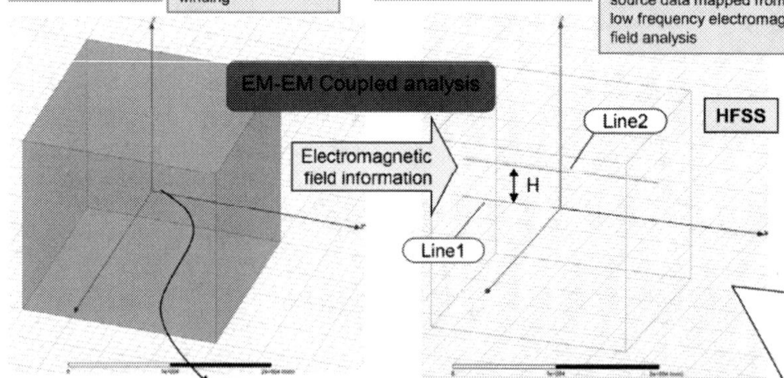


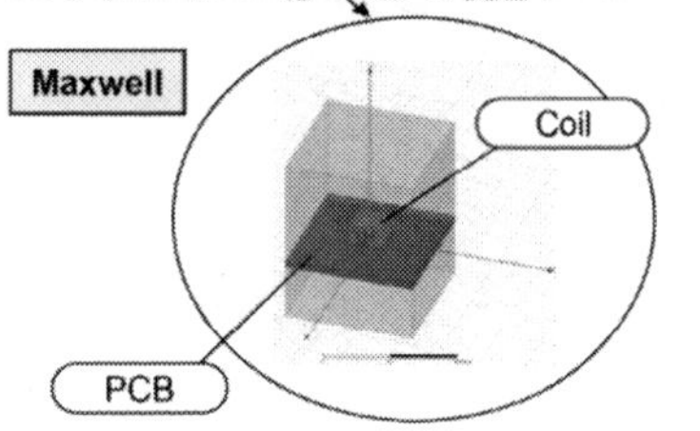


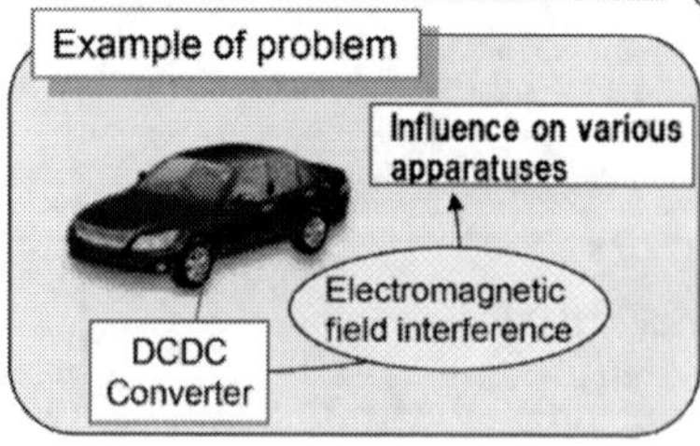


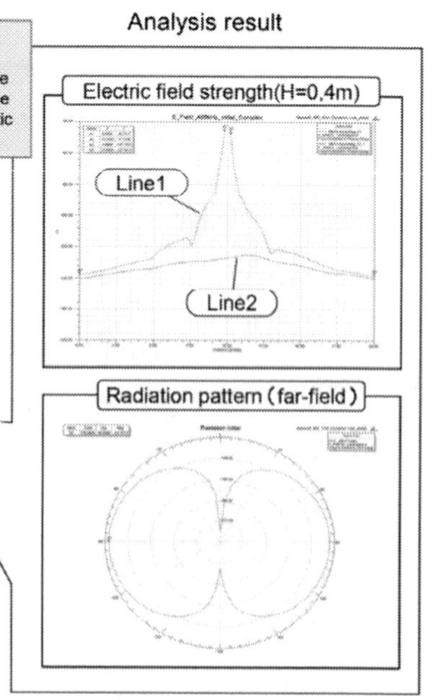


24

## 2–(5). Measure the radiation noise ~ Electric field strength analysis around the coil ~

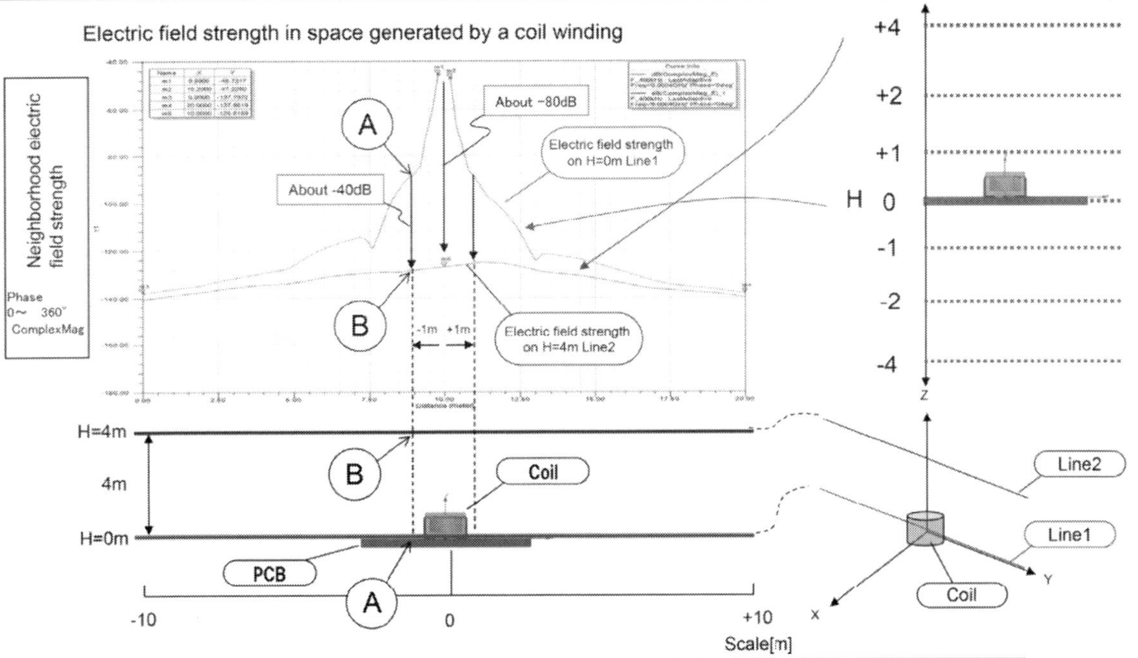


Electric field strength is the highest along the coil basement (at 0m). There is about 40dB difference between 4m(B) and 0m(A) in height at +-1m position from the coil center.

25

## 2–(5). Measure the radiation noise ~ Electric field strength analysis around the coil ~

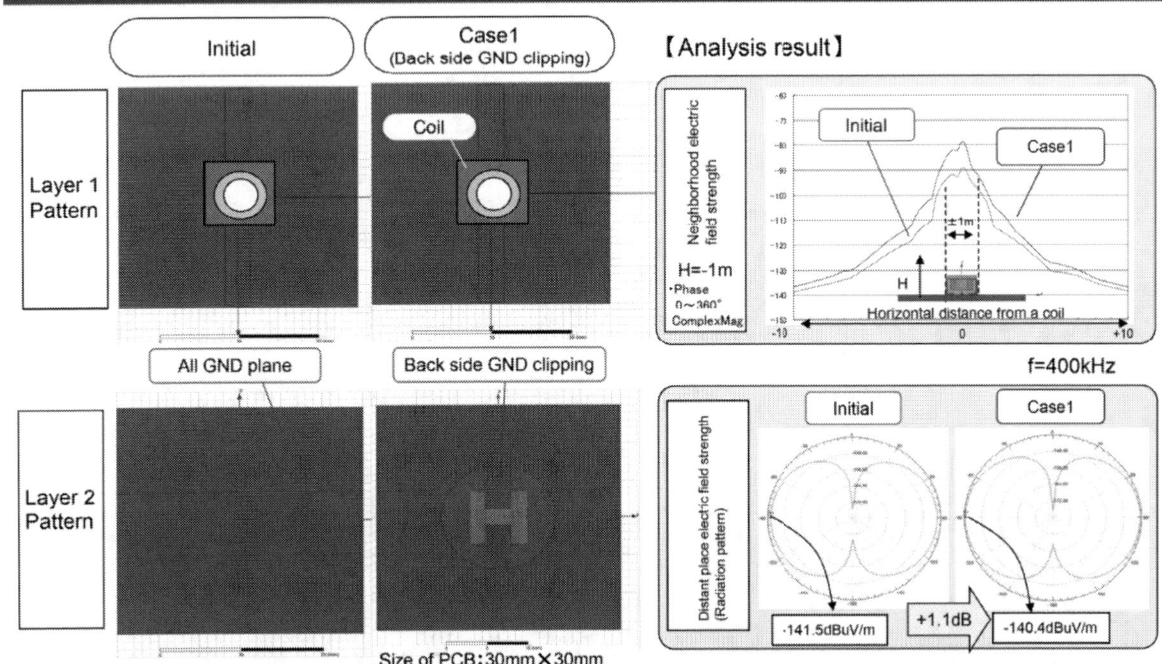


Simulation result replicates the measurement data that shows emitted electric field becomes higher for the Case1 model (GND plane under the coil is clipped) than for the initial model.

26

# Agenda

1 Summary of the noise solution activity

2 Example 1) Radiation noise

3 Example 2) Conduction noise

4 Example 3) Immunity (static electricity)

5 Example 4) Immunity (Radio irradiation)

27

## 3. Conduction noise measures ~ Noise source measures by the conduction noise reduction ~

Power system platform

Simplorer | ANSYS Designer | PlanarEM

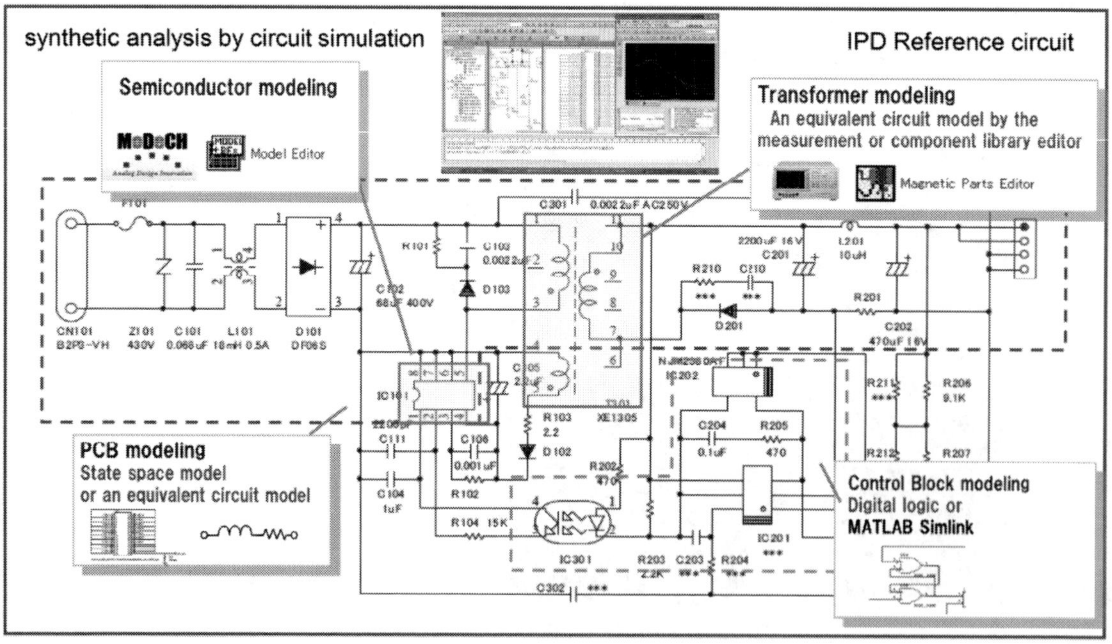


28

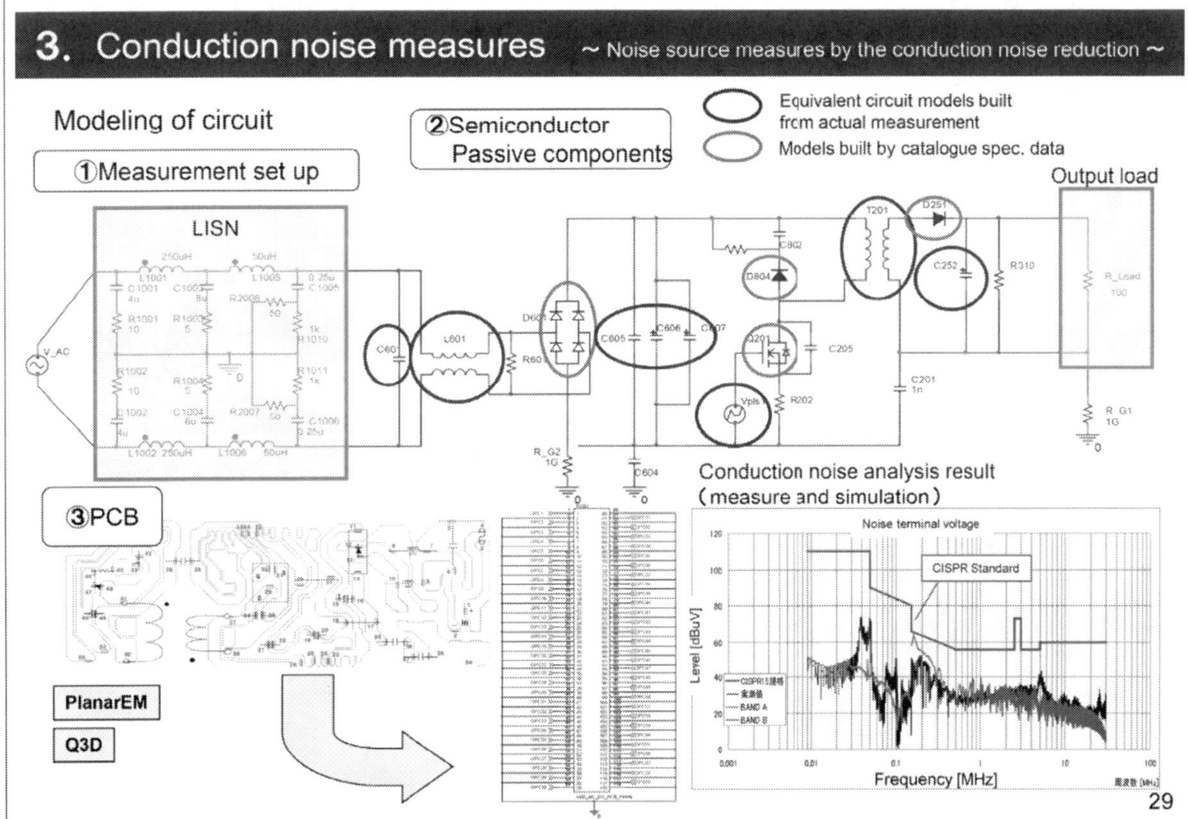
3. Conduction noise measures ~ Noise source measures by the conduction noise reduction ~
Modeling of circuit
①Measurement set up
②Semiconductor
Passive components
Equivalent circuit models built from actual measurement
Models built by catalogue spec. data
Output load
LISN
③PCB
PlanarEM
Q3D
Conduction noise analysis result
(measure and simulation)
Noise terminal voltage
CISPR Standard
Level [dBuV]
Frequency [MHz]
29

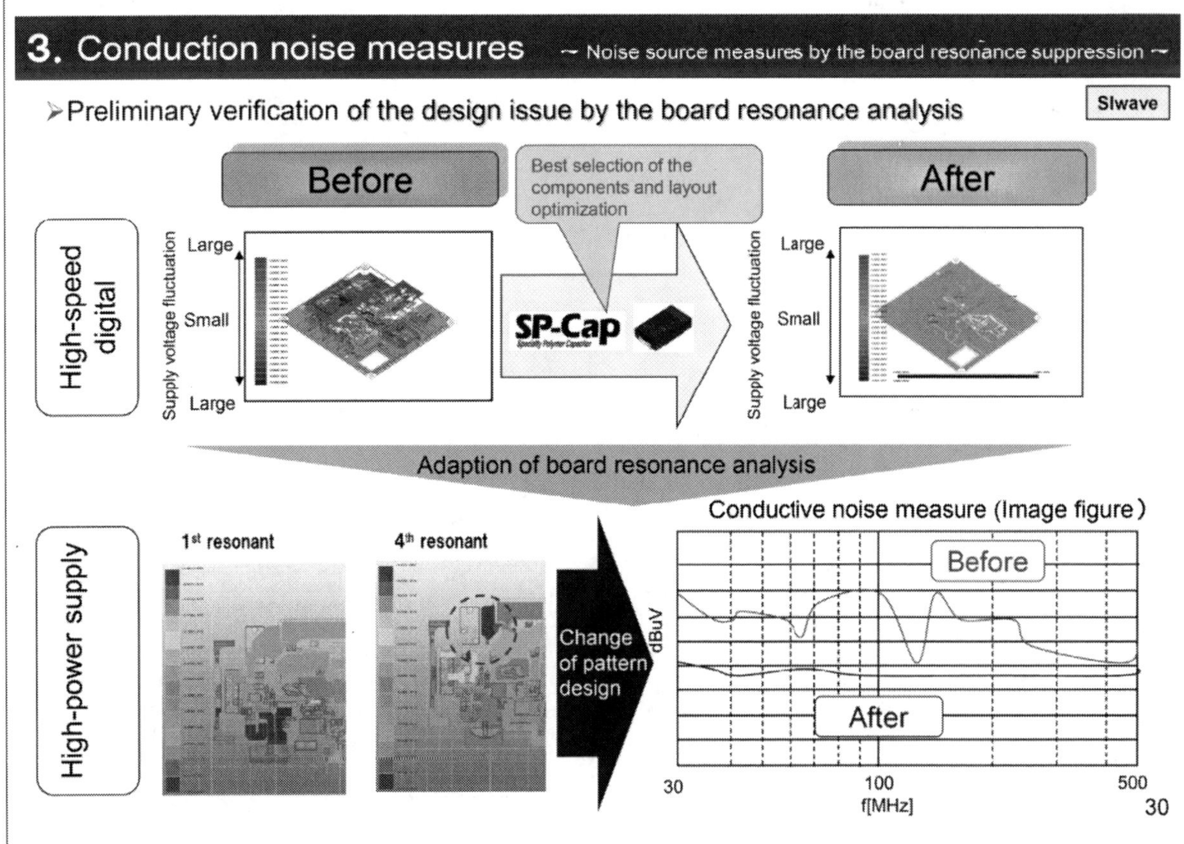
3. Conduction noise measures ~ Noise source measures by the board resonance suppression ~
➢Preliminary verification of the design issue by the board resonance analysis
SIwave
Before
Best selection of the components and layout optimization
After
High-speed digital
Supply voltage fluctuation
Large
Small
Large
SP-Cap
Adaption of board resonance analysis
High-power supply
1st resonant
4th resonant
Change of pattern design
Conductive noise measure (Image figure)
Before
After
dBuV
30
100
500
f[MHz]
30

## 3. Conduction noise measures

~ Consideration of the influence of the spatial coupling of the magnetic component ~

**Without spatial coupling**
<magnetic component (2D characteristics)>

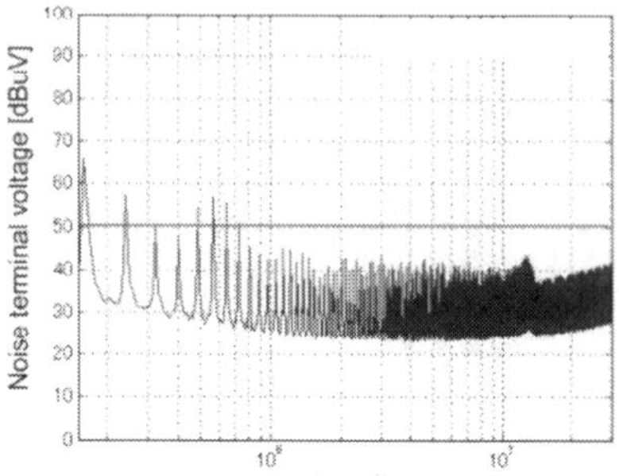


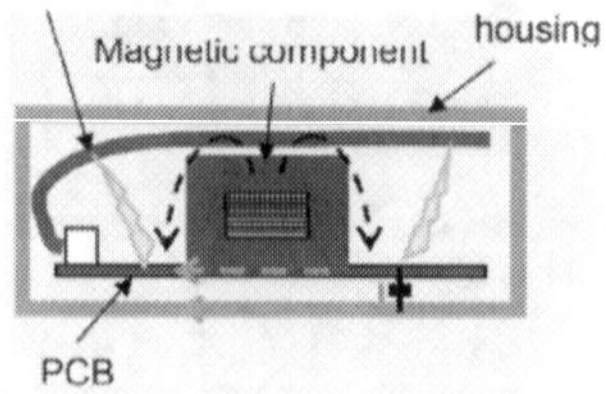


Modeling in consideration of three-dimensional spatial electromagnetic field coupling

**With spatial coupling**
<magnetic component (3D characteristics)>

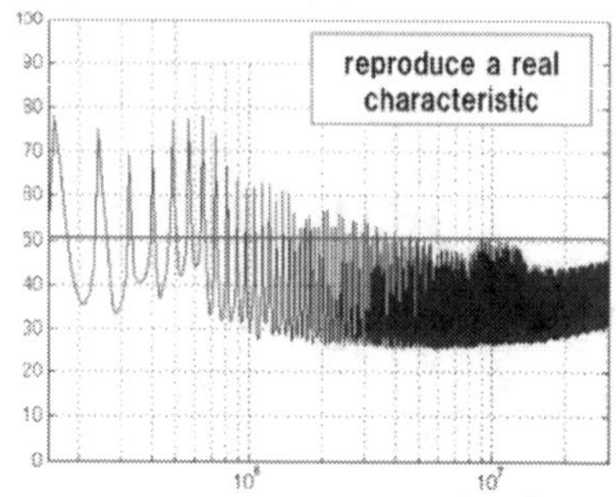


Important point of the EMC analysis

- Simplification of the simulation model
- Trade-off of accuracy versus time for analysis simulation

Equivalent circuit model

High fidelity model

Electromagnetic field model

① Magnetic field, electric field combination with the magnetic component and PCB wiring
② Combination with a housing and PCE (Common mode noise)

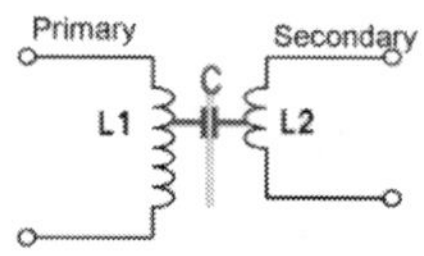


Simplorer

ANSYS Designer

Maxwell

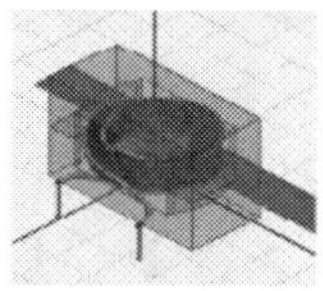

31

## 3. Conduction noise measures

~ Conduction noise problem measures by the spatial coupling of the magnetic component ~

◎ Reduction of the coupling effect inside transformer

Eliminating the placement position dependency of the electrostatic shield type transformer

**Before**

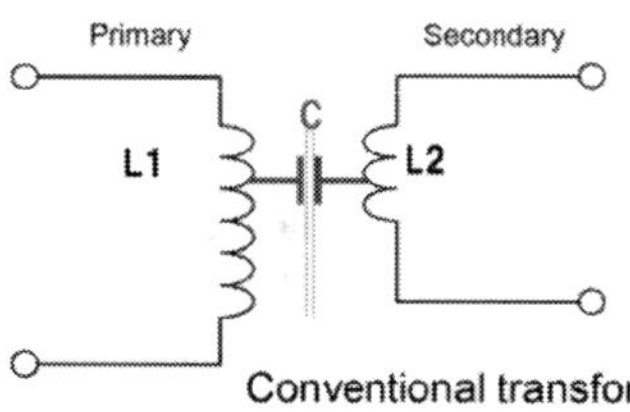


Conventional transformer

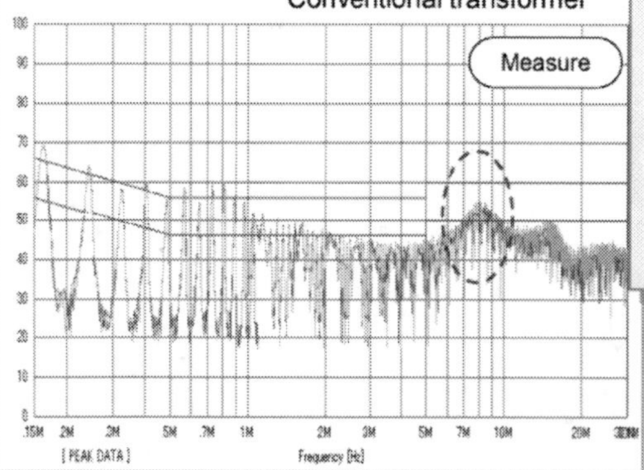


**After**

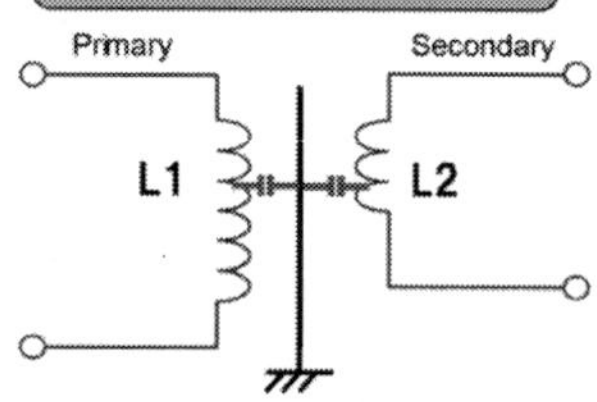


Electrostatic shield type transformer

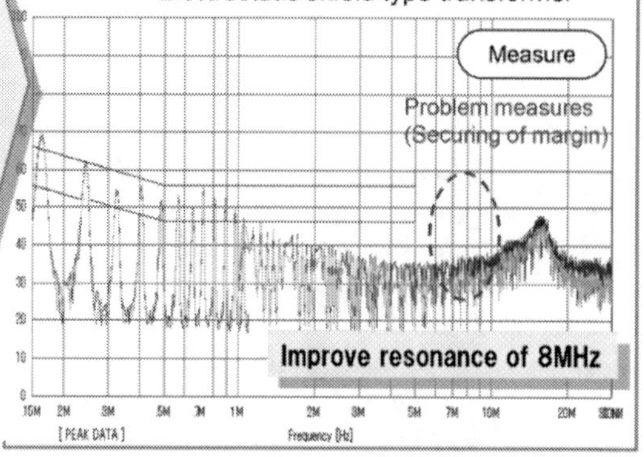


**Improve resonance of 8MHz**

32

# Agenda

1 Summary of the noise solution activity

2 Example 1) Radiation noise

3 Example 2) Conduction noise

4 Example 3) Immunity (static electricity)

5 Example 4) Immunity (Radio irradiation)

33

## 4. Immunity (static electricity) measures ~ measure for electrostatic discharge at PCB GND ~

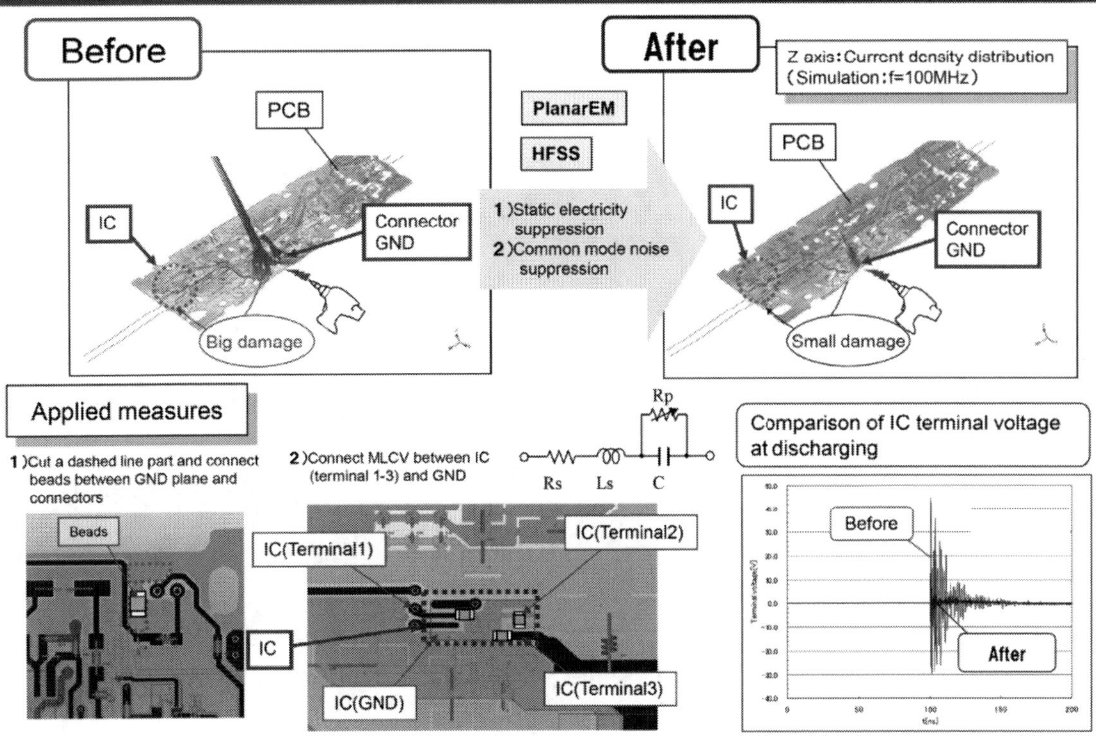


34

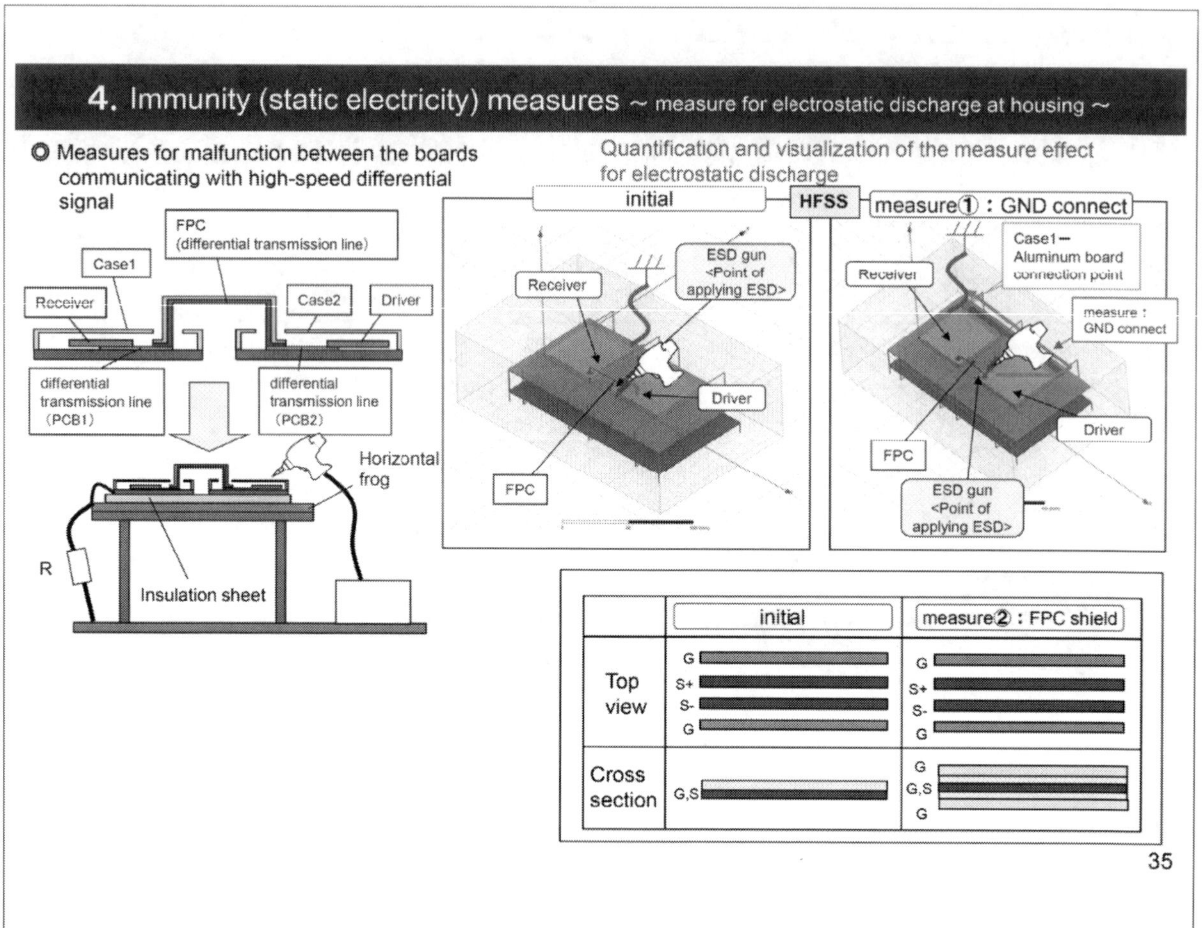
4. Immunity (static electricity) measures ~ measure for electrostatic discharge at housing ~
Measures for malfunction between the boards communicating with high-speed differential signal
FPC (differential transmission line)
Case1
Receiver
Case2
Driver
differential transmission line (PCB1)
differential transmission line (PCB2)
Horizontal frog
R
Insulation sheet
Quantification and visualization of the measure effect for electrostatic discharge
initial
HFSS
measure① : GND connect
ESD gun <Point of applying ESD>
Receiver
Driver
FPC
Case1 – Aluminum board connection point
measure : GND connect
initial
measure② : FPC shield
Top view
G
S+
S-
G
Cross section
G,S
G
G,S
G
35

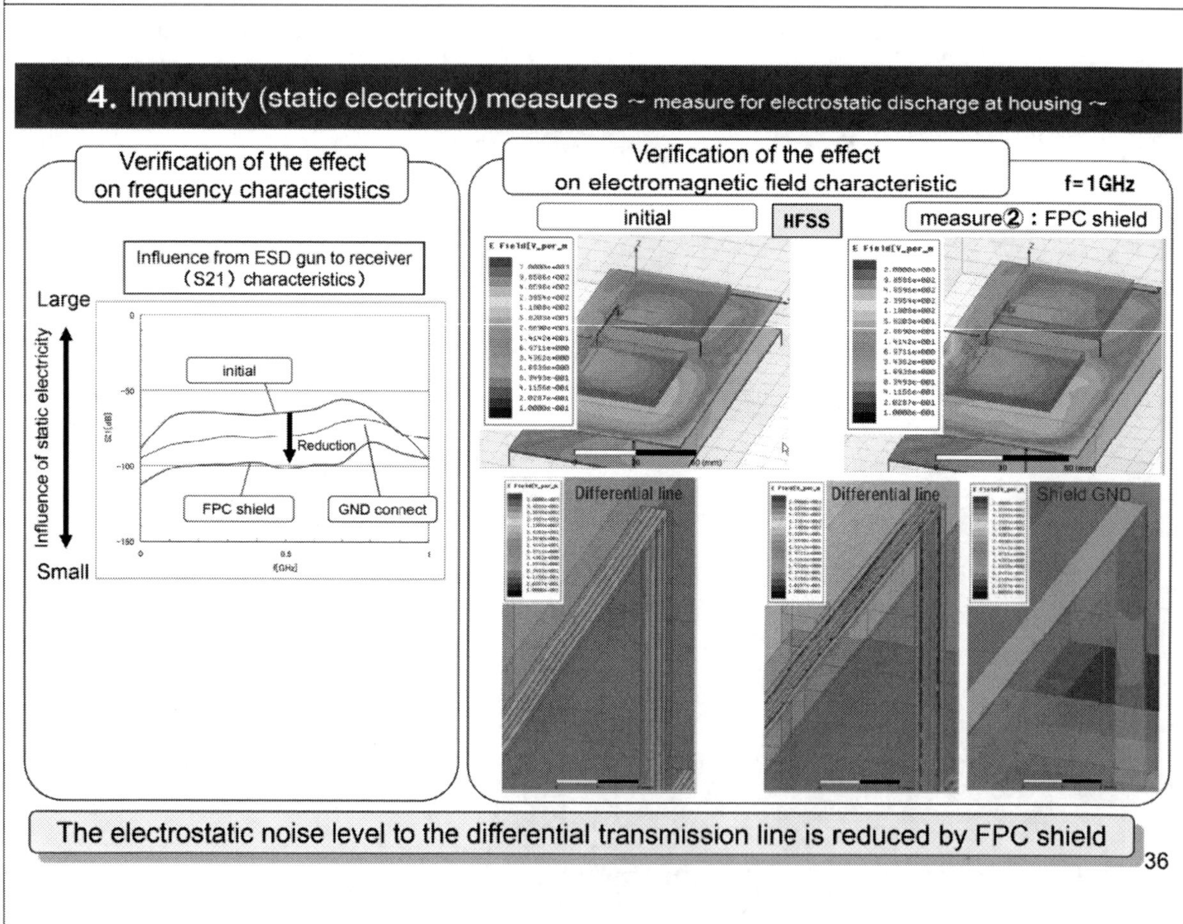
4. Immunity (static electricity) measures ~ measure for electrostatic discharge at housing ~
Verification of the effect on frequency characteristics
Influence from ESD gun to receiver (S21) characteristics)
Large
Influence of static electricity
Small
initial
Reduction
FPC shield
GND connect
Verification of the effect on electromagnetic field characteristic
f=1GHz
initial
HFSS
measure② : FPC shield
Differential line
Differential line
Shield GND
The electrostatic noise level to the differential transmission line is reduced by FPC shield
36

## 4. Immunity (static electricity) measures ~ measure for electrostatic discharge at housing ~

◎Visualization of physical phenomena by HFSS Transient analysis

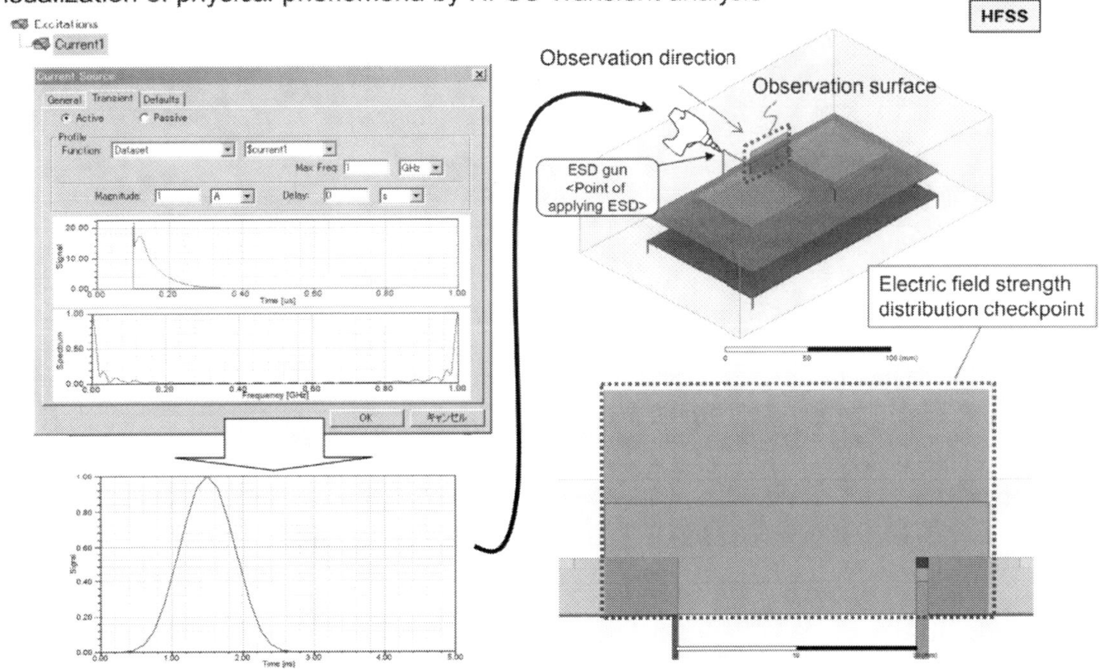


Fast inspection by using substitute wave pattern of the electrostatic gun wave

37

## 4. Immunity (static electricity) measures ~ measure for electrostatic discharge at housing ~

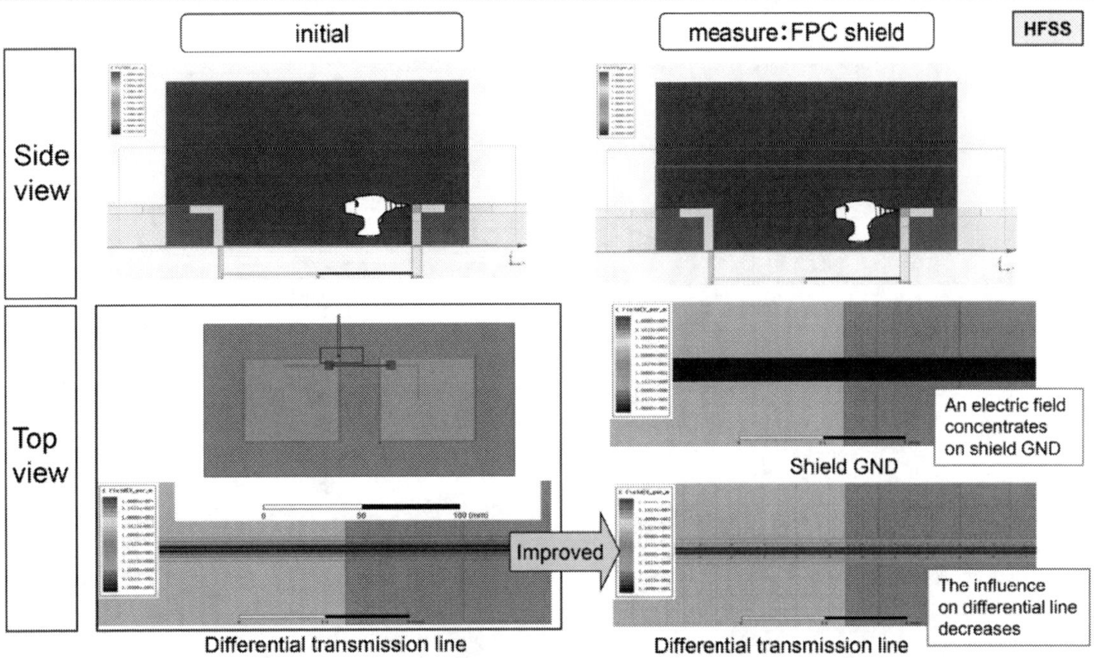


In addition to frequency domain analysis at a single frequency point , visualization by the transient analysis is effective to comprehend the realistic electromagnetic phenomenon

38

# Agenda

1 Summary of the noise solution activity

2 Example 1) Radiation noise

3 Example 2) Conduction noise

4 Example 3) Immunity (static electricity)

**5 Example 4) Immunity (Radio irradiation)**

39

## 5−(1). Immunity (Radio irradiation) measures

~ Verification of simulation by simple model test and Application to the real-life designs~

**Approach (1)** Simulation accuracy verification

Evaluation (G-TEM Cell) ~ Simple coupon examination ~

Comparison between measurement and simulation for the simple model

- TEM wave irradiation condition

Electric field E=100V/m
Frequency:100MHz-1GHz

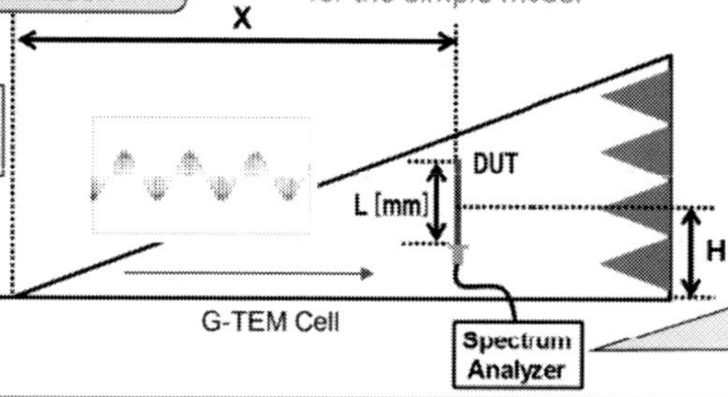


observes the voltage level of the DUT by spectrum analyzers in applying an electric field

**Approach (2)** Adoption of simulation to analyze the factors and measures

Simulation( Radio irradiation ) ~Set PCB and module~

HFSS

Validation of the measure, verification of its effect for the real set structure with PCB and module

- TEM wave irradiation condition

-applying each xy polarized wave 200V/m in Z direction

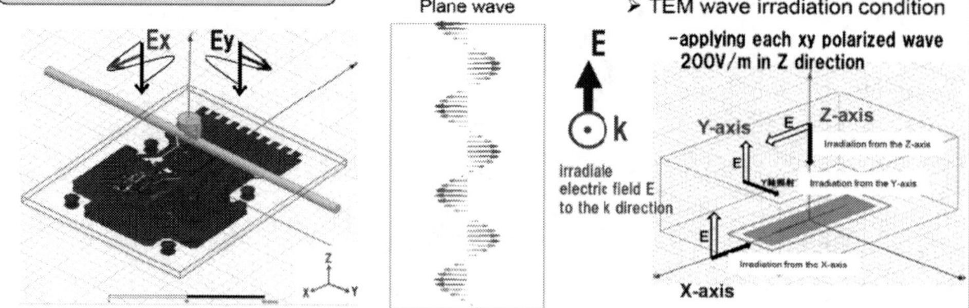


40

## 5−(1). Immunity (Radio irradiation) measures

~ The influence of the housing shape and effect of the noise measure components ~

◎Countermeasure for differential line malfunction under the radio irradiation

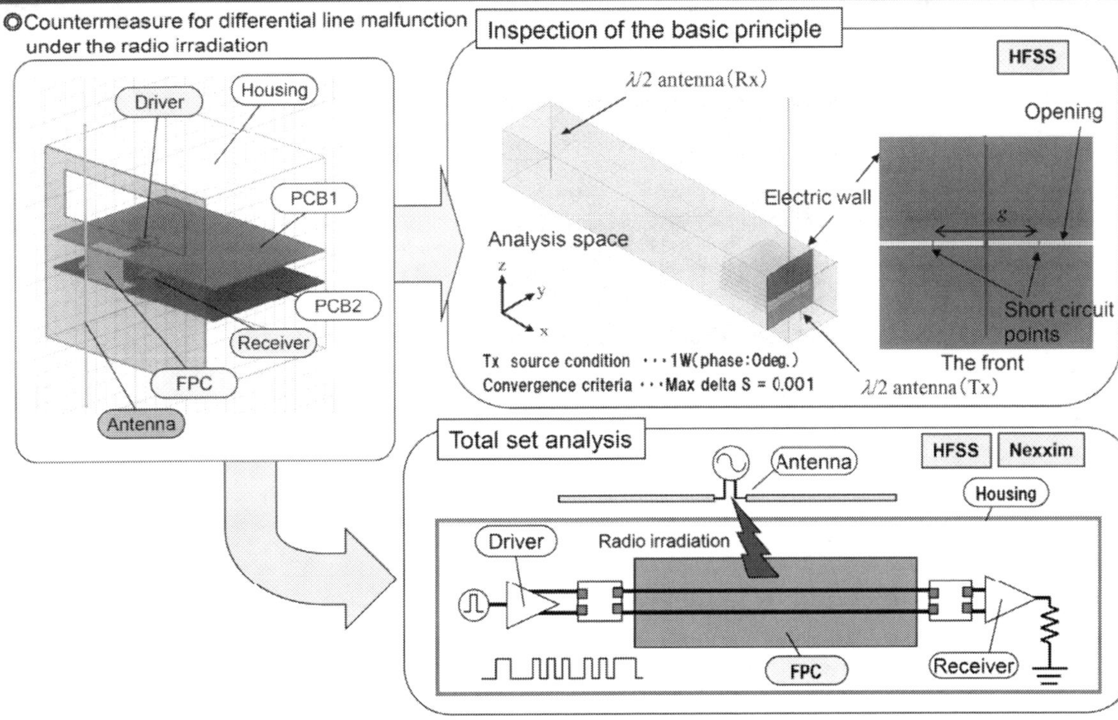


41

## 5−(1). Immunity (Radio irradiation) measures

~ Basic examination of electromagnetic wave propagation mechanism in the housing ~

Magnetic field distribution vs short circuit conditions of the opening aperture

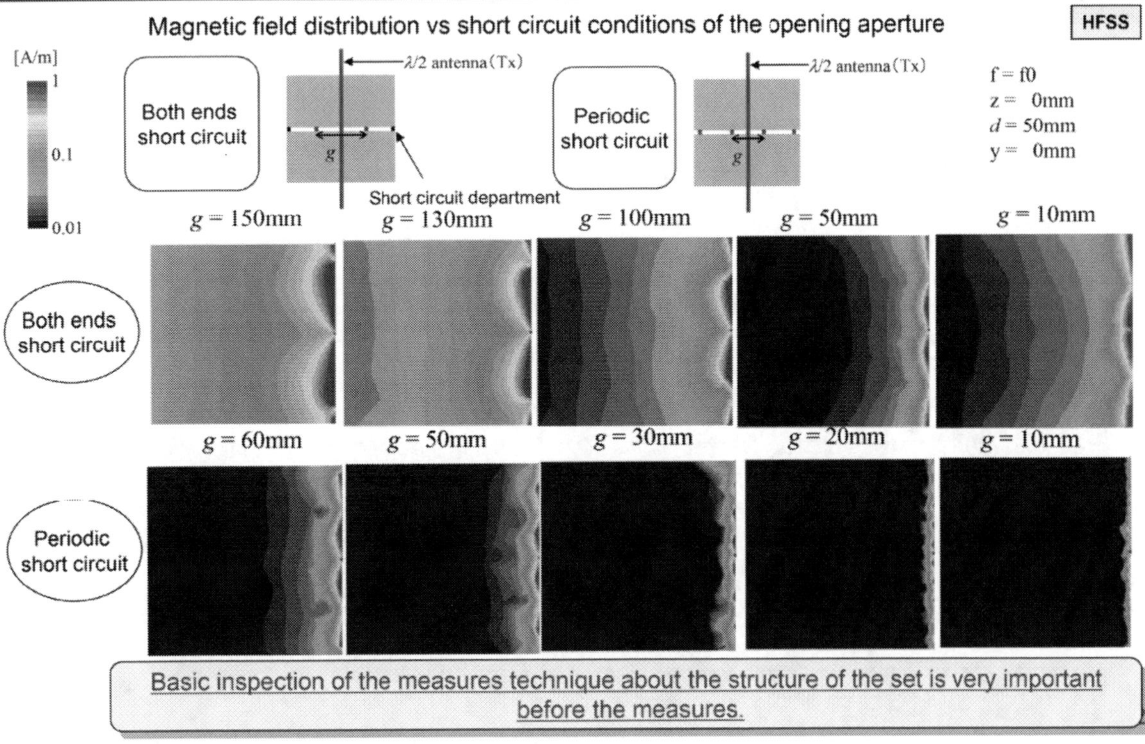


Basic inspection of the measures technique about the structure of the set is very important before the measures.

## 5−(1). Immunity (Radio irradiation) measures

~ Solve propagation path of the radio irradiation by analyzing total structure ~

HFSS

### ~Set total analysis~

**Analysis model**

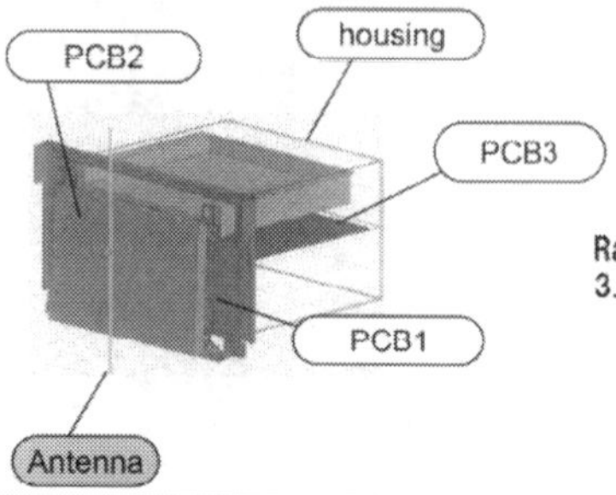


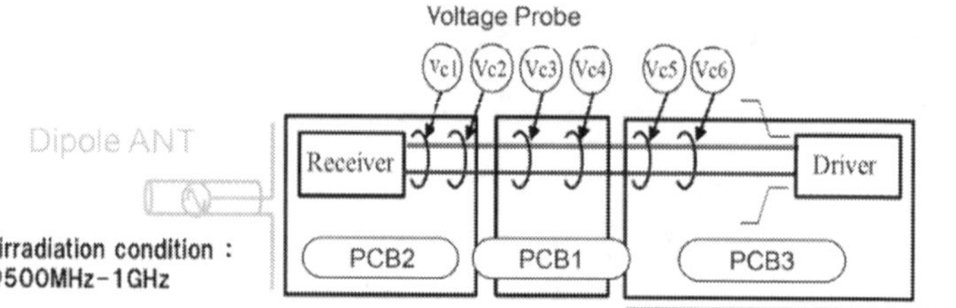


Radio irradiation condition : 3.5W@500MHz-1GHz

※ High-speed differential line : 100 ohm

500mV@
RiseFall Time = 1ns
Pulse Width = 3ns

**Radio irradiation antenna position**

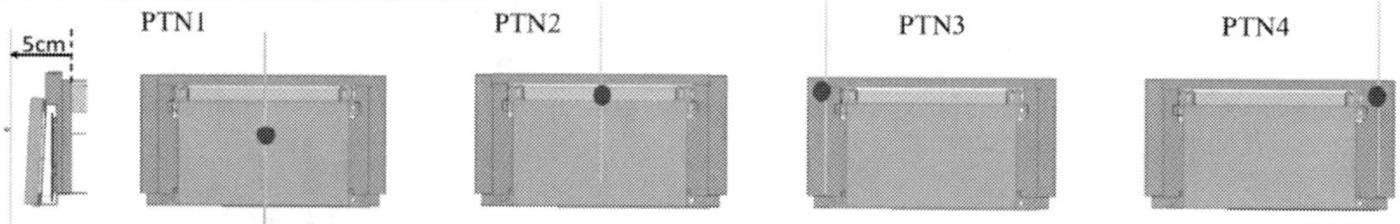


**※ Evaluate the propagation path of the radio irradiation on both vertical and horizontal conditions of the dipole antenna.**

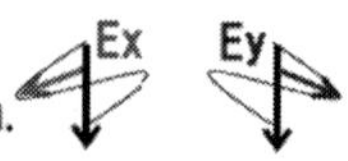


43

## 5−(1). Immunity (Radio irradiation) measures

~ Solve propagation path of the radio irradiation ~

◎ Investigation of the effect by the antenna placement

HFSS

Common mode voltage
(Observation point : Vc1~Vc6)

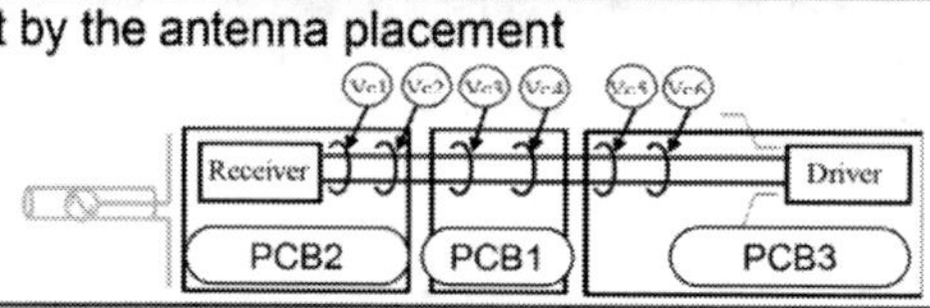


— PTN1
— PTN2
— PTN3
— PTN4

**※Frequency 1 or 2**
**500MHz-1GHz**

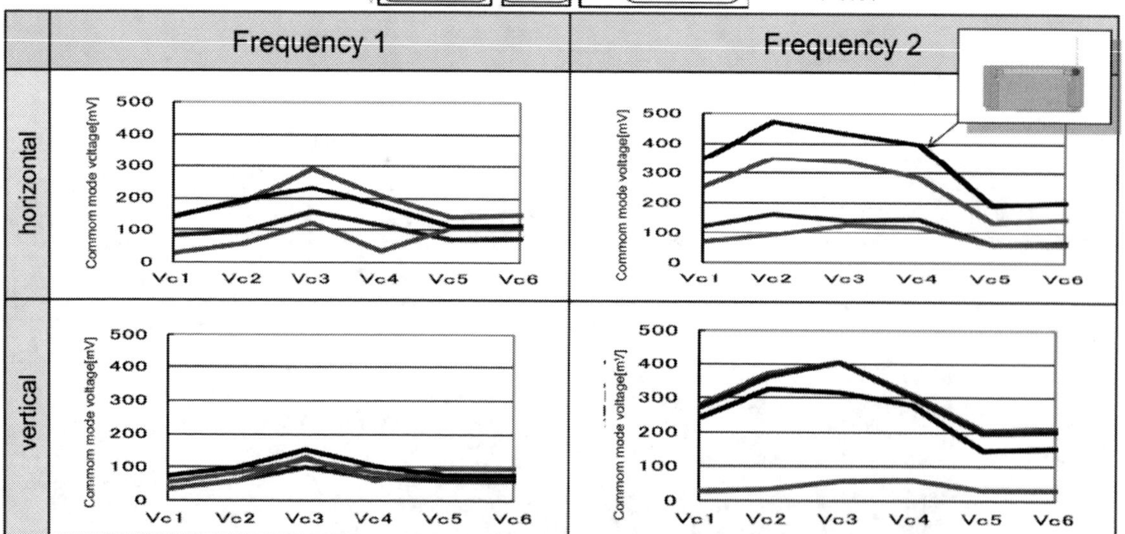


Common mode voltage tends to become higher for the antenna placed on the condition of pattern 4.

44

Thank you for your kind attention.

Panasonic

45